Understanding and Troubleshooting Cisco Catalyst 9800 Series Wireless Controllers

Simone Arena

Francisco Sedano Crippa, CCIE No. 14859

Nicolas Darchis, CCIE No. 25344

Sudha Katgeri, CCIE No. 45857

Cisco Press

Understanding and Troubleshooting Cisco Catalyst 9800 Series Wireless Controllers

Simone Arena
Francisco Sedano Crippa, CCIE #14859
Nicolas Darchis, CCIE #25344
Sudha Katgeri, CCIE # 45857

Published by:
Cisco Press

ScoutAutomatedPrintCode

Library of Congress Control Number: 2022906015

ISBN-13: 978-0-13-749232-9

ISBN-10: 0-13-749232-4

Warning and Disclaimer

This book is about deploying and troubleshooting a wireless network with the next generation Catalyst 9800 Wireless Controller. It covers the software and hardware architecture, the design and deployment aspects and provides useful troubleshooting tools. Every effort has been made to make this book as complete and as accurate as possible, but no warranty or fitness is implied.

The information is provided on an "as is" basis. The authors, Cisco Press, and Cisco Systems, Inc. shall have neither liability nor responsibility to any person or entity with respect to any loss or damages arising from the information contained in this book or from the use of the discs or programs that may accompany it.

The opinions expressed in this book belong to the authors and are not necessarily those of Cisco Systems, Inc.

Trademark Acknowledgments

All terms mentioned in this book that are known to be trademarks or service marks have been appropriately capitalized. Cisco Press or Cisco Systems, Inc., cannot attest to the accuracy of this information. Use of a term in this book should not be regarded as affecting the validity of any trademark or service mark.

Special Sales

For information about buying this title in bulk quantities, or for special sales opportunities (which may include electronic versions; custom cover designs; and content particular to your business, training goals, marketing focus, or branding interests), please contact our corporate sales department at corpsales@pearsoned.com or (800) 382-3419.

For government sales inquiries, please contact governmentsales@pearsoned.com.

For questions about sales outside the U.S., please contact intlcs@pearson.com.

Feedback Information

At Cisco Press, our goal is to create in-depth technical books of the highest quality and value. Each book is crafted with care and precision, undergoing rigorous development that involves the unique expertise of members from the professional technical community.

Readers' feedback is a natural continuation of this process. If you have any comments regarding how we could improve the quality of this book, or otherwise alter it to better suit your needs, you can contact us through email at feedback@ciscopress.com. Please make sure to include the book title and ISBN in your message.

We greatly appreciate your assistance.

Editor-in-Chief: Mark Taub

Alliances Manager, Cisco Press: Arezou Gol

Director, ITP Product Management: Brett Bartow

Executive Editor: Nancy Davis

Managing Editor: Sandra Schroeder

Development Editor: Ellie C. Bru

Project Editor: Mandie Frank

Copy Editor: Chuck Hutchinson

Technical Editors: Shobhit, Flavio Correa

Editorial Assistant: Cindy Teeters

Designer: Chuti Prasertsith

Composition: codeMantra

Indexer: Timothy Wright

Proofreader: Barbara Mack

Americas Headquarters
Cisco Systems, Inc.
San Jose, CA

Asia Pacific Headquarters
Cisco Systems (USA) Pte. Ltd.
Singapore

Europe Headquarters
Cisco Systems International BV
Amsterdam, The Netherlands

Cisco has more than 200 offices worldwide. Addresses, phone numbers, and fax numbers are listed on the Cisco Website at **www.cisco.com/go/offices.**

Figure	Credit/Attribution
Figure 3-12, Figure 12-22, Figure 12-34, Figure 13-16, Figure 13-19, Figure 13-20, Figure A-1, Figure A-3, Figure A-5, Figure A-11, Figure A-12, Figure A-16 Figure A-25, Figure A-28	Microsoft Corporation
Figure 6-2-Figure 6-4, Figure 9-5, Figure 6-11, Figure 6-12	Wireshark
Figure 7-4	nostal6ie/Shutterstock
Figure 7-5	Terry Leung/Pearson Education Asia Limited
Figure 7-6	Monkey Business Images/Shutterstock
Figure 11-2	Bluepixel Technologies
Figure 12-11-Figure 12-13, Figure 12-20, Figure 12-21, Figure 12-24, Figure 13-7, Figure 13-8	Internet Society
Figure 12-23, Figure 12-26, Figure 12-27, Figure 12-30, Figure 12-31, Figure 12-32	Postman, Inc
Figure 13-3, Figure 13-10, Figure A-17-Figure A-20, Figure A-29, Figure A-30	GitHub, Inc
Figure 13-26-Figure 13-30	Grafana Labs
Figure A-6-Figure A-10, Figure A-13-Figure A-15, Figure A-23, Figure A-24, Figure A-26, Figure A-27, Figure A-31, Figure A-32, Figure A-37	Apple, Inc

About the Authors

Simone Arena is a principal technical marketing engineer (TME) within the Cisco Enterprise Networking & Cloud group and is primarily focused on enterprise network architecture and on all things related to wireless and mobility. Simone is based in Italy and is a Cisco veteran, having joined Cisco in 1999. Throughout the years, Simone has covered multiple roles at Cisco, starting as a software engineer working with Catalyst switching platforms, to consulting system engineer in the field, to TME within different teams (Enterprise Solution Engineering, Wireless Business Unit, Enterprise Networking and Cloud, and now Networking Experiences Group). Today Simone is the lead TME architect for Catalyst Wireless, and his time is split between helping customers and partners design the best solution that fits their needs and engineering and product management, trying to evolve and improve the products and solutions. Simone is a Distinguished Speaker at Cisco Live and has spoken at Cisco Live events all over the world for several years. Besides wireless and networking, Simone has two passions: his family, with his two daughters Viola and Anita; and Fiorentina, the best soccer team in the world...no question. In his spare time, Simone enjoys listening to music, especially through his new tube amplifier (simply awesome!).

Francisco Sedano Crippa, CCIE No. 14859, joined Cisco in 2006. After some years at TAC supporting voice solutions and as a system engineer working with service providers, he moved to the development side, where he worked on routing, datacenter and, during the past 10 years, as a technical leader on the Wireless Controller development team, focused in serviceability, location services, programmability, and cloud. He's a Cisco Live speaker and is passionate about DevOps and automation, and he is now working on architecting next-generation cloud-based lab services. When not working, he spends his time building a full-size Boeing 737 simulator in his basement and enjoying his other passion: his daughter, Scarlett, and son, Marco, and his wife, Isabel.

Nicolas Darchis, CCIE Wireless No. 25344, joined the Wireless and AAA Cisco TAC team in Belgium in 2007, where his main focus was troubleshooting wireless networks, wireless management tools, and security products. Since 2016, Nicolas has been working as a technical leader for wireless at the same technical assistance center in Brussels; he has shifted a big part of his focus to improving product serviceability of new and upcoming products, as well as new software releases. He is also a major contributor to online documentation of Cisco wireless products and has participated in many of the wireless "Ask the Expert" sessions run by the Cisco support community. Nicolas has been a CCIE Wireless No. 25344 since 2009 and, more recently, he has achieved CWNE No. 208.

Sudha Katgeri, CCIE No. 45857, is a technical leader in services for Enterprise Wireless and has been with Cisco since 2006. Besides supporting customer escalations, Sudha collaborates with Customer Experience (CX), Enterprise Networking (ENB) Escalation, Engineering, and Product Management to improve product quality and serviceability in the next-generation Catalyst wireless stack. Sudha has a CCIE in Wireless (#45857) and is an author and contributor to Wireless TAC Innovation Tools like Wireless Config Converter, CLI Analyzer, and several documents on cisco.com.

About the Technical Reviewers

Shobhit is a principal engineer in Enterprise Wireless Engineering at Cisco Systems. He has over 15 years of experience working on Enterprise Wireless LANs and Mobility, across a range of products, and has been involved with the Catalyst 9800 wireless LAN controllers since they were conceptualized. He has extensively worked on the architecture, design, and implementation of the software, which runs on Catalyst 9800 WLCs, and he continues to be passionately involved with Catalyst 9800 adoption.

Shobhit lives in Bangalore, India, with his wife, Shweta, and their eight-year-old daughter, Snehi.

Flavio Correa, CCIE Wireless No. 38913, joined Cisco in 2008 and is based in Brazil. He is a lead technical solutions architect for Latin America with more than 20 years of experience designing and deploying wireless technologies. He holds a bachelor's degree in electrical and electronics engineering from Mackenzie University and an MBA in data science from the University of São Paulo.

Dedications

I would like to dedicate this book to my father; I know he is watching me from up there, and he would be very proud. I also want to thank my family (all of them, all members of my big Italian family!) for pushing me to be a better man every day and cheering for me when I need it. Final mention to my two dogs, Barney and Mia, and the two cats, Harry and Ginny, that keep me company in my long hours in the office.

—Simone Arena

I would like to first dedicate this book to my parents. When I started my studies, I often went with them to buy books to prepare for CCIE, and we dreamed about writing one someday. It has been possible, thanks to all their support. I'm also lucky to have the best friends ever, especially Abel, Oscar, and Jaci. We shared so many long study nights and some unforgettable memories. And my wife, Isabel, who has supported me in every (usually crazy) idea I have. And, my son, Marco, and daughter, Scarlett. Without them, the book would have probably been published earlier, but the world would be less awesome.

—Francisco Sedano Crippa

I would like to dedicate this book to my family, who have supported me on every step of this long journey: Caroline, my wife, and Maxime, Emeline, and Capucine, my children who accepted I spent time writing instead of playing with them. I would like to particularly thank all the people in Cisco Engineering who love the product they work on and are always happy to provide answers, help, and confirmation about certain behaviors and are always eager to hear feedback about how we could improve the product. Giving names would mean I probably would forget some, so I prefer to thank them privately myself and collectively here.

—Nicolas Darchis

I dedicate this book to my family, my parents for inspiring me to follow my dreams, and my husband and kids for their unwavering love and support, without which I could not have fulfilled my goals or been part of this book. I also want to thank my coauthors for their understanding and encouragement throughout this undertaking.

—Sudha Katgeri

Acknowledgments

Special thanks to our awesome technical reviewers: Shobhit and Flavio Correa. They both helped us tremendously throughout this book journey not only with their corrections and attention to details but also with their suggestions on the content, identifying missing use cases and hence improving the quality of the book.

We would like to thank our management leadership in Cisco for supporting us during the lengthy process of writing a book. Thank you to the EMEA and US CX leadership: Matthew Batson, Kathy Ferguson, Wes Moss; and EN Product Management leadership: Greg Dorai, Chandan Mehndiratta, and Muhammad Imman.

This book couldn't have been possible without the support of many people on the Cisco Press team. Thanks to Nancy Davis, executive editor; her enthusiasm and dedication were instrumental in getting the book done. Eleanor Bru, development editor, did an amazing job in the technical review cycle, and it has been an absolute pleasure working with you. Also, many thanks to the numerous unknown soldiers at Cisco Press working behind the scenes to make this book happen.

Contents at a Glance

Contents

Reader Services

Register your copy at www.ciscopress.com/title/ISBN for convenient access to downloads, updates, and corrections as they become available. To start the registration process, go to www.ciscopress.com/register and log in or create an account*. Enter the product ISBN 9780137492329 and click Submit. When the process is complete, you will find any available bonus content under Registered Products.

*Be sure to check the box that you would like to hear from us to receive exclusive discounts on future editions of this product.

Icons Used in This Book

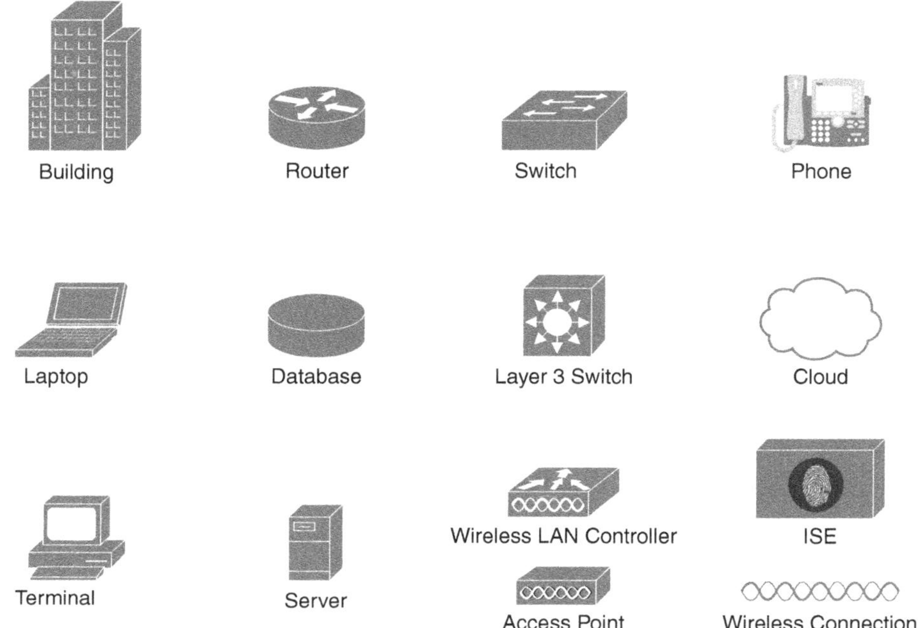

Command Syntax Conventions

The conventions used to present command syntax in this book are the same conventions used in Cisco's Command Reference. The Command Reference describes these conventions as follows:

- **Boldface** indicates commands and keywords that are entered literally as shown. In actual configuration examples and output (not general command syntax), boldface indicates commands that are manually input by the user (such as a **show** command).
- *Italics* indicate arguments for which you supply actual values.
- Vertical bars (|) separate alternative, mutually exclusive elements.
- Square brackets [ ] indicate optional elements.
- Braces { } indicate a required choice.
- Braces within brackets [{ }] indicate a required choice within an optional element.

Introduction

Wireless networks are continuously evolving: the technology is changing at a very fast pace from Wi-Fi 4/5 to Wi-Fi 6/6E and to Wi-Fi 7, which is not too far away on the horizon. The applications and the services enabled on the Wi-Fi network are evolving as well, becoming more and more business critical with stringent requirements on performance and latency. The number of devices that needs to be supported is increasing, and high-density deployments are becoming the norm.

To meet these fast-changing requirements, you need a wireless network that can quickly adapt, that is secure and can perform and scale to higher standards, and that can be easily and programmatically managed. This is the reason behind the introduction of the next-generation wireless controller, the Catalyst 9800 (C9800), which is a critical element of the new Catalyst wireless stack.

The C9800 is based on a new modern and secure operating system, Cisco IOS-XE, and it's built from the ground up for Cisco intent-based networks to deliver on the next wave of wireless innovations.

Goals and Methods

The goal of this book is to educate wireless professionals on how to design, deploy, and manage a Catalyst wireless network built using the new Catalyst 9800 wireless LAN controller, providing practical tips and recommendations.

Who Should Read This Book?

This book is for all readers who are passionate about wireless and want to learn how to design, deploy, and manage a Cisco Catalyst wireless network. The book has been written with three groups of readers in mind:

Technical Decision Makers (TDM): The book familiarizes TDMs with a deep technical view of the Catalyst 9800 wireless controller and helps you in addressing your specific business needs.

Network architects: The book provides technical details on the software and hardware architecture of the Catalyst wireless solution so that you can better understand how to design your own network and implement the features and functionalities you need.

Network operators and IT professionals: The book provides useful tools and tips for setting up, configuring, and monitoring and troubleshooting the network, making your work as a network operator easier.

How This Book Is Organized

The book layout follows the lifecycle of the design and deployment of a Catalyst wireless network using a C9800 wireless controller and hence is divided into three parts:

1. Designing and bringing up the network: Day Zero
2. Configuring and deploying: Day One
3. Monitoring and troubleshooting: Day Two/Day N

Most of the theory is laid out in advance so you can understand the concepts before you see the practical aspects of each topic. We recommend you go through the chapters sequentially to get the full benefit of the book.

Book Structure

In part one, you find the following chapters:

Chapter 1, Cisco C9800 Series: This chapter introduces the Catalyst 9800 wireless controller, its benefits, and its main characteristics. The chapter clarifies the reasons for a new wireless controller for intent-based networking and the role it plays in a cloud-based management network. The chapter also describes the flexible options to manage the C9800, the licensing model, and how this new wireless controller fits into the next-generation wireless stack.

Chapter 2, Hardware and Software Architecture of the C9800: This chapter covers the software architecture of the Catalyst 9800, the reasons behind using a unified IOS-XE for all Cisco enterprise platforms, and the benefits of a scalable architecture.

Part two has the following chapters:

Chapter 3, C9800 Configuration Model: This chapter introduces the new configuration model available for the Catalyst 9800 wireless controller. Important design considerations and best practices are illustrated to get the best performance and stability out of your C9800-based wireless network.

Chapter 4, C9800 Deployment and Installation: This chapter illustrates the different deployment modes supported for the private and public cloud and describes how to bring up your C9800 to an operational state. It then suggests different methods to easily configure your first WLAN on a Catalyst wireless network.

Chapter 5, Security: This chapter covers all the security aspects of the Catalyst 9800 controller, focusing on AAA operations, the use of access control lists (ACLs) to restrict client traffic or to protect the controller management plane, and rogue detection and the Wireless Intrusion Prevention System (WIPS).

Chapter 6, Mobility and Client Roaming: This chapter explains in detail the concept of seamless client roaming in an enterprise deployment. It also describes optimizations such as Fast Secure Roaming, 802.11k, 802.11v, and 802.11r-FT. Deploying mobility and related roaming optimizations on the C9800, and in co-existence with existing AireOS WLCs, is key to enabling a successful adoption of the C9800.

Chapter 7, RF Deployment and Guidelines: This chapter covers basic antenna concepts and radio resource management (RRM) features and functionalities and provides the tuning and recommendation details you need to have your wireless network running like clockwork.

Chapter 8, Multicast and Multicast Domain Name System (mDNS): This chapter describes the configuration and optimizations available on the Catalyst 9800 wireless controller to deliver broadcast, multicast, and mDNS traffic effectively and efficiently.

Chapter 9, Quality of Service (QoS): The C9800 is based on IOS-XE and leverages the Cisco Modular QoS CLI (MQC) to define and implement an end-to-end architecture for QoS. The chapter describes the design and deployment of the C9800 "trust DSCP" QoS model and provides best practices.

Chapter 10, C9800 High Availability: This chapter describes high availability capabilities of the Catalyst 9800 wireless controller. The C9800 provides stateful switchover (SSO) to allow subsecond failover for access points (APs) and clients. The chapter delves into the design and deployment of the network to leverage redundancy features for the C9800.

Part three features the following chapters:

Chapter 11, Cisco DNA Spaces Integration and IoT: This chapter covers at a high level the value-added services that the wireless network can provide, mainly via Cisco DNA Spaces. The chapter explains the integration of DNAS with the Catalyst 9800 and APs through the Cisco DNA Spaces connector, the protocols involved, and the solutions offered.

Chapter 12, Network Programmability: This chapter describes network programmability concepts and the protocols used in modern network programmability, such as NETCONF and RESTCONF. It describes the YANG data models used in the Cisco 9800. It also presents examples illustrating how Python can be used to query the Cisco 9800 operational model and how to create site tags using RESTCONF.

Chapter 13, Model-Driven Telemetry: This chapter describes how you can benefit from model-driven telemetry on the Catalyst 9800 wireless controller and how to integrate the rich data models available with open-source tools.

Chapter 14, Cisco DNA Center/Assurance Integration: This chapter gives an overview of how Cisco DNA Center Assurance works and how it can help you in troubleshooting your Catalyst wireless networks. The chapter focuses on the Catalyst 9800 integration with Cisco DNA Center.

Chapter 15, Backing Up, Restoring, and Upgrading Your C9800: This chapter covers different operational aspects of the C9800, including how to save the configuration on the controller, how to back it up to an external location, and how to restore the configurations in case of problems on the device or in case of complete device replacement.

Chapter 16, Troubleshooting: This chapter explains the various monitoring, debugging, tracing, and packet-capturing features that are native to the C9800 and introduces tools that are available on-the-box and offline that help with data collection and analysis to determine the root cause of problem scenarios.

Appendix A, Setting Up a Development Environment: In this extra hands-on appendix, you learn how to install and use the tools needed to work with C9800 programmability. This appendix offers an overview of a modern development environment, including Git and container-based environments.

Chapter 1

Cisco C9800 Series

The Cisco Catalyst 9800 series is the next generation of wireless controllers for the enterprise. Based on the Cisco IOS-XE operating system, the Catalyst 9800 is built from the ground up for intent-based networking to deliver on the next wave of wireless innovations and to address new requirements coming from emerging standards like Wi-Fi 6, Wi-Fi 6E, and Wi-Fi 7 in the near future.

Cisco Catalyst 9800 series wireless controllers integrate over 20 years' worth of Cisco radio frequency (RF) excellence with a modern, scalable, and programmable operating system to create a best-in-class wireless network. Together with Catalyst Access Point, Cisco DNA Center, and Cisco DNA Spaces, it provides the next generation of wireless experience and addresses the enterprise's evolving and growing digitization needs.

Cisco Catalyst 9800 series wireless controllers are feature-rich and enterprise-ready to power your business-critical operations and transform end-customer experiences. The main advantages are as follows:

- **High availability:** In-Service Software Upgrade (ISSU), seamless software upgrades and updates enabled by hot and cold patching, and stateful switchover (SSO) keep your clients and services always on during both planned and unplanned events.
- **Security:** You are able to secure the air, devices, and users with the Cisco Catalyst 9800 series. The wireless infrastructure becomes the strongest first line of defense with Cisco Encrypted Traffic Analytics and software-defined access (SDA). The controllers come with built-in security: secure boot, runtime defenses, image signing, integrity verification, and hardware authenticity.
- **Flexibility:** You can deploy anywhere to enable wireless connectivity everywhere. Whether on-premises, in a public or private cloud, or embedded on a switch or access point, the Cisco Catalyst 9800 has multiple deployment and scale options to best meet your organization's needs.
- **Open and programmable framework:** Built on a modular operating system, Catalyst 9800 controllers feature open and programmable APIs that enable automation of

your Day Zero (Day 0) to Day N network operations. Model-driven streaming telemetry provides deep insights into the health of your network and clients.

Cisco Catalyst 9800 series wireless controllers are available in multiple form factors:

- **Cisco Catalyst 9800-L:** This compact wireless controller appliance is perfect for small- to medium-sized network deployment. Data ports can operate in 1 GE and 10 GE mode, supporting different SFP/SFP+ transceivers and up to 5 Gbps of throughput (up to 10 Gbps with the optional performance license). It scales to 500 access points and 10,000 clients with a performance license.
- **Cisco Catalyst 9800-40:** This one rack unit fixed wireless controller appliance scales from medium to large deployments. Data ports can operate in 1 GE and 10 GE mode, supporting different SFP/SFP+ transceivers and up to 40 Gbps of throughput. It scales to 2000 access points and 32,000 clients.
- **Cisco Catalyst 9800-80:** Wireless controller appliance modular uplinks provide flexible connectivity options supporting 10 GE, 40 GE, and 100 GE QSFP hot-swappable. Fixed data ports can operate in 1 GE and 10 GE mode, supporting different SFP/SFP+ transceivers and up to 80 Gbps of throughput. It scales to 6000 access points and 64,000 clients.
- **Cisco Catalyst 9800-CL for Private Cloud:** This cloud controller with deployment flexibility offers the hypervisor of your choice (KVM, VMware ESXi, Microsoft Hyper-V, or Cisco ENCS). It scales from 1000 to 6000 access points and up to 64,000 clients.
- **Cisco Catalyst 9800-CL for Public Cloud:** This cloud wireless controller is used for the public cloud of your choice (Amazon Web Services, Google, or Azure Cloud) and your Infrastructure-as-a-Service (IaaS) deployments. It scales from 1000 to 6000 access points and up to 64,000 clients. It is available for FlexConnect deployments only.
- **Cisco Catalyst 9800 Embedded Wireless Controllers (EWC) on Catalyst Access Points:** This embedded wireless controller for the Cisco Catalyst 9100 access point can be positioned for small standalone deployments. It scales to 100 access points and 2000 clients. It is available for FlexConnect deployments only.
- **Cisco Catalyst 9800 Embedded Wireless Controllers (EWC) on Catalyst Switches:** This embedded wireless controller is used with the Cisco Catalyst 9000 series switches for software-defined access deployments. It scales up to 200 access points.

Why Cisco C9800?

You might be wondering why Cisco decided to invest in a new generation of wireless local-area network (LAN) controllers, the Catalyst 9800. The AireOS-based controllers have provided an industry-first enterprise class solution for the last 15 years, innovating and driving customers through many iterations of the Wi-Fi standards: from the early days of 802.11n, to 802.11ac, waves 1 and 2, to the most recent 802.11ax standard. So why C9800 and, most importantly, why now? To answer this fundamental question, we need to introduce intent-based networking as the Cisco strategy for the enterprise and understand how Catalyst 9800 fits into it.

Intent-Based Networking (IBN)

Enterprise customers need to deal with a continuously evolving technology landscape. The pace of change is fast and driven primarily by global trends like mobility, Internet of Things (IoT), and cloud adoption. As a result, networks are becoming more complex, need to change frequently, and must respond to new and sophisticated cybersecurity threats.

Customers need a simpler way to design, deploy, and operate their network to be agile in this evolving digital economy. Cisco's answer to these trends is intent-based networking (IBN).

IBN builds on software-defined networking (SDN) by using a network controller that acts as a central control point for network activity. Such controllers are crucial for network abstraction and automation that lets IT treat the network as an integrated whole. Controller-led networks in all domains (including access, WAN, data center, and cloud) collaborate and extend their benefits throughout the enterprise and help make digital transformation a reality.

The controller relies on an intelligent network that can be programmed through open and secure interfaces so that you can express "intent" (what you want the network to do), and this gets translated and instantiated automatically into a network device optimized for configuration. This is IBN automation. On the other side, the intelligent network needs to continuously capture relevant analytics and efficiently stream it, leveraging streaming telemetry, back to the controller to be analyzed, interpreted, and correlated. This is IBN assurance, which allows you to realize a closed-loop system and verify whether the actual intent has been met. The controller for the enterprise is Cisco DNA Center.

As shown in Figure 1-1, the closed-loop system of Cisco DNA Center leverages two important building blocks: the IOS-XE network operating system and Cisco programmable hardware.

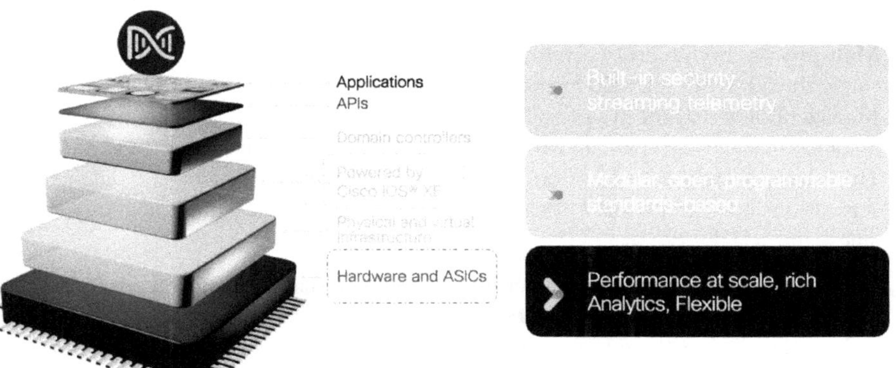

Figure 1-1 *The principles of intent-based networking*

These two pillars of the enterprise architecture allow Cisco to continuously deliver innovation at a solution/application level and bring value to customers. And it's on these two foundation elements that the next generation of Catalyst 9800 wireless LAN controller (WLC) has been built.

Flexible Software

One of the pillars of IBN is Cisco IOS-XE, a modern and secure network operating system embedded in all Catalyst products: wireless, routing, and switching devices. This gives Cisco a great advantage: a network that is not made by silo products but instead built as one network, with a common operating system, a common policy structure, a common automation and assurance framework, with common programmable interfaces. This operating system hugely simplifies how customers consume networking services.

Figure 1-2 lists some of the key characteristic of a modern networking operating system.

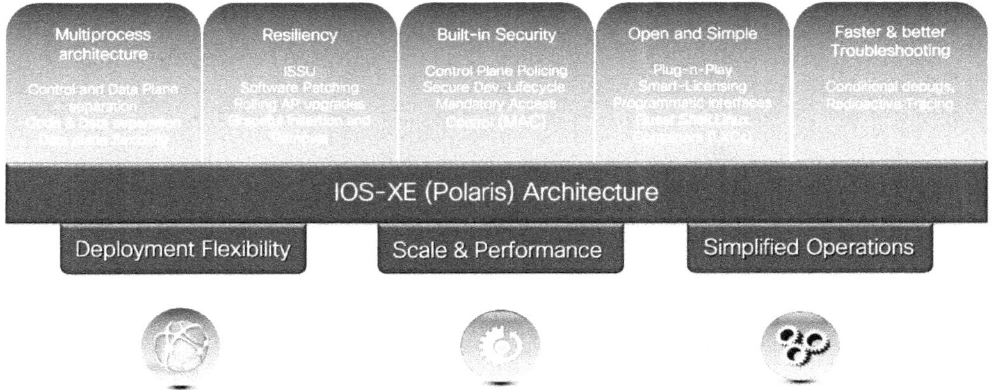

Figure 1-2 *IOS-XE: A modern networking operating system*

Let's look closely at the advantages that Cisco IOS-XE brings to Catalyst 9800:

- **Multiprocess architecture:** Every important function in C9800 is a single process, single thread with separated memory and fault domain: the AP and client Session Manager (known as the Wireless Network Controller Daemon, or WNCd), radio resource management (RRM), Mobility Manager, Rogue Management, and so on. This software architecture not only allows for management, control, and data plane separation and segregation, but also provides a framework for more flexibility, resiliency, and scalability.
- **Resiliency:** A multiprocess architecture and data externalization (both configuration and operational data) allow IOS-XE to provide a much more resilient software architecture. This is the foundation for process restart, process patching, and for In-Service Software Upgrades (ISSUs) and rolling AP upgrades, an RF-based intel-

ligent AP software upgrade mechanism. All of these are important innovations that deliver unprecedented resiliency to your wireless network.

- **Built-in security:** C9800 can also leverage the key security features embedded in IOS-XE, such as control plane policing, secure development lifecycle, and mandatory access control (MAC).
- **Open, simple, and programmable framework:** Features include Plug and Play, zero-touch provisioning (ZTP), Embedded Event Manager, Guest Shell, smart licensing, programmatic interfaces with NETCONF/RESTCONF and Yang models. These are some of the features directly inherited from IOS-XE and now available to Catalyst 9800 customers.
- **Faster and easier troubleshooting:** Conditional debugs and radioactive tracing are two of the main innovations for troubleshooting and serviceability that provide the next level of network insights and deep visibility into users, applications, and device behaviors. When an issue happens and you need to identify the root cause, this information is available "on box," but it's also sent via streaming telemetry to Cisco DNA Center Assurance, which provides a single pane of glass for monitoring your entire network.

As shown in Figure 1-2, all these IOS-XE innovations translate directly into tangible customer advantages: from Day Zero deployment flexibility and ease of use; to Day One scale, performance, and resiliency; to Day Two operation efficiency. Across the entire lifecycle, this network is easy to design, deploy, and operate.

Flexible Hardware

> *People who are really serious about software should make their own hardware.*
> —Alan Kay

This famous quote is from Alan Kay, a computer scientist and pioneer in work on object-oriented programming and graphical user interface (GUI) design. In other words, the intersection of flexible hardware and flexible software really brings value to an organization, especially in the current time of rapid technology and organizational change. Flexible hardware allows organizations to adapt their network over time and adopt new functions, new protocols, and new solutions without compromising on performance (thanks to the hardware acceleration) and without replacing the network hardware. This provides investment protection and maximizes the return on investment (ROI) in their infrastructure.

As illustrated in Figure 1-3, the Catalyst 9800 is built on a programmable application-specific integrated circuit (ASIC), as with the Cisco Quantum Flow Processor (QFP) in the C8000-80 and 9800-40 or the Unified Access Data Plane (UADP) chip in the C9800 embedded in the Catalyst 9000 switches. Flexible silicon in the Catalyst 9800 allows customers to take advantage of the latest networking innovations and at the same time provides industry highest performances in terms of throughput and scale.

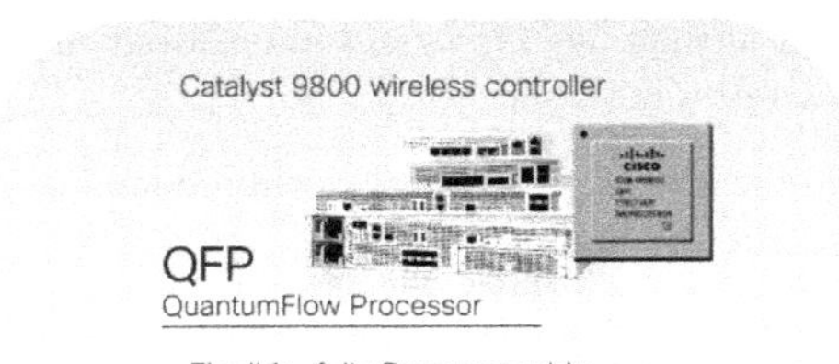
Catalyst 9800 wireless controller
QFP
QuantumFlow Processor
• Flexible, fully Programmable
• Feature-Rich
• Scalable
• Advanced on-chip QoS
• Secure
• Extensible Architecture

C9800 embedded in Catalyst 9000 switches

UADP
Unified Access Data Plane
• Flexible, Programmable
• High-Performance
• Scalable
• Advanced on-chip QoS
• Secure
• Extensible Architecture
Unlocking the power of Catalyst wireless at hardware speeds

Figure 1-3 *Catalyst 9800 is built on programmable silicon*

With Cisco Catalyst Wireless, the advantages of programmable ASIC are not limited to C9800 but are also extended to the new Wi-Fi 6 Catalyst AP, as shown in Figure 1-4.

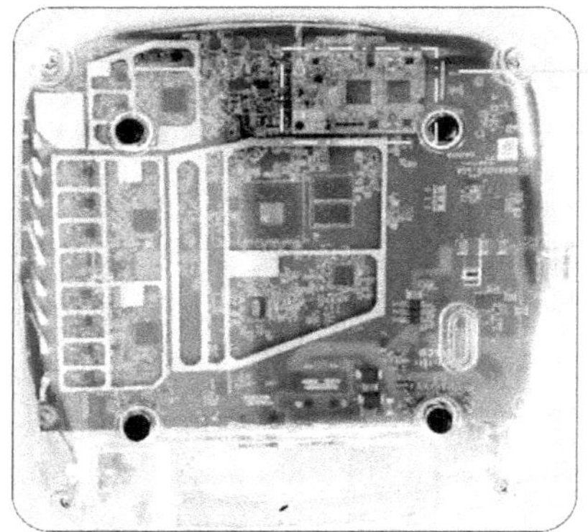

Cisco RF ASIC

Figure 1-4 *Cisco RF ASIC: Software-defined radio using a Mini-PCIe slot*

The Cisco RF ASIC is a software-defined "radio on a module" whose main purpose is to analyze a large range of frequencies and convert the RF baseband to data, which is then analyzed by the CleanAir engine. Capable of extremely high resolution of 78.125 kHz (at least four times better than the nearest competitor), the module is embedded in the Catalyst 9130, 9120, and 9124 access points. It provides Cisco's RF excellence-related features:

- **CleanAir:** Monitors the spectrum for and identifies non–Wi-Fi sources of interference
- **Zero Wait Dynamic Frequency Selection (DFS):** Allows a channel availability check of the DFS channel; allows immediate use without 60s penalty

- **Dual Filter Dynamic Frequency Selection (DFS):** Provides dedicated radar detection to augment the radio vendor detection algorithms; reduces false detection by 99.9 percent
- **FastLocate:** Provides consistent and fast location updates, without requiring dedicated monitor hardware to capture data traffic
- **Off-Channel RRM:** Provides zero client impact off-channel monitoring for RF management, leaving the client serving on radio 100% of the time on-channel availability

As illustrated in Figure 1-5, the 15 years of wireless innovations and RF excellence that customers have enjoyed from AireOS-based wireless controllers are not lost; everything that has been learned has been brought over to the Catalyst 9800, augmented with the advantages you get from a modern, scalable operating system like IOS-XE and combined with Cisco flexible hardware.

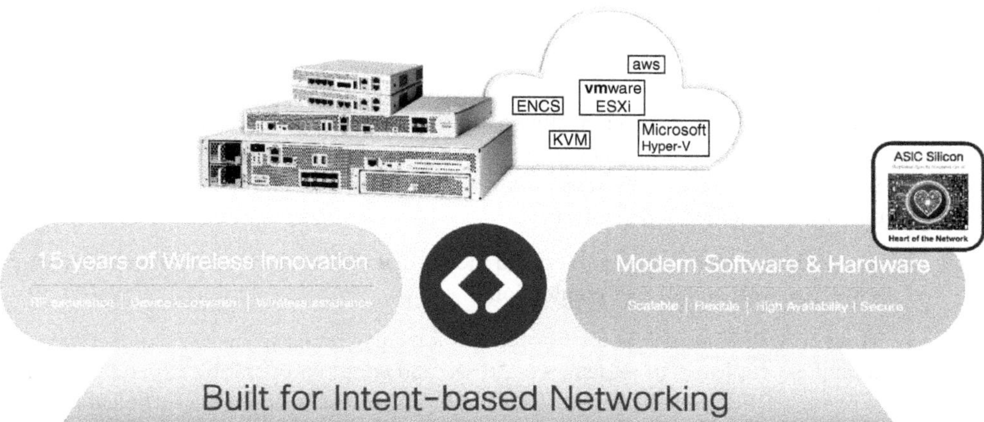

Figure 1-5 *Catalyst 9800 built for intent-based networking*

The Catalyst 9800 has been built from the ground up according to the principles of intent-based networking to deliver on the next wave of wireless standards like Wi-Fi 6, Wi-Fi 6E, and Wi-Fi 7.

The Role of the Wireless Controller in a Cloud Era

The adoption of the cloud is one of the most relevant trends in IT, and networking is no different. Cloud networking is defined as a technology that allows customers to build their networks leveraging cloud-based services. These software and infrastructure services include but are not limited to network management, policy management, and analytics and assurance functions.

The *cloud* here refers to both the public and private cloud. With the public cloud, customers don't own the infrastructure (computing, storage, networking, databases, and so on), and resources are consumed "as a service", usually through a graphical interface or application programming interface (API) provided by the cloud provider. With the private cloud, customers have exclusive access to dedicated virtualized or physical resources. The resources are either located at on-premises data centers or hosted by a colocation provider.

Here are the main reasons for customers to adopt the cloud and an "as a service" model:

- **Scale:** You have the ability to scale up and down on demand when needed without planning for it up front. This flexibility usually helps to optimize resources.
- **Agility:** This capability usually means bringing new features to market more rapidly because cloud services are usually built independently, often relying on microservices architecture.
- **Up-to-date services:** Services are managed directly by the cloud provider, which will keep them updated as new features are added or issues are fixed.
- **High availability:** Cloud services are redundant in nature because they are deployed across multiple data centers or zones, and all are managed by the provider, so customers don't have to worry about backups and redundancy configuration.
- **Simplicity:** Even if not technically related to the cloud, services in the cloud are usually designed around simplicity because they have to be used by multiple customers. Sometimes this simplicity happens at the cost of customization and control for specific functions that may be desired.
- **Accessibility:** Deploying in the cloud means that the services can be reached from anywhere with an Internet connection.
- **Move to OPEX model:** Some IT organizations may prefer to be less exposed on capital expenditures or expenses (CAPEX) and adopt an operating expenditures or expenses (OPEX)–based model. This is a business reason and not a technical one.

There are no doubts that the cloud can bring a lot of benefits to IT and networking, especially in simplifying operations. Does this mean that everything will move to the cloud and the on-premises infrastructure will be commoditized and not relevant anymore? What is the role of the wireless LAN controller in a cloud-managed network?

To answer these questions, we need to first analyze the functions that a WLC has performed traditionally and see how they should evolve to fit in a cloud management era.

Cisco introduced the AireOS-based WLC in 2005 mainly to address the requirements for scale in rapidly growing wireless networks, and since then, the WLC functions have continuously evolved and become richer. Here is a list of the key functions:

- Access point (AP) image management (no more updating each single AP)
- Radio resource management (RRM; defining and implementing an RF channel and transmitting a power plan for the whole network)
- Configuration management
- Network access server (NAS) for client authentication and authorization

- Data plane termination and anchor point for mobility (to allow roaming at scale and across different subnets, or Layer 3 roaming)
- Consistent client mobility database
- Single point of policy definition
- Single point of ingress to the wired network
- Single serviceability and troubleshooting point
- Communication with external services (AAA, DHCP, NetFlow Collector, and so on)
- Telemetry collection, aggregation, and optimization

These functions can be classified as belonging to the management, control, and data planes and were all initially included in the WLC "box," either a physical or virtual appliance.

From a network architecture perspective, different functions or services have different requirements. Management functions (for example, AP image and configuration management, analytics and assurance) and non–real-time control plane services (for example, RRM, rogue management, policy definition) need to be designed for scale, ease of use, and role-based access. On the other hand, real-time control plane functions (for example, client authentication, roaming management) and data plane services (for example, data plane termination, policy enforcement) should be designed around low latency, data consistency, data optimization, and high density.

As a general wireless architecture principle, it is important to understand where these functions may belong: management and non–real-time control plane functions may be centralized either in the cloud or in a central location where it's easier to deploy at scale. Real-time functions are better deployed on-premises, close to the client devices to fulfill the stringent requirements on latency.

In the design of the next-generation Catalyst 9800, these architecture requirements were taken into consideration, which resulted in a platform that is very flexible and can be deployed according to customer requirements for scale and performance and also according to the type of consumption model—an "as a service" or a "do it yourself" on-premises model.

Here are the main characteristics of the Catalyst 9800 that make it a wireless controller built for the cloud era:

- **Network function virtualization:** With C9800, the wireless LAN controller should not be considered a box anymore, but more a collection of wireless network services that can be deployed in different places on the network: in an appliance (physical or virtual) to have hardware-accelerated, ASIC-based, data plane termination that scales to 80 Gbps of throughput and 6000 access points and 64,000 clients; or embedded in a network device like Catalyst 9000 switches, where the control plane runs on the switch and the data plane is distributed to the wired network. Similarly, for the C9800 embedded in the Catalyst AP for small deployments.
- **API-driven:** Every single configuration for the Catalyst 9800 is available through programmatic interfaces and open configuration models (Yang and OpenConfig models).

- **Model-driven telemetry:** Deep analytics information is captured and streamed efficiently and at scale thanks to streaming telemetry protocols like gRPC/gNMI.
- **Deployable "as a Service":** The Catalyst 9800 is an industry-first wireless controller that can be deployed on a public cloud with an Infrastructure-as-a-Service (IaaS) model. The C9800-CL (where *-CL* stands for cloud) allows you to choose the public cloud of choice—Amazon Web Services (AWS), Google Cloud Platform (GCP), or Microsoft Azure—and enjoy the benefits of the cloud for the compute, storage, and networking resources. At the same time, it allows you to retain full control of your wireless controller services' software releases and configuration as if it were deployed on-premises.

Customer adoption of cloud services varies based on the specific requirements and needs. That's why it's important to provide a wireless architecture that is flexible and can adopt to different deployment scenarios. For example, some customers may require high throughput and fast secure roaming across Layer 3 boundaries, in large roaming domains like the ones found in hospitals and in manufacturing plants or large warehouses. They call for real-time controller plane mobility functions and data plane termination functions to be located on-premises close to the user devices to reduce latency. Other customers may need to reduce the data center footprint and move all possible workloads in the public cloud; this would be the case for a retailer with multiple small branches with few access points that do not actually need the WLC functions to be onsite and hence could benefit from AWS, GCP, or Azure installation. The Catalyst 9800 allows you to design for a flexible architecture that best meets your needs.

Throughout the rest of the book, you learn a lot of Catalyst 9800 functions, what they do, and how to configure and troubleshoot. Before that, however, let's focus on how to manage this platform.

Managing the Cisco C9800

The Catalyst 9800 wireless controller has been built with flexibility in mind, and this is also reflected in the different options that you have to manage the platform, as illustrated in Figure 1-6. Let's analyze each option in more detail.

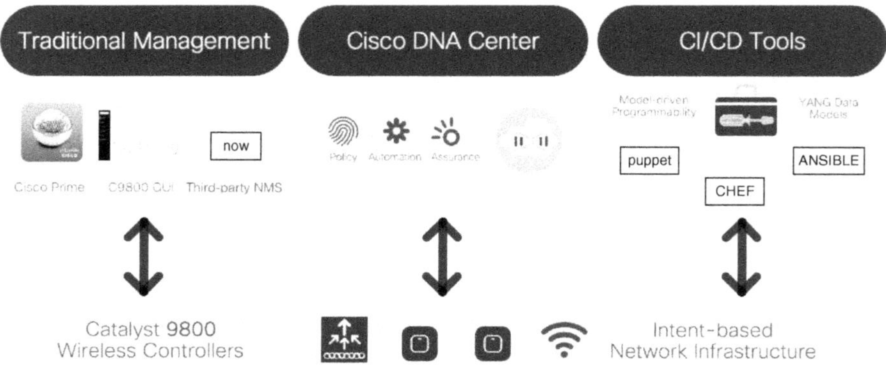

Figure 1-6 *Catalyst 9800 flexible management options*

Traditional Management Tools

You may decide to utilize traditional SNMP-based third-party network management systems (NMSs) to manage the Catalyst 9800. These systems are fully supported, and you can look up the available management information bases (MIBs) in the Cisco Feature Navigator tool and select the desired platform, as illustrated in Figure 1-7.

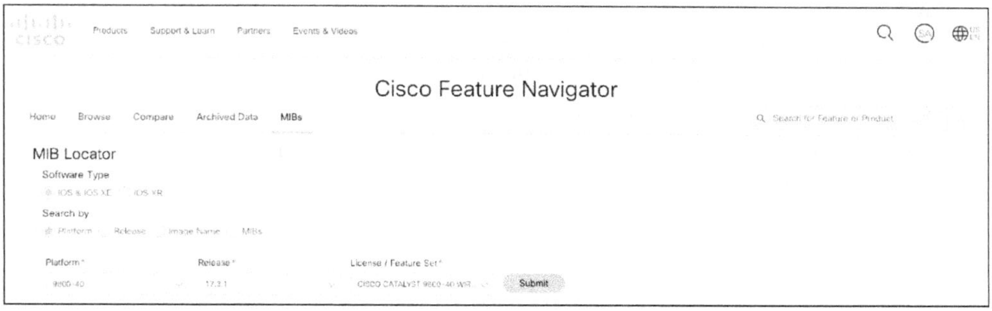

Figure 1-7 *Cisco Feature Navigator*

This method offers only the base functionalities and doesn't really exploit the full capabilities of the Catalyst 9800, when compared with other Cisco traditional management tools like the integrated graphical user interface and Cisco Prime Infrastructure.

"On Box" Management

There are two ways to manage the Catalyst 9800 directly "on box":

- Text-based command-line interface (CLI)
- Web-based graphical user interface or WebUI

If you know the IOS-XE CLI, now the same interface is available on the wireless controller as well. Some people find the CLI easier to use than WebUI because it allows you to simply copy and paste, to complete commands, to set alias commands, and so on. As with Cisco routers and switches, there are three ways to access the CLI: the local console, remote Telnet, and Secure Shell (SSH).

To log in to the box and configure it remotely, you first need to define a local username with the right privileges. The following CLI shows the minimal recommended configuration to create an administrator user (with privilege 15) and enable SSH:

```
username admin privilege 15 secret 0 password
hostname name
ip domain name domain-name
crypto key generate rsa modulus 2048 label label-name
ip ssh rsa keypair-name label-name
line vty 0 4
   transport input ssh
   login local
```

Starting with release 17.4, the Catalyst 9800 provides a startup wizard via the console CLI to take you through a step-by-step flow and configure the basic settings all at one time. This capability is available only for an "out-of-the-box" scenario where the C9800 has never been configured before or after a factory reset.

If you want a more graphical approach to management, Cisco has also invested in the design of a brand-new graphical user interface for the Catalyst 9800, making it intuitive and easy to use. The new GUI is built on modern front-end web technology to make it light and fast. It features key tools for making a configuration intuitive and for troubleshooting and gathering important insights on your network.

If you're new to Catalyst 9800, the Web interface features a few setup tools to help you get started with the configuration of the most common and important settings you might need to use. You have these different options:

- You can use the WLAN Wizard. Setting an SSID is definitely one of the first tasks you will have to do to get your wireless network up and running. From release 17.5 onward, the C9800 GUI provides a step-by-step flow to configure the most common WLAN profiles and its associated policy for different deployment modes (Local, FlexConnect, and so on). Figure 1-8 shows how you can set up a PSK SSID. Check out the CLI preview, which is very useful.

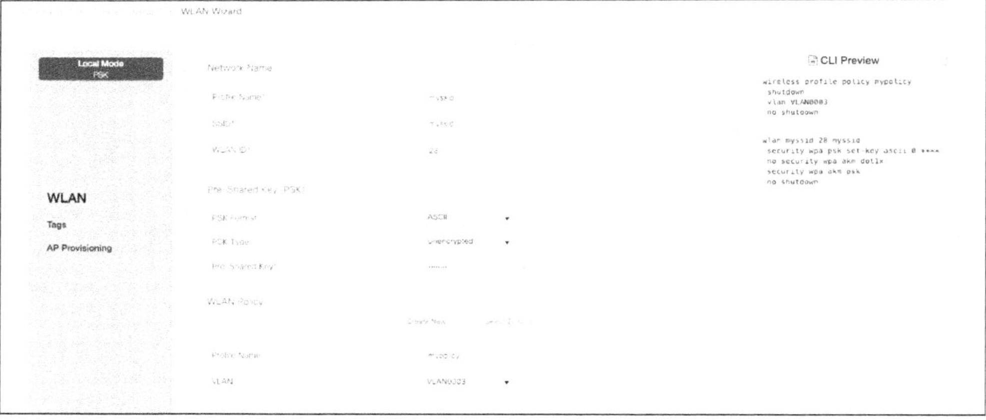

Figure 1-8 *Setting up a PSK SSID with the WLAN Wizard*

- If you need to understand more about the configuration constructs used in C9800, then it's recommended you go through the Advanced Setup Wizard because it introduces you to profiles and tags and takes you through the explicit steps for configuring a wireless network, as illustrated in Figure 1-9. You can find it under **Configuration > Wireless Setup > Advanced.**

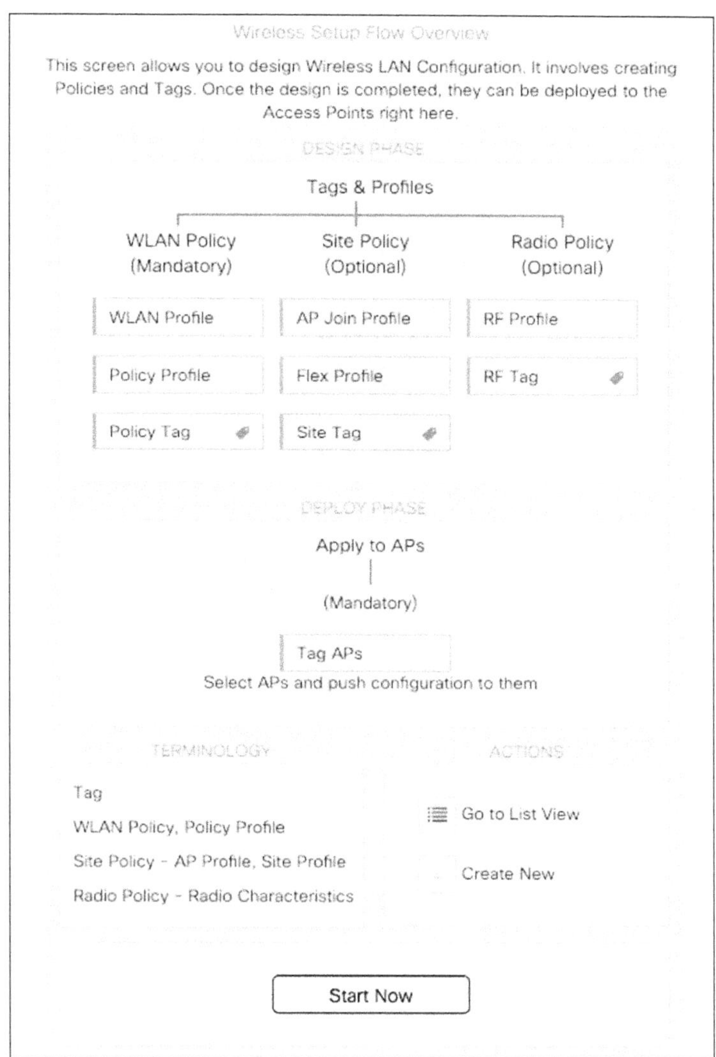

Figure 1-9 *The Advanced Setup Wizard*

- Last but not least, if you have any configuration concerns, you can rely on the Guided Assistance tool. On any page of the Catalyst 9800 GUI, you can click the small banner in the bottom-right corner to launch the tool. At that point, you can type what you're looking for and follow the online instructions, as illustrated in Figure 1-10. This tool also provides an explanation of the different fields you have to configure.

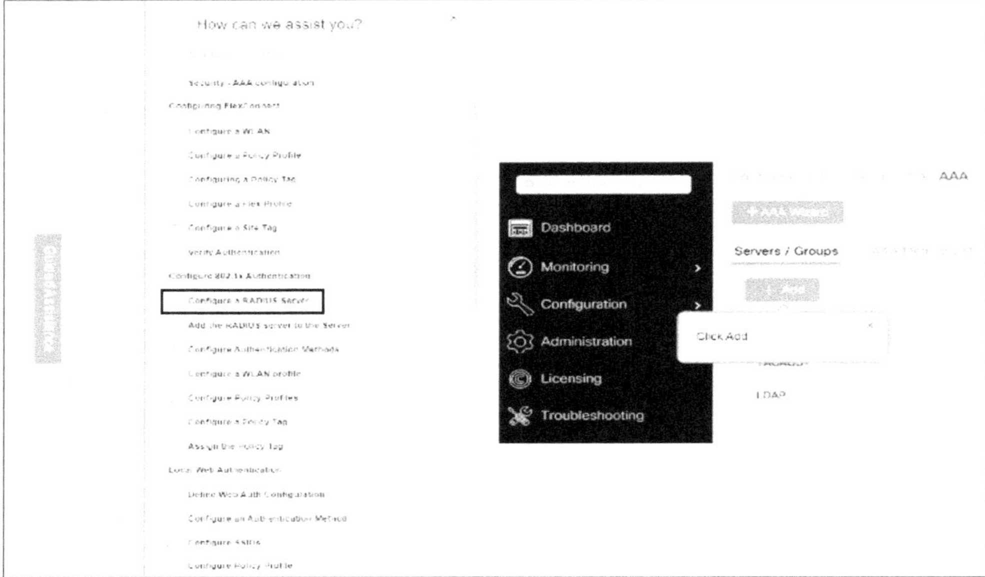

Figure 1-10 *Guided Assistance tool*

If your intent is to dig deeper into an issue that you might have on the network, the Catalyst 9800 WebUI also offers a new and useful troubleshooting tool in the Troubleshooting section of the dashboard. For example, the Packet Capture tool allows you to capture traffic on any interface (wireless management interface, port channel, or physical ports) in an easy and quick way. Just go to the Troubleshooting tab, select **Packet Capture**, and define your capture, as shown in Figure 1-11.

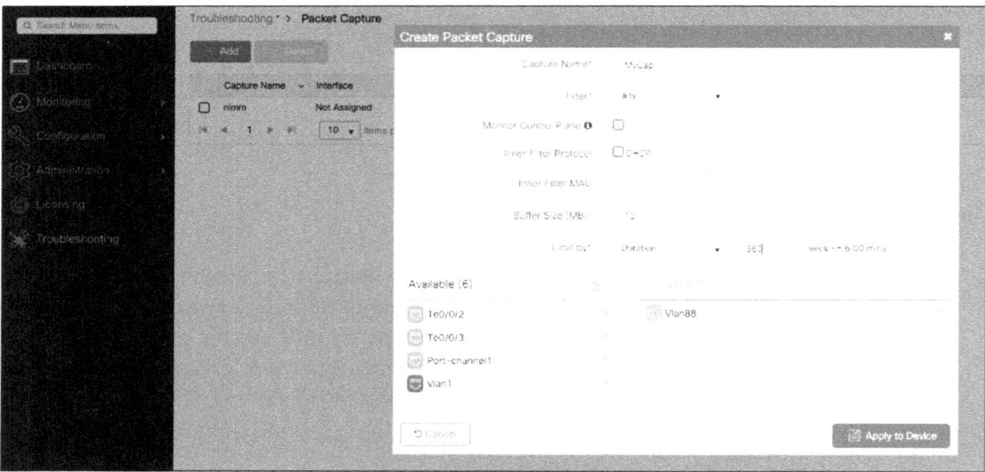

Figure 1-11 *Defining a Packet Capture on C9800's GUI*

Packet Capture on the C9800 allows you to look at the control plane traffic as well, so you can see the messages sent to and from the access point that are DTLS encrypted and hence not visible with a standard Packet Capture on a switch. You can start and stop the capture directly on the GUI and then download the capture file directly on your desktop.

There are some important best practices that you need to keep in mind as you utilize the GUI. The Catalyst 9800's Web GUI leverages Virtual Teletype (VTY) lines for processing HTTP requests. At times, when multiple connections are open, the 15 VTY lines set by the device (the default number) might get exhausted. Therefore, it is strongly recommended that you increase the number of VTY lines to 50. Use the following configuration commands to do this:

```
C9800(config)# line vty 16-50
```

Another useful recommendation is to configure the service tcp-keepalives to monitor the TCP connection to the box. In this case, use the following commands:

```
C9800(config)# service tcp-keepalives-in
C9800(config)# service tcp-keepalives-out
```

Starting with release 17.3, you are able to configure HTTP/HTTPs independently for WebUI access and for portal redirection of client web authentications. For more secure access to the box, it is recommended you disable HTTP for WebUI access. You can go to **Administration > Management > HTTP/HTTPs/NETCONF** and enable it, as shown in Figure 1-12. On the same page, it's also important to explicitly associate a trustpoint to be used for HTTPs connections.

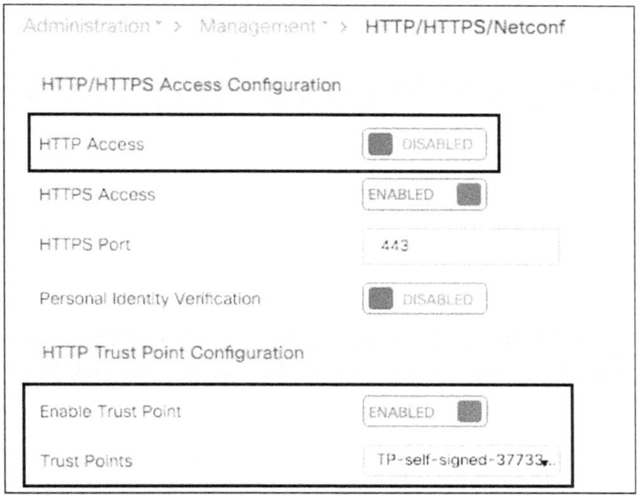

Figure 1-12 *Disabling HTTP web access and selecting a specific trustpoint for HTTPs*

You can find other important recommendations in the Cisco Catalyst 9800 Series Configuration Best Practices online (see the "References" section for details).

Cisco Prime Infrastructure

Cisco Prime Infrastructure (PI) has evolved throughout the years to become a complete management system for enterprise customers, covering the branch to the data center. PI simplifies the management of wireless and wired networks and offers Day 0 and Day 1 provisioning, as well as Day N assurance. Figure 1-13 provides a view of the Prime dashboard.

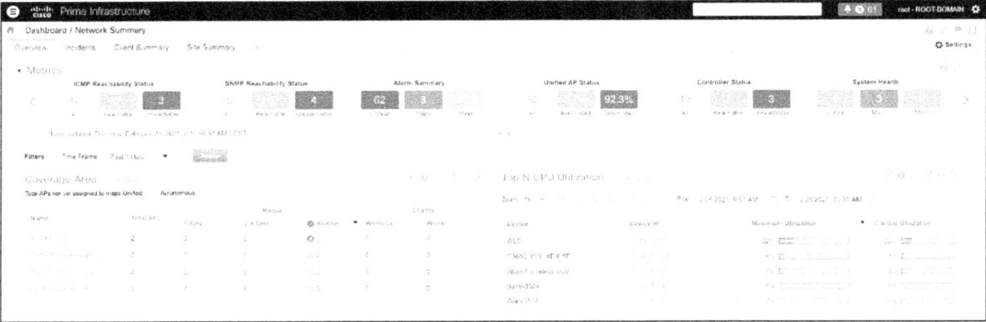

Figure 1-13 *Prime Infrastructure dashboard*

Prime Infrastructure leverages CLI and SNMP information to verify reachability and to add the device to its inventory. It also uses SNMP to push configuration templates as well as support SNMP traps for access points and client events. However, for PI to gather access points and client statistics, NETCONF is leveraged because it's more efficient and allows telemetry to scale better.

For Prime Infrastructure to configure, manage, and monitor the Catalyst 9800, PI needs to be able to access it via SSH, SNMP, and NETCONF. This means that you need to do some preliminary configuration on C9800 itself before PI can manage it. These are the protocols and ports being used in the communication:

- SSH access (recommended over Telnet) is on TCP port 22.
- All configurations are pushed via SNMP templates using UDP port 161 or CLI over SSH/Telnet.
- Operational data for the C9800 box itself is obtained via SNMP UDP port 162.
- AP and client operational data leverages streaming telemetry.

- Prime Infrastructure to C9800 direction: TCP port 830 is used to push telemetry configuration using NETCONF.
- C9800 to Prime Infrastructure direction: TCP port 20830.

Next, you need to follow these steps:

Step 1. Add SNMP configuration: enable SNMP and add community strings for SNMPv2 (a read and write community is required). You can find this on the GUI under **Administration > Management > SNMP**, as shown in Figure 1-14.

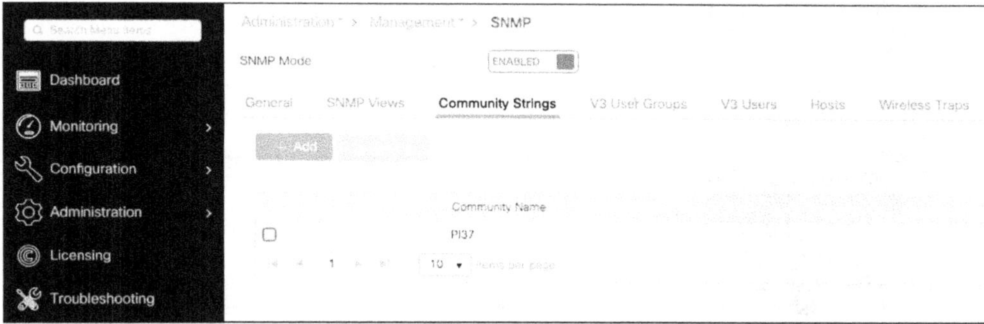

Figure 1-14 *SNMP configuration of the C9800*

You also can configure SNMP via the CLI. Here is an example for SNMPv3 configuration:

```
C9800(config)#snmp-server view view-name iso included
C9800(config)#snmp-server group group-name v3 auth write
view-name
C9800(config)#snmp-server user username group-name v3 auth
{md5 | sha} password priv {3des | aes | des} {optional for
aes 128 | 192| 256} privacy-password
```

Step 2. Enable NETCONF under **Administration > Management > HTTP/HTTPs**, as illustrated in Figure 1-15.

If you already have authentication and authorization settings on the box (aaa new-model is configured), you also need to make sure you specify the repository (local or remote) where the credentials are validated, as shown in the following commands:

```
C9800(config)#aaa authorization exec default local|radius|
tacacs group
C9800(config)#aaa authentication login default local|radius|
tacacs group
```

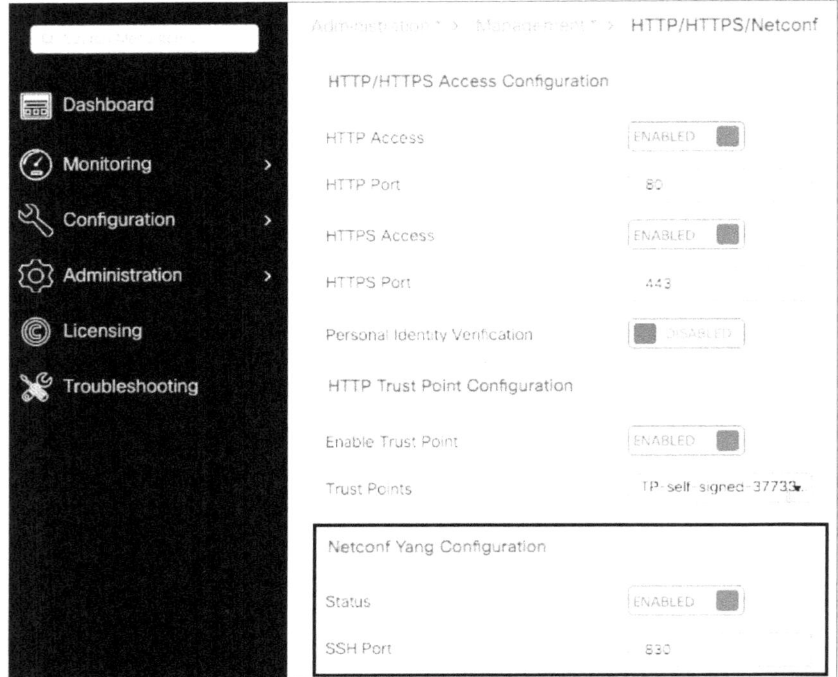

Figure 1-15 *Enabling NETCONF on the Catalyst 9800 GUI*

NETCONF on C9800, by design, leverages the aaa default method (and it cannot be changed) for both authentication and authorization. In case you want to define a different method for SSH connections, you can do so under the "**line vty**" command line. But NETCONF keeps using the default exec authorization method.

Step 3. Grab the Wireless Management Interface (WMI) IP address and the administrator password that you have already configured on the box.

One thing to be careful of when managing the C9800 wireless controller with Prime: when adding to inventory, Prime overwrites the **aaa authentication login default** and **aaa authorization exec default** methods if already configured on the box to point them to the local database. This means if you are using TACACS, you lose CLI access after adding the C9800 to Prime. Keep this in mind and revert those aaa configurations back and make them point to TACACS if that is your preference.

At this point, you are ready to go to Prime Infrastructure, add the C9800 using the WMI interface as the reference IP address, and start managing it. This discussion is beyond the scope of this book, but you can check the configuration guide as a starting point.

Cisco DNA Center

Cisco DNA Center is the enterprise orchestrator and network management platform for the Cisco intent-based network, as illustrated in Figure 1-16.

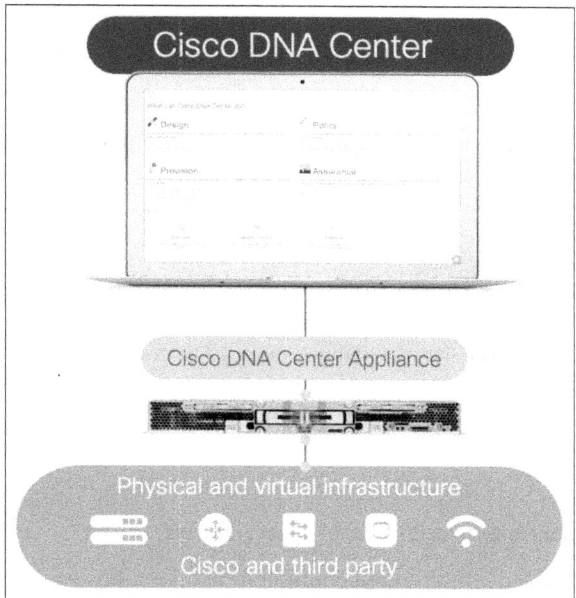

Figure 1-16 *Cisco DNA Center*

DNA Center represents a new and radically different approach to network management; it provides an intuitive single pane of glass for your entire network and embeds automation, analytics, and assurance into one single platform. DNA Center helps you solve IT challenges by

- **Simplifying operations:** DNA Automation allows you to onboard new network devices in minutes, without onsite support visits. This is key when you have thousands of access points and hundreds of switches to deploy. You can also automate additions, changes, policy provisioning, and software upgrades and make sure your network is compliant.
- **Optimizing the user experience:** DNA Assurance provides a single dashboard to monitor and troubleshoot your networks. It leverages the analytics received from the network and uses advanced artificial intelligence (AI) and machine learning (ML) to proactively monitor, troubleshoot, and optimize your network. For wireless networks, this capability is particularly important because a lot of information can be gathered from wireless clients and access points.

- **Embedding security everywhere:** A unified policy for wired and wireless and network segmentation is easily automated with Cisco software-defined access (SDA). Thanks to the integration with Cisco Identity Services Engine (ISE), Cisco Umbrella, and Cisco Secure Network Analytics (formerly Cisco Stealthwatch), Cisco DNA Center ensures comprehensive security and trusted access everywhere on your network. DNA Policy helps you create policies to scale and control access, route traffic, and prioritize applications, and enforce them consistently across your campus, WAN, and multicloud data center network domains.

A full description of Cisco DNA Center and how to manage the Catalyst 9800 is beyond the scope of this book, but it's important to clarify here what is needed before C9800 can be successfully discovered and managed.

C9800 Prerequisites for Cisco DNA Center

The first step to manage the C9800 with DNA Center is to add it to the inventory. You can do this via the Discovery tool. Here is the list of settings you need to configure on the C9800 before successfully discovering it:

- Configure an admin user with privilege 15.
- Enable SSH on the box.
- Configure a Wireless Management Interface and a route to reach DNA Center.
- Make sure that NETCONF is enabled. The required configuration commands for NETCONF and AAA authorization are

  ```
  c9800(config)#netconf-yang
  c9800(config)#aaa new-model
  c9800(config)#aaa authorization exec default local
  ```

  If you're using the AAA server to authenticate the user credentials, make sure that the NETCONF user returned from AAA is defined with privilege 15.
- For discovery, it is recommended you use an IP range with just one IP address (the one for the Wireless Management Interface of the C9800). Using the WMI is preferred and recommended over the service port (SP).
- If you are discovering a C9800 stateful switchover pair, make sure you don't use the Redundancy Manager Interface.

It's also important to keep in mind that DNA Center pushes its own self-signed certificate to the managed devices; the default certificate is **sdn-network-infra-iwan**. When the Catalyst 9800 has more than one certificate configured on the box (for example the self-generated trustpoint and the one pushed by DNA Center), it is strongly recommended you specify the certificate to be used for HTTPs access to the device. Not doing so may result in the Catalyst 9800 picking the wrong one and breaking access to the graphical user interface. In this case, use the following CLI command:

```
c9800(config)#ip http secure-trustpoint trustpoint-name
```

Alternatively, you can configure this in the GUI by going to the **Administration > Management > HTTP/HTTPS/Netconf** page and then selecting the specific certificate in the HTTP Trust Point Configuration section.

CI/CD Tools

Continuous integration and continuous development (CI/CD) are fundamental DevOps best practices for software development that emphasize automation to build and test software, focusing on achieving a software-defined lifecycle.

The Catalyst 9800 wireless controller is built on IOS-XE software, which provides open, standard-based programmable interfaces that support the integration of various CI/CD tools available in the market or allow you, as a network admin, to build your own.

Chapter 12, "Network Programmability," provides a technical description of the data models, protocols, and interfaces that allow C9800 to be programmatically managed.

Licensing

Cisco Smart Software Licensing makes it easier to buy, deploy, track, and renew Cisco software licenses for your Catalyst 9800. It allows you to create a pool of license resources that can be shared across multiple C9800 wireless controllers by removing older device-level entitlement and enforcement.

No licenses are required to boot up a C9800 wireless controller. However, each access point requires two licenses to be entitled to connect: one AIR Network License and one AIR DNA License. Both of these licenses can be configured to be either Essential or Advantage level. If there are not sufficient Cisco DNA licenses to cover all the access points connected to a Cisco Catalyst controller, an out-of-compliance message is displayed. This out-of-compliance message is purely informational and does not impact the functionality of the wireless deployment.

Starting with software release 17.3.2a, C9800 supports the Smart Licensing Using Policy, which is an enhanced version of Smart Licensing, with the overarching objective to further simplify the licensing solution:

- **Seamless Day 0 operations:** After a license is ordered, no preliminary steps, such as registration or generation of keys, are required unless you use an export-controlled or enforced license. There are no export-controlled or enforced licenses on Cisco Catalyst wireless controllers, and features can be configured on the device right away. There are no more evaluation licenses either.
- **Consistency in Cisco IOS-XE:** Campus and industrial Ethernet switching, routing, and wireless devices that run Cisco IOS-XE software have a uniform licensing experience.
- **Visibility and manageability:** Tools, telemetry, and product tagging are on the customer Smart Licensing Portal.
- **Flexible time-series reporting to remain compliant:** Easy reporting options are available, whether you are directly or indirectly connected to Cisco Smart Software Manager (CSSM), or in an air-gapped network.

For more information about licensing for the C9800, take a look at the frequently asked questions link found in the "References" section.

Cisco Next-Generation Wireless Stack

The Catalyst 9800 wireless controller is a key component of the next-generation wireless stack, as shown in Figure 1-17. Together with Cisco Catalyst Access Points, Cisco DNA Center, and DNA Spaces, it provides a complete end-to-end wireless solution for companies of all sizes, to manage the growing number of connected wireless devices and fulfill the evolving and growing digitization needs.

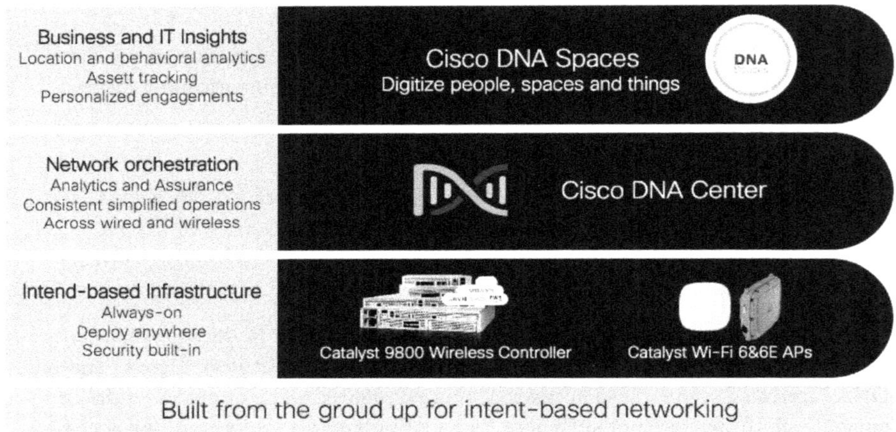

Figure 1-17 *Cisco next-generation wireless stack*

The Cisco next-generation wireless stack introduces wireless innovations at each and every layer, as illustrated in Figure 1-18.

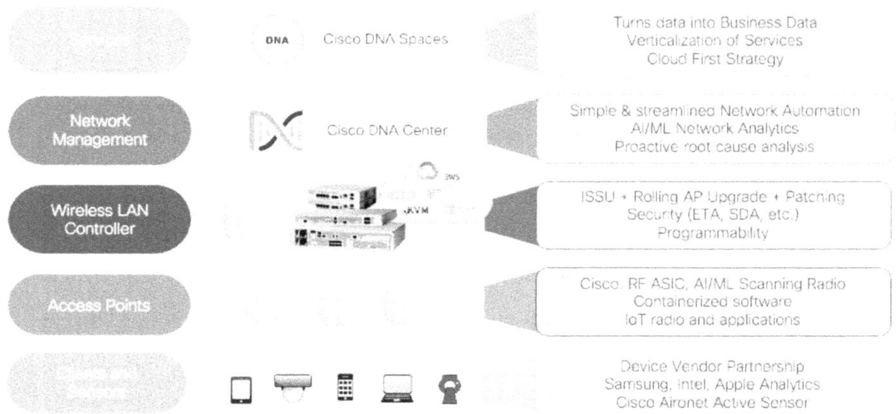

Figure 1-18 *Wireless innovation at each later of the stack*

Wireless innovations start at the device level with key partnerships with vendors like Apple, Samsung, and Intel; they allow the Cisco infrastructure to gain additional insights (iOS, Samsung, and Intel analytics) on device behavior, and provide additional quality of experience (for example, Apple Fastlane and Apple Fastlane+) so that these devices behave better on a Cisco wireless network. The Catalyst access points—thanks to Cisco RF ASIC, the IoT radio, and the IOx framework—go beyond the latest Wi-Fi 6 (802.11ax) standard and are ready for growing user expectations, IoT, and next-generation cloud-driven applications.

You previously learned about the advantages and innovations introduced by the Catalyst 9800 and Cisco DNA Center, so the last element is Cisco DNA Spaces. Whether it's learning more about visitors to your organization, your employees, or your things, such as assets and sensors, Cisco DNA Spaces digitizes your physical space. It does so by synthesizing location data across your sites to deliver location-based services at scale. This information can be used to enhance the customer experience, improve business operations and efficiencies (and reduce costs), realize industry-specific business outcomes, and much more.

Cisco DNA Spaces is a cloud-based platform for location: it leverages information from Wi-Fi to BLE tags, beacons, and other IoT sensors, and with gateway-enabled Cisco Wi-Fi 6 access points, it can easily scale advanced use cases while lowering total cost of ownership (TCO).

Summary

The Cisco Catalyst 9800 Series is the next generation of wireless LAN controllers from Cisco. It combines radio frequency (RF) excellence, gained over 20 years of leading the wireless industry, with Cisco IOS-XE software, a modern, modular, scalable, and secure operating system. This chapter introduces the Catalyst 9800, its benefits, and main characteristics, and clarifies the reasons for a new wireless controller for intent-based networking and the role it plays in a cloud-based management network. The chapter also describes the flexible options to manage the C9800, the licensing model, and how this new wireless controller fits into the next-generation wireless stack.

References

Feature Navigator tool: https://cfnng.cisco.com/mibs

Cisco Catalyst 9800 Series Configuration Best Practices: https://www.cisco.com/c/en/us/products/collateral/wireless/catalyst-9800-series-wireless-controllers/guide-c07-743627.html

Prime Infrastructure configuration guide: https://www.cisco.com/c/en/us/support/cloud-systems-management/prime-infrastructure/products-installation-and-configuration-guides-list.html

Licensing: https://cisco.com/go/smartlicensing

https://www.cisco.com/c/en/us/td/docs/wireless/controller/9800/17-3/config-guide/b_wl_17_3_cg/m-sl-using-policy.html

C9800 frequently asked questions: https://www.cisco.com/c/en/us/products/collateral/wireless/catalyst-9800-series-wireless-controllers/nb-06-cat9800-ser-wirel-faq-ctp-en.html

Cisco IOx framework in the Cisco IOS-XE and Linux OS integration: https://www.cisco.com/c/en/us/products/cloud-systems-management/iox/index.html

Chapter 2

Hardware and Software Architecture of the C9800

Hardware is often overlooked by software companies and vice versa. However, computer scientist Alan Kay said, "People who are really serious about making software should make their own hardware." This is a principle dear to Cisco as well to other successful companies, such as Apple, that can leverage unique software capabilities thanks to their hardware. This chapter does not simply list numbers, cite specifications, and throw complex terminology at you but explains the important ideas that are the foundation of the Catalyst 9800 hardware and software architecture and its unique strengths. Let's start with a quick introduction to the overall controller protocols architecture, which hasn't changed much conceptually from the days of Cisco Unified Wireless.

General CAPWAP Split MAC Architecture

To this day, the access point (AP) software is still referenced on cisco.com as lightweight, and this makes a direct reference to the split from MAC architecture explained later in this section. There is no more "fat" (that's really old naming) or "autonomous" software for wireless access points anymore (for 11ac Wave 2 APs and later), and they are always expected to join a wireless LAN controller (WLC) to operate. However, there are other modes of operation, such as Embedded Wireless Controller on AP (for Catalyst APs), where the AP acts both as an AP and a WLC (so it is autonomous in some way, but it still runs the lightweight software on the AP side of things), or the Workgroup Bridge mode, where the AP acts as a client to the infrastructure.

CAPWAP stands for Control and Provisioning of Wireless Access Points and is an IETF standard protocol. It builds two tunnels between an access point and the WLC: one for control and one for data. The control channel uses UDP 5246 port, and the data channel uses UDP 5247 on the WLC side (it's the destination port when the AP sends traffic to the WLC, and the source port when the WLC sends traffic to the APs), in IPv4 or IPv6 (UDP Lite is used in IPv6 to save on processing of the checksum). APs use a random source port to send CAPWAP traffic to the WLC, and the WLC uses that same port to reply to APs.

The control protocol allows the WLC to centrally manage and configure all the access points and make sure they are always on the same software versions that correspond to the WLC version. It leverages Datagram Transport Layer Security (DTLS) for security and uses encryption of every packet.

The data protocol is an optionally encrypted way of tunneling back all the client data (including the initial onboarding and authentication, which are considered to be control-plane traffic from a controller handling perspective) to the controller to simplify the topology. Benefits of tunneling the traffic back to the controller are

- You don't need to span many VLANs to all the wired infrastructure hosting the APs. APs are connected on access ports in a management VLAN (typically dedicated to the APs).
- The WLC can host the client policies such as QoS or ACLs.
- The WLC is a central and unique point of contact for RADIUS and AAA authentication.
- Wireless clients' MAC and IP point of presence is behind the WLC data port from a wired network standpoint, which simplifies roaming by limiting MAC move events and lookups.

The split MAC architecture gets its name from the fact that the MAC layer responsibilities are split between the WLC and APs, as shown in Table 2-1.

Table 2-1 *Split MAC Architecture Responsibilities*

Access Point MAC Functions	Controller MAC Functions
802.11 beacons and probe responses (although probes are forwarded to the WLC as well)	802.11 associations requests, management and action frames
802.11 frame transmission and acknowledgments (including client power save handling and buffering)	802.11 QoS resource reservation
802.11 QoS frame queuing and packet prioritization	Client authentication in general
802.11 MAC layer data encryption and decryption	Client data traffic forwarding
Monitoring RF environment and scanning other channels	

Basically, all this boils down to a latency concern. It wouldn't be practical and efficient to send every frame to the controller to be decrypted, nor can the controller effectively manage the AP airtime and transmissions on the medium in real time. For everything else where there are benefits in centralizing the task on the controller and if latency is less of a concern, the task is best handled by the WLC.

This description also explains why there is a FlexConnect mode for APs in remote branches where the latency might be a bit higher and where you can optionally move some of the MAC functions back to the AP (like client authentication or data traffic forwarding).

DTLS is an IETF protocol based on TLS but targeted for use on UDP. All Cisco APs and appliance controllers are shipped with a manufacturer installed certificate (MIC), which is used to establish the CAPWAP DTLS tunnel and build the control tunnel encryption key generation and mutual authentication. A locally significant certificate can also be generated and used for this purpose. CAPWAP control encryption is always used after the DTLS handshake occurs, but CAPWAP data packet encryption is an optional setting (automatically turned on in OfficeExtend mode). DTLS data encryption can have a performance impact on the global throughput numbers forwarded by the WLC, so enabling it when the data is transported over unsecured networks is advised. The data over the air is encrypted with the L2 security defined in the WLAN, if any (for example, WPA2-AES), and is always decrypted at the AP. At that stage, the AP encapsulates the wireless client frame in a UDP CAPWAP packet to send to the WLC. This client data could be perfectly readable and decodable in Wireshark if someone has access to the AP management network and collects a sniffer capture there. With CAPWAP data packet DTLS encryption, this CAPWAP payload would be encrypted, and the only thing someone capturing traffic would see is an encrypted CAPWAP data packet sent from a specific AP to the WLC without any clue with regards to what it contains.

The Controller Control Plane Architecture Elasticity

In the early codenames of the Catalyst 9800 during its inception, the word *elastic* was very present. Understanding why is the first step to understanding the whole architecture and its benefits.

IOS-XE Software Architecture

IOS-XE is based on a UNIX system named binOS internally (or Polaris as of its 16 and 17.x versions) and is a Cisco-modified version of UNIX. IOS (the now legacy one) was a monolithic operating system, a single process with a single memory space and fault domain. IOS-XE moves away from this architecture by adopting a multiprocess, modular, and scalable approach, separating the operating system (binOS) from the network tasks that are now managed by a process called IOSd. IOSd still takes care of the routing and interface configuration, but more specific tasks (like wireless tasks) are separated into dedicated processes. The management and config replication (in case of high availability) is also separated from the IOSd process. Figure 2-1 depicts the various processes in the IOS-XE architecture. Each of them is covered later in this chapter.

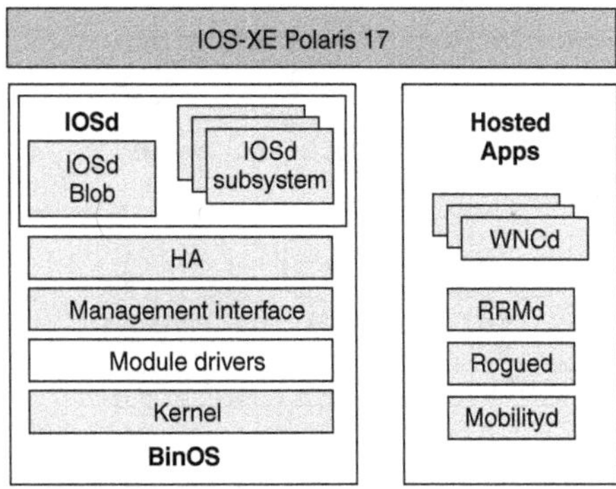

Figure 2-1 *IOS-XE general software architecture*

WNCd: The Heart of the Wireless Controller Control Plane

The Catalyst 9800 was designed with single-threaded event-driven processes in mind. The idea is to have as many processes as possible for the task while keeping those processes as independent as possible for session management purposes. Of course, processes can communicate with each other and they do, but vital activities around a specific given task are best handled by a unique process to avoid synchronization locks and wait times. There is no dynamic spawning of processes; they are statically spawned depending on the appliance hardware specifications. Similarly, multiple databases duplicate information for contention-free access: each process has its own database to which it has exclusive, contention-free access. These databases are duplicated into a central, consolidated database used for manageability and high availability replication. Thus, manageability does not impact session management and vice versa.

The databases are based on a versioned data model. This is a huge step forward compared to previous architectures, and this change (among a few others) is at the source of the programmability and the patchability of the problem. In Service Software Upgrade (ISSU), the new hitless upgrade system where you can upgrade your WLC high availability pair without any noticeable downtime in the network, is allowed by the fact that the WLC is based on modeled and versioned data. Having separate processes with their own databases and only minimal synchronization between processes and databases compared to the previous architecture, where all the threads were accessing the same central databases, avoids mutual exclusions and locks that prevent further scalability.

As depicted in Figure 2-2, the previous architecture relied on a single process running many threads fighting for a single memory space in a single fault domain for centralized data, causing contention. The Catalyst Wireless architecture, on the other hand, relies on several processes that are single-threaded and nonblocking. The number of Wireless Network Control daemon (WNCd) processes can be easily increased for scale, and there is not a single fault domain anymore (for example, memory separation). Although the increase in databases may seem like extra complexity, it is the conclusion of the platform being data model driven and able to easily externalize any data very quickly, boosting programmability in an efficient manner. The whole platform was also built with process patchability and restartability in mind.

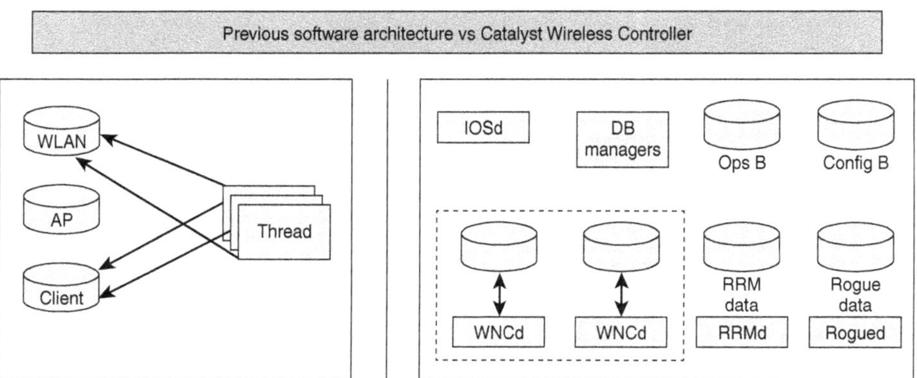

Figure 2-2 *The old wireless architecture on the left compared to the Catalyst Wireless architecture on the right*

The key wireless process is called Wireless Network Control daemon. The number of WNCd processes varies depending on the hardware that 9800 is running on (details are given in the "Hardware Overview" section). The WNCd process is a critical process managing APs and client sessions. Each WNCd process handles a specific set of access points and all the clients present on those access points; this maintains the approach where each process has everything it needs to handle a specific client session and keeps interprocess communication and replication to a minimum. The WNCd process is a single point for receiving and sending packets to the APs it manages but also implements a few other AP-facing capabilities like RRM client or probe handling.

This approach gives the vision of scalability of future Catalyst wireless platforms that will support an ever-increasing number of APs and clients simply by running more WNCd processes on more CPU cores. To oversee these processes, the WNCMgrd process manages the load-balancing of APs to WNCd instances. This means that the WNCMgrd is the one handling CAPWAP discoveries for the whole controller and assigns each new AP to a specific WNCd process (according to rules detailed later). The WNCMgrd is also in charge of centralizing information from each WNCd process in order to have a single go-to place for the "show wireless" CLI commands providing all the APs and clients information regardless of the number of WNCd processes. It has read-only access to the Centralized Wireless operational Database (CWDB), which contains all the real-time operational data. It can then consolidate information from each WNCd process and perform AP load-balancing and CLI information centralization tasks.

WNCd is a large process that is the center pillar of the Catalyst Wireless architecture (specific to the C9800) and contains many libraries inside it. You may hear a lot about SANET and SISF and see references to them in the logs and believe they are processes, but they are libraries inside the WNCd process. As a matter of fact, the general IOS-XE SANET library (in charge of AAA) has been copied (although modified) inside the WNCd process to manage the AAA authentication of wireless clients and their session management within the same process. Figure 2-3 depicts the various responsibilities of the SANET library that can be found inside a WNCd process.

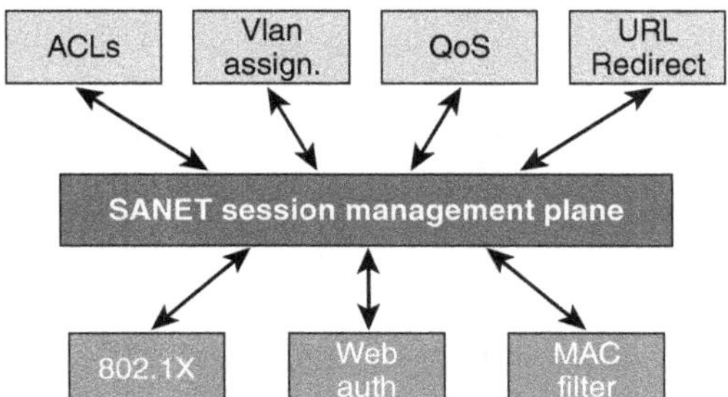

Figure 2-3 *SANET library responsibilities*

On top of that modified SANET library within WNCd, the Catalyst 9800 still has a SANET library inside the Session Manager Daemon (SMD) process just like other IOS-XE devices. That one handles wired session management (not really used in the Catalyst

9800) but also central Change of Authorization processing. Figure 2-4 shows the "classical" SANET library related to the SMD process in relation to the new SANET libraries inside the wireless processes.

Figure 2-4 *WNCd processes and the AAA task split between the existing SMD process and the new SANET libraries inside WNCds*

Similarly, the Switch Integrated Security Features (SISF) library is integrated in the WNCd process to handle the wireless client DHCP or IP tracking process, among other things. SISF originated on the Catalyst switches to provide an integrated framework to process IPv6 control traffic as well as act on IPv6 data traffic. SISF responsibilities on the Catalyst 9800 include

- IPv6 NDP inspection (barring bogus NDP messages)
- NDP address gleaning: populating the binding table with information snooped in NDP traffic
- Device tracking
- IPv4 address gleaning: ARP and DHCP messages snooping
- DHCP relay with configured helper address
- NDP and ARP multicast suppression: unicasting NDP or ARP messages or responding on behalf of targets to save on broadcast/multicast traffic
- DAD proxy: duplicate address detection
- DHCP requirement: making sure the IP can only be learned through DHCP process

If you are accustomed to legacy Cisco Unified Wireless controllers, you might expect the Catalyst 9800 to do ARP proxy by default, but that is not the case. ARP proxy is a feature configurable in the policy profiles since IOS-XE 17.3 but is not enabled out of the box. The default behavior is for the 9800 to transform broadcast ARP messages destined to the wireless clients into a unicast message for the specific MAC address. This saves on airtime because the message does not have to be sent on all access points at the same time. It also brings efficiency because the destination client of the ARP request also learns about the source MAC at the same time. In other words, it saves one ARP request-reply exchange in the case of two wireless clients having to contact each other. However, it may cause spam or battery drain if a device on the network is constantly ARPing for one or more destinations. Enabling ARP proxy then allows the WLC to reply to the ARP request on behalf of the destination client without having bothered that client at all in the process (as long as it's a known and registered wireless client with the WLC).

A mobility agent library also takes care of the client roaming from inside the WNCd process. This strategy greatly reduces interaction with any other centralized process like IOSd over the interprocess communication (IPC) channel. Mobility is ensured no matter where the clients are hosted. A client being handled by an AP on one WNCd process can roam to an AP managed by another WNCd process without any noticeable delay; this is mostly a matter of internal process and resource optimization.

Other Wireless Processes

While WNCd is the biggest wireless process by far (because it contains many libraries as SANET and SISF, which were covered previously), there are smaller and more specific wireless-oriented processes, each with a role:

- Mobility daemon (Mobilityd) is a 9800-specific process handling all communications with mobility peers. It also maintains a PMK cache for roaming clients to have a centralized repository for those within IOS-XE. It knows a client's presence within the WLC to answer mobility peers accordingly during mobility announce messages.
- Radio Resource Manager daemon (RRMgrd) is a process that handles communications with Radio Resource Management (RRM) group members. It runs the RF grouping, TPC, DCA, and FRA algorithms. It talks to each RRM-client process in each WNCd.
- Rogued handles rogues, Wireless Intrusion Prevention System (WIPS), and Rogue Location Discovery Protocol (RLDP) functionalities. It gets rogue reports from WNCd processes and classifies the rogue APs and clients.
- NMSPd is a process handling all communications toward CMX or the DNA Spaces connector.
- Wstatsd handles Netflow (FNF) packets and maintains a database of AVC statistics.

Contrary to WNCd, all the preceding processes are single instances.

DBM is the process that handles the configuration database (which is used for WebUI querying), while ODM is the process that manages the CWDB (operational database). PubD is in charge of centralizing the information for model-driven access. Those are not C9800-specific processes but have important tasks in a daily C9800 operation.

Wireless Client State Machine

This discussion is all theoretical so far. A good way to see this information in action is to follow a client onboarding. Wireless clients associated with an AP are serviced by the WNCd instance that is also servicing that specific access point. This means there is a minimal dependency on other processes to complete a client join process. The client authentication is done by the SANET library inside the WNCd process; the client IP learning and DHCP phase are handled by the SISF library of the same WNCd process, and a mobility agent within WNCd handles any roaming event. In case the client moves to an AP that is handled by another WNCd process (which should ideally not really happen; more on this in Chapter 3, "C9800 Configuration Model"), the mobility agent hands over the client to the other WNCd instance. There is no particular restriction to a client roaming between different WNCd processes. Client-related data is stored in separate database tables inside the respective WNCd process. A master table is available in the WNCmgrd to provide global tracking of clients wherever they are hosted or moved to. Each client is tracked by a finite state machine. The onboarding of a client is considered to be a Control Plane activity—that is, a task where traffic is analyzed and tracked by the Control Plane, and when the client moves to the final forwarding state of that state machine (called the RUN state), the traffic is forwarded by the datapath without mobilizing the Control Plane anymore. The client state machine's information is replicated between WLCs in a high availability pair of WLCs to guarantee a smooth failover.

The client state machine starts when an 802.11 management frame is received from the client (association request, reassociation request, 802.11 authentication frame). The client goes through various states until the RUN state where it is fully plumbed to the datapath. A client is deleted from the state machine based on certain timers (idle timeout, session timeout) or state timeout (after a certain time stuck in a transition state, the client is deleted). For example, if a client associates with a web authentication SSID (which is typically open) and does not log in on the web portal, the client entry is deleted after 5 minutes by default. This does not prevent the client from associating and trying again, but this really limits the overload and spam from clients who passively try to connect but do not finish the work and use resources (including IP address) for nothing.

Here is an example of an 802.1X client join state machine processing in a centrally switched WLAN configuration:

- The wireless client sends an association request, which is encapsulated in CAPWAP by the AP over the data tunnel to the controller.
- The WNCd (managing that AP) event library receives a socket data event.
- The CAPWAP data library inside the WNCd parses the CAPWAP data payload, classifies the payload as an 802.11 management payload, and invokes the registered handler function.
- The client orchestrator looks up an existing client state machine for this specific client based on the MAC address as unique identifier (which means that clients using random MAC and failing to associate and rotating their MAC address after each association attempt will create a lot of temporary client entries). If no existing state machine is found, it instantiates one and passes on the event to the 802.11 library to process the association request.
- The 802.11 library processes the association request and reads the BSSID operational table to validate the 802.11 parameters. It then sends an association response back. It invokes an internal API to send the association response over to the CAPWAP data tunnel.

- The client orchestrator sends an "add mobile" payload to the AP over the CAPWAP control tunnel to create the client entry on the AP itself. This is ACKed by the AP.
- The client orchestrator verifies the client L2 authentication policy configured. Because it is 802.1X authentication, it triggers the authentication interface state machine to start the Layer 2 authentication. A call to the SANET library is made.
- SANET takes care of the 802.1X exchange and forwards accordingly the packets encapsulated inside RADIUS to the RADIUS server up until an authentication failure or success.
- The SANET process writes the client policies, which may have been overridden by RADIUS if AAA override is configured. Those client policies are kept within SANET and available through internal API to other libraries.
- An IPC call is made to Mobilityd to update the PMK Cache key. It is maintained in Mobilityd at all times, even if there is no roaming to other WLCs. Mobilityd publishes the PMK key to the mobility group if any.
- The authentication interface triggers the four-way key exchange to generate the PTK, GTK, and IGTK. The authentication interface gets back to the client orchestrator with the result of the Layer 2 authentication.
- The client orchestrator then sends the transient keys to the AP in another "Add mobile" payload inside CAPWAP control.
- If there are controllers in the mobility group, the Mobility interface starts the processing by announcing the client and finds out if the client already exists on another WLC (and is therefore roaming between different WLCs). The WLCs then negotiate the exchange of the client information.
- The client orchestrator transitions to the "IP learn" state and triggers the IP learn library (called SISF) inside WNCd to learn the IP address of the client.
- The IP learn interface state machine plumbs a client entry into the datapath. This allows the datapath to learn about the client MAC and punt all the DHCP payloads in the client VLAN to the WNCd process (that is, the DHCP packets are taken out of the dataplane and sent to the CPU for further processing and learning).
- When the client receives its IP from the DHCP server, the SISF library identifies the IP address, creates an event, and returns the information to the client orchestrator.
- The client orchestrator drives the client policy bind to the datapath by providing all the policies such as QoS, ACL, VLAN, and IP.
- The client orchestrator sends the final "add mobile" payload with the QoS attributes to the AP and then transitions the client to RUN state.

- From this point onward, the client traffic is forwarded according to policies and will be until a timeout (session or idle for example) is hit or until another client management event is triggered (roaming, new association, and so on).

The controller needs to learn about at least one IP address used by the client but supports up to eight addresses learned for the same MAC (useful in IPv6). The preceding is a detailed explanation of the state machine changes in plain English. It may help to see a diagram showing the state transitions. Figures 2-5 and 2-6 depict the same overall client state machine upon connecting a new wireless client to a WPA2 Enterprise SSID until the client is placed in the RUN state. The client states are written as they are found in RadioActive tracing logs (for more details, see Chapter 16, "Troubleshooting").

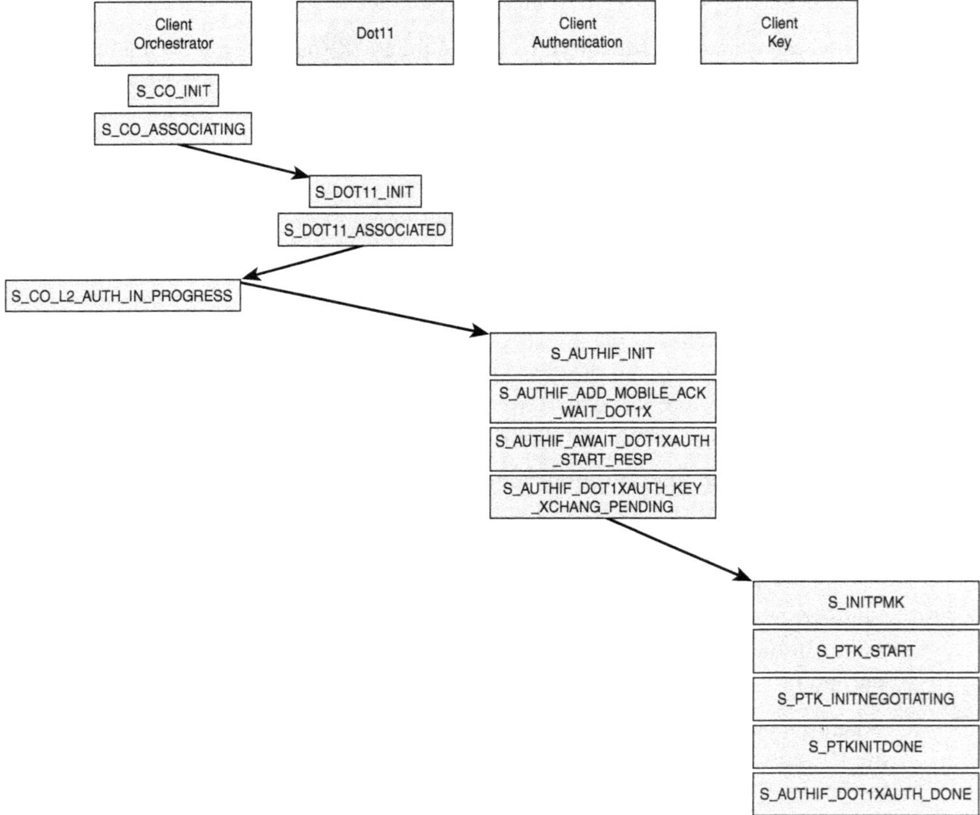

Figure 2-5 *Client state machine on a WPA2 Enterprise SSID*

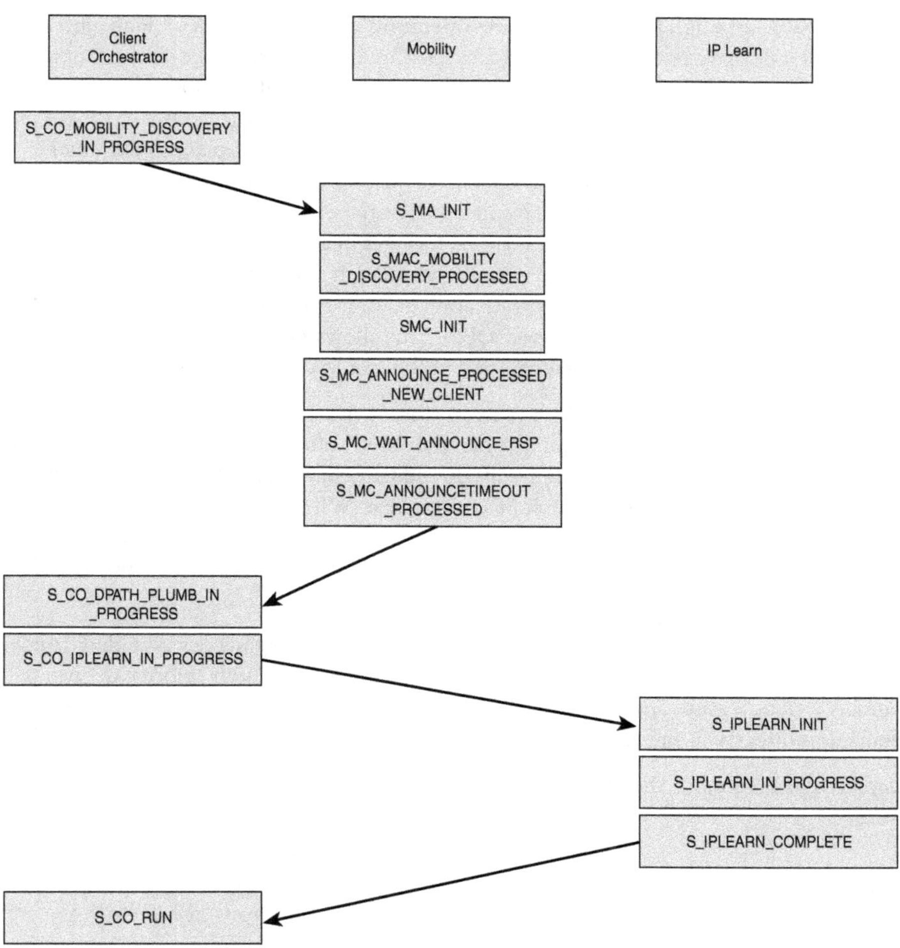

Figure 2-6 *Client state machine on a WPA2 Enterprise SSID (continued)*

One Dataplane to Rule Them All (or Three at the Maximum)

One of the great achievements with IOS-XE 17 (although it started in the IOS-XE 16 train) is to unify the software for routers, switches, and now wireless controllers too. The goal is to run the same IOS-XE on all the enterprise network devices. It would be easy to say that the Catalyst 9800 is very similar to the rest of the Catalyst switches, but that is not the case although they share the IOS-XE operating system and a somewhat similar process structure. After all, access points are also part of the Catalyst family now, and they don't run the same operating system at all. The dataplane is one big difference, and this is why you cannot use the same software file to install on a Catalyst 9300 and on a Catalyst 9800: although they run the same operating system and Control Plane infrastructure, they do have dataplane differences. The Catalyst switches rely on the Doppler chipset and the Forwarding Engine Driver (FED) dataplane. Catalyst 9800 was meant to run on various hardware from the start and therefore adopted the Cisco Packet Processor (CPP) complex dataplane that is also used in the ASR1000. This dataplane is extremely flexible and written in a way where it can be implemented completely on software or through a hardware-accelerated chip. This means that the exact same dataplane is leveraged in a 9800CL (the virtual machine form factor) and in a 9800-40 (an appliance form factor), for example. In the case of the virtual controller, it is run as a process on specific CPU cores, whereas in the case of hardware appliances, it runs in a Cisco programmable chip called the Quantum Flow Processor (QFP). This allows you to have a unique code for all platforms, although the implementation will differ at compilation time.

The title of this section is justified by the exception of the Embedded Wireless Controller (EWC on switch and on AP). When the 9800 software runs on a Catalyst 9000 series switch, it runs its Control Plane freely on the CPU but must rely on the platform dataplane and therefore uses the switch FED dataplane. When the 9800 is running on a Catalyst 9100 AP, it is the AP OS that provides its dataplane to the controller code running in a container. This is why, for example, the EWC on AP cannot use more than one SVI.

The QFP is a 64-core Cisco packet processing chip with four threads per core dedicated to processing packets. It is fully programmable and implemented with priority queues in mind as well as feature arrays so that ACLs, AVC, Netflow, and other features can be implemented directly in the packet processing workflow.

As depicted in Figure 2-7, the life of a packet inside the 9800 dataplane is quite straightforward:

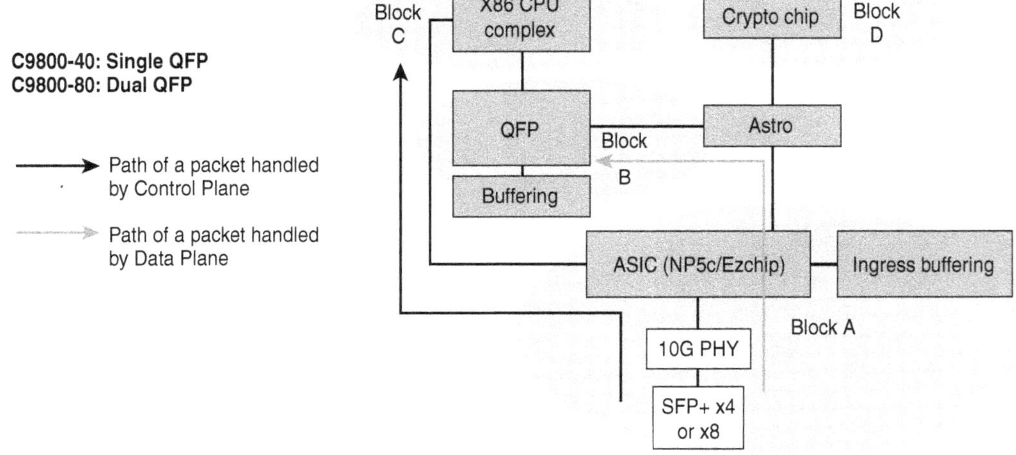

Figure 2-7 *Block diagram of the life of a packet in the 9800 appliance's dataplane*

Packets arrive on the data interfaces (block A) through an SFP and then PHY driver. The ASIC is in charge of buffering and sorting out the packets. If the packet is for the Control Plane (for example, destined to one of the WLC IP addresses, or simply a packet to be intercepted, such as for web authentication), the ASIC immediately sends it to the CPU complex (block C). If the packet is for fast forwarding on the dataplane (typically a wireless client traffic), it is handed over to the QFP dataplane (block B) through a bus (called Astro in the picture). There might be a small detour through the crypto chip (block D) if the packet is encrypted to have it decrypted for the dataplane processing. This means that all packets go through the dataplane (even briefly), but some take a quick shortcut to the CPU for further processing while others go through the dataplane set of packet processing features and don't involve the CPU.

Figure 2-8 logically belongs to the previous sections of this chapter because it describes the path of a packet inside the Control Plane, but it can better be understood after we have explained the processes, and it is a logical continuation of Figure 2-7 after a packet is handed over to the CPU complex.

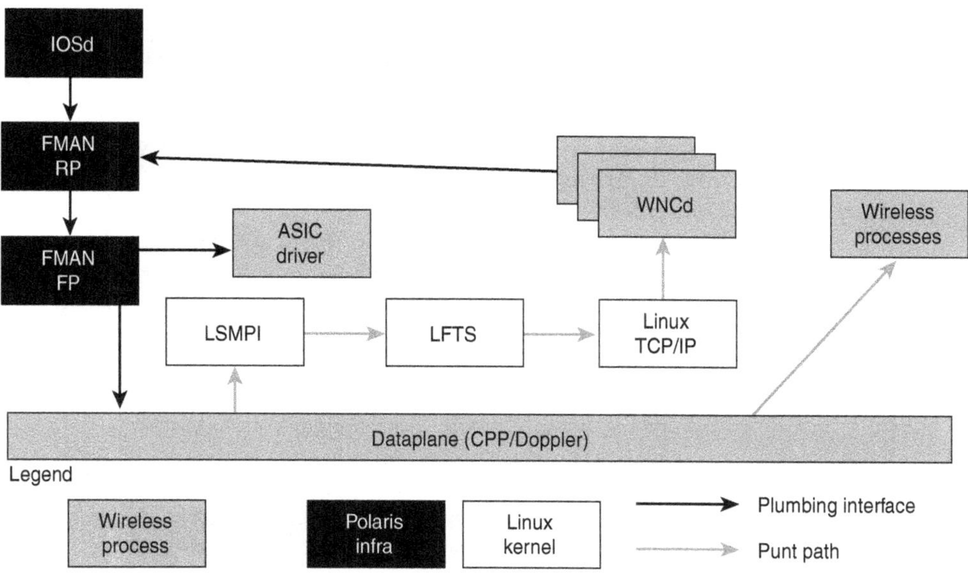

Figure 2-8 *Control Plane packet processing*

After the packet is handed over by the dataplane to the Control Plane, it goes through the Linux kernel (Linux Shared Memory Pool Interface, or LSMPI, and Linux Forwarding Transport Service, or LFTS). These kernel libraries identify if the packet is for the IOSd process (critical routing task), and if not, hand over to the TCP/IP stack and then to the right process (one of the WNCds, for example, if it's an AP CAPWAP control packet).

Hardware Overview

We briefly introduced the various controller models in the introduction of this book. As already hinted at in the introduction, the differentiating factor of hardware controllers is to have a dataplane acceleration that allows them to support higher throughput of centrally switched traffic. Let's now dig deeper into the specifics of each.

C9800-40 and C9800-80

The 9800-40 and 9800-80 continue a long Cisco wireless tradition of specifying the aggregate maximum throughput in the last digits of the part number (just like the 5520 and 8540). The two big brothers of the Cisco wireless family are similar, just like their aforementioned predecessors, so you can consider them identical except for a few points covered in this section.

While the first one touts a maximum of 40 Gbps of aggregate throughput (that is, not 40 Gbps upload and download at the same time, but 40 Gbps in one direction or 20 Gbps simultaneously in full duplex in each direction) through 4 TenGigabitEthernet ports and the second a maximum of 80 Gbps through 8 TenGigabitEthernet ports, the first thing to underline is that these numbers are maximum throughput numbers. Throughput depends on a lot of factors, but the most important one is the packet's length. Those numbers are measured with all network packets being 1500 bytes, which optimizes the dataplane processing capability. If you have only 64 bytes-long packets in your network (which is an exaggeration in the other direction), the throughput numbers are much lower because the dataplane spends a lot of processing cycles on very small packets and would literally have to treat 20 times the number of packets to reach similar throughput numbers. On top of this, the presence of heavy access lists or application visibility (AVC) being enabled can also slightly impact the throughput numbers. All these factors are the reason why you cannot predict a specific maximum throughput number for your network and have to monitor dataplane utilization. Typically, the maximum number of APs/clients is reached before the maximum dataplane capacity is reached. The reason is that wireless is a shared medium, and the more contention, the less overall throughput you observe as air time is lost in collisions and management frames. Clients typically only send bursts of traffic and don't all download large files at the same time; therefore, it is quite uncommon to hit the maximum throughput limit of the WLC, and other limits (such as number of APs, number of clients, Internet bandwidth) are hit first. Figure 2-9 shows a 9800-40 WLC.

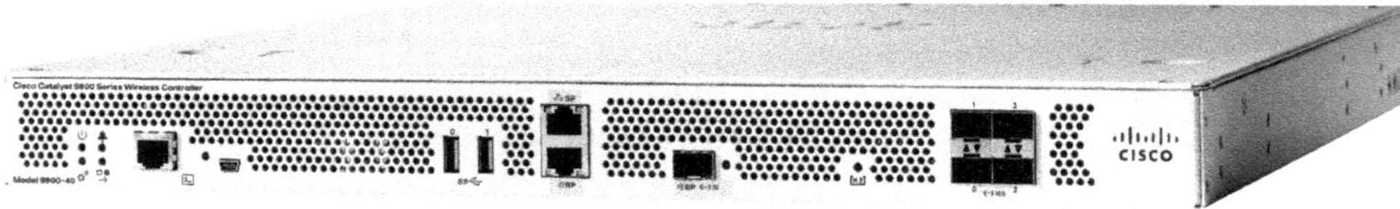

Figure 2-9 *A 9800-40 appliance*

The 9800-40 and 9800-80 benefit from the QFP dataplane chip (which is also present in other Cisco products like the ASR1k). The 9800-80 has two of those QFP dataplane chips (which explains why its max throughput is doubled) but also has a stronger CPU and memory, which help increase the maximum AP and client count supported on it to 6000 APs and 64,000 clients (2000 APs and 32,000 clients for the 9800-40). The 9800-40 has five WNCd processes, whereas the 9800-80 has eight WNCd processes, using most of the 12-core CPU of the 9800-80 appliance.

The 9800-40 and -80 are also the controllers of choice for Encrypted Traffic Analytics, a feature in combination with Cisco Stealthwatch (or Secure Network Analytics) where the controller sends (when traffic is centrally switched) detailed Netflow statistics about encrypted traffic streams so that Stealthwatch is able to identify threats and malware even on encrypted traffic.

The 9800-80 final major differentiation is the presence of an extension bay (which is empty by default) where you plug a variety of modules if you prefer to have, for example, two FortyGigEthernet ports or even a single 100GigEthernet port. As soon as an extension bay is connected, the eight default TenGigabit ports are disabled in favor of the port on the extension bay. They cannot all be used at the same time. You can see this bay on the bottom right of Figure 2-10.

Figure 2-10 *A 9800-80 appliance*

Apart from the differences highlighted until now, the two controllers are nearly identical and can run the same software files (even if on cisco.com they would have different filenames, they are the same type of software). Both appliances have a 1 Gbps RJ45 and a 1 Gbps SFP redundancy port (only one of them works at any given time), giving you the choice of connectivity for redundancy in your HA pair. They also have a 1 Gbps management service port. This service port (SP) named "GigabitEthernet0" in IOS-XE is hard-coded to be part of the **mgmt-intf** VRF and therefore has separate routing from the rest of the appliance ports. It is wired directly to the appliance CPU and is only able to transmit and receive control plane traffic. Both appliances have 32 GB of bootflash memory for IOS-XE operations as well as a 240 GB SSD **harddisk** partition (where logs and backup images can be stored). On top of the hardware QFP dataplane chip, both appliances have a crypto chip allowing for hardware acceleration of encryption/decryption of the dataplane traffic because no CPU could handle the encryption at such data rates.

C9800-L

The smallest appliance in the 9800 portfolio is only small in size and makes no compromise in its feature set. It exists as 9800-L-F, which has two SFP ports for flexible connectivity or, as depicted in Figure 2-11, a 9800-L-C with all copper RJ45 ports.

Figure 2-11 *A 9800-L-C appliance*

It is similar to the AireOS-based 3504 wireless controller that it replaces in the Cisco wireless portfolio, with the motto "smaller than the bigger appliances, but still a full feature parity with the biggest controller." From a size standpoint, you can fit two 9800-L within the same rack unit (RU), but it is advised to leave some room on top of it because the thermal dissipation holes are on top of the units.

The 9800-L has four mGig ports capable of 1 or 2.5 Gbps ports that can be used in a LAG and two extra ports that go up to 10Gbps. It is available in two variations: the 9800-L-C or the 9800-L-F, where the two extra data ports are SFP+ (and therefore support fiber) supporting up to 10 Gbps. It does not leverage a QFP chipset but instead uses dedicated x86 CPU cores for its dataplane. It enjoys a maximum total forwarding throughput of 5 Gbps, which can be pushed to 10 Gbps by installing a performance license. Aside from these connectivity-related differences and different maximum throughput numbers or maximum supported APs and client numbers, the 9800-L does not have any particular restrictions or limitations: it supports the same feature set as the bigger appliances. It does require a specific software file, which differs from the bigger appliances though.

The 9800-L only has a bootflash and no hard disk. It supports a maximum of 250 APs and 5000 clients and double these numbers if installing the Performance License.

C9800-CL

The 9800-CL is not hardware at all, but the software version that is able to run on a private or public cloud. Its architecture differs only from a dataplane implementation perspective (as it is virtualized) and therefore typically has lower maximum throughput numbers than the appliances. It is therefore typically advised and recommended for FlexConnect local switching deployment or deployments that do not send too much centrally switched traffic.

The 9800-CL is available in several footprints. The bigger the virtual machine footprint, the more APs and clients are supported through the use of more WNCd processes. Although less scalable from a data throughput standpoint, the 9800-CL scales up to the size of the 9800-80 in terms of supported AP and client count. On the dataplane topic, the 9800-CL has a fixed maximum throughput (around 2 to 3 Gbps, depending on certain conditions), but this throughput can be doubled through the use of SR-IOV NICs. Using specific compatible NIC cards for the virtual machine and SR-IOV compatible drivers, the 9800-CL can more or less double the maximum data throughput available by allowing more CPUs to the dataplane task. Separate SR-IOV virtual machine images labeled "high throughput" are available should you want to deploy them.

There are no specific restrictions with regards to 9800-CL compared to appliances: they support all the same feature sets. It might be interesting to underline a couple of points that pertain to the virtual machines:

- LAG is supported starting IOS-XE 17.5 on 9800-CL SR-IOV images, while NIC teaming is the recommended way to achieve redundancy and higher throughput for other situations.
- When deployed in the public cloud, your wireless management interface has to be one of the routed L3 ports. While 9800-CL supports central switching to some extent, the public cloud deployments do not support central switching at all, nor the features that revolve around it (such as sniffer mode AP).
- If your virtual host system does not support MAC learning (such as VMWare ESXi by default and before release 6.7), you need to enable *promiscuous mode* as well as *forged transmits* in the port group configuration. All the wireless client MAC addresses appear as coming from behind the WLC switchport, and if MAC learning is not enabled in your platform, the host system does not know to which VM to hand over a specific packet because it has not learned the wireless client MACs. Enabling promiscuous mode and forged transmits allows sourcing packets with a wireless client MAC as well as distributes all unknown MAC traffic to all your 9800-CL instances in the same VLAN/port group.
- VM snapshots are supported, but be aware that such operations typically tend to "freeze" the VM for a couple of seconds and increase the load on the host appliance, which can lead to WLC crashes or HA failovers.

Summary

This chapter covers software architecture of the Catalyst 9800, the reasons behind using a unified IOS-XE for all Enterprise platforms, and the benefits of the scalable architecture of the Catalyst 9800 controller. It covers a client onboarding example through the state machine, which is a good illustration of the software components coming together and something you will be able to see when you troubleshoot your wireless client through radioactive traces (which is covered in the troubleshooting chapter). Finally, this chapter also covers the different hardware available that can run the Catalyst 9800, their differences, and which one is best suited for each type of deployment.

Chapter 3

C9800 Configuration Model

This chapter describes the new configuration model available on the Catalyst 9800 wireless controller series. Understanding how the new configuration constructs work is important because they represent the basis of the C9800 design, deployment, and troubleshooting that you will learn in the rest of the book.

In designing the Catalyst 9800 from scratch, Cisco decided to build a new configuration model to overcome some of the limitations of AireOS: Most of the changes in AireOS configuration would have a global network impact; AireOS doesn't have a concept of location (physical or virtual), so it is not possible to limit configuration changes to a group of APs; the AireOS configuration is not reusable, and the user has to repeat the same configuration multiple times, even if, all it needs is a single parameter change.

The new configuration model of the Catalyst 9800 is based on the design principles of reusability, simplified provisioning, enhanced flexibility, and modularization to help simplify the design, implementation, and management of wireless networks at scale and to cope with always-changing business and IT requirements.

C9800 New Configuration Model

The Cisco Catalyst 9800 series' new configuration model is based on two main constructs: profiles and tags. Profiles group a set of features and functionalities, and tags allow you to assign these features and functionalities to the access points (APs).

There are five types of profiles:

- **AP Join profile (or AP profile):** Contains general AP settings such as Control and Provisioning of Wireless Access Points (CAPWAP) timers, 802.1X supplicant, SSH/Telnet settings, and many more. These settings in AireOS are usually at a global configuration level and apply to all the APs.

- **WLAN profile:** Defines the SSID name and profile and all the security settings.
- **Policy profile:** Contains the policy to be associated with the WLAN. It specifies the settings for client VLAN; authentication, authorization, and accounting (AAA); access control lists (ACLs); session and idle timeout settings; and so on.
- **FlexConnect profile:** Groups all settings to be assigned to a FlexConnect AP or FlexConnect group: native VLAN, ACL mapping, and so on.
- **RF profile:** Defines the radio frequency (RF) characteristics of each band, as in AireOS.

Tags allow you to bind the settings defined in the profiles to the access point. There are three types of tags:

- **Policy tag:** Ties together the policy profile and the WLAN profile.
- **Site tag:** Assigns the AP Join profile settings to the AP and determines if the site is a local site, in which case the APs will be in local mode, or not a local site, in which case the APs will be in Cisco FlexConnect mode. Additionally, when Enabled Local Site is unchecked, the site tag allows you to select and assign a FlexConnect profile (the default-flex-profile is chosen if nothing is specified).
- **RF tag:** Binds the 6 GHz, 5 GHz, and 2.4 GHz profiles to the AP.

As shown in Figure 3-1, an access point is always assigned three tags, one for each type.

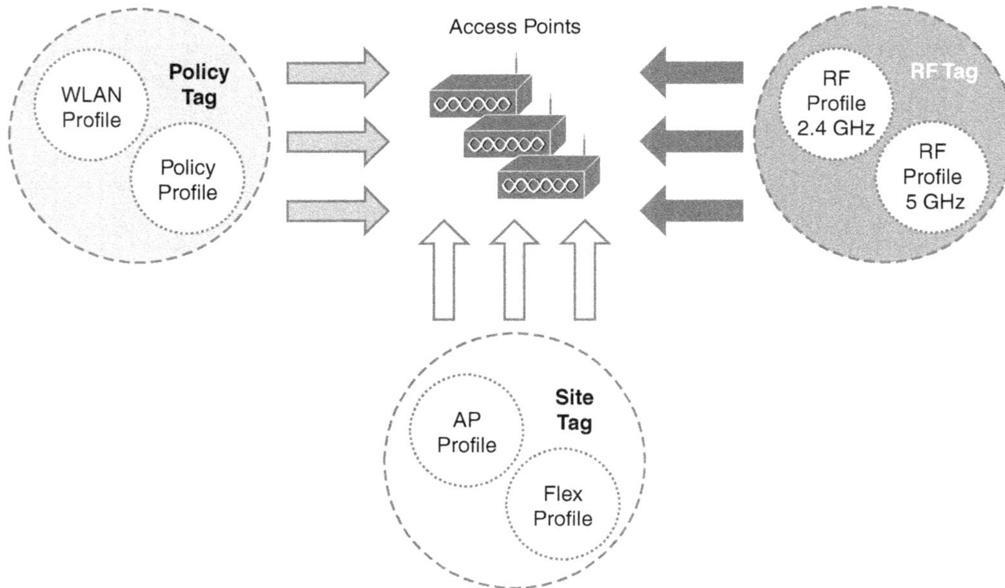

Figure 3-1 *Profiles and tags and AP assignment*

If the three tags and related profiles are not explicitly assigned, the AP gets the default values automatically (default-policy-tag, default-site-tag, and default-rf-tag) with the related default profiles upon joining the C9800.

To understand a bit more about tags, let's start with the easier one: the RF tag. For each AP, it specifies the RF profile for the 2.4 GHz, 5 GHz, and 6 GHz radio, and it's similar to the concept of the RF profile in AireOS. The default RF tag takes the 5 GHz and 2.4 GHz RF profiles settings directly from the global configuration unless you create a named RF profile, with custom configurations, and assign it to the default RF tag.

The policy tag defines the list of WLANs/SSIDs to be broadcasted on the AP and the associated policies. For this reason, it defines the broadcast domain and can be considered equivalent to the AP group in AireOS (you see later that the access group has many more settings than the ones contained in the policy tag, but this comparison can help you understand it better). Regarding the default-policy-tag, of course, there is no such thing as a default WLAN profile; you would always have to define your WLAN settings. Also, starting with release 16.12.2s, for the SSID to be broadcasted, you must explicitly map any WLAN profile (no matter the WLAN ID number) to a policy profile within a policy tag. This is also true for the default-policy-profile and default-policy-tag. In other words, no SSID is broadcasted by default. This is a different behavior from previous releases where the first 16 WLANs were broadcasted by default.

The site tag is a new concept introduced in the C9800, because it defines settings at a site level. These configurations are contained in the AP Join profile that is associated with the site tag. There is no AireOS equivalent of the site tag because most of these settings (CAPWAP timers, AP dot1x credentials, Backup primary/secondary, and so on) are global in AireOS. The site tag is also the place where you may define a deployment as FlexConnect.

In AireOS, FlexConnect is a configuration mode at the AP level; in C9800, it is defined as a site configuration and hence at the site tag level. If you consider a FlexConnect location, the site tag can be considered as equivalent to Flex groups in AireOS. The Flex profile contains all the FlexConnect-specific settings, and it also defines the roaming domain. As with FlexConnect group, if fast and secure roaming is a requirement, then the maximum number of APs per site tag is equal to 100.

The C9800 configuration model allows you to have much more flexibility in tweaking the configuration to fit a specific wireless deployment. Let's consider a few examples: you have multiple remote locations, and many of them have a different characteristics of the transport network, which calls for different and optimized **CAPWAP heartbeat timers** and **TCP MSS Adjust** settings on the APs at each location. In AireOS, these are global configurations; if you have a single controller managing all the sites, the same settings are applied to all the APs. There is no per site configuration; either you apply the changes to all the APs, or you don't.

With the new C9800 configuration model, values such as **TCP MSS Adjust** and **CAPWAP timers** are set at the AP Join profile level, so you can evaluate the transport network at each site and decide the value that is best for a specific group of APs. This flexibility allows for a better network design.

Another practical example: say you have to design the RF coverage of a large warehouse. Most of the requirements are straightforward, and the coverage area is uniform, but you have a refrigerator area where you need to provide coverage inside the fridge cells. This calls for an external antenna AP model so that the APs can be mounted outside and only the antennas placed inside the cold area. Given the specific RF requirements of the metal cells, you conclude that some special RF settings are needed just for this small group of APs. All the other settings (SSIDs, security and network policies, and so on) are the same: all you need to do is assign the same policy and site tags for all the APs and just configure and associate a different RF tag (with related custom RF profile) only to this group of APs.

Profiles and tags offer the following advantages:

- Modular and reusable configuration constructs. You define your WLAN/SSID once, and then you can have a different policy (VLAN, QOS) at each location.
- Flexibility in assigning a configuration to just a group of APs.
- Ease of management of site-specific configuration across geodistributed locations.
- No need to reboot when applying config changes via tags.

The last advantage is very important from an operation point of view. Say you decide to broadcast only a subset of SSIDs on the APs in a certain location. With AireOS, you would likely need to create an AP group, add the SSIDs you want, and then associate the APs with the AP group. This results in the APs rebooting. With C9800, all you have to do is to create a new policy tag with the desired SSIDs and change the policy tag assignment to the APs. This triggers a CAPWAP restart, but not a reboot. The difference? There is about a 20–25-second downtime versus a 3-minute reboot cycle in AireOS.

One important clarification: in this example, a CAPWAP restart is needed because you are changing the tag (policy tag) associated with the AP. If your configuration change affects only a setting in an already-assigned profile and/or tag, no CAPWAP restart is needed, and the change doesn't result in a CAPWAP disconnect.

What Does My AireOS AP Group Migrate To?

To better understand the new C9800 configuration constructs, let's compare C9800 with AireOS and see how the settings present in the AireOS AP group would migrate to C9800 profiles and tags.

We mentioned previously that because the policy tag defines the SSIDs broadcasted by the AP, it represents the broadcast domain, and it could be considered equivalent to the AP group. The AP group contains more settings than the SSIDs and associated interfaces in AireOS. It allows you to assign different types of parameters such as the RF profiles, RLAN, Hyperlocation, Intelligent Capture (iCap), and some general settings. As shown in Figure 3-2, when mapped to C9800, these configurations result in a combination of settings present in all three different tags. As you can see, the configuration in C9800 is modularized, and you can decide to change some settings and not others.

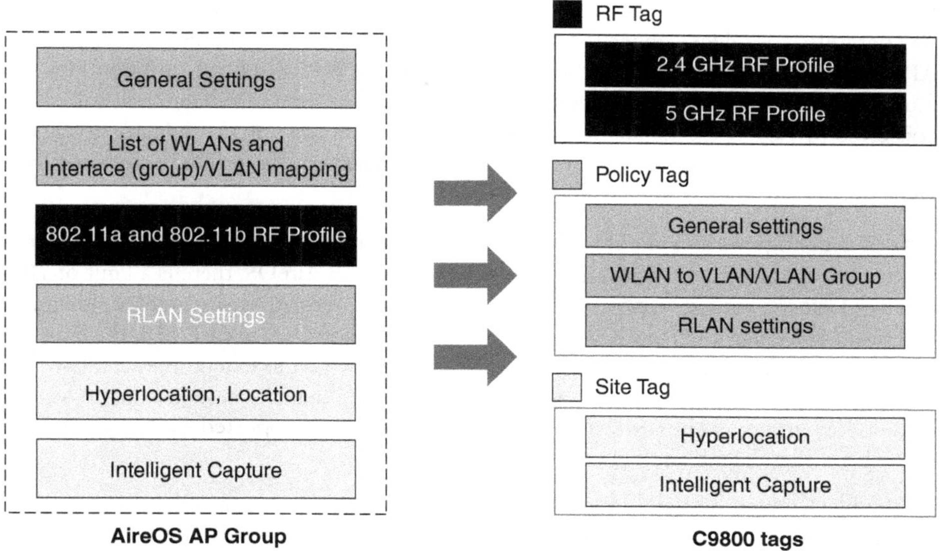

Figure 3-2 *AireOS AP Group to C9800 site tags mapping*

What About FlexConnect?

In the new C9800's configuration model, FlexConnect is not an AP mode configured on the AP itself. Instead, it's a site-level configuration, and therefore, you can find it in the site tag settings. The idea behind it is that FlexConnect is usually deployed at a remote location, and you would have multiple APs to be configured in this mode. With C9800, all you have to do is assign the AP to a site tag configured for FlexConnect, and all the APs assigned to that site tag are automatically converted to Flex APs. This scenario is illustrated in Figure 3-3.

Figure 3-3 *Setting the site tag as FlexConnect = nonlocal site*

By default, the site is local, meaning that APs assigned to the site tag are in local mode. To configure it as a FlexConnect site, just uncheck the **Enable Local Site** setting, and the APs already joined to that site tag will go through a CAPWAP reset and turn into Flex mode. All new APs joining the C9800 and assigned to that site tag will automatically be converted to Flex. If the **Enable Local Site** setting is unchecked on the default-site-tag, all the APs joining the WLC with the site tag set as default-site-tag will be automatically converted to Flex.

After it is configured as a remote site (or better, not as a local site), the site tag is similar to the concept of a FlexConnect group in AireOS. As with AireOS, there is a limit of 100 APs per Flex site tag for supporting seamless roaming, and roaming across Flex site tags will result in a client full reauthentication. Starting with software release 17.8, the limit has been extended to 300 APs per site tag. This will allow you to extend the seamless roaming domain at the site. One important note regarding default-site-tag: as with the default Flex group in AireOS, fast and secure roaming is not supported for FlexConnect APs when using default-site-tag.

All the FlexConnect settings are configured under the FlexConnect profile that you select and assign to the site tag after it is configured as a remote site, as shown in Figure 3-3.

Starting with release 17.3.3, C9800 supports clients with overlapping IP addresses in different Flex locations. There are two requirements for this to happen: the policy profile mapped to the SSID needs to be configured for FlexConnect local switching, and each location needs to be assigned its own named site tag, which is used as an additional identifier to handle two clients having the same IP addresses in different sites. At the time of writing, client IP address overlapping is not supported for policy profiles configured for central switching, for APs either in local or FlexConnect mode.

Cisco C9800 Series Profile and Tag Considerations

As just described, with the Catalyst 9800 wireless controller's profiles and tags model, some configurations are done differently when compared to AireOS with the intent of making the settings more flexible and easier to use. Functionalities that you are used to in AireOS wireless controllers are also supported in the C9800, but you need to get familiar with the configuration model because they are done differently. Plus, the new configuration model is made to be extended to the new differentiating features supported by the C9800.

Next, we describe what you need to know about profiles and tags and give some tips and best practices on how to best use them.

Assigning Tags

Each access point needs to be assigned three unique tags: one policy, one site, and one RF tag. By default, when an AP joins the C9800 wireless controller, it gets the default value of each tag, namely default-policy-tag, default-site-tag, and default-rf-tag. You can make changes to the default tags or create custom tags. To know what tag has been configured on each AP, you can go to the GUI under **Configuration > Wireless > Access Point**, as illustrated in Figure 3-4.

Configuration > Wireless > Access Points

All Access Points

Number of AP(s): 3

AP Name	AP Model	Slots	Admin Status	IP Address	Base Radio MAC	AP Mode	Operation Status	Configuration Status	Policy Tag	Site Tag	RF Tag	Tag Source
AP-3800I-SJ	AIR-AP3802I-B-K9	2	✓	172.16.11.10	006b.f19c.5060	Local	Registered	Healthy	PT_US-WE_SJC-2_Floor2_6d981	building24	TYPICAL	Static
AP3700_f9a0-SJ	AIR-CAP3702E-A-K9	3	✓	172.16.11.11	5897.bd37.f9a0	Local	Registered	Healthy	PT_US-WE_SJC-2_Floor3_59278	building24	TYPICAL	Static
AP2800-f960-SJ	AIR-AP2802I-B-K9	2	✓	172.16.11.12	f80c.cbd6.f960	Local	Registered	Healthy	PT_US-WE_SJC-2_Floor3_59278	building24	TYPICAL	Static

Figure 3-4 *AP tags in the GUI*

In release 16.12.2s and later, you can also get more details by clicking the icon (2329_03ICON01.png) next to the AP, and a popup window opens, as shown in Figure 3-5.

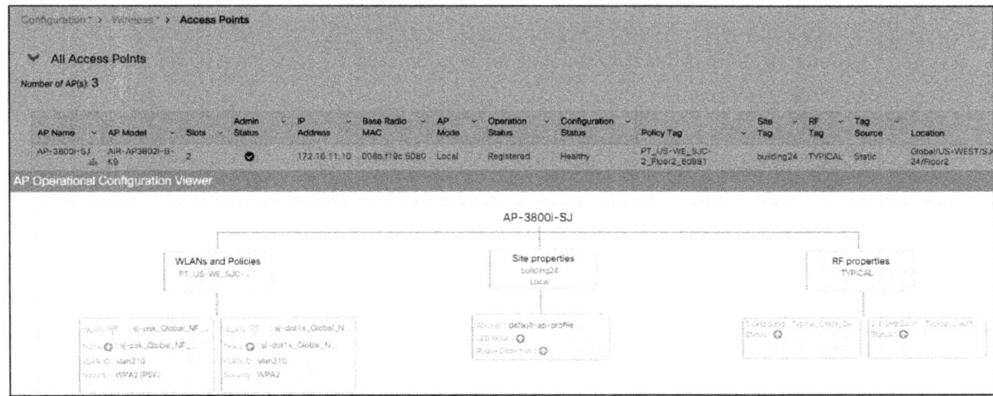

Figure 3-5 *AP tags details in the GUI*

This icon shows whether the SSID is being broadcasted or not (it will be gray and not green). The (2329_03ICON01.png) icon turns red if there is a tag misconfiguration.

With the CLI, you can use the **show ap tag summary** command. The output is shown in Figure 3-6.

This command clearly indicates whether there is a misconfiguration involving tags and profiles under the Misconfigured column. A typical example of tag misconfiguration is assigning the same WLAN to two different policy profiles with different Application Visibility and Control (AVC) settings. In this case the **show avc status <WLAN name>** command flags it as an error, with the related explanation.

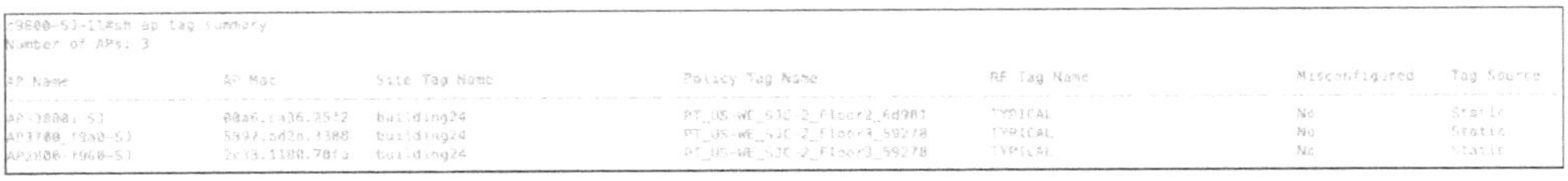

```
c9800-SJ-11#sh ap tag summary
Number of APs: 3

AP Name              AP Mac          Site Tag Name        Policy Tag Name                  RF Tag Name          Misconfigured   Tag Source
--------------------------------------------------------------------------------------------------------------------------------------------
AP-3800i-SJ          00a6.ca36.2512  building24           PT_US-WE_SJC-2_Floor2_6d981      TYPICAL              No              Static
AP3700_f9a0-SJ       5897.bd26.3388  building24           PT_US-WE_SJC-2_Floor3_59278      TYPICAL              No              Static
AP2800-f960-SJ       2c33.1180.70fa  building24           PT_US-WE_SJC-2_Floor3_59278      TYPICAL              No              Static
```

Figure 3-6 *AP tag summary*

Notice the Tag Source column in the output of the **show ap tag summary** command; it tells you how the AP got the tags. The possible sources, in order of priority, are

- **Static:** As network admin, you select the AP and assign it specific tags. The configuration is saved on the controller based on the AP's Ethernet MAC address. When an AP joins that specific controller, it is always assigned the specified tags.
- **Location:** This is a configuration construct internal to the C9800 (it's not the AP location that you can configure on each AP), and it's used primarily in the Basic Setup flow on the WebUI. A location allows you to create a group of three tags (policy, site, and RF) and assign APs to it. You can ignore it unless you are using the Basic Setup flow.
- **Filter:** You can configure the C9800 to use a regular expression (regex) to assign tags to APs as they join the controller. As of today, you can set a filter based only on AP name, so this method cannot be used for out-of-the-box APs but is very useful in a migration from an AireOS-based network or a brownfield scenario where the APs already have a name.
- **AP:** The AP itself carries the tag info learned through Plug and Play (PnP) or pushed from the controller and saved in the AP memory.
- **Default:** This is the default tag source; no explicit assignment is done.

The first two sources (static and location) are user static mapping configurations to assign APs to tags and hence have the highest priorities. The filter allows you to define a dynamic mapping of APs to tags based on regex expressions. When the source is the AP, it means that this information is saved on the AP itself and is presented to the controller when the AP joins. Finally, if there is no explicit tag mapping configuration on the C9800, and if the APs doesn't carry any tag information, the AP is assigned the default tags. Hence, the default option has the lowest priority.

If you don't use Cisco DNA Center or Cisco Prime Infrastructure, a simple way to assign multiple APs to a set of tags is to leverage the programmable interfaces of C9800 and write a small software program that does exactly the job. An example of such programs is explained in Chapter 12, "Network Programmability."

Another simple but less automated way is to use the web interface under the Advanced setup in the GUI (accessed through **Configuration > Wireless Setup > Advanced**). Click **Start Now** on the main page and then go to the **Apply** section and click the icon to display the AP list, as highlighted in Figure 3-7.

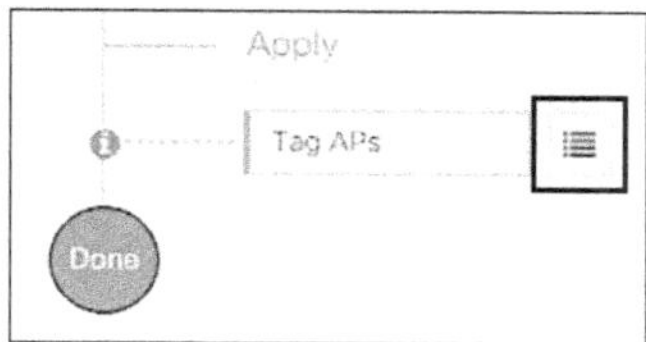

Figure 3-7 *Click the menu icon to select APs and assign tags*

On the following page, select the APs you want and click **+ Tag APs** blue icon; then assign the tags in the popup window, as shown in Figure 3-8.

Starting with software release 17.6, the tags can be automatically saved on APs leveraging the **AP tag persistency** feature. This feature is enabled globally on the controller with the following CLI command:

```
C9800(config)#ap tag persistency enable
```

In release 17.6, the feature is disabled by default for backward compatibility with previous releases, but Cisco recommends enabling it. When the tag persistency feature is enabled, APs joining a C9800 wireless controller have the configured tags saved on the AP automatically.

Before AP tag persistency was introduced, to push and save the tags to the AP, you had to use a CLI command in exec mode, per single AP:

```
c9800-1#ap name <APname> write tag-config
```

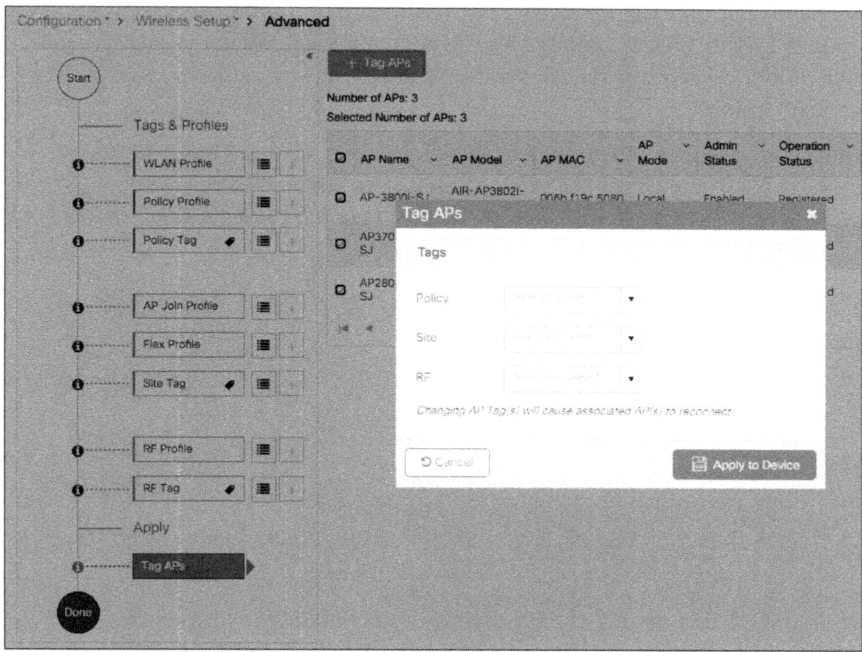

Figure 3-8 *Assign the tags to the selected APs*

This method could be very tedious because the command is done per AP on the CLI. Since release 17.4.1, the same command is also available in the GUI under each AP configuration, as shown in Figure 3-9.

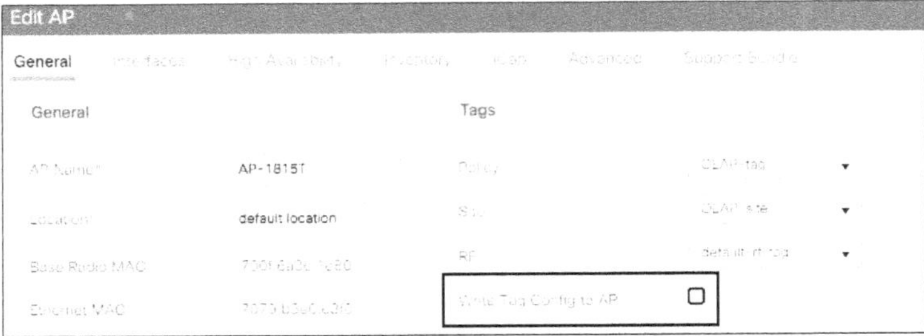

Figure 3-9 *Starting with release 17.4.1, the write tag is available in the GUI*

The operational advantages of the AP tag persistency feature are clear when you need to move APs between wireless controllers. This can be in the context of AP migration or in a primary/secondary (N+1) high availability deployment. Because the tags are saved on the AP, when the AP joins the second WLC, it presents the tags; as long as these tags are configured on the controller, the mapping is honored. Of course, the tag source priorities still apply, and the AP tag source is considered only if no static or filter-based mapping is present for that AP.

Another way to preserve tag assignment when moving APs from one controller to the other is to use an AP tag filter. Let's say you want to move a group of access points between C9800 controllers; more specifically, you want to move APs that are on floor 1 from WLC1 to WLC2. Let's also assume that you have named the APs accordingly as APx_floor1, where *x* is the AP number. In this case, you can easily use the filter tag assignment option.

First, you need to configure the desired tags on both controllers and then, on WLC2, configure a filter rule to match any joining AP name that ends with floor1 and assign it to the desired tag combination. Go to **Configuration > Tags & Profiles > Tags**, click the **AP** tab, and then click the **Filter** tab, as shown in Figure 3-10.

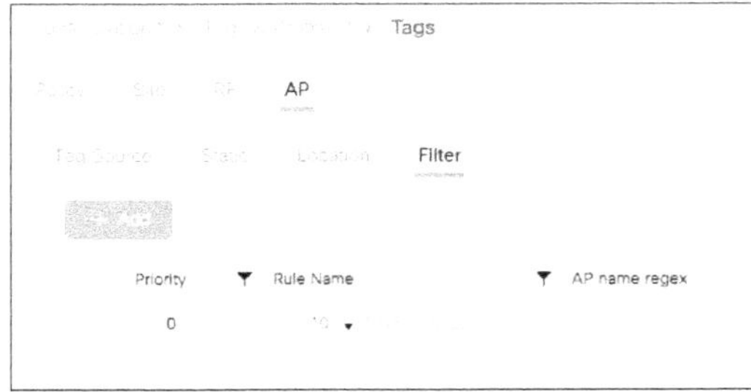

Figure 3-10 *Go to the Filter tab to define the tag filter*

You can add a new rule by clicking **+Add**, shown in Figure 3-10. Figure 3-11 shows an example of a rule that matches any AP name ending with floor1.

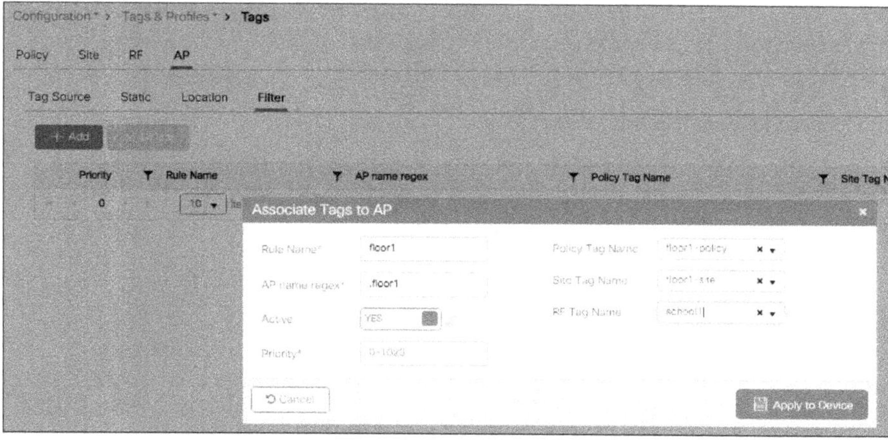

Figure 3-11 *A tag filter*

Another way you can ensure that the AP is assigned to the right tags when joining the C9800 wireless controller is by preconfiguring the AP to tag mapping by loading a CSV file directly in the GUI. This is easily done in two steps:

Step 1. Create the CSV file first. It can be a simple text file, but it needs to be in a specific format: *AP Ethernet MAC, policy tag name, site tag name, RF tag name*. An example is shown in Figure 3-12.

AP-list - Notepad

File Edit Format View Help

```
80e8.6fd8.61e0,OEAP-policy-tag,OEAP-site,default-rf-tag
c4f7.d54d.0b7c,OEAP-policy-tag,OEAP-site,default-rf-tag
```

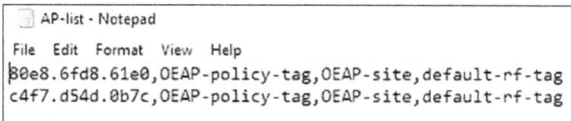

Figure 3-12 *CSV import file format example*

Step 2. Load the CSV file on the WLC by going to **Configuration > Tags & Profiles> Tags,** clicking the **AP** tab, and then clicking the **Static** tab (as indicated in Figure 3-13) and clicking **Select File.**

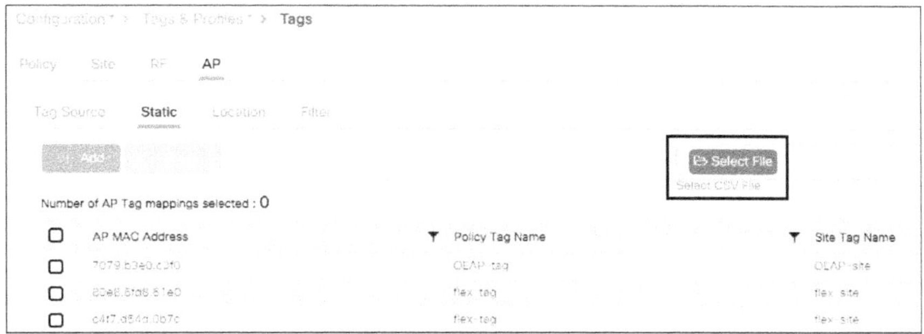

Figure 3-13 *Selecting the file to load the AP to tag mapping*

Because you can modify the existing tags, create new ones, and attach them to the APs in different ways, it's recommended that you validate the tag configuration using the following command in exec mode to catch any inconsistencies:

```
C9800#wireless config validate
```

Moving APs Between Wireless Controllers and Preserving Tags

The preceding section described how the C9800 handles the mapping of tags to APs. Given this information, you should consider the following when moving APs between two C9800 wireless controllers (C9800-1 and C9800-2):

- If the AP on C9800-1 doesn't hold any tag information (either via the **ap tag persistency** feature or via the exec command **ap name <APname> write tag-config**) and there is no explicit mapping configured on C9800-2, the AP is assigned default tags upon joining C9800-2. This means that the AP may start broadcasting different SSIDs or no SSID at all.
- The AP retains the tag information when moving between the controllers if both WLCs have the same mapping of AP to tags configured. This can be done via static configuration, by assigning the AP to a location via basic setup, or via tag filters.
- The AP also retains its tag when moved between the two controllers if the tags are saved on to the AP itself (either via the **ap tag persistency** feature or via the command **ap name <APname> write tag-config**), the tags are defined on both controllers, and there is no higher priority mapping defined (that is, the AP is assigned another set of tags on C9800-2 via static configuration).
- If the AP has saved tags assigned and joins a controller where those tags are not present, it is assigned to the default tags (assuming no other mapping is configured on the controller for that AP).
- In all cases, if the AP retains its tag name assignment but the settings within the tag are different on the two controllers, the AP is configured based on the settings present on the newly joined controller.

Note The previous list applies, for example, to APs moving from the Primary controller to the Secondary upon a Primary failover using N+1 redundancy.

When you're moving an AP from an AireOS wireless controller to a C9800, since the AP doesn't carry any tag information from AireOS, it is mapped to the default tags; this is true unless a static or dynamic tag preassignment is done on the C9800 controller, as explained previously.

Roaming Between Policy Tags

Policy tags are used to decide which SSID is being broadcasted by which AP and with what policy, so they define the broadcast domain for a group of APs. In this, the policy tag is similar to the concept of the AP group in AireOS.

A client roaming between APs configured to broadcast the same SSID but with different associated policy profiles results in a slow roam. In other words, roaming across two different policy tags (same SSID but different policy profile names) forces the client to go through a full authentication and DHCP process to renew its IP address. This behavior is by design on the C9800: if you, as an admin, associate a different client policy with different groups of APs, it would be correct to reevaluate that client policy when the client roams across so that the client has to go through a full reauthentication, and the new policy can be applied. This is true if doing intracontroller roaming, so roaming between APs on the same C9800. When roaming between wireless controllers (intercontroller roaming), you need to set up a mobility group, and we address that in Chapter 6, "Mobility and Client Roaming."

Note If the policy profile associated with the SSID is the same (same name and content) even if using different policy tags, roaming for that SSID is seamless. The slow roam happens if there is a change in the policy profile associated with the same SSID.

In some scenarios, this behavior may not be desirable; hence, it needs to be considered when designing your wireless network with the C9800. Consider the following customer use case: a university has a rule to use /22 subnets across the campus. It uses one networkwide faculty SSID, and because it has more than 1022 users (the max number of IP addresses allocated to a /22 subnet), it needs to assign multiple client subnets to the SSID.

In C9800, there are three common ways of implementing this use case:

1. Using a VLAN override from the AAA server to assign different groups of users to different subnet/VLANs.
2. Using the vlan group feature to map multiple client subnets to the same SSID and assign clients in a round-robin fashion to the available VLANs in the group. This is the equivalent of VLAN Select (a.k.a. the interface group feature) in AireOS.
3. Using policy profiles and tags to map a specific VLAN to the same SSID for each group of APs. This also allows the user to know deterministically which IP subnet the client belongs to as it joins that location (group of APs). In AireOS, this would be achieved using AP groups.

The VLAN AAA override is common and fully supported with C9800. You can also use option 2 by using a feature like AireOS's VLAN Select, which on a Catalyst wireless controller is called VLAN group. Recall that the Cisco Catalyst 9800 doesn't need a Layer 3 interface associated with the client VLAN/subnet, so you can group the Layer 2 VLANs. In the GUI, go to **Configuration > Layer2 > VLAN**, select the **VLAN Group** tab, configure the VLAN group first, and assign the VLANs (VLANs 210 and 211), as shown in Figure 3-14.

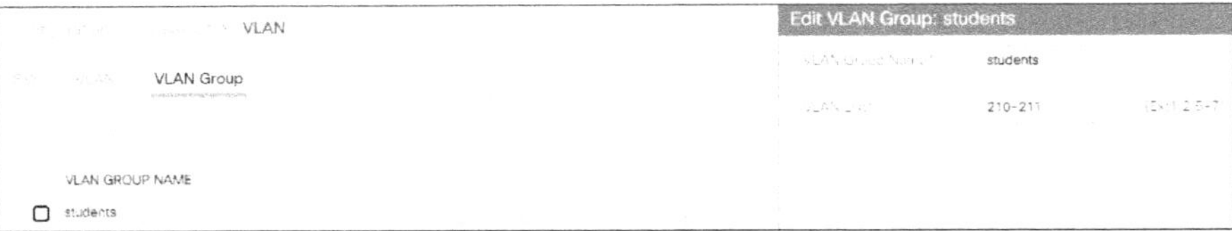

Figure 3-14 *VLAN group definition*

Then configure the policy profile to map the SSID to the defined VLAN group, as shown in Figure 3-15.

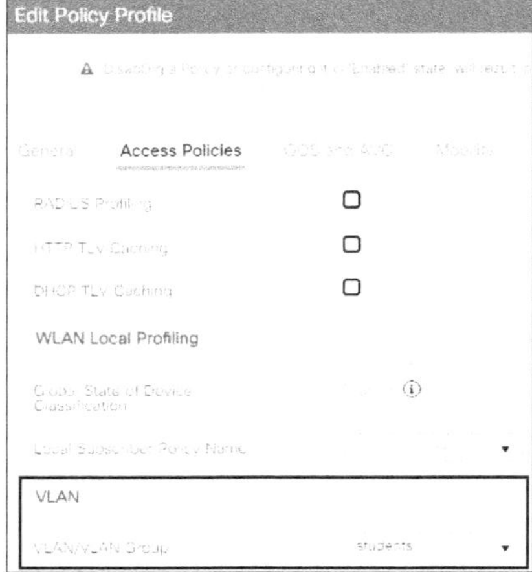

Figure 3-15 *VLAN group assignment within the policy profile*

Then assign all the APs to the same policy tag where the SSID is mapped to this policy.

For the third option, to do something like the AP Group solution in AireOS, you would have to define two policy profiles, one with VLAN 210 and one with VLAN 211, and map them to the same SSID using a different policy tag. Then you apply the different policy tags to the different groups of APs.

In this case, you need to consider the limitation of slow roaming across policy tags mentioned earlier: if the two locations are separated and have an air gap, there is no problem because the client will have to disconnect anyway. But if the locations are in the same roaming domain, you need to consider that the client will go through a full reauthorization as it roams across the two policy tags with different VLANs. This is different from AireOS behavior: an AireOS WLC would allow seamless intra-WLC roaming across two AP groups mapped to different client VLANs.

The good news is that starting with Cisco IOS XE Release 17.3, if the policy profiles differ only for certain parameters (VLAN and ACL being the most important), seamless roaming is also allowed across policy profiles (and related policy tags). To configure the feature, enter the following command in global config mode:

```
c9800(config)#wireless client vlan-persistent
```

Even if the command only mentions **VLAN**, many other parameters can differ between the two policy profiles and still result in a seamless roam. For a complete list of these attributes, see the configuration guide link in the "References" section.

The recommendation is to consider this behavior as you design your policy tag assignment: all APs in the same roaming domain should have the same policy profile. If you need to assign different policies, we recommend you deploy release 17.3 and newer and use the **wireless client VLAN-persistent** feature.

Designing with Site Tags in Mind (Local Mode APs)

As illustrated in Chapter 2, "Hardware and Software Architecture of the C9800", the Catalyst 9800 has a multiprocess software architecture. The main process responsible for AP and client sessions is called the Wireless Network Controller daemon (WNCd); the number of WNCd processes varies from platform to platform, as you can see in Table 3-1.

Table 3-1 *Number of WNCd Processes per Platform*

Platform	# of WNCd Instances
EWC (on AP or C9k switch)	1
C9800-L	1
C9800-CL (small)	1
C9800-CL (medium)	3
C9800-40	5
C9800-CL (large)	7
C9800-80	8

You can verify the number of WNCds in your platform by using the following CLI command:

```
C9800#show processes platform | include wncd
```

When APs join the C9800, they are distributed among the available WNCds (if multiple processes are present). As you can imagine, load-balancing APs (and the related clients) among the available WNCd processes results in improved scale and performance because it better exploits the available resources on the C9800. At the time of writing, AP distribution is based on site tags: APs associated with the same site tag join and hence are managed by the same WNCd.

As you design your Cisco Catalyst wireless network, it becomes important to understand how to best assign APs to site tags. The considerations are different for deployments with APs in local mode or in FlexConnect mode. Let's examine the two separately and start with the general recommendations that you need to keep in mind to get the best performance out of your Catalyst 9800 wireless controller with local mode APs:

1. Use custom site tags and not the default-site-tag when roaming and fast roaming are requirements.
2. Assign the same site tag to all the APs in the same roaming domain.
3. Limit the number of APs you assign to a single site tag (a value of 500 APs per site tag is recommended for best performance).
4. Whenever possible, do not exceed the maximum number of access points per single site tag shown in Table 3-2.

Table 3-2 *Recommended Max Number of APs per Site Tag*

Platform	Maximum Number of APs per Site Tag*
C9800-80, C9800-CL (medium and large)	1600
C9800-40	800
Any other C9800 platform	Equal to the maximum number of APs supported

*This is true for local mode APs. For FlexConnect APs and related remote site tags, if seamless roaming is required, as of 17.6 release, the limit is still 100 APs per site tag (the same as for AireOS).

5. If you're dealing with large deployments or high-density scenarios, use the recommended number of site tags per platform and evenly distribute APs among them.

These recommendations are just that—suggestions. For example, if you have more than 500 APs in the same site tag, things will still work, but you would probably not get the best performance out of your network.

Let's analyze the listed suggestions, one by one.

The first recommendation tells you not to use the default-site-tag. This suggestion would help improve the way the resources are used internally on the C9800, optimizing interprocess communication. Let's try to understand why: if using the default-site-tag, the APs are distributed among the available WNCd processes in a round-robin fashion. Consider a local site (APs in local mode) that you are bringing up, connecting one AP after the other. Suppose you are using the C9800-CL (medium), and therefore, you deal with three WNCd processes. As illustrated in Figure 3-16, the first AP joins WNCd1, the second AP joins WNCd2, the third AP joins WNCd3, the fourth AP is again assigned to WNCd1, and so on, in round-robin fashion. As clients roam from one AP to the other at the site, there is a lot of interprocess (inter-WNCd) roaming, which is not ideal. Will it work? Will fast roaming and 802.11r work? Yes, absolutely; but the design is not optimal, and therefore, it's not recommended.

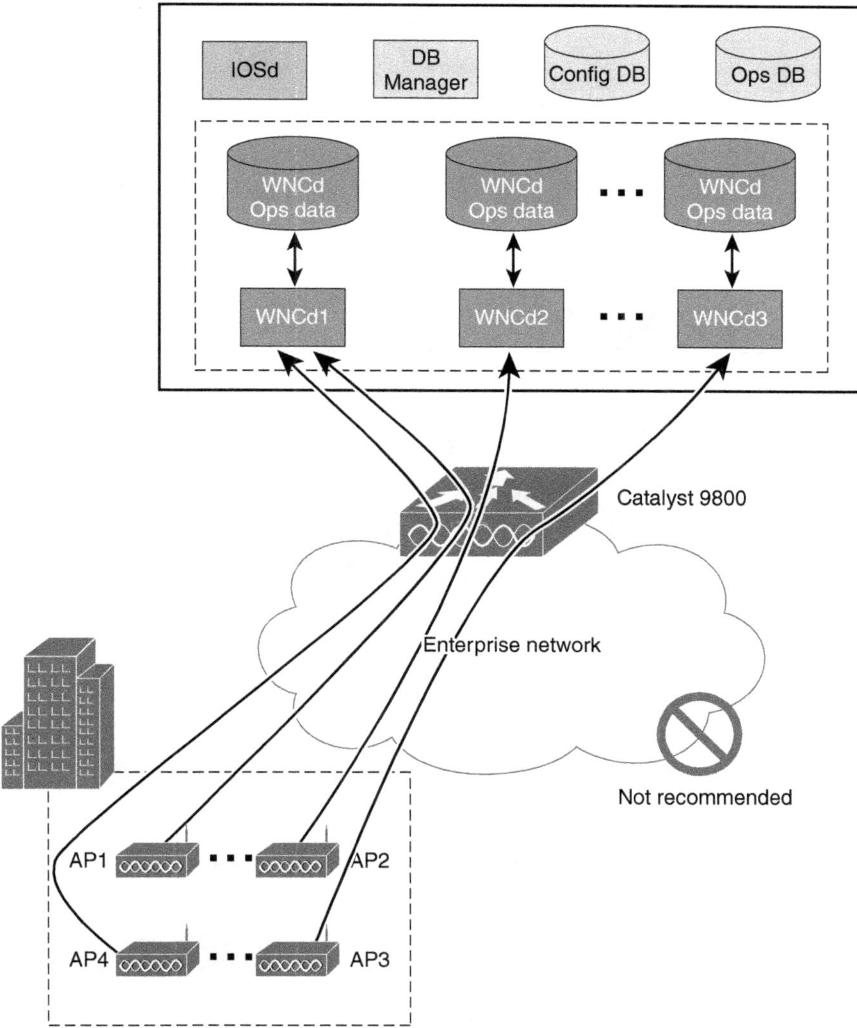

Figure 3-16 *APs to WNCd distribution using the default-site-tag*

In this case, the recommendation is to use a custom/named site tag and assign all the APs at the site to that named tag so that they are all assigned to the same WNCd, and all the roaming triggers intraprocess (intra-WNCd) communication. This scenario is shown in Figure 3-17.

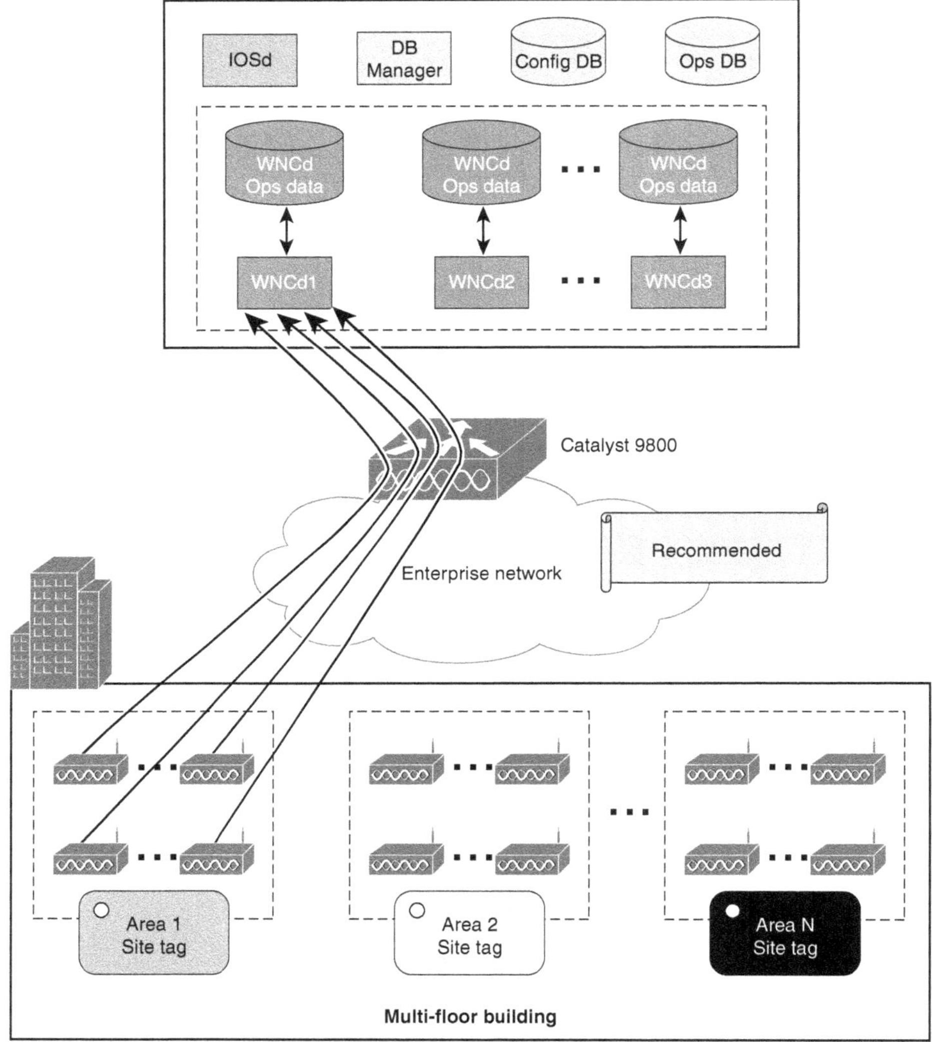

Figure 3-17 *APs to WNCd distribution using custom site tag per roaming domain*

Note Again, here we are talking about APs in local mode. As stated earlier in the case of FlexConnect, if using the default-site-tag, fast roaming is not possible because the clients' keys, derived from the first authentication, are not redistributed among the APs assigned to the default-site-tag. For this reason, using a custom tag is a must for Flex if fast roaming is a requirement.

In releases before 17.6, assigning the same site tag to all the APs in the same roaming domain is also particularly important if you require *optimized* fast roaming for applications that are delay sensitive, such as voice-over WLAN. *Optimized* here means that C9800 would leverage protocols such as 80211k/v to pass additional information to the client and assist the roaming process; for example, the list of neighbor APs the client could roam to is provided via the 802.11k neighbor list. What is a roaming domain? It's considered the group of APs where the majority of client roaming happens: it can be an area within a floor, multiple floors, an entire building, and so on.

When roaming between two APs in different site tags and hence across WNCd processes, the AP neighbor information is lost, and therefore protocols such as 802.11v and 802.11k that rely on this information are not optimized. The recommendation is to assign all the APs in the same roaming domain (where seamless and fast roaming is needed) to the same site tag. This affects only 802.11k/v and doesn't affect fast and seamless roaming, which is supported across site tags.

This limitation is removed starting with release 17.6.1, so clients roaming across site tags can benefit from 802.11k/v. In this release, you have to manually check if the SSID is enabled on the neighboring APs by making sure that it's included in the policy tag. Starting with release 17.7.1, the check is automatic.

Note For APs in local mode (for central association SSIDs), seamless roaming with 802.11r, Cisco Centralized Key Management, and opportunistic key caching (OKC) work across site tags. No limit there.

Why not assign all the APs in one single site tag and get it over with? Here is where the third suggestion applies to optimize the resources internally: For the best performance, you should limit the number of APs per site tag and hence to the WNCd. The number Cisco recommends is around 500 APs per site tag. This is just a reference number that can be used for all the different Catalyst 9800 form factors. Let's be clear: nothing will break if you assign more than 500 APs per site tag if you stay within the limits that have been tested and hence officially supported and that are specified in Table 3-2. Going beyond the maximum limits is not recommended, and you will start seeing some undesired performance effects: the client roamings per second may decrease, and the same for the authentications per second. Customers may also see syslog events indicating an overload in a WNCd process. These are all effects of overloading a single WNCd process. Imagine if C9800 was a car's engine and the WNCd processes its cylinders: if you ran your car with just one cylinder, the results would not be great and optimized, right?

The last suggestion on the list is about high-density (HD) deployments where, to optimize the performance of C9800, it's recommended that you choose the number of site tags according to the specific platform, as listed in Table 3-3.

Table 3-3 *Platforms and Corresponding Site Tag Recommendations*

Platform	Recommended Number of Site Tags
C9800-80	8 or a multiple (16, 24, ...)
C9800-CL (large)	7 or a multiple (14, 21, ...)
C9800-40	5 or a multiple (10, 15, ...)
C9800-CL (medium)	3 or a multiple (6, 9, ...)

After you select the number of custom tags, you also need to try to evenly distribute APs across these site tags. Again, remember that the site tag doesn't have to correspond to a physical site, but you would have to create virtual areas where you group APs.

Let's consider some examples to put all the recommendations together to understand how to implement them:

- You have to design for a very large venue (such as a stadium) with 3000 APs and a large number of clients. Seamless roaming is required everywhere, so this is indeed a large roaming domain. You have selected a 9800-80 to manage this deployment. The recommendation is to identify at least eight (or multiple) virtual roaming areas (grouping sectors in the stadium, for example) where you know that the majority of roaming will happen and define a site tag for each one. Say you chose eight site tags; then you would distribute the 3000 APs across these tags. In this case, it would be 375 APs per site tag. Of course, it doesn't have to be a precise cut, but the recommendation is to have an equal distribution of APs and to avoid overloading a few site tags, even if it would make sense from a physical location/site point of view. On the other hand, if you have small areas (such as the ticketing areas) where you have few APs, merge them with other APs to get to a site tag size that is close to the recommended one, 375 in this case.
- You have a small campus with three buildings with a total of 600 APs on a C9800-40. Most of the time there would be no Wi-Fi coverage (air gap) between the buildings, and there is no roaming across. In this case, you can configure three site tags, one per building. This means 200 APs per site tag, which is well within the recommended settings.
- You have a large campus and multiple buildings for a total of 1200 APs on a C9800-40, and this time roaming needs to be across the entire campus (such as a hospital campus). Because 1200 exceeds the maximum number of APs per site tag, and this is a large roaming domain, it is recommended that you use five site tags (grouping buildings together in five virtual areas). In this case, you would have a very good balanced system with 240 APs per site tag. Remember that seamless roaming is fully supported across the site tags; it's only 802.11k/v that will not be leveraged at the time of writing when you cross the border areas between site tags.

If using Cisco DNA Center, you do not have to deal with APs to site tag assignment at all, unless you want to. Cisco DNA Center hides the implementation details and just presents you with a simple provisioning task: indicate where the APs are located. Starting with release 2.1.x, Cisco DNA Center automatically assigns APs to site tags, including the 500 APs per site tag recommendation, and automatically uses multiple site tags when necessary. The logic is simple: APs assigned to floors within the same building are part of the same site tag; if the number of APs exceeds 500, Cisco DNA Center automatically creates an additional site tag with the same name of the existing one and an increased index.

Starting with release 2.2.2, you can even overcome this rule and assign a custom site tag and put as many APs as you want within a single site tag. A use case for this would be to control which APs are assigned to which site tag and place, say, 600 APs in the same site tag across multiple buildings.

Designing with Site Tags in Mind (FlexConnect Mode APs)

For FlexConnect deployments, a site tag identifies the fast-roaming domain because key caching and distribution only happen within a single Flex site tag. Here are the recommendations for Flex:

- The default-site-tag is a no-go for Flex deployments where fast roaming is a requirement, so using custom site tags is always recommended. The reason is that the client key is not distributed among the FlexConnect APs in default-site-tag. You should configure at least one site tag per Flex site.
- If support for fast seamless roaming (802.11r, CCKM, OKC) is needed, the max number of APs per site tag for a Flex site is 100. This has been extended to 300 APs per site tag starting with release 17.8.
- If you are running release 17.6 and the branch has more than 100 APs, define at least two site tags and design APs to site tag assignment so that each site tag has fewer than 100 APs. Roaming across the two site tags is not a fast roam (the client needs to go through a full reauthentication), the same as AireOS and Flex groups.
- Don't use the same site tag name across multiple FlexConnect sites (this includes default-site-tag). The C9800 doesn't know about physical locations, and there is no point in distributing client keys across APs in different physical locations because roaming will never happen. Also, different site tag names are a requirement to support client overlapping IP addresses across FlexConnect sites for local switching SSIDs.

To better understand this last topic, let's consider a real-life scenario where the customer requirement is to migrate from AireOS to C9800 and at the same time simplify the configuration and hence the number of configured AP tags and profiles.

As a customer, you have hundreds of branches geographically distributed, with a small number of APs for each site. The settings (SSIDs, associated policies, and so on) are the same across sites, and today a FlexConnect group is configured at each site. You have the following design question when migrating to C9800: If I have 5 APs at each location, can I use the same site tag for 20 of these sites? The logic here is that 5 APs multiplied by 20 branches (5 × 20 = 100) gives you a total of 100 APs for each site tag, and that is exactly the limit for seamless roaming in FlexConnect. If fast secure roaming is a requirement, it's recommended to have one site tag at each location, for two main reasons:

1. If you're using overlapping IP addresses across branches, site tags need to be unique.

2. Having APs in the same site tag across geolocations would create unnecessary overhead because the WLC distributes client roaming keys across all the APs in the same site tag even if in different branches. This also unnecessarily consumes AP memory because the AP has to keep info about clients that will never roam to that site.

It's true that you still have as many site tags as branches to configure, but migrating to C9800 allows you to simplify and optimize the configuration. The policy tag and RF tag can be the same for all branches, and the AP Join profile can be modified if some parameters need to be different for each location; otherwise, they also can be the same. If you're not interested in providing fast roaming within the site and there is no need to support overlapping IP addresses across sites, you could use just one site tag, the default-site-tag. As noted earlier, C9800 never distributes client keys among APs in the default-site-tag, so you don't have that key distribution overhead.

One last consideration about site tag design: what if you are forced to have a mix of large and small site tags and you cannot distribute the APs evenly as recommended? This would be the case where you have a deployment with a campus (with local mode APs) and many small remote sites (with FlexConnect APs). As explained earlier, for FlexConnect, every site should be its own site tag because it defines the fast secure roaming domain, so you don't have much choice around the number of tags. In this case, to have the best load-balanced system and follow the recommendations for local mode APs, it's probably best to have two WLCs: one to manage the campus APs and a different one dedicated to the branches, maybe using a 9800-CL to optimize costs.

Summary

This chapter introduces the new configuration model available for the Catalyst 9800 wireless controller. The model leverages the concepts of configuration profiles and AP tags to provide a modular, flexible, and reusable configuration for the C9800. The chapter also introduces important design considerations and best practices to get the best performance and stability out of your C9800-based wireless network. It's important for you to understand these new configuration constructs because they will be used in the remainder of the book.

References

Configuration guide: "vlan-persistent" feature: https://www.cisco.com/c/en/us/td/docs/wireless/controller/9800/17-3/config-guide/b_wl_17_3_cg/m_client_roaming_policy_profile.html

Cisco Catalyst 9800 Series Configuration Best Practices: https://www.cisco.com/c/en/us/products/collateral/wireless/catalyst-9800-series-wireless-controllers/guide-c07-743627.html

Regular expressions in IOS XE: https://www.cisco.com/c/en/us/td/docs/ios/termserv/configuration/guide/xe_16/tsv-xe-book/tsv_reg_express.html

Chapter 4

C9800 Deployment and Installation

In this chapter, you will learn how to install your Catalyst 9800 wireless LAN controller (WLC) and set up your first Catalyst wireless network.

One of the advantages of the C9800 is its deployment flexibility. Private or public cloud, physical or virtual appliance, embedded in a Catalyst switch or access point—you can really deploy the C9800 the way you want. You can think of the C9800 as a network function that you can choose where to run—whether in a dedicated appliance, embedded in a network device, or even in the public cloud. Not only do you have the same look and feel, but given a chosen deployment mode (local mode, FlexConnect, software-defined access [SDA], and so on), the features are the same across all the different form factors.

The first part of this chapter illustrates the different deployment models, giving you important information and useful tips to successfully install the Catalyst 9800. The second part is dedicated to the basic instructions to set up a WLAN and start seeing your WLC in action.

C9800 Deployment Models

Are you considering adopting the public cloud as a key enabler for your digital transformation strategy and leveraging a public cloud-based solution for your wireless access network? Or, given your data protection and security policy, do you still want to have full control of your network and hence want to manage your wireless LAN controller on-premises in a private cloud? The Catalyst 9800 gives you the flexibility to choose the deployment that best fits your requirements.

C9800 for Private Cloud

Private cloud refers to the deployment model where the customer has unique access to dedicated, virtualized or physical, compute and storage resources.

When deployed on-premises in a private cloud, the Catalyst 9800 is usually connected in a centralized location in the company core network or data center, or hosted by a third-party colocation provider. In all cases the device is on a private network.

C9800 Physical Appliance

Customers who want full control of their wireless controller services usually choose the physical appliance form factor. This model provides dedicated, purpose-built, high-performance hardware with Cisco-embedded software (in this case, the Cisco IOS-XE operating system). You have one single box to manage, one single place to go to troubleshoot the network; a lot of customers love this model. Based on your scalability and throughput requirements, you can choose among the following models, as you saw in Chapter 2, “Hardware and Software Architecture of the C9800”:

- **Cisco Catalyst 9800-80:** This model supports up to 6000 access points and 64,000 clients, with a line rate throughput of 80 Gbps.
- **Cisco Catalyst 9800-40:** This model supports up to 2000 access points and 32,000 clients, with a line rate throughput of 40 Gbps.
- **Cisco Catalyst 9800-L:** This model supports up to 250 access points and 5000 clients, with a line rate throughput of 5 Gbps. The scale and performance numbers double with the performance license.

In terms of C9800 physical appliance installation, after racking, powering, and cabling the box following the best practices found in the installation guide, you would have to go through the bootstrap configuration. This is the minimal configuration that allows you to connect securely to the box (via SSH or HTTPs) to finish the configuration. There are two ways to bootstrap the C9800 appliance:

- Manually, through the console port using the IOS-XE CLI
- Automatically, using the Plug and Play (PnP) protocol Zero Touch Provisioning (ZTP), Pre-boot Execution Environment (PXE) or other automation methods

The C9800 physical appliance provides both an RJ-45 port and a USB serial port to attach a console terminal. The RJ-45 console is an asynchronous serial (EIA/TIA-232) port labeled CON on the front panel. The USB serial console port connects directly to the USB connector of a PC using a USB Type A to 5-pin mini-USB Type-B cable. The default parameters for the console port are 9600 baud, 8 data bits, no parity, and 1 stop bit.

Starting with release 17.4.1, when you connect to the console port for the first time, the C9800 presents a setup wizard to guide you through the box configuration. This setup is covered later in this chapter. For previous releases, you would have to enter the minimal configuration manually. The minimal bootstrap configuration would be

- Hostname
- User login credentials (username and password) with a privilege level of 15 to give the admin access to the box
- Enable password

- IP address and route to reach the box (an SVI interface is preferred)
- SSH configuration

The Catalyst C9800 offers the possibility to push the bootstrap configuration via the PnP protocol. This is supported starting with release 17.3.1 on all the physical appliances, and you may leverage Cisco DNA Center as the PnP server. Cisco DNA Center has a dedicated PnP provision workflow, as represented in Figure 4-1, that greatly simplifies the C9800 onboarding operations.

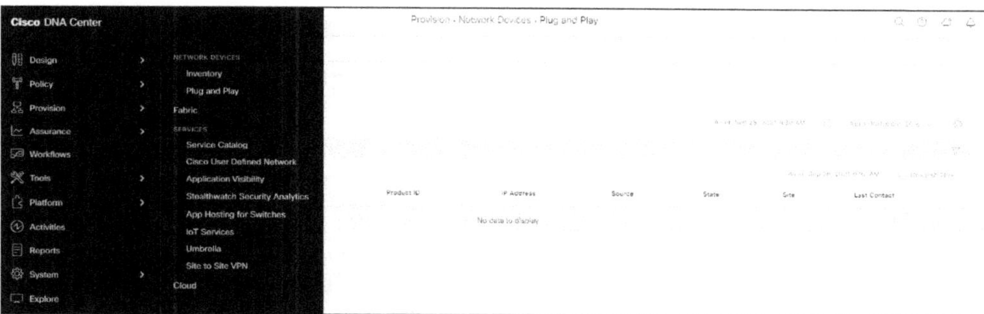

Figure 4-1 *Plug and Play provision flow*

The PnP workflow allows you to claim the device, assign it to a site, and push the basic configuration (wireless management interface, credentials, NTP, country code, and time zone) to make the wireless controller operational in a few clicks.

How does a Catalyst C9800, out of the box, find the PnP server to connect to? There are three different methods:

- **DHCP option 43:** Add the option 43 string into your DHCP server configuration to include your Cisco DNA Center's IP. The PnP agent running on the C9800 receives that string via DHCP and starts the PnP connection process. Here is a sample configuration for the DHCP server running on an IOS-XE router:

```
ip dhcp pool vlan10
network 10.100.10.0 255.255.255.0
default-router 10.100.10.1
option 43 ascii 5A1N;B2;K4;I<IP address of PnP server>;J80
```

  In this case, VLAN10 is the VLAN where C9800 would get its IP address

- **DNS discovery:** Map the pnpserver.localdomain hostname to Cisco DNA Center's IP in the locally administered DNS server. You need to pass the DNS domain and DNS server IP information via DHCP to C9800 for the resolution to work.

- **Plug and Play Connect service:** This method relies on the public cloud Cisco PNP redirection service. When C9800 doesn't receive any PnP info via DHCP or DNS, it resolves the DEVICEHELPER.cisco.com fully qualified domain name (FQDN) and discovers and connects to the Cisco PnP Connect service. There you would have to preconfigure a profile to map the C9800 device (serial#) to a Cisco DNA Center IP address or hostname.

When C9800 discovers and reaches the PnP server on Cisco DNA Center, it appears in the Plug and Play dashboard, ready to be claimed. This is shown in Figure 4-2.

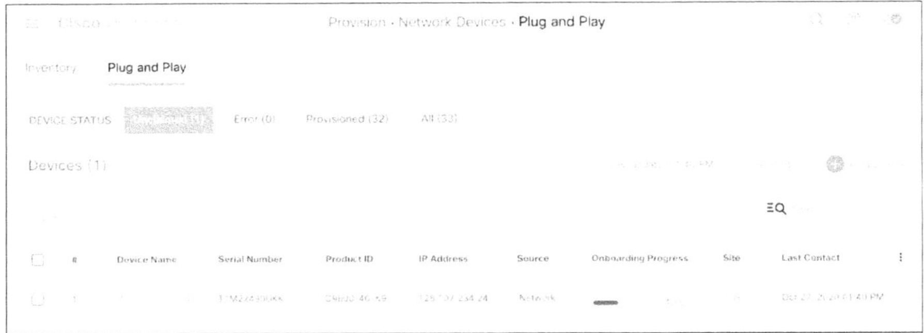

Figure 4-2 *C9800 ready to be claimed in Plug and Play dashboard*

Remember that for a PnP agent to run on the C9800, there should be no startup configuration at all, and you should not type anything on the command-line interface (CLI). If you want to watch PnP in action on an already-configured C9800, you can issue the following command:

```
C9800# pnpa service reset
PnP Reset command will erase config and reload the device.
Are you sure you want to continue? [yes/no]: yes
```

This command resets the box to the factory default and restarts the PnP process.

One important detail: The PnP protocol is assumed to use VLAN 1 as a default to receive the DHCP IP address. If VLAN 1 is not configured on the switch and you want to use another VLAN to allow it to reach the PnP server, you have to issue the following command on the upstream switch where C9800 is connected:

```
switch(config)#pnp startup-vlan ?
  <2-4094> PNP vlan id; default vlan is 1
```

Usually, C9800 is connected to the upstream switch via multiple ports in an EtherChannel or a Link Aggregation Group (LAG) configuration for better throughput and resiliency. Here are the steps that you need to keep in mind in this case:

1. An upstream switch must be configured in LACP ACTIVE mode (PAGP is not supported).

2. An upstream switch can have one or more links in the LAG, but C9800 adds only one link into port-channel 1 that will be created automatically.

3. This autoconfiguration of port-channel 1 allows DHCP to get the IP address on either vlan1 or PnP? startup-vlan (if configured on the upstream switch).

4. When C9800 reaches DNAC, the template push can add remaining links to port-channel 1 or create a new port-channel <xyz> and move all upstream links into it.

At the time of writing, the PnP flow on Cisco DNA Center creates the LAG configuration but only with a primary uplink interface; the rest of the ports can be added outside of the PnP process—for example, using a configuration template during provisioning. Also, you need to make sure that the PnP VLAN and the VLAN where the wireless management interface is configured are the same.

The physical appliance provides a dedicated management Ethernet port, called the service port (SP). Just like your Catalyst routers and switches, this interface is internally named GigabitEthernet0 and usually connected to an out-of-band (OOB) network. The purpose of this interface is to allow administrative users to perform management tasks on the controller connecting via HTTP/HTTPS or SSH/Telnet. The interface is most useful in troubleshooting scenarios when other data plane forwarding interfaces are inactive or unreachable. The management interface also provides a complete control plane and data plane separation because it is part of its own internal Virtual Route Forwarding (VRF), called Mgmt-intf. The VRF is preconfigured and cannot be changed.

If the box is unpacked on a desk in the lab, you may decide to access it directly via a browser through one of the front-panel 1G/10G ports or via the service port. C9800, out of the box, acts as a DHCP server for clients connected through the uplink ports, so you can use these steps to connect:

1. Connect the laptop to the front panel port with SFP using Ethernet cable.
2. The laptop is assigned an IP in the 192.168.1.x pool automatically, via DHCP.
3. Access the Day 0 GUI at http://192.168.1.1.
4. Use **webui** as the username and **appliance serial number** as the password (or **admin/admin**).

To receive a DHCP address from the box, you need to set the DHCP client-id = webui on your laptop:

- For MacOS laptops, go to **System Preferences > Network > Advanced > TCP/IP**, as shown in Figure 4-3.
- For Windows laptops, go to **Registry Edit > HKEY_LOCAL_MACHINE > SYSTEM > CurrentControlSet > Services > TCP/IP > Interfaces > Add DhcpClientIdentifier.**

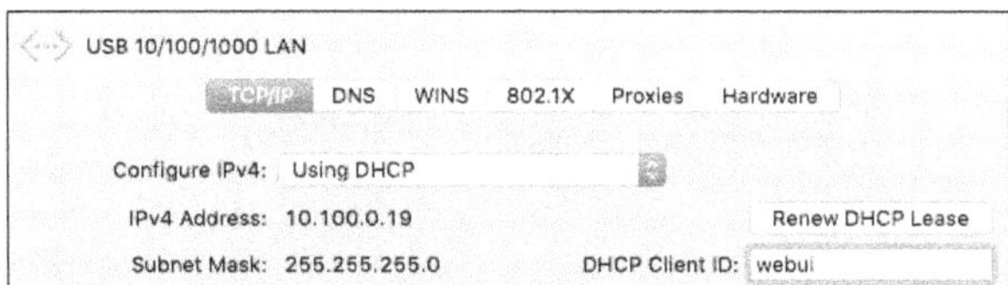

Figure 4-3 *DHCP client ID setting on Mac*

If you want to connect using the SP port, the procedure is a bit different because you do not get a DHCP IP through this port:

1. Connect your laptop to the SP port using Ethernet cable.
2. Assign a static IP to the service port via the CLI, connecting through the console port.
3. Access the GUI at http://<SP IP address>.
4. Use **webui** as the username and **appliance serial number** as the password (or **admin/admin**).

What if you want to reset the appliance to factory defaults? This would include configuration, additional downloaded images, additional files, boot variables, and all. A use case for this would be if you must return the appliance and want to make sure it has nothing on it. The command for this is

```
C9800# factory-reset {all | config |boot-vars}
```

Use the **all** option to erase all the content from the NVRAM, all the Cisco IOS images except the current boot image, boot variables, startup and running configuration data, and user data. Use the **config** option to reset the startup configurations. Use the **boot-vars** option to reset the user-added boot variables. This command is supported only for physical appliances.

C9800 Virtual Appliance

The Cisco Catalyst 9800-CL is the next generation of enterprise-class virtual wireless controllers for the cloud. Specifically for the private cloud, the C9800-CL is a virtual machine that you can deploy using the hypervisor and server hardware of your choice in your data center.

A virtual machine (VM) is a software implementation of a computing environment in which an application with its own operating system can be installed. The hypervisor enables multiple applications to share a single host machine. It does this by emulating a physical computing environment: although each operating system appears to have the dedicated use of the host's processor, memory, and network resources, the hypervisor controls and allocates only the required resources to each application and ensures that the operating systems (VMs) do not disrupt each other. This allows the C9800 VM to provide the following benefits:

- **Hardware independence:** Because the controller runs on a VM, it can be supported on any x86 hardware that the virtualization platform supports.
- **Sharing of resources:** The resources used by the wireless controller are managed by the hypervisor; these resources can be shared among VMs.
- **Flexibility in deployment:** You can easily move the C9800 VM from one server to another.

Figure 4-4 illustrates the different supported options for C9800-CL.

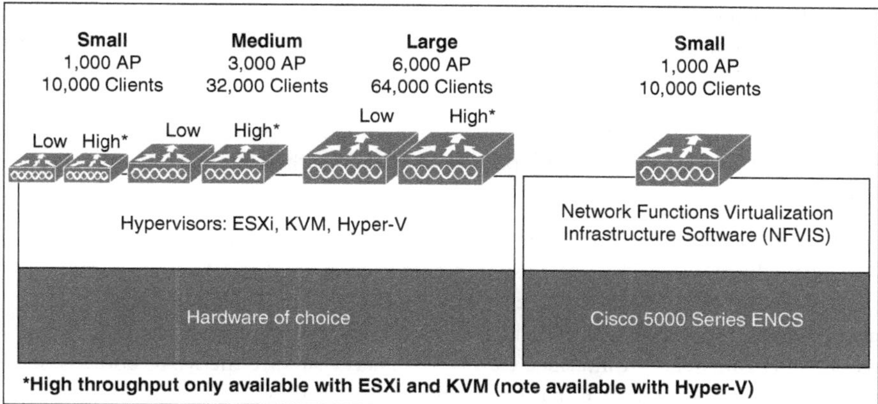

Figure 4-4 *Cisco Catalyst 9800-CL for private cloud*

C9800-CL comes in multiple scale and throughput profiles with a single deployment package to best meet your organization's needs. The profile is selected at installation time, so you don't have to download a different image.

It can be deployed on the most common hypervisors: VMware ESXi, KVM, Hyper-V, and Cisco NFVIS (on ENCS). For the latest on the supported hypervisor versions, see the Cisco Catalyst 9800-CL Wireless Controller for Cloud Deployment Guide (see the link in the "References" section).

C9800-CL supports up to 2 Gbps of throughput in a centralized wireless deployment. This is with the low throughput profile and using Internet Mix (IMIX) traffic; with larger packets, you would see more gigabits per second. With a high (enhanced) throughput profile, it can go up to 5 Gbps on ESXi and KVM, with the right set of network cards and resources (SR-IOV-enabled NIC card).

The C9800-CL has deployment mode parity with the physical appliance; it supports centralized, FlexConnect, Mesh, and Fabric (SD-Access) AP deployment modes with the same features as the physical form factors. This means, for example, that the C9800-CL can be deployed at full AP and client scale, it supports stateful switchover (SSO) and In Service Software Upgrade (ISSU) high availability, it can be configured to be a guest anchor, it can be managed via Cisco DNA Center or through programmable interfaces, and so on; all this is not available in the AireOS-based virtual WLC.

By default, the Catalyst 9800-CL boots up with three network interfaces. The possible interface mapping may be as follows:

- **GigabitEthernet1:** Device management interface. This may be mapped to the out-of-band management network. This is the equivalent of the service port on the physical appliance.
- **GigabitEthernet2:** Wireless management interface. Map it to your network to reach APs and services. Usually, this interface is configured as a trunk to carry multiple VLANs (wireless management and client traffic).

- **GigabitEthernet3:** High availability interface. Map it to a separate network for peer-to-peer communication for HA SSO. This is the equivalent of the RP port. This port is not needed if HA SSO is not going to be configured.

Do not connect two interfaces to a single virtual network because that may cause network loops. When a trunk port is used, pruning the VLANs that can reach the wireless controller is recommended.

One other important factor to keep in mind is the security settings in the virtual switch (vswitch) on the hypervisor. Usually, a vswitch port is connected to one application, which means one MAC address. In the case of C9800-CL, even if deployed with a single interface, you have more than a MAC address behind that port: you have the MAC address of the virtual network interface card (vNIC), and you have the MAC address of the wireless management interface that is usually associated to an SVI.

By default, most hypervisors like VMware prevent MAC addresses other than the one of the vNIC to be transmitted or received on the vswitch port. For C9800-CL to have network connectivity, you need to change these settings. Let's look specifically at VMware ESXi: as illustrated in Figure 4-5, you need to change the security settings on the port group and set both Promiscuous mode and Forged transmits to Accept.

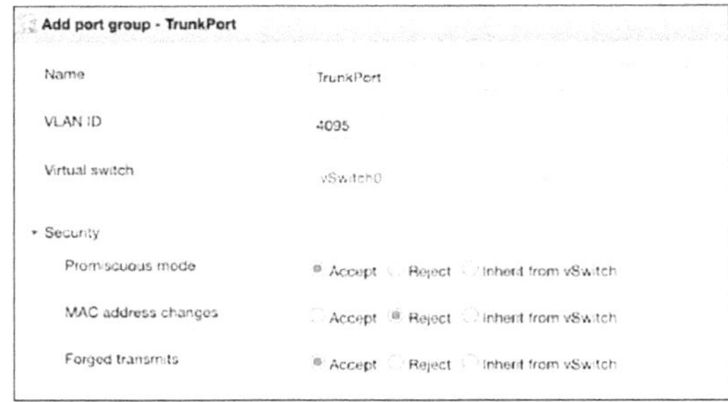

Figure 4-5 *Security settings on VMware port group*

Promiscuous mode allows MAC addresses other than the one of the vNIC to be delivered to the C9800-CL (so it's used for the incoming traffic); Forged transmit is used for the outgoing traffic because it slows the source MAC to be different from the one on the vNIC.

These settings are needed for both trunk and nontrunk connections. You can set the security settings at the vswitch level or at the port group within the vswitch; if you do it at the switch level, changes affect all the VMs on that vswitch. The recommendation is to dedicate a port group to the C9800 and enable Promiscuous mode and Forged transmit there; if the VLANs are available only at that specific port group level, these settings do not affect other VMs connected to other port groups.

Here is a list of additional best practices and recommendations you should keep in mind when deploying C9800-CL in a private cloud:

- Allocate CPU and memory resources as reported in the data sheet and deployment guide.

- Hyperthreading is not supported and needs to be disabled. This means that vCPUs cannot be shared with other VMs and need to be equal to the numbers of physical cores present.
- VMware vMotion is supported with caveats when deploying C9800-CL in an SSO configuration: Due to a limitation in ESXi for Virtual Guest Tagging (VGT Mode), upon switchover, there might be an extended data outage if no packets originate from 9800-CL. As a workaround, you need to initiate a continuous ping from the 9800-CL to update the MAC address in the right port of the physical switch. With a standalone C9800CL, VMware vMotion works without caveats.
- Operations like VMware DRS, Snapshot, vMotion, and vNIC teaming are not supported with SR-IOV interfaces.
- Cloning from snapshots is not supported.

It is common for customers using virtualization to leverage automation tools to automatically spawn and configure virtual machines. But there might be a case where the customer wants to bootstrap the C9800-CL using the PnP protocol. The PnP process was described in the previous section, "Catalyst 9800 Physical Appliance," for the physical appliance, and it applies for C9800-CL as well. As you recall, the PnP agent runs only when there is zero configuration on the device. When you are deploying using VMware vCenter and an .ova file, there is an OVF template that embeds some default configs:

```
<ovf:Property ovf:key="enable-scp-server" ovf:type="boolean"
ovf:userConfigurable="false" ovf:value="true" />
<ovf:Property ovf:key="enable-ssh-server" ovf:type="boolean"
ovf:userConfigurable="false" ovf:value="true" />
<ovf:Property ovf:key="ios-config-0001" ovf:type="string"
ovf:userConfigurable="false" ovf:value="ip http server" />
<ovf:Property ovf:key="ios-config-0002" ovf:type="string"
ovf:userConfigurable="false" ovf:value="ip http secure-server" />
<ovf:Property ovf:key="ios-config-0003" ovf:type="string"
ovf:userConfigurable="false" ovf:value="netconf-yang" />
```

The VMware orchestrator pushes these useful configurations to automatically enable SSH, SCP, HTTP and HTTPs, and Netconf. But these settings prevent PnP from running when the VM boots up. The same thing happens if you do a configuration reset and reload the box; when the VM boots up again, the basic configuration is pushed.

If you are using vCenter but you still want to use PnP, here is the procedure to make sure that the VM starts with no startup configuration:

1. Deploy the VM from an .ova file as you normally do.
2. When you get to the Customize Template section, skip filling out the additional configuration.
3. After you finish deploying the OVA and the VM is copied to the ESXI host, go to Edit settings before powering on the VM.
4. In this section there are two CD/DVD drives. Drive #1 contains the ISO datastore with the 9800-CL image. Drive #2 contains a datastore ISO with the bootstrap

configs. Delete Drive 2 and save the settings (it's not enough to disconnect the disk; you need to delete it).

5. Power on the VM. There is no startup configuration, and now the PnP agent starts.

If customers want the advantage of virtualization but don't want to own the physical infrastructure, they can leverage the C9800-CL as a service in the public cloud for deployments where FlexConnect local switching fits the customer requirements. This scenario is described later in this chapter.

Embedded Wireless Controller on Catalyst AP and Switch

If you are deploying wireless at a branch or small campus location and you don't want to install, power, configure, and maintain an appliance (either physical or virtual), you can use the Cisco Embedded Wireless Controller (EWC) in the Catalyst 9100 access point or in a Catalyst 9000 switch. In this mode, the wireless LAN controller becomes a service that you embed in the network infrastructure.

The Cisco Embedded Wireless Controller on Catalyst Access Points (EWC-AP) is the next-generation Wi-Fi solution, combining the most advanced enterprise features with the latest Wi-Fi 6 access points, creating a best-in-class wireless experience for your controller-less deployment. The EWC-AP advantages are shown in Figure 4-6.

If considering this deployment mode, consider that EWC-AP is ideal for small deployments (up to 100 APs and 2000 clients ... well, not really small) where the customer cares about ease of deployment. You can manage it via Cisco DNA Center if you have many sites; or you can decide to use the GUI and/or the EWC mobile application if it is a standalone deployment. EWC-AP is all about ease of provisioning, and when the AP boots up in the EWC mode, it broadcasts a provisioning SSID ending with the last digits of the MAC address. You can connect to the provisioning SSID using the PSK-based password **password** and when opening a browser, you are redirected to mywifi.cisco.com, which takes you to the AP web UI. Enter the username as **webui** and the password as **cisco**.

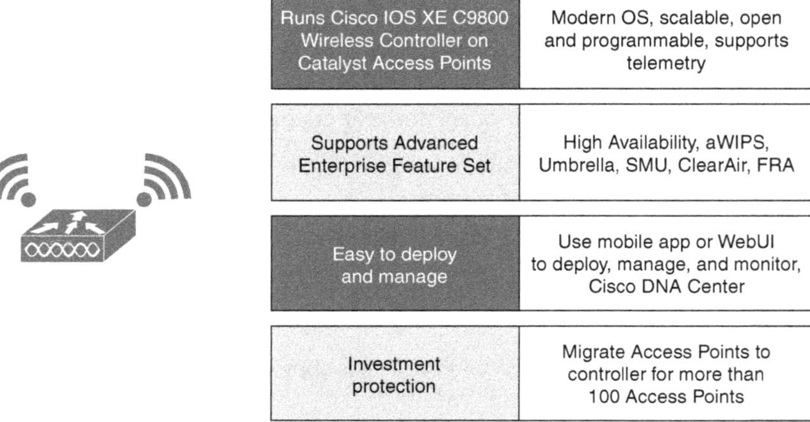

Runs Cisco IOS XE C9800 Wireless Controller on Catalyst Access Points
Modern OS, scalable, open and programmable, supports telemetry
Supports Advanced Enterprise Feature Set
High Availability, aWIPS, Umbrella, SMU, ClearAir, FRA
Easy to deploy and manage
Use mobile app or WebUI to deploy, manage, and monitor, Cisco DNA Center
Investment protection
Migrate Access Points to controller for more than 100 Access Points

Figure 4-6 *Cisco EWC-AP characteristics*

All Cisco Catalyst 9100 access points (except for Catalyst 9105AXW) can run in master mode where the wireless controller functions run, and all the Catalyst APs and the Cisco Aironet 802.11ac Wave 2 access points can operate in client serving mode.

The other model of the embedded Catalyst 9800 wireless controller is the one on the Catalyst 9000 switches. Although it is supported for non-SDA deployments in the 17.3 release (and only in this release), it's critical to use this mode only in the context of software-defined access (SDA) and hence deployed and managed using Cisco DNA Center, as illustrated in Figure 4-7.

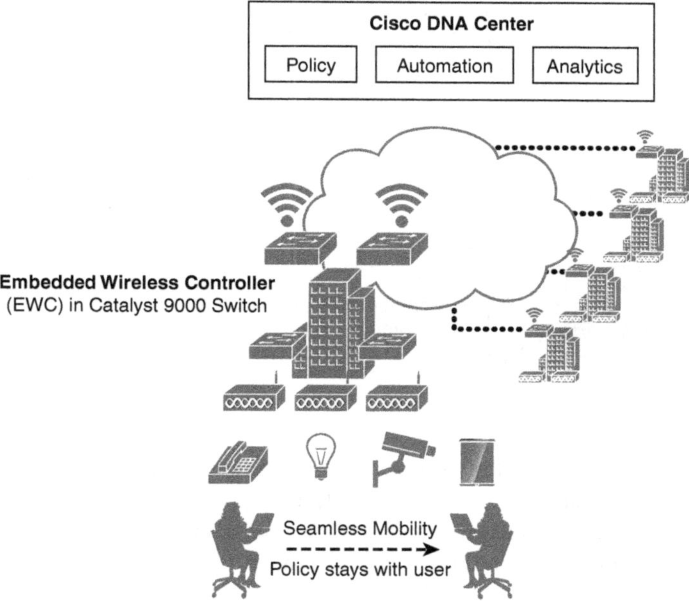

Figure 4-7 *Cisco Embedded Wireless Controller (EWC) in Catalyst 9000 switches for SDA*

SDA wireless is outside the scope of this book. Refer to the SDA Wireless deployment guide for more information.

C9800 for Public Cloud

What does "for public cloud" mean for a wireless LAN controller service? Let's try to understand this first and then see how C9800-CL can be deployed in the public cloud.

According to the National Institute of Standards and Technology (NIST) special publication 800-145, three cloud service models are commonly available:

- Software as a Service (SaaS)
- Platform as a Service (PaaS)
- Infrastructure as a Service (IaaS)

Figure 4-8 shows a representation of these service models, highlighting the most important stack components and the areas of responsibility between the cloud provider and the consumer that is using these services.

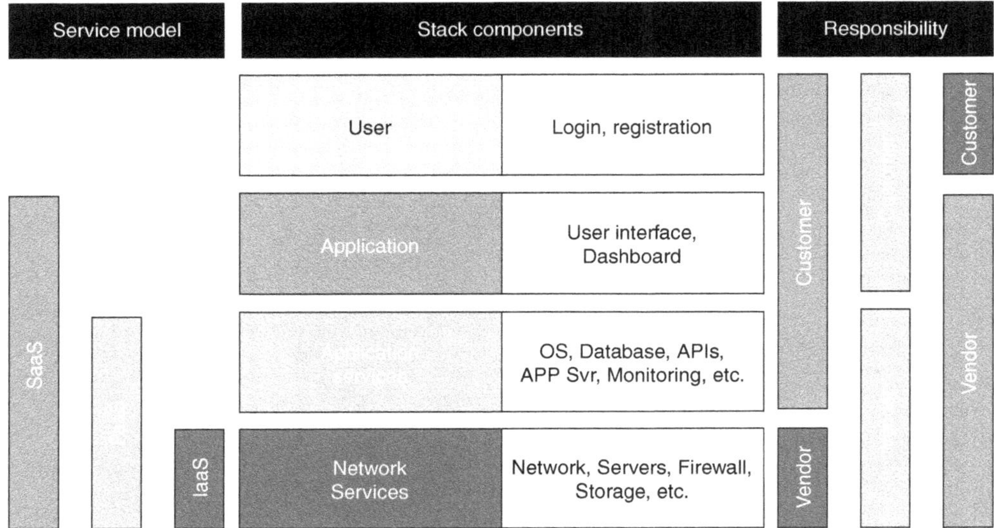

Figure 4-8 *Public cloud service models*

The SaaS model provides the highest level of abstraction and simplification. The whole "service" stack, from the infrastructure to the application level, is provided by the cloud service provider. As the customer, you directly access the service through a user web interface or an application programming interface (API). On the other side, the consumer does not manage or control the underlying cloud infrastructure. Cisco Meraki leverages this model.

In the PaaS model, the cloud service provider provides all the components of the platform needed to run your own applications. This includes the underlying cloud infrastructure networking and security plus additional services like databases and monitoring tools.

Finally, the IaaS model provides the computing resources, along with the storage and networking components (such as routers and firewalls), and leaves the customer to deploy applications and services. This model gives the customer the highest level of control and flexibility, but some integration work is required.

The IaaS model is the one chosen for the Catalyst 9800 wireless controller for the public cloud. This means that you can rely on the public cloud vendor to provide the networking, the computing, and the security infrastructure, but then you will be fully managing the C9800 as a virtual machine in the cloud.

Customers should look at deploying C9800-CL in the cloud with an IaaS model for two main reasons:

- Retaining the customization and control they usually have with the controller on-premises in the private cloud
- Exploiting the advantages of the public cloud

There are many advantages in adopting the public cloud. The most significant ones for the C9800 are as follows:

- **Agility:** It takes a few minutes to spawn a C9800-CL instance in AWS, CGP, or Azure; it becomes extremely easy to launch a wireless controller to test some new feature or functionality and terminate it when done.
- **Scalability:** There are no physical limits in the public cloud, so you can add new instances as the requirements for additional APs or clients increase.
- **Global footprint:** This is important for latency but also for security and privacy policies. The public cloud providers have a global footprint, so from any location where you install APs, you should be able to reach a C9800-CL instance in the cloud in less than 50 ms. Some customers have a strict security policy dictating that user data and traffic need to stay within the region; the public cloud providers have a data center in every geographical region.
- **OPEX model:** You shift from a capital expenditure (buying up front) model to an operational expenditure (pay-as-you-go) model. It may also reduce the data center footprint and infrastructure costs.

At the same time, the customer has full control of the configuration of the Catalyst wireless controller: what software image to run, what functionalities to turn on or off, what configuration tweaks to apply. For the deployment modes supported, there is no difference in deploying a wireless controller in a private cloud or public cloud. On the other side, compared to SaaS, using an IaaS model and having the flexibility to deploy and manage your own instance in the cloud mean that more work is on the customer to integrate the C9800-CL with the cloud infrastructure.

Cisco Catalyst 9800-CL is available as an Infrastructure-as-a-Service (IaaS) solution on the most important cloud providers' marketplaces: Amazon Web Services (AWS), Google Cloud Platform (GCP), and Microsoft Azure. Two deployment modes are supported:

- **Managed VPN deployment mode:** A VPN connection is established between the virtual private cloud (VPC), or equivalent, where the C9800-CL is installed and the on-premises sites where the APs are located. This could be a VPN tunnel between a router on-premises and a cloud VPN router service, or it could be a direct private connection between the customer data center and the cloud provider (for example, AWS Direct Connect). Figure 4-9 shows an example of this deployment mode for AWS.

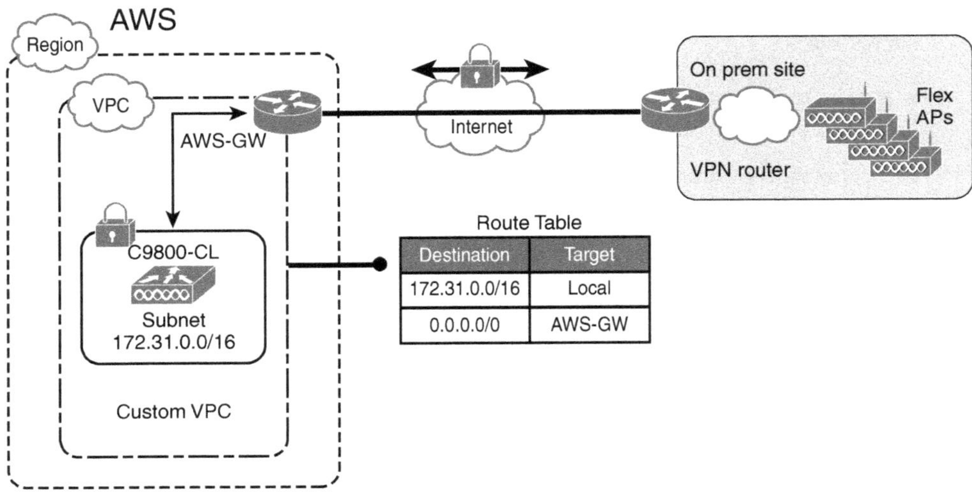

Figure 4-9 *Managed VPN deployment mode on the public cloud*

- **C9800-CL with public IP:** This mode is supported starting with 17.3.2 and allows the APs to connect directly to the WLC in the public cloud, without VPN.

Exposing C9800-CL with a public IP means that you need to implement some sort of filtering to make sure that only your own APs join the cloud instance. There are multiple ways of doing that filtering on IP subnet, AP MAC address, and starting from release IOS-XE 17.3.2, AP's serial number. This is usually preferred because the serial numbers are available at order time, and the C9800-CL can easily be provisioned in advance to accept APs only from a specific list. Figure 4-10 shows how to configure this in the C9800-CL GUI under **Configuration > Security > AAA > AAA Advanced > AP Policy.**

AAA | Show Me How >

AAA Advanced

Global Config
RADIUS Fallback
Attribute List Name
Device Authentication
Password Policy
AAA Interface

Authorize APs against MAC — DISABLED
Authorize APs against Serial Number — ENABLED
Authorization Method List — default

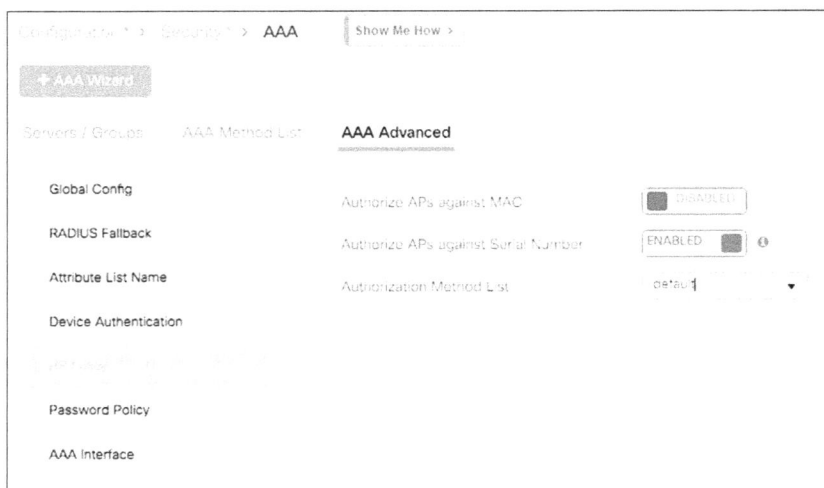

Figure 4-10 *Authorize APs against serial number*

A detailed description of C9800-CL deployment in the public cloud is not within the scope of this book (refer to the related deployment guide for that), but here are some important considerations you should keep in mind when using this option:

- Because APs are on-premises and 9800-CL is in the cloud, only the FlexConnect deployment is supported. The AP join profile is configured to convert the joining APs to Flex mode upon registration.
- Specifically, using Cisco FlexConnect central authentication and local switching is recommended so that traffic stays local to the site. AP local authentication can be configured as a fallback when the AP loses connection to the cloud.
- In the public cloud, there is no concept of a VLAN or Layer 2 broadcast domain; hence, the C9800-CL is deployed with only one Layer 3 interface, the wireless management interface. This means that the device management and wireless management interfaces are the same, and there is no SP port equivalent.
- Because the wireless management interface is a Layer 3 port (management IP assigned to an L3 physical or loopback interface), the following features are not supported: AP sniffer mode, HyperLocation, and Multicast.
- For high availability, C9800-CL in the public cloud supports N+1 deployments with a primary and a secondary WLC. It doesn't support stateful switchover. If you use more than one instance, deploying them in multiple availability zones is recommended to leverage the built-in resilience of the cloud provider infrastructure.
- Cisco DNA Center is not officially supported for two reasons: DNA-C would reside on-premises, and any traffic from the public cloud is charged (for example, assurance traffic). Cisco DNA Center cannot manage a device that is behind a NAT router. If you have a private dedicated connection to the cloud provider VPC, it's really like a private deployment and therefore it would be supported.
- In terms of cloud latency and bandwidth requirements, refer to the requirements for a FlexConnect deployment guide.

Setting Up Your First Catalyst Wireless Network

Now you have all the elements to set up your first Catalyst wireless network. To do this, at a minimum, you must perform the following three tasks:

1. Configure the Catalyst 9800 with the basic settings to make it operational, which is to allow AP and clients to join. This is referred to as the *Day Zero (Day 0) setup*.
2. Allow the AP to discover and join the C9800.
3. Configure an SSID and have a client connect to it.

Next, let's look at how to achieve these tasks. The focus is not on the configuration itself—for that, you have the configuration guides—but rather on the most effective ways you can achieve it.

C9800 Initial Setup

For a Catalyst 9800 wireless LAN controller to be operational, accepting AP and client joins, you need some basic settings on top of what you have already configured during bootstrap, as mentioned in the previous section.

This configuration, at a minimum, includes the following:

- Hostname, login credentials, enable password, IP address, and route to reach the box (This is what was considered part of the bootstrap configuration.)
- Mode of operation—standalone or part of the SSO pair (active or standby)
- NTP and time zone settings
- Country code within a regulatory domain
- Wireless Management Interface (WMI)
- Trustpoint certificate for APs to join (This certificate is needed only on the C9800-CL because the physical appliances have the manufacturer installed certificate, or MIC.)
- WLAN (Optional; this could be configured as part of the Day 1 configuration.)

The Day 0 configuration can be done in multiple ways; the following are the most common:

- IOS-XE CLI setup wizard, also known as Cisco Setup Command Facility
- Day 0 web interface
- Cisco DNA Center provisioning

Starting in release 17.4.1, Catalyst 9800 supports a wireless setup wizard through the CLI interface. This is a step-by-step guided procedure to enter the Day 0 configuration. Before this release, the setup facility was generic for an IOS-XE device, and you could not configure any wireless controller settings.

When connecting to the console on an out-of-the-box C9800, with no startup configuration, you get this banner:

```
**********************************************************************
This is a Wireless LAN Controller (WLC) step wizard. This wizard gives
you the option to configure a Device management interface, a.k.a.
Service Port. If a separated Service Port is not desired, the device
can also be managed using the same interface which is used for wire-
less management. For such a case, please select [no] in the prompt
below
**********************************************************************
```

```
Configure device management interface?[yes]:
```

If you select **yes**, then you are asked to enter the SP port and related IP address information, plus the user login credentials. When done, you get another banner:

```
*******************************************************************

Basic management setup is now complete. At this point, it is possible
to save the above and continue wireless setup using the webUI (for
this, choose 'no' below)

*******************************************************************

Would you like to continue with the wireless setup?[yes]
```

At this point you have the choice to break from the wizard and complete the configuration directly on the web interface through the Day 0 dashboard, which is covered later in this section. If you select **no** at the prompt, you see

```
Configure device management interface?[yes]: no
```

This means you want to set up the box with only one interface for device and wireless management, and the setup facility prompts you to enter all the information to configure the WMI. This includes a VLAN, IP address, route and gateway info, and login credentials. You are prompted with the same banner again, and again you can decide whether you want to proceed with the Day 0 GUI or finish the wireless setup on the CLI wizard. If you answer **yes** or simply click **Enter** to this prompt, you see

```
Would you like to continue with the wireless setup?[yes]
```

Then you are prompted to enter all the Day 0 settings on CLI.

If you enter **no** here or if you go through all the setup, at the end, you get to review the configuration to be applied and choose one of the following options:

```
[0] Go to the IOS prompt without saving this config
[1] Return back to the setup without saving this config
[2] Save this config to nvram and exit
Enter your selection: 2
```

If you select **0**, you discard all the settings done via the setup wizard and you go to the CLI IOS prompt where you can configure the box manually. If you type **1**, you start all over with the setup wizard. If you enter **2**, you save the changes.

If you exit the setup wizard after configuring the basic settings to reach the box, you can finish the configuration via the Day 0 GUI. Just point your browser to the IP address you configured (either the one of the SP or the WMI interface) and use the admin credentials to access it. The Day 0 GUI is a simplified dashboard to guide you through the additional configuration that you need to enter, as you can see in Figure 4-11.

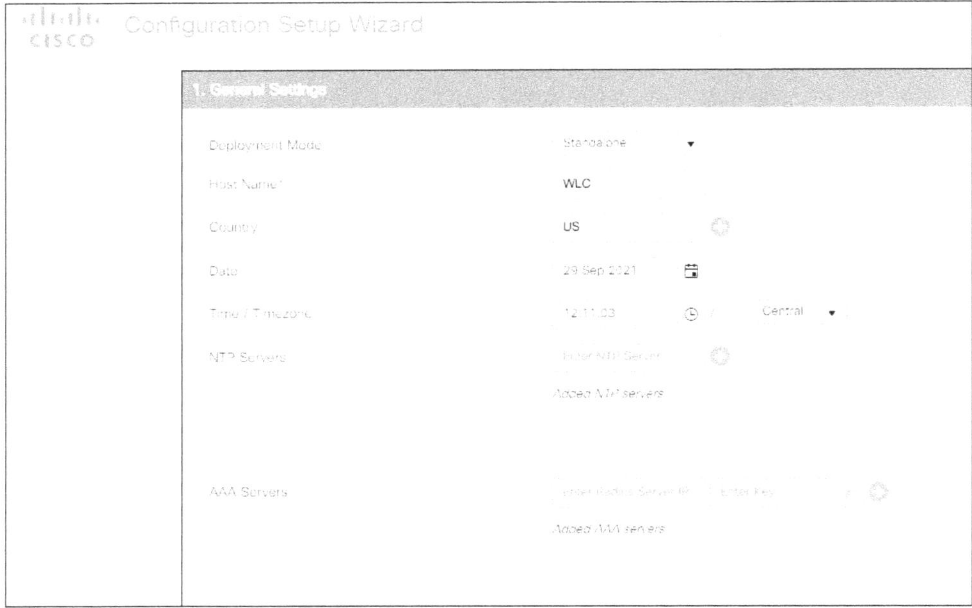

Figure 4-11 *Day 0 Configuration Setup Wizard*

When you're done, the configuration is pushed to the box, and you are prompted to log in again. This time, you are connected to the main Day 1 dashboard, as shown in Figure 4-12. Now your Catalyst 9800 wireless LAN controller is operational.

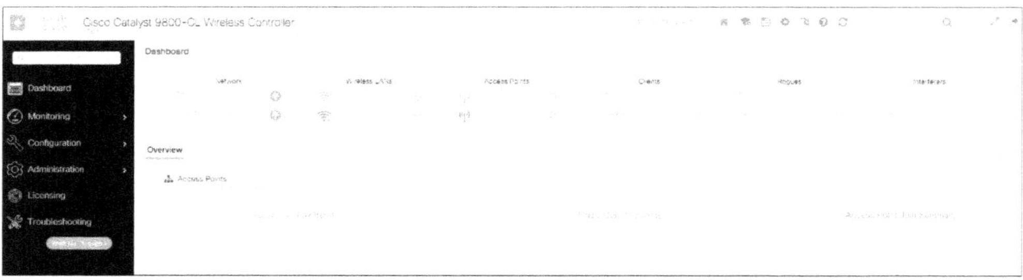

Figure 4-12 *Catalyst 9800 wireless LAN controller main dashboard*

Note When you're browsing to the box, it's the country code setting that determines if you get the configuration wizard or the main dashboard; without the country code, the WLC cannot be operational as the access points cannot join.

In the "References" section, you can find a link to a document that contains all the recommended initial configuration for the Catalyst 9800; it's called "Cooking recipe: Minimum bootstrap CLI configuration for Catalyst 9800" and it could be really useful if you start from scratch.

Access Point Join

Now you have an operational Catalyst 9800 wireless LAN controller. At this point you need to connect an access point and make it join. The joining process is the same you experienced with the AireOS controller; nothing changed. This means that, out of the box, the AP needs to learn the IP address on the C9800 to be able to send a CAPWAP discovery and then join.

How does the AP find out about the C9800? Again, nothing new here; the methods are essentially the following:

- **Static configuration via the AP console:** Manually enter the WLC information via the **AP#capwap ap primary-base <name> <IP address>** command.
- **CAPWAP broadcast:** AP sends a CAPWAP broadcast packet to find out whether WLC is on its same Layer 2 broadcast domain/VLAN.
- **DHCP option 43:** Configure the DHCP server to pass the IP address information via option 43. Here is a sample configuration from the IOS-XE DHCP server:

  ```
  ip dhcp pool <AP pool name>
           network IP <Network> <Netmask>
           default-router <default router hostname/IP>
           dns-server <DNS Server hostname/IP >
     option 43 hex <hex string>
  ```

  The hexadecimal (hex) string is assembled by concatenating the TLV values as shown:

  Type + Length + Value

  Type is always f1 (in hexadecimal). *Length* is the number of wireless controller management IP addresses times four in hex. *Value* is the IP address of the controller listed sequentially in hex.

  For example, suppose that two controllers have management interface IP addresses, 10.126.126.2 and 10.127.127.2. The type is f1(hex). The length is 2 * 4 = 8 = 08 (hex). The IP addresses translate to 0a7e7e02 and 0a7f7f02. Assembling the string then yields f1080a7e7e020a7f7f02. The resulting Cisco IOS command added to the DHCP scope is as follows:

  ```
  option 43 hex f1080a7e7e020a7f7f02
  ```

- **DNS name resolution:** Configure your DNS to return controller IP addresses in response to the CISCO-CAPWAP-CONTROLLER.localdomain query, where local-domain is the access point domain name learned through DHCP.
- **Previously joined WLCs:** Locally stored IP addresses from previously joined WLCs.
- **WLC in the same mobility group:** If a new WLC is placed in the same mobility group as the WLC where the AP is currently joined, the AP learns the new WLC's IP and tries to discover it.

- **Plug and Play with Cisco DNA Center:** The PnP option can really simplify your network operations when used with Cisco DNA Center. As illustrated in Figure 4-13, the onboarding process can be completely automated from purchasing to device provisioning:

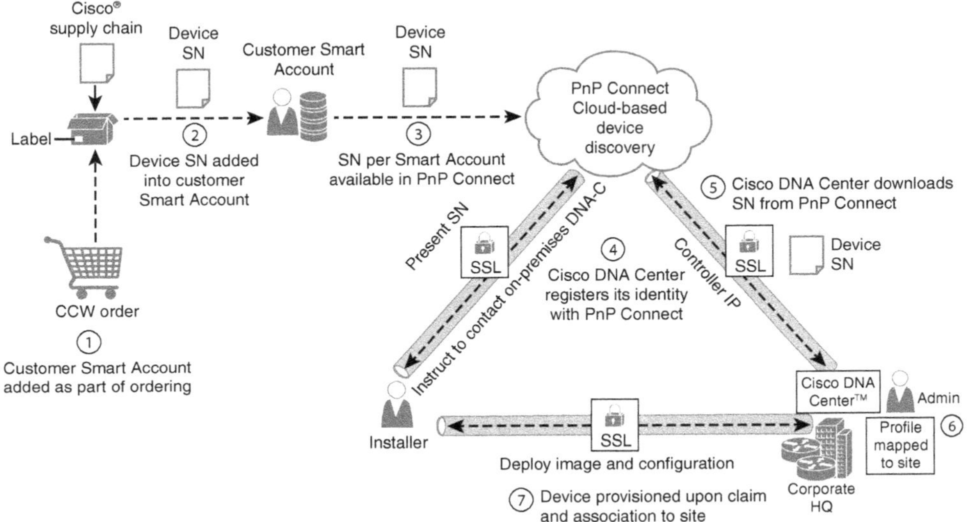

Figure 4-13 *AP end-to-end onboarding process with PnP*

In this case, the serial number of the purchased access point can be automatically transferred from the customer Cisco Smart Account to the Cisco cloud-based PnP redirection service and from there to the on-premises Cisco DNA Center. When the AP boots out of the box without any configuration, it uses DNS resolution to connect to the PnP cloud service, and then it is redirected to Cisco DNA Center, where you can claim it in the Plug and Play flow. As part of the claiming workflow, you assign the AP to a site; the AP is configured with the IP address of the C9800 managing that site and joins the WLC. As explained in the previous section, "Catalyst 9800 Physical Appliance," for the C9800 appliance, there are other methods for AP to be onboarded via PnP. You can configure the PnP server in the DHCP address scope using the option 43 string **"5A1N;B2;K4;I<IP address of PnP server>;J80"** or you can leverage DNS resolution of pnpserver.localdomain.

When the AP is configured with the IP address of the C9800, it starts the CAPWAP discovery process and eventually joins the WLC. As explained in Chapter 3, "C9800 Configuration Model," to be operational, the AP needs to be assigned to three tags: site, policy, and RF tags. If nothing is configured, the AP is associated with the default tags.

Note Ensure that the Control and Provisioning of Wireless Access Points (CAPWAP) UDP ports 5246 and 5247 are not blocked by an intermediate device between the AP and WLC locations.

Configuring WLAN and Connecting a Client

The last step to have your catalyst network operational is to configure a wireless LAN (WLAN) network so that the AP can broadcast an SSID, and clients can join.

Here again the intent is not to present you with the several configuration steps; you can easily find them in the configuration guides; rather, the focus of this section is to explain the different options to simply achieve the task.

As you learned in the previous chapter, C9800 leverages a different configuration model based on profiles and tags. When you understand them, they provide a simple yet powerful and flexible way to configure your wireless network. But what if you have not mastered the new configuration model yet? Where to start?

One way is to leverage Cisco DNA Center as it abstracts the configuration details from the user and presents an intuitive provisioning flow.

If you want to directly configure the C9800, the other way is to log in to the C9800 dashboard, open the configuration guide on the side, and go from there. This is probably the best way to learn all the concepts. To make it easier for you, the C9800 dashboard provides two interesting, assisted configuration methods:

- WLAN Wizard
- Walk Me tool

The WLAN Wizard was introduced in release 17.6.1 and, as the name suggests, is focused on WLAN configuration. In the GUI, it can be found under **Configuration > Wireless Setup > WLAN Wizard**, as shown in Figure 4-14.

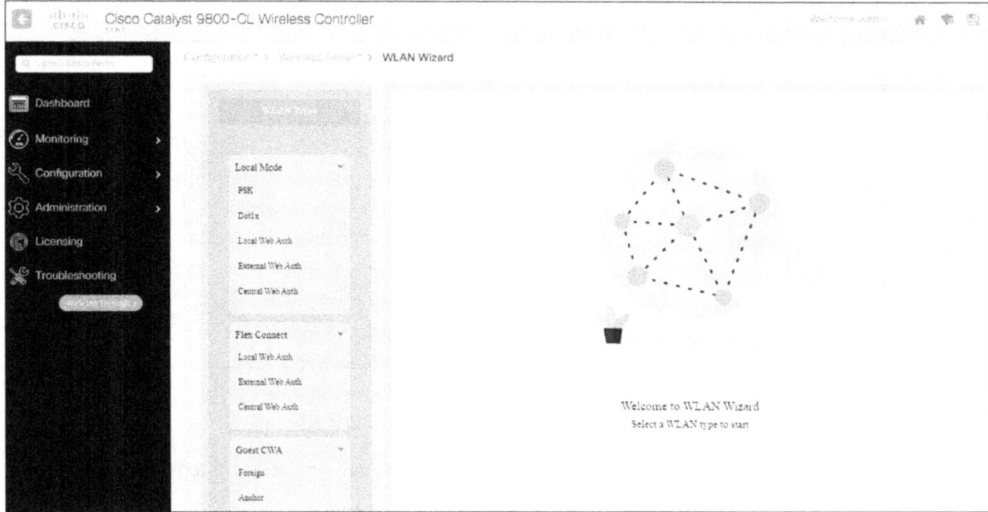

Figure 4-14 *WLAN Wizard*

The wizard not only guides you step-by-step to define a WLAN (SSID) and the related configurations like the WLAN policy, AAA configuration, ACLs, and URL filters as applicable to the WLAN deployment type selected (PSK, dot1x, or Web Auth). It also allows you to associate the WLAN to tags and apply the tags to the APs.

The other great advantage of this tool is that it provides a preview of the CLI commands for each configuration step. What a great way to get familiar with the new IOS-XE–based configuration! Figure 4-15 shows an example of a PSK SSID configuration.

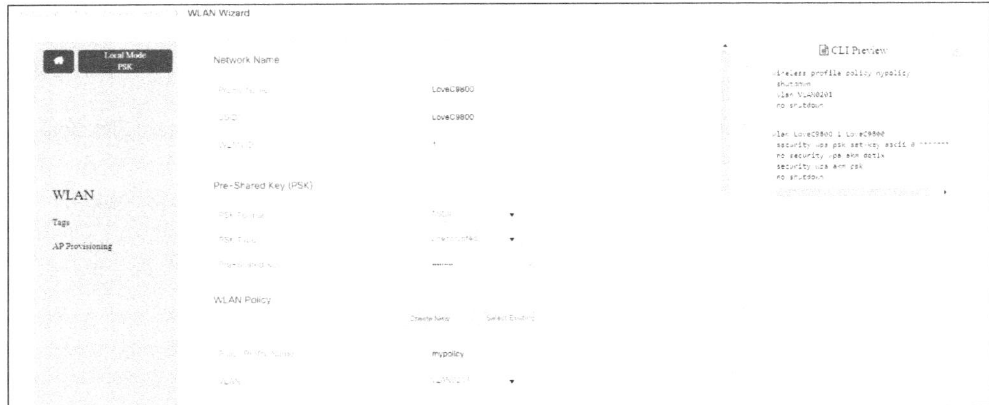

Figure 4-15 *WLAN Wizard, CLI preview*

The Walk Me tool is much more than WLAN; it provides an assisted and interactive configuration tool for any configuration you can deploy on the C9800. You can find the Walk Me blue icon on each webpage. Just click it and you are asked how you can be assisted. Type what you need. In this case, type **Configure SSID** or just **SSID**, as shown in Figure 4-16.

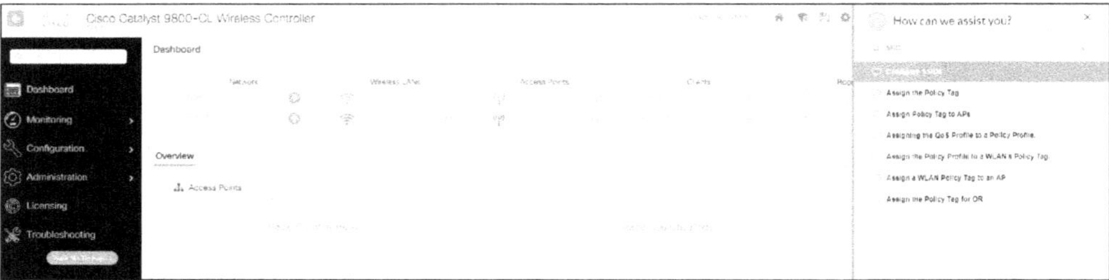

Figure 4-16 *Walk Me configuration tool*

When you click the first link, Configure WLAN, you can start adding a WLAN profile and then the related policy, as shown in Figure 4-17.

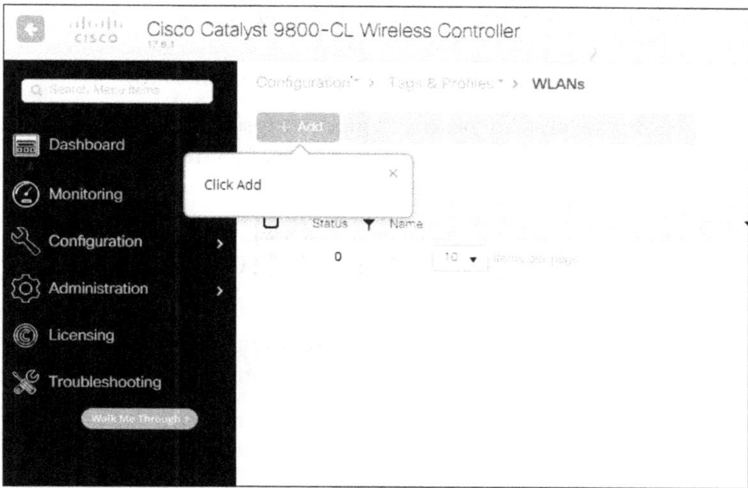

Figure 4-17 *WLAN creation in Walk Me*

In the case of Walk Me, after the WLAN is configured, the tool walks you through what you need to do next. In this case, you would have to configure the policy tag and then associate that policy tag to the AP.

After you assign the three tags to the AP, in whatever way you have picked, your SSID is broadcasted, and the client joins your Catalyst wireless network.

One important thing to remember: the Walk Me tool is embedded in the browser of your own laptop and communicates directly to the Walk Me server; this means that your laptop needs to be able to reach the Internet. On the other hand, it's not installed in the WLC itself.

Summary

The Catalyst 9800 wireless LAN controller is a flexible platform. This chapter illustrates the different deployment modes supported for the private and public cloud and describes how to bring up your C9800 to an operational state. It then suggests different methods to easily configure your first WLAN on a Catalyst wireless network.

References

Catalyst 9800 installation guides: https://www.cisco.com/c/en/us/support/wireless/catalyst-9800-series-wireless-controllers/products-installation-guides-list.html

Catalyst 9800 configuration guides: https://www.cisco.com/c/en/us/support/wireless/catalyst-9800-series-wireless-controllers/products-installation-and-configuration-guides-list.html

Cooking recipe: Minimum bootstrap CLI configuration for Catalyst 9800: https://www.cisco.com/c/en/us/support/docs/wireless/catalyst-9800-series-wireless-controllers/217368-cooking-recipe-minimum-bootstrap-cli-co.html#anc8

Catalyst 9800-CL on private cloud deployment guide: https://www.cisco.com/c/en/us/products/collateral/wireless/catalyst-9800-cl-wireless-controller-cloud/nb-06-cat9800-cl-wirel-cloud-dep-guide-cte-en.html

Catalyst 9800-CL in GPC deployment guide: https://www.cisco.com/c/dam/en/us/td/docs/wireless/controller/9800/9800-cloud/deployment/c9800-cl-gcp-deployment-guide.pdf

Catalyst EWC-AP deployment guide: https://www.cisco.com/c/en/us/td/docs/wireless/controller/ewc/16-12/config-guide/ewc_cg_16_12.html

Cisco SDA wireless deployment guide: https://www.cisco.com/c/dam/en/us/td/docs/cloud-systems-management/network-automation-and-management/dna-center/deploy-guide/cisco-dna-center-sd-access-wl-dg.pdf

Cisco DNA Center guides: https://www.cisco.com/c/en/us/support/cloud-systems-management/dna-center/products-installation-guides-list.html

Plug and Play Connect Portal: https://software.cisco.com/software/csws/ws/platform/home?locale=en_US#module/pnp

Chapter 5

Security

Everyone typically agrees that wireless always had a reputation of being less secure than cables. This belief is due to the inherent nature of an open medium where an attacker can place attacks without any physical contact and without cutting into cables, as well as the ever-increasing mobile workforce, which makes identifying and tracking connected devices more difficult. Security has always been important and has become even more important over wireless due to the open nature of the medium. With the ever-increasing power of brute-force cloud resources and continuously discovered weaknesses in existing security protocols, security will continue to increase in importance.

Cisco Identity Services Engine (ISE) is the platform of choice for identity and access control and policy management. It enables organizations to enforce wireless access controls by implementing 802.1X, MAC authentication, web authentication, and administrative user authentication. It also has many other features in terms of visibility and monitoring, but these topics are covered in other books. This chapter covers all of the security aspects, one by one, even if each section references others because security is very closely integrated into all these areas.

Network Security Fundamentals

Wireless security uses many concepts and services familiar to the wired security world. As a matter of fact, a good part of the security infrastructure sits on the wired network. It is therefore essential to have a good grasp of concepts like access control lists, authentication servers, and certificates.

Access Control Lists (ACLs)

Access control lists are an important part of the security of a platform. Conceptually, ACLs allow you to match specific traffic and take specific actions on that traffic. The traffic is matched by giving a network address and adding a wildcard mask, which is a kind of inverted subnet mask. For example, an ACL covering 192.168.1.0 with mask 0.0.0.255 affects all IP addresses from 192.168.1.0 to 192.168.1.255. The binary 1s of the wildcard mask define which bits can vary ("don't care" bits), and the 0s define which bits have to statically match.

An access control entry (ACE) is a single statement in this ACL; it covers a network address, a wildcard mask combination, and a permit or deny statement. An ACL is made up of one or more ACEs; because an ACE can cover only a contiguous range of addresses covered by a wildcard mask, the ACL can combine several ACEs to deny or permit unrelated address ranges (and sometimes port combination too). The order in which the ACEs are defined matters because it is the order in which the network device compares the traffic and takes the corresponding action as soon as one ACE matches. The rest of the ACEs are not even considered as soon as a match is found.

An ACL defines a specific set of traffic to be matched and does not take any action on that traffic until you apply it somewhere. Where the ACL is applied is the key and determines exactly what action is taken. Too often, network administrators link ACLs with dropping or allowing traffic, and although these are important use cases, they are not the only ones. ACLs can define where to apply specify quality of service (QoS) actions in the traffic; ACLs also can define the traffic to intercept for a web portal and many other things.

ACLs use computing power intensively in the way they work because traffic evaluated against the ACL has to be evaluated against each statement of the ACL. If the statement does not match, the Catalyst 9800 moves to the next statement until a match is found. If the end of the ACL is reached and no statement matched, a default action is taken, which is typically to deny the traffic, but the action depends on the type of ACL. Luckily, the 9800 appliances are hardware-accelerated (as discussed in Chapter 2, "Hardware and Software Architecture of the C9800"). ACLs have little impact overall on the platform performance, but it is always good to consider writing the most efficient ACLs with as few lines as possible to meet the security criteria.

Defining ACLs

IOS-XE inherits several types of ACLs that can be configured, but from a practical viewpoint, extended ACLs are the only ones you should be concerned about. They are like standard ACLs but allow you to identify the traffic more granularly. There is no reason to define a standard ACL on a Catalyst 9800 controller because the extended ACLs supersede them in every way.

An exception to the previous statement is the fact that the Catalyst 9800 supports Cisco TrustSec and Scalable Group ACLs (SGACLs), also known as role-based ACLs, and is part of the TrustSec solution as well as the SD-Access solution.

The command-line interface (CLI) allows you to define ACLs by name or by an identifying number. It's much easier to use names in general:

```
c9800-CL(config)#ip access-list extended ?
  <100-199>     Extended IP access-list number
  <2000-2699>   Extended IP access-list number (expanded range)
  WORD          Access-list name
```

The structure of a CLI ACL statement or ACE is as follows:

```
<sequence number> [permit/deny] <protocol> <address or any> eq <port
number> <subnet> <wildcard>
```

For example:

```
1 permit tcp any eq www 192.168.1.0 0.0.0.255
```

The sequence number allows you to specify where in the ACL order of ACEs to insert the ACE. It is usually a best practice to define your statements with the sequence 10, 20, 30, 40, and so on. This way, you can "insert" a statement later without deleting half of the ACL by using sequence 5 or 15, for example, in the statement you want to insert between two existing statements.

The WebUI allows you to write a complete ACL much more easily by going to the **Configuration > Security > ACL** page, as shown in Figure 5-1. You can then see a list of protocols to pick from and make changes to an existing ACL much more conveniently (rather than having to use the "no" form of a statement to remove it and reconfigure it differently afterward).

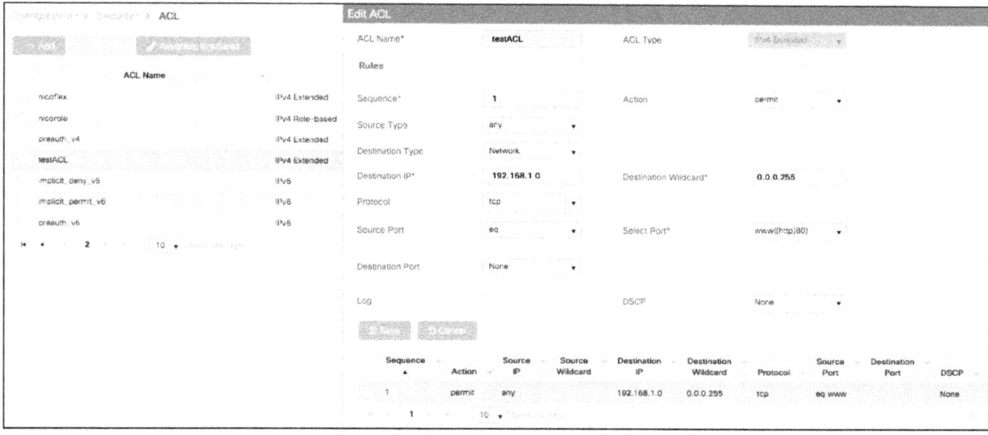

Figure 5-1 *Creating an ACL on the Catalyst 9800*

At the end of an ACL statement, you can add the **log** keyword; this has the effect that a syslog message will be thrown when that specific statement is matched. This option is extremely verbose on the controller terminal and will typically overwhelm it. Considering the syslog is thrown on the controller, it also only works with centrally switched ACLs and does not work when the ACL is applied by the FlexConnect AP.

Applying ACLs

An ACL might be called by different names depending on where it is applied and its function. In all cases, the ACL is defined the same way on the controller (as an extended

IP ACL). There are several ways of applying an ACL, but two stand out and are named differently because they have a different impact on the traffic:

- **Security ACL:** This name refers to an ACL in its most popular understanding, that is, defining what traffic is allowed through the device and which type of traffic is blocked and dropped.
- **Punt ACL** or **redirect ACL:** This name refers to an ACL that specifies which traffic is sent to the CPU (instead of its normal expected handling by the data plane) for further processing. A good example is the central web authentication (CWA) redirect ACL, which defines which traffic is intercepted and redirected to the web login portal. The ACL does not define any traffic to be dropped or allowed through but simply what follows the regular processing or forwarding rules and what is sent to the CPU for interception. A redirect ACL has an invisible last statement that is an implicit deny. This implicit deny is applied as a security access list entry (and therefore drops traffic that isn't explicitly allowed through or sent to the CPU).

You may have heard of DNS ACLs (or URL filters) or downloadable ACLs; we touch on those later. There are other ways of applying an ACL, but they typically refer to a specific feature (for example, an ACL in a QoS policy specifies traffic where this QoS policy will apply). On top of the "how," there is also the "where." A security ACL permits or drops traffic, as we have already stated, but where the ACL is applied also plays a role, so let's clarify that point.

A security ACL can be applied as follows:

- **On an SVI interface:** The ACL is evaluated only against the traffic that is routed through that interface. Using an SVI as a default gateway for a wireless client is not supported by the Catalyst 9800 at the time of this writing, so this only leaves the scenario where you have SVIs for management purposes and the ACLs applied on those SVIs apply to traffic destined to those interfaces.

```
myc9800-CL(config)#interface Vlan<number>
myc9800-CL(config-if)#ip access-group myACL in/out
```

- **On a physical interface of the controller:** The ACL is evaluated against all traffic that passes through that interface. Along with applying ACLs on the SVI, this is your other option for restricting traffic hitting the Catalyst 9800 management plane.

```
myc9800-CL(config)#interface GigabitEthernet1
myc9800-CL(config-if)#ip access-group myACL in/out
```

- **In a wireless policy profile or WLAN:** This vague category encompasses several places where you can configure an ACL that will be applied to the wireless client traffic both in the case of central switching or local switching of traffic. Such ACLs are supported only in the inbound direction.
- **On the AP itself:** In the case of FlexConnect local switching, the ACL is still configured and applied from the policy profile on the controller, but there is an extra step

of downloading this ACL to the AP through the Flex profile. ACLs must be downloaded to the AP before they can be applied at all. As an exception, fabric mode APs (in the case of software-defined access) also use Flex ACLs even though the AP is not operating in Flex mode.

Applying Wireless ACLs on the WLC

When editing a policy profile, you can specify an IPv4 and a separate IPv6 ACL (see Figure 5-2) as WLAN ACL. This is the most straightforward way of applying an ACL to all traffic to and from clients that are fully authenticated (that is, in the RUN state).

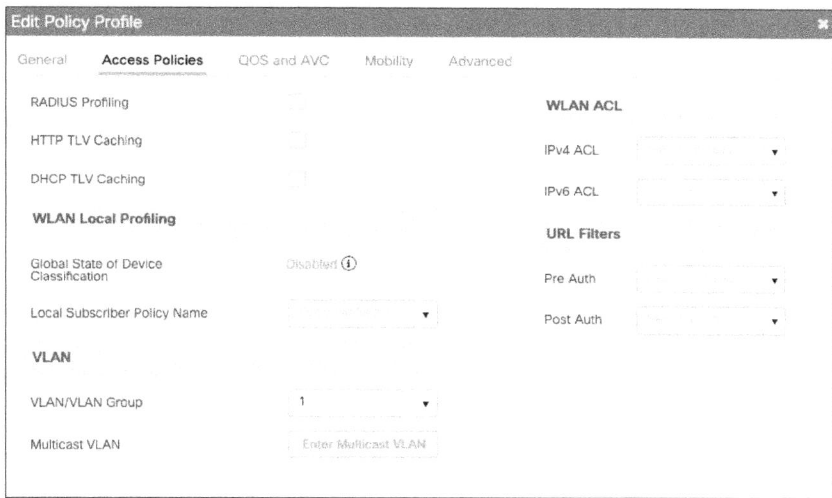

Figure 5-2 *Access policies of the policy profile on the C9800*

The explicit mention of clients in the RUN state is deliberate. Are there cases where you want to apply ACLs to clients that have not hit the RUN state yet? Apart from getting an IP address, clients are not supposed to be sending traffic before[el]except in the case of a web authentication WLAN. In that scenario, they can send and receive the traffic necessary to submit their credentials on the web login page before being in the final RUN state of the client state machine (for more details, see the "Wireless Security Fundamentals" section of this chapter). In some cases, you might want to allow some traffic before it is authenticated. For example, the external portal should be allowed, or you might want to also allow specific internal resources to be consulted before being authenticated. This requires a preauthentication ACL. In the WLAN edit section, illustrated by Figure 5-3, you can see a preauthentication ACL section (again doubled between IPv4 and IPv6) that allows you to define traffic to be allowed before the RUN state. A preauthentication ACL should mostly contain permit statements. Whatever does not match a statement in the preauthentication ACL automatically matches a default deny because the client is in a WEBAUTH_REQD state where all traffic is blocked apart from HTTP (that is intercepted).

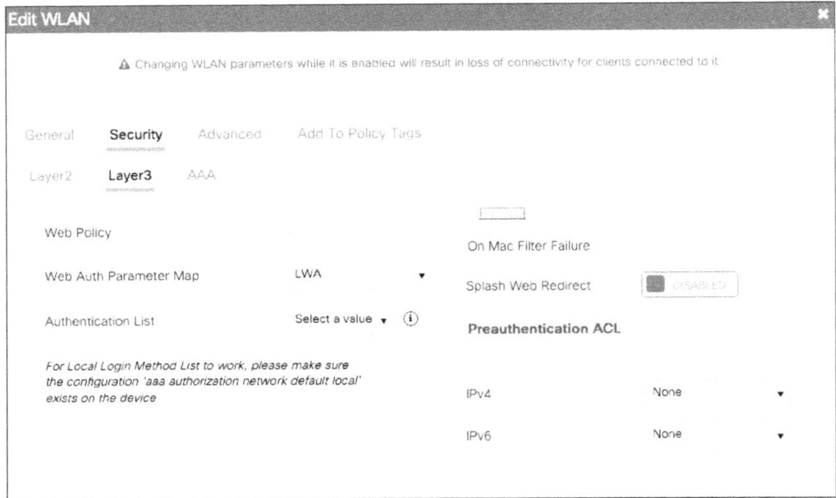

Figure 5-3 *Layer3 WLAN security settings where the preauthentication ACL can be configured*

These examples cover the static definition of ACLs. The Catalyst 9800 wireless LAN controller (WLC) supports the dynamic assignment of ACLs from a RADIUS server through the use of the Airespace-ACL-Name. This is covered with an example in the later "RADIUS" section. In this case, it requires enabling AAA override on the policy profile.

FlexConnect ACLs on the AP

If your AP is running as FlexConnect and locally switching traffic, there is no client traffic hitting the WLC data plane at all, and therefore, the WLC cannot physically apply any policing to the client traffic. This means the AP has to do the job itself. Defining an ACL for use on a FlexConnect AP is the exact same thing as covered before: you define a regular ACL on the Catalyst 9800. The specificity is that you have an extra step of predownloading it to the AP. The term *predownload* is more accurate than *apply* because the latter term implies that traffic might be matched against it already. But there are basically three different steps in the configuration for a local switching scenario:

- You define the ACL on the WLC.
- You predownload it on the AP(s).
- You apply the ACL somewhere in the configuration.

The predownload step is simple. Navigate to **Configuration > Tags and Profiles > Flex** and select the **Policy ACL** tab of a given Flex profile (see Figure 5-4). There, selecting **Add** means predownloading an ACL to the AP. You are able to add any ACL that was configured on the controller, and when you click **Save**, the ACL statements are downloaded on the AP and visible in a **Show ip access-list** on the AP itself (as displayed in Example 5-1).

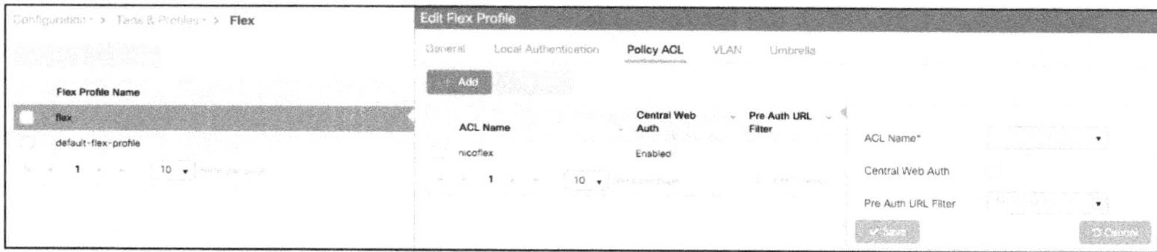

Figure 5-4 *The Flex profile Policy ACL tab allows you to download ACLs to the APs*

Example 5-1 *Output of show ip access-list on an AP*

```
9120-AP#show ip access-list
Extended IP access list nicoflex
    1 deny icmp any 1.1.1.1 0.0.0.0
    2 deny icmp any 2.2.2.2 0.0.0.0
    3 permit icmp any any
    4 deny icmp any 3.3.3.3 0.0.0.0
```

On that page, you select the **Central Web Auth** check box if your ACL is meant to be a redirect/punt ACL for use in a central web auth scenario. Its effect is to invert the permit and deny statements because the redirect ACL has inverted meaning on the AP operating system versus IOS-XE. (The preauth URL filter is covered in the next section.)

When the provisioning of ACLs on the APs is done, if the ACL is called out in the WLAN or policy profile, as demonstrated before, it is properly applied given that the ACL is now defined on the AP as well.

The Case of Downloadable ACLs (DACLs)

A downloadable ACL is an access list that is not predefined on the network device but returned dynamically, including all its statements, to the device via RADIUS during a network authentication and authorization. This type of ACL is popular with wired 802.1X deployments because it's much easier to centralize all the ACL definitions on the RADIUS servers than it is to make sure they are deployed on all network switches. The RADIUS server returns an authorization result containing a specific AV-pair mentioning the DACL name. The network device is then expected to start a new RADIUS request of a special type to download all the statements from that ACL. Therefore, the DACL download shows up as a separate successful authentication after the client authentication in the RADIUS logs. Wireless deployments have less need for DACL because they use a centralized controller most of the time. The 9800 controller benefits from the DACL implementation inside IOS-XE but is not adapted to the wireless workflow yet, and therefore, at the time of this writing in IOS-XE 17.7.1, DACLs are not supported on the Catalyst 9800. There is a good chance that support will come in future releases though, so keep an eye on the document titled *List of IOS-XE Wireless Features per Release* for it or release notes of each version.

URL Filters (a.k.a. DNS-Based ACLs)

Allowing access based on IP addresses is fine when covering your internal network, but what if you want to allow or restrict access to public resources on the Internet? IP addresses can change at any time without you knowing, and maintaining an IP-based access list covering Internet resources is just not practical. This is where URL filters come in. They work like ACLs in the sense that they permit or deny access, but their statements include URLs instead of IP addresses.

URL filters don't require you to have DNS configured on your wireless controller because they snoop the wireless client DNS traffic in real time. When traffic is centrally switched through the WLC, it is the controller that does this DNS snooping, whereas the AP can do the job for fabric or locally switched WLANs.

Before using an app on a smartphone or before visiting a website on a browser, the client device typically does a DNS lookup to get an IP address resolution of the app or website domain and then sends traffic to the specific IP addresses returned. Those IPs can be load-balanced and can change based on which DNS server you ask or your geographic location. When the wireless client is dynamically sending this DNS request, the WLC or the AP snoops and takes a peek at the DNS response. It looks at the first A record in the DNS reply and notes the IP address linked to it if the URL is a match to anything configured in the URL filter on the controller. It then adds this IP address dynamically in an IP cache table for this wireless client. It is therefore some kind of IP-based access list that is specific to clients and changing in real time based on the DNS requests made by the clients.

There are two types of URL filters: standard and enhanced. Standard URL filters can be applied before client authentication (preauth) or after a successful client authentication (postauth). Preauth filters are extremely useful in the case of external web authentication to allow access to the external login page, as well as potentially some internal websites before authentication takes place. Postauth, they can work to block specific websites or allow only very specific Internet websites while all the rest is blocked by default, but this type of URL filtering postauth is better handled by using Cisco DNS Layer Security (formerly known as Umbrella) for a lot more flexibility. The standard URL filters apply the same action (permit or deny) for the whole list of URLs: it's either all permit or all deny. Standard URL filters work on both local mode APs and FlexConnect APs.

Enhanced URL filters allow specification of a different action (deny or permit) for each URL inside the list and have per-URL hit counters (see Figure 5-5). They are supported only on FlexConnect APs in local switching (or fabric APs).

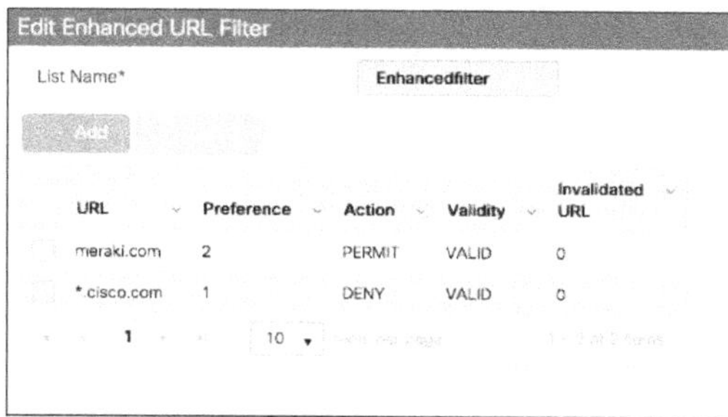

Figure 5-5 *URL filter configuration*

In both types of URL filters, you are allowed to use a wildcard subdomain such as *.cisco.com. URL filters are standalone but always applied along with an IP-based ACL. As of release 17.6, a maximum of 20 URLs is supported in a given URL filter. Considering that one URL can resolve to multiple IP addresses, it is interesting to note that only up to 40 resolved IP addresses can be tracked for each client. Another important limitation to note is that only DNS A records are tracked by URL filters. The WLCs or APs do not track the resolved IP address of a URL if the DNS answer uses a CNAME alias record.

Certificates and Trustpoints

Explaining public key infrastructure (PKI) in detail is beyond the scope of this book. However, certificates are a very important part of any network device and especially for the Catalyst 9800. Certificate-based authentication is a method to identify a user, device, or machine before it can be granted access to a network. A wireless network, comprising a wireless LAN controller (hereafter referred to as WLC), access points (APs), and clients, commonly uses certificate-based authentication to validate the identities of peer devices when participating in services such as AP join, device management access, and web authentication. Each service can use different sets of client and server certificates.

But how do devices get their digital identities?

To begin with, each participating device (controller, access point, or client) has its own device certificate and a certificate authority (CA) certificate that validates its authenticity. A closer look at the certificates available on the Catalyst 9800 controller shows the following types:

- **Cisco-installed manufacturer installed certificate (MIC):** On physical appliances (Catalyst 9800-40, Catalyst 9800-80, Catalyst 9800-L), these are, by default, factory installed and widely known as the Cisco-installed MIC or Secure Unique Device Identifier (SUDI) device certificate. In addition, controllers and access points have a Cisco manufacturing certificate authority (CA) certificate that is used to sign and validate device certificates.

- **Wireless LAN controller self-signed certificate for virtual controller:** The Catalyst 9800-CL (the virtual instance of the controller) does not come with any manufacturing certificate. In the absence of an identity certificate, it relies on the self-signed certificate that has to be generated by the Day 0 wizard or manually using a script and validated by the local Cisco IOS certificate authority (which is a self-signed local CA and is not the same as the manufacturing Cisco CA certificate). This acts as the Catalyst 9800-CL's local identity certificate and is used for AP joins, mobility connections, and Network Mobility Services Protocol-Connected Mobile Experience (NMSP-CMX) connections.

- **IOS-XE device self-signed certificate:** The default self-signed certificate is autogenerated during the controller's initial startup if any HTTPS, SSH, or NETCONF service is configured on the controller.

The listed default certificates provide an easy and out-of-the-box method of early trust between peer devices. However, if you want to provide better security, you can consider using third-party validated certificates, including locally significant certificates (LSCs).

Third-party certificates require a PKI framework that enables encryption of public keys and digital certificates. Along with different authentication protocols, the PKI model works with certificate authorities, root certificates, and asymmetric key encryption to ensure that the digital certificates are securely exchanged over encrypted tunnels during a client and server exchange.

On Catalyst 9800 controllers, these digital certificates are configured and held in containers called *trustpoints* and used when the devices initiate a secure communication with other network devices or network clients. A trustpoint is one of the most important configuration entities for a PKI client. A trustpoint includes the identity certificate of the CA that signed the device certificate, CA-specific trustpoint configuration parameters, and an association with an enrolled identity certificate.

Trustpoints provide a mapping between the identity certificate and the application or service that needs the certificate. For example, for the SSL/HTTPS server functionality, the **ip http secure-trustpoint <*trustpoint name*>** command tells the controller what identity certificate to present to an SSL client. Depending on your requirement, you can configure many trustpoints.

A Case for Trustpoints

Identity validation using certificates spans across a range of functions and protocols in the Catalyst 9800 wireless environment. Certificates are primarily used for authentication when an access point joins the controller using CAPWAP with DTLS, for web administration and web authentication using HTTP with TLS, and for local EAP authentication. Certificates are also used when the controller communicates with Cisco Connected Mobile Experience (CMX), Cisco Digital Network Architecture Center (DNA Center), and Digital Network Architecture Spaces (DNA Spaces). Some of these exchanges require additional configuration, whereas others do not require any action from your side.

How to Add a Certificate on the Controller

To add a certificate on the controller, there are basically two things you can do: either start a certificate request on the WLC (as shown in Figure 5-6) or import a ready-for-use certificate to the WLC (as shown in Figure 5-7), get it signed, and install the resulting certificate. You can do various activities outside of the Catalyst 9800 device to end up with the same result. A certificate has to contain a private key and be linked to a given certificate authority.

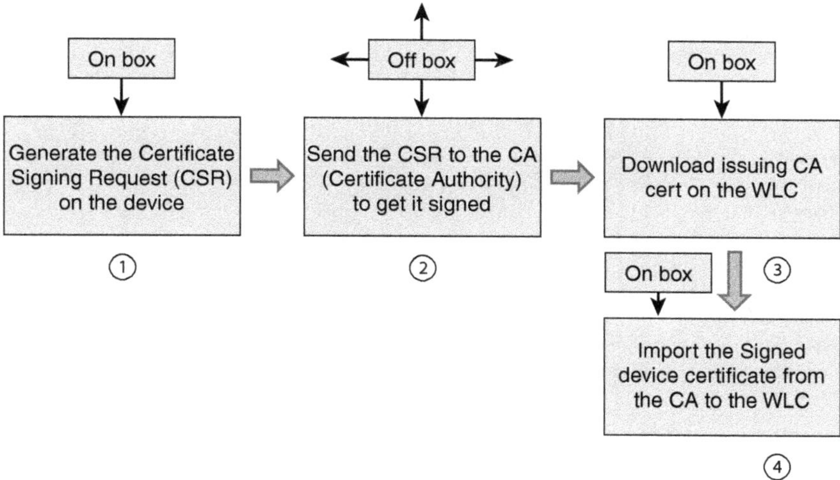

Figure 5-6 *Adding a certificate by generating the question on the controller*

Generating the certificate on the WLC itself is the most secure because the private key never leaves the device. However, it does not allow for configuring the SAN field at this time. The workflow is shown here and can mostly happen in the **Configuration > Security > PKI Management > Add Certificate** page of the WebUI:

Step 1. Go to the **Key Pair Generation** tab of the PKI Management page, as depicted in Figure 5-7. Click **Add**, enter a key name, choose **RSA**, choose a Modulus of 4096, make sure it's exportable, and then click **Generate**.

Configuration > Security > **PKI Management**

Trustpoints | CA Server | **Key Pair Generation** | Add Certificate

+ Add

Key Name	Key Type	Key Exportable	Zeroise Key
ssh-key	RSA	No	Zeroise
TP-self-signed-899645931	RSA	No	Zeroise
WLC_CA	RSA	No	Zeroise
vasa.p12	RSA	No	Zeroise
myc9800-CL_WLC_TP	RSA	Yes	Zeroise
paolokey	RSA	Yes	Zeroise
bigchain.pfx	RSA	No	Zeroise
SLA-KeyPair2	RSA	Yes	Zeroise
SLA-KeyPair	RSA	Yes	Zeroise
ssh-key.server	RSA	No	Zeroise

1 | 10

Key Name* Enter Key Name
Key Type* RSA Key EC Key
Modulus Size* 4096
Key Exportable*
Cancel | Generate

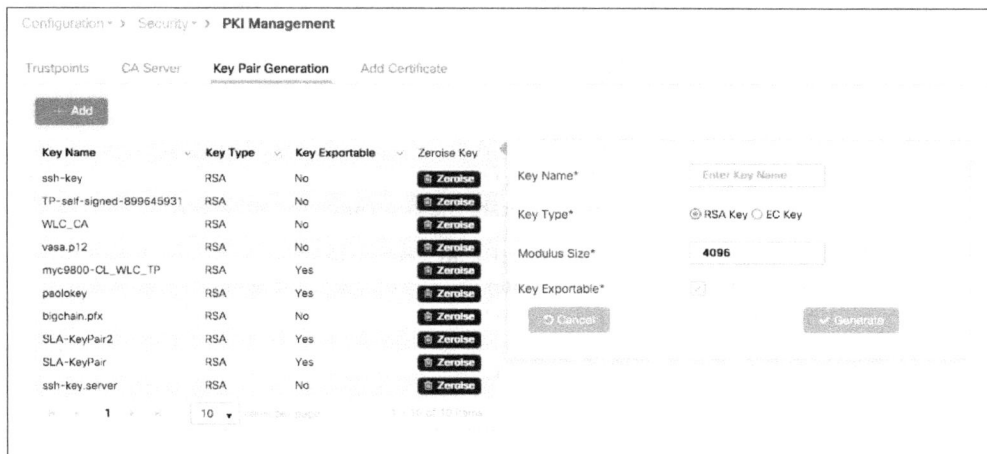

Figure 5-7 *Creating a key pair for use with a certificate*

Step 2. To generate the CSR on the WLC, go to **Generate Certificate Signing Request**, as shown in Figure 5-8, and enter the desired fields that will appear on your certificate after it is signed.

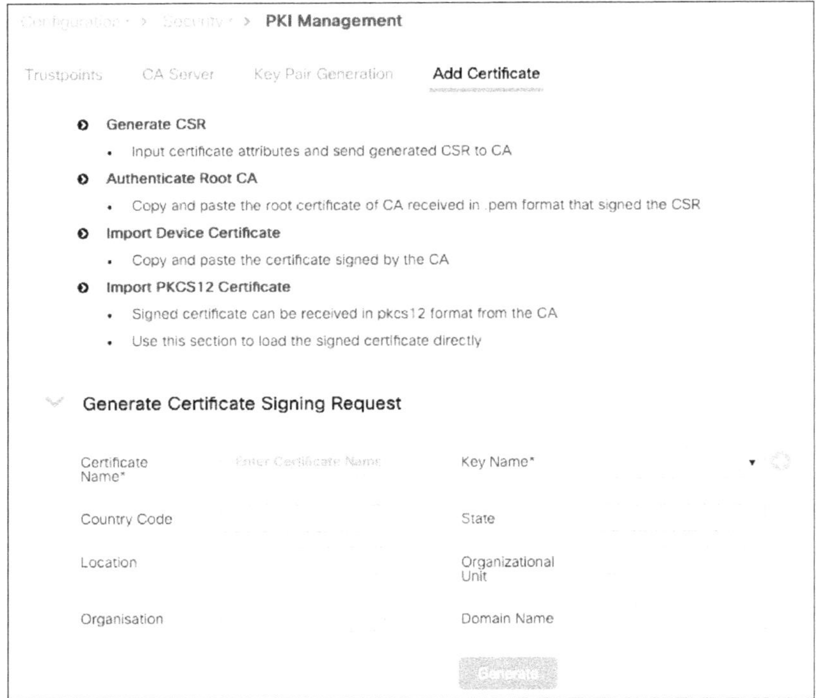

Figure 5-8 *Generate a certificate signing request from the WebUI*

Step 3. Get the certificate signed by the certificate authority of your choice. The details of this step vary greatly depending on which certificate authority you choose. Many publicly trusted CAs have a website where you can easily get your certificate signed.

Step 4. Go to the **Authenticate Root CA** section of the **PKI Management** page, as depicted in Figure 5-9. Enter a trustpoint name (choose the name you want to use for your certificate) and paste the content of the PEM file.

Step 5. Go to the **Import Device Certificate** section of the **PKI Management** page, as depicted in Figure 5-10. Enter the same trustpoint name as in step 3 and paste the content of the PEM file of the device signed certificate you received from your CA.

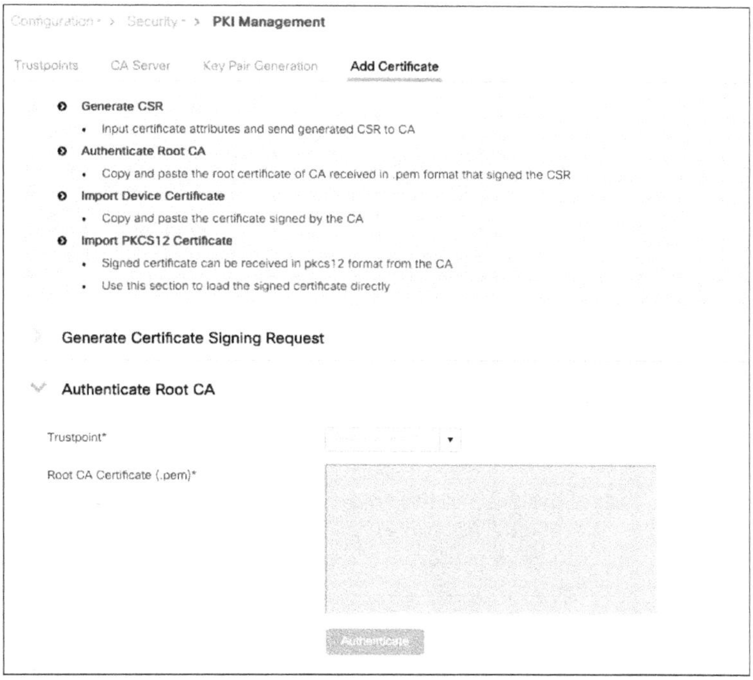

Figure 5-9 *Authenticating the CA that issued your device certificate*

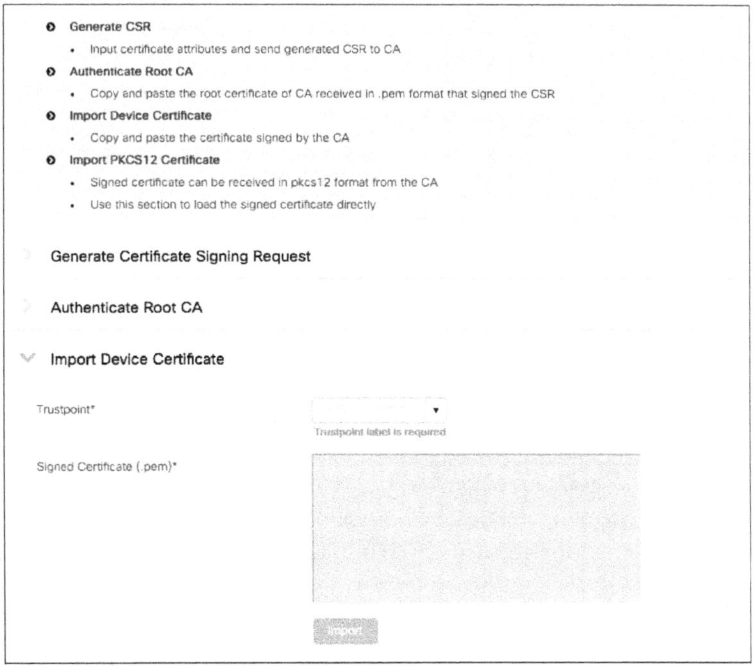

Figure 5-10 *Adding the device signed certificate received from the CA*

Generating the CSR outside of the 9800 is a possibility (depicted in Figure 5-11). In this case, you typically have to bundle your private key along with your certificate in a PKCS12 file format and download everything at one time to the WLC:

1. The first step would be to use a tool such as OpenSSL to generate your CSR as well as a private key. Some public CAs also offer a web page where you can easily generate a key and CSR for them to sign as well.
2. Have your CA sign your certificate.

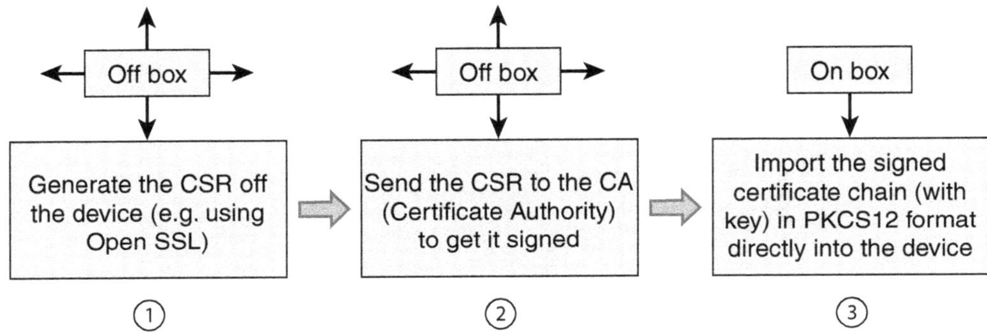

Figure 5-11 *Adding a certificate when the private key was generated outside of the 9800*

3. Obtain the PKCS12 formatted file containing the certificate chain (your device certificate and the CA chain), along with your private key. If your CA does not provide this information, you can combine the pieces you received together with OpenSSL to obtain the PKCS12 file. As depicted in Figure 5-12, go to the **Import PKCS12 Certificate** section of the **PKI Management** page in the WebUI and enter the file location and private key password.

When you're importing PKCS12 files, it is possible to import a complete chain if your PKCS12 contains a chain of certificates up to a root CA (if the CA that signed your device certificate is not a root CA but an intermediate CA, for example). Be aware that the Catalyst 9800 controller does not send the whole chain of certificates when a client connects to the web interface, and therefore, the client should have most of the chain imported rather than only the top root CA. The **9800 Web UI PKI Management** page contains a guided workflow for certificate import to simplify all these steps. The exact workflow is explained in great detail in the guide titled *Configuring Trustpoints on Cisco Catalyst 9800 Series Controllers*. Commands and some tips are also covered in *Generate CSR for Third-Party Certificates and Download Chained Certificates to Catalyst 9800 Wireless Controllers*.

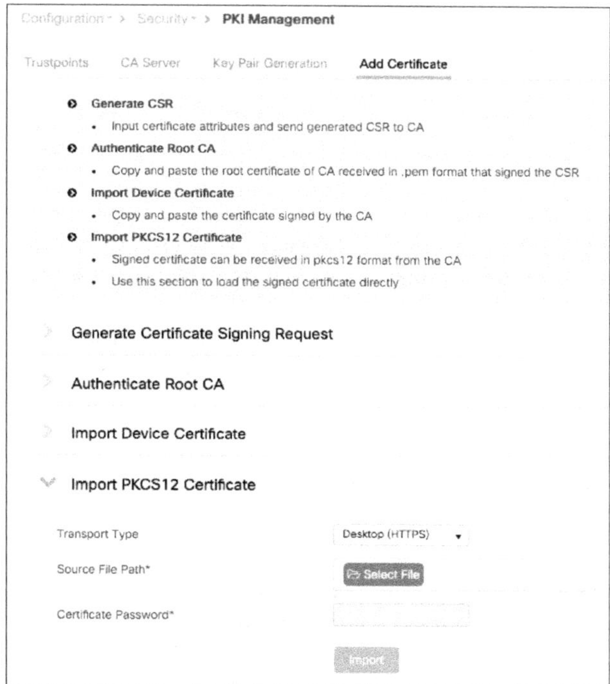

Figure 5-12 *Adding a certificate chain in PKCS12 format to the WLC*

AAA

Network authentication, authorization, and accounting (AAA) provide the means of getting answers to the identity of a person or device requesting access, what resources are being accessed, if this person has specific rights toward that resource, and logging activities of the person or device during a session. The term *AAA* is used in a relaxed manner to depict any part of the access control process, but each letter stands for a specific phase of this flow:

- **Authentication:** This term is the most used, even to depict the whole flow. Technically, the authentication phase consists of verifying the identity of a device or a person. This is done by requesting user credentials, which can be a username and password pair or even something else such as a certificate.

- **Authorization:** This phase consists of checking the privilege or authorization level the client has. The client may ask for permission to access every single resource uniquely, or this can happen in a one-time operation where the authentication server returns the list of privileges and accesses that the client can benefit from, and it's then up to the network access server to apply those privilege levels to the user.

- **Accounting:** This often-optional phase happens after the authentication is complete and allows you to track the activity of the user or device on the network. It is also useful to know when the session ends, that is, when the device is no longer connected to the network.

RADIUS

The Remote Authentication Dial-In User Service (RADIUS) is a protocol used to control network access. It is quite old and, as its name implies, dates from the dial-up era, which explains why some of the terms and attributes sound quite outdated and unrelated to wireless, but it still does the job and is the main protocol used for wireless enterprise class security. RADIUS operates in a client/server model where the client is the wireless controller and the server is the RADIUS authentication server (ISE in a Cisco environment). The exact terms are

- **Network access server (NAS) or network access device (NAD):** This device is responsible for passing the user information to the RADIUS server and acting on the response that is returned.
- **RADIUS server:** This device is responsible for processing the access request and returning all required parameters for the NAS to provide access to the user.

RADIUS works over UDP on port 1812 for authentication and authorization and on UDP port 1813 for accounting.

There are four authentication packet types:

- **Access-request:** The NAS sends this type of packet to provide the server with information required to proceed with the authentication. It is basically used for any RADIUS packet sent to the RADIUS server.
- **Access-challenge:** This is the type used by the RADIUS server to send data to the NAS to proceed with the authentication. You can guess that these first two types are repeated a certain number of times (depending on the authentication type and details), and the value is in the attributes and data they carry rather than the message type. The request and challenge RADIUS packets carry the EAP authentication frame in one of the attributes (called EAP payload).
- **Access-accept:** The RADIUS server sends this packet type to tell the NAS the authentication has been successful. The access-accept contains all the necessary attribute fields to provide the authorization details to the NAS. This is a specificity of the RADIUS protocol where authorization is not a separate phase but is embedded as a one-time operation on the final authentication packet.

- **Access-reject:** The RADIUS server uses this packet type to notify the NAS that the authentication failed. It rarely contains attribute fields other than the username in the case of 802.1X.

There are two extra packet types for accounting:

- **Accounting-request:** The NAS sends these packets to deliver information about the session to the server.
- **Accounting-response:** These packets work as a kind of acknowledgment that the server has received and processed the accounting request successfully.

RADIUS Attributes

We have already hinted at the fact that the authentication server can return attributes that involve privilege levels and specific information about accessing or not accessing specific resources. RADIUS attribute-value pairs (AVPs), often called *AV-pairs*, are a major part of the success of RADIUS as a protocol. They are used to exchange information between the NAS device and the authentication server. They can provide extra information about the request being made and provide various parameters for the network device to enforce. They contain a type, a length, and a value that defines the attribute format and how to read it (shown in Figure 5-13) and are extremely flexible, which is the reason RADIUS was able to stay relevant and survive so many technology changes. Because the field is an octet, up to 255 attributes can be used. Some of them are predefined and well known, whereas others leave room for vendors to communicate vendor-specific information by encapsulating their own extended attributes inside it.

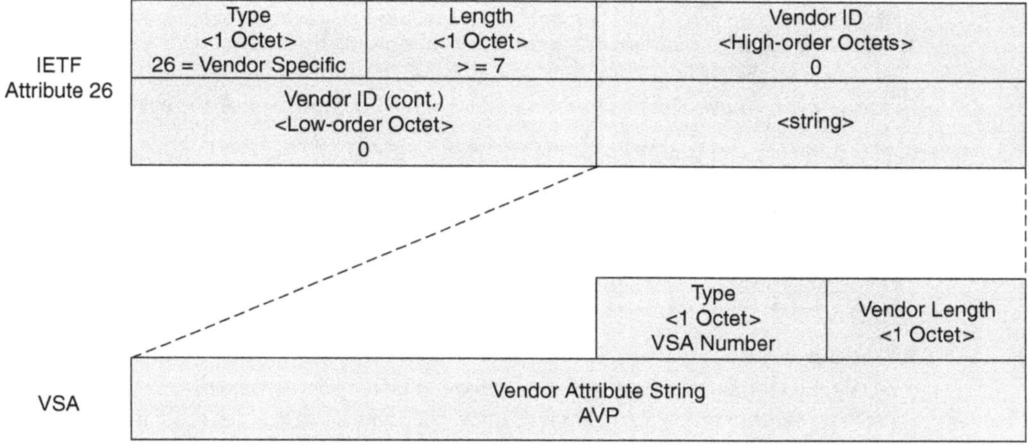

Figure 5-13 *RADIUS AV-pair format*

RADIUS Sequence Example

Now that we've covered the protocol, let's examine, at a high level, the operational sequence that shows how all the components work together during an 802.1X authentication. The EAP exchange is illustrated in Figure 5-14.

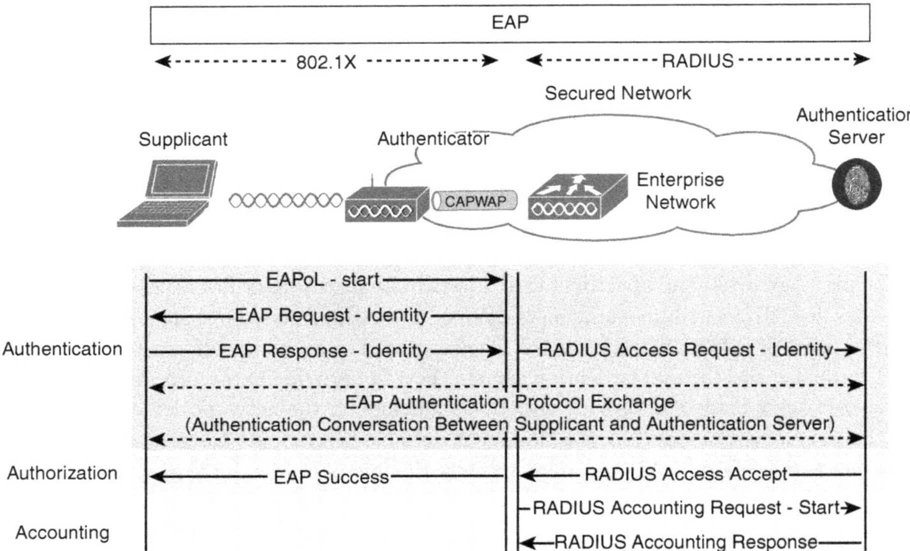

Figure 5-14 *Workflow of an EAP authentication between a supplicant and authentication server*

1. The supplicant can initiate the authentication process by sending an EAPoL-Start message to the authenticator, but this is optional. Both switches and access points know when a new client is connecting and can be the initiator of the authentication.

2. The authenticator (the WLC or the AP) then sends an EAP identity request frame asking for the client's identity. The WLC is the authenticator from the point of view of initiating the RADIUS traffic in the case of central authentication, but the AP can be the authenticator in the case of FlexConnect with local authentication. When the WLC is the authenticator, it is the one originating the EAP messages sent to the client, and AP is just transmitting those messages over the air. That is part of the split MAC model.

3. The supplicant submits its identity for the authenticator to forward to the authentication server. First of all, this identity does not have to be the real username (and never a certificate). Because many EAP methods use a TLS tunnel, the identity of the first authentication method is not so useful, and many clients send **anonymous** as the identity. This is called the *outer identity* (in the case of a tunnel EAP method like PEAP). This outer identity is the only thing the authenticator (that is, the WLC) is able to use to refer to the user. It is not able to read the inner identity sent inside the encrypted tunnel because it does not have the keys to decrypt this tunnel and is merely forwarding data between the supplicant and the authentication server. The WLC can sometimes learn the supplicant

identity if the authentication server reveals it in RADIUS attributes like it sometimes does by appending the username RADIUS attribute with the access-accept. This identity is carried to the authentication server through RADIUS encapsulation, and the authenticator appends a number of RADIUS AVPs to give extra detail about the authentication, such as the AP MAC address (as Called-station-id), the client MAC address (as calling-station-id), the SSID details, and so on.

4. The rest of the authentication is defined by the specific EAP method in use. The choice of the method is determined by the authentication server proposing EAP methods one by one to the client until the client acknowledges the use of a method. So, the RADIUS server suggests a method, and the supplicant is free to accept or request another one (without knowing what will be offered next).

5. The EAP authentication goes on, and the authenticator forwards each frame/packet back and forth between the supplicant and the authentication server. When the authentication is successful, the server sends a RADIUS access-accept message that may contain extra RADIUS AVPs to define the type of authorization granted (a privilege level, a VLAN assigned, an access list to apply, a session timeout, and so on). The authenticator relays this as an EAP success message.

6. On the supplicant side, WPA keys typically are exchanged then. On the infrastructure side, the authenticator starts sending accounting packets with informative details about the session. Depending on the configuration, the authenticator can keep sending regular accounting packets during the session (which is useful for tracking bandwidth consumption).

RADIUS Change of Authorization (CoA)

In a standard RADIUS interaction, the network device is the one initiating all the communications. Those communications are either a new authentication request or some accounting data. There is no means for the authentication server to spontaneously send data to an NAS or to terminate a current session (that is, a client that has been authorized before to access the network). RFC 5176 adds this flexibility to the RADIUS protocol by defining enhancements that allow the authentication server to dynamically modify the authorization of a user session via Change of Authorization (CoA) messages.

CoA messages are transported over UDP using 3799 as the destination port; however, Cisco devices use port 1700 for this task. The NAS replies with a CoA acknowledgment if it can successfully change the authorization for the user session or a nonacknowledgment if it is unsuccessful. Here are a few examples of what the authentication server can do:

- Terminating a user session
- Requesting the NAS to reauthenticate the user/device from scratch

Practical examples vary a lot:

- The ISE can decide to reauthenticate a client after new profiling data is made available (the ISE obtains new DHCP packets or intercepts new HTTP packets that refine the profiling information), and this potentially results in a different policy applied for that client.

- The ISE can decide to terminate a specific session if a security threat is automatically or manually detected.
- The ISE uses CoA in regular workflows such as central web authentication (this topic is covered in the section titled, "*Central Web Authentication*").

To do so, the authentication server has to attach several attributes to the CoA request, such as the accounting session ID, the audit session ID (a Cisco vendor-specific attribute), and a calling session ID, which is basically the host MAC address. With this information, the NAS is able to execute on the authentication server request.

CoA is not enabled by default on many Cisco NAS or WLCs, and you need to specifically configure that you will tolerate the authentication server sending CoA messages that affect user sessions.

RADIUS Configuration and Load Balancing

Authentication of the wireless client involves defining a RADIUS server, optionally creating a server group, defining AAA methods that use the server you created, and calling these methods from the WLAN configuration, for example (or someplace else if talking about a form of authentication other than wireless client authentication).

Configuring RADIUS Servers

To configure a RADIUS server, you can use the WebUI in **Configuration > Security > AAA**, as depicted in Figure 5-15. Under the CLI, you would enter the following:

```
radius server <servername>
 address ipv4 8.60.0.252 auth-port 1812 acct-port 1813
key  <key>
```

You can choose the server name, and it is only for reference. The address can be configured with IPv4 or IPv6 or a hostname (that requires configuring a DNS server for the 9800 to use). It is advised you enter the key in clear text. The encrypted options are mostly used in CLI when you are pasting the key that is already encrypted from another configuration file. The timeout defines the number of seconds the 9800 will wait before declaring the RADIUS packet lost (and send retries), and the retry count is the number of times a RADIUS request will be retried until declared over.

Configuring RADIUS Server Groups

If you are planning to use several RADIUS servers for redundancy or load balancing, creating a server group makes things easier. It basically consists of naming a specific list of servers. The RADIUS servers need to be created on the Catalyst 9800 configuration first before you can assign them to a server group. It is easily defined in the WebUI under **Configuration > Security > AAA**, where you have a **Server Groups** tab, as shown in Figure 5-16, or in the CLI:

```
Aaa group server radius <group name>
   Server name <server1>
   Server name <server2>
```

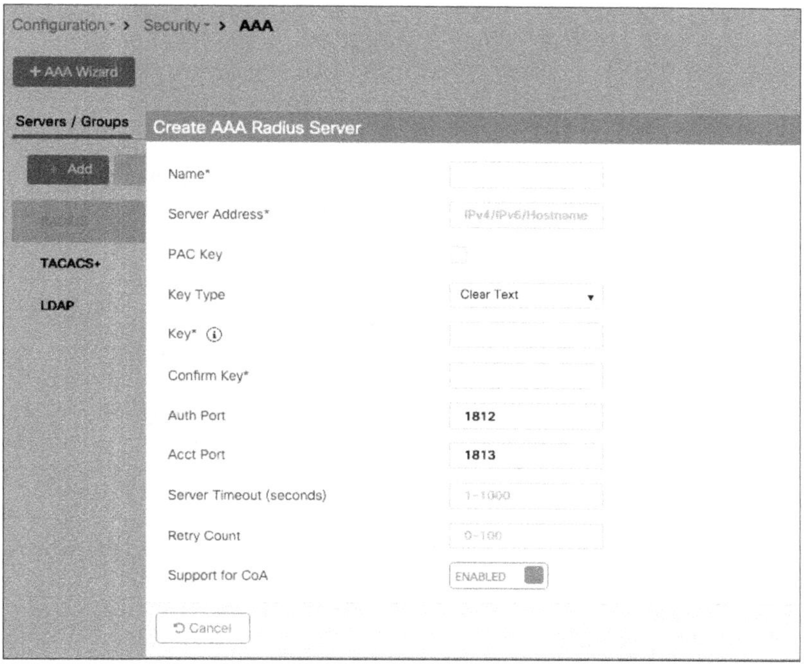

Figure 5-15 *RADIUS server creation on the C9800 WebUI*

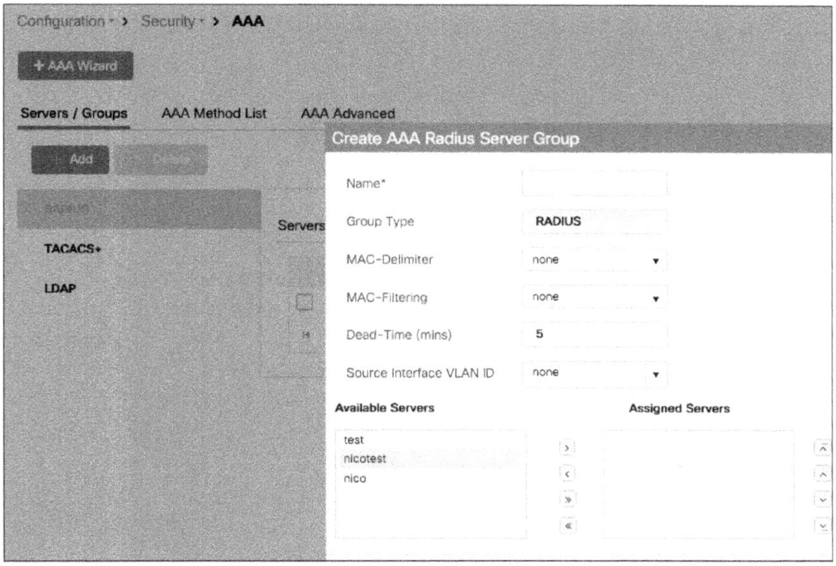

Figure 5-16 *RADIUS server group configuration*

The server group is also a specific place to configure extra RADIUS settings:

- In the case of MAC filtering authentication, for example, you can define what the MAC address delimiter will be (hyphen, colon, and so on) because different RADIUS servers might expect different formats.
- Similarly, the MAC filtering attribute allows you to define what value will be used in the password field. Some RADIUS servers expect the MAC address to be repeated there, whereas others expect the RADIUS shared secret to be used.
- The dead-time is a critical setting to configure how long a RADIUS server will be dead before it can be attempted again. If you don't configure this setting, the RADIUS server is instantly marked back alive after being declared dead, and you basically never fail over. Although newer software versions configure 5 minutes by default, you may still have unconfigured dead time if you are upgrading from an earlier release and kept your configuration. A RADIUS server is considered dead when it does not reply after a configurable number of retries and a configurable timeout to RADIUS requests (configuration illustrated in Figure 5-15) sent by the C9800.
- The source VLAN ID allows you to overwrite the typical interface. The 9800 uses its routing table to determine the interface or SVI to use to send RADIUS traffic to the server. You can override this globally with the **ip radius source-interface <int>** command or on a per AAA server group basis with this source VLAN ID. This is similar to the **RADIUS interface overwrite** feature in AireOS controllers except that, because the 9800 does not have dynamic interface, this setting is on a per group basis and allows you to choose a specific interface to source RADIUS traffic for each group.

The order in which you define the servers in the group is the order in which they are used. It is not possible to change this order after creation unless you delete and re-create the server group with the servers in another order.

Any setting or timer that is set under the specific RADIUS server configuration supersedes the same setting configured under the AAA server group, if present.

The global setting **radius-server dead-criteria time <seconds> tries <number>** defines the criteria to declare a RADIUS server dead. Basically, when the 9800 does not get a reply to a RADIUS request after the configured server timeout, it retries sending that packet the number of times configured in the retry settings. If there is still no reply, that specific RADIUS request is dropped on the 9800, but the RADIUS server is not marked dead until the dead-criteria is met (in both time and number of tries of different RADIUS requests). The dead time configured in the RADIUS server group supersedes the globally configured setting if there is one. More specific configuration, in general, supersedes more global configuration if they represent the same setting.

RADIUS Server Fallback

We have explained the conditions under which the controller moves from the primary RADIUS server in the list to the next one and so on. Let's bring a bit of clarity on when the controller falls back to using the primary RADIUS server.

The equivalent of the passive fallback setting in AireOS is achieved by using the **dead-time** command previously mentioned. This means that when you enter **radius-server deadtime <minutes>**, this command defines the number of minutes for which the RADIUS server is considered dead, and therefore, the WLC uses the next RADIUS servers in the list. After this time has passed, the WLC automatically starts using the primary RADIUS server again with the next authentication that it has to perform.

If you add the command **automate-tester username <dummy username> probe-on** to the RADIUS server configuration section, test RADIUS authentications (using the dummy username you entered) are sent to the RADIUS server only when it is marked dead to see if it is back alive. If you configure **automate-tester username <dummy user> idle-time <minutes>**, the controller sends the test authentication every "idle-time" period even when the server is alive (which can be useful to detect whether it goes dead when there are no authentications ongoing). The automate tester considers the server to be alive if it receives any reply from the server; the tester does not need to receive a successful authentication result (especially because no password was configured). Just make sure the RADIUS server does not ignore such a plaintext PAP authentication, which can sometimes be the case with a default configuration.

Somewhat similarly, but in a purely manual fashion, you can test an authentication against a RADIUS server or group with the exec mode command **test aaa group radius <user> <password> new-code**. This command also sends a PAP authentication request to the RADIUS server list with the given username and password.

RADIUS Load Balancing

Load balancing helps in handling authentications during a server failure but can also be used to balance the load between several servers and make sure no RADIUS servers get overwhelmed with a storm of authentication requests. By default, configuring several servers in a radius server group allows for redundancy, but only one server is used at a time; the controller uses the next RADIUS only if the current server does not respond and gets declared dead. You can also configure several RADIUS server groups in your AAA authentication methods, but the effect is the same: the second RADIUS server group is used only if the first one is completely dead.

If you add the **load-balance method least-outstanding** command to the RADIUS server group (or globally with **radius-server load-balance method least-outstanding**), the WLC rotates RADIUS servers even when they are alive. For a new batch of RADIUS requests, it checks which server has the least number of transactions and uses it.

RADIUS Accounting

A similar configuration can be done to add a RADIUS accounting server. Periodic interim accounting records can be sent to the accounting server (otherwise, a START is sent when the session starts and a STOP when the client is deleted) if you enable **aaa accounting update [newinfo] [periodic] number** in global configuration mode.

AAA Methods

The command **aaa new-model** is required to configure anything pertaining to AAA that is described here, and it unlocks many other commands. You can create various types of AAA methods from there:

- **Authentication dot1x:** The **aaa authentication dot1x <method name> <aaa server group>** command allows you to define the authentication server(s) for 802.1X authentication for both wireless clients or mesh APs (if applicable).
- **Authentication login:** The **aaa authentication login <method> <server group>** command allows you to define methods for logging in the device (GUI or CLI).
- **Authorization network:** The **aaa authorization network <method name> <server group>** command is a kind of multipurpose method. Its main use is to allow the override of the AAA attribute, that is, to accept and apply RADIUS or TACACS attributes received from the server during an authorization result. For example, it is required for the controller to apply a dynamic RADIUS VLAN assignment or session timeout or ACL. This method type is also used for MAC address authorization (client or APs).
- **Authorization credential-download:** The **aaa authorization credential-download <method name> <server group>** command allows you to define the server(s) that will be used for verifying credentials of an authentication happening locally. It is used in the case of the local EAP (when the 9800 acts as RADIUS server) or LDAP authentication.
- **Authorization exec:** The **aaa authorization exec <name> <server group>** command is used to determine which server is used to authorize users in starting an EXEC shell on the device.
- **Accounting identity:** The **aaa accounting identity <name> start-stop <aaa server group>** command allows you to define RADIUS accounting servers that will receive session statistics for network users.

When doing a local authentication (as in the case of local EAP) on the 9800, you need to define which methods are used with

```
9800(config)#aaa local authentication <method> authorization <method>
```

Only named AAA methods have been mentioned so far. Every method also has the "default" keyword that can replace the name. A named method has no effect until you call it from a specific part of the configuration. The default method is automatically applied for the said use case. For example, if you configure the **aaa authentication login default** method, you impact the way users authenticate when accessing the 9800 via the CLI (unless you configured a named method in the VTY line configuration already). They are sometimes required (for example, the NETCONF authentication cannot be configured with named methods at the time of this writing and will always use the default one), but in general you should try not to use them unless you know what you are doing exactly.

AAA methods can point to RADIUS server groups or TACACS server groups (except for dot1x) but can also use the "local" destination. An important point is that if you select local first and then add a server group, the method first checks in the local database and consults the external server if the user is not found locally. However, the opposite is not true: when defining a server group first and then local as fallback, the controller checks only in the local database if the server group is completely down (and not if the user is not found in the server group because the controller has no means of knowing why an authentication has failed or if the user was not found or has an invalid password).

When you're using a local database of users, it is possible to define a set of AAA attributes for them. Defining these attributes could be useful if you are doing local EAP—that is, local authentication of 802.1X users on the WLC, and you want to return a specific VLAN or ACL. It can also be leveraged when you're doing local MAC address authentication. You can define an AAA attribute list this way:

```
9800(config)#Aaa attribute list user1_list
9800(config-attr-list)#Attribute type ssid "test-ssid"
```

You can then map this list to the user in the local database:

```
9800(config)#User-name user1
9800(config-user-name)#Aaa attribute list user1_list
```

Or if you have MAC address users, you can use the following:

```
9800(config)#User-name <MAC> mac aaa attribute list <attribute-list-
name>
9800(config-user-name)#Aaa attribute list user1_list
```

You need to set the **aaa authorization network** method to point to local to accept these attributes. Don't forget to enable AAA override on the policy profile as well.

Possible and popular attributes that you can use are

- **SSID:** You can restrict the SSID that the user can connect to.
- **VLAN:** You can assign a VLAN depending on the username.
- **QoS:** You can assign QoS policies specifically for the user.
- **Session-timeout:** You can override the WLAN session timeout on a per-user basis.
- **ACL:** This attribute helps assigns a predefined ACL to the user.

Local EAP

Often confusingly shortened as LEAP (which is a very old and insecure EAP type), *local EAP* refers to the capability of the C9800 to act as a RADIUS server for wireless clients only. It supports PEAP, EAP-TLS, and EAP-FAST (let's not even mention *LEAP* anymore, which still shows up as an option in the configuration). It is meant for use in limited-scale deployments or as backup when the RADIUS server is not reachable (in branches where you have a controller, for example). Local EAP is explained in detail in a configuration example on Cisco.com and referenced at the end of this chapter, but let's still look at the basic concepts around it.

Local EAP requires you to

- Configure a local user, or "network" type of user, with **username <username> password 0 <password>**, for example.
- Configure a dot1x authentication method pointing locally.
- Configure a credential download authorization method pointing locally.
- Define the methods name under **aaa local authentication.**
- Configure an EAP profile where you define the EAP types supported as well as point to a trustpoint for the 9800 to use a specific certificate.
- Define the local EAP authentication profile in the WLAN settings.

TACACS+

Terminal Access Controller Access Control System (TACACS) is a protocol from 1984 that got enhanced by Cisco under the name TACACS+ in 1993. TACACS+ has core differences with RADIUS, despite the fact both are authentication protocols covering all three *A*s of the AAA acronym. Here are a few of the core differences of TACACS+ compared to RADIUS:

- TACACS+ uses TCP instead of UDP. TCP is reliable and offers a connection-oriented transaction system. TCP covers the retransmission problems, whereas with RADIUS you need to specify timers and handle retransmissions in the application layer itself. TCP also allows you to know if the server is still dead or simply ignoring the actual authentication request for some reason.
- RADIUS only encrypts the password in the access-request packet while all the rest of the RADIUS packet is unencrypted. TACACS+ encrypts the entire body of the packet but leaves a standard TACACS+ header. This protects the username as well as other network attributes present in the packet.
- RADIUS combines the authorization phase with the authentication phase by returning all the required privileges and attributes in the access-accept payload. TACACS+ separates all the phases, and the authentication phase is fully separate and only identifies the user without covering access privileges. The NAS can then immediately do an authorization request to request access to a specific resource. The TACACS+ server can reply to this authorization request without reauthenticating the client. This allows you to force the NAS to request authorization for every specific resource and therefore allows much more granularity.

The separation of the authorization phase, combined with the increased number of packets of the whole process (as seen in Figure 5-17), makes TACACS+ a protocol better suited for network device management than end user authentication (like 802.1X).

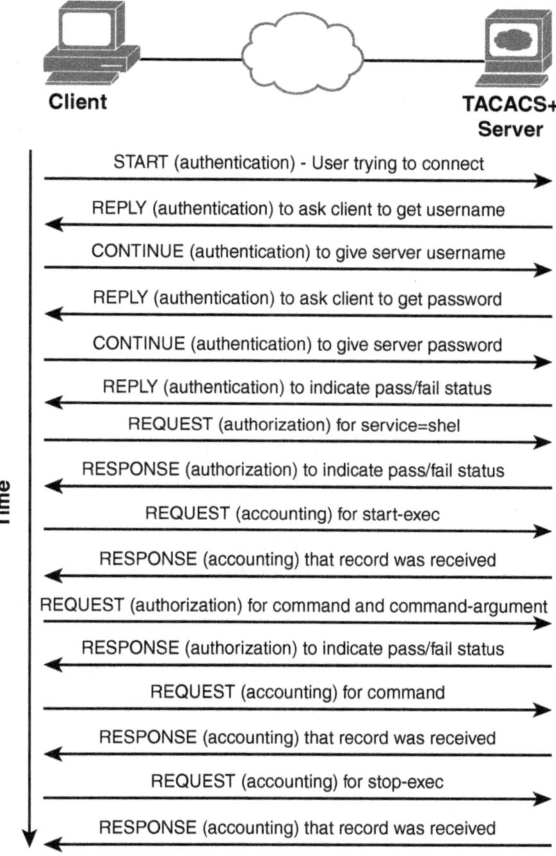

Figure 5-17 *TACACS workflow*

As a matter of fact, TACACS+ does not support encapsulating an 802.1X authentication, which means RADIUS is your one and only option to transport 802.1X on your wireless network. To secure and control access to your network devices, such as the wireless controller, on the other hand, you have the choice between RADIUS and TACACS+. Using RADIUS would seem like a logical option because this means installing only one type of authentication server (although ISE does both RADIUS and TACACS+) and using a single protocol in your network. However, the drawback of using RADIUS for network device management is that you can return only a one-time authorization set of attributes, which typically are an IOS privilege level, and possibly some specific attributes (like defining the user as a lobby admin and not a regular admin user). TACACS+, because it allows you to force an authorization for every specific access, allows you to define granular CLIs or web pages that can be accessed. The NAS then asks for authorization for every single command typed on the CLI, and the authentication server accepts or denies each of them. TACACS+ command authorization obviously can add latency if you are going to paste many commands, but it allows for very granular security where specific users are allowed only a very specific subset of commands.

LDAP

Lightweight Directory Access Protocol has remained popular over the years, although it is not a very secure end-to-end authentication solution. LDAP is a network protocol for accessing user directories. It does not take care of the authorization part of things (no network access level defined or custom attributes returned) but simply validates user credentials. It can be used as a back-end user database for web authentication or for local EAP. There are many limitations around LDAP (limited use cases, insecurity of the passwords, impossibility to use it for PEAP-MSCHAPV2 authentication, and so on), so this book does not cover it. There is a document about it on cisco.com if you seek to learn more; check the "References" section at the end of this chapter for the link.

As you have read in the previous sections, the AAA override check box allows the WLC to honor the attributes received from the RADIUS server and that are attached to the RADIUS access-accept packet. Some of the attributes don't have any particular prerequisites and can be honored on the spot, such as the session timeout (it overrides the session timeout configured in the policy profile). On the other hand, some attributes, such as ACL or VLAN assignment, require the VLAN or ACL to be preconfigured on the WLC for the WLC to simply have to apply the setting. In FlexConnect, those settings need to exist on the AP itself, where they have to be applied. This means that you need to precreate specific VLANs and/or ACLs on the APs before being able to return them via the AAA server. This can be done in the Flex profile where you can add policy ACLs (which download the ACL to the AP and not necessarily apply it yet because it is dynamically applied), and VLANs in the VLAN-ACL mapping (where you do not necessarily have to assign an ACL but can just add VLANs).

Wireless Security Fundamentals

Next, we cover in enough detail all the security settings that can be applied on an SSID so that you can feel comfortable understanding what you are configuring on a Catalyst 9800 WLAN.

Wired Equivalent Privacy (WEP)

Wired Equivalent Privacy is not supported by the Catalyst 9800. WEP is not considered security and has been deprecated by the IEEE for many years. History books are available to learn more about WEP.

Wi-Fi Protected Access (WPA)

Wi-Fi Protected Access is the Wi-Fi Alliance certification based on the 802.11i amendment to the standard. The first version was released as a stopgap for devices that did not have the hardware to implement the final 802.11i specifications, and therefore, WPA1 implemented the Temporal Key Integrity Protocol (TKIP), which was based on the same RC4 algorithm used by WEP. WPA2 uses only the Advanced Encryption Standard (AES) algorithm, which is still considered secure to this day.

WPA2 originally combines AES with Counter Mode with Cipher Block Chaining Message Authentication Code Protocol (CCMP), which is, to make things simple, the way that the AES encryption algorithm is applied to the data.

WPA uses the existing legacy open system authentication, which basically consists of the client and AP sending each other empty successful authentication frames before being able to send the association request. Although this approach seems weird, the reason is that the only other authentication frame type at the time referred to share key WEP authentication and was very unsecure. Therefore, the core of the WPA exchange happens after the association. The association request and response are the opportunity for both the client and the AP to exchange their security capabilities. The Robust Security Network (RSN) Information Element (IE) is there to specify which authentication and encryption methods and ciphers can be used by both parties.

WPA comes in two flavors (aside from the WPA1 prestandard certification and WPA2 final certification): Personal and Enterprise. The encryption of the traffic between the wireless client and the AP is unique (that is, it's different for every session) and securely created from the user authentication phase. It is important to note that only data frames are encrypted by WPA, and all the management frames are sent in clear text. WPA Personal is the most popular flavor out there due to the simplicity to deploy it: the wireless network (or SSID or WLAN) uses a preshared key that is known by all authorized clients and that serves as authentication (if you know it, you must be in on the secret and therefore allowed to join the network) as well as a seed for the unique encryption key. This means that every wireless client, in reality, uses a different key for sending and receiving traffic to and from the AP, but all those keys are generated from the same master secret. WPA Enterprise requires the use of a RADIUS server and relies on the 802.1X/EAP authentication framework (which typically works based on username/password pairs but can also accommodate certificates, tokens, smartcards, and the like) to generate the key seed. The benefits of WPA Enterprise are its use of different credentials for each user, and the fact that each user encryption key is based on those unique credentials rather than shared credentials. The 802.1X/EAP can come with stronger security during the authentication phase, but that largely depends on the EAP method chosen and the way it is implemented. The cost of WPA Enterprise includes setting up a RADIUS server (which can be more or less of a hassle depending on what you choose) and the management of unique credentials for every client (although they could technically still use identical credentials if you choose that for simplicity of management). IoT devices (printers, scanners, and so on) typically implement WPA Personal only for the simplicity of implementation and do not allow WPA Enterprise to be configured.

The security of both WPA flavors lies mainly in the credentials chosen. Even with the most secure EAP method for WPA Enterprise, if a user password is **P@ssw0rd** (even with that @ and 0 included), the most basic tools can guess or brute-force it very quickly. The same goes for PSK: if you choose a complex preshared key that is long enough, your network is very secure to brute-forcing, and the main weakness resides in the fact that all clients share this same key. Therefore, someone could leak it to an attacker, and all the clients are then compromised instead of just one. There are multiple proprietary implementations of PSK, which allow for having more than one PSK (up to having a different key for every device if you wanted to). The methods supported by the Catalyst 9800 are covered in the section "Preshared Key for WPA Personal."

Protected Management Frames (PMF) is a standard amendment, compatible with WPA2 but ratified much later, that encrypts unicast management frames and provides protection against forgery to broadcast management frames as well. Thanks to PMF, it is not possible anymore for an attacker to impersonate the AP in an existing client-to-AP–established encrypted relationship. Probes and association frames, for example, are still in clear text because they happen before a key handshake can occur.

The encryption key is derived from the seeding material through what is often called the WPA *four-way handshake*. In reality, several keys are created during this exchange because there exists a separate key for broadcast traffic called the Groupwise Transient Key (GTK) and a few others that are beyond the scope of this chapter. This four-way exchange is always the last step before traffic can be sent (typically the DHCP DISCOVER would be the first traffic sent by the client) in an encrypted manner on the network. This four-way handshake has to happen between every client and every AP the clients connect to: because the encryption key is unique to the client-AP session, a different key has to be negotiated when the client roams to another AP. This handshake is a ping-pong between the AP and the client exchanging nonces (that is, random and arbitrary numbers used to see the cryptographic communication) to derive the same encryption key without transmitting that key over the network and considering both parties know a shared secret (either the PSK or the result of the 802.1X/EAP authentication that occurred just before). Those messages are called EAPoL Message 1 to 4 or EAPoL M1, M2, M3, or M4.

Those four messages are enough to exchange all the needed session keys between the AP and the client. Because the broadcast traffic is encrypted by a separate group key that is identical between every device connected to the same AP, that group key may need to be rotated regularly or when clients join or leave the WLAN. The AP can sometimes send an M5 message to which clients reply with an M6 message to derive a new group key for broadcast traffic.

WPA3 recently became the latest cool thing in the domain of wireless security and is covered later. It was created to improve on the weaknesses of WPA2. Despite the fact that WPA2 has still not been fully cracked yet to this day, it still showed vulnerabilities that required improvements. The main issues of WPA2 are as follows:

- The passphrase/password is used for both authentication and encryption.
- It is susceptible to offline attacks if the four-way handshake is captured. The attacker can try a dictionary of keys or passphrases until they are able to verify message 3 and 4 with it. With the public cloud offering a lot of compute power at a rental cost, this is making dictionary or brute-force attacks possible for anyone, even on a budget.
- There is no forward secrecy, which means if you know the password or passphrase, you are able to derive all the session keys for all past, present, and future sessions.
- By default, only the data frames are encrypted, which means that all the control and management frames are in clear text, and this basically opens the door for imperson-

ation and for attackers to pretend to be someone else and disconnect other clients remotely. PMF was released to take care of this issue but didn't get much traction from a Wi-Fi device manufacturer's standpoint and so was not widely implemented.

802.1X for WPA Enterprise

802.1X-2010 (yes the *X* is capital) is a protocol framework for Layer 2 access control. This implies that the network device controls the user identity before providing an IP address to it and permits or denies network connectivity based on the identity of the user or device requesting access. Built around a wired port concept, it still translates perfectly as a virtual port concept for wireless users. 802.1X allows you to authenticate wired and wireless devices similarly, as shown in Figure 5-18. The controlled port blocks all data except for 802.1X-related authentication frames (called EAPoL frames) until authentication has completed successfully. At that point, the port forwards all or some traffic only (based on the authentication result). Over wireless, each association between a client and access point generates a unique pair of 802.1X virtual ports. The only difference with 802.1X over wireless is that the authentication results in the generation of cryptographic keys that are used for traffic encryption (which is not necessarily the case over a wired connection).

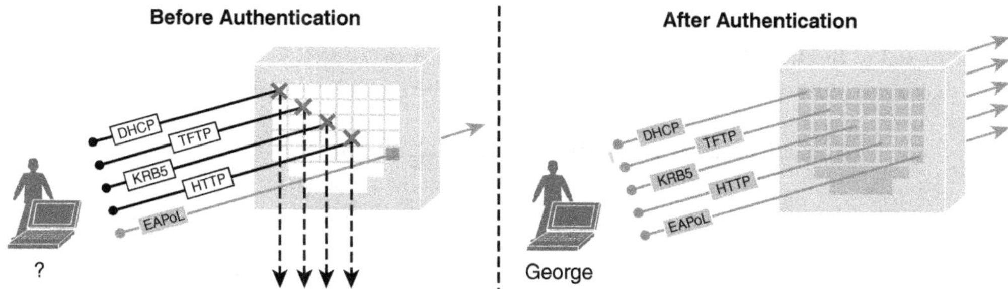

Figure 5-18 *802.1X port-based authentication system*

802.1X Components

As specified in the standard, and as shown in the Figure 5-19, a few terms depict the various roles of devices taking part in an 802.1X authentication:

- **The supplicant:** The client requesting access to a network resource and submitting credentials for authentication.
- **The authenticator:** A network device controlling the access to the network and facilitating the authentication process by relaying the credentials of the supplicant to the authentication server. This typically is a switch or, in this case, a WLC.
- **The authentication server:** A server responsible for validating the credentials sent by the supplicant and determining the network access level to grant to the user. In this case, this is the ISE server.

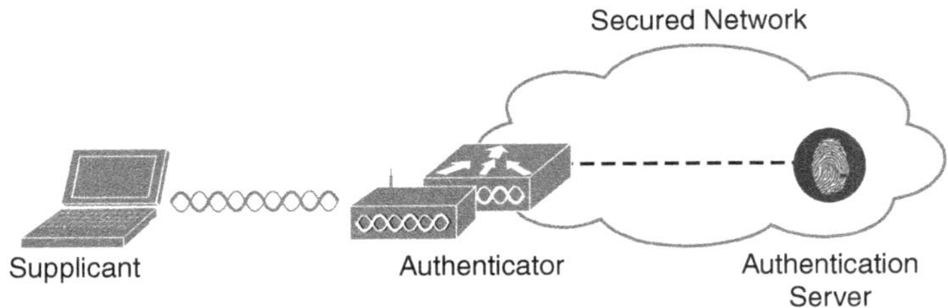

Figure 5-19 *802.1X components*

EAP

The Extensible Authentication Protocol (EAP) is at the heart of the 802.1X authentication process and provides a flexible framework and protocol, which supports multiple methods that are negotiated between the client and the authentication server and used to mutually authenticate them. EAP messages between the supplicant and the authenticator run directly over the data link using EAP over LAN (EAPoL) encapsulation (as shown in Figure 5-20). The authenticator can then extract the EAP payload and forward it to the authentication server over the wired network, typically inside a RADIUS packet. There is a clear demarcation between the EAP exchange over EAPoL between the client and the authenticator, which uses no addressing because the link is point to point and is supposed to be the only traffic allowed, and the transport of this EAP exchange over the IP network to the authentication server, which can be much further away on the network.

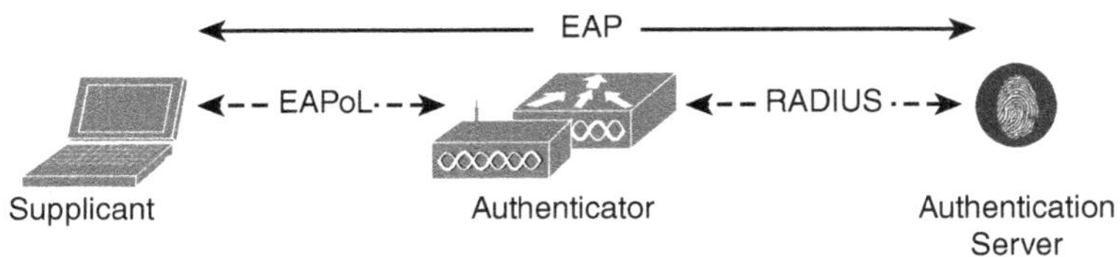

Figure 5-20 *EAP authentication workflow*

EAP is a request-response protocol consisting of only four EAP type packets identified in the "code" field:

- **Request:** This packet is sent by the authenticator to the supplicant and uses the type field to indicate what is requested (for example, identity, or EAP method to use).
- **Response:** This packet is sent by the supplicant in response to a valid request from the authenticator.
- **Success:** This packet is sent by the authenticator to signify a successful completion of an authentication.
- **Failure:** This packet is used by the authenticator if the authentication ends in a failure.

EAPoL frames also have several types and can be an EAPoL START (to signify one party wants to start a new authentication) or an EAPoL key (which transports a key).

EAP Methods

EAP is a framework protocol allowing for many different authentication methods (each having a unique ID) that define how credentials are exchanged. Although many exist, only a handful are used for authentication over wireless, and they are covered here.

- **PEAP (Protected EAP):** As depicted in Figure 5-21, this method relies on server-side TLS authentication to create an encrypted tunnel using a valid certificate that the authentication server sends to the supplicant at the beginning of the handshake. Just like during HTTPS web browsing, a TLS tunnel is built, and the server side always has to provide a certificate to validate its identity. The client doesn't have to provide anything. Historically, many operating systems allowed the supplicant to not enforce validation of the server certificate, which means you could run a bogus certificate on your ISE server without issues, but more and more operating systems started to enforce this validation with no option to disable it manually. This means it is becoming a requirement to have a certificate on your ISE server that the client has in its trust store. Some clients allow you to manually import the ISE certificate in real time, whereas others require you to preload it. When this TLS tunnel is established, a new EAP negotiation using a new/another EAP method takes place inside the tunnel.

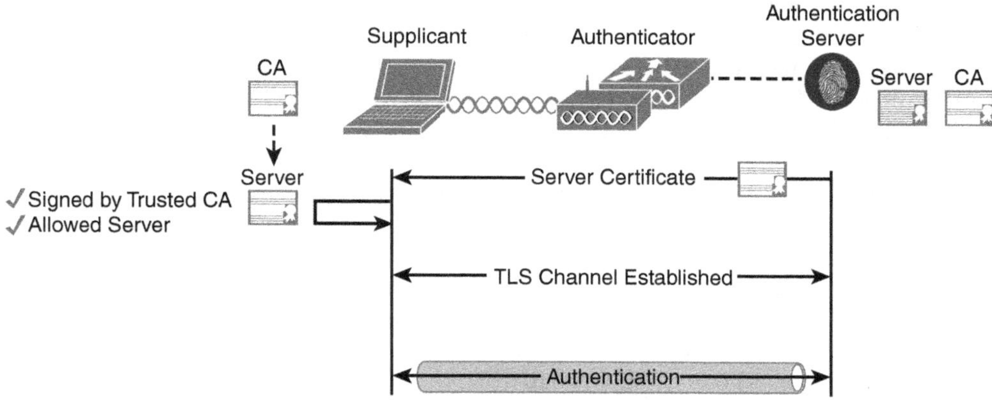

Figure 5-21 *PEAP authentication high-level workflow*

Therefore, the goal of PEAP is only to provide an encrypted tunnel to securely use another (even less secure normally) EAP method inside it. That new inner EAP method is typically EAP-GTC, MS-CHAPv2, or EAP-TLS.

- **PEAP with inner EAP-GTC:** This EAP method was originally designed for one-time password authentication or token authentication. This means sending the password in clear text (but it's inside the PEAP original TLS tunnel, so it's not visible to an attacker/eavesdropper). This method is required if you are using a third-party user credentials database like LDAP or such, as is typically the case when using one-time passwords. The external user database needs to get the password in clear text to verify it because it is not able to verify the MS-CHAPv2 encrypted version of it.

- **PEAP with inner EAP-MS-CHAPv2:** This is the most popular EAP method so far among Windows PCs and smartphones. EAP-MS-CHAPv2 is a way to exchange a username and password in encrypted form (it's encrypted twice because CHAP already encrypts the password and it's inside the PEAP TLS tunnel as well).

- **PEAP with inner EAP-TLS:** This is maybe the most secure method because another TLS authentication happens inside the encrypted TLS tunnel. This inner TLS authentication, this time, requires both server- and client-side certificate authentication. This means typically running an Enterprise PKI (something ISE can do for you with BYOD workflows) and issuing your own certificates to your clients. The client certificate is encrypted to any attacker/eavesdropper, so it is impossible for anyone other than ISE to know the certificate fields such as the username. Considering it involves two TLS handshakes, this method is quite slow if the client has to reauthenticate fully every time it roams to another access point. Fast roaming is advised if you want to deploy it.

- **EAP-FAST (EAP-Flexible Authentication via Secure Tunneling):** Depicted in Figure 5-22, developed by Cisco, and published as IETF RFC 4851, EAP-FAST enables mutual authentication by using a shared secret, called the Protected Access Credential (PAC), to establish a TLS tunnel that secures the exchange of user authentication messages. It has similarities to PEAP in this regard, but the PAC is at the core of EAP-FAST and allows for more deployment options. The PAC can be manually distributed to the client out-of-band if you want to ensure the most secure deployment. EAP-FAST has three phrases. Phase 0 is required only if you opted out of manually distributing the PAC out of band. During that phase, the PAC is provisioned in-band on the supplicant using either an anonymous TLS handshake (using the Diffie-Hellman protocol) or an authenticated tunnel using the server certificate. Concretely, this means you can avoid using a valid and trusted certificate on the authentication server in case you are doing manual PAC out-of-band provisioning or if you provision in-band but decided to use an anonymous TLS tunnel. Phase 1 is the TLS tunnel establishment using the PAC, which takes care of mutual authentication (if both sides have the PAC, they must know each other). Finally in phase 2, user authentication credentials are passed within the TLS tunnel created in phase 1. Either EAP-MS-CHAPv2 or EAP-TLS is available as the inner method, just like for PEAP. EAP-FAST also allows for proprietary features such as EAP-chaining where two identities (the device itself and the user on it) can be provided in the same authentication flow. The security of EAP-FAST depends greatly on your deployment options and the inner method used. Most people use it with anonymous in-band provisioning, which typically requires a first failed authentication to deliver the PAC before authentications can succeed after the client has installed the PAC correctly.

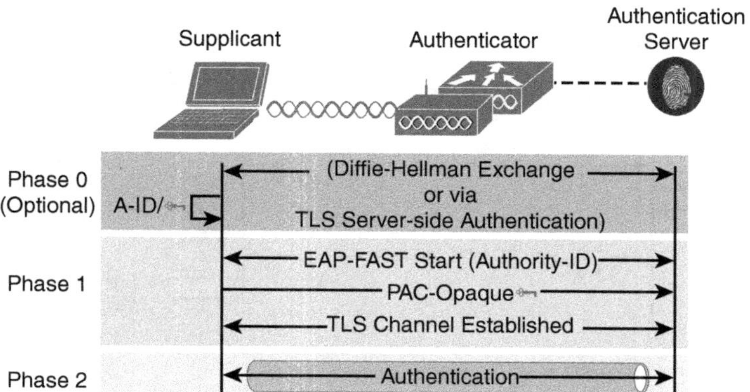

Figure 5-22 *EAP-FAST high-level overview*

- **EAP-TLS:** Depicted in Figure 5-23 and described in IETF RFC 5216, this method was covered previously as an inner method to PEAP or EAP-FAST. It is also available as a standalone method. A TLS tunnel is then built based on both a server-side as well as a client-side certificate. The tunnel itself is the authentication, and no other inner method occurs. It is therefore possible for an eavesdropper to read certificate details (which is not necessarily a major security concern) during the authentication. For an EAP-TLS authentication to succeed, the client device should have the root CA that issued the authentication server certificate chain in its trusted root CA store, and the authentication server should similarly have the root CA on top of the client certification chain in its trusted root CA store. This way, both parties can authenticate the other party certificate as issued by a trusted chain of CAs. It is interesting to note that there is technically no username in EAP-TLS authentication: a certificate is provided, and it's either trusted by the other side or it is not. The certificate is not issued to a username but to specific fields such as Common Name (CN), Country (C), and Organization (OU). To be able to identify the certificate uniquely to return a specific set of privilege, the authentication server can be configured to consider any of these certificate fields (or a combination of fields) as a username.

The validity of a certificate in an 802.1X context consists of verifying whether the certificate is still in its validity period and if it is issued by a trusted root CA on top of the chain and optionally if the certificate hasn't been revoked in a certificate revocation list (CRL). Some authentication servers can proceed in making a binary comparison of the certificate; that is, they compare bit by bit that the certificate presented is the same one as the one stored in the authentication server database, but that is uncommon. Being issued by a trusted source is typically enough. There is no verification of the name on the certificate because users don't really map to domain names like web servers do. The client also can hardly verify the authentication server DNS FQDN because 802.1X is a Layer 2 authentication method happening before any sort of IP connectivity. The client only relies on having the trusted CA installed in its store as well as the validity period. On its side, the ISE can be configured to read one field of the certificate (typically the CN) to obtain a username out of the certificate if that can be useful for matching authorization policies.

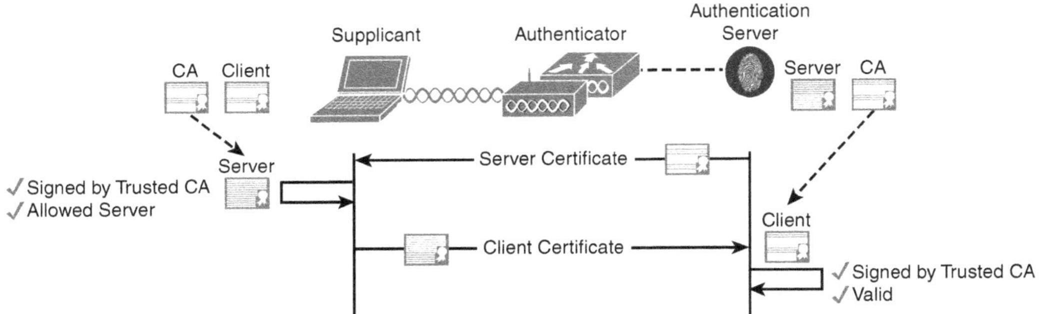

Figure 5-23 *EAP-TLS high-level overview*

WPA3 Enterprise

WPA3 also comes with a revision of the Enterprise flavor of WPA. There is no major advertised change, but the main thing is that WPA is also an interoperability certification, and this certification was revised to take care of many previous flaws and attack vectors around its implementation. More notably, WPA3 Enterprise brings a 192-bit cryptographic strength aimed at sensitive government and finance sectors, while the 128-bit security remains the default. The list of supported and allowed cryptographic protocols is revised and updated in WPA3, and a more restrictive option called SuiteB brings a uniform list of very secure protocols for the most demanding customers. PMF is now a requirement for WPA3, which means that clients supporting WPA3 support PMF, and on the administrator side, it is impossible to configure a WPA3 SSID without PMF being enabled.

A transition mode exists to allow WPA3 clients to use PMF while WPA2 clients can still connect. WPA transition mode basically downgrades the security of your SSID to a regular WPA2 SSID though.

Fast Transition (FT), the standard for fast and secure roaming (covered in more detail in the next chapter, along with client mobility), is not supported with 192-bit WPA3 Enterprise at this time; this would need a new AKM for 192 bits + FT and would need clients to support it, so it's an industrywide limitation. Also, Catalyst 9115 and 9120 access points do not support WPA3 SuiteB GCMP256 ciphers (AES256 is okay).

Preshared Key for WPA Personal

When you're configuring WPA Personal with a preshared key, what you typically think of as "the key" is actually the passphrase. It is not what is used to encrypt the traffic and not even what is used to generate the actual encryption key. Details are beyond the scope of this book, but the passphrase is converted to a 256-bit key, called a Pairwise Master Key (PMK), using a standard conversion method. That 256-bit key is used to generate the actual final encryption key. This allows you to have a variable-length passphrase (imagine having to remember a 64-character passphrase, for example), which is easier to remember. The requirements for the passphrase are that it must contain between 8 and 63 printable ASCII characters. Eight characters is really not a lot and should definitely not be used in production because it is a reasonable length to brute-force by using dictionaries of commonly used passphrases. Using a longer passphrase (including fancier characters) randomizes the 256-bit final key better and protects you from brute-force attacks.

One of the inherent weaknesses of the PSK authentication and key management protocol is that even if the encryption/session key is unique for each client, these keys are derived from the same seeding material (the passphrase). In practice, this means that if an attacker starts to listen to over-the-air traffic and knows the passphrase, it is not able to decrypt the traffic because it cannot derive the same session key. However, if the attacker is listening when a given client is associating and the attacker hears the four-way handshake, the attacker is able to derive the same session key as the client because they know the passphrase.

WPA3 SAE

WPA3 Personal is also called SAE for Simultaneous Authentication of Equals and brings much more than just a change of name. Contrary to its predecessors (WPA1 and 2), SAE does not rely on the open authentication mechanism but adds a new SAE authentication frame type, as shown in Figure 5-24. This adds two frames to the overall frame exchange count because there are four SAE authentication frames exchanged in total (two back and forth between the client and the AP), then the association and the typical four-way handshake.

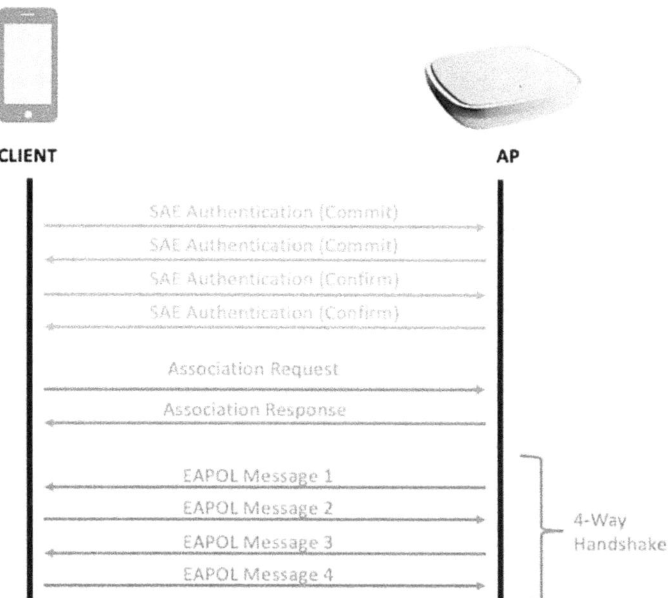

Figure 5-24 *WPA3 SAE workflow*

The SAE authentication frame exchange consists of four frames because it is a kind of key exchange by itself. Finishing the SAE initial four-frame exchange results in deriving a Pairwise Master Key (PMK) already. The concept of SAE is to protect the actual four-way key exchange by having it encrypted thanks to an SAE exchange done first,which allows you to exchange anonymously session keys. This process is similar in principle to the PEAP concept of building a secure tunnel to exchange credentials more securely. The effect of this is that even if an attacker knows the passphrase (which is still the weakest attack vector to get on to your network), the attacker is not able to decrypt the traffic from any wireless client, and that is a strong advantage over WPA2. In summary, SAE strongly enhances privacy while authentication security stays related to the strength of your passphrase and how you share it with your users. The SAE handshake also prevents the attacker from doing offline brute-force attacks on a captured handshake because that attacker has to be online and try one key at a time, which drastically reduces the speed of brute-forcing (and allows you to detect the attack).

WPA3 SAE requires PMF to be enabled (as for all WPA3 deployment types) and does not support Fast Transition (FT) at the time of this writing.

MPSK

MPSK (Mulitple PreShared Key) is an option you can configure in the WLAN settings (as shown in Figure 5-25); it allows you to configure up to five preshared keys for a given WLAN. This allows you to rotate keys or to use specific keys for specific devices. Keep in mind, however, that there is no authentication: anyone can connect using any of the keys entered.

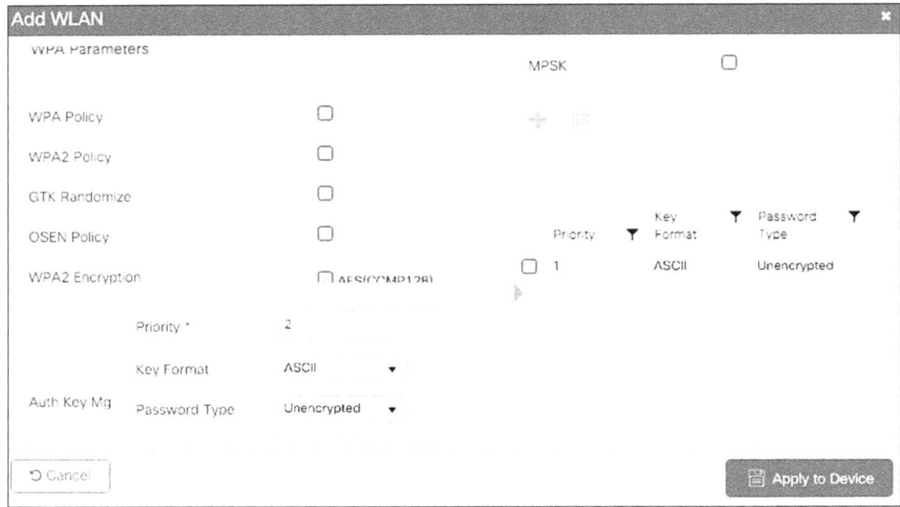

Figure 5-25 *MPSK configuration in WLAN settings*

Due to the way SAE works, MPSK is not supported with WPA3.

Identity PSK (iPSK)

Identity PSK is a standard PSK SSID where the administrator configures MAC filtering to have a pretext to reach out to a RADIUS server (you don't necessarily want to validate the MAC addresses). The RADIUS server can then return, through specific attributes, a (possibly) unique key for use with the wireless client. It is assumed the wireless client knows the right key to use. Although many people think this means they have to add wireless device MAC addresses in the RADIUS server and return specific PSKs depending on the MAC used for authentication, this is not the only possibility. As a matter of fact, the RADIUS server can use any RADIUS attribute in its authorization policy to determine which key to return: the location of the AP/WLC, the type of client device used (through profiling), or any creative combination based on the RADIUS attributes present in the MAC filtering authentication request. An example is shown later, but first let's look at MAC filtering itself and RADIUS concepts.

MAC Filtering

If you configure MAC filtering (also sometimes called MAC bypass) on a WLAN, the controller basically verifies the client MAC address authorization after the client sends the association request frame and before sending the association response. The wireless controller can either check locally in its local database or send the request to an external authentication server. The association response status code sent back to the client then depends on the authentication result: the association is rejected if the MAC address is not authorized to access the network.

There are several concerns with this system:

- It is not a security measure. A MAC address can be spoofed, and MAC addresses can be easily discovered by eavesdropping because they are always in clear text, and therefore it cannot be considered as a proper security measure.
- The fact that more and more devices use a randomized MAC address to connect to networks makes this measure completely unpractical and unmanageable. The last remaining use case could be IoT devices that do not perform MAC randomization or devices where the administrator manually configured the wireless adapter to not change the MAC address. But again, it cannot be considered secure because a MAC can be spoofed.
- It can be used as a pretext to trigger a RADIUS authentication to the external authentication server, as is the case with central web authentication (covered later).
- The MAC filtering authentication happens between the association request and the association response, which means that if the wireless controller has to verify the MAC address with an external RADIUS server, this can add a great deal of latency, and the association response sent back to the client can be delayed. There is no standard recommended maximum latency, but after more than 50 ms, most clients do not wait for an association response anymore and try to connect to other APs, possibly on other channels. MAC filtering has to be implemented with a low-latency nearby MAC address database.

Note At the time of this writing, the only MAC address format tolerated in the WLC local database for authorizing clients and APs via MAC filtering is aaaabbbbcccc (without any separators).

Enhanced Open

Not part of the WPA3 specification but released at the same time, Enhanced Open brings more privacy to open networks by encrypting all the guest traffic without requiring any key to be shared. Guest networks are here to stay, and although some administrators configured a preshared key on their guest WLAN, it is clear that this approach is neither secure nor practical: you have to share the key with your guests one way or another, and this often leads to anyone being able to figure out what the key is. This means that anyone can get on your network without having to type in anything. Enhanced Open does not bring any security in the authentication sense of the word; anyone can still onboard the network without any form of verification of identity. However, based on Opportunistic Wireless Encryption (OWE), the client and AP securely derive an encryption session key pretty much out of thin air (that is, without having a shared secret known beforehand from both parties).

Once connected, their traffic is private, meaning no attacker is able to decrypt their traffic (offline dictionary attacks are rendered impossible). Enhanced Open also requires PMF, which means it is not possible anymore for an attacker to remotely deauthenticate anyone else or impersonate certain management frames. Interestingly, Enhanced Open even beats WPA2-PSK from a privacy standpoint because when the attacker knows the WPA2-PSK, they can decode everyone's traffic, which is not the case with Enhanced Open, because every client uses a different master key that differs at each session.

Enhanced Open comes with a great transition mode because client support for Enhanced Open is still limited at this time. When using Enhanced Open transition mode, the controller creates two SSIDs in fact. One is broadcasted and is a regular open SSID, and the second is hidden (that is, its name is not present in beacons) and uses Enhanced Open as security. The regular open SSID beacons contain an information element advertising the presence of a hidden Enhanced Open BSS. This behavior is standardized, which means it is not a Cisco hack, and any client supporting Enhanced Open is aware of this technique. The result is that the legacy clients not supporting Enhanced Open simply see a regular open guest SSID and connect to it without seeing anything different (and do not benefit from any improved privacy, of course), whereas clients supporting Enhanced Open automatically connect to the Enhanced Open SSID even when the user selected the regular one on their device.

Securing the Air

Many configuration examples are available on cisco.com, showing very specific configuration. Here, we highlight the usual security combinations and the configuration logic, not the detailed steps.

WPA2 Personal

When adding a WLAN, as depicted in Figure 5-26, simply choose WPA+WPA2 as the Layer 2 security mode. Fast Transition is possible to configure with a PSK but does not bring a lot of advantages to the table.

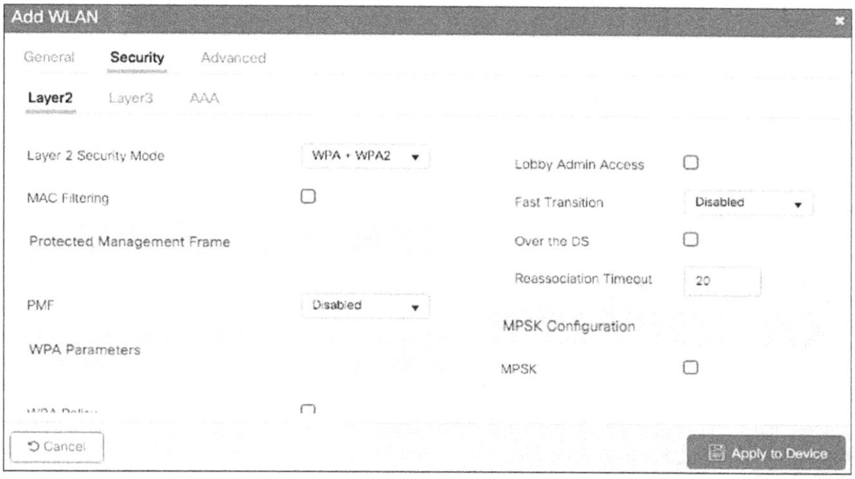

Figure 5-26 *WPA 2 PSK SSID settings*

The standard configuration, as shown in Figures 5-27 and 5-28, is PSK as AKM with AES128. GCMP is a protocol that allows for fast encryption of very high throughputs of data but is not required at typical Wi-Fi 6 speeds yet, and 256 bits of encryption is typically used in conjunction with WPA3 (although not necessarily, because not all clients support it).

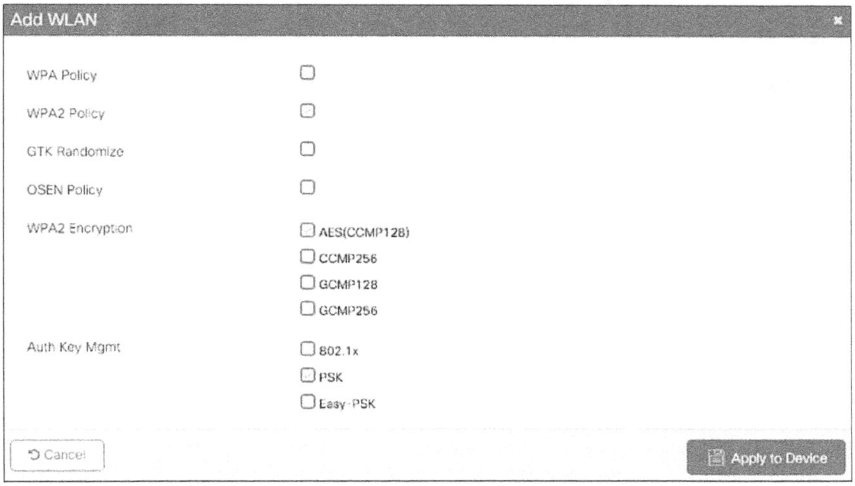

Figure 5-27 *WPA2 PSK Layer2 security settings*

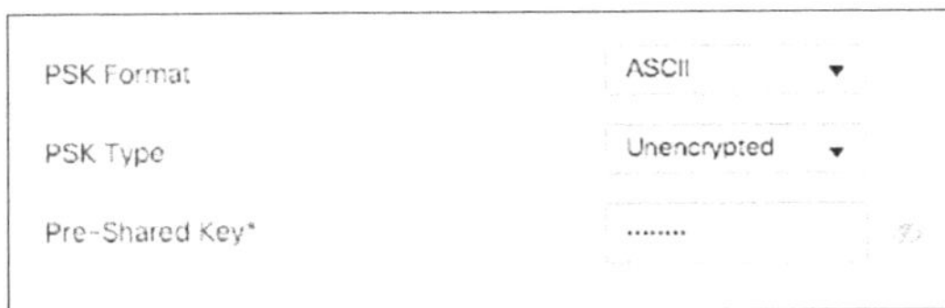

Figure 5-28 *WPA2 PSK SSID key configuration*

Enter the key as unencrypted if you are typing it manually. You would choose an AES key if the passphrase you were pasting is already AES-encrypted (if you were copying from an existing encrypted configuration file, for example).

Figure 5-29 shows the Robust Security Network (RSN) information element (a field from the beacon frame of the AP) of the WPA2 PSK SSID illustrated in Figure 5-26.

```
▼ Tag: RSN Information
    Tag Number: RSN Information (48)
    Tag length: 20
    RSN Version: 1
  ▼ Group Cipher Suite: 00:0f:ac (Ieee 802.11) AES (CCM)
      Group Cipher Suite OUI: 00:0f:ac (Ieee 802.11)
      Group Cipher Suite type: AES (CCM) (4)
    Pairwise Cipher Suite Count: 1
  ▼ Pairwise Cipher Suite List 00:0f:ac (Ieee 802.11) AES (CCM)
    ▶ Pairwise Cipher Suite: 00:0f:ac (Ieee 802.11) AES (CCM)
    Auth Key Management (AKM) Suite Count: 1
  ▼ Auth Key Management (AKM) List 00:0f:ac (Ieee 802.11) PSK
    ▶ Auth Key Management (AKM) Suite: 00:0f:ac (Ieee 802.11) PSK
  ▶ RSN Capabilities: 0x0028
```

Figure 5-29 *RSN information element of a beacon frame of a WPA PSK SSID*

WPA3 SAE

For configuring SAE, select **WPA3** (starting with 17.6 IOS-XE) or **WPA2+WPA3** as the security mode, as shown in Figure 5-30. As a reminder, FT is not supported with SAE at this time.

Figure 5-30 *SAE WLAN example configuration*

Choose **AES-128** and **SAE** for the AKM, as shown in Figure 5-31. WPA3 offers the same 128-bit encryption as was used with WPA2 but allows you to turn it up to 256 bits and offers the choice between the same CCMP mode of operation or to use Galois Counter Mode Protocol (GCMP). GCMP was meant to be used with 802.11ac wave 2 because it is better suited for very high throughput encryption, but it did not get traction in the field back then, especially on the client side. Many WPA3-certified devices now support 256-bit AES and GCMP.

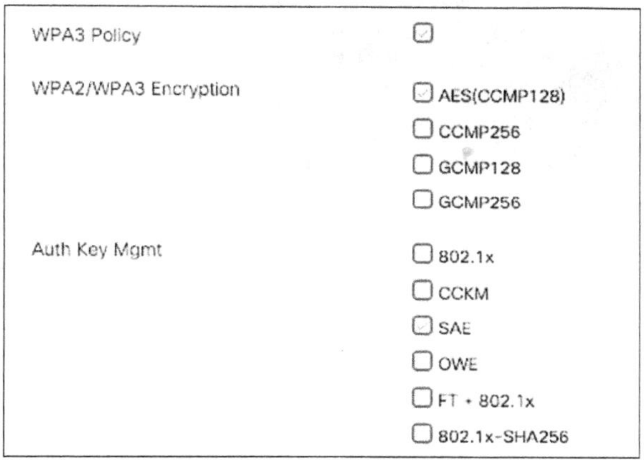

WPA3 Policy	☑
WPA2/WPA3 Encryption	☑ AES(CCMP128)
	☐ CCMP256
	☐ GCMP128
	☐ GCMP256
Auth Key Mgmt	☐ 802.1x
	☐ CCKM
	☑ SAE
	☐ OWE
	☐ FT + 802.1x
	☐ 802.1x-SHA256

Figure 5-31 *Ciphers option example for SAE*

Enter your preshared key unencrypted, as depicted in Figure 5-32, and you should be good to go.

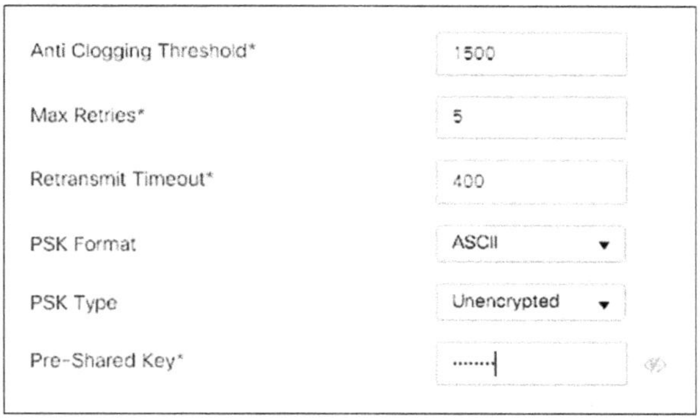

Anti Clogging Threshold*	1500
Max Retries*	5
Retransmit Timeout*	400
PSK Format	ASCII ▾
PSK Type	Unencrypted ▾
Pre-Shared Key*	....

Figure 5-32 *PSK configuration for SAE*

If you take a wireless capture and look at the beacon RSN IE, you can see SAE advertised as the authentication key management method (see Figure 5-33).

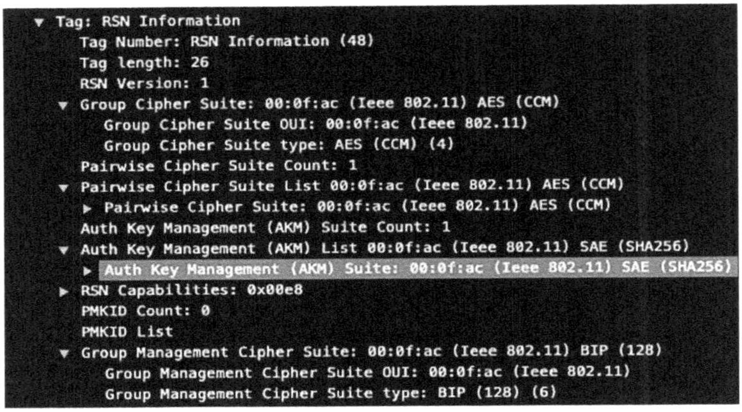
Tag: RSN Information
Tag Number: RSN Information (48)
Tag length: 26
RSN Version: 1
Group Cipher Suite: 00:0f:ac (Ieee 802.11) AES (CCM)
Group Cipher Suite OUI: 00:0f:ac (Ieee 802.11)
Group Cipher Suite type: AES (CCM) (4)
Pairwise Cipher Suite Count: 1
Pairwise Cipher Suite List 00:0f:ac (Ieee 802.11) AES (CCM)
Pairwise Cipher Suite: 00:0f:ac (Ieee 802.11) AES (CCM)
Auth Key Management (AKM) Suite Count: 1
Auth Key Management (AKM) List 00:0f:ac (Ieee 802.11) SAE (SHA256)
Auth Key Management (AKM) Suite: 00:0f:ac (Ieee 802.11) SAE (SHA256)
RSN Capabilities: 0x00e8
PMKID Count: 0
PMKID List
Group Management Cipher Suite: 00:0f:ac (Ieee 802.11) BIP (128)
Group Management Cipher Suite OUI: 00:0f:ac (Ieee 802.11)
Group Management Cipher Suite type: BIP (128) (6)

Figure 5-33 *SAE advertised in the SSID beacon RSN IE*

WPA2 with iPSK

The concept of Identity PSK is to use a different WPA preshared key depending on the identity of the client connecting. This differs from MPSK, where several keys are available, and anyone can use any key. In this method, you rely on the AAA server to return a specific key, depending on the authentication result. iPSK is covered in detail in *Configure Catalyst 9800 WLC iPSK With Cisco ISE* on cisco.com, but let's tackle the important points and use the opportunity to explain the concepts of RADIUS attributes. On the WLC side, you need to create a WPA PSK WLAN and add MAC filtering, as illustrated in Figure 5-34. The MAC filtering is basically an excuse to reach out to the AAA server and initiate a RADIUS conversation; the intent is not necessarily to authorize the MAC address.

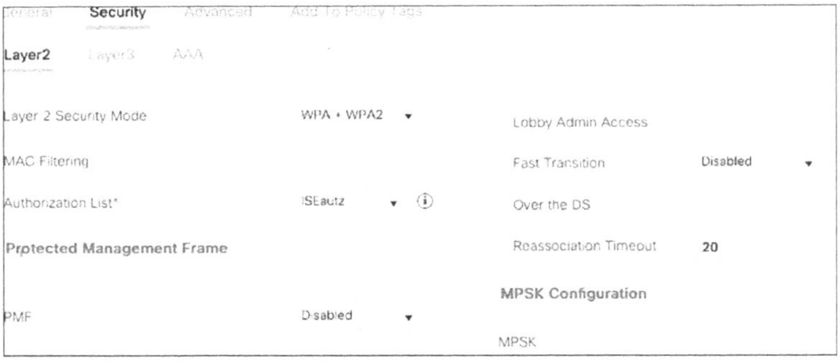
Security
Layer2
Layer 2 Security Mode
WPA + WPA2
Lobby Admin Access
MAC Filtering
Fast Transition
Disabled
Authorization List*
Over the DS
Protected Management Frame
Reassociation Timeout
20
MPSK Configuration
PMF
Disabled
MPSK

Figure 5-34 *iPSK SSID security settings part 1*

As you can see in Figure 5-35, you still need to configure a PSK that is the default key, used when the AAA server returns only a permit-access and not specifying any extra key.

WPA2 Encryption	AES(CCMP128)
	CCMP256
	GCMP128
	GCMP256
Auth Key Mgmt	802.1x
	PSK
	Easy-PSK
	CCKM
	FT + 802.1x
	FT + PSK
	802.1x-SHA256
	PSK-SHA256
PSK Format	ASCII
PSK Type	Unencrypted
Pre-Shared Key*	•••••••

Figure 5-35 *iPSK SSID security settings part 2*

The AAA method used in MAC filtering is a "aaa authorization network" type. The policy profile used also needs to allow AAA override for the WLC to accept RADIUS attributes returned by the AAA server.

On the Cisco ISE side, you have many options. You can enter specific devices' MAC addresses in the endpoint database so that MAC addresses are authenticated. You could also configure your authentication policy to proceed and continue with the authorization despite the authentication failing if the user is not found. Figure 5-36 illustrates an example of authorization rules you could configure. The first rule returns an access-accept along with the PSK "Cisco123" that the client has to use. The second rule illustrates the fact that you do not necessarily have to enter specific MAC addresses in your RADIUS server database but can use any of the RADIUS attributes present in the authentication request.

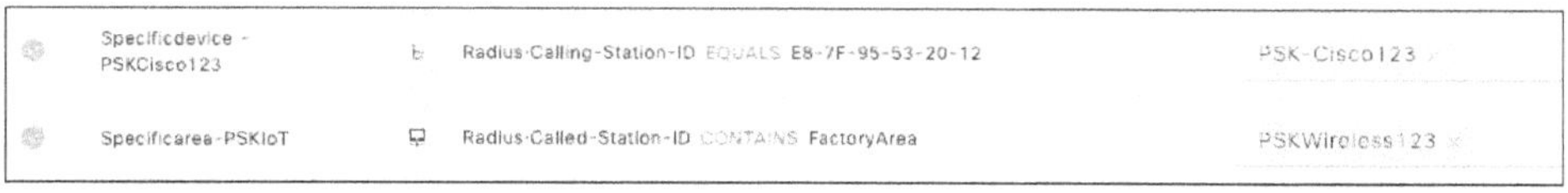

Specificdevice - PSKCisco123	Radius-Calling-Station-ID EQUALS E8-7F-95-53-20-12	PSK-Cisco123
Specificarea-PSKIoT	Radius-Called-Station-ID CONTAINS FactoryArea	PSKWireless123

Figure 5-36 *ISE authorization policies for IPSK*

This example uses the called-station-id and relies on the fact that the 9800 is configured to specify the AP locations in that attribute. You can configure it under **Configuration > Security > AAA > AAA Advanced > Global Config > Show Advanced Settings,** as shown in Figure 5-37. The same page also allows you to specify the MAC address format that will be handy if you are using a third-party RADIUS server such as FreeRadius You can therefore define a specific key to use in a specific area of the network, regardless of who the client is. It could also be an option to write a rule based on the device group the client MAC belongs to (either through static group membership in the ISE endpoint database or through profiling dynamic rules): your creativity and the existing RADIUS attributes set are the limit.

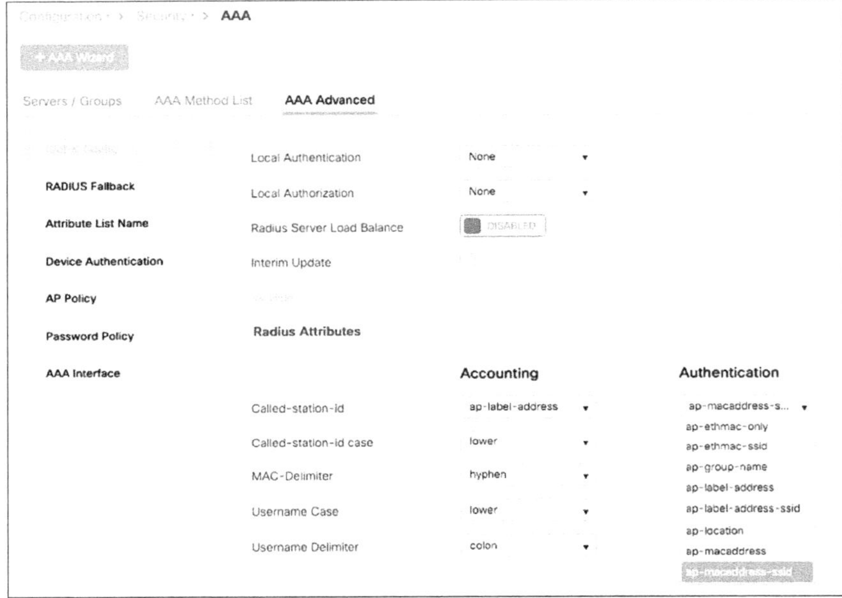

Figure 5-37 *Customization of the called-station-id RADIUS field*

The custom authorization results are straightforward and only need two attributes that are Cisco proprietary and called cisco-av-pair (they are shown in Figure 5-38). The first attribute is cisco-av-pair, and the value is **psk=<actual key you want to use>**, whereas the other is an identical cisco-av-pair and its value is **ascii** because you probably want to use your WPA key as an ASCII key.

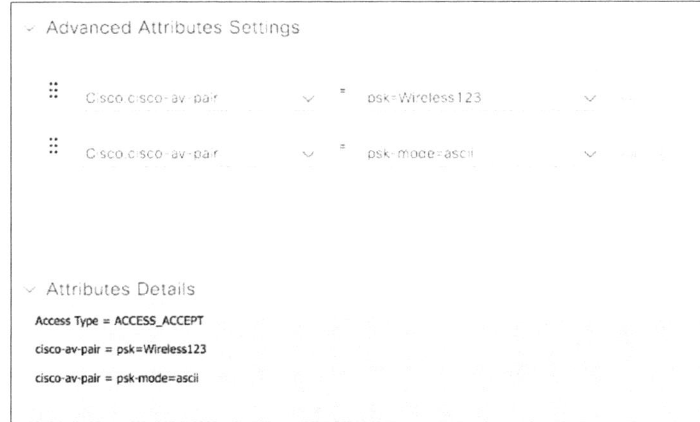

Figure 5-38 *ISE authorization result*

iPSK works because the MAC filtering is an authentication that happens at the 802.11 association phase. The AP or WLC authenticates the MAC address before replying with an association response, and therefore, this is a time-sensitive process where the client can give up on waiting for this association response and move to another channel if your RADIUS server is not quick enough.

Let's take a look at the RADIUS exchange from the RadioActive Trace output taken from the Catalyst 9800. The exchange is simple: the controller sends an access-request mentioning the MAC address of the client as both username and password. The RADIUS server is expected to reply with access-accept or access-reject. Example 5-2 shows the request, as sent by the 9800 for an iPSK SSID.

Example 5-2 Always-On Logs Corresponding to a RADIUS Access-Request Packet Being Sent

```
wncd_x_R0-0{1}: [radius] [25734]: (info): RADIUS: Send Access-Request to
192.168.1.99:1812 id 0/59, len 384
wncd_x_R0-0{1}: [radius] [25734]: (info): RADIUS: authenticator 9b 9e 12
1c 3b d9 d2 b3 - 53 4d f5 f0 2b 63 ae 1c
wncd_x_R0-0{1}: [radius] [25734]: (info): RADIUS: User-Name        [1]
14  "e87f95532012"
wncd_x_R0-0{1}: [radius] [25734]: (info): RADIUS: User-Password    [2]
18  *
wncd_x_R0-0{1}: [radius] [25734]: (info): RADIUS: Service-Type     [6]
6  Call Check                 [10]
wncd_x_R0-0{1}: [radius] [25734]: (info): RADIUS: Vendor, Cisco    [26]
31
wncd_x_R0-0{1}: [radius] [25734]: (info): RADIUS:  Cisco AVpair    [1]
25  "service-type=Call Check"
wncd_x_R0-0{1}: [radius] [25734]: (info): RADIUS: Framed-MTU       [12]
6  1485
wncd_x_R0-0{1}: [radius] [25734]: (info): RADIUS: Message-
Authenticator[80]     18  ...
wncd_x_R0-0{1}: [radius] [25734]: (info): RADIUS: EAP-Key-Name     [102]
2  *
wncd_x_R0-0{1}: [radius] [25734]: (info): RADIUS: Vendor, Cisco    [26]
49
wncd_x_R0-0{1}: [radius] [25734]: (info): RADIUS:  Cisco AVpair    [1]
43  "audit-session-id=8501A8C00000004ECC162224"
wncd_x_R0-0{1}: [radius] [25734]: (info): RADIUS: Vendor, Cisco    [26]
18
wncd_x_R0-0{1}: [radius] [25734]: (info): RADIUS:  Cisco AVpair    [1]
12  "method=mab"
wncd_x_R0-0{1}: [radius] [25734]: (info): RADIUS: Vendor, Cisco    [26]
30
wncd_x_R0-0{1}: [radius] [25734]: (info): RADIUS:  Cisco AVpair    [1]
24  "client-iif-id=50334480"
wncd_x_R0-0{1}: [radius] [25734]: (info): RADIUS: Vendor, Cisco    [26]
17
```

```
wncd_x_R0-0{1}: [radius] [25734]: (info): RADIUS:  Cisco AVpair    [1]
11  "vlan-id=1"
wncd_x_R0-0{1}: [radius] [25734]: (info): RADIUS: NAS-IP-Address   [4]
6  192.168.1.133
wncd_x_R0-0{1}: [radius] [25734]: (info): RADIUS: NAS-Port-Id      [87]
17  "capwap_9000001b"
wncd_x_R0-0{1}: [radius] [25734]: (info): RADIUS: NAS-Port-Type    [61]
6  802.11 wireless            [19]
wncd_x_R0-0{1}: [radius] [25734]: (info): RADIUS: NAS-Port         [5]
6  116
wncd_x_R0-0{1}: [radius] [25734]: (info): RADIUS: Vendor, Cisco    [26]
28
wncd_x_R0-0{1}: [radius] [25734]: (info): RADIUS:  Cisco AVpair    [1]
22  "cisco-wlan-ssid=iPSK"
wncd_x_R0-0{1}: [radius] [25734]: (info): RADIUS: Vendor, Cisco    [26]
30
wncd_x_R0-0{1}: [radius] [25734]: (info): RADIUS:  Cisco AVpair    [1]
24  "wlan-profile-name=iPSK"
wncd_x_R0-0{1}: [radius] [25734]: (info): RADIUS: Called-Station-Id
[30]     24  "d4-ad-bd-a2-8f-20:iPSK"
wncd_x_R0-0{1}: [radius] [25734]: (info): RADIUS: Calling-Station-Id
[31]     19  "e8-7f-95-53-20-12"
wncd_x_R0-0{1}: [radius] [25734]: (info): RADIUS: Vendor, Airespace
[26]     12
wncd_x_R0-0{1}: [radius] [25734]: (info): RADIUS:  Airespace-WLAN-ID  [1]
6  5
wncd_x_R0-0{1}: [radius] [25734]: (info): RADIUS: Nas-Identifier
[32]      7  "Katar"
```

Note a couple of interesting RADIUS attributes:

- **User-name:** In this case, the client didn't initiate any authentication and therefore does not provide any username. It's the 9800 that decides to fill the User-name field with the MAC address using a specific format.
- **User-password:** This is one of the few fields that is encrypted with the RADIUS shared secret.
- **Service-type:** Although the naming of the possible options doesn't match current usages (they go back to the origin of RADIUS for dial-up communications), they depict whether the authentication is a MAC authentication or an 802.1X authentication, for example ("Call check" in the case of MAC, "Framed" in case of 802.1X).
- **Framed MTU:** The controller expresses the MTU it can honor for RADIUS. It is interesting to note that the RADIUS server is not required to follow this value, and Cisco ISE doesn't follow it; therefore, fragmentation of RADIUS packets can occur.
- **Message authenticator:** This field is encrypted using the shared secret, to make sure the WLC is allowed to communicate with the RADIUS server.

- **Cisco AVPair** : A couple of Cisco AVPairs mention some additional and optional information such as the SSID name, the WLAN profile name, the VLAN, or the fact that it's a MAC address filtering authentication ("method=mab").
- **Audit-session-id** : It is created by the WLC and serves as session ID (for example, if you want to terminate the session via CoA from ISE) even when accounting is not in use.
- **Airespace-WLAN-ID** : The Catalyst 9800 keeps using the same Airespace proprietary attributes introduced by the former generation of WLCs, which specifies the WLAN ID the client is connecting to.
- **Calling-station-id** : This attribute is always the MAC address of the client.
- **Called-station-id** : This attribute is, by default, the MAC address of the AP but can be customized to append other values. Here, the SSID name is appended to it.

All these attributes can be used to create authentication and authorization rules in the ISE, which means it is possible to authenticate the client based on where it is trying to connect, on which SSID it is trying to connect, its MAC address, and so on.

After this request, the RADIUS server replies with an access-accept, authorizing the client on the network and also providing the PSK that this client should be using, as shown in Example 5-3.

Example 5-3 Always-On Logs Corresponding to a RADIUS Access-Accept Packet Being Received

```
wncd_x_R0-0{1}: [radius] [25734]: (info): RADIUS: Received from id
1812/59 192.168.1.99:0, Access-Accept, len 185
wncd_x_R0-0{1}: [radius] [25734]: (info): RADIUS: authenticator 2c d3 14
54 c5 e4 f2 7d - e1 4a b3 ee 80 ba 3f 2f
wncd_x_R0-0{1}: [radius] [25734]: (info): RADIUS: User-Name           [1]
19  "E8-7F-95-53-20-12"
wncd_x_R0-0{1}: [radius] [25734]: (info): RADIUS: Class
[25]    50  ...
wncd_x_R0-0{1}: [radius] [25734]: (info): RADIUS: Message-
Authenticator[80]    18  ...
wncd_x_R0-0{1}: [radius] [25734]: (info): RADIUS: Vendor, Cisco
[26]    23
wncd_x_R0-0{1}: [radius] [25734]: (info): *
wncd_x_R0-0{1}: [radius] [25734]: (info): RADIUS:  Cisco AVpair       [1]
17
wncd_x_R0-0{1}: [radius] [25734]: (info): RADIUS: Vendor, Cisco
[26]    22
wncd_x_R0-0{1}: [radius] [25734]: (info): RADIUS:  Cisco AVpair       [1]
16  "psk-mode=ascii"
wncd_x_R0-0{1}: [radius] [25734]: (info): RADIUS: Vendor, Cisco
[26]    33
wncd_x_R0-0{1}: [radius] [25734]: (info): RADIUS:  Cisco AVpair       [1]
27  "profile-name=Apple-Device"
```

iPSK becomes a bit trickier to use for clients such as smartphones that now typically use a private MAC address by default (that is subject to change) but stays relevant for IoT devices that typically have a fixed MAC address because privacy is not their concern.

iPSK is supported in FlexConnect local switching mode. The P2P blocking action configuration drop-down in the WLAN advanced settings allows for an "allow private group" intriguing setting. This, when combined with iPSK, allows users using the same PSK to talk to each other but not talk to users using a different PSK.

Enhanced Open

Let's cover a typical use case where you, as administrator, want to configure Enhanced Open but still allow older clients to be able to connect to the guest SSID. Figure 5-39 shows the final result: one WLAN is configured for WPA3 Enhanced Open (called OWE), and the other is a fully open SSID. Only the fully open SSID called "Open" has its name broadcasted in the beacons while Enhanced Open is hidden.

	EnhancedOpen	3	EnhancedOpen	[WPA3][OWE][AES]
	Open	4	Open	[open]

Figure 5-39 *A pair of Open/Enhanced Open SSIDs for Enhanced Open Transition mode*

For this example, create the first SSID, call it **EnhancedOpen** (in reality, choose a name that is relevant for your company), and make sure it is hidden by disabling **Broadcast SSID**, as depicted in Figure 5-40.

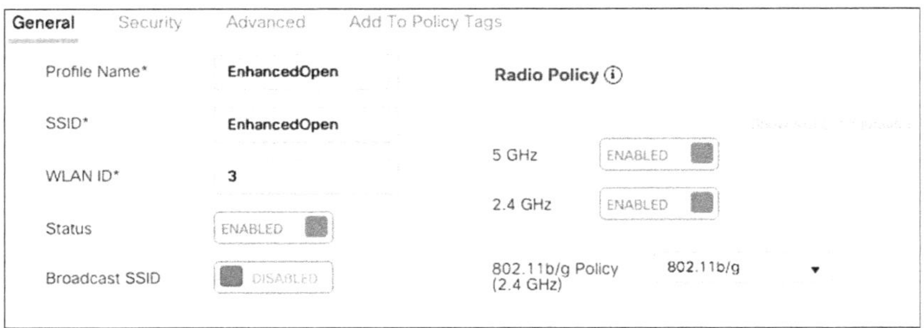

Figure 5-40 *Enhanced Open SSID general settings*

Configure the security settings to be WPA3 with OWE as AKM and choose a cipher (AES 128 is a good conservative choice for a guest SSID), as shown in Figures 5-41 and 5-42. The last field asks you for the transition mode WLAN ID. This basically asks you for the "other SSID in the pair." So, when configuring the Enhanced Open SSID, enter the WLAN ID of the open SSID (**4** in this example).

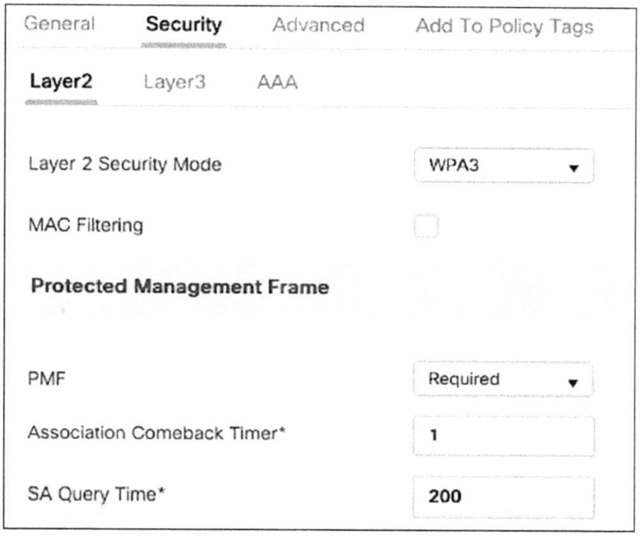

General Security Advanced Add To Policy Tags

Layer2 Layer3 AAA

Layer 2 Security Mode: WPA3

MAC Filtering

Protected Management Frame

PMF: Required

Association Comeback Timer*: 1

SA Query Time*: 200

Figure 5-41 *Enhanced Open security settings part 1*

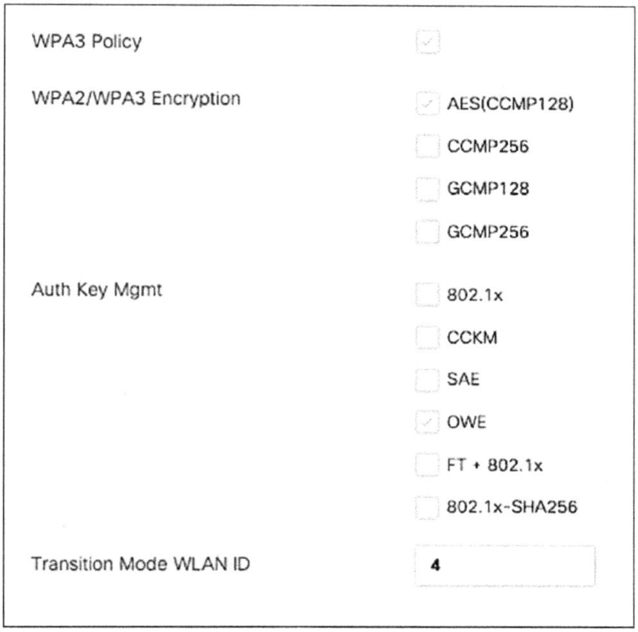

WPA3 Policy

WPA2/WPA3 Encryption: AES(CCMP128), CCMP256, GCMP128, GCMP256

Auth Key Mgmt: 802.1x, CCKM, SAE, OWE, FT + 802.1x, 802.1x-SHA256

Transition Mode WLAN ID: 4

Figure 5-42 *Enhanced Open SSID security settings part 2*

Proceed with the creation of the regular open SSID, which is broadcasted and with no particular security settings other than enabling OWE transition mode and specifying the WLAN ID of the Enhanced Open SSID, as shown in Figure 5-43.

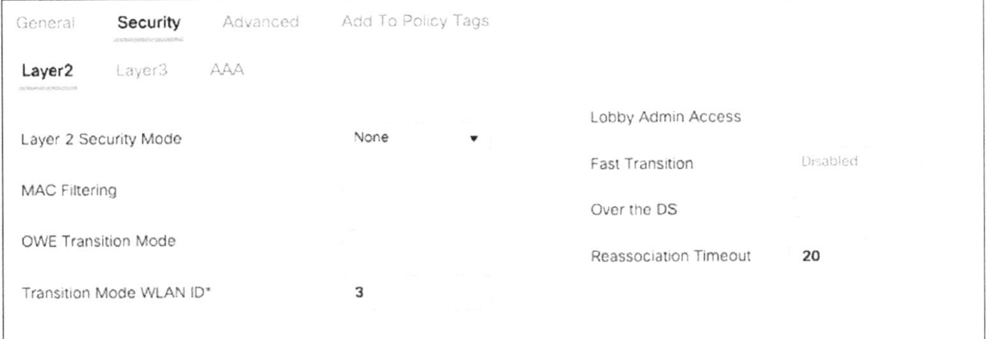

Figure 5-43 *Regular open SSID linked with transition mode to an Enhanced Open SSID*

When you're done, if you look at the scan list of SSIDs for a wireless client present in the area where the WLANs are active, no matter what it supports, it shows only the Open SSID (because it is the only one broadcasted). If your client does not support Enhanced Open, it connects to the Open SSID, and it is business as usual. If your client supports Enhanced Open, even if you click the "Open" SSID in the list of heard SSIDs, in reality you are connected to the secure SSID. You can see from the **Monitor > Clients** page in the WebUI that the client is, in fact, using encryption, as depicted in Figure 5-44.

WLAN Profile Name	EnhancedOpen
Wireless LAN Network Name (SSID)	EnhancedOpen
Client Entry Create Time	21 seconds
Policy Type	WPA3
Encryption Cipher	CCMP (AES)
Authentication Key Management	OWE

Figure 5-44 *Security details of a client connected to an Enhanced Open SSID*

(Local) Web Authentication

Cisco.com offers a great guide titled *Configuring Web-based Authentication on Cisco Catalyst 9800 Series Controllers* covering the various types of web authentication available on the Catalyst 9800 as well as recommendations and limitations.

Web authentication is a Layer 3 security policy (contrary to central web authentication, which is covered in the next section) and is sometimes called local web authentication because the ACL and portal URL are locally configured on the WLC. It does not mean that the portal cannot be hosted on an external server but only that the configuration of these details has to appear in the WLC configuration. The naming convention can quickly become confusing around this feature; therefore, it is better and easier to fully refer to each component. The authentication part of the web authentication can be hosted locally on the controller or distributed to an LDAP database or a RADIUS server. The portal page can also be hosted on the WLC (and even customized) or on an external web server (that only has a web server role and does not authenticate the user). For example, saying you need to configure "Local web authentication with a local user database and an external login page" is easier to understand (although much longer to articulate) than saying you are configuring external web authentication. Figure 5-45 depicts the workflow of web authentication, although many different examples would be needed to cover all the possibilities, depending on where each part is hosted, but the concepts stay the same.

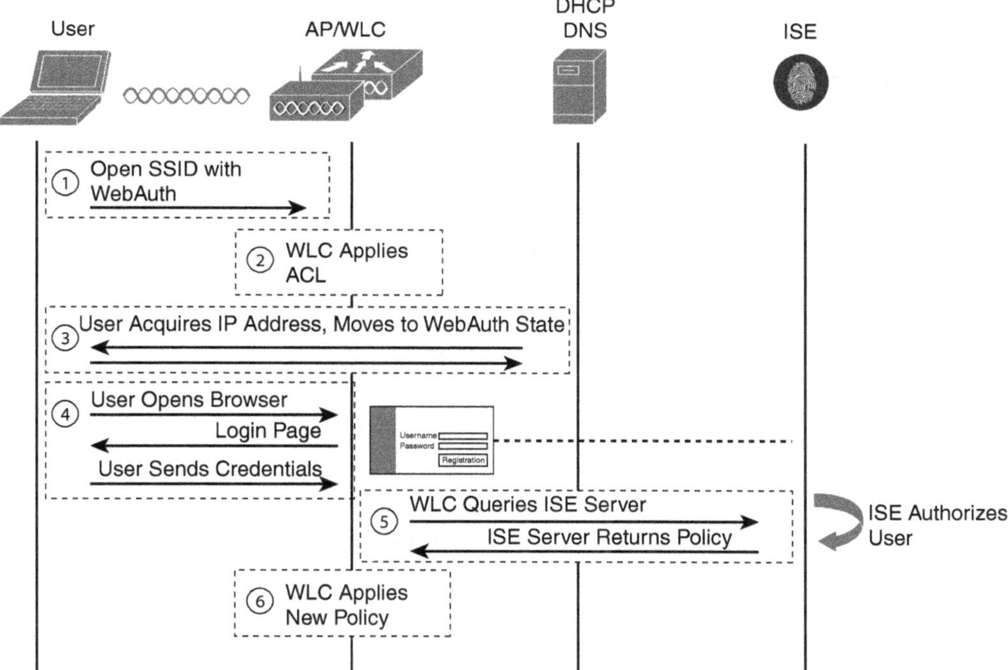

Figure 5-45 *Local web authentication workflow*

Web authentication can be configured on top of an existing preshared key or even 802.1X SSID. What defines a local web authentication SSID is the fact that the Layer 3 security tab of the WLAN configuration shows one type of web authentication configured. When a client connects to a web authentication SSID, a hidden preauthentication ACL is applied to it; it blocks all traffic except for DHCP and DNS while HTTP traffic is intercepted and redirected for web authentication. The wireless client can obtain an IP address (step 3) and detect the web portal by sending HTTP packets (step 4). The WLC spoofs whatever IP address was the destination of this HTTP packet and replies back with a redirection to the login page (whether it is hosted on the WLC or not). The login page can potentially be external and is only an HTML repository. The authentication process is completely separate, and the WLC either authenticates the user credentials locally or sends them via LDAP or RADIUS to an external server for validation. When the authentication is complete, the WLC moves the client to the RUN state, and the preauthentication ACL is removed.

A very focused reader wonders how the WLC can know the user credentials if they are entered in a web page that is not necessarily hosted on it. The controller has to use a trick for this. When the page is hosted on the WLC, the controller uses its virtual IP (a nonroutable IP like 192.0.2.1 typically) to serve it. The user credentials are then submitted to it, and this is how the WLC knows the credentials. A similar system is used even if the page is hosted externally: the web redirection sends the client first to the virtual IP anyway, which then sends the user again to the external login page while it adds arguments to the URL, such as the location of the virtual IP. Even when the page is hosted externally, the user still submits its credentials to the virtual IP thanks to this trick so that the WLC can receive the user credentials and authenticate them. If you are to write your own login page, refer to the HTML example provided in the Catalyst 9800 wireless controller download section on cisco.com to see how to have the credentials properly submitted to the WLC.

When we talk about web authentication, it is important to talk about the certificate validation. For a good web authentication experience, it is crucial to install a trusted certificate on the controller and apply it for use with web authentication. There are multiple criteria for a browser to validate a certificate; the main ones are

- The certificate must be within its validity period.
- The certificate must have a Common Name (CN) field that corresponds to the URL visited.
- The certificate must be issued by trusted certificate authorities (CAs).

The second criterion is also related to the fact that you typically cannot issue certificates to IP addresses. You must configure a virtual hostname linked to the virtual IP you are using. Instead of being redirected to the virtual IP, the clients are then redirected to the hostname entered. That's the hostname you need to issue your web authentication portal certificate to. It is also expected that the DNS used by the wireless client redirects requests to the virtual hostname toward the virtual IP.

The third one does not necessarily mean that you need to have a public third-party-signed certificate, but if you use an enterprise PKI, you need to make sure that the root CA is installed on every wireless client (which can be challenging in the case of a guest SSID).

The catch is that if you are hosting your login page on an external web server, the client is presented with two certificates during the login experience: the WLC virtual IP certificate and the external login page certificate.

There are other slight variations of web authentication:

- **Webconsent:** Sometimes also called *web passthrough*, this variation removes the authentication phase. The page typically displays only the terms and conditions and asks the user to agree before letting them through on the network. The page can optionally ask the user to leave an email address, but no validation is performed. This type is used by many external portals relying on social network login credentials because the controller is not able to perform the authentication of the credentials. It is then the login page that is expected to perform the credentials validation, but in the back end, the controller has no means of verifying it.
- **Web authentication on MAC filter failure:** This variation is a MAC filtering SSID where the client is put in a web authentication pending state if the MAC authentication fails and is fully allowed on the network if it is successful.

Central Web Authentication

Central web authentication (CWA) is named this way because the configuration of the web authentication details (such as the ACL and the redirect URL) are configured centrally on the RADIUS servers. This means that the SSID on the WLC is configured with MAC filtering enabled (to have a pretext to send RADIUS requests to the ISE server) and optionally a preshared key or Enhanced Open. The ISE server then replies with an access-accept, along with a redirect URL and redirect ACL, as depicted in step 3 in Figure 5-46. The redirect ACL is a punt ACL that needs to be predefined on the WLC (or the AP in the case of FlexConnect local switching). The ISE server just returns the name of the ACL and not its definition. The redirect ACL defines traffic (matching "deny" statements because it denies redirection for it) that is allowed through on the data plane and traffic (matching "permit" statements) that is sent to the control plane toward the CPU for further processing (that is, web interception and redirection in this case). The ACL has implicit (that is, invisible) statements allowing DHCP and DNS traffic toward all IPs, just like the case with LWA. It also ends with an implicit deny statement, which is acting as security ACL implicit deny (which means it drops traffic). This means that the last invisible statement of the redirect ACL drops all the traffic that does not match any previous punt statement. An example of a redirect ACL would be (considering 10.48.39.28 is the ISE IP address):

```
ip access-list extended REDIRECT
 deny ip any host 10.48.39.28
 deny ip host 10.48.39.28 any
 permit tcp any any eq 80
```

The first two statements allow traffic toward ISE to not be redirected and follow regular processing. The last statement sends all HTTP traffic toward the CPU for URL redirection, and an implicit security ACL ACE "deny" is present afterward to drop all the other types of traffic.

The redirect URL simply defines toward which URL the WLC redirects the guest clients.

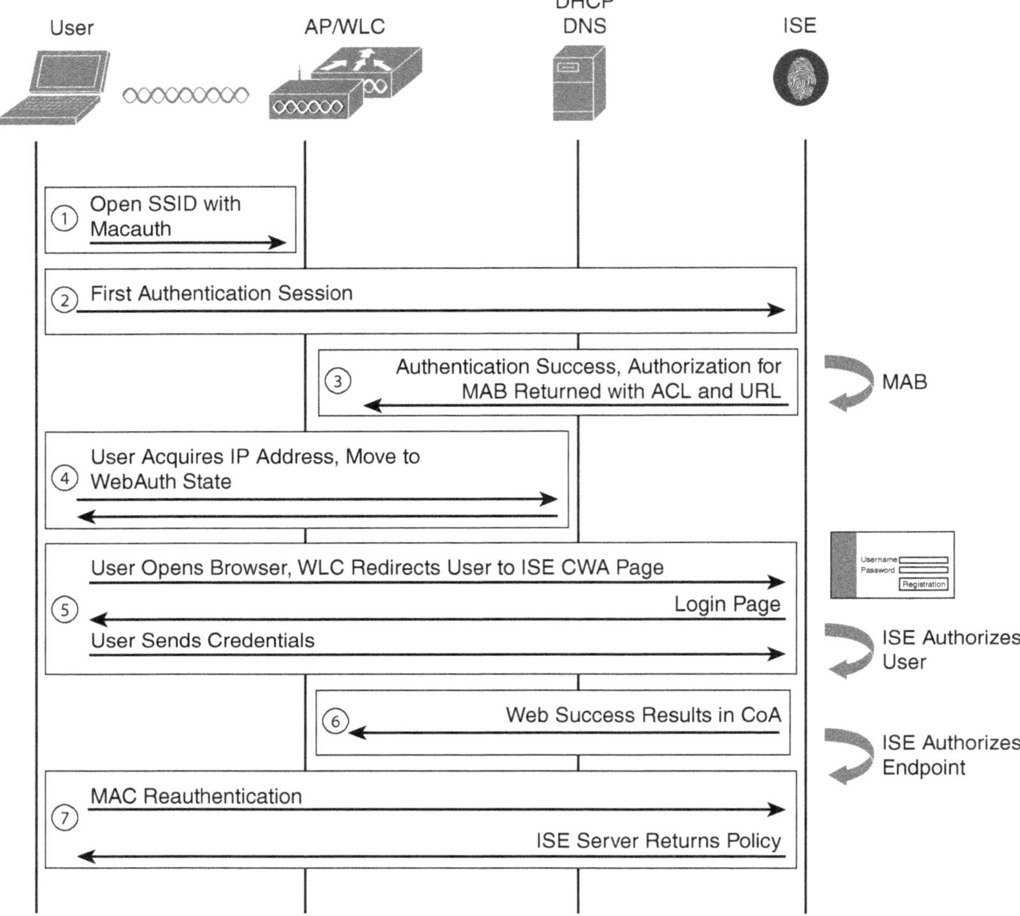

Figure 5-46 *Central Web Authentication workflow*

The Wi-Fi client can then obtain an IP address (DHCP and DNS traffic are allowed by default, HTTP traffic is intercepted for redirection, and the rest is dropped). All modern operating systems have a portal detection mechanism that works by sending HTTP probe packets to specific test URLs. If the device obtains the requested page, the device must have Internet access. If the device does not obtain the page, that's where you get the famous "limited or no connectivity" sign or any of its variations on your device. However, if the HTTP request gets a reply that is a redirection, it means the device is behind a captive portal. The behavior then varies based on operating systems and even from version to version, but overall, the operating system either shows a pop-up that the user needs to log in to get Internet access or automatically opens a browser window showing the login page. This means that you probably do not have to open your web browser manually to test for a login portal.

The fifth step of the workflow includes all the login details (depending on the type of portal you choose, you can self-register or enter an authorization code) on the portal page(s) until the user has submitted their credentials and received the success message. Something unique about CWA is that after a successful login, the user is basically reauthenticated.

In step 6, the ISE sends a RADIUS Change of Authorization message to the WLC, containing the user session ID (from the audit-session-id attribute, it does not require accounting necessarily) and a reauthentication action. The WLC then restarts all the client state machines back to square one.

Step 7 is then a repetition of step 2 (but with a twist): the WLC sends a MAC authentication request to the ISE, which still simply contains the MAC address of the wireless client to the ISE. However, the ISE remembers that this client just logged in and appends an internal attribute containing the client username entered on the portal page to the authentication workflow. The client authentication can then hit authorization rules that are based on the client user identity and return any other authorization attribute: a VLAN (even if it is not recommended to change the VLAN of a client after it already obtained an IP address), a session timeout, an ACL, an SGT, and so on. This "trick" is done because the portal authentication happens via HTTPS, and the WLC knows nothing from the credentials that the user typed in the login page; therefore, no RADIUS authentication can happen at that stage containing the portal page username. This way, it is still possible to return various authorization results depending on the client login page identity, and they are applied during the Layer 2 authentication phase. The key is that the second MAC authentication taking place returns an access-accept that can contain various authorization attributes but should not contain any redirect URL/ACL again; otherwise, the portal page is shown again to the user in a loop.

One benefit of the CWA workflow is that the WLC Virtual IP is not used, and therefore, the WLC does not have to present its certificate to the user. Only the ISE login portal page gives its certificate to the client, so this reduces some of the administrative burden. Another benefit is that there is a load balancing of the portal pages taking place naturally: the redirect URL points to the portal page of the specific ISE policy node against which the client authenticated previously. This means that your portal load balancing is, in fact, following your RADIUS load-balancing strategy.

Web Authentication Best Practices

There are many mistakes when it comes to web authentication that lead to a poor user experience or simply a nonworking wireless network. Let's look at the most common points.

HTTPS Redirection

HTTPS redirection is bad and should never be configured. When doing an HTTP redirection, the wireless client triggers the TCP handshake (on port 80) with whatever website it requested, but in reality, the WLC is intercepting this traffic, spoofing the website IP, and doing the TCP handshake. The client then sends its HTTP GET to request whatever page it intended to, and the WLC replies with a redirection to the login page. Intercepting HTTP sessions is not too hard for the controller, and there are no obstacles to task.

When doing an HTTPS redirection, the wireless client still triggers the TCP handshake (on port 443), and the controller still has an easy time spoofing the website IP address and completing the TCP handshake. But before the client can send a GET request, the TLS handshake of HTTPS must happen, and the server side (the controller here) is supposed to send a certificate to prove its identity. Even if the controller has a valid certificate installed, this certificate is issued to the controller hostname and definitely not to whatever website the client was requesting. This is the whole point of HTTPS—being able to verify the identity of the site you are visiting—and therefore, any other certificate presented depicts a session hijacking (which web authentication is, in fact).

There are no good solutions to this situation because this is the whole point of the protocol and why it is secure in the first place. The controller sends its certificate to complete the TLS handshake. The client browser always throws at least an error on this, and most of the time it is even impossible to ignore or avoid. Some browsers do not even show the page or do not show any error at all. This is why HTTPS redirection is a broken concept to begin with. On top of that, simply enabling it is extremely resource intensive for the controller.

The rationale behind this feature comes from people noticing that the Internet is now mostly HTTPS, and therefore, when a user opens up a browser, there is a good chance that an HTTPS URL is entered or that the home page is HTTPS, even if the user does not type **HTTPS** in the address bar. Therefore, for a time, this was indeed an issue, and the only way to have people notice there is a portal page was to enable HTTPS redirection. However, as explained in the previous section on CWA, most operating systems today have a portal detection mechanism relying on HTTP probes. Even for operating systems not benefiting from this feature, web browsers themselves implement this detection too and send an HTTP probe when opened to warn the user whether a captive portal is in place. There are therefore little to no reasons to still need to have HTTPS redirection enabled.

Some networks have configured all their laptops to use a web proxy to access the Internet or even an intranet, which means the clients never send their packets on ports 80 or 443 but typically on a higher port used by the web proxy. This issue is once again solved by relying on the operating system HTTP probe that is sent on port 80 no matter what the proxy configuration in the browser is.

Captive Portal Bypass

Captive portal bypass is a feature that administrators often enable out of a trial-and-error process when something is not working right with web authentication. It is important to understand its origin and purpose. Apple was the first company to implement the captive portal detection system on its company phones. What is now considered a great feature started with a few concerns because the captive portal page is not opened in the favorite or default browser but in a pseudo-browser called Apple Captive Network Assistant. The problem was that this pseudo-browser did not support dynamic pages very well and would display a blank page on many bring-your-own-device (BYOD) portals. The captive portal bypass feature on the WLC then made the WLC spoof an "HTTP 200 OK" reply from the captive detection URL, pushing the client to believe it had full Internet connectivity and therefore not show any pop-up at all. The user was then expected to open their browser and get redirected to the captive portal BYOD page displayed on this full-blown browser. The problem, as mentioned previously, is that nowadays most websites are HTTPS, so this workflow would typically not work at all. But this is not a concern because both the Apple CNA pseudo-browser solved some of its limitations, and the ISE BYOD portal page got adapted to fully display on those pseudo-browsers. Android now also uses a pseudo-browser, and no problems are reported with the vast majority of captive login pages. It is therefore recommended to disable captive portal bypass in the webauth parameter map, as shown in Figure 5-47, and to fully rely on the captive portal detection system. One limitation of those pseudo-browsers, still, is their inability to do the client provisioning flow from ISE.

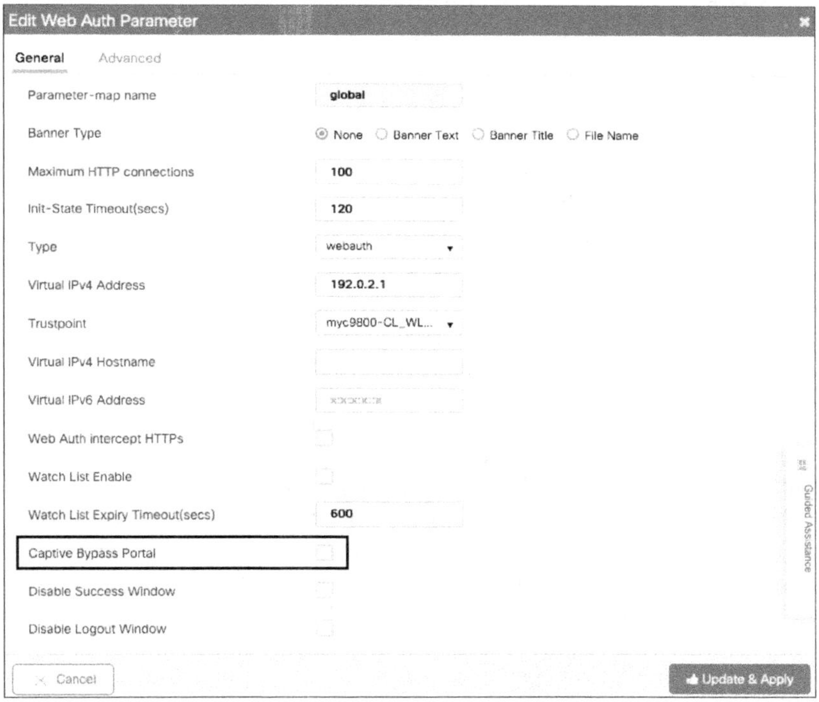

Figure 5-47 *Do not enable Captive Bypass Portal unless you really have a good reason*

Web Authentication Takeaways

It is important to summarize the key points of web authentication because they will save you a lot of headaches when deploying your configuration:

- Do not enable HTTPS redirection; it's pointless.
- Do not enable captive portal bypass.
- Make sure you have valid certificates on your portal page (always) and on the WLC (if you are not using CWA).

- You need a working DNS. Really.
- Rely on the client devices' captive portal detection pop-up.
- Configure both an IPv4 and an IPv6 virtual IP; there's little reason not to.

RFC 8910 is also something to watch out for because it is a new standard for detecting captive portals. Whether it becomes popular will depend on its client adoption rate.

Rogue Detection and WIPS

Rogue detection, WIPS, and more generally protecting against wireless attacks are covered in Chapter 7, "RF Deployment and Guidelines."

Securing Your Access Points

Securing your wireless networks doesn't mean much if your APs are not secured as well. If anyone can join an access point to your network and start broadcasting your WLANs in unwanted places, you have a problem. If your wired network simply has strict security guidelines, you need to set some policies in place for the wireless network to fit in with the existing policies.

AP Authorization

If any AP plugged to your network is allowed to join your WLC and broadcast the configured SSIDs, that configuration is considered quite insecure. Even if you don't consider this a concern because somehow your wired network is physically very secure, you may have several WLCs in your network and may still look for a way to make sure which WLC an AP can join and which WLC it can't. Some people think DHCP Option 43 covers this requirement, but Option 43 only allows the AP to learn about WLC IP addresses. If the AP learns another WLC IP address (via broadcast or via the mobility group), it tries to discover it too. Enabling AP authorization on a WLC means that globally, no AP is allowed to join until its MAC address is verified by the WLC. MAC addresses have to be entered in the **Device Authentication** page under the **AAA Advanced** section in the *aabbccddeeff* format, that is, without any separator.

The **AP Policy** page under **AAA Advanced** allows you to enable the authorization of APs against MAC addresses globally and to pick the AAA method that authenticates them (you can use a list on the WLC itself or use a AAA server). The AAA method you need to select refers to the command **aaa authorization credential-download**.

An option to authorize APs against their serial number has also been available since IOS-XE 17.3.2.

Here's a summary of the CLI commands required:

```
9800# config t
9800(config)# aaa new-model
```

```
9800(config)# aaa authorization credential-download <method name>
local
9800(config)# ap auth-list authorize-mac
9800(config)# ap auth-list method-list <method name>
9800(config)# username <aaaabbbbcccc> mac
```

This topic is covered in more detail in the configuration example titled *Catalyst 9800 Wireless Controllers AP Authorization List* on cisco.com.

Whether or not you have configured AP authorization, any AP that is in Bridge mode has to be added to the device authentication list. This explains the confusion people experience when they order an outdoor AP like a 1562 or a 9124, receive it as a local mode AP, and have no problems joining it to the WLC. Then, when converting the AP to Bridge mode or Flex+Bridge mode, the AP does not join anymore. If checking the wireless join statistics (from the widget in the WLC home page that tells you some APs are disjoined), you see they are stuck in "AP auth pending" status.

On top of having their Ethernet MAC address verified when they join, the Bridge mode APs need to authenticate themselves either through PSK or through 802.1X to the controller. The reason is that anyone could be joining a Mesh mode AP to your backhaul and joining your WLC, so some form of authorization is always required.

The configuration example titled *Configuring Mesh on Catalyst 9800 Wireless LAN Controllers* is a good resource to learn more about configuring Bridge mode APs.

AP 802.1X Authentication

A secure wired network often means implementing security at the switchport level. Having good physical security on your premises is one thing, but even your own employees could be plugging unwanted devices in the network. And if you think about it, if the only thing it takes to get access to sensitive resources or a management VLAN is to find the right switchport and connect to it physically, you can understand that this is vastly unsecure by today's standards. Many networks implement 802.1X authentication on most wired ports accessible to access devices. There are often exceptions for server ports or IoT devices that cannot perform 802.1X authentication, but that's another story. Considering how easy it is to climb on a chair and remove an AP from the ceiling and use its network cable, securing the AP switchport is definitely good sense. Enabling 802.1X on the switchports where APs are connected means that when the cable gets plugged in, the only traffic that is allowed to pass is the EAP authentication protocol until the authentication succeeds, and only then can the AP get an IP address and send some traffic. Not much is required to configure on the wireless to achieve this: the main thing required is to configure credentials on the access points. You obviously need to configure those credentials somewhere in the user database that your RADIUS server uses and need to configure that RADIUS server appropriately from an authorization policy standpoint.

You can configure 802.1X credentials for the access points in the AP join profile under the **Management > Credentials** tab (not globally anymore, as was the case in AireOS controllers). You simply have to set a username and password, and the AP tries to trigger an EAP-FAST authentication with those credentials when connected to the network. There is no harm in configuring the 802.1X credentials even if the AP is connected to a non-802.1X switchport. Now the problem is that this only works for APs that join the WLC once to get this configuration provisioned on the AP. That can work if your APs are staged or provisioned on an unsecured port before being placed in the secure network. Alternatively, you can enter the credentials directly on the AP console if it is plugged directly to an 802.1X-enabled port. The command is **capwap ap dot1x username <user> password <password>**.

The 802.1X is typically used on access switchports to authenticate hosts, but an access point can also be in FlexConnect mode and therefore connected on a trunk port. Having the AP perform 802.1X authentication on a trunk port is also supported, provided that the switchport authentication is set to multihost mode. First of all, this allows several MAC addresses to be seen on the switchport (802.1X normally limits every switchport to communicate with only one MAC), and only the first MAC address (logically the AP) is authenticated. All the extra MAC addresses learned afterward are applied the same authorization result as the AP.

Using 802.1X to authenticate access points is also a great way of assigning a specific VLAN dynamically to the AP. This means you do not need to have static VLAN configurations on your ports, and depending on the username authenticating, the right VLAN (a user VLAN or the AP management VLAN) is dynamically assigned. Another way to achieve a similar goal is to use smartport macros that detect whether an AP is connected via CDP learning and is able to apply more switchport configuration lines. However, the CDP detection mechanism is probably a bit less secure than a full-blown 802.1X authentication.

One important point for AP 802.1X to work on trunk ports is that if you configure periodic reauthentication on your switchport, you need to make sure to configure **Termination action** (RADIUS attribute 29) to be **Radius-request** (a value of 1). If left to the default, it means that the AP loses network connectivity for the time of the reauthentication. When you set this attribute, the AP is able to keep sending traffic during the reauthentication until the reauthentication completes successfully.

Securing the AP Join Process Using Locally Significant Certificates

The concept of configuring LSC is simple: you decide to issue certificates yourself to each and every access point you own, and the WLC allows only APs with such a certificate to join, therefore securely preventing any new APs from joining without your authorization. LSC requires owning an enterprise PKI that automatically issues a certificate to each access point. This certificate distribution happens during a provisioning phase that you define where any AP that joins is automatically provided with a custom certificate signed by your enterprise CA. When you determine that provisioning phase to be over, only APs with a valid enterprise-signed certificate are allowed to join. This certificate supersedes the typical Cisco-manufactured MIC certificate (which isn't deleted but stays as backup) for establishing the DTLS connection.

This solution is much more secure than relying on a MAC address as well as much more scalable, but it requires an enterprise PKI supporting Simple Certificate Enrollment Protocol (SCEP). This procedure is covered in the document titled *Configure SCEP for Locally Significant Certificate Provisioning on 9800 WLC* on cisco.com.

Until 17.5.1, there was a limitation where the enterprise CA used for SCEP had to be a root CA, but intermediate CAs are now supported as well.

Securing Your Wireless Controller

Network devices must be secured because they are a target of choice for attackers. When an attacker can access one network device, that attacker can typically easily access more network resources. The controller is an excellent target because it manages the configuration of the whole wireless infrastructure. Let's review a couple of ways to protect unintended administrative access to the C9800.

Securing Administrator Access

One of the first measures to secure your wireless controller is to restrict administrator access to it. Leveraging local users stored on the controller can work only for small teams. In large companies, where more administrators have access to the controller and even just for simplicity of password management (having your credentials stored securely in one place is better than having them all over the network on each device, especially if you have to change your password), it becomes critical to use RADIUS or TACACS to authenticate users accessing your device as a network administrator.

Using TACACS+

A configuration example titled *Configure RADIUS and TACACS+ for GUI and CLI Authentication on 9800 Wireless LAN controllers* is available on cisco.com to better illustrate the process of using TACACS+. Let's cover it from a conceptual perspective nonetheless.

The first step is to configure the TACACS+ server on the WLC. You can easily do this in the **Configuration > Security > AAA > Servers/Groups** page. You need to specify an IP address and a shared secret.

The second step is to configure AAA methods. You need an **aaa authentication login <methodname> group <tacacs server group>** and an **aaa authorization exec <methodname> group <tacacs server group> method.** The login method takes care of authentication, and the exec authorization method is required to allow the user to enter any command (and basically do anything on the device). Because these are named methods, they do not have any effect until they are called somewhere else in the configuration.

To use the TACACS+ server for CLI authentication, add your method in the VTY configuration (it is also configurable under **Administration > Management>HTTP/HTTPS/ Netconf/VTY** in the WebUI after 17.6):

```
9800(config)#line vty 0 15
9800(config-line)#login authentication <aaa login method name>
9800(config-line)#authorization exec <aaa exec method name>
```

To use it for WebUI:

```
9800(config)#ip http authentication aaa login-authentication <aaa
login method name>
9800(config)#ip http authentication aaa exec-authorization <aaa exec
method name>
```

You can also configure this from the WebUI in the **Administration > HTTP/Netconf** page, as shown in Figure 5-48.

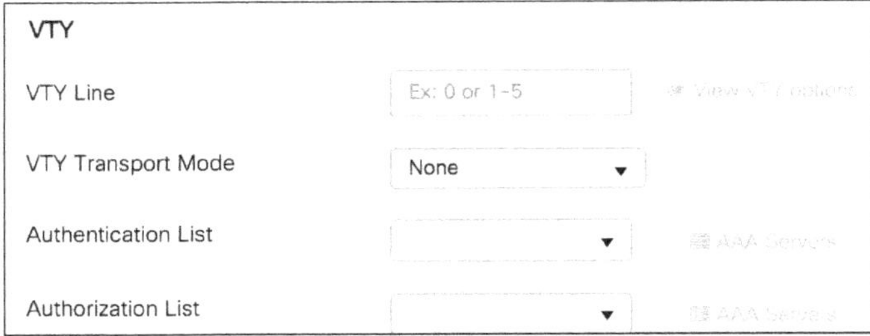

Figure 5-48 *VTY configuration for CLI access in the HTTP/Netconf web page*

For the settings to take effect, you might need to restart the web server by typing **no ip http server** followed by **ip http server**.

Some extra configuration is required on ISE (refer to the configuration example for more), but in a nutshell, you need to return two things on top of a successful authorization result:

- A TACACS+ shell profile allows you to set some attributes, and the most important for this situation is the default privilege level. This is the privilege level at which the user ends up after connecting. Although it ranges from 0 to 15, there are no differences by default between all the values ranging from 1 to 14. A privilege level of 0 allows you only some basic commands, and a privilege level of 15 allows full access.
- On top of a TACACS profile, you can also assign a command set from ISE, which is a list of CLI commands that the user is allowed to run. For each command, the WLC asks for authorization to run the specific commands. You can use wildcards or allow all but a subset of commands if you want.

Using RADIUS

Restricting WLC administrator access through RADIUS follows the exact same concept as with TACACS+. Obviously, you define a RADIUS server in lieu of a TACACS server, but the AAA methods and other commands stay the same. One behavioral difference with RADIUS is that it does not ask for authorization every time a CLI is entered: no shell profiles and command sets can be used. Instead, you need to edit your RADIUS authorization result and return a cisco-av-pair with the value **shell:priv-lvl=15**.

Guest Users

Guest users are created by the administrator or the lobby ambassador, which we discuss in more detail next. They have no device administration access (SSH or GUI). They are only able to connect to the network and access the Internet. They can also be temporary and be deleted after a set timer.

The Lobby Ambassador Type of User

There is an extra type of administrative user who is not simply defined by the privilege level. The lobby ambassador is a type of user who can be created on the WLC (either on WebUI or CLI) and whose only purpose is to create guest users. When you log in as a lobby ambassador, your only privilege is to create guests, and you get a special WebUI dedicated to this activity. In previous releases, it was tricky to authenticate lobby ambassadors via TACACS because it required configuring a command on the WLC for each ambassador username. After release 17.5, the only thing you require from ISE is to return the mandatory **user-type** attribute with value **lobby-admin** as a custom attribute in the shell profile. Without this, your lobby ambassador gets full administrator privilege on the controller, which you probably want to avoid. This process is covered in detail in the configuration example titled *Configure 9800 WLC Lobby Ambassador with RADIUS and TACACS+ Authentication*.

NETCONF

NETCONF is what Cisco DNA Center mostly uses to configure the Catalyst 9800 and a protocol you can use yourself as well, with custom scripts or third-party tools even. At the time of this writing, NETCONF security is ruled only by the default login authentication and exec authorization methods. This means that you need to set **aaa authentication login default local** and **aaa authorization exec default local** to have it working. It is also possible to point these methods to a RADIUS or TACACS server for authentication if you want. A common catch is that the management stations (Prime Infrastructure or Cisco DNA Center) tend to overwrite these config lines and point them back to local, so you may want to have a backup administrative user on the WLC to keep accessing it in case of trouble. The NETCONF implementation of the 9800 WLC uses SSH over port 830 and currently does not allow you to manually select ciphers or the certificate used for authentication although this may change in the future.

Granularity of WebUI Access

The WebUI is an interface that uses a hybrid data model in the background. For some pages, it actually opens a VTY line and enters CLI commands (which are authorized for each of them if you are using TACACS, for example), whereas some other pages use a TDL programmability model and do not use a CLI back end at all. This means that the WebUI is affected by changes you make to the CLI authorization method. At this time, there is no way to restrict a WebUI user from using certain pages and not others: a user either has privilege level 15 and has full access to the WebUI or has privilege level 1 and has access only to the dashboard and some monitoring subpages.

Connect to the WebUI Using Certificates

In the HTTP(s) administration page on the WLC, you can enable Personal Identity Validation. You then have to point to a trustpoint. This feature restricts the WebUI connection to only devices that can present a client certificate that will be trusted according to the designed trustpoint on the WLC. HTTPS websites always require the server to present its certificate, but optionally the client can do the same. This means you could get rid of credentials and require a smart card or certificate to be able to access the 9800 WebUI.

Securing Traffic

Another key to securing your controller is to make sure it is not receiving unwanted traffic and that only the right people connecting from the right locations can access it. By default, your wireless clients are not allowed to access the WLC command line or WebUI. To lift that restriction, you need to enable **Management Via Wireless** under **Configuration > Wireless > Wireless Global.**

A Catalyst 9800 controller can be managed, using any of its SVI IP addresses. It is recommended that you minimize the number of SVIs you configure on the 9800 for this reason, unless you have specific needs for SVI (in the case of mDNS proxy, for example, SVIs in wireless client VLANs are required). You can restrict access by applying ACLs on the 9800 SVIs.

Catalyst 9800 wireless controller appliances provide you with a service port to manage the WLC. The specificity of this SP is to be really out-of-band; that is, the port is physically connected to the WLC CPU and separated from the other data ports and the data plane. From a software perspective, IOS-XE places this port in a hard-coded VRF called mgmt-intf. You cannot change this VRF configuration on the service port. You can add routing for the service port, but it has to be part of the mgmt-intf VRF.

Encrypted Traffic Analytics

The fact that more and more network protocols are encrypted means it is harder and harder to analyze the traffic going through the network. Encrypted Traffic Analytics (ETA) is a solution from Cisco (its flow is illustrated in Figure 5-49) where network devices all report information from the network traffic to a Cisco Secure Network Analytics (formerly Stealthwatch) appliance that runs a machine learning algorithm to be able to identify threats and attacks, not based on the traffic content but on the traffic patterns. The Catalyst 9800 can be a part of this solution, leveraging the Flexible Netflow configuration directly on the C9800 (not on the FlexConnect APs themselves). You can find more information in the *Encrypted Traffic Analytics White Paper* and the configuration guide.

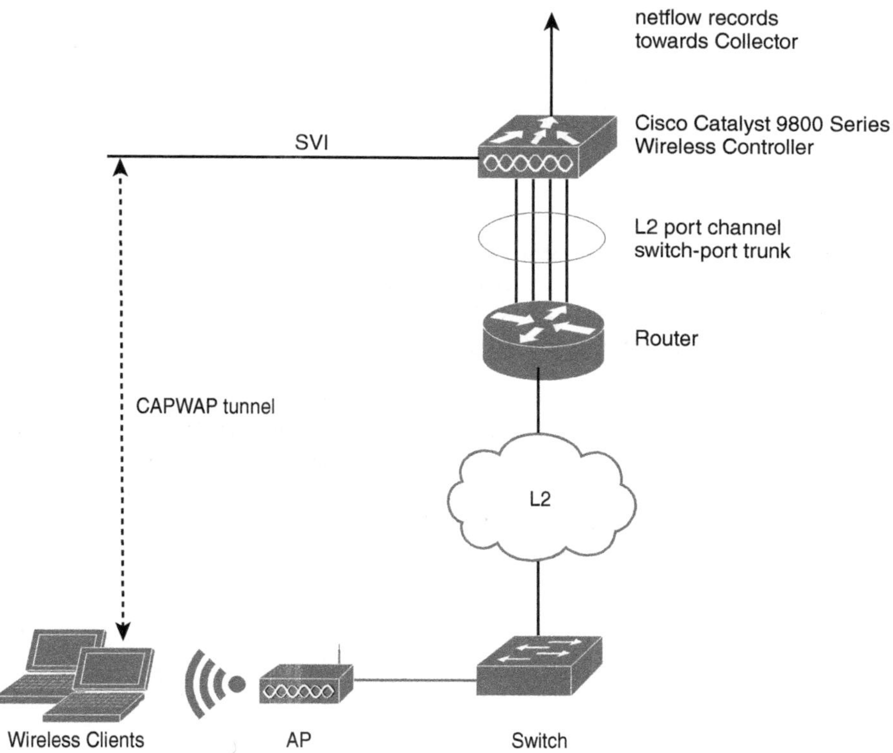

Figure 5-49 *ETA workflow topology*

Cisco Umbrella

Cisco Umbrella is a cloud DNS service. The first question that may come to your mind is "What is the relation with wireless?" Well, many administrators try to secure network access by restricting the content that wireless users can access. This can be done with AVC if you want to target specific applications, but websites are more complicated because you would need to do URL filtering. Solutions exist in the shape of web proxies, firewalls, or content engines, but those indeed have little integration with wireless. First of all, Cisco Umbrella is a very fast DNS server that helps guarantee a very low latency of DNS resolutions and accelerates web browsing. On top of that, because it is cloud managed, it is constantly updated with the latest safety against malware and helps you to not resolve malicious domain names. Simply configuring a client to use Umbrella as a DNS server is enough to provide some protection. On top of that,

Umbrella offers an administrator dashboard that allows you to restrict what URLs can be resolved in your network. Without personally choosing to block website X or Y, you can decide to ban all adult-content websites or all websites promoting violence, for example. You can also, of course, add specific sites to be allowed or blocked, as well as add personal URLs that you want resolved (as if it were your private DNS server). Rather than relying on antivirus software analyzing the threats downloaded to your clients, Umbrella prevents your clients from accessing the malicious content, an approach that is much more efficient and secure.

How does Umbrella relate to wireless though? The integration allows you to do two things:

- Make sure that your clients use Umbrella as a DNS server.
- Allow you to mark the DNS traffic coming from your network toward Umbrella to use specific DNS policies on Umbrella. For example, you may want to allow more websites on your secure employee SSID but block a lot more things on your guest SSID (such as illegal software, movies, and VoD websites).

You must first register your controller to the Umbrella cloud by going to the **Configuration > Security > Threat Defense > DNS Layer Security/Umbrella** page (as shown in Figure 5-50), where you can create a global umbrella parameter-map in your configuration. You can obtain the token from the Umbrella cloud account and enter it on the controller. You can then enable DNS packet encryption (which means the WLC encrypts the packets it sends to Umbrella) and create parameter maps. Those are just names that you are able to apply to a policy profile to identify requests on the Umbrella cloud site.

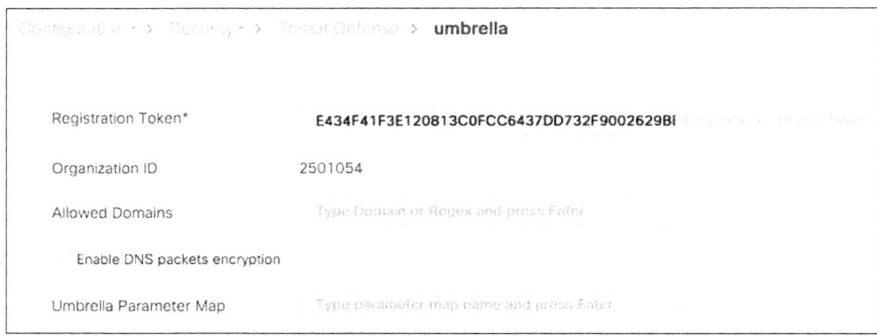

Figure 5-50 *Umbrella global configuration on the 9800*

On the policy profile advanced settings section, you can choose one of the parameter maps you created to segregate DNS policies on the Umbrella cloud. By default, the controller injects the Umbrella DNS IP address into DHCP option 6 (which points to the DNS servers) when the clients receive their IP address so that you don't even have to configure the Umbrella IPs in your DHCP pools. The other settings depicted in Figure 5-51 allow you to also inject the DHCP option 6 setting in Flex APs when the traffic is locally switched and to force the DNS traffic redirection. This means that even if the client then manually configures a custom DNS server, the WLC intercepts the DNS queries anyway, sends them to OpenDNS, and spoofs the original DNS IP back so that the client doesn't notice that it's another DNS server that replied.

Figure 5-51 *DNS Layer Security (formerly called Umbrella) section of a policy profile.*

Cisco Secure Development Lifecycle (CSDL)

Cisco Secure Development Lifecycle is at the heart of new Cisco development, and the Catalyst 9800 controller fully embraces the CSDL concepts. These requirements are broad and range from code validation to vulnerability and penetration testing, to default settings on the devices or the use of third-party modules in the code. Security is a never-ending journey, and the Catalyst 9800 controller keeps implementing stricter and stricter security guidelines and principles with every release.

Summary

This chapter covers the security aspects of the Catalyst 9800 controller. The biggest aspect revolves around AAA, which is either locally handled by the controller or delegated to an external RADIUS or TACACS server. Security also consists of ACLs to restrict the traffic that clients can pass or to protect the controller management plane from undesired access. Encrypted Traffic Analytics, rogue detection and WIPS, and Cisco Umbrella are other security components that can help secure your overall solution. Security happens at every layer of the OSI model and is an all-encompassing topic.

References

Trustpoints on 9800: https://www.cisco.com/c/en/us/td/docs/wireless/controller/9800/config-guide/trustpoints/b-configuring-trustpoints-on-cisco-catalyst-9800-series-controllers/m-overview-of-trustpoints-on-catalyst-9800.html

Generate and download CSR on Catalyst 9800: https://www.cisco.com/c/en/us/support/docs/wireless/catalyst-9800-series-wireless-controllers/213917-generate-csr-for-third-party-certificate.html

Configuring WLC with LDAP for 802.1X and web-auth: https://www.cisco.com/c/en/us/support/docs/wireless/catalyst-9800-series-wireless-controllers/216744-configuring-catalyst-9800-wlc-with-ldap.html

Local EAP configuration on 9800: https://www.cisco.com/c/en/us/support/docs/wireless/catalyst-9800-series-wireless-controllers/215026-local-eap-authentication-on-catalyst-980.html

List of IOS-XE wireless features per release: https://www.cisco.com/c/en/us/support/docs/wireless/catalyst-9800-series-wireless-controllers/214855-ios-xe-wireless-feature-list-per-release.html

Wireless controller AP authorization list: https://www.cisco.com/c/en/us/support/docs/wireless/catalyst-9800-series-wireless-controllers/213916-catalyst-9800-wireless-controllers-ap-au.html

Configure SCEP for LSC certificate on 9800 WLC: https://www.cisco.com/c/en/us/support/docs/wireless-mobility/wireless-lan-management/215557-configure-scep-for-locally-significant-c.html

Configure RADIUS and TACACS+ for GUI and CLI authentication on 9800 wireless LAN controllers: https://www.cisco.com/c/en/us/support/docs/wireless/catalyst-9800-series-wireless-controllers/214490-configure-radius-and-tacacs-for-gui-and.html

Configure Lobby ambassador with RADIUS and TACACS authentication: https://www.cisco.com/c/en/us/support/docs/wireless/catalyst-9800-series-wireless-controllers/215552-9800-wlc-lobby-ambassador-with-radius-an.html

Cisco CSDL: https://www.cisco.com/c/dam/en_us/about/doing_business/trust-center/docs/cisco-secure-development-lifecycle.pdf

Chapter 6

Mobility and Client Roaming

The word *mobile*, derived from the Latin word *mōbilis*, means movable. In wireless networking, *mobility*, or *roaming*, refers to the ability of a wireless client to move from one access point (AP) to another while maintaining a wireless connection. When a client is onboarded to a Service Set Identifier (SSID), it goes through association, authentication, and IP addressing phases before it can pass traffic. Wireless clients can be predominantly stationary like wireless printers or truly mobile like smartphones. The time taken to onboard a client varies with the security method used and the network latency, and it ranges from a few seconds to several tens of seconds. As the client is onboarded, a set of contextual information is associated with the client that defines the access and service level that the client traffic will receive. This information includes the type of encryption that needs to be used for all the client traffic, the services that the client should have access to, and any preferential treatment, Quality of Service (QoS), to be given to specific traffic types sourcing from the client.

When the client roams, all client traffic, including delay-sensitive traffic like voice, is expected to roam without service impact. This means that the client needs to be onboarded on the roam-to AP, within milliseconds. All the contextual information associated with the client needs to be carried through at each roam to ensure the same type and level of service is experienced by the client.

In this chapter, you learn about

- The difference between slow, fast, and secure roaming
- Intra- and inter-controller roaming
- How to deploy mobility on the Catalyst 9800
- Inter-Release Controller Mobility (IRCM) with the AireOS wireless LAN controller (WLC)

802.11 Roaming

In a wireless deployment, the decision to roam is made by the wireless client. A wireless client usually roams in the following circumstances:

- When the connection to the current AP is bad. This is deemed bad if the client misses consecutive beacons from the AP or if the client repeatedly retransmits and does not receive 802.11 acknowledgments or when the received signal strength indication (RSSI) from the current AP drops below a certain threshold.
- If there is a better AP. Another AP is considered better when RSSI from the AP is higher than the AP that the client is associated to. Clients may also prefer APs for a specific band (5 GHz over 2.4 GHz) if they are set accordingly.

The specific value for number of missed beacons, maximum retries, RSSI values, and RSSI differentials that drives clients to roam is client-vendor specific. The wireless infrastructure can assist and optimize the client transition between APs, but the final decision to roam is always with the client. For a client to roam successfully, a set of steps needs to be executed:

1. **RF scanning:** First and foremost, a client has to identify the APs that it can potentially roam to. A wireless client connected to an access point on a specific channel periodically goes off-channel and collects information about the available wireless local-area networks (WLANs), the APs broadcasting those networks, and the RSSI at which the client hears the APs. This scanning can be passive, where the client listens to the beacons sent by the access point, or active, where the client sends probe requests and listens for the probe responses from the APs. In either case, the client builds a list of candidate APs where each entry can include MAC address, RSSI, signal-to-noise ratio (SNR), load, channel, and so on for the AP. The RSSI at which the wireless clients look to roam is usually below the range of –70 to –75 dBm. The exact value varies from vendor to vendor.

   When the client decides to roam, the client uses the collected heuristics and, based on band, RSSI, SNR, and load, among others, determines the best AP among the candidate AP list. Then it initiates an 802.11 authentication to the AP, and this is the next step in the client onboarding or roaming process.

2. **802.11 authentication:** To roam, the client sends an 802.11 authentication frame to the neighboring AP. When the AP responds back with an authentication response frame, it establishes the physical layer communication between the client and AP over which the rest of the frames will be exchanged. This exchange is done with open authentication, and no encryption is involved at this stage. Note that the client can exchange these 802.11 authentication frames with one AP indefinitely or can try with multiple APs if one fails to respond.

3. **802.11 (re)association:** When 802.11 authentication is complete to this AP, the client sends an 802.11 reassociation request, which includes information about WLAN and the AP it is roaming from. The AP responds back to the client with a

reassociation response and allocates an association identifier (AID) corresponding to this client join.

4. **L2 authentication:** For the simplest scenario of no Layer 2 security, there is no context information that needs handing off. When Layer 2 security is in use, the security method used determines the context and time taken to onboard the client both at a fresh join to the network and at roam. This scenario is discussed in further detail in subsequent sections.

5. **Client access policies handoff:** VLAN, IP address, access control list (ACL), and quality of service (QoS) are some additional details that need to be carried across when the client is handed off to continue to deliver consistent service to all client traffic.

Each step in this process takes time and needs to be considered for the total client roaming time. The L2 authentication (depending on security methods used) takes the longest part in the roaming process. Different approaches have been defined to optimize the time taken at each step, which include

- Techniques to reduce time taken for L2 authentication
- Roaming optimizations like 802.11k and 802.11v to reduce time taken for RF scanning

Full-Auth Roaming (or Slow Roam)

As explained in Chapter 5, "Security," wireless enterprise networks usually have some type of Layer 2 security enabled on their SSIDs. The two common security methods are

- **Wi-Fi Protected Access (WPA)/WPA2-Personal/WPA3-Personal:** After the client associates, a preshared key (PSK) is used to authenticate clients.
- **WPA2-Enterprise/WPA3-Enterprise:** The Extensible Authentication Protocol (EAP) method and 802.1X are leveraged to validate a client's credentials (username and password), certificates, or tokens against an authentication, authorization, and accounting (AAA) server. This involves an exchange of several additional frames as compared to PSK and therefore takes longer.

In addition to the security type, factors like RF conditions resulting in retransmissions, number of clients, and round-trip time between the Catalyst 9800 Wireless LAN Controller and AAA server all lead to an increase in the time taken to onboard the client.

In both security methods, post PSK or EAP authentication, a WPA/WPA2/WPA3 four-way handshake occurs between the client and the AP to negotiate the encryption key. Figure 6-1 illustrates the steps involved in the WPA/WPA2 key generation and distribution process:

1. From PSK or EAP authentication, a Master Session Key (MSK) is derived, and this serves as root for deriving other keys.

2. From this MSK, the client and AAA server derive the Pairwise Master Key (PMK). The AAA server sends the PMK to C9800/AP.

3. The C9800/AP derives a Group Master Key (GMK).

4. The C9800/AP initiates the WPA/WPA2 four-way handshake consisting of four messages (M1, M2, M3, and M4). During this exchange, the client and C9800/AP derive actual encryption keys.

5. The client and C9800/AP derive the Pairwise Transient Key (PTK) from PMK, which is used to encrypt unicast frames. The C9800/AP derives the Group Transient Key (GTK) from the GMK, which is used to encrypt multicast/broadcast on this specific SSID/AP.

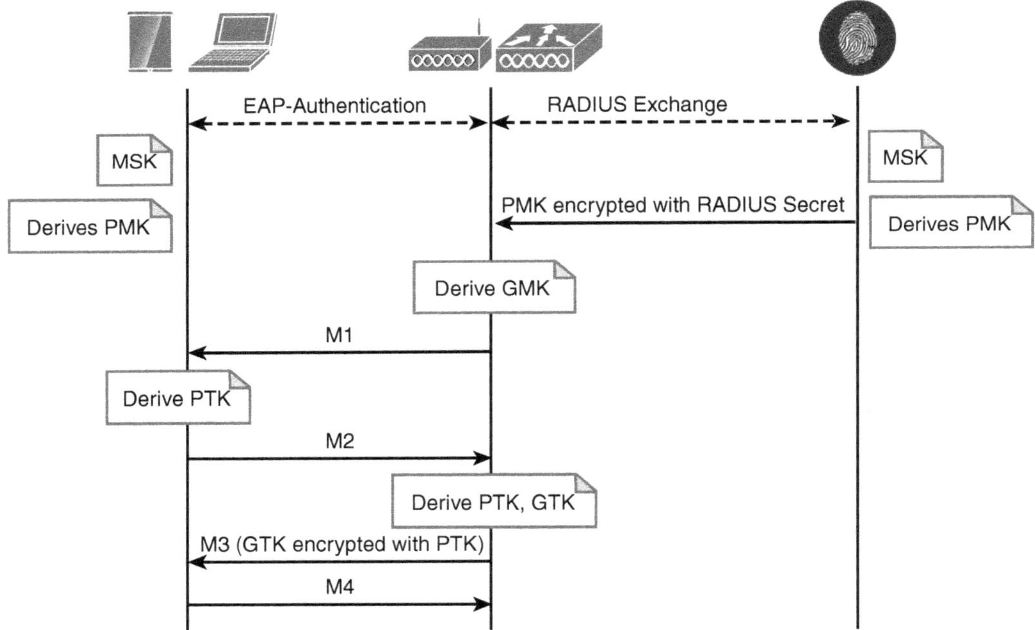

Figure 6-1 *802.11 key management and distribution*

Until the keys are derived, no data traffic is allowed, including Dynamic Host Configuration Protocol (DHCP). The derived key is used to encrypt a DHCP exchange between the wireless client and AP. The encryption key is valid for a limited period (the length of the session typically) and unique between the client and AP. In other words, every AP needs to negotiate a key with the client as the client roams.

Figure 6-2 shows the packet capture of a client roaming when WPA/WPA2-PSK is in use, and Figure 6-3 shows the packet capture of a client onboarding when WPA/WPA2-EAP is used.

No.	Time	Source	Destination	Length	Info
1716	0.022097	e2:2e:04:18:04:24	Cisco_9f:c3:2a	108	Authentication, SN=3813, FN=0, Flags=........C
1718	0.001447	Cisco_9f:c3:2a	e2:2e:04:18:04:24	108	Authentication, SN=0, FN=0, Flags=........C
1720	0.001086	e2:2e:04:18:04:24	Cisco_9f:c3:2a	361	Reassociation Request, SN=3814, FN=0, Flags=........C, SSID=slowroampsk
1723	0.002113	Cisco_9f:c3:2a	e2:2e:04:18:04:24	287	Reassociation Response, SN=1, FN=0, Flags=........C
1725	0.001293	Cisco_9f:c3:2a	e2:2e:04:18:04:24	221	Key (Message 1 of 4)
1731	0.020722	e2:2e:04:18:04:24	Cisco_9f:c3:2a	221	Key (Message 2 of 4)
1733	0.000607	Cisco_9f:c3:2a	e2:2e:04:18:04:24	255	Key (Message 3 of 4)
1735	0.012708	e2:2e:04:18:04:24	Cisco_9f:c3:2a	199	Key (Message 4 of 4)
1743	0.000052	e2:2e:04:18:04:24	Cisco_b1:6c:fd	144	QoS Data, SN=0, FN=0, Flags=.p.....TC

Figure 6-2 *WPA/WPA2 PSK reassociation (slow roam)*

No.	Time	Source	Destination	Protocol	Length	Info
4713	16:54:43.246686	Apple_64:5d:e5	68:7d:b4:5e:43:8f	802.11	69	Authentication, SN=2755, FN=0, Flags=........C
4715	16:54:43.246163	68:7d:b4:5e:43:8f	Apple_64:5d:e5	802.11	69	Authentication, SN=0, FN=0, Flags=........C
4717	16:54:43.248853	Apple_64:5d:e5	68:7d:b4:5e:43:8f	802.11	211	Reassociation Request, SN=2756, FN=0, Flags=........C, SSID=cvoice
4719	16:54:43.263550	68:7d:b4:5e:43:8f	Apple_64:5d:e5	802.11	183	Reassociation Response, SN=1, FN=0, Flags=........C
4721	16:54:43.266930	68:7d:b4:5e:43:8f	Apple_64:5d:e5	EAP	72	Request, Identity
4723	16:54:43.269581	Apple_64:5d:e5	68:7d:b4:5e:43:8f	EAP	76	Response, Identity
4725	16:54:43.276763	68:7d:b4:5e:43:8f	Apple_64:5d:e5	EAP	73	Request, TLS EAP (EAP-TLS)
4727	16:54:43.278065	Apple_64:5d:e5	68:7d:b4:5e:43:8f	EAP	75	Response, Legacy Nak (Response Only)
4729	16:54:43.284187	68:7d:b4:5e:43:8f	Apple_64:5d:e5	EAP	73	Request, Protected EAP (EAP-PEAP)
4731	16:54:43.285117	Apple_64:5d:e5	68:7d:b4:5e:43:8f	TLSv1.2	260	Client Hello
4736	16:54:43.302692	68:7d:b4:5e:43:8f	Apple_64:5d:e5	EAP	1079	Request, Protected EAP (EAP-PEAP)
4738	16:54:43.303598	Apple_64:5d:e5	68:7d:b4:5e:43:8f	EAP	73	Response, Protected EAP (EAP-PEAP)
4740	16:54:43.307753	68:7d:b4:5e:43:8f	Apple_64:5d:e5	TLSv1.2	351	Server Hello, Certificate, Server Key Exchange, Server Hello Done
4742	16:54:43.317178	Apple_64:5d:e5	68:7d:b4:5e:43:8f	TLSv1.2	203	Client Key Exchange, Change Cipher Spec, Encrypted Handshake Message
4744	16:54:43.323470	68:7d:b4:5e:43:8f	Apple_64:5d:e5	TLSv1.2	124	Change Cipher Spec, Encrypted Handshake Message
4746	16:54:43.324119	Apple_64:5d:e5	68:7d:b4:5e:43:8f	EAP	73	Response, Protected EAP (EAP-PEAP)
4748	16:54:43.328433	68:7d:b4:5e:43:8f	Apple_64:5d:e5	TLSv1.2	107	Application Data
4750	16:54:43.329524	Apple_64:5d:e5	68:7d:b4:5e:43:8f	TLSv1.2	111	Application Data
4752	16:54:43.333737	68:7d:b4:5e:43:8f	Apple_64:5d:e5	TLSv1.2	137	Application Data
4754	16:54:43.334653	Apple_64:5d:e5	68:7d:b4:5e:43:8f	TLSv1.2	165	Application Data
4756	16:54:43.346834	68:7d:b4:5e:43:8f	Apple_64:5d:e5	TLSv1.2	153	Application Data
4758	16:54:43.347552	Apple_64:5d:e5	68:7d:b4:5e:43:8f	TLSv1.2	108	Application Data
4760	16:54:43.351714	68:7d:b4:5e:43:8f	Apple_64:5d:e5	TLSv1.2	106	Application Data
4762	16:54:43.352310	Apple_64:5d:e5	68:7d:b4:5e:43:8f	EAP	73	Response, Protected EAP (EAP-PEAP)
4764	16:54:43.366697	68:7d:b4:5e:43:8f	Apple_64:5d:e5	EAP	71	Success
4766	16:54:43.367407	68:7d:b4:5e:43:8f	Apple_64:5d:e5	EAPOL	184	Key (Message 1 of 4)
4768	16:54:43.368210	Apple_64:5d:e5	68:7d:b4:5e:43:8f	EAPOL	202	Key (Message 2 of 4)
4770	16:54:43.370090	68:7d:b4:5e:43:8f	Apple_64:5d:e5	EAPOL	274	Key (Message 3 of 4)
4772	16:54:43.370773	Apple_64:5d:e5	68:7d:b4:5e:43:8f	EAPOL	162	Key (Message 4 of 4)

Figure 6-3 *WPA/WPA2 EAP reassociation (slow roam)*

When a client does a regular roam without using any RF or network optimizations, the authentication and four-way handshake process must be repeated at each roam. This has a particular impact when using WPA2/3 Enterprise, where the communication with the AAA server needs to happen at every roam. The AAA server might be centralized, and there could be a delay introduced by the network. Therefore, this is called a *full authentication roam* or *slow roam*. Applications that are delay-sensitive, such as voice or applications that do not tolerate longer delays, see a significant user experience impact such as audio gaps or application disconnects. To work around this issue, several optimizations have been introduced in the 802.11 standard to accelerate the roaming process and make it fast and secure.

Fast Secure Roaming

The higher latency seen in EAP authentication compared to PSK authentication is introduced by the EAP frames exchanged with the AAA server to accomplish the EAP authentication process itself. When using fast secure roaming, the client goes through a full 802.1X/EAP authentication only the first time it associates to the network; on subsequent roams to neighboring APs, the EAP exchange is skipped. Then the only remaining part is the four-way handshake, which is relatively fast because the AAA server is not involved. Table 6-1 lists the different fast secure roaming methods supported by the C9800 in both local mode and FlexConnect deployments.

Table 6-1 *Fast Secure Roaming Support on the C9800 for Central and FlexConnect Deployments*

Method	Local Mode	FlexConnect (Connected Mode; Central Authentication)	FlexConnect (Connected Mode—Local Authentication)	FlexConnect (Standalone)
Sticky Key Caching (SKC)	Not Supported	Not Supported	Not Supported	Not Supported
Opportunistic Key Caching (OKC)	Supported	Supported	Supported	Supported (Until the client session expires)
Cisco Centralized Key Management (CCKM)	Supported	Supported	Supported	Not Supported
FT-802.11r/ Adaptive 11r (Over the Air)	Supported	Supported	Supported	Supported (Until the client session expires)
FT-802.11r/ Adaptive 11r (Over the DS)	Supported	Supported	Not Supported	Not Supported

In the case of FlexConnect deployments, irrespective of central or local switching configuration, fast secure roaming is supported across APs that belong to the same site tag. This is true for all user-defined site tags but not for the default-site-tag: the client key is not distributed to the Flex APs belonging to default-site-tag, so fast roaming is not supported.

PMKID Caching (Sticky Key Caching)

Pairwise Master Key ID (PMKID) caching, or Sticky Key Caching (SKC), is the first and optional fast secure roaming method suggested by WPA2. With SKC, the infrastructure (WLC and AP) and wireless client cache the PMKs derived during the full 802.1X/EAP authentication for some time after the initial connection. This PMKID is used to derive session encryption keys. Upon roam, the wireless client sends PMKID in the reassociation request frame. If the client roams back to an AP to which it previously associated and the previous session has not timed out, then AP will use cached PMK. EAP exchange is skipped and only the four-way handshake occurs.

- **Pros with SKC:** The client experiences a fast roam (less than 100 ms), and there is a reduced load on the AAA server.
- **Cons with SKC:** SKC applies only to roams to the APs to which the client has previously connected. Every roaming toward a new AP involves a full authentication. Also, there is a limited number of PMKIDs that each AP would have to store for clients that are not connected anymore. These limitations are the reason SKC is not widely adopted and not supported at all on C9800.

OKC

Opportunistic Key Caching (OKC) is an enhancement over PMKID caching but is not part of the 802.11 standard. In this method, the initial association of a client into the WLAN uses full authentication. The wireless client and the C9800 cache this original PMK. In case of local mode deployment, the C9800 distributes the PMK to other WLCs in the mobility group and makes the PMK available to APs that need to derive encryption keys. In the case of FlexConnect mode deployment, the C9800 shares the cached PMK to all APs within a custom site tag. PMKID is hashed from the client MAC address, AP MAC address (BSSID of the WLAN), and original PMK. With PMK cached for the specific client, the only variable part is the AP MAC address.

At each roam, the client adds PMKID on the WPA2 Robust Security Network Information Element (RSN IE) to inform the AP that a cached PMK is used for fast secure roaming. The PMKID hash is generated from the client's own MAC, cached PMK, and MAC address of roamed-to AP. When the AP receives this request, if the infrastructure does not support OKC, the PMKID sent by the client is ignored and a full authentication occurs. But if the infrastructure supports OKC for the local mode, the C9800 hashes the PMKID using the cached PMK, client MAC, and its own AP MAC address and sends a successful reassociation response confirming the PMKID match. In the case of FlexConnect deployment, this hash computation is done by the AP. With this, 802.1X/EAP exchange can be avoided, and only a four-way handshake occurs to derive the new encryption keys. In Figure 6-4, a packet capture shows the OKC roam, and Figure 6-5 shows an over-the-air packet capture snippet of the OKC roam for a local mode deployment.

No.	Time	Source	Destination	Length	Info
554	0.002172	92:4c:d7:78:89:f8	Cisco_23:c6:4f	108	Authentication, SN=1823, FN=0, Flags=........C
556	0.000992	Cisco_23:c6:4f	92:4c:d7:78:89:f8	108	Authentication, SN=3128, FN=0, Flags=........C
558	0.000985	92:4c:d7:78:89:f8	Cisco_23:c6:4f	347	Association Request, SN=1824, FN=0, Flags=........C, SSID=OKC
559	0.000136	92:4c:d7:78:89:f8	Cisco_23:c6:4f	347	Association Request, SN=1824, FN=0, Flags=....R...C, SSID=OKC
561	0.004662	Cisco_23:c6:4f	92:4c:d7:78:89:f8	269	Association Response, SN=3129, FN=0, Flags=........C
563	0.027169	Cisco_23:c6:4f	92:4c:d7:78:89:f8	109	Request, Identity
569	0.008778	92:4c:d7:78:89:f8	Cisco_23:c6:4f	115	Response, Identity
573	0.021785	Cisco_23:c6:4f	92:4c:d7:78:89:f8	110	Request, Protected EAP (EAP-PEAP)
575	0.006260	92:4c:d7:78:89:f8	Cisco_23:c6:4f	241	Client Hello
599	0.000000	Cisco_23:c6:4f	92:4c:d7:78:89:f8	101	Request, Protected EAP (EAP-PEAP)
602	0.001996	92:4c:d7:78:89:f8	Cisco_23:c6:4f	110	Response, Protected EAP (EAP-PEAP)
604	0.003478	Cisco_23:c6:4f	92:4c:d7:78:89:f8	1321	Server Hello, Certificate, Server Key Exchange, Server Hello Done
606	0.007491	92:4c:d7:78:89:f8	Cisco_23:c6:4f	236	Client Key Exchange, Change Cipher Spec, Encrypted Handshake Message
608	0.001755	Cisco_23:c6:4f	92:4c:d7:78:89:f8	165	Change Cipher Spec, Encrypted Handshake Message
610	0.002876	92:4c:d7:78:89:f8	Cisco_23:c6:4f	110	Response, Protected EAP (EAP-PEAP)
612	0.001465	Cisco_23:c6:4f	92:4c:d7:78:89:f8	144	Application Data
614	0.002854	92:4c:d7:78:89:f8	Cisco_23:c6:4f	150	Application Data
616	0.001802	Cisco_23:c6:4f	92:4c:d7:78:89:f8	144	Application Data
618	0.002206	92:4c:d7:78:89:f8	Cisco_23:c6:4f	152	Application Data
620	0.001553	Cisco_23:c6:4f	92:4c:d7:78:89:f8	150	Application Data
622	0.002853	92:4c:d7:78:89:f8	Cisco_23:c6:4f	150	Application Data
624	0.001847	Cisco_23:c6:4f	92:4c:d7:78:89:f8	108	Success
626	0.000466	Cisco_23:c6:4f	92:4c:d7:78:89:f8	221	Key (Message 1 of 4)
628	0.007728	92:4c:d7:78:89:f8	Cisco_23:c6:4f	221	Key (Message 2 of 4)
630	0.001235	Cisco_23:c6:4f	92:4c:d7:78:89:f8	255	Key (Message 3 of 4)
633	0.000585	92:4c:d7:78:89:f8	Cisco_23:c6:4f	199	Key (Message 4 of 4)
634	0.000000	92:4c:d7:78:89:f8	Cisco_23:c6:4f	199	Key (Message 4 of 4)
2026	0.022106	92:4c:d7:78:89:f8	Cisco_9f:c3:2f	108	Authentication, SN=2190, FN=0, Flags=........C
2028	0.001362	Cisco_9f:c3:2f	92:4c:d7:78:89:f8	108	Authentication, SN=0, FN=0, Flags=........C
2030	0.001320	92:4c:d7:78:89:f8	Cisco_9f:c3:2f	371	Reassociation Request, SN=2191, FN=0, Flags=........C, SSID=OKC
2033	0.003157	Cisco_9f:c3:2f	92:4c:d7:78:89:f8	287	Reassociation Response, SN=1, FN=0, Flags=........C
2035	0.001324	Cisco_9f:c3:2f	92:4c:d7:78:89:f8	221	Key (Message 1 of 4)
2043	0.034286	92:4c:d7:78:89:f8	Cisco_9f:c3:2f	239	Key (Message 2 of 4)
2045	0.001272	Cisco_9f:c3:2f	92:4c:d7:78:89:f8	255	Key (Message 3 of 4)
2048	0.000248	92:4c:d7:78:89:f8	Cisco_9f:c3:2f	199	Key (Message 4 of 4)

Figure 6-4 *Packet capture: OKC fast roam in local mode deployment*

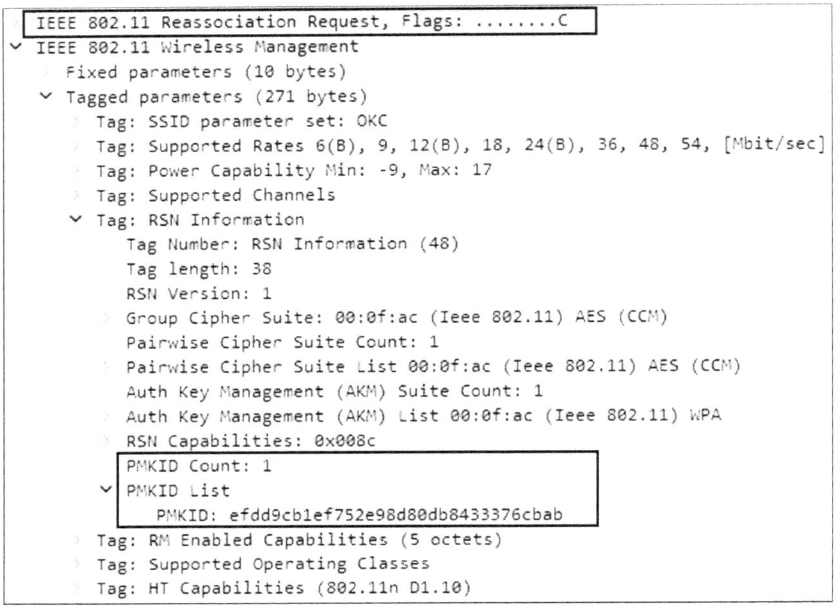

```
  IEEE 802.11 Reassociation Request, Flags: ........C
v IEEE 802.11 Wireless Management
  > Fixed parameters (10 bytes)
  v Tagged parameters (271 bytes)
    > Tag: SSID parameter set: OKC
    > Tag: Supported Rates 6(B), 9, 12(B), 18, 24(B), 36, 48, 54, [Mbit/sec]
    > Tag: Power Capability Min: -9, Max: 17
    > Tag: Supported Channels
    v Tag: RSN Information
        Tag Number: RSN Information (48)
        Tag length: 38
        RSN Version: 1
      > Group Cipher Suite: 00:0f:ac (Ieee 802.11) AES (CCM)
        Pairwise Cipher Suite Count: 1
      > Pairwise Cipher Suite List 00:0f:ac (Ieee 802.11) AES (CCM)
        Auth Key Management (AKM) Suite Count: 1
      > Auth Key Management (AKM) List 00:0f:ac (Ieee 802.11) WPA
      > RSN Capabilities: 0x008c
        PMKID Count: 1
      v PMKID List
          PMKID: efdd9cb1ef752e98d80db8433376cbab
    > Tag: RM Enabled Capabilities (5 octets)
    > Tag: Supported Operating Classes
    > Tag: HT Capabilities (802.11n D1.10)
```

Figure 6-5 *OKC fast roam RSN IE*

OKC is enabled by default on the C9800 and cannot be disabled for central authentication either for local or FlexConnect deployment. It can only be disabled for FlexConnect local authentication. To do this, on the C9800 GUI, navigate to **Configuration > Tags and Profiles > WLANs > Advanced** and uncheck **OKC**, as shown in Figure 6-6.

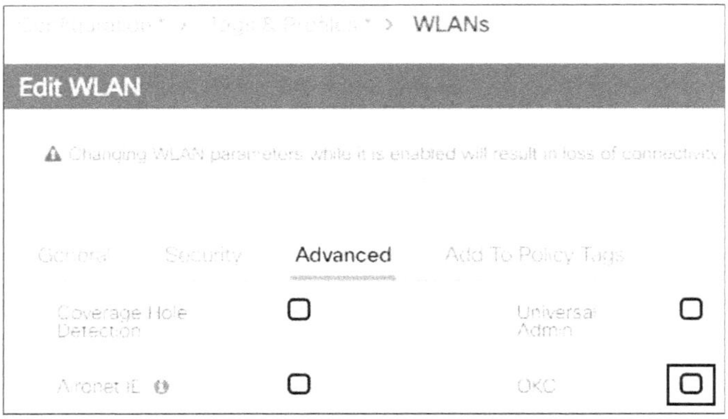

WLANs

Edit WLAN

Advanced

Figure 6-6 *Disabling OKC for FlexConnect local authorization*

The type of client that roams using OKC is reflected as 802.11i. On the C9800 GUI, navigate to **Monitoring > Wireless > Clients > Mobility History**, as shown in Figure 6-7.

The equivalent command-line interface (CLI) is **#show wireless client mac-address <mac-address> mobility history.**

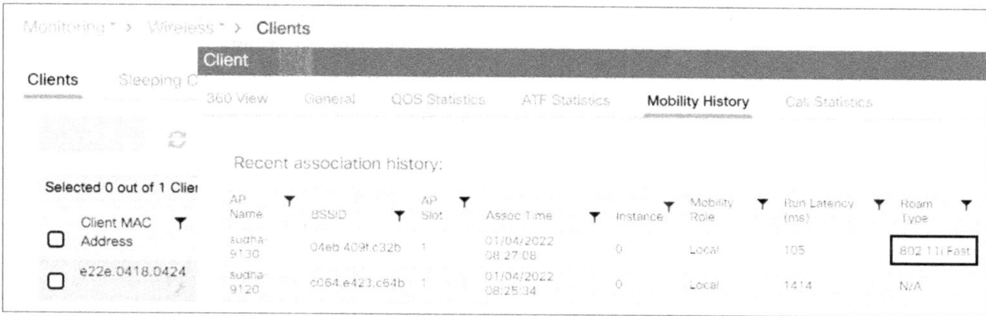

Figure 6-7 *Monitoring the client roam type on the C9800*

- **Pros with OKC:** The client requires only one full 802.1X/EAP authentication and can fast-secure roam to APs that it has not previously associated to. Also, the wireless client and infrastructure don't need to store multiple PMKIDs but only the original PMK from the initial 802.1X/EAP authentication, making this a scalable method.
- **Cons with OKC:** This method is not an 802.11 standard, so client support may not be there. However, this is one of the popular alternatives to 802.11r or Fast Transition (FT) method.

CCKM

Cisco Centralized Key Management (CCKM) is a Cisco proprietary fast-secure roaming method only supported by Cisco wireless infrastructure devices and wireless clients from different vendors that are Cisco Compatible Extension (CCX)-compatible. Support for CCKM is indicated in the Auth Key Management (AKM) Suite of RSN IE sent in beacon, probe response, and association request frames.

With CCKM, like regular WPA/WPA2, an MSK (also known as the Network Session Key, or NSK) is mutually derived between the client and the AAA server. This master key is sent from the AAA server to the WLC after a successful authentication. The WLC and the client derive the seed information from NSK and go through a four-way handshake like WPA/WPA2, to derive the unicast (PTK) and multicast/broadcast (GTK) encryption keys with the first AP.

When the CCKM client roams, it sends a single reassociation request frame to the AP/WLC, including a message integrity check (MIC) and a sequentially incrementing random number. This information, combined with BSSID of the roam-to AP, is used to derive PTK by the client as well as the wireless infrastructure. Because PTK can be derived with reassociation request itself, a four-way handshake, in addition to EAP authentication, is skipped. AP responds with a reassociation response, and the client can continue to forward data.

CCKM is supported with central authentication for local mode and FlexConnect deployments. With FlexConnect local authentication, in connected mode, the cache is distributed from the AP to the C9800 and then to other APs in the FlexConnect site tag. In standalone mode, only cached entries work; new authentications do not work. Enabling CCKM requires you to explicitly disable the Fast Transition setting, as shown in Figure 6-8.

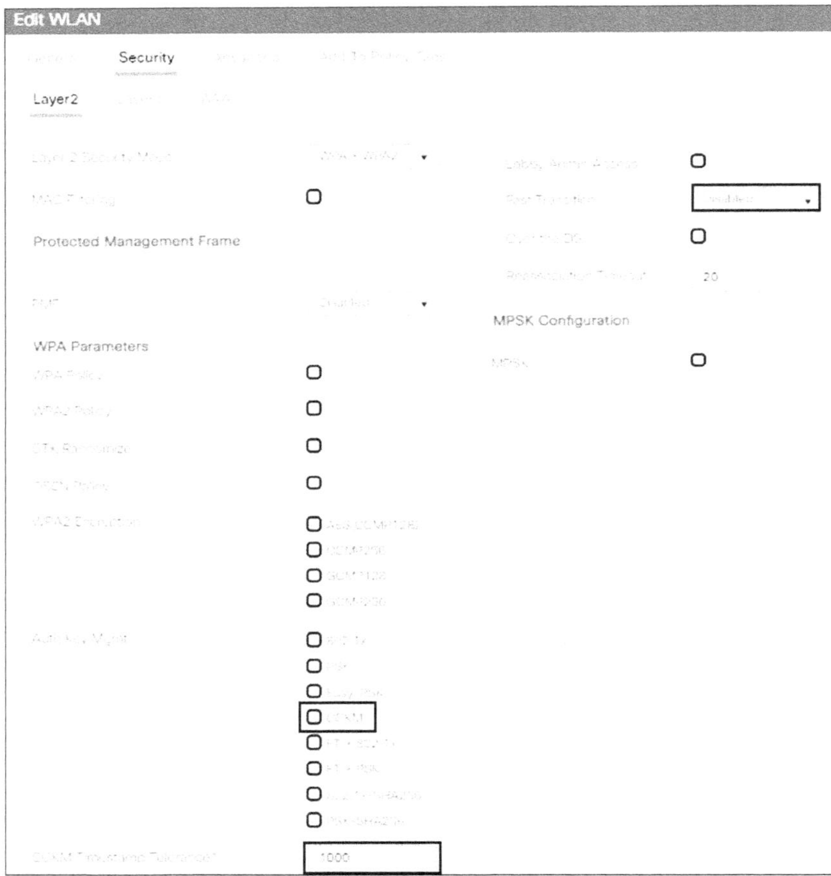

Figure 6-8 *Configuring CCKM on the C9800*

- **Pros with CCKM:** CCKM is faster than SKC and OKC because a four-way handshake and an EAP/802.1X exchange are avoided. It supports all the encryption methods within the 802.11 standard: Wired Equivalent Privacy (WEP), Temporal Key Integrity Protocol (TKIP), and Advanced Encryption Standard (AES).
- **Cons with CCKM:** Because CCKM is proprietary, it is supported only on Cisco wireless infrastructure and CCX wireless clients. Lack of adoption means CCKM is essentially obsolete and is only relevant in Cisco work group bridge (WGB) deployments, which only support CCKM.

Fast Transition (802.11r)

The 802.11r Fast Transition (FT) or Fast Basic Service Set (BSS) Transition is the IEEE standard for fast secure roaming. In Fast BSS Transition, the initial handshake with the new AP occurs before the client roams to it, thus reducing the roam time. During this initial handshake itself, PTK is derived both by client and wireless infrastructure, and data forwarding begins immediately after reassociation completes. The 802.11r adds a new Authentication Key Management (AKM) suite under WLAN profile configuration called FT.

In FT deployment, the first client association to a wireless network goes through the normal full authentication process, including 802.11 open authentication, association, EAP exchange, and four-way key handshake to let the client forward data. Beyond that, the 802.11 management frames all carry additional 802.11r information to complete FT negotiation as described next.

Beacon and Probe Responses

The 802.11r information carried in the 802.11 beacons and probe responses sent from the AP is shown in Figure 6-9 and Figure 6-10. The 802.11r fields contained within these 802.11 management frames are as follows:

1. Mobility Domain Information Element (MDIE), which includes
   - Element ID 54.
   - A Mobility Domain Identifier (MDID) that identifies groups of APs capable of FT. This MDID can be verified from the C9800 CLI with **#show wireless mobility summary**.
   - The FT capability set to 1.
   - The FT Over-the-DS field, which indicates how the client exchanges messages with the target AP to complete the FT roam.
     - If it is set to 1, the client does FT roaming to target the AP via the current AP over the wired network, referred to as FT over the distribution system (DS).
     - If it is set to 0, the client does FT roaming to target the AP directly over 802.11. On C9800, all WLANs perform FT key exchange over-the-air by default.
2. The Robust Security Network Information Element (RSN IE) in the beacon is determined by the FT mode in use.
   - **FT only:** Includes the FT AKM suite.
   - **FT Mixed Mode:** Includes two AKM suites: PSK/802.1X and PSK/802.1X FT over IEEE.

- **FT Adaptive:** Includes one AKM suite: PSK or 802.1X.

The difference between these modes is discussed further in the next subsection.

```
IEEE 802.11 Beacon frame, Flags: ........C
IEEE 802.11 Wireless Management
  Fixed parameters (12 bytes)
  Tagged parameters (383 bytes)
    Tag: SSID parameter set: 11renable
    Tag: Supported Rates 6(B), 9, 12(B), 18, 24(B), 36, 48, 54, [Mbit/sec]
    Tag: DS Parameter set: Current Channel: 100
    Tag: Traffic Indication Map (TIM): DTIM 1 of 1 bitmap
    Tag: Country Information: Country Code US, Environment Unknown (0x04)
    Tag: Power Constraint: 3
    Tag: TPC Report Transmit Power: 0, Link Margin: 0
    Tag: RSN Information
    Tag: Mobility Domain
      Tag Number: Mobility Domain (54)
      Tag length: 3
      Mobility Domain Identifier: 0x34ac
      FT Capability and Policy: 0x00
      .... ...0 = Fast BSS Transition over DS: 0x0
      .... ..0. = Resource Request Protocol Capability: 0x0
    Tag: QBSS Load Element 802.11e CCA Version
```

Figure 6-9 *802.11r FT Beacon MDIE*

```
IEEE 802.11 Beacon frame, Flags: ........C
IEEE 802.11 Wireless Management
  Fixed parameters (12 bytes)
  Tagged parameters (383 bytes)
    Tag: SSID parameter set: 11renable
    Tag: Supported Rates 6(B), 9, 12(B), 18, 24(B), 36, 48, 54, [Mbit/sec]
    Tag: DS Parameter set: Current Channel: 100
    Tag: Traffic Indication Map (TIM): DTIM 1 of 1 bitmap
    Tag: Country Information: Country Code US, Environment Unknown (0x04)
    Tag: Power Constraint: 3
    Tag: TPC Report Transmit Power: 0, Link Margin: 0
    Tag: RSN Information
      Tag Number: RSN Information (48)
      Tag length: 20
      RSN Version: 1
      Group Cipher Suite: 00:0f:ac (Ieee 802.11) AES (CCM)
      Pairwise Cipher Suite Count: 1
      Pairwise Cipher Suite List 00:0f:ac (Ieee 802.11) AES (CCM)
      Auth Key Management (AKM) Suite Count: 1
      Auth Key Management (AKM) List 00:0f:ac (Ieee 802.11) FT over IEEE 802.1X
        Auth Key Management (AKM) Suite: 00:0f:ac (Ieee 802.11) FT over IEEE 802.1X
          Auth Key Management (AKM) OUI: 00:0f:ac (Ieee 802.11)
          Auth Key Management (AKM) type: FT over IEEE 802.1X (3)
      RSN Capabilities: 0x0028
    Tag: Mobility Domain
```

Figure 6-10 *802.11r FT Beacon RSN IE AKM*

Association Request

The association request is sourced from a wireless client and includes MDIE and FT over IEEE 802.1X AKM suite in RSN IE.

Association Response

The association response from the C9800/AP shown in Figure 6-11 includes the Fast BSS Transition Information Element (FTIE) in addition to MDIE. The FTIE includes information such as the MIC, nonces, PMK-R0 key holder identifier (R0KH-ID), and PMK-R1 key holder identifier (R1KH-ID), which are needed to perform the FT authentication sequence during a Fast BSS Transition.

```
> IEEE 802.11 Association Response, Flags: ........C
v IEEE 802.11 Wireless Management
  > Fixed parameters (6 bytes)
  v Tagged parameters (276 bytes)
    > Tag: Supported Rates 6(B), 9, 12(B), 18, 24(B), 36, 48, 54, [Mbit/sec]
    > Tag: Vendor Specific: Microsoft Corp.: WMM/WME: Parameter Element
    > Tag: HT Capabilities (802.11n D1.10)
    > Tag: HT Information (802.11n D1.10)
    > Tag: Extended Capabilities (10 octets)
    > Tag: VHT Capabilities
    > Tag: VHT Operation
    > Tag: Mobility Domain
    v Tag: Fast BSS Transition
        Tag Number: Fast BSS Transition (55)
        Tag length: 96
        MIC Control: 0x0000
        0000 0000 .... .... = Element Count: 0
        MIC: 00000000000000000000000000000000
        ANonce: 0000000000000000000000000000000000000000000000000000000000000000
        SNonce: 0000000000000000000000000000000000000000000000000000000000000000
        Subelement ID: PMK-R1 key holder identifier (R1KH-ID) (1)
        Length: 6
        PMK-R1 key holder identifier (R1KH-ID): 84e87e81d09a
        Subelement ID: PMK-R0 key holder identifier (R0KH-ID) (3)
        Length: 4
        PMK-R0 key holder identifier (R0KH-ID): 33457d3f
```

Figure 6-11 *802.11r FT association response*

Authentication

After association, PSK or 802.1X authentication kicks off. As with the other methods, MSK is derived from PSK or 802.1X/EAP authentication, both by the client and the AAA server. The MSK is used as a seed for the FT key hierarchy.

After authentication, an FT four-way handshake occurs. FT specifies three layers of key hierarchy that are cached:

1. **PMK-R0:** The Pairwise Master Key (PMK-R0) is the first level PMK and is derived from the MSK. The key holders for this PMK are the WLC (R0KH) and the client (S0KH).

2. **PMK-R1:** The Pairwise Master Key (PMK-R1) is the second level PMK and is derived from the PMK-R0 by the WLC. The WLC uses a secure channel to transmit unique PMK-R1 to each AP in the mobility domain. The key holders of PMK-R1 are the APs (R1KH) managed by the WLC and the client (S1KH).

3. **PTK:** The Pairwise Transient Key (PTK) is the third and final level key in the hierarchy. PTK is derived from PMK-R1 and is used to encrypt the data frames by the APs and the clients.

Figure 6-12 shows packet capture of the initial association of the client and the roaming of the client to target AP on an FT-enabled SSID.

No.	Time	Source	Destination	Length	Info
285	0.002251	MurataMa_60:29:3e	Cisco_23:c6:4c	108	Authentication, SN=271, FN=0, Flags=........C
287	0.001038	Cisco_23:c6:4c	MurataMa_60:29:3e	108	Authentication, SN=938, FN=0, Flags=........C
289	0.001673	MurataMa_60:29:3e	Cisco_23:c6:4c	358	Association Request, SN=272, FN=0, Flags=........C, SSID=11renable
291	0.004228	Cisco_23:c6:4c	MurataMa_60:29:3e	372	Association Response, SN=939, FN=0, Flags=........C
293	0.004293	Cisco_23:c6:4c	MurataMa_60:29:3e	109	Request, Identity
295	0.024719	MurataMa_60:29:3e	Cisco_23:c6:4c	115	Response, Identity
299	0.088306	Cisco_23:c6:4c	MurataMa_60:29:3e	110	Request, Protected EAP (EAP-PEAP)
301	0.003845	MurataMa_60:29:3e	Cisco_23:c6:4c	241	Client Hello
308	0.000000	Cisco_23:c6:4c	MurataMa_60:29:3e	101	Request, Protected EAP (EAP-PEAP)
310	0.003464	MurataMa_60:29:3e	Cisco_23:c6:4c	110	Response, Protected EAP (EAP-PEAP)
312	0.003413	Cisco_23:c6:4c	MurataMa_60:29:3e	1321	Server Hello, Certificate, Server Key Exchange, Server Hello Done
314	0.010654	MurataMa_60:29:3e	Cisco_23:c6:4c	236	Client Key Exchange, Change Cipher Spec, Encrypted Handshake Message
316	0.002056	Cisco_23:c6:4c	MurataMa_60:29:3e	165	Change Cipher Spec, Encrypted Handshake Message
318	0.002469	MurataMa_60:29:3e	Cisco_23:c6:4c	110	Response, Protected EAP (EAP-PEAP)
320	0.002267	Cisco_23:c6:4c	MurataMa_60:29:3e	144	Application Data
322	0.002323	MurataMa_60:29:3e	Cisco_23:c6:4c	150	Application Data
324	0.001642	Cisco_23:c6:4c	MurataMa_60:29:3e	144	Application Data
326	0.003633	MurataMa_60:29:3e	Cisco_23:c6:4c	152	Application Data
328	0.001511	Cisco_23:c6:4c	MurataMa_60:29:3e	150	Application Data
330	0.002657	MurataMa_60:29:3e	Cisco_23:c6:4c	150	Application Data
332	0.001764	Cisco_23:c6:4c	MurataMa_60:29:3e	108	Success
334	0.000000	Cisco_23:c6:4c	MurataMa_60:29:3e	221	Key (Message 1 of 4)
336	0.020090	MurataMa_60:29:3e	Cisco_23:c6:4c	342	Key (Message 2 of 4)
338	0.002610	Cisco_23:c6:4c	MurataMa_60:29:3e	391	Key (Message 3 of 4)
341	0.000000	MurataMa_60:29:3e	Cisco_23:c6:4c	199	Key (Message 4 of 4)
342	0.000000	MurataMa_60:29:3e	Cisco_23:c6:4c	199	Key (Message 4 of 4)
367	0.046878	MurataMa_60:29:3e	Broadcast	462	QoS Data, SN=2, FN=0, Flags=.p.....TC
397	0.024374	MurataMa_60:29:3e	Broadcast	144	QoS Data, SN=3, FN=0, Flags=.p.....TC
412	0.002264	MurataMa_60:29:3e	10.6.1.12	274	QoS Data, SN=4, FN=0, Flags=.p.....TC
868	0.022004	MurataMa_60:29:3e	Cisco_9f:c3:2c	243	Authentication, SN=466, FN=0, Flags=........C
870	0.001272	Cisco_9f:c3:2c	MurataMa_60:29:3e	239	Authentication, SN=0, FN=0, Flags=........C
872	0.001386	MurataMa_60:29:3e	Cisco_9f:c3:2c	480	Reassociation Request, SN=467, FN=0, Flags=........C, SSID=11renable
875	0.002663	Cisco_9f:c3:2c	MurataMa_60:29:3e	467	Reassociation Response, SN=1, FN=0, Flags=........C
899	0.012668	MurataMa_60:29:3e	Cisco_b1:6c:fd	144	QoS Data, SN=1, FN=0, Flags=.p.....TC
901	0.007196	MurataMa_60:29:3e	Broadcast	144	QoS Data, SN=2, FN=0, Flags=.p.....TC

Figure 6-12 *802.11r (FT) initial association and roam (over-the-air)*

During this initial association to the mobility domain (Frames 285, 287), an FT four-way handshake occurs to derive the PTK (Frames 334-341). However, in the case of an FT roam (Frame 868 onwards), in addition to 802.1X/EAP authentication, the four-way handshake exchange is also avoided by piggybacking FT information into 802.11 open authentication or FT action frames and 802.11 authentication frames. When the client wants to roam to the target AP using FT, it has two options to communicate with the target AP: over-the-air or over-the-ds.

FT Over-the-Air

In this case, when the client decides to FT roam, it initiates 802.11 communication directly with the target AP. FT information is piggybacked on 802.11 authentication and reassociation frames, which allows for the PTK and GTK to be generated at reassociation itself. In short, both an 802.1X/EAP exchange and a four-way handshake needed to derive keys are avoided, and the client starts forwarding traffic as soon as a reassociation response is processed, as shown in Figure 6-13.

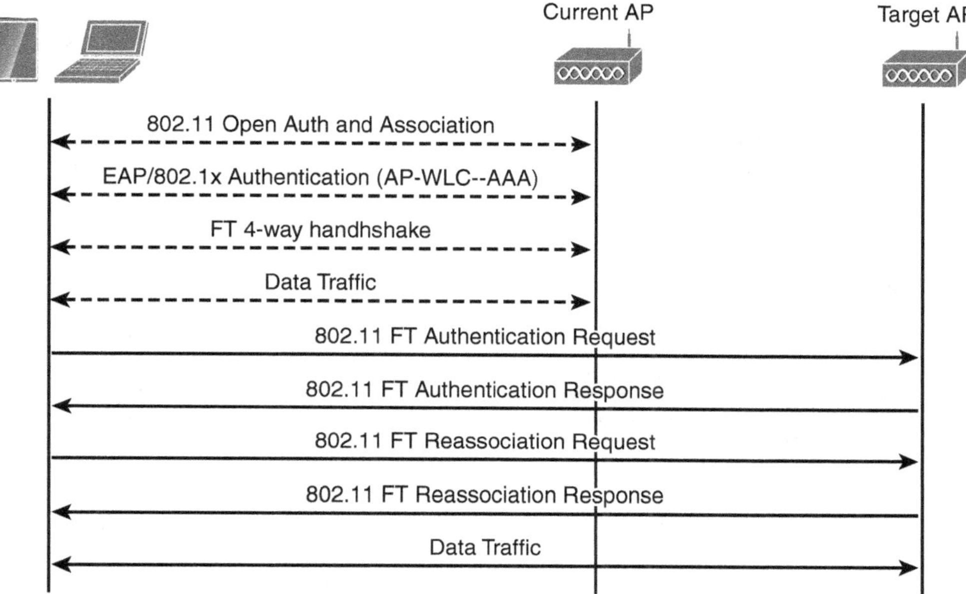

Figure 6-13 *FT roam over-the-air*

Over-the-DS

DS refers to distribution system, the wired infrastructure, to which APs are connected. In this case, when a client decides to FT roam, the client initiates FT request action frames to its current connected AP, including the BSSID (MAC address) of the AP it wants to roam to. The current AP forwards this request action frame over the wired infrastructure to the target AP. The target AP responds with an FT response action frame over the DS to the current AP. When action frames have been successfully exchanged, the client sends a reassociation request to the AP over-the-air and receives a reassociation response to complete the roam and derive the final keys to encrypt data from the client, as shown in Figure 6-14.

FT over-the-DS has not been adopted by many client vendors and is therefore not recommended. On the C9800, FT over-the-DS needs explicit configuration, with FT over-the-air being the system default.

When 802.11r first came out, clients that were not FT-capable could not process new FT authentication frames and failed to connect to the FT-enabled SSID, thereby requiring two SSIDs to be configured: one for FT and one for non-FT. To work around this problem and have all clients coexist on one SSID, two solutions came into play:

- **Mixed-mode FT:** With mixed-mode FT, two AKM suites are advertised in the RSN IE: one is the regular PSK or 802.1X, and the other is FT. The expectation was that non-FT clients would use regular PSK and 802.1X, whereas FT-capable clients could avail FT. However, a lot of legacy client drivers struggled to deal with multiple AKMs and could not connect to SSID. To configure mixed mode on a WLAN profile, navigate to **Configuration > Tags and Profiles > WLANs.** Then create or select a WLAN. On the following page, navigate to the **WLAN > Security > Layer 2 >** under **Auth Key Management**, select both **802.1X** and **FT+ 802.1x** OR both **PSK** and **FT+ PSK**. Figure 6-15 show the RSN IE with two security methods: regular WPA 802.1X and FT over 802.1X. This is the bit that can confuse some older wireless clients. MDIE is present in beacon and probe response frames. Figure 6-16 shows the Aironet IE when mixed-mode FT is in use. This shows up even when Aironet IE is disabled under the WLAN.

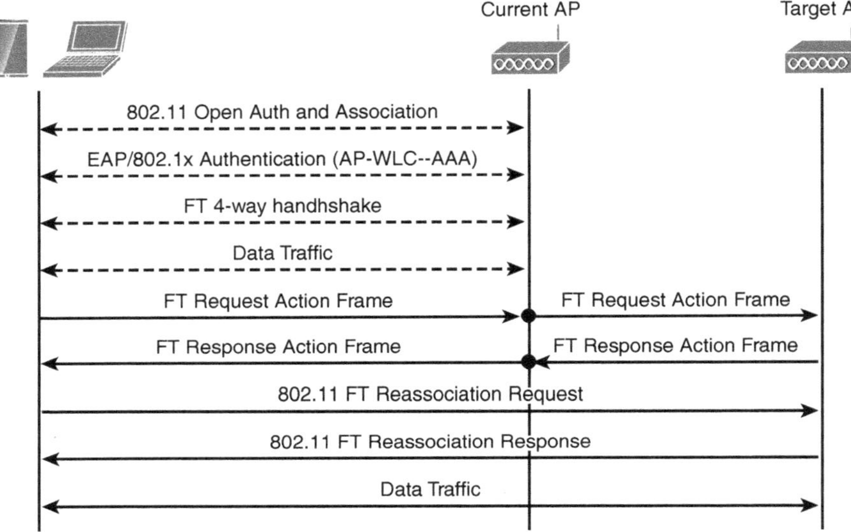

Figure 6-14 *FT roam over-the-DS*

```
IEEE 802.11 Beacon frame, Flags: ........C
IEEE 802.11 Wireless Management
  Fixed parameters (12 bytes)
  Tagged parameters (360 bytes)
    Tag: SSID parameter set: 11rmixed
    Tag: Supported Rates 6(B), 9, 12(B), 18, 24(B), 36, 48, 54, [Mbit/sec]
    Tag: Traffic Indication Map (TIM): DTIM 1 of 1 bitmap
    Tag: Country Information: Country Code US, Environment Unknown (0x04)
    Tag: Power Constraint: 0
    Tag: RSN Information
      Tag Number: RSN Information (48)
      Tag length: 24
      RSN Version: 1
      Group Cipher Suite: 00:0f:ac (Ieee 802.11) AES (CCM)
      Pairwise Cipher Suite Count: 1
      Pairwise Cipher Suite List 00:0f:ac (Ieee 802.11) AES (CCM)
      Auth Key Management (AKM) Suite Count: 2
      Auth Key Management (AKM) List 00:0f:ac (Ieee 802.11) WPA 00:0f:ac (Ieee 802.11) FT over IEEE 802.1X
        Auth Key Management (AKM) Suite: 00:0f:ac (Ieee 802.11) WPA
          Auth Key Management (AKM) OUI: 00:0f:ac (Ieee 802.11)
          Auth Key Management (AKM) type: WPA (1)
        Auth Key Management (AKM) Suite: 00:0f:ac (Ieee 802.11) FT over IEEE 802.1X
          Auth Key Management (AKM) OUI: 00:0f:ac (Ieee 802.11)
          Auth Key Management (AKM) type: FT over IEEE 802.1X (3)
      RSN Capabilities: 0x0028
    Tag: QBSS Load Element 802.11e CCA Version
    Tag: RM Enabled Capabilities (5 octets)
    Tag: Mobility Domain
```

Figure 6-15 *AKM in RSN IE in beacon of mixed-mode FT WLAN*

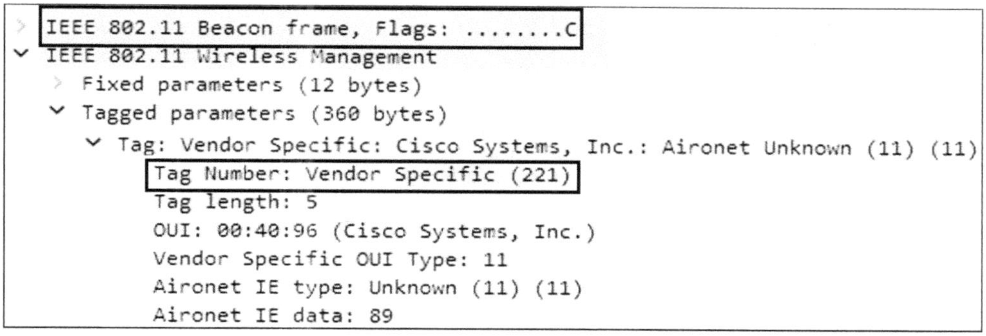

```
> IEEE 802.11 Beacon frame, Flags: ........C
v IEEE 802.11 Wireless Management
  > Fixed parameters (12 bytes)
  v Tagged parameters (360 bytes)
    v Tag: Vendor Specific: Cisco Systems, Inc.: Aironet Unknown (11) (11)
        Tag Number: Vendor Specific (221)
        Tag length: 5
        OUI: 00:40:96 (Cisco Systems, Inc.)
        Vendor Specific OUI Type: 11
        Aironet IE type: Unknown (11) (11)
        Aironet IE data: 89
```

Figure 6-16 *Aironet IE in beacon of mixed-mode FT WLAN*

- **Adaptive FT**: Adaptive FT is a unique solution offered by Cisco where Authentication Key Management in RSN IE does not advertise FT methods but only includes 802.11r MDIE in beacon and probe response frames. In addition, a specific bit on Aironet IE is set to indicate FT support. Legacy clients see only regular PSK or dot1x SSID advertised and can connect to the SSID. Some FT-capable clients like Samsung S10 devices and above, Apple devices running iOS 10 and later, iPadOS, and macOS 12 detect FT support even without FT AKM advertised in RSN IE and do a proprietary handshake to establish an 802.11r association.

  This mode allows for maximum compatibility for FT and non-FT clients while enabling FT-enabled clients to take advantage of FT fast secure roaming. The drawback is that FT-capable devices that are not Samsung or Apple and depend on AKM in RSN IE to do FT are forced to use legacy fast roaming methods. Adaptive FT is the recommended configuration on WPA SSIDs.

  Figure 6-17 shows the content of a beacon frame transmitted by an AP configured for a WLAN set to adaptive FT and where the security settings are only set to 802.1X AKM. The RSN IE shows only WPA security; MDIE is seen to indicate 802.11r support without claiming FT is configured on the SSID. Figure 6-18 shows Aironet IE in the same beacon frame set to a value that will change when WLAN is not configured for adaptive FT.

  The type of the client that roams using FT is reflected as 802.11R. On the C9800 GUI, navigate to **Monitoring > Wireless > Clients > Mobility History**, as shown in Figure 6-19. The equivalent CLI is **#show wireless client mac-address <mac-address> mobility history**.

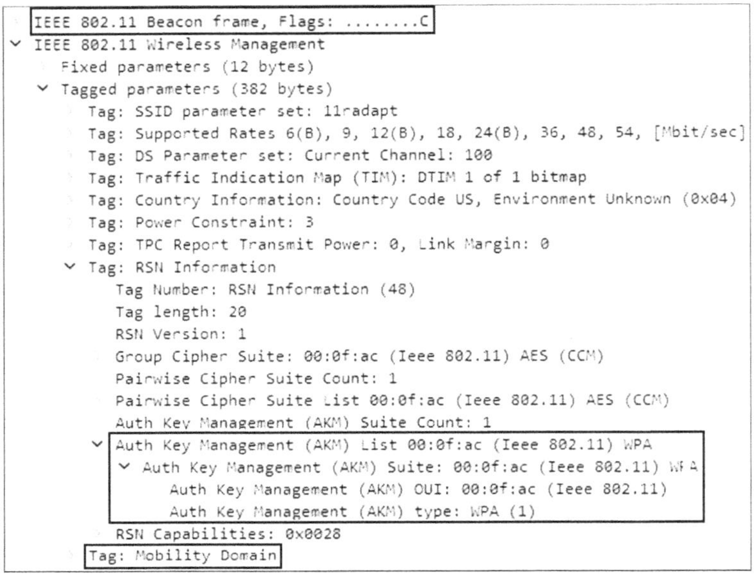

```
> IEEE 802.11 Beacon frame, Flags: ........C
v IEEE 802.11 Wireless Management
  > Fixed parameters (12 bytes)
  v Tagged parameters (382 bytes)
    > Tag: SSID parameter set: 11radapt
    > Tag: Supported Rates 6(B), 9, 12(B), 18, 24(B), 36, 48, 54, [Mbit/sec]
    > Tag: DS Parameter set: Current Channel: 100
    > Tag: Traffic Indication Map (TIM): DTIM 1 of 1 bitmap
    > Tag: Country Information: Country Code US, Environment Unknown (0x04)
    > Tag: Power Constraint: 3
    > Tag: TPC Report Transmit Power: 0, Link Margin: 0
    v Tag: RSN Information
        Tag Number: RSN Information (48)
        Tag length: 20
        RSN Version: 1
      > Group Cipher Suite: 00:0f:ac (Ieee 802.11) AES (CCM)
        Pairwise Cipher Suite Count: 1
      > Pairwise Cipher Suite List 00:0f:ac (Ieee 802.11) AES (CCM)
        Auth Key Management (AKM) Suite Count: 1
      v Auth Key Management (AKM) List 00:0f:ac (Ieee 802.11) WPA
        v Auth Key Management (AKM) Suite: 00:0f:ac (Ieee 802.11) WPA
            Auth Key Management (AKM) OUI: 00:0f:ac (Ieee 802.11)
            Auth Key Management (AKM) type: WPA (1)
      > RSN Capabilities: 0x0028
    > Tag: Mobility Domain
```

Figure 6-17 *AKM in RSN IE of beacon of adaptive FT WLAN*

```
> IEEE 802.11 Beacon frame, Flags: ........C
v IEEE 802.11 Wireless Management
  > Fixed parameters (12 bytes)
  v Tagged parameters (382 bytes)
    > Tag: SSID parameter set: 11radapt
    > Tag: Supported Rates 6(B), 9, 12(B), 18, 24(B), 36, 48, 54, [Mbit/sec]
    v Tag: Vendor Specific: Cisco Systems, Inc.: Aironet Unknown (11) (11)
        Tag Number: Vendor Specific (221)
        Tag length: 5
        OUI: 00:40:96 (Cisco Systems, Inc.)
        Vendor Specific OUI Type: 11
        Aironet IE type: Unknown (11) (11)
        Aironet IE data: c9
```

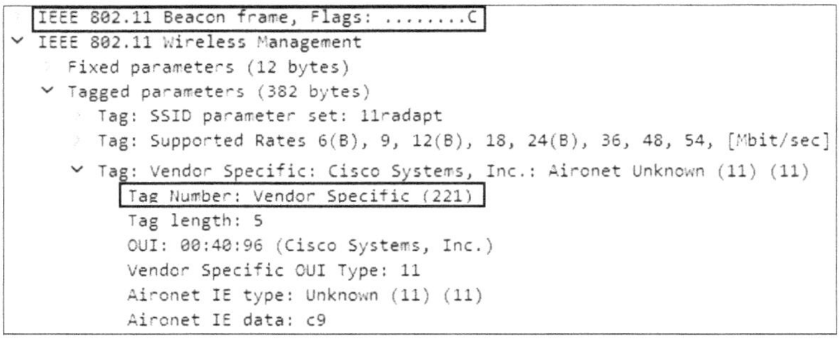

Figure 6-18 *Aironet IE in beacon of adaptive FT WLAN*

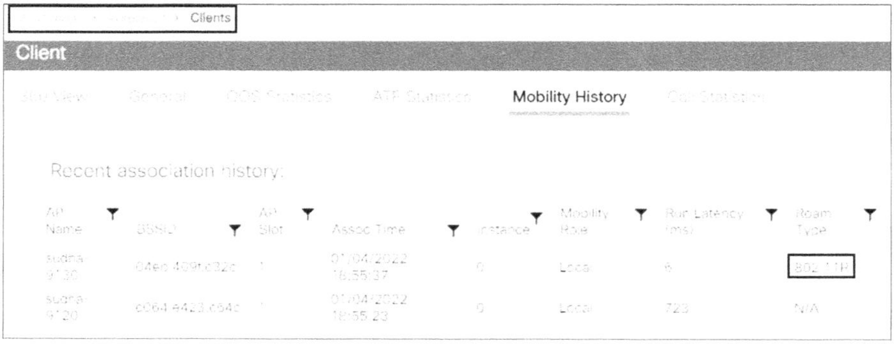

Figure 6-19 *Monitoring the mobility history of a client on the C9800 GUI*

Roaming Optimizations

The 802.11 standard has a couple of amendments whose goal is to optimize the client roaming experience. Specifically, with 802.11k, the network provides the client with a list of neighboring APs to roam to, reducing the need to perform off-channel scanning and hence the roaming time. The 802.11v standard has multiple components, but with 802.11v BSS Transition Management, the infrastructure assists the client in the roaming decision by sending it solicited or unsolicited roaming.

802.11k

The 802.11k enables 11k-capable clients to request a neighbor report containing information about APs that would be good candidates for roaming within the service set. This enables clients to avoid passive or active RF scanning before roaming. The C9800 supports a feature called *11k assisted roaming*, which creates and delivers an optimized neighbor list to the 802.11k clients. The 802.11k neighbor list is generated on-demand and can be different for two clients on different APs because the WLC would consider the individual client's RF relationship with the surrounding APs.

An AP configured for 802.11k broadcasts three information elements (IEs) in its beacon and probe responses. The presence of all three IEs indicates that the SSID is configured for 11k:

- **Country element:** This IE allows a client to identify the regulatory domain in which it is located and configure its PHY for operating in that regulatory domain.
- **Power Constraint element:** This IE allows the client to determine the local maximum transmit power in the current channel.
- **Radio Management (RM) Enable Capabilities element:** This IE is 5 octets long. In a beacon or probe response, bit 1 informs of the APs' 11k capability that it can provide a neighbor list. In an association request, this same bit 1 signifies the client's request for a neighbor list.

Steps involved in 11k assisted roaming are as follows:

1. When an 11k client associates with an AP advertising 11k capability, it sends an 802.11 action request frame with Category IE set to Radio Measurement (5) and Action IE set to Neighbor Report Request (4).
2. The C9800 searches its Radio Resources Management (RRM) neighbor list for a list of RF neighbors on the same band as the AP to which the client is currently associated.
3. The C9800 uses RSSI to determine the top six neighbors after adjusting RSSI based on whether the client was seen by a neighboring AP and accounting for the adjustable floor bias for the neighboring AP on same floor as the current associated AP. Note that the default floor bias is set to 15 dBm but can be configured up to 25 dBm.
4. By default, the neighbor list contains neighbors only in the same band as the client, but a dual-list configuration option on C9800 allows the 11k response to return neighbors in both bands.

5. The returned neighbor list includes the BSSID, channel, and RSSI of the neighboring radios that facilitate roaming.

For non-11k clients that do not send neighbor list requests, you can still avail this roaming optimization by enabling prediction optimization. When it is enabled, after every client association/reassociation, a prediction-based neighbor list is generated using the same criteria as described in step 3 for each non-11k client, and a resulting neighbor list is stored in the mobile station software data structure on the C9800. The number of neighbors in the prediction list is set to 3 by default and can be configured up to 6.

The clients send probes prior to any association or reassociation, and different APs see probes with different RSSI values. To accurately predict the AP that the client will roam to, a neighbor list needs to be created with the latest probe data and is therefore run at each association or reassociation for each non-11k client. Further, the C9800 discourages association to APs that are not in the neighbor list by denying association requests to such APs. By default, the maximum number of associations denied is 5 but can be configured up to 10.

To enable 802.11k assisted roaming, along with dual list, which returns preferred neighbors from both bands and enables prediction roaming on WLAN, navigate on the C9800 GUI to **Configuration > Tags and Profiles > WLANs Profile > Select WLANs > Advanced > Assisted Roaming (11k)** and select appropriate fields, as shown in Figure 6-20.
To configure the parameters such as floor bias, maximum associations denied, and minimum number of entries in the prediction-based neighbor list on the C9800 GUI, navigate to **Configuration > Wireless > Wireless Global > Assisted Roaming**, as shown in Figure 6-21.

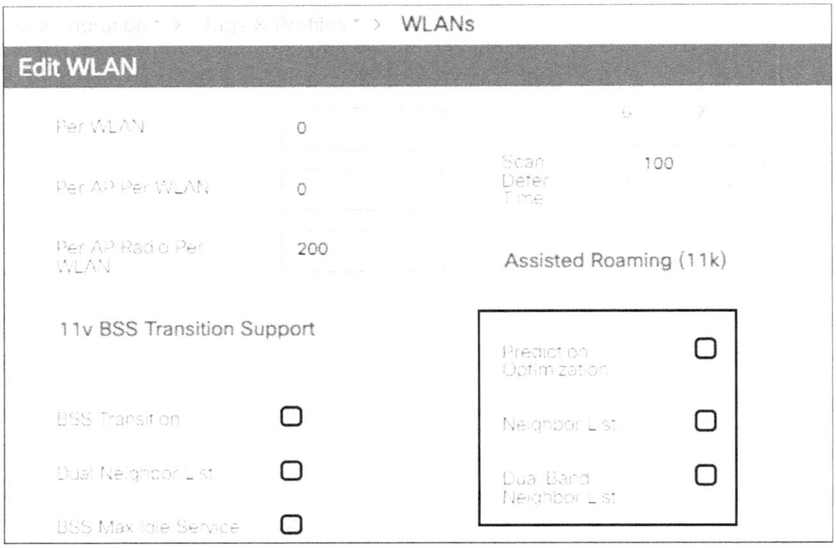

Figure 6-20 *Configuring an 802.11k on the C9800 WLAN Profile*

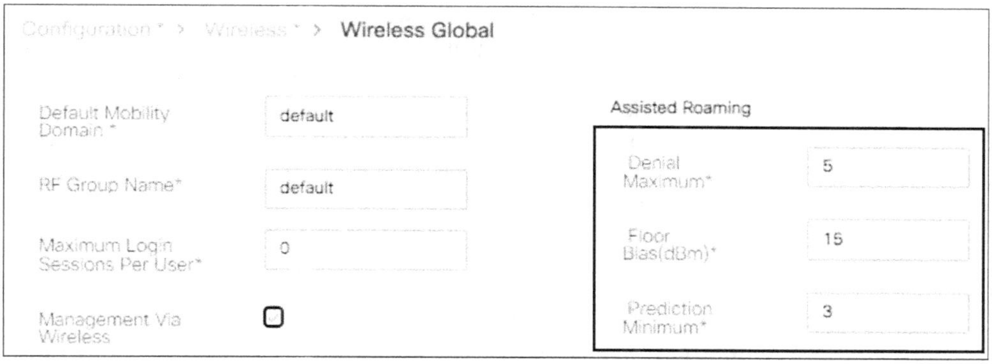

Figure 6-21 *Configuring assisted roaming parameters on the C9800*

802.11v BSS Transition

The 802.11v standard is an amendment of the IEEE 802.11 standards that incorporate two main enhancements to the wireless network management:

- **Network-assisted power savings:** Improves battery life of clients by defining the max idle period, which is the amount of time that a client can sleep without receiving any frame. A client is informed of this max idle period via association and disassociation frames.
- **Network-assisted roaming:** Enables the wireless infrastructure to suggest the client roam away from its current AP and provides the client the list of APs that it can roam to in the same extended service set (ESS).

The 802.11v BSS Transition Management messages are exchanged in three scenarios to get the client to roam:

- **Solicited request:** The client can send an 802.11v BSS Transition Management Query before roaming to seek out better options of AP to reassociate with. The C9800 and AP respond with a BSS transition management request, providing the list of candidate APs to roam to.
- **Unsolicited load-balancing request:** Load balancing is a feature that allows AP to load balances clients across APs on the same controller to avoid one AP being overloaded. When the client counts exceed the configured load-balancing threshold for an AP, any new clients trying to associate with the AP are denied with an association response with status 17 (AP busy). Typically, the denied clients keep trying to associate to the same loaded AP if, from the RSSI perspective, that AP is their best option. For example, consider 40 users in a conference room serviced by one AP. With an 802.11v BSS Transition Management query, a load-balancing failure can be handled more smoothly by providing clients with a list of candidate APs to roam to.

- **Unsolicited optimized roaming request:** The wireless clients are expected to scan RF and roam to AP with the highest signal. However, some clients have displayed a sticky behavior where they stay with the AP to which they are associated, even when a neighboring AP provides a stronger signal. This is referred to as a sticky client problem. To address this problem, the C9800 supports a feature called *optimized roaming* where the RSSI of the client data packets and data rate are monitored, and the client is proactively disassociated. The 802.11v BSS Transition Management Request enhances optimized roaming by informing the client of an imminent disassociation and providing a list of APs to roam to.

To configure 802.11v, navigate on the C9800 GUI to **Configuration > Tags and Profiles > WLAN Profile > Advanced > 802.11v** and select **BSS Transition,** as shown in Figure 6-22. By default, only AP neighbors in the band to which a client is associated are provided. However, you can configure the C9800 to gather a candidate AP list from both bands and share with the client.

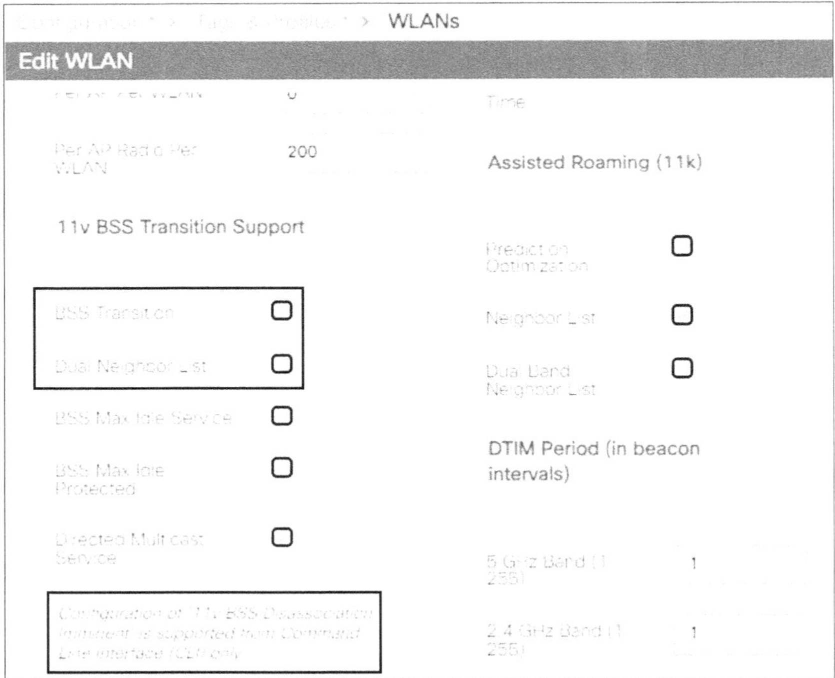

Figure 6-22 *Configuring an 802.11v BSS transition on the C9800*

An 802.11v BSS Transition Management Request sent by an AP to a client is only a suggestion. Clients can honor the suggestion or discard it. The C9800 provides a configuration option called *Imminent Disassociation* for you to force the clients to disassociate if the client does not reassociate with another AP within a defined window of time. Clients are informed of imminent disassociation via the 802.11v BSS Transition Management Request frame. Note that the request frame sent by AP in solicited request use case cannot leverage imminent disassociation. On the C9800, imminent disassociation can be configured only from CLI using **#bss-transition disassociation-imminent** under a specific WLAN profile.

Types of Client Roaming

When a wireless client associates and authenticates to an access point, the C9800 managing the AP creates an entry for the client in its client database. This entry includes the client's MAC and IP addresses, security context, quality of service (QoS) context, wireless LAN (WLAN) information, and details of the AP to which the client is associated. The C9800 uses this information to forward frames and manage traffic to and from the wireless client. At each client roam, this context of the client needs to be maintained.
So far in this chapter, the methods we have discussed look to reduce the time taken for the client to identify the target AP and to get the security context, including authentication and encryption. The last step in the onboarding process is the client getting an IP address so that it can start forwarding traffic. This can happen dynamically via the Dynamic Host Configuration Protocol (DHCP) or with a static IP configuration, in which the client sends an Address Resolution Protocol (ARP) packet claiming its own IP address. Here, you learn about the distinct types of roaming that can happen in a deployment and how the client IP context is maintained across roams.

When a client roams across two APs, depending on whether the APs are connected to the same or different WLCs, you can have two types of roaming:

- **Intra-controller roaming:** When the two APs are registered to the same C9800
- **Inter-controller roaming:** When the two APs are registered to two different C9800s

Intra-Controller Roaming

In intra-controller roaming, the client roam is limited to APs on one C9800. However, depending on tags and profiles applied to the APs, three types of intra-controller roaming can occur:

- Intra-WLC, Intra-WNCd roaming
- Intra-WLC, Inter-WNCd roaming
- Intra-WLC, Inter-policy profile roaming

Intra-WNCd Roaming (Same Site Tag, Same Policy Profile)

On the C9800, the Wireless Network Control daemon (WNCd) is the process responsible for terminating Control and Provisioning of Wireless Access Points (CAPWAP) tunnels from AP and processing all control/management traffic. C9800 form factors like C9800-CL (medium, large), C9800-40, and C9800-80 have multiple WNCd instances. APs are load-balanced across multiple WNCd instances based on the site tag to which they belong. Site tags can be user-defined custom tags or the system-defined default-site-tag. It is recommended to configure custom site tags, intentionally grouping RF-proximate APs. Each custom site tag with an AP on it is mapped to a single WNCd instance. So, when a client roams between two APs that are RF neighbors, it is likely a roam within the same WNCd instance, as shown in Figure 6-23. This roam across APs on the same WNCd instance or within the same site tag is referred to as Intra-WNCd roam. The exception to this rule is when APs are assigned to the default-site-tag. In this case, the APs are distributed across the multiple WNCd instances in round-robin fashion, and client roaming within the default-site-tag will be across multiple WNCds and hence not efficient.

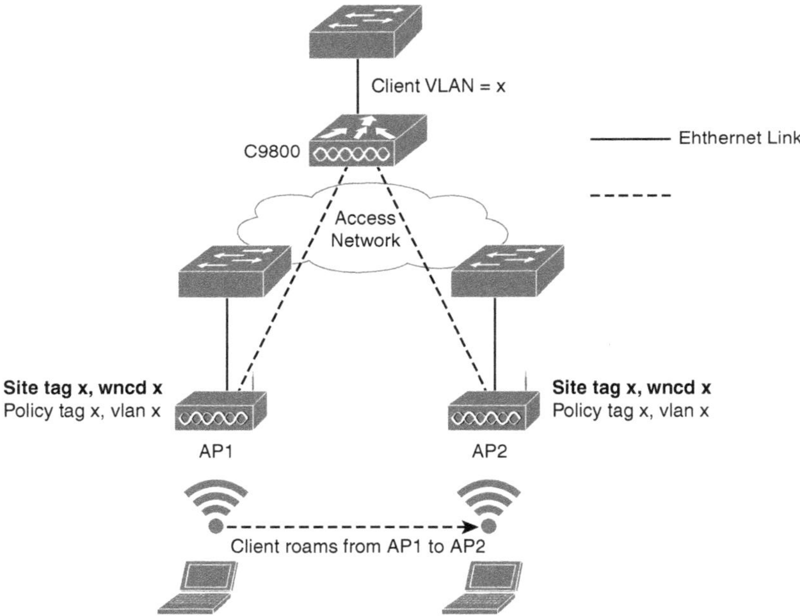

Figure 6-23 *Intra-controller Intra-WNCd roaming on the C9800*

In the case of Intra-WNCd roaming, when the client roams, the C9800 simply updates the client database with the information of the new AP to which the client associated. The client retains its IP address, and the C9800 continues to forward traffic for the client. The 802.11r and all fast secure roaming optimizations discussed, including 11k and 11v, are supported with this roam. For C9800 form factors like C9800-CL (small) and C9800-L, all roams are intra-WNCd roams.

Inter-WNCd Roam (Different Site Tags, Same Policy Profile)

When you're grouping RF-proximate APs into one custom site tag, bear in mind that the maximum number recommended on each site tag for local mode deployment is about

500 APs. In the case of FlexConnect deployment, the maximum number of APs allowed in the site tag is 100 if secure fast roaming is a requirement.

Depending on the size of deployment, it is possible to have APs that are RF neighbors but belong to different custom site tags. When a client roams across two APs belonging to different custom site tags, the roam is across two WNCd instances, as shown in Figure 6-24. This is referred to as *Inter-WNCd roam*. If you use the default-site-tag, the APs are distributed across WNCd instances in a round-robin fashion. So, at site boundaries, roams across RF neighbors are more likely to be inter-WNCd roams. With inter-WNCd roaming, fast secure roaming is supported because the PMK cache is shared across WNCd instances. Starting with release 17.6.1, 802.11k and 802.11v are also supported when roaming across WNCd instances because the AP neighbors are calculated independently from the WNCd instance.

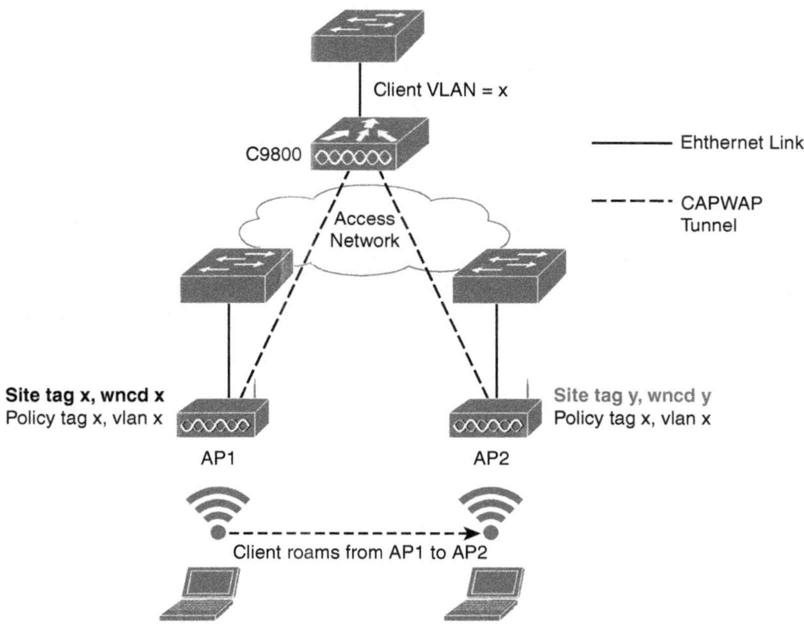

Figure 6-24 *Intra-controller inter-WNCd roaming on the C9800*

Intra-WLC Roam (Same Site Tag, Different Policy Profile)

Besides the site tag, APs are also mapped with policy tags. A policy tag maps WLAN profiles to policy profiles. A WLAN profile defines SSID and its security settings, whereas the policy profile is responsible for the policies that apply to the SSID, including client VLAN, QoS policy, access control lists (ACLs), and many more. The same WLAN can be mapped to a different policy profile via a different policy tag. Different APs can be mapped to two different policy tags on the same C9800, irrespective of their site-tag configuration. This flexibility comes in handy to customize WLANs and policies independent of site boundaries.

When a client roams within a WLAN across two APs mapped to different policy tags, two scenarios may occur:

- **Same WLAN profile, same policy profile (name and content), but different policy tags:** Consider a deployment with two buildings that need some SSIDs to be broadcast in both buildings but restrict some SSIDs to individual buildings. This configuration is achieved by creating two policy tags, one per each building. The roaming of a client on the WLAN profile mapped under two different policy tags but the same policy profile, both in terms of name and content, will be seamless. In other words, tag name or tag string does not influence roaming.
- **Same WLAN profile, different policy profile (either name or content):** Consider a deployment that broadcasts the same SSID on two floors of a building but maps different VLANs to this SSID on each floor to manage the desired subnet size. The most common configuration approach to meet this design requirement is to create two separate policy profiles, one per floor, each using a client VLAN for that building. Then these policy profiles are mapped to the same SSID using two different policy tags, and the applicable policy tags are applied to the group of APs deployed in the specific floor. Figure 6-25 depicts this type of inter-policy roaming setup. When a client roams across two policy profiles with two different VLANs, the client goes through a full authentication and DHCP. The reason is that, by design, a different policy profile configuration usually means that the client policy needs to be reevaluated. Starting with Cisco internetwork operating system (IOS-XE) release 17.3, this same roam scenario across policy profiles can be made seamless by configuring **#wireless client vlan-persistent.** In addition to VLAN, 17 other parameters can be different on the two policy profiles but would still result in seamless fast secure roaming.

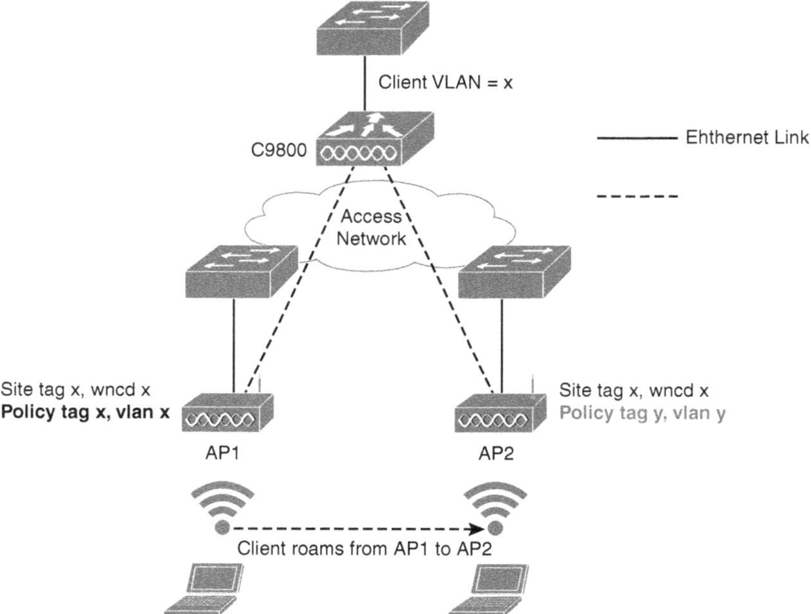

Figure 6-25 *Intra-controller inter-policy profile roaming on the C9800*

Inter-Controller Roaming

In a local mode deployment of multiple C9800s, clients can roam across APs that are registered to different C9800s; this is called *inter-controller roaming*. For clients to be able to roam across controllers, the C9800s must be aware of each other. This is referred to as the *mobility domain* created by configuring each wireless controller as a mobility peer in a full-mesh configuration. For each mobility peer, you need to enter the MAC address, the IP address of the Wireless Management Interface (WMI), and mobility group name. On a C9800 local mode deployment, a mobility group defines the boundary for sharing security context and PMK cache of the wireless clients between controllers. In other words, fast secure roaming occurs within a mobility group for local mode. Clients can also roam between two local mode C9800s in different mobility groups, provided they are defined in each other's mobility domain or list. However, this is a case of inter-mobility-group roaming, and the client undergoes a full authentication while keeping its IP address.

Depending on the client VLAN to which the client belongs, two types of inter-controller roaming can occur. Note that the operational VLAN can be assigned to the client either statically via C9800 configuration in the policy profile or dynamically via AAA override.

Layer 2 Roaming

If the C9800 that the client is roaming to supports the same operational VLAN as the C9800 that the client is roaming from, the client roam is Layer 2. Note that you need to make sure the client VLAN is forwarded on the C9800 uplink and supported by the upstream wired networks to ensure traffic does not blackhole. In this case, when the client roams to an AP registered to a new C9800 other than the one it is on, the new C9800 exchanges mobility messages with the original C9800, and the client database entry, with all its context, is moved to the new controller and deleted from the original WLC, as shown in Figure 6-26. In the example note the different policy tag names on the two C9800s, but this is irrelevant in the context of inter-controller mobility. Any new security context and associations are established as needed, and a client database entry on a new C9800 is updated to point to the new AP. The roam is seamless, and the end user is unaware of the roam.

Note that the C9800 uses the VLAN ID only to determine the type of roam. If a deployment has two C9800s configured with the same VLAN ID but the VLAN is mapped to two different subnets, then the roam across those two C9800s is still treated as Layer 2 roam. The client context is moved with the client holding onto its IP address, and all client traffic post-roam gets blackholed until an idle timeout or a session timeout or the user triggers client deletion and fresh association to get a new IP address. Because this defeats the purpose of seamless roaming, such a deployment is not recommended.

Layer 3 Roaming

If the C9800 that the client is roaming to supports a different operational VLAN than the C9800 it is roaming from, the client roam is Layer 3. Mobility messages are exchanged between the two C9800s when the client roams, as is the case with Layer 2 roaming. However, the client entry is not moved to the new C9800, but a copy of the client entry is created on the roam to C9800. Further, the client entry is marked as "foreign" on the new roam to C9800 (also called *foreign WLC*) and as "anchor" in the roam from WLC (called *anchor WLC*). The client retains the IP address it received when it was associated with the original C9800. All the client traffic is tunneled from the Foreign C9800 to the Anchor C9800, which then forwards it to the wired infrastructure, as shown in Figure 6-27.

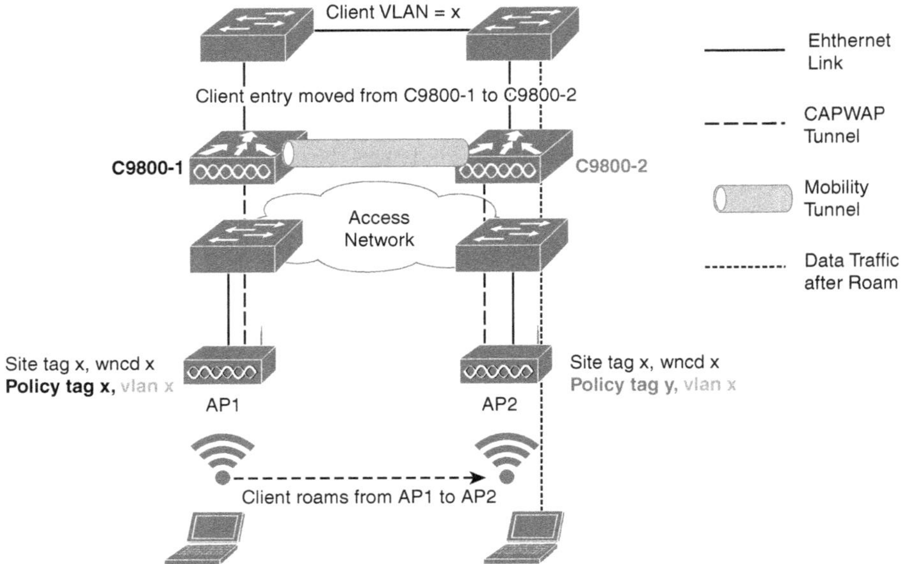

Figure 6-26 *Layer 2 CAPWAP Mobility Tunnel C9800*

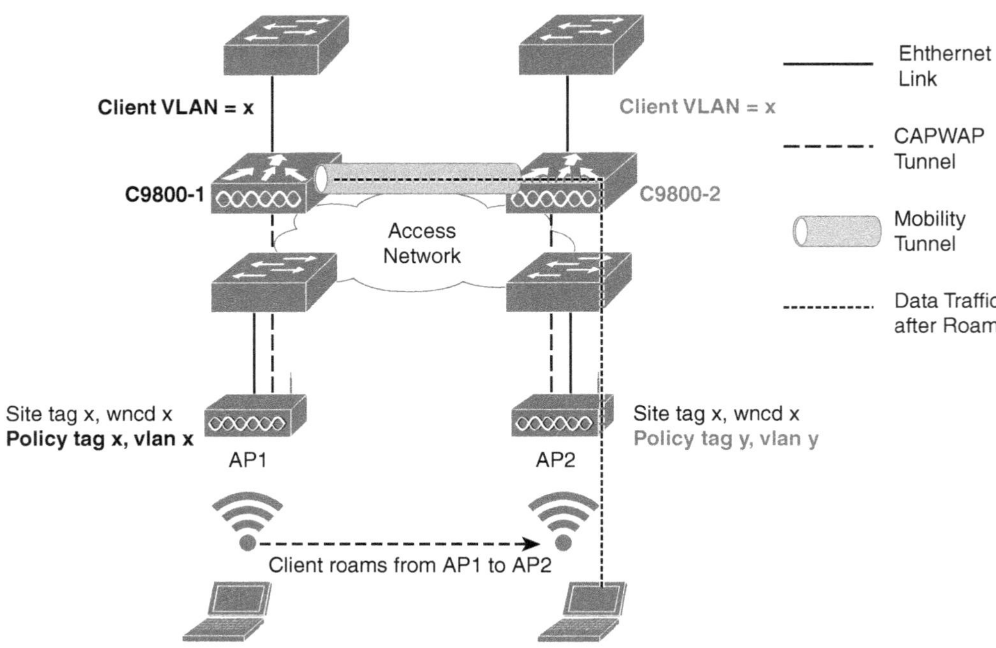

Figure 6-27 *Layer 3 inter-controller roaming on the C9800*

Note that in the case of Layer 3 roaming, if ACL and QoS profiles are in use, either due to configuration on a policy profile or obtained via AAA override, the QoS policy gets applied on the Foreign C9800, whereas the ACL policy gets applied on the Anchor C9800.

Static IP Client Mobility

If clients are assigned static IP addresses and they roam across C9800s or across policy tags where the same subnet as the static IP is not present, the clients are not able to connect to the wireless network. Enabling this feature anchors the static IP client traffic back to the controller whose wired infrastructure supports the subnet of the static IP.

This feature works only in centrally switched SSIDs and when the client is associated with Open, PSK, or dot1x SSID. It mandates that the feature be enabled on all the C9800s in the mobility list. It is mutually exclusive with the auto anchoring feature and IRCM with AireOS WLC. To configure, on the C9800 GUI, navigate to **Configuration > Tags and Profiles > Policy Profile > Mobility** and select **Static IP Mobility**, as shown in Figure 6-28.

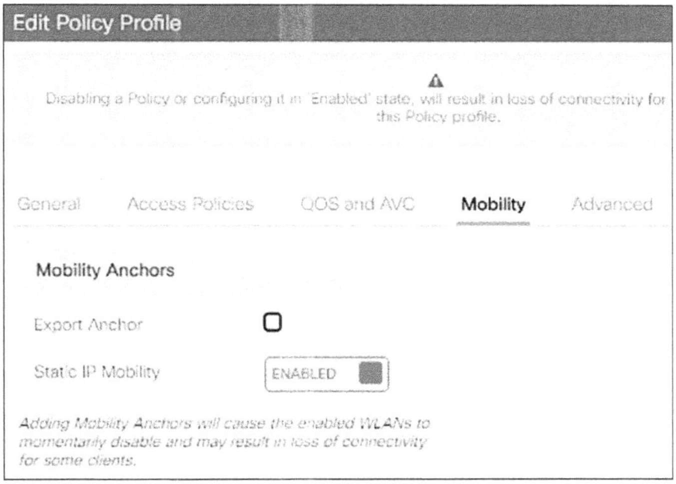

Figure 6-28 *Configuring Static IP Mobility on the C9800*

Auto-Anchor Mobility (Guest Tunnel)

When you need to provide wireless guest access in an enterprise deployment, one of the major design requirements is to provide segregation of guest traffic from all other VLANs. Auto anchoring refers to an intentional way of tunneling clients on a WLAN to a specific anchor WLC, which is then responsible for forwarding the traffic out to the wired infrastructure. In the case of guest access, this anchor WLC is usually located in the DMZ area of the network, thereby achieving the desired segmentation. This feature is sometimes referred to as *guest tunneling*, which is a misnomer. Wireless guest access is the more common use case for auto anchoring, but the feature is not limited to wireless guest auth SSID. Auto anchoring requires that these parameters match between foreign and anchor WLC:

- WLAN profile > Security settings
- Policy profile >DHCP settings
- Webauth parameter-map type

Auto anchoring can be enabled as shown in Figure 6-29 from the C9800 GUI by navigating to **Configuration > Tags and Profiles > Policy Profile > Mobility**, then selecting **Export Anchor**, and finally selecting **Anchor IP** from the available list. Note that the foreign WLC points to the anchor WLC wireless management IP, whereas the anchor WLC points to itself.

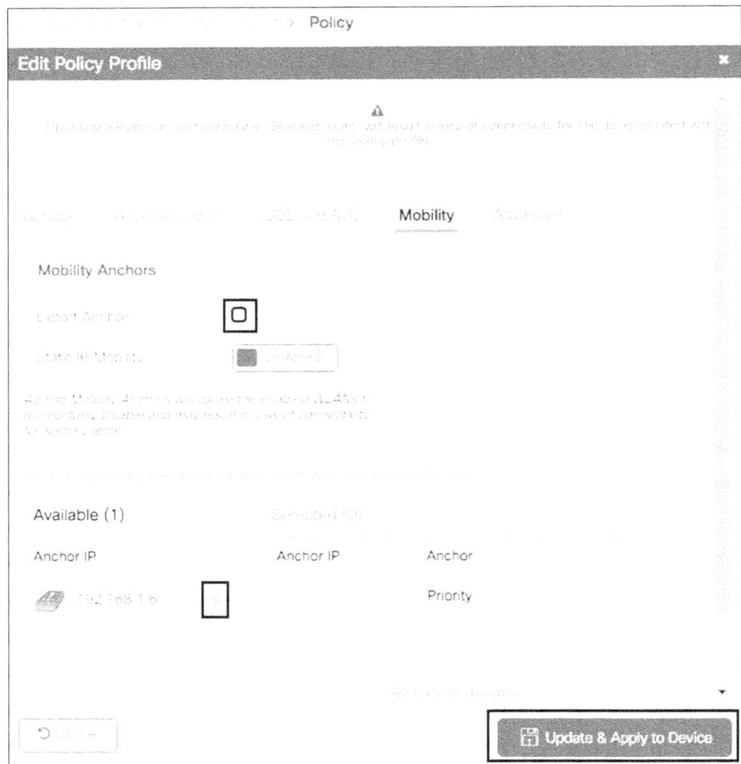

Figure 6-29 *Configuring auto anchoring on the C9800*

Configuring Secure Mobility Tunneling on a C9800

When two C9800s are defined in each other's mobility list, they communicate using CAPWAP over User Datagram Protocol (UDP) ports 16666 and 16667 to establish a secure tunnel between each other. The control channel of the tunnel is always Dynamic

Transport Layer Security (DTLS) encrypted, whereas the data channel can be optionally encrypted through explicit configuration. This is called *secure mobility*, and this tunnel is used to exchange mobility messages like mobility announce, mobility handoff, and many more to provide seamless client roaming. Each C9800 can have a maximum of 72 members in its mobility domain, across all mobility groups. However, within a mobility group, only 24 WLCs are allowed. On the C9800, the mobility group is identified by a mobility group name that is part of the Day 0 configuration wizard. If no custom mobility group name is defined, the C9800 boots up with the mobility group name "default." You can edit the mobility group name anytime on the C9800 GUI by navigating to **Configuration > Wireless > Mobility > Define Mobility Group Name**, as shown in Figure 6-30.

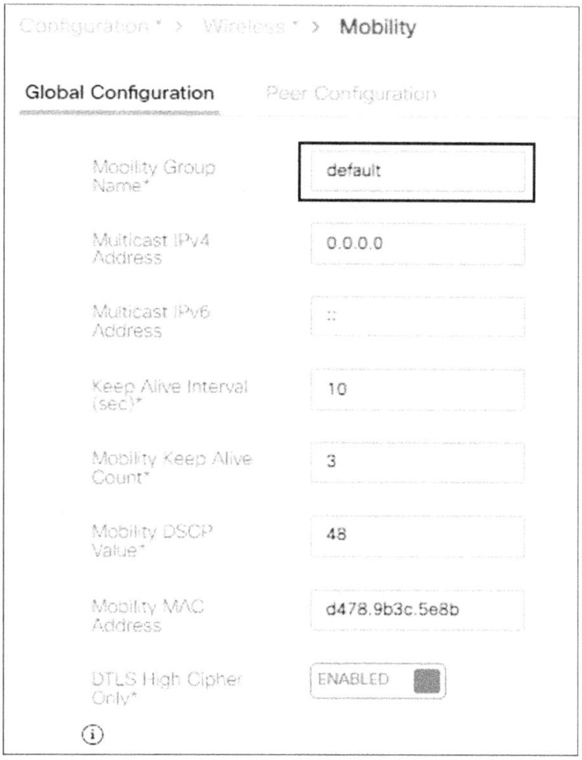

Figure 6-30 *Configuring secure mobility on the C9800*

The following configuration fields are relevant:

- **Mobility Group Name:** It can be a custom string or "default."
- **Multicast IPv4 Address:** By default, the mobility tunnel uses CAPWAP unicast so the C9800 must create multiple copies of mobility announce messages matching the number of mobility peers. However, mobility messages can be exchanged over the CAPWAP multicast tunnel by defining the multicast group IPv4 address to which all the C9800s subscribe. You need to enable it on all the WLCs in the mobility domain.

Each C9800 subscribed to the multicast group acts as both sender and receiver for this mobility multicast group IP.

- **Multicast IPv6 Address:** The mobility tunnel uses IPv6 CAPWAP multicast tunnels.
- **Keep Alive Interval:** Keepalives are exchanged between mobility peers to monitor the health of the tunnel. This field defines time between keepalives sent to peers. By default, this is set to 10 seconds but supports a range of 1 to 30 seconds.
- **Mobility Keepalive Count:** This field dictates the number of keepalive retries before a peer is declared DOWN.
- **Mobility DSCP Value:** Mobility keepalives on the C9800 are sent with a Differentiated Services Code Point (DSCP) of 48, which prioritizes the mobility traffic but can be modified to any value from 0 to 63.
- **Mobility MAC Address:** This field is populated, by default, to the burned-in MAC address of the WMI. It is mandatory to configure this field on a C9800 pair configured for stateful switchover (SSO) to ensure mobility tunnels do not drop in case of a failover.
- **DTLS High Cipher Option Only:** By default, the DTLS mobility tunnel supports Elliptic Curve Diffie-Hellman Ephemeral (ECDHE)/Galois Counter Mode (GCM) along with the AES128-SHA cipher suite. This field allows you to disable weaker cipher suites like AES128-SHA.

To define mobility peers for the C9800 in its mobility domain or list, navigate to **Configuration > Wireless > Mobility > Peer Configuration** and enter the MAC address, IPv4/IPv6 address, and Mobility Group Name of the peer C9800 obtained from the **Mobility > Global Configuration** tab of the peer, as shown in Figure 6-31. This configuration needs to be done on both the local C9800 and peer C9800 and repeated for all the peers in the mobility domain.

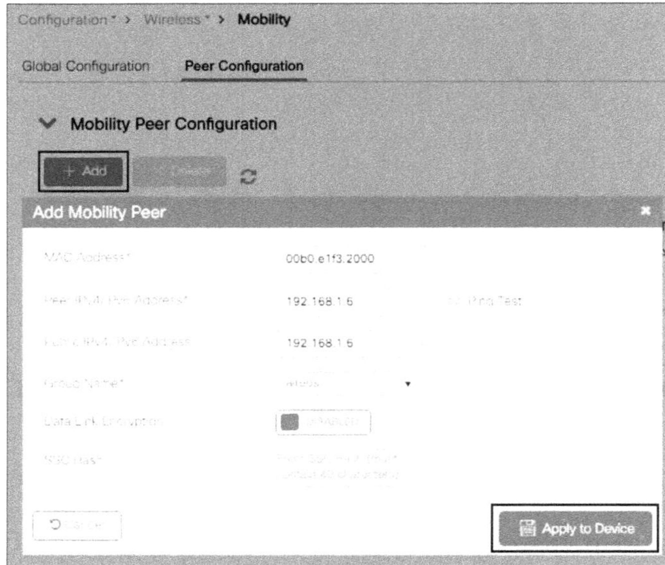

Figure 6-31 *Configuring mobility peers on the C9800*

Note that two other configuration fields are available:

- **Data Link Encryption:** Only the mobility control tunnel is DTLS encrypted by default. This field allows you to enable DTLS encryption for the mobility data tunnel. This configuration needs to be enabled on both the local and peer for encryption and decryption to occur.
- **SSC Hash:** The DTLS tunnel uses a manufacturer installed certificate (MIC) in the case of hardware form factors of C9800. However, virtual form factor C9800-CL has a self-signed certificate (SSC). The SSC hash is used as an additional validation when peering with the C9800-CL.

You can verify mobility tunnel status by navigating to **Configuration > Wireless > Mobility > Peer Configuration**, as shown in Figure 6-32. The equivalent CLI is **#show wireless mobility summary.**

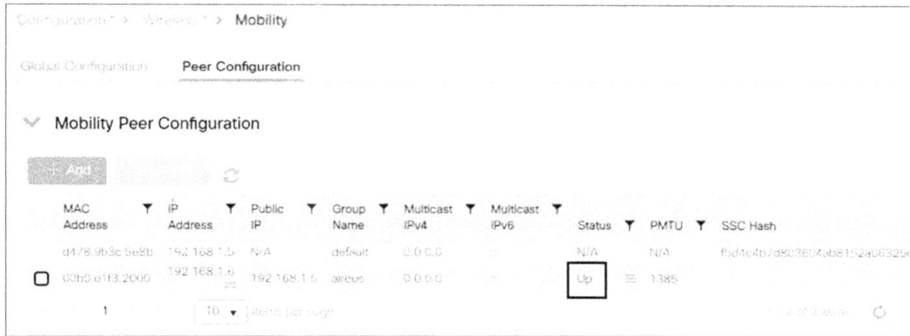

Figure 6-32 *Monitoring mobility tunnel status on the C9800*

You can view mobility tunnel statistics by navigating to **Monitoring > Wireless > Mobility.** The CLI equivalent to see the statistics is **#show wireless stats mobility** and **#show wireless stats mobility messages**

C9800 to AireOS Inter-Release Controller Mobility (IRCM)

A C9800 running IOS-XE is termed a *next-generation WLC*. By this definition, the term *legacy WLCs* refers to *AireOS WLCs*. It is a common requirement to have AireOS and C9800 WLCs coexist because the network is migrated from AireOS to IOS-XE. In this hybrid deployment, it is expected to have clients be able to seamlessly roam across these WLCs that differ in hardware and software. This is referred to as Inter-Release Controller Mobility.

AireOS also supports mobility and client roaming. AireOS WLCs use UDP port 16666 for the control tunnel but use IP protocol 97 (that is, Ethernet over IP, or EoIP, tunnels) for the mobility data tunnel. To peer with a C9800, the AireOS WLC needs to support the secure mobility feature, available starting with AireOS release 8.8.111.0. The AireOS WLCs that are supported in these releases are Cisco 3504, Cisco 5520, and Cisco 8540.

However, a large deployment base exists for Cisco 5508 and Cisco 8510. To accommodate these deployments, a special 8.5 IRCM code is available. Note that other AireOS WLCs like Cisco 2504, Cisco WiSM2, and Cisco Flex7510 do not support the IRCM release and hence cannot create a mobility tunnel with C9800.

The important point that you need to remember is that when a client performs IRCM roaming, it is always Layer 3, irrespective of the client VLAN configured on either AireOS WLC or C9800. When the client first connects to the network, whichever WLC it lands on serves as the anchor WLC. This means that the client gets an IP address from the VLAN/subnet associated to the WLC that it first connects to. If the client VLAN/subnet is the same on both the WLCs, this works well. But if the VLAN/subnet is different on the two WLCs, you need to make sure that you don't have clients with static IP because that IP would be valid on only one of the WLCs. When guest tunnels are used with anchor and foreign WLCs, it is also possible to have an AireOS WLC serve as an anchor to the C9800 foreign, and vice versa. To define the AireOS WLC and C9800 as mobility peers, you need to use the same steps described for Figure 6-31. The only exception is that the AireOS WLC does not have secure mobility enabled by default and needs explicit configuration, as shown in Figure 6-33. Also note that AireOS WLCs mark mobility tunnel traffic with DSCP 0. However, this mismatch does not prevent a tunnel from coming up.

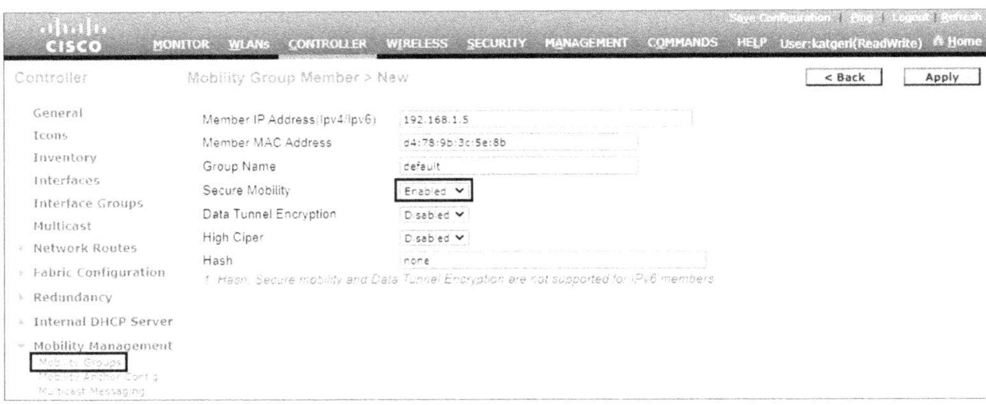

Figure 6-33 *Configuring secure mobility on AireOS WLC*

Summary

This chapter explains in detail the concept of seamless client roaming in an enterprise deployment. It describes the optimizations like fast secure roaming, 802.11k, 802.11v, and 802.11r-FT, which allow roaming to occur within milliseconds, thereby providing uninterrupted service, even for delay-sensitive traffic like voice. Deploying mobility and related roaming optimizations on the C9800, along with the coexistence of the C9800 and AireOS WLCs, is key to enable a successful adoption of C9800.

References

9800 Best Practices Guide: https://www.cisco.com/c/en/us/products/collateral/wireless/catalyst-9800-series-wireless-controllers/guide-c07-743627.html

IRCM Deployment Guide: https://www.cisco.com/c/en/us/td/docs/wireless/controller/technotes/8-8/b_c9800_wireless_controller-aireos_ircm_dg.html

Mobility Configuration Guide: https://www.cisco.com/c/en/us/td/docs/wireless/controller/9800/config-guide/b_wl_16_10_cg/mobility.html

Chapter 7

RF Deployment and Guidelines

Wireless networking relies a lot on the characteristics of the physical layer, the transmit medium, the so-called radio frequency (RF) layer. Other technologies do as well, but for example, in the case of wired Ethernet networks, the cable is an isolated and dedicated piece of wire and hence often considered to be reliable (and even if not, it becomes reliable if replaced) and to perform consistently. It also connects only two devices although this was not necessarily the case in the past. Wi-Fi relies on a shared medium: it's a half-duplex technology, and this brings many complexities. It definitely does not have consistent operations and performance even when considered in the same physical area because other wireless networks or even unrelated devices occupying the same frequency space would alter the propagation. Worried? You should not be. This chapter, although cannot cover all the theory of 802.11 wireless communications, does cover what you need to understand about the relevant features and settings of the Catalyst 9800 so that you can operate your network in the most efficient way.

Radio Resources Management (RRM) Concepts and Components

In a somewhat lazy abuse of language, wireless engineers often refer to radio frequency (RF) or radio resources management (RRM) issues when referring to anything pertaining to the Layer 1 signal transmission, reception, and decoding aspect. It is essential to understand all the pieces interacting together before making any configuration changes in this domain because the effects of a change are not always easy to measure directly.

Antennas and Signal Propagation

Wireless signal power is measured in decibel-milliwatt (dBm) or in milliwatts. While the first one is a logarithmic scale, the latter is a more usual scale with a linear progression. They are really equivalent. Take the measurement unit dBm: *dB* stands for decibel, which is a logarithmic scale, and the *m* is added when the scale is referenced to 1 milliwatt. You can use one or the other based on the context because the decibel scale makes it easier to calculate antenna gains and signal attenuation. Each time 3 decibels are added, it means the power in milliwatts is doubled. If you add 6 decibels, you quadruple the power and so on. Each time 10 decibels are added, the power is multiplied by 10 (which is why the scale is not linear but logarithmic). Remember that 0 dBm equals 1 mW of power; from there, thanks to the previous tricks, you can easily calculate that 10 dBm equals 10 mW and 20 dBm equals 100 mW.

The antenna gain is often mentioned in dBi units, which have a relative meaning (hence, the *i* referring to a comparison against an identical perfectly isotropic antenna) and not an absolute one like all the other scales. This has the advantage that it can be added and subtracted from the dBm scale. A gain expressed in dBi does not have any power equivalent per se because an antenna is a passive device requiring no electricity, however, an antenna can add gain, measured in dB. The decibel-based scale is much easier to use because the typical receiver signal strength would have to be expressed in billionths of a milliwatt, and it is also easier to use in calculations like free space path loss or antenna gain and connector loss calculations, which become simple additions and subtractions.

For example, if an AP transmits at 20 dBm (100 mW), and you are using an antenna providing 3 dBi of gain, the AP actually is said to have 23 dBm (200 mW) of equivalent isotropically radiated power (EIRP) in the direction where the antenna provides the best of its gain. This means that, because antennas do not perfectly radiate equally in all directions, it is as if the AP were transmitting at 23 dBm with an imaginary perfect antenna radiating equally in all directions. The reason is that antennas always provide a gain in some direction and less power in other directions. Antenna gain is considered passive: the signal is not dynamically boosted (which would require a power input) but simply radiated unequally from a directional standpoint, and therefore, some directions get a more powerful signal than others. Let's use another example to explain the concept of EIRP: if you had the perfect antenna (that is, isotropical radiator) radiating your signal equally in all directions so that the energy leaving the antenna is 20 dBm and then you replaced the antenna with one that focuses the signal in one direction and not the other, what would happen? In one direction, you would have no signal at all (real life is not so black and white, but this is an example), and in the other direction you would have double the amount of energy in a given location because all the same amount of energy (20 dBm) is radiated in fewer directions so that's more energy on the same area. While your transmit power has not changed and is still 20 dBm, you would have 23 dBm of EIRP in the direction where the signal is going, considering the antenna gave a 3 dBi gain in that direction.

Understanding the previous example is key to understanding the AP and antenna configuration. On the Catalyst 9800, the APs have transmit power (often called *txpower* and that refers to the power it sends to the antenna connector) and antenna gains configured and shown separately. You therefore need to add the two to figure out the actual EIRP over the air. The controller takes care of meeting regulations for you, and there is no risk of having your APs radiate too much power, provided that you configured the antennas correctly. The antenna gain can be configured for each AP on the 9800, and it does not have any particular effect apart from modifying the power levels to make sure not to radiate too much power. Indeed, the more gain you configure, the lower (in dBm) the value of each power level will be to keep meeting the maximum EIRP allowed in the country.

The AP has different power levels. Power level 1 is the maximum power the AP can use; the actual transmit power depends on the antenna and antenna gain configured. Power level 2 then sets itself to be half of the power level 1 value (which means 3 dBm less because it is a logarithmic scale). Each power level then is half of the previous power level and basically is 3 dBm less each time. If you look at the power level of the AP, it does not tell you how strong the signal will be for the clients (this requires knowing the antenna gain and direction), but if the AP is transmitting at its maximum or if it has got some margin from it. The power at which the client receives the signal depends on the distance of the client from the AP (it degrades gradually with the distance using inverse-square law) as well as the presence of any obstacle on the way that causes attenuation (that is, absorbs some of the signal energy) or reflects the signal in other directions.

Antennas provide a passive gain (otherwise, they would be called amplifiers rather than antennas) and are, typically "dumb" devices in the sense that the access point is not able to talk with the antenna and figure out its characteristics. However, Cisco has released a number of antennas called *self-identifying antennas (SIAs)* that contain an EEPROM that can be read by the AP to automatically configure the antenna type and gain. SIA antennas exist with RP-TNC connectors as well as DART4 or DART8 connectors. The SIA antennas help you, as administrator, by automatically configuring the right antenna gain and make sure the AP EIRP stays within legal boundaries. SIA antennas can be identified by their part numbers ending with *S*, such as AIR-ANT-2566P4W-RS. In 2021, Cisco pushed the smart antenna concept even further by releasing a stadium antenna with a dynamically configurable beamwidth so that each AP can configure its antenna to radiate the most optimal way.

Countries and Domains

Originally, there were only a handful of regulatory domains, or areas of the globe that agreed on similar requirements for wireless devices. The U.S. Federal Communications Commission (FCC) ruled that the -A domain applied in most of the Americas, and the European Telecommunications Standards Institute (ETSI) ruled that the -E domain be used in EMEA and even many countries in Asia, for example. Over the years, countries have started deviating from common regulatory domain specifications for channels and power. It is thus essential for the APs to broadcast their configured country code in their beacons to make it clear to the clients what channels and power they can use and under which conditions.

In the **Configuration > Wireless > Access Points** page, you can enable one or more countries in the **Country** tab. When the AP joins, its regulatory domain must be covered by at least one of the countries configured. In Figure 7-1, the administrator configured South Africa, Belgium, Canada, and Qatar, which enables the domain's -A, -B, -E, -M, and -N APs to join the controller.

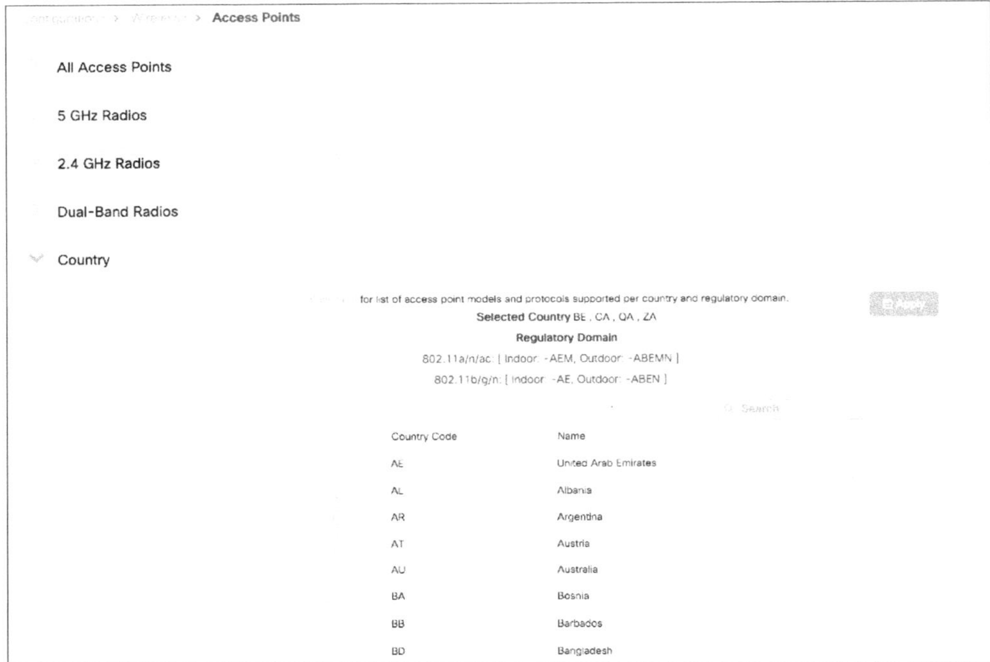

Figure 7-1 *The country configuration page*

The AP itself can be set to any of the configured countries that match its regulatory domain. As shown in Figure 7-2, you can set the AP to Belgium or South Africa because they both match the fact that the AP is -E, covering ETSI countries (C9130AXI-E in this example). The AP cannot be set to Canada because Canada corresponds to -A or -B domains.

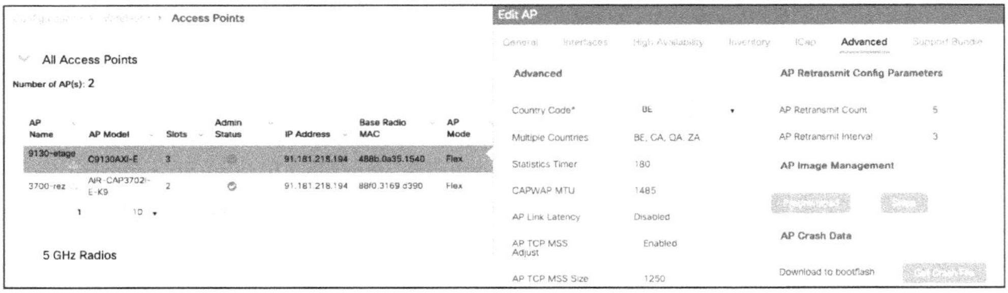

Figure 7-2 *Country selection for each AP*

The CLI to configure the country codes has recently (in 17.3) changed to

```
(config)#wireless country BE
(config)#wireless country CA
```

Notice that each country gets its own command-line statement, which makes it easier to add or remove a country compared to the previous command, **ap country**, which took a list of up to 20 country codes as arguments. The Catalyst 9800 now supports 200 country codes to be configured, which should be sufficient even for an international deployment. Country codes configured globally on the wireless LAN controller represent all the possible countries its APs can be assigned to. The WLC country list is thus a superset of the countries its joined APs are in.

The country code is also something that you can configure in an AP join profile to assign the same specific country to a group of APs belonging to the same site. A given AP can be assigned only one country (because it can logically be present in only one country at a time). This data allows the AP to advertise the right regulations, as can be seen in the beacons illustrated in Figure 7-3, where a country information element defines not only the country code but also allowed channels and power. With this information, the clients know what powers and channel plan they are supposed to use when connecting to this network.

```
▶ Tag: SSID parameter set: Darchis
▶ Tag: Supported Rates 18, 24(B), 36, 48, 54, [Mbit/sec]
▶ Tag: DS Parameter set: Current Channel: 64
▶ Tag: Traffic Indication Map (TIM): DTIM 1 of 1 bitmap
▼ Tag: Country Information: Country Code BE, Environment Unknown (0x04)
    Tag Number: Country Information (7)
    Tag length: 72
    Code: BE
    Environment: Unknown (0x04)
  ▶ Country Info: First Channel Number: 36, Number of Channels: 1, Maximum Transmit Power Level: 23 dBm
  ▶ Country Info: First Channel Number: 40, Number of Channels: 1, Maximum Transmit Power Level: 23 dBm
  ▶ Country Info: First Channel Number: 44, Number of Channels: 1, Maximum Transmit Power Level: 23 dBm
  ▶ Country Info: First Channel Number: 48, Number of Channels: 1, Maximum Transmit Power Level: 23 dBm
  ▶ Country Info: First Channel Number: 52, Number of Channels: 1, Maximum Transmit Power Level: 23 dBm
  ▶ Country Info: First Channel Number: 56, Number of Channels: 1, Maximum Transmit Power Level: 23 dBm
  ▶ Country Info: First Channel Number: 60, Number of Channels: 1, Maximum Transmit Power Level: 23 dBm
  ▶ Country Info: First Channel Number: 64, Number of Channels: 1, Maximum Transmit Power Level: 23 dBm
  ▶ Country Info: First Channel Number: 100, Number of Channels: 1, Maximum Transmit Power Level: 30 dBm
  ▶ Country Info: First Channel Number: 104, Number of Channels: 1, Maximum Transmit Power Level: 30 dBm
  ▶ Country Info: First Channel Number: 108, Number of Channels: 1, Maximum Transmit Power Level: 30 dBm
  ▶ Country Info: First Channel Number: 112, Number of Channels: 1, Maximum Transmit Power Level: 30 dBm
  ▶ Country Info: First Channel Number: 116, Number of Channels: 1, Maximum Transmit Power Level: 30 dBm
```

Figure 7-3 *Country information element in an AP beacon*

Cisco is now releasing APs belonging to the -ROW (Rest of World) domain, which allows for much more simplicity because the AP is physically the same and adapts its channel and power settings to the configured country. A lot of former regulatory domains are now folded into the -ROW domain and can therefore be software-configured when you assign the country to the AP. It is essential to configure the right country code on each of those APs for them to respect the proper regulations. The -ROW domain is supported starting with IOS-XE 17.6 and will progressively include more countries and be available on more AP models.

Challenging RF Environments

What is a challenging environment from a wireless perspective? A lot of environments could qualify for the term *challenging*: designing a network for a very high-density deployment where you may have hundreds of clients under the coverage of a single AP is indeed challenging. A completely different type of challenge would be designing a network in a factory with very few clients but a lot of metal surfaces around or dealing with a large open space and interflow coverage like you would find in a mall. Here, we highlight a few specific use cases of wireless network designs and some of the most typical traps and tricks that you need to be aware of and that may have an effect on your Catalyst 9800 wireless controller configuration.

Metal-Heavy Areas

Steel is probably one of the worst enemies of a wireless engineer. If you survey a factory filled with steel machinery (such as depicted in Figure 7-4), you may not see a lot of impact on your data if you are looking only for signal strength as a metric. Indeed, in a place filled with steel, the signal faces a lot of multipath effects as it rebounds differently on different steel surfaces. Many mobile clients still have just one antenna, or two at the most, and are therefore subject to multipath effects. Reflection is somewhat of an invisible adversary because you cannot prove it through RSSI (the signal is not weaker in strength due to the presence of metal reflections); you cannot see it in a spectrum analysis but you simply experience a lot of corrupted frames. The best workaround is to place the APs (and especially antennas) as far away as possible from pillars, corners, metallic surfaces, and other obstacles.

Figure 7-4 *A steel-heavy industrial environment where reflections cause problems*

High-Density Crowd Areas

Major events, whether they are in a conference center or an outdoor venue like a stadium, typically go for a high-density coverage because they expect of lot of attendees proportionally to the area surface. The problem rookie designers might forget is that bodies create a lot of attenuation: an empty venue behaves very differently from a filled venue. You will probably have to configure RF profiles to blast a transmit power that would be considered too high when the place is empty but perfectly fine when the place is filled with attendees.

Shielded Doors and Sudden Turns

The problems caused by shielded doors and sudden turns can happen in very different types of environments, but using the example of a shielded door makes it maybe a bit more obvious. Imagine a mobile device such as a smartphone that is located in a room, with an AP nearby, and this room is closed with a shielded door that creates a very strong attenuation (see Figure 7-5). The device therefore does not hear (or at least not at a good signal level) any APs located behind that door. Suddenly, a user decides to walk out and, within 2 seconds, opens the door and closes it behind. At this point, the Wi-Fi client cannot get a good signal to the AP located inside the room where it was connected previously, and it must scan in emergency to find out on which channel it can find APs on this side of the door. Depending on the algorithm and the channels on which the APs are located, this scan could take any amount of time from a few seconds to close to a minute. Beyond shielded doors, this type of situation happens in a corridor with thick walls when the corridor makes 90-degree turns and basically any time where a wall of furniture creates a higher-than-usual attenuation.

Figure 7-5 *A heavy door that can cause a lot of signal attenuation and typically breaks roaming*

The workarounds to this problem are to set APs around that area on non-DFS channels (that are typically scanned more quickly by client devices), to enable 802.11k support (it helps only if the APs are able to hear each other, which is not necessarily the case depending on the physical configuration of the space), and to add an AP in the problematic transition area.

Uneven Ceilings

Places like supermarkets, warehouses, or convention centers are typically huge indoor buildings with a natural ceiling that is very high, but often there are trusses or fake lower ceilings to place lighting, smoke detectors, and other devices (such as access points sometimes) but also to provide a more pleasant ceiling to the human eye, at a more reasonable height. Given these surroundings, technicians might place APs at different heights from each other, or simply very high compared to the ground. If APs are on a 30-foot (10-meter) high ceiling, they actually hear each other louder than they hear their clients, and the RRM algorithm does not perform as expected because it mostly focuses on how APs hear each other. If APs are at varying heights, depending on their antenna patterns, you may face similar RRM algorithm glitches because one AP might hear its neighbors (the ones broadcasting from above itself), but the opposite would not necessarily be true. The AP height entered on the map is currently not taken into account in the RRM algorithm, so the solution is to use RF profiles for groups of APs that are at similar height to better control their behavior.

Atriums

In the words of a lot of wireless engineers: "Atriums are the worst!" These areas (illustrated in Figure 7-6) are inside a building where many floors communicate through some kind of well (for light, air circulation, or simply visual effect), which means that the AP signal leaks between floors. This setup is especially problematic for location tracking because clients can connect to an AP from another floor very easily. Even just from a signal propagation standpoint, the atrium is a place of big adjacent channel interference. There is no one-size-fits-all solution, and placing APs far away from the atrium while maintaining coverage might be a solution to avoid too many APs being heard in the atrium area. For location tracking, APs in monitor mode may help.

Figure 7-6 *An atrium is an open space with no attenuation connecting several floors, a Wi-Fi interference nightmare*

Radio Resources Management (RRM)

Radio Resources Management is Cisco's state-of-the-art radio frequency management system that provides a systemwide view of your entire wireless RF environment. It does so by collecting a lot of data from all the APs and processing centrally through a series of intelligent algorithms. Each AP spends time listening within its environment and collecting some data on utilization statistics. This information drives several algorithms, and not only the ones directly related to RRM. APs gather information about neighbors and channel conditions. This information helps to define groups of APs that can hear one another, called *RF neighborhoods*. They are centrally managed by the RF Group leader WLC, which is the elected WLC in the network currently taking care of RRM settings and analytics. Next, we describe the basics of the RRM algorithms and how the system works.

Data Collection

An AP operates on a given channel, which may change over time but has to stay stable for a good while for the sake of client stability. While on the channel, the AP listens to the medium whenever it is not transmitting, and during these times it is very easy for the AP to collect statistics on the current channel it sits on without any effort. On top of that, it scans other channels very briefly (to not disrupt their currently connected clients) with the objective of figuring out which APs are nearby and what are the statistics of other channels (from a load or noise standpoint). This is the monitoring task of an AP (monitor mode APs do this full time).

APs also send Neighbor Discovery Protocol (NDP) messages when they are on other channels, to help other APs locate them in their neighborhood. These NDP messages are managed centrally from the WLC. They are sent to the special multicast address 01:0B:85:00:00:00 that all Cisco APs monitor, at the highest power allowed for the channel and at the lowest data rate supported in the band. This is to allow APs to figure out which other managed APs are around them, regardless of the power and data rate configuration on the WLC. An NDP packet contains the antenna details of the sending radio, the power the message was sent at, the channel it was sent on, the operating channel of the AP, optional encryption details, the IP address of the sender AP's RRM group leader WLC, the hashed RF group name, the radio slot ID, and a group ID. When an AP hears an NDP message while operating on its usual channel, it validates that the message comes from a member of its RF group (via the hash) and, if so, forwards the message along with the received channel and RSSI to the controller. Each WLC keeps a list of up to 24 neighbors for each AP radio, and this data is forwarded to the RF group leader regularly. The WLC can then compute out of this data for each target AP:

- **RX neighbors:** How this radio hears other radios
- **TX neighbors:** How other radios can hear this radio.

The NDP message exchange between APs basically allows the WLC to calculate the free space path loss (including walls and obstacles) between all the APs. The **RRM > General** pages (available for each band), as depicted by Figure 7-7, allow some configuration of this data collection. The Monitoring Channels section allows you to configure which channels are covered by the off-channel scan. You can either configure all channels (but NDPs are sent only on country-allowed channels, and the AP listens only on country unallowed channels), country channels, or only the channels configured in DCA. The latter can be useful for 2.4 GHz where you might want your APs to scan only channels 1, 6, and 11 (which are the commonly configured DCA channels in 2.4 GHz) and never spend any time on the channels in between because they probably have no value and little activity. The RRM Neighbor Discovery type field allows you to either have NDP packets sent as clear (the "transparent" setting) or encrypted (the "protected" setting).

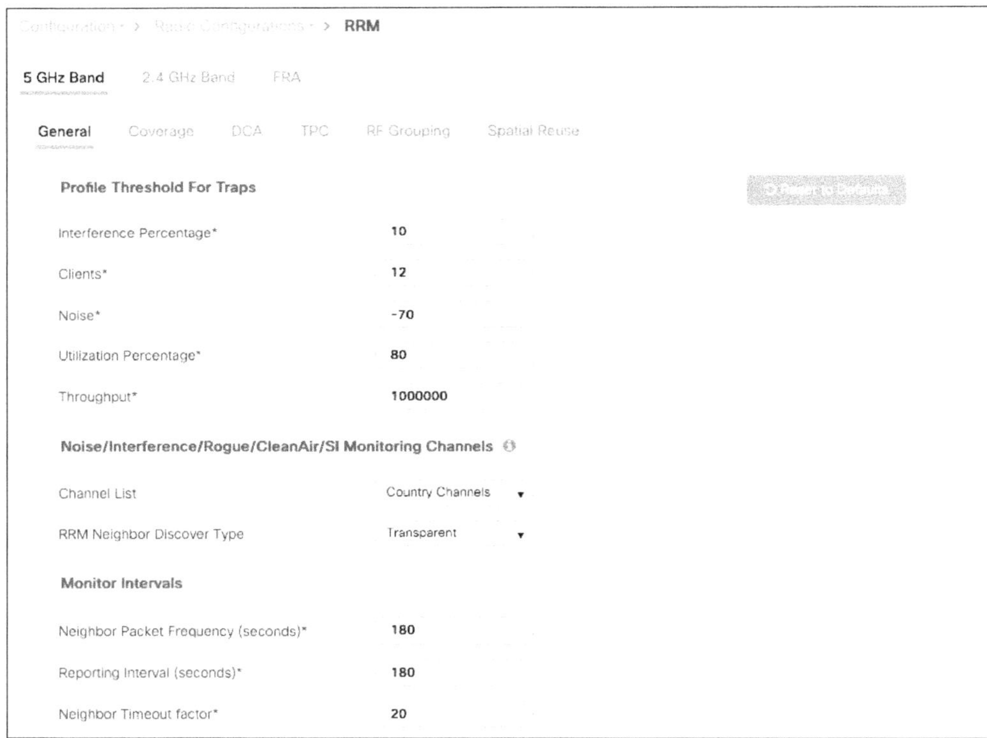

Figure 7-7 *Configuring RRM data collection settings*

The intervals at which the AP scans and sends NDP can also be configured. The default Neighbor Packet Frequency is 180 seconds, which means the AP goes over all the channels to monitor every 180 seconds, and depending on the number of channels you allowed, this defines the interval between two scans. If you configured 2.4 GHz to monitor only DCA channels and configured those to be only 1, 6, and 11, it means your APs go off-channel every minute to scan one of them. If your 5 GHz band is set to monitor country channels (say 20 channels for this example), that means your APs go off-channel every 9 seconds (180 seconds divided by 20). The reporting interval defines the cycle after which the AP has sent an NDP frame to every channel it has to monitor. In the given example of the default 180-second values, it means the AP goes off-channel twice (once for 50 ms for listening and once just for sending an NDP frame) for every channel configured in the monitor list. The timeout factor is the number of reporting intervals after which an AP will delete a given neighbor AP from its table if it didn't hear about it anymore (20 would mean the neighboring AP wasn't heard in the last 20 reporting periods, that is, 20 × 180 seconds = 1 hour). There are valid reasons why this timeout factor has to be relatively high like that: a busy network or a configured voice SSID (where off-channel scan defer is enabled) decreases the opportunities for the AP to go off-channel, and the AP might skip cycles and not scan a specific channel at all in the usual 3-minute interval. You typically do not want a neighbor to disappear from the list of AP neighbors in such a situation. The only reason an AP should be deleted from the neighbor list is if that AP went down or if the RF environment changed (for example, obstacles) and the two APs cannot physically hear each other anymore. It is also important to note than an AP cannot send an NDP frame on another DFS channel than the one it is on (and that it cleared from radar) unless a beacon or probe response from another AP is heard within the last 5 seconds before the transmission, which can cause some APs to not send any NDP frames at all on other DFS channels.

On the same page, the Profile Threshold for Traps section defines thresholds and conditions for sending SNMP traps. They do not have any operational effect on RRM apart from sending an SNMP message.

NDP forms the foundation of the understanding of the RF propagation in the specific deployment. It is very important for

- RF grouping
- Transmit Power Control (TPC)
- Flexible Radio Assignment (FRA)
- Rogue detection
- Dynamic Channel Assignment (DCA)
- CleanAir merging of interferers (based on which APs are close to each other and could hear the same interferer)
- CMX/DNA Spaces calculation of AP RF distance and pathloss measurements

It is possible to analyze what other APs are hearing a specific AP by going to the **Monitoring > Wireless > Radio Statistics** page on the WLC, as illustrated in Figure 7-8, clicking a specific access point, and going to the **Client/AP** tab.

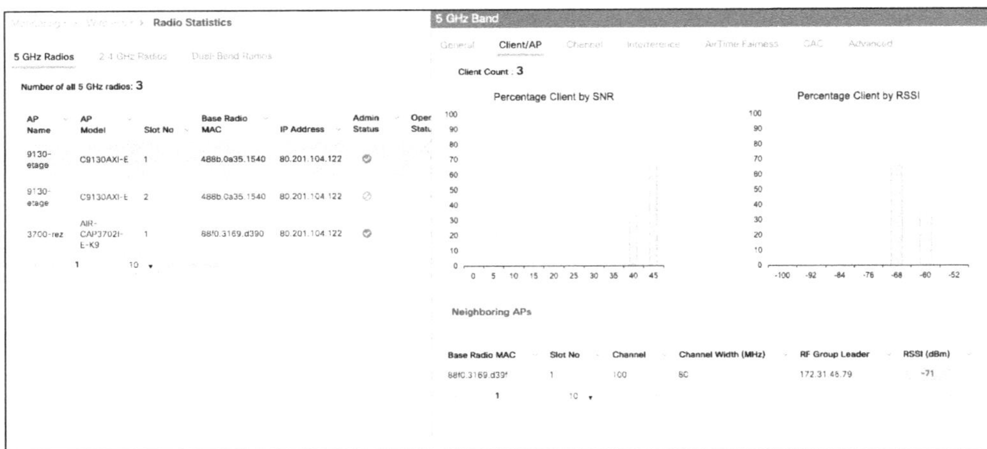

Figure 7-8 *Monitoring AP neighbors*

RF Grouping

An RF group is defined by the RF group name you define on your wireless LAN controller in the initial configuration. It can be changed at any time in the **Configuration > Wireless >Wireless Global** configuration page. It is a string-based name that you can assign to all the WLCs managing the same RF space or, in other words, with APs that can hear each other. This means that if you configure a different RF group name on all your WLCs, their respective APs do not work with each other to map and figure out the common RF environment. Using a different RF group name also results in one WLC reporting the other APs (heard above an RSSI threshold) as rogue because these APs belong to a different RF management domain.

Because more than one WLC can share the same RF group name, an RF group leader needs to be elected and put in charge of running most of the RRM algorithms on behalf of the whole RF group. When a WLC initializes, it has to assume that it is alone in the RF group and therefore acts at least temporarily as RF group leader, creates a unique group ID, and instructs its APs to use the RF group name in their NDP messages. NDP messages received from APs belonging to another WLC mention their RF group leader IP address as well as RF group name. If they share the same RF group name but have another leader, the WLC contacts this other WLC to add to the RF group list. The one with the highest group ID takes the role of leader. If a new leader is selected, the information is sent to all the APs, and APs start sending to the newly elected RF group leader IP in their NDP advertisements. The RF group leader can then start to assemble an RF neighborhood, that is, a group of APs that hear each other at a signal better than –80 dBm. If the signal of an AP suddenly goes lower than –85, it is deleted from the neighborhood. The WLC sends Hello messages every 10 seconds to every other WLC in the currently formed RF group.

RF Grouping Modes

RRM starts in automatic grouping mode, which basically behaves as explained in the previous section. It is also possible to turn off grouping to have each WLC act as standalone. A **Leader** setting, as shown in Figure 7-9, in **Configuration>Radio Configurations > RRM**, makes the grouping static and defines the WLC where this setting is enabled on an automatic leader. On top of this, each WLC platform model has a specific static priority number added to the equation, in order to avoid having small WLCs acting as leader for large WLCs. When this mode is set, you can manually add the other WLCs to the RF group one by one. A more deterministic approach allows you to choose a specific bigger controller to act as RF leader. It is interesting to note that there is a maximum number of APs for which a WLC can act as leader in an RF group, as described in Table 7-1, so it is not realistic to keep one big campus with tens of thousands of APs in the same RF group because that will go over the limit of the strongest controller hardware available. If too many APs have to be managed by an RF group leader, another WLC is elected to be a second leader, and the APs are split among them, so there is no impact there.

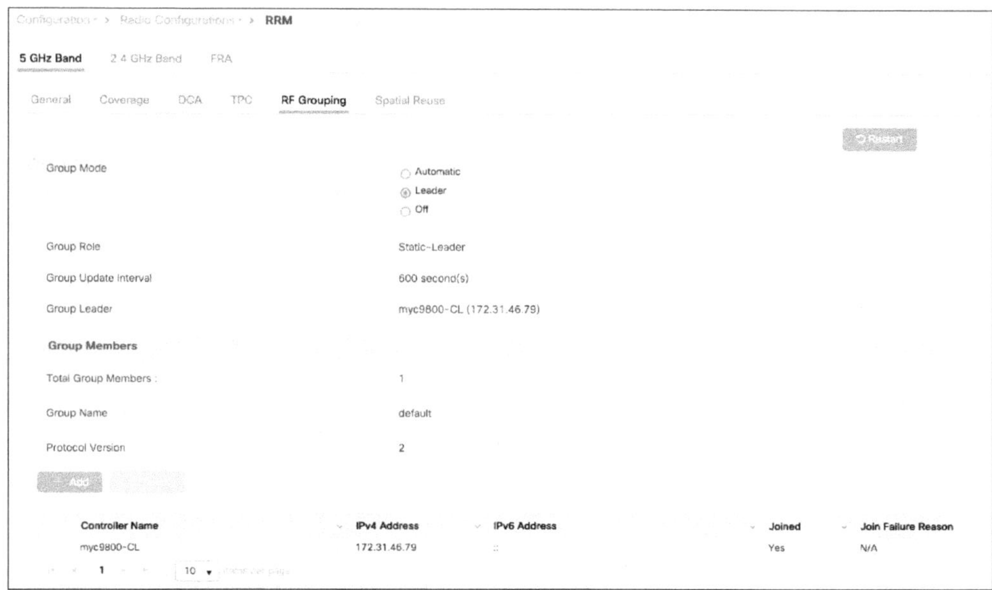

Figure 7-9 *RF grouping monitoring and configuration page*

Table 7-1 *RF Grouping Maximum AP Limits*

Group Leader WLC	Maximum APs	Maximum APs per RF group
9800-L	250 (500 with performance license)	500
5508	500	1000
9800-CL Small	1000	2000

Group Leader WLC	Maximum APs	Maximum APs per RF group
9800-40	2000	4000
9800-CL Medium	3000	6000
5520	1500	3000
8510/8540	6000	6000
9800-CL Large	6000	12,000
9800-80	6000	12,000

Because the RF group leader is running the RRM algorithms, only the RRM configuration on the leader WLC matters when it comes to RRM. It is a classic mistake to change RRM settings on a given WLC and be surprised that there are no effects when the RF group leader was in fact another WLC. It is safer to make sure RRM settings are identical on all WLCs susceptible to form an RF group to avoid any surprises.

TPC

Transmit Power Control (TPC) is the name of the RRM algorithm that focuses on lowering the transmit power of APs if needed. The only case where TPC can increase the transmit power is covered in the coverage hole detection section. The objective of a good wireless deployment is to have sufficient coverage but without having too much signal overlap, which would cause co-channel interference.

TPC Overview

TPC runs on the RF group leader WLC and, contrary to TPC on AireOS controllers, exists in only one flavor, formerly known as TPC version 1. The TPC algorithm can run automatically every 10 minutes (that's the default setting), can also run on demand at the click of a button only, or not run and freeze the power settings until instructed otherwise.

APs out of the box, after joining the WLC, transmit at their maximum allowed power level; after set by TPC, APs retain their last transmit power setting across reboots. TPC runs separately for each AP in the RF group and determines whether the transmit power of the AP should be lowered based on the AP neighbor details. Based on the power threshold (which is –70 dBm by default, as illustrated in Figure 7-10), if three APs are heard louder than this threshold (no matter which channel they operate on), the algorithm starts to consider lowering the power. TPC gives a particular focus on the worst-performing APs but also considers cascading effects, so the exact formula is a bit more complicated. But the point is that if you raise the power threshold closer to the maximum of –50 or if a specific AP is isolated and does not hear many neighbors, it may be expected to run at full power. When a power change is recommended, the AP decreases its power by one level (that is, 3 dB). TPC does not take channels into account by default because DCA might change them at the same time TPC runs anyway. It is, however, advised to enable **TPC Channel Aware** for the 5 GHz band, which typically allows for a lot more channels and could tolerate much better APs hearing several neighbors on different channels. It is better to leave **Channel Aware** disabled for the 2.4 GHz band.

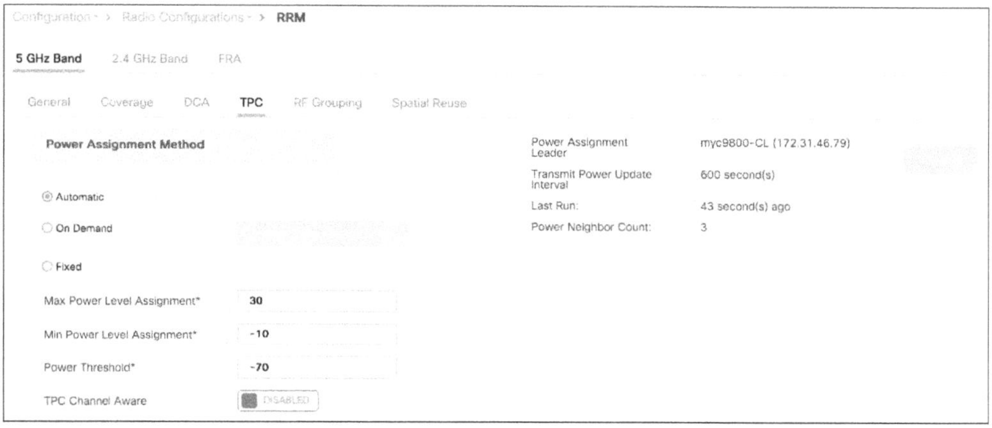

Configuration > Radio Configurations > **RRM**

5 GHz Band 2.4 GHz Band FRA

General Coverage DCA **TPC** RF Grouping Spatial Reuse

Power Assignment Method

Automatic

On Demand

Fixed

Max Power Level Assignment* 30

Min Power Level Assignment* -10

Power Threshold* -70

TPC Channel Aware DISABLED

Power Assignment Leader myc9800-CL (172.31.46.79)

Transmit Power Update Interval 600 second(s)

Last Run: 43 second(s) ago

Power Neighbor Count: 3

Figure 7-10 *TPC settings*

The same TPC web interface page shows you which WLC is the current RF group leader, what time the algorithm interval is set to, and when it ran the last time.

While TPC mostly lowers the transmit power, it can also increase in case of sudden AP failure to compensate for the coverage gap.

TPC Minimum and Maximum

The minimum and maximum power setting you configure runs on every controller. It is designed as a safety to prevent using power that you know will be too high or too low for the deployment. For example, this might be the case in a situation where you could not afford to place your APs in the best ideal position but have to deal with "all the APs in the hallway" in a university or hospital deployment. If you let the TPC algorithm run without guidelines, it would lower the power severely because APs can hear each other very well with a clear line of sight in the hallways. However, coverage in the rooms would probably suffer, and some places would not have a good coverage. You probably would be ready to accept a bit of co-channel interference in the hallways if that means a proper coverage of the places where devices will effectively be used: this can be achieved by increasing the TPC minimum power. Similarly, if you're mounting APs very high on the ceiling in warehouses or similar venues, the AP signal not might even reach the ground if it is under a certain level. If you don't set a TPC minimum, there is a good chance that the APs would go below this threshold because they probably hear each other very well on the ceiling as they are closer to each other than they are to the clients. Decreasing the maximum TPC power could make sense in situations where APs cannot hear each other too well because of distance or antenna orientation, but clients can hear all APs very well.

It may seem a bit confusing, but the TPC minimum and maximum fields expect a value in dBm and not in power level. This value does not take any antenna gain into account and is the transmit power as shown by an AP. This means the value is absolute and is comparable regardless of AP and antenna model or channel used. The latter may be extra confusing because the maximum allowed power may vary depending on the channel, so this is something to take into account when designing your plan and settings.

Coverage Hole Detection

The coverage hole detection algorithm is different from the TPC algorithm because it is in charge of increasing the transmit power only if a coverage hole is detected. It focuses on clients and is able to discriminate between clients that genuinely have a low signal strength because they are in an area of poor coverage that can be remediated and clients that might have a static lower transmit power and where any remediation would not help.

Coverage hole detection can be enabled globally in the RRM settings. You can configure a long list of thresholds for coverage holes to trigger, as shown in Figure 7-11.

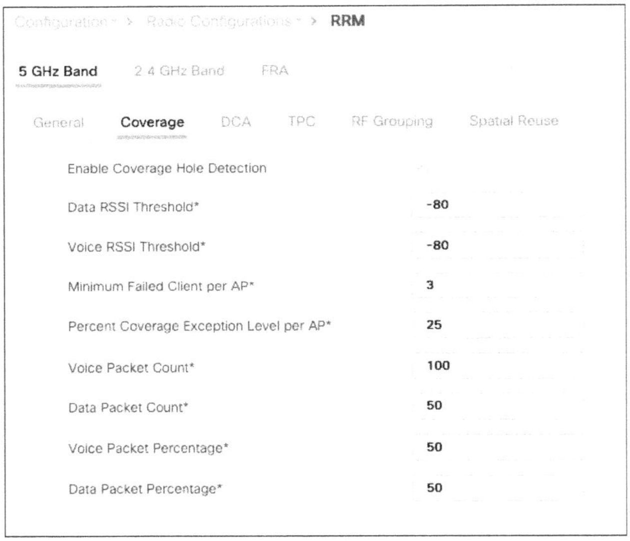

Configuration > Radio Configurations > RRM

5 GHz Band | 2.4 GHz Band | FRA

General | Coverage | DCA | TPC | RF Grouping | Spatial Reuse

Enable Coverage Hole Detection	
Data RSSI Threshold*	-80
Voice RSSI Threshold*	-80
Minimum Failed Client per AP*	3
Percent Coverage Exception Level per AP*	25
Voice Packet Count*	100
Data Packet Count*	50
Voice Packet Percentage*	50
Data Packet Percentage*	50

Figure 7-11 *Coverage hole detection algorithm settings*

If the WLC detects a single client that keeps showing a signal worse than –80 dBm (by default) for more than 5 seconds without roaming to another access point, it logs a pre-coverage hole event. This is a syslog and trap sent to the management platform, but no action is taken because a single client may not be representative of the situation. The type of client that holds on to an access point without roaming, ignoring the possibility of a better connection through a different AP, is called a "sticky" client (although if these clients are in a real coverage hole, they should not be called that because any client in such an area is forced to stay connected at low RSSI). The signal threshold can be configured for both data and voice frames (it can make sense to have a louder, more sensitive threshold for voice-tagged frames because those clients are more critical, although the packet count is higher because voice packets are less frequent). You can also configure the number of clients (three by default) required to trigger an actual coverage hole event (instead of a simple precoverage hole alert). On top of that, the **Percent Coverage Exception Level per AP Setting** must be met for the remediation to occur. For example, if it is set to 25 percent, it requires that the three sticky clients represent at least 25 percent of the client count of the AP (which means the AP should have 12 clients or fewer). When all conditions are met, the WLC raises the power level of the AP hearing those clients by one level.

Because the AP reports this data every 90 seconds to the WLC, it means that it requires several clients to stay connected at a poor signal strength to the same AP for a bit of time before any action happens. This condition helps to avoid false positives and avoid having your APs climb to max power all the time (before TPC calms things down) only because of a couple of clients with poor roaming logic. Extra settings such as packet count or packet percentage also allow you to require a certain amount of traffic to be received at a poor signal before something is done (to avoid triggering for clients that are connected but barely passing any traffic).

Coverage hole detection can be enabled on a per-WLAN basis, in the WLAN settings. This means clients of that WLAN are used for the coverage hole detection algorithm. WLANs in which the feature is disabled do not have their client participate in the coverage hole detection algorithm, but it is important to understand that if enough clients in a given WLAN with CHD enabled are in a coverage hole, the power is increased for all clients on the AP. The feature only segregates which clients count or do not count for matching a coverage hole trigger condition. It can be useful to exclude a WLAN where legacy devices (which may have a poor roaming algorithm) connect to avoid spurious coverage hole events.

DCA

The Dynamic Channel Assignment (DCA) algorithm also runs on the RF group leader and runs on a per AP basis. It oversees the determination of the best channel for the AP to operate on. DCA calculates an RSSI-based custom metric that allows you to compare channels with each other. It takes several factors into consideration when computing this metric:

- **Same channel contention:** The impact of managed APs and client communication on the channel served by the AP.
- **Foreign channel or rogue:** The impact of nonmanaged APs and client operating on the same channel as the target AP.
- **Noise:** Non–Wi-Fi communications that might interfere with the target AP.
- **Channel load:** Actual usage (through the QBSS element) of the channel is taken into account. The QBSS Information Element is present in quality of service–enabled BSS (when you enabled WMM basically) and advertises the amount of airtime the AP is busy receiving or transmitting useful signals.

A sensitivity threshold can also be specified to choose the difference margin that will be required for DCA to consider moving an AP to another channel. A "high" sensitivity means the algorithm needs a difference of only 5 dB between two channels to move the AP to the better channels, whereas a "low" setting means a difference of 20 dB is required to change the channel. DCA runs by default every 10 minutes, which can sometimes be a bit too aggressive. However, it does not mean that DCA will change your channels on every AP each 10 minutes. But it will run at that interval and evaluate for each AP if it is worth changing the channel. Changing the channel causes a tiny interruption of service because the clients need to understand where the AP went, but operating on a channel that is impacted by interference is also very disruptive. Some administrators prefer that the DCA algorithm runs once or twice per day only, which can be achieved by increasing the interval. You can even specify an anchor time that will specify the time of day at which it will run the first time (and it will run again after one interval). Having DCA run only a few times per day works great in environments that are somewhat stable in terms of interference or foreign APs. The complete opposite administrative decision is to have the APs react to every noticeable interference and try to switch to a better channel. This effect can be achieved with a shorter DCA interval but also by enabling event-driven RRM (ED-RRM). This feature, when combined with CleanAir-enabled access points, has the APs react in real time (outside of the DCA interval) to interferers that are considered severe enough to cause an impact. This mechanism can be very helpful in environments where people can fire up any kind of interferer at any time and you do not want your channels to be completely unusable for a long time. Figure 7-12 illustrates all the settings mentioned previously as well as some channels settings.

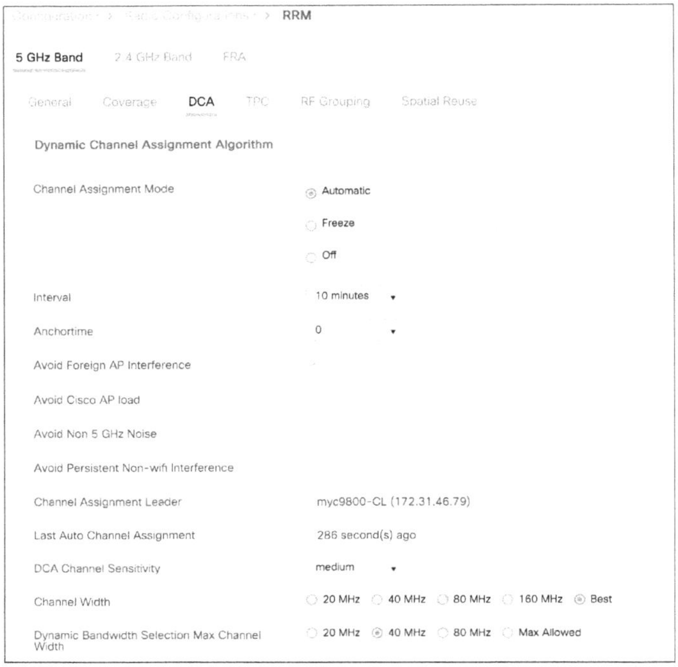

RRM
5 GHz Band | 2.4 GHz Band | FRA
General | Coverage | DCA | TPC | RF Grouping | Spatial Reuse
Dynamic Channel Assignment Algorithm
Channel Assignment Mode: Automatic / Freeze / Off
Interval: 10 minutes
Anchortime: 0
Avoid Foreign AP Interference
Avoid Cisco AP load
Avoid Non 5 GHz Noise
Avoid Persistent Non-wifi Interference
Channel Assignment Leader: myc9800-CL (172.31.46.79)
Last Auto Channel Assignment: 286 second(s) ago
DCA Channel Sensitivity: medium
Channel Width: 20 MHz / 40 MHz / 80 MHz / 160 MHz / Best
Dynamic Bandwidth Selection Max Channel Width: 20 MHz / 40 MHz / 80 MHz / Max Allowed

Figure 7-12 *DCA settings*

DCA allows you to choose the channel width you want DCA to assign to your APs. The setting "Best" tries to use larger channels (up to 80 MHz) wherever possible and where 802.11ac or Wi-Fi 6 clients are detected. It may choose to use 40 MHz or 20 MHz channels if it notices that APs are very close to each other, if there are interferences, or simply if it thinks the network would perform better with smaller channels. One inconvenience of the "Best" setting is that some clients always try to connect to the largest channel available before looking at the best RSSI. (Apple has documented this issue in its client roaming behavior; other clients don't often publish their roaming logic, so it's hard to know the real ratio.) Having APs with varying channel width would then be detrimental because clients would not necessarily pick the closest AP to them but the AP with the largest channel around. It is possible to also configure a sort of "ceiling" by defining the **Dynamic Bandwidth Selection Max Channel Width**. Enabling this setting can be useful if you know that 80 MHz will not be efficient at all in your network and that you want to aim at 40 MHz whenever possible, but some areas could benefit by using 20 MHz channels due to their density. FlexDFS is an automatic feature (that is, there is nothing to configure for it) where if an AP is set to use a 40 or 80 MHz channel and radar is detected in a 20 MHz subchannel, the AP can reduce its bandwidth to 20 or 40 MHz to avoid the problematic 20 MHz range. This may explain why some APs keep using smaller channels even if you set DCA to use larger ones.

The same configuration page allows you to select the DCA channel list, which is the list of channels the algorithm will choose from and possibly the list of channels that the APs will be monitoring completely. This page allows you to completely remove some channels if you are aware of specific static interferers.

Similarly to the transmit power level, APs use their previous channel when they join any WLC. However, out of the box, on their initial configuration, they use the first channel of the band. When a new standalone WLC starts or reboots, or when a new RF leader is elected in a group, DCA enters a "startup mode" for 100 minutes. This consists of 10 runs of the DCA algorithm (every 10 minutes) regardless of your interval configuration. These runs use the high sensitivity setting to help shuffle the APs and create a channel plan. After this startup mode is finished, the DCA algorithm uses the configured settings and interval.

Overlapping Basic Service Set (BSS)

To keep interoperability with legacy versions of the Wi-Fi standard, larger channel widths keep using a primary 20 MHz channel where management frames are still sent at legacy mandatory data rates (which is not the case anymore in the 6 GHz band, which drops legacy support). When setting a channel statically—for example, by going to the radio list for a specific band in **Configuration->Wireless>Access Points** and choosing 80 MHz and channel 100 as shown in Figure 7-13 with the 3700 AP—the AP chooses the configured channel as the primary 20 MHz and the channels above to complete the 80 MHz width.

Access Points

All Access Points

5 GHz Radios

Number of AP(s): 3

AP Name	Slot No	Base Radio MAC	Admin Status	Operation Status	Policy Tag	Site Tag	RF Tag	Channel	Power Level
9130-etage	1	488b.0a35.1540			NicoHouseEtage	Nicohouse	NicehouseRF	(64,60,52,56)	*1/7 (16 dBm)
9130-etage	2	488b.0a35.1540			NicoHouseEtage	Nicohouse	NicehouseRF	(36)*	*8/8 (-4 dBm)
3700-rez	1	88f0.3169.d390			NicoHouseprod	Nicohouse	NicehouseRF	(100,104,108,112)	*1/6 (17 dBm)

1 10

Figure 7-13 *Manual channel configuration result*

The DCA algorithm also prioritizes this behavior, and if it has to place APs on the same channel, it tries to assign the primary 20 MHz to be the same so that APs can hear each other's management frames and act nicely. However, that is not always the best option, and DCA might choose to assign the primary channel differently. In Figure 7-13, notice that the 9130 AP uses channels 52 through 64 (which makes for 80 MHz of width) but channel 64 is the primary 20 MHz. This setup can complicate sniffer captures because sniffers typically offer to sniff channels only above the primary channel. It is more efficient at first sight to use another primary channel for another AP that would operate on overlapping channels because potentially both APs could transmit at the same time management frames (since they are on different channels), but they still compete for the whole channel width when sending data frames. This situation gets even more complicated if you consider varying channel sizes (APs with 20 MHz, some with 40, some with 80, and some on 160 MHz) as shown in Figure 7-14, because on top of a primary 20 MHz, there is also a primary 40 MHz and 80 MHz (for stations not supporting the whole width). Different Wi-Fi versions of the protocol have different rules and thresholds to access the medium, and shuffling primary channels like this could cause some stations to forbid access to the medium by other stations. Long story short, it is often better to have APs use the same primary channel for all their clients to play nicely together, but in a busier environment, the DCA algorithm may use overlapping channels with a different primary 20 MHz if it feels justified from a load or interference standpoint. If you see this happening, the WLC is trying to squeeze the most efficiency out of the network but is lacking some channel reutilization. You can inspect whether rogue wireless networks are active on some of your channels, study the possibility of adding more channels, or hunt for interferers.

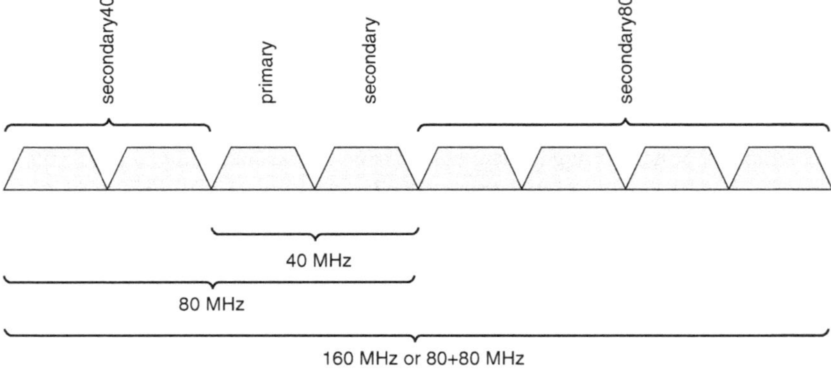

Figure 7-14 *Complex primary and secondary channel plans for various channel widths*

Cloud-Based RRM

By the time you read this, Cisco will have released a cloud-based RRM solution. The data collection and measurements stay the same, but moving the RRM algorithm "brain" to the cloud brings a few advantages. The first is scale because the cloud can compute the best RRM for your network regardless of the deployment size, whereas the current RRM solution might elect different RF leaders if there are many APs to take into account.

RRM automates a lot of tasks for you but currently works based on a set of thresholds that are configurable and where the defaults should work fine for most deployments. A cloud RRM solution brings the power of AI and gets rid of thresholds. Patterns are identified based on your real environment assurance data, and the RRM decisions are made accordingly. Decisions are not only snapshot-based (that is, based on the data at a given fixed time of the day) but can be taken based on usage patterns, identified peak hours, and historical data. The algorithm can also self-improve because it is able to observe the results of previous RRM decisions on assurance data and identify what works best.

Finally, an RRM control center gives you comprehensive data about the network current status, the RRM changes, and settings applied.

RF Profiles

An AP gets its RF settings based on the RF tag it is assigned to. The RF tag is configured with RF profiles, one for each frequency band. An AP that is not configured with any specific RF profile uses the default RF settings policy globally configured on the WLC. You can define them in **Configuration > Tags and Profiles > RF/Radio**, as shown in Figure 7-15. You can configure the name, radio band, and NDP mode. The basic NDP mode is auto, where the AP uses its serving radio to send the NDP messages. When set to off-channel, the AP uses its software monitor radio chipset (only on Catalyst 9120 and above) to send the NDP messages (they are heard with the main serving radio on the current channel mostly), completely freeing the client-serving radio of this task. Off-channel is therefore a better setting, but not available in all APs and also not necessarily optimal if you have a dual 5 GHz radio with external antennas where the software radio with the CleanAir chipset then does not have the same propagation pattern as the data traffic. This can be the case on a 9130 with external antennas where you may be using very directional antennas for the 5 GHz radios with different orientations between the two 5 GHz radios. The internal software-defined radio is not connected to all the external antenna chains and has a different view of the RF space.

The 802.11 tab allows you to define the data rates supported, as shown in Figure 7-16. To this day, it is still a good idea to offer support for legacy data rates (that is, up to 54 Mbps) because many clients expect this support. On top of disabling or enabling support for certain data rates, you can also define (for the legacy rates) which rates are mandatory. The lowest mandatory data rate is the rate at which the beacons and management frames are sent by the AP. Professionals often consider this as defining the coverage area of a given radio cell. Although it is a little abuse of language (because the signal energy travels the same distance regardless of the data rate configuration), it does define the usable coverage area, that is, the maximum distance at which beacons are heard and a client can send an association frame that will be decodable by the AP. Defining more than one mandatory rate helps with multicast traffic, which can be sent at a higher mandatory data rate. There is no particular effect in disabling certain 802.11n MCS data rates apart from making sure the clients downshift rates a bit faster.

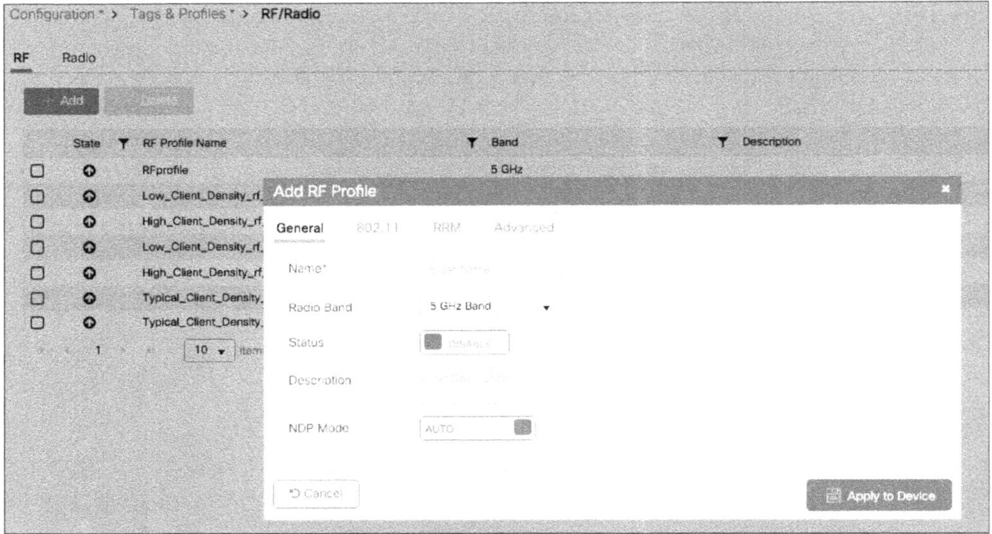

Figure 7-15 *RF profile general configuration*

Figure 7-16 *RF profile configuration*

RF profiles allow you to configure the typical coverage hole and TPC settings specifically for the APs assigned to this RF profile. The DCA tab offers some (not all) of the global DCA configuration knob and is especially handy in letting you configure the channel width and channel list for the specific group of APs, as shown in Figure 7-17. It also allows you to configure certain 802.11v settings for high-speed client roaming.

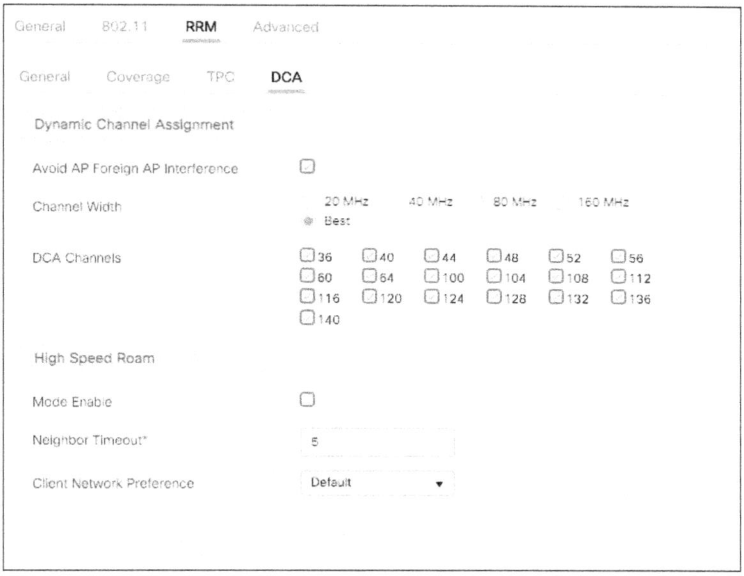

Figure 7-17 *RF profile DCA configuration*

The Advanced Settings tab offers configuration for High Density (such as Receiver Start of Packet, or RxSoP), aggressive client load balancing, airtime fairness, FRA, and Wi-Fi 6 OBSS-PD, as illustrated in Figure 7-18.

RxSoP is an advanced feature that should be enabled only in high density environments and with adequate testing. High density means that you want to have small cells and have clients roam to the nearest AP as soon as possible to maintain the best connectivity and data rate. However, some clients might be slow to roam and stick to an access point that is not the best and nearest to them. Even if you select directional antennas and do the best RF deployment possible, clients might still hear the AP at a low signal level, such as –78 dBm or –80 dBm, and decide to stick to that AP, taking precious AP airtime because the communication would be slower. They might even cause retries and use low data rates, impacting the whole cell. RxSoP can be set to a custom value between –85 and –60 dBm. Below this threshold, the AP completely ignores the received frame and considers it as background energy (noise). This means that a client probe received below the configured RxSoP RSSI level is ignored and the AP does not waste resources for this communication. This frees up the AP to work only with closer clients and not waste time with sticky clients. Configuring RxSoP settings requires you to have a dense coverage and to be sure that the client will be able to roam easily. On top of the custom values, it is possible to use predefined value thresholds.

Figure 7-18 *RF profile advanced configuration*

The same Advanced tab allows you to configure the multicast data rate, which by default is the highest mandatory rate, but thanks to RF profiles can be configured to be a specific data rate. That data rate chosen in the RF profile should still be configured as a mandatory rate to make sure all clients support it. The Advanced tab also allows you to configure Overlapping Basic Service Set Preamble Detection (OBSS-PD), which is covered in the Wi-Fi 6 features section later in this chapter.

Since the 17.6 version of IOS-XE, you can also configure a Radio profile, illustrated in Figure 7-19, which allows you to configure the radiation pattern of the C-ANT9104 antenna and the C9130AXE-STA AP, which is a stadium antenna that can adapt its radiation pattern dynamically through configuration. This Radio profile can be assigned along with the RF profile in the RF tag, as shown in Figure 7-20.

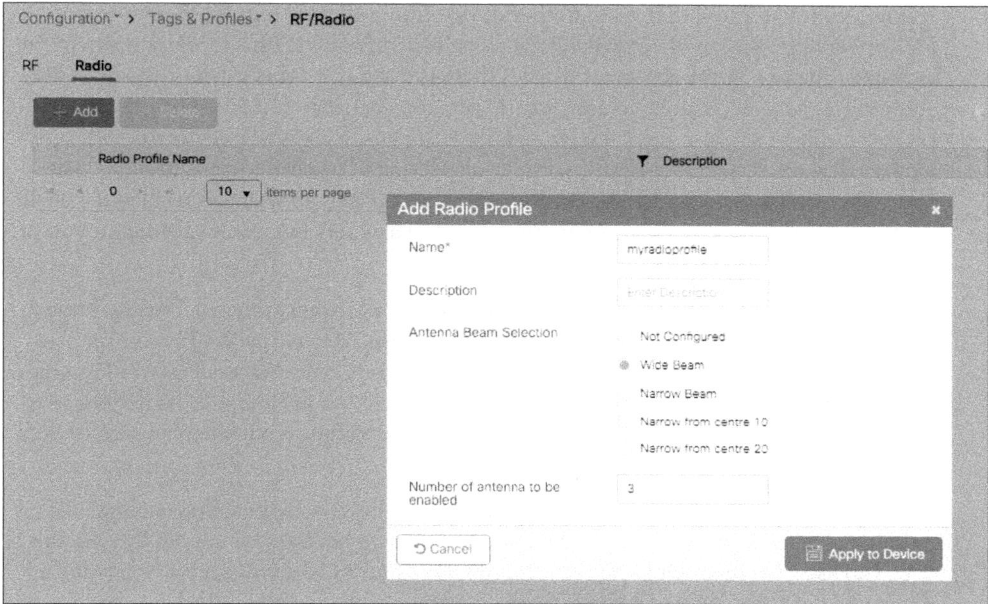

Figure 7-19 *Radio profile configuration*

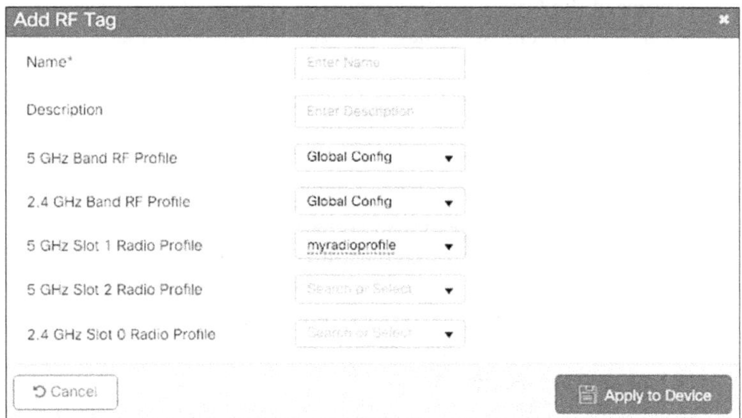

Figure 7-20 *RF tag configuration*

Spectrum Intelligence and CleanAir

When you're operating a wireless network, especially one covering one of the challenging environments described, it is essential to have some form of visibility over the wireless medium. Wi-Fi is a form of free-for-all where there are rules to access the medium, but it is not possible to fully prevent another device from transmitting. All Wi-Fi devices are expected to play by the rules and get along together, but if this does not happen (for example, due to a faulty device or driver), there is not much other devices can do. This situation is made even more complicated if you consider that 802.11 is just one of the protocols allowed to use the 2.4 GHz and 5 GHz bands. Non Wi-Fi devices do not operate by the same rules and transmit without respecting your clients' transmissions. The least you can do, considering the band is unlicensed and free for use, is to identify devices that are actively using the same frequency space as yours but do not belong to you or your wireless network.

This activity of detecting and recognizing other types of transmitters is broadly known as spectrum intelligence (SI). This implies that the Cisco AP use its Wi-Fi radio chipset to get as much data as possible on those transmissions it cannot decode as Wi-Fi signals. From those patterns (for example, the bandwidth of the transmission or its hopping pattern in the case of frequency hopping), it may be able to figure out what type of device/protocol is in action.

Higher-end Cisco Catalyst access points embed a dedicated, software-programmable, radio chipset to do the job. On Wi-Fi 4 and 5 access points (802.11n and 802.11ac), this is called a Cisco CleanAir chipset. The Catalyst 9120 and 9130 series access points include a separate software-defined radio of a new generation that performs the CleanAir duty. These dedicated radio chipsets include a full-blown spectrum analyzer that is specialized in this task.

Although the RF-ASIC present in the Catalyst access points is capable of much more than the previous generation CleanAir chipsets, they still both work in the same way with regards to the interference detection process. The spectrum analysis radio constantly listens to the medium and scans the whole frequency range more or less every second. CleanAir can analyze the spectrum only when the AP is not transmitting (because the AP transmissions would be way too loud and cover everything else). When the AP is listening to the medium (which is basically all the time it is not transmitting), if a Wi-Fi frame is received, it is processed directly by the Wi-Fi radio, but any other non–Wi-Fi signal is processed and analyzed by the CleanAir chipset. It listens to the signals received by the Wi-Fi radio on its current channel. Both chips share the same antenna chain, but the CleanAir chip has a very high sampling rate and dedicated hardware. This allows the CleanAir chip to detect different Bluetooth transmitters that hop at the same time to different neighboring 1 MHz-spaced frequencies.

Both spectrum intelligence (available on the 1800 series and 9105/9115 series APs) and CleanAir focus on detecting non–Wi-Fi interferers. The latter do it with better resolution and detect more types of interferers. The former have a small performance impact because the AP needs to spend a bit more time off-channel to perform the detection, whereas CleanAir has no performance toll whatsoever (except only if you enable BLE detection by CleanAir, which causes a 10 percent packet loss on the 2.4 GHz band).

CleanAir provides interference device reports (IDRs), illustrated in Figure 7-21, back to the controller that can be consulted in real-time under **Monitoring > Wireless > CleanAir Statistics.** The information contained in the report includes

- The AP that is closest to the interference.
- The type of interferer if it was recognized.
- The Wi-Fi channels that will be affected by it (because a non–Wi-Fi interferer can affect multiple channels).
- The duty cycle, which is a percentage representing the amount of airtime blocked by the interferer.
- The severity, which is a number between 1 and 100 representing the severity of the impact. It is majorly influenced by the duty cycle (the more airtime wasted, the higher the impact) but also by the proximity/loudness of the interference (if an interferer transmits 100 percent of the time but is of very low signal strength because it is far away, its severity is less).
- The RSSI at which it is heard.
- A device ID that uniquely identifies the interferer and differentiates separate interferers that are in the same physical space.
- A Cluster ID that allows you to cross-identify a unique interferer heard by two separate reporting APs so that it shows up as a single interferer and not two different entries. Clustering is done by client-serving APs on the basis of their list of neighboring APs (and comparing the interferer-heard RSSI, among other things). Monitor mode APs do not participate in this activity because they do not transmit neighboring messages.

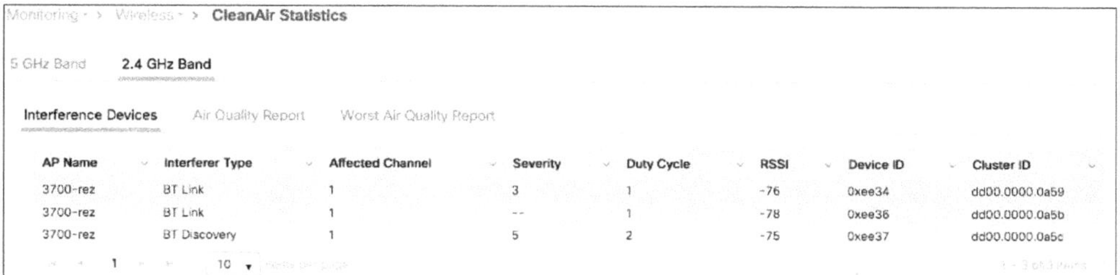

Monitoring > Wireless > **CleanAir Statistics**

5 GHz Band **2.4 GHz Band**

Interference Devices Air Quality Report Worst Air Quality Report

AP Name	Interferer Type	Affected Channel	Severity	Duty Cycle	RSSI	Device ID	Cluster ID
3700-rez	BT Link	1	3	1	-76	0xee34	dd00.0000.0a59
3700-rez	BT Link	1	--	1	-78	0xee36	dd00.0000.0a5b
3700-rez	BT Discovery	1	5	2	-75	0xee37	dd00.0000.0a5c

1 10

Figure 7-21 *2.4 GHz CleanAir Interferer device report*

Another thing reported by CleanAir is the air quality (AQ). It is reported by a virtual index between 0 and 100, where 100 represents perfection. CleanAir takes a rolling average of the number or severity of the interferers present on a given channel to compute this metric. This metric helps giving an easy overview of the channel when interferers may come and go and not be constantly present in the real-time interferer list.

Configuring CleanAir

In the **Configuration > Radio Configurations > CleanAir** page illustrated in Figure 7-22, you can enable CleanAir globally. You then can check the CleanAir status (up or down) for each AP depending on its support for CleanAir. The same location allows you to enable spectrum intelligence for the non–CleanAir-capable APs. There is little to no reason to disable CleanAir because it does not bring a performance impact. It is, however, recommended to keep spectrum intelligence disabled if your network is sensitive to performance or a small packet loss due to the increased off-channel activity.

You can enable the reporting of interferers (otherwise, only air quality indexes are calculated) and select which interferer is reported. The BLE beacon is the only one excluded by default because it has some performance impact, whereas all other interferer types do not have any performance toll.

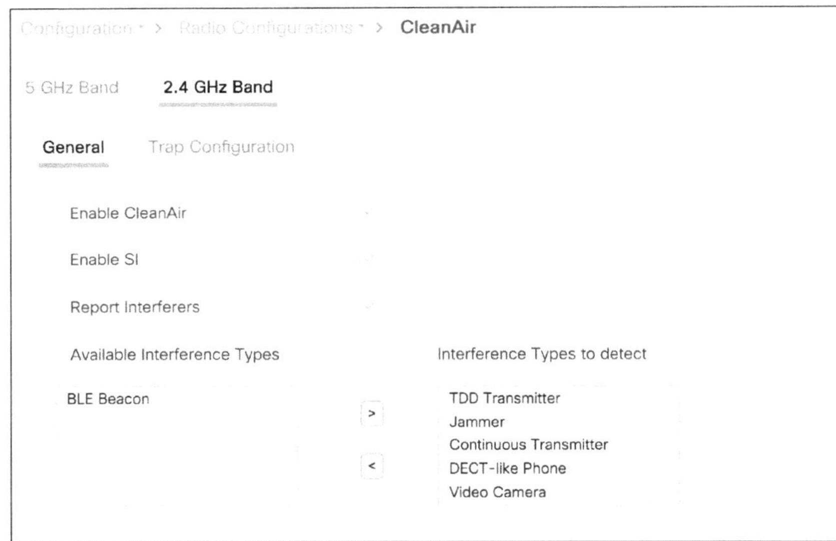

Figure 7-22 *The CleanAir configuration page*

Monitoring the Spectrum Live

APs with the CleanAir capability also allow you to connect to them and see the live view of their spectrum. This basically means having as many spectrum analyzers onsite as you have APs, and all of them are accessible remotely without any effort. In the past, having this capability required putting the AP in a special mode called SE-connect, but this mode is not required anymore. Simply go to **Configure > Access Points** and click a CleanAir-enabled AP. In the General section of the settings, you can find the CleanAir NSI key, which you need to connect to the AP, as illustrated on Figure 7-23.

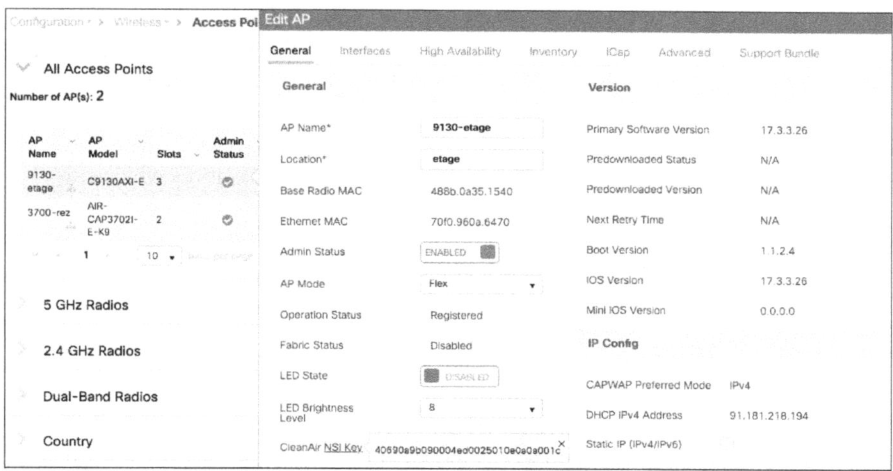

Figure 7-23 *Obtaining the CleanAir key for a given AP*

To connect to the AP spectrum analyzer, you need either Cisco Spectrum Expert or some third-party tool that supports CleanAir APs. You can then connect to the AP and enter the NSI key to see the live spectrum. Cisco DNA Center also allows you to view this within your web browser and without entering any key; to do so, click **Spectrum Analysis** in the AP 360 Assurance view, as shown in Figure 7-24.

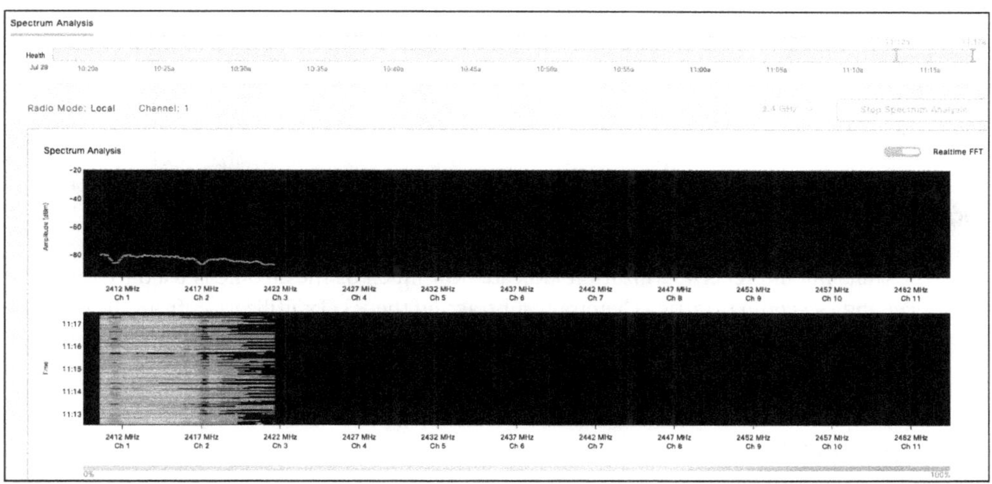

Figure 7-24 *Cisco DNA Center spectrum live view*

Interferer Location Tracking

If you have Cisco CMX or DNA Spaces, it is possible to locate interferers on the map. This capability, however, suffers from several issues that impact its accuracy. While Wi-Fi devices have an expected and typically similar transmit power, any other type of interferer could have very different and even varying transmit power throughout its transmission. CleanAir mostly works on the current Wi-Fi channel, and therefore, it is typically harder to have at least three of your APs hearing the same interferer because it would require all these APs to operate on a channel that overlaps with the interfering device. This requirement can be countered by having CleanAir-capable APs in monitor mode, which can scan all the frequencies rapidly. Most regulatory agencies agreed to open up the 5 GHz frequency range only if the new devices (Wi-Fi clients and APs) operated in total respect of the incumbents (the radars).

Monitoring the RF Space

Going to **Monitoring > Wireless > AP Statistics**, you can have a 360-degree view of the AP RF status. The main metric is **Channel Utilization**, as seen in Figure 7-25, which corresponds to the logical airtime consumed by signals above the clear channel assessment line. This includes the AP utilization as well as other APs on the same channel and non–Wi-Fi interferers.

	Slot 0 (2.4 GHz)	Slot 1 (5 GHz)
Radio Type	802.11ax - 2.4 GHz	802.11ax - 5 GHz
Radio Role	Remote	Remote
Admin Status	Enabled	Enabled
Number of Clients	3	2
Current Channel	6	64
Power Level	*3/8 (9 dBm)	*1/7 (16 dBm)
Channel Utilization	15%	1%
Transmit Utilization	0%	1%
Receive Utilization	0%	0%

Figure 7-25 *Radio statistics of an AP from the C9800 web interface*

The transmit and receive utilization indicate how much of the airtime is used by the AP receiving or sending traffic. Therefore, subtracting the Rx/Tx utilization from the channel utilization allows you to see how much of the airtime is consumed by other devices and interferers.

Advanced RF Features

Next, we cover a number of RF-related features that allow you to tweak certain client or RF-related behaviors:

- Band select
- Client load balancing

- Off-channel scan defer
- Airtime fairness
- Wi-Fi-6–specific features

Optimized roaming could have a place here but is discussed in greater detail in Chapter 6, "Mobility/Client Roaming."

Band Select

The 5 GHz band is generally much better than the 2.4 GHz band because it provides more channels and less interference (at least in most cases). If your WLAN network is dual-band (that is, available in both bands), you could face several issues such as the clients sometimes going to the 2.4 GHz band (because it typically has a stronger RSSI) and facing worse performance. You could also have the clients going back and forth between the 2.4 GHz and the 5 GHz bands, which can cause longer roaming times because roaming between different frequency bands can require some time for a client adapter to achieve. The Band Select feature pushes the AP to ignore all the client 2.4 GHz probing requests if it has seen that the client is 5 GHz capable. This means that upon the first probe request on 2.4 GHz received by a client, the WLC does not answer for a little while, waiting to see if the client is also 5 GHz capable. If the client shows dual-band capability, the WLC answers only the 5 GHz probe requests. If the client is seen probing only in 2.4 GHz, the WLC replies to the 2.4 GHz probe requests. When a client is classified as dual-band, there is no negative impact at all caused by Band Select (apart from the client believing there is no 2.4 GHz network at all), but this dual-band classification does delay the first probes from the client. This impact may become greater because lately many Wi-Fi devices (especially smartphones) use randomized MAC addresses to probe, meaning the WLC cannot relate one probe request with another from the same client if the client changed the MAC address in-between. This may lead the WLC to introduce delay to all probe responses toward such a client, which is why this feature is particularly not recommended for SSIDs carrying voice applications.

Band Select can be enabled or disabled on a per-WLAN basis in the WLAN settings.

Global Band Select settings can be configured in **Configuration > Wireless > Advanced** web page, as shown in Figure 7-26, but should be left at the default most of the time. Those settings give you an idea of the delays in question though. Clients that hear better than –80 dBm are considered for Band Select. When they are declared to be dual-band, for 60 seconds the 2.4 GHz probes are suppressed. The WLC, upon seeing a new 2.4 GHz probe request, waits to see if the client also probes on 5 GHz. If it sees two 2.4 GHz probes without corresponding 5 GHz probes in the 200 ms that follow, it declares the client single band and keeps answering the 2.4 GHz probe requests.

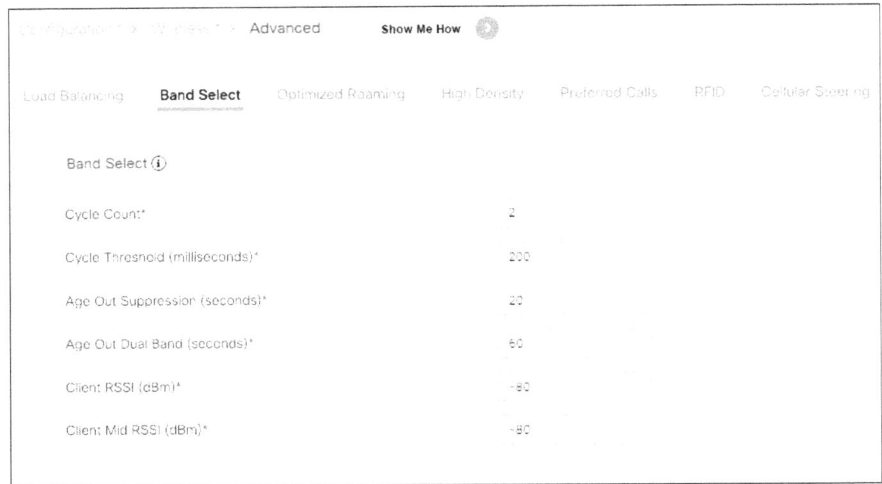

Figure 7-26 *Band Select global configuration*

Aggressive Client Load Balancing

Seeing all the clients on one AP while other APs around it have little to no clients may be upsetting sometimes. This situation may have a variety of possible causes: typically, the first AP in the entrance of a larger area attracts the client initially, and it may not be in a hurry to move to another AP slightly further away. This can happen in large reception areas or in large auditoriums, for example. Aggressive load balancing is a way to remediate this situation by having a busy AP push clients to connect to another AP. When load balancing is enabled in the WLAN advanced settings, an AP that has too many clients (we cover what "too many" means in a moment) starts to reject new association requests with the proper association response code 17 ("AP busy"). The hope is that the client will look for another AP to join. Some client drivers are not so well programmed and do not understand this response code, whereas other clients may have understood perfectly but consider there is no other decent candidate AP to join, so there is a safety mechanism, and the AP ends up accepting the association request after a couple of attempts. The load balancing window that you can configure in the load balancing settings (**Configuration > Wireless Advanced > Load Balancing** tab) is the acceptable client count delta between the AP and the neighbor APs.

If a neighboring AP has three clients and load balancing is configured with a window of five, the AP with eight clients starts rejecting the ninth client, hoping to see it connect to the AP with three clients, for example.

The load balancing window and attempt count can be configured in the RF profile as well as global in the advanced wireless settings.

Off-Channel Scanning Defer

Off-channel scanning defer is a per-WLAN setting (**Configuration > Tags & Profiles > WLANs > Advanced** tab) and has a profound impact on your QoS applications and the RF algorithms as well. We already covered in detail how each AP goes off-channel to scan and to send NDPs, which is the source of data for many algorithms, including RRM. This off-channel scan is short and does not impact data transfer but can be noticed on WLANs that are used for critical real-time voice applications. The AP going off-channel for around 50 ms or a bit more should ideally not be noticed if buffering is in place, but combined with other events, it could lead to small disruptions. You can enable off-channel scanning defer per UP category. By default, as shown in Figure 7-27, it is enabled for UP 5 and 6 (the voice category basically) for a duration of 100 ms. This means the AP postpones any off-channel activity for 100 ms upon receiving a frame tagged with UP 5 or 6. This process can repeat, and as long as voice frames are sent, the AP keeps postponing its off-channel scanning duty and stays on-channel to serve the clients. Although this process is great for the client traffic, it means the AP does not participate in rogue detection or metrics collection for RRM. It can be interesting to also enable UP 7, which is used for 802.1X authentication: you probably prefer your AP not to move away to another channel while a client is starting a long EAP-TLS authentication, for example. If you enable all UPs, it effectively disables all off-channel scanning while there is any traffic ongoing.

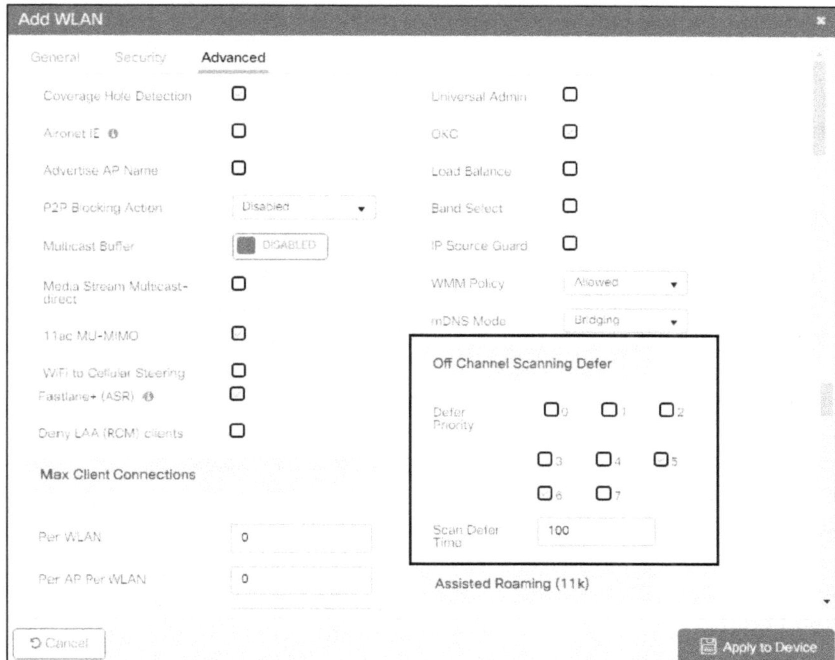

Figure 7-27 *Off-Channel Scanning Defer setting*

Airtime Fairness (ATF)

Traditional methods for reaching user fairness like rate limiting do not cater well to the bandwidth consumption over the air, giving rise to the need for airtime fairness. Rate limiting is sometimes required due to business reasons but is never an efficient way to restrict bandwidth because packets are being dropped on the wired network with no restrictions whatsoever taking place over the wireless network, where clients can still monopolize the airtime with retries, low data rates, and so on. ATF happens mostly on the AP, which takes note of every client's airtime utilization. It makes more sense to restrict clients to a percentage of usable airtime because that airtime can depend on many things: load and interference caused by other APs on the same channel, low data rates, or retries taking place. When the AP receives a frame to transmit downstream to the client, it verifies whether the client still has some airtime available in its bucket. If not, it defers the frame and leaves it in the client priority queue for a while until the client has some airtime available in its configured bucket and potentially drops the frame if it cannot send it for some time.

Airtime fairness allows you to configure profiles that restrict the percentage of airtime available for a given SSID (compared to other SSIDs) or for each client in the SSID. It applies only on data frames (not management or control frames) and in the downstream direction only. ATF is configured in the **Configuration > Wireless > Air Time Fairness** page, where it can be enabled globally for each band. You can then create ATF profiles with a specific weight that you are able to assign in the policy profile you attach to SSIDs.

Airtime fairness is supported only on 802.11ac access points at the time of this writing and not yet on Catalyst access points.

Wi-Fi 6 Features

Wi-Fi 6 is the Wi-Fi Alliance name for 802.11ax and was designed to improve the efficiency of the protocol rather than pure throughput like previous amendments of the standard, which is why it is called High Efficiency (HE) in the protocol itself. The major change to the protocol itself is the move from orthogonal frequency-division multiplexing (OFDM) to orthogonal frequency-division multiple access (OFDMA). Two extra more complex modulations are added (MCS 10 and 11) for increased throughput at very close range. Less visible but still important changes were done to the modulation rules such as the four times increase in subcarriers, while at the same time the intersymbol guard time has been drastically increased. These last two changes to the modulation rules do not change the overall throughput (because both measures somewhat balance each other out) but are meant to bring more resiliency, especially outdoors. Here, we cover the biggest Wi-Fi 6 features and their impact.

Wi-Fi 6E starts to be referenced in IOS-XE 17.7.1 and gets full support in later releases. Wi-Fi 6E is Wi-Fi 6 but in the 6 GHz band. In 2021, many countries worldwide started to release this new band, which typically ranges from 500 to 1200 MHz of spectrum depending on the country. The first and largest advantage of Wi-Fi 6E is more channels for more capacity and less interference. While Wi-Fi 5 (802.11ac) already offered 160 MHz channels, they have been rarely usable in the enterprise setting so far due to the fragmented bandwidth available in the 5 GHz band. Wi-Fi 6E is the same protocol but gives a better chance to 80 MHz and 160 MHz channels to work in the enterprise segment. It also helps support high-density venues like stadiums or concerts by providing less channel reuse and allowing more cells to cover an area. There are other benefits like the fact that the 6 GHz band is not opened to legacy Wi-Fi devices, making it a "Wi-Fi 6 and above only" band. The 6 GHz band is not covered by DFS but by automated frequency coordination (AFC) to make sure Wi-Fi does not interfere with incumbents on the same frequency. Due to the sheer number of possible 20 MHz channels in the 6 GHz band, discovering APs also happens differently for Wi-Fi 6E. Although scanning specific "primary" channels in 6 GHz is possible, it is expected that clients discover 6 GHz radios through existing 5 GHz radios and specific advertisement mechanisms. Last but not least, Wi-Fi 6E only allows WPA3 security or Enhanced Open (that is, no unencrypted networks).

OFDMA

The new OFDMA access rules allow the AP to subdivide the channel into several resource units (RUs), which are basically smaller channels. This means that for a given transmission opportunity a specific client can use 2 MHz out of the channel width to send its data, whereas other clients are assigned other subchannel widths, and everyone can then transmit their data at the same time. This capability increases efficiency drastically, reduces latency (because everyone can transmit at the same time), reduces collision probability (because clients have their own dedicated subchannel bandwidth), and allows deterministic performance where the AP can prioritize clients that have smaller real-time data frames to send. Multi-user OFDMA transmissions can happen both in uplink and downlink.

Multi User–Multiple Input Multiple Output (MU-MIMO)

Already present in 802.11ac, MU-MIMO is taken to the next level in Wi-Fi 6 because it becomes bidirectional and is more efficient with APs having more transmit and receive chains (like the 8×8 9130 AP). MU-MIMO allows up to eight users to send or receive data in parallel on a 106 RU or higher. Acknowledgments (ACKs) and clear-to-send (CTS) frames can be processed for all clients simultaneously. The AP is in charge of maintaining a channel matrix with device locations and moving devices between user groups; they move around so that the AP always knows which clients can receive traffic at the same time through MU-MIMO because it requires destination clients to be physically separated by some distance for the beamforming to work. Each MU-MIMO client can have a different MCS rate. All these improvements combined with a better support on the client side for MU-MIMO promise a better adoption of this feature than 802.11ac wave 2 brought us.

Target Wake Time (TWT)

Target wake time is targeted at increasing battery life and reducing power-save–related management overhead. Until now, devices could sleep and wake up every few beacons to check if there was buffered traffic. Devices could hardly sleep for more than 1 second without facing the risk of being deauthenticated. Thanks to TWT, devices can communicate with the AP and go for very long sleep durations, which is essential for IoT devices that need to run on low power.

All the Wi-Fi 6 features can be enabled or disabled granularly in the Advanced tab of the WLAN configuration page, as shown in Figure 7-28.

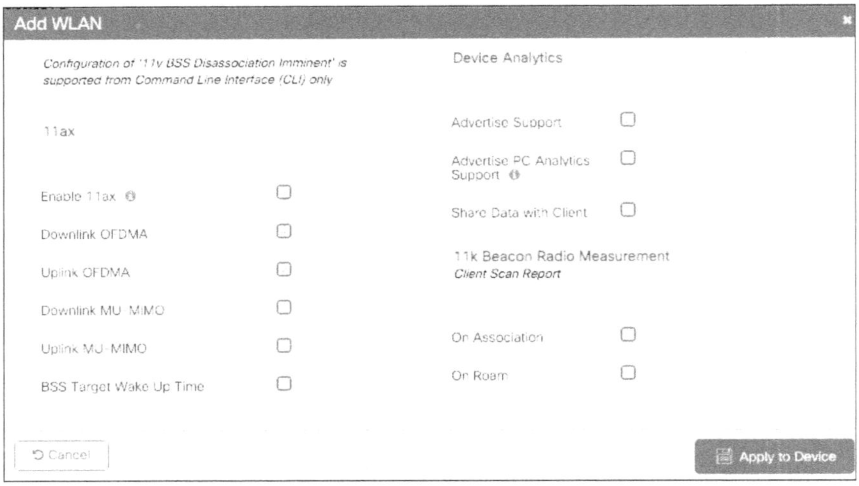

Figure 7-28 *WLAN 11ax features configuration*

You can configure other Wi-Fi 6 settings in the global network configuration page under **Configuration> Radio configuration>Parameters**, as shown in Figure 7-29.

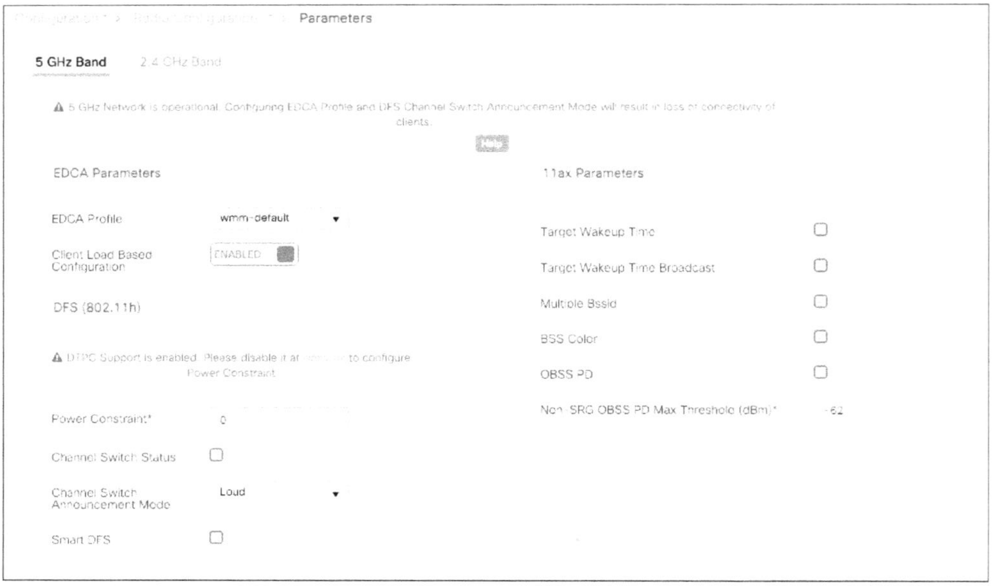

Figure 7-29 *Global 11ax features configuration*

BSS Coloring

The idea behind BSS coloring is that in the real world, in a high-density Wi-Fi deployment, you can have APs on the same channel, and their respective clients could still be transmitting at the same time toward their own APs without really interfering with the other ones. Before Wi-Fi 6, if a client heard a Wi-Fi transmission on its channel louder than –82 dBm, it had to stay silent. However, if clients are decently close to their APs, they can potentially talk at the same time, provided that they each talk to a different AP, and the APs would hear their client just fine. This allows for a lot of channel reuse between floors or in large warehouses where APs hear each other but the clients on the ground don't necessarily hear other APs. The way to achieve this is to mark each BSS with a "color," which is basically a unique identifier between 0 and 63. If a client hears another transmission on its channel and using its color (that is, belonging to the same BSS), it stays silent if it is louder than –82dBm. However, if the client hears a transmission on another BSS (using another color), it uses a much higher RSSI threshold (for example, –62 dBm) to determine whether it can talk at the same time. The color is advertised by the BSS in the beacons, and the client has to use the color after associating to the BSS.

It is required for devices to understand and use the BSS color (which is included in the Wi-Fi 6 certification), but some devices do not use the differentiated RSSI level for overlapping BSS transmission (called OBSS-PD) and keep operating like before and simply use the color advertised by the BSS, therefore not really getting any benefit from the feature. So as with previous certifications, the Wi-Fi Alliance ensures devices can interoperate with each other, but not every device pulls every possible benefit from the IEEE standard.

BSS coloring PD with a custom threshold (that is, the RSSI above which even a signal from another BSS is considered to be too loud for clients to transmit at the same time) can be configured in the RF Profiles Advanced tab.

BSS coloring as a whole (assigning colors to each AP/BSS) can be toggled in the RRM menu, as shown in Figure 7-30.

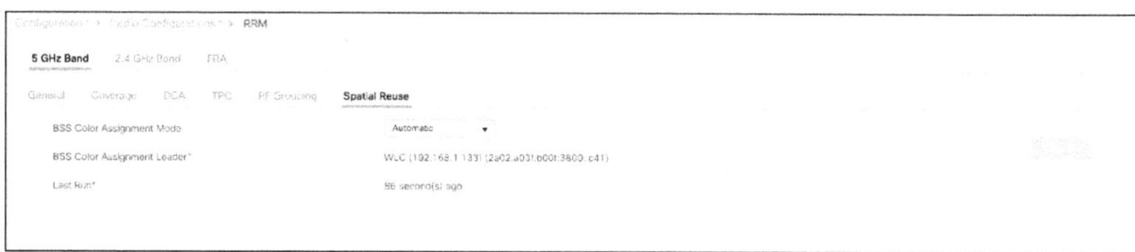

Figure 7-30 *OBSS coloring configuration*

On each AP radio, you can configure whether the AP follows the global BSS Color RRM settings or if it uses an AP-specific configuration, as shown in Figure 7-31. In this case, you can even set the color manually. It is generally a better idea to let the RRM algorithm choose the colors itself though.

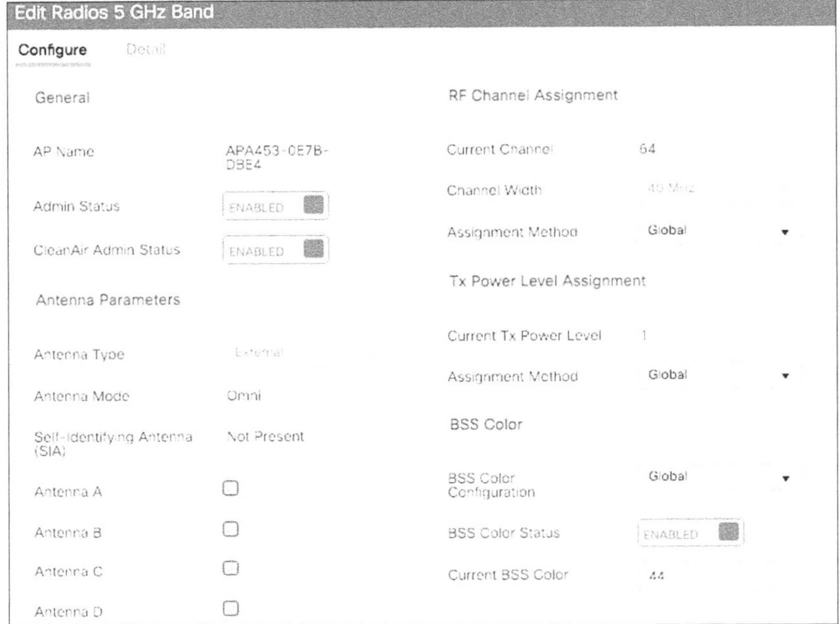

Figure 7-31 *AP-specific BSS color options*

Channel Width

Similar to 802.11ac wave 2, Wi-Fi 6 offers 20 MHz, 40 MHz, 80 MHz, 160 MHz, and 80+80 MHz channels. Each doubling of the channel width comes with a 3 dB penalty to the SNR because of the physics of wider channels, and therefore, it makes it harder to reach higher data rate modulations on wider channels at the same distance. An exception to this rule is 6 GHz Wi-Fi where the rules allow a 3 dB power increase each time the channel doubles in width to compensate for this effect. An AP with more MIMO chains can compensate the drop in signal to noise ratio when using larger channels a little bit by using the combined gain on all transmit and receive antennas when using 8×8 MIMO.

Dynamic Frequency Selection (DFS)

Dynamic Frequency Selection sounds like the name of a fancy feature but is in reality the required access-to-medium protocol in certain frequency ranges in certain parts of the world. In the past, radars used to be the only devices operating in their frequency ranges. But over time, more and more radars have become present in many regions (ships, airports, and also weather radars), and more and more devices are asked to operate on the same frequency range as radar devices.

DFS Overview

The general behavior of a device complying with the DFS protocol is to be able to detect when a radar occupies the channel, then to stop using that occupied channel for a long period of time, to monitor another channel and jump on it only if it is clear from radars.

The process of detecting a radar is complicated and left to radio chipset vendor implementation. Regulatory domain agencies (ETSI, FCC, OFCOM, and so on) have specific tests set up, and as long as the wireless device detects specific types of radar and does not interfere with it, it is approved. Nothing around radar detection is configurable, and each vendor takes the responsibility of fine-tuning the algorithm to avoid false detections (the AP backing off from a channel despite no radars being present) as much as possible. This detection accuracy also depends on the radio chipset used because a radar pulse has very different characteristics from a Wi-Fi signal; it is hard to catch accurately on the lower-end chipsets. DFS was required early on for ETSI devices working in the European Union (and countries following ETSI regulations) in the ETSI 5 GHz band. It is not necessarily mandatory in other parts of the world and also depends on the frequency range. The FCC has now made it mandatory for the UNII-2 and UNII-2 extended frequency range like ETSI.

Radars may be fixed (often civilian airport or military base, but also weather radar) or mobile (ships). A radar station transmits a set of powerful pulses periodically and observes the reflections. Because the energy reflected back to the radar is much weaker than the original signal, the radar has to transmit a very powerful signal. Also, because the energy reflected back to the radar is very weak, the radar could confuse that energy with other radio signals (like a wireless LAN, for example).

Because the 2.4 GHz band is free of radar, the DFS rules apply only to the 5.250–5.725 GHz band.

When the radio detects a radar, it must stop using the channel for 30 minutes at least to protect that service. It then monitors another channel for at least 1 minute and can start using it if no radar is detected during that period.

The burden of DFS compliance falls mostly on the AP. It can hardly be feasible to require the same detection accuracy from all the mobile clients out there, and it's much simpler for regulatory bodies to certify the APs instead. The clients can use a channel if an AP is broadcasting beacons on it, while an AP has to perform a silent detection on a new channel before being able to transmit anything on it. This has important implications. An AP has to listen carefully for 1 minute, checking whether any radar is heard, before being able to send beacons on a given DFS-regulated channel, which severely impacts the ability of the AP to change channel in case of interference (it still can but typically has to face a 1-minute blackout at least or more if it has to scan several DFS channels). On the other side, the client, when moving to a new DFS channel, cannot send a probe request until it has heard a probe response or a beacon from an AP: at that moment, it can safely assume that the AP did the proper verifications and that the channel is clear of radars. However, this has more importance than it may first seem because clients scan other channels much more frequently than APs do. While an AP might decide after a while to consider moving to another channel, which may offer less interference or less congestion, a client has to constantly scan all channels to find an appropriate AP to roam to when it is moving away from the current AP (or even when standing still if the RF conditions change). A client would have to spend only a few milliseconds on a non-DFS channel, basically the time to set its radio to the new channel, send a probe request, wait for the response, and move back to the original channel, causing very little disturbance or absence from its current channel where the APs can be buffering traffic in the meantime. On the contrary, scanning a DFS channel implies moving to that channel, then waiting for the next beacon, and then changing the channel again, which often means a scan time longer than 100 ms. Because of this impact to operations, a lot of client devices prioritize scanning the non-DFS channels over the DFS channels. Considering the typically high number of DFS channels combined with the 100 ms complete traffic loss, client devices can scan a given DFS channel only once every 30 seconds or once every minute.

It is possible that APs falsely report radar events when no radars are around. Cisco has a clear competitive advantage with the APs that have a CleanAir or RF ASIC chipset that can participate along with the Wi-Fi radio to detect radar with a much higher accuracy than any classical wireless radio. A radar event is declared only if both chipsets have heard the radar, which drastically reduces false positives to close to zero.

Some channels, like channels 120, 124, and 128, have even stricter DFS rules because they are used for Terminal Doppler Weather Radar (TDWR), which is of critical importance. The scan time for an AP to be able to use these channels goes up to 10 minutes instead of one. Not all AP models decide to use these channels. The 2800, 3800, and 4800 do support these three channels, for example. Support among AP models is subject to change because APs are able to get certified.

Transmit Power Control (TPC), not to be confused with the RRM feature of the same name, is a system that got defined in the 802.11 standard along with DFS and that allows for the AP to set a transmit power maximum (that is lower or equal to the country maximum) for all stations to respect, in case you want to force your clients to use a lower transmit power.

DFS in the C9800

The **Configuration > Radio Configurations > Parameters** page contains several settings related to the DFS behavior.

Smart DFS is a feature you can enable on the WLC. Without it, each time a radar is heard on a given channel, the AP blocklists this channel for 30 minutes, as per the regulation requirement. The idea is that if there is a constant radar using a channel, you probably do not want your APs to keep trying to use that channel all the time, so Smart DFS, when enabled, doubles the blocklist time each time the channel is blocklisted. As an example, the first time radar is heard on a given channel, that channel is blocked for use on that AP for 30 minutes. If, some time later, radar is heard again on that channel, it is blocked for 1 hour, then for 2, and so on with a maximum of 24 hours' block time.

The Power Constraint field becomes available if you disable DTPC support in the Network page. You can then specify in dBm what is the maximum power clients should be allowed to transmit at.

Enabling the channel switch status means that your outdoor APs (indoor APs typically do not hear radar), when hearing radar on their channel, announce that radar was heard and that the AP will go offline on this particular channel. In loud mode, the AP announces what channel it plans on going toward, but if that new channel is a DFS-regulated one, the AP has to observe the 1-minute scan time when moving to it before being able to transmit beacons. In quiet mode, the AP orders all stations (including clients) that heard the channel switch announcement frame to stop any transmission immediately, whereas loud mode allows for a couple more beacons to be transmitted to make sure everyone heard the change. Some specific clients support the channel switch announcement frame, but so far it is not a majority of the mainstream clients unfortunately. There is no harm in enabling it though. This capability is extremely handy in the case of a mesh network where other outdoor APs are connected to your root AP and not all access points may have heard the radar directly. This way, all the APs go offline at the same time if one of the APs detected radar.

Flexible Radio Assignment (FRA)

Flexible Radio Assignment is a feature focused on APs with dual 5 GHz capability like the Aironet 2800, 3800, 4800 or the Catalyst 9120, 9124, or 9130. Those APs have a radio that can either operate on 2.4 GHz or 5 GHz via configuration while the other radio slot is a static 5 GHz radio. Having a dual 5GHz capability is becoming more and more of a need because the 2.4 GHz space is overcrowded and very limited in terms of number of channels; therefore, adding more APs typically does not add more capacity. Having APs with two 5 GHz radios allows you to add more density and more capacity to the network without overcrowding the RF space, thanks to the typically larger number of 5 GHz channels available in most countries. Having dual 5 GHz radios on an AP is possible only thanks to hardware filtering but also an enforced separation of at least 100 MHz between the two 5 GHz radios and sometimes a difference in transmit power. For internal antenna AP models, when both radios are set to 5 GHz, one of them can reach the maximum transmit power while the other faces a limitation of transmit power to avoid signal bleeding between the two radios. These are called *macro* and *micro cells* (literally, big and small cells). When you use external antennas, thanks to a different set of antennas and a bigger physical separation between the transmitters, no transmit power limitations are in place at all.

FRA refers to the algorithm that automatically defines the best role for a flexible radio. It is possible to statically assign that flexible radio to be monitoring or to be client serving in 2.4 GHz or in 5 GHz, but you could also rely on FRA to determine the best course of action. FRA calculates the redundancy level of the 2.4 GHz coverage and computes the Coverage Overlap Factor (COF). This calculation can even run on APs that do not have a flexible radio. When the COF is above a certain level, FRA moves some flexible radio to client-serving 5 GHz. If the 5 GHz coverage is already dense enough, flexible radios can be moved into the monitoring state and can scan both bands continuously.

FRA can be configured globally in **Configuration > Radio Configuration > RRM**, as shown in Figure 7-32. When enabled, FRA runs at every FRA interval (which needs to be higher or equal to the DCA interval). If the COF calculated for an AP is higher than the sensitivity threshold (100 for low, 95 for medium, and 90 for high), the AP flexible radio role is changed.

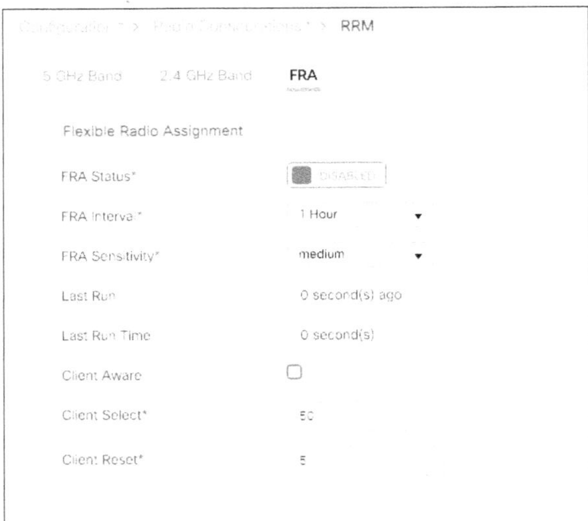

Figure 7-32 *FRA settings on the Catalyst 9800 controller*

You can verify the COF factor calculated for your APs with the command **show ap fra**. The algorithm also mentions the suggested mode for the radio (5 GHz client serving or monitoring). One problem with the algorithm configured as shown in Figure 7-32 is that once the radios are moved to 5 GHz, they do not scan 2.4 GHz coverage at all. So, with every flexible radio moved to 5 GHz operation, the WLC loses a bit of visibility in the 2.4 GHz coverage. Enabling Client Aware FRA allows you to move between 5 GHz monitor and client serving modes when the load on the other 5 GHz radio requires it.

One point to keep in mind is that the RF group leader should ideally be the same WLC between 2.4 GHz and 5 GHz when FRA is in use. To have consistent COF calculations, it is advised to statically set the leader to avoid any problems. This note raises the fact that indeed you could potentially have different WLCs acting as the RF leader for 2.4 GHz and 5 GHz, which sounds a bit weird but is fine. The reason is that 2.4 GHz has a different coverage range from 5 GHz, and therefore, the number of APs in the RF neighborhood is different, and possibly the list of WLCs involved is not the same between both bands.

Tri-radio

Certain AP models such as the Catalyst 9130 and the 9124 are capable of tri-radio operations. The flexible slot is not the 2.4 GHz slot anymore, but thanks to the 8×8 5 GHz radio, they can divide their 5 GHz radio (slot 1) into two separate 5 GHz radios (slot 1 and slot 2), each capable of 4×4 operations. This makes for three Wi-Fi radios (one 2.4 GHz and two 5 GHz) on top of the Bluetooth and IoT radios that can still be operational at the same time and the RF-ASIC, which could also count as an extra radio.

By default, such APs operate with a single 5 GHz radio in 8×8. Enabling tri-radio must first be done globally; otherwise, the AP-specific option stays inactive. The global setting is in **Configuration > Radio Configuration > Network**, as shown in Figure 7-33.

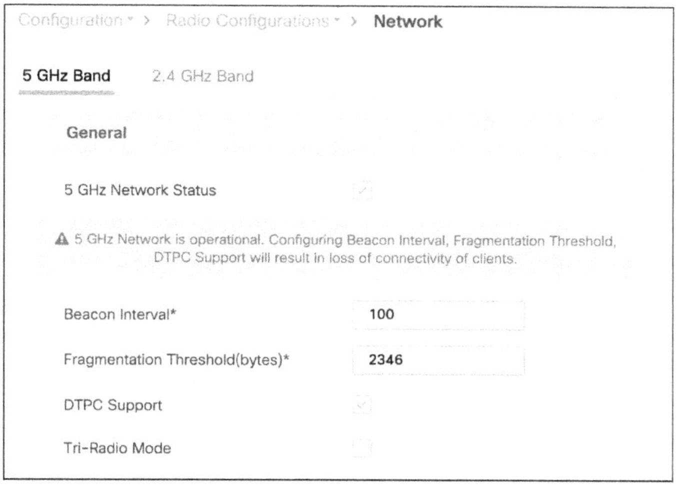

Figure 7-33 Global tri-radio setting

By default, these tri-radio–capable APs show the slot 2 radio, but it stays disabled and the 5 GHz radio still operates in 8×8 MIMO mode. Each tri-radio–capable AP is then, by default, in "auto" radio role. These radios can be statically enabled (to get a dual 4×4 5 GHz radio) or disabled (to keep the 8×8 single radio). If they are enabled, the second slot becomes operational and can be configured like an independent radio. When they are left to auto, FRA is in charge of determining the radio role (either one 8×8 radio or two 4×4 radios). FRA can decide to keep a single 5 GHz radio, use both, or have one in monitoring mode and the other in 4×4 client serving mode.

Figure 7-34 illustrates the slot 1 configuration, which allows you to set the modes to auto or manual. Notice the **Global Tri-Radio Mode** is displayed and gives you a warning if it is disabled.

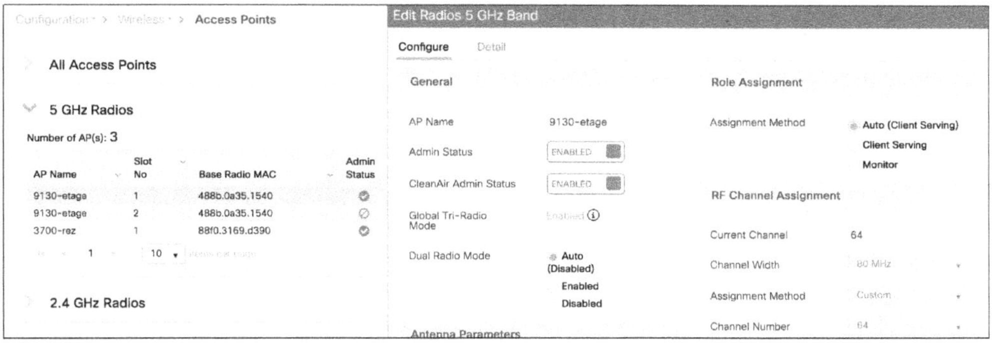

Figure 7-34 Global tri-radio setting

An important note is that 160 MHz operations require both radios to operate together in 8×8 and are not available in tri-radio mode.

Wireless Intrusion Prevention System (WIPS) and Rogue Detection

Wireless brings a new level of possible problems and attack vectors: even if your network is very secure with the latest encryption, you can still be subject to a lot of impersonation or Layer 1 attacks. Rogue detection (not *rouge* as can often be read on the Internet) refers to the detection of unwanted access points in your physical area. A Wireless Intrusion Prevention System refers to the detection and possibly mitigation of low-level (Layer 1 and Layer 2) attacks against your network. A rogue client is defined as a client that connects to a rogue AP. Although a rogue AP is always, by definition, an unwanted device in your network, a rogue client can be one of your own legitimate clients that got tricked into connecting to a rogue AP or could also be an unwanted client.

Rogue AP Detection and Classification

A rogue access point is defined as an access point that you do not manage. The definition could even be extended to "an access point that you do not manage through the same management system." This term can refer to a malicious access point that you probably want to have removed but can also apply to an AP from the neighboring company that can still be heard from your premises or even one of your own APs that is joined to a lab WLC. This means classification of rogues is important so that you know what actions to take (or not) against them.

Detecting a Rogue Access Point

We have already established that APs scan other channels regularly (in a 3-minute interval by default) and briefly listen on those channels for around 50 ms. This gives over a 50 percent chance to hear a beacon transmitted by any AP present on that other channel (considering the most default beaconing interval of 100 ms). All these APs heard are considered rogues by default, but from this list, you need to subtract APs that are managed by the same WLC or other WLCs from the same managed domain. If the AP has heard an NDP frame from the candidate rogue AP and this NDP could be decoded (that is, if the two APs belong to the same RF group), it is not a rogue but an AP from the same deployment. APs that did not send an NDP frame (non-Cisco APs) or whose NDP frame could not be decoded (they belong to another RF group name) are unclassified rogue access points.

A few more facts on rogue detection:

- Monitor mode APs, because they spend all of their time scanning all channels and not serving clients at all, are much more efficient at detecting rogue APs.
- Rogue detection is disabled on OfficeExtend access points because typically there is little use in detecting rogues in domestic areas at each remote office worker location.
- APs with an RF-ASIC (9120 and 9130 series) benefit from the extra software-defined radio doing the scanning for rogue detection; the client serving radio is offloaded from this task. This is also true for aWIPS signatures (discussed later).

The **Configuration > Security > Wireless Protection Policies** page offers some configuration settings for rogue detection and classification, as shown in Figure 7-35. The Low, High, and Critical settings are an easy way to automatically configure the rest of the settings. You can choose any level and see what it configures, or choose **Custom** and define settings yourself. Table 7-2 details the settings configured by each level.

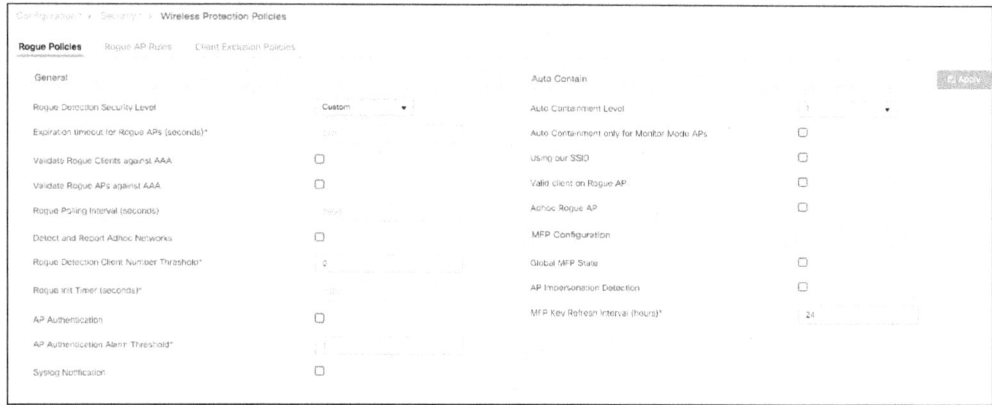

Figure 7-35 *Rogue detection settings*

Table 7-2 *Rogue Severity Level Settings*

Parameter	Critical	High	Low
Cleanup Timer	3600	1200	240
AAA Validate Clients	Disabled	Disabled	Disabled
Adhoc Reporting	Enabled	Enabled	Enabled
Monitor-Mode Report Interval	10 seconds	30 seconds	60 seconds
Minimum RSSI	–128 dBm	–80 dBm	–80 dBm
Transient Interval	600 seconds	300 seconds	120 seconds
Auto Contain Works only on Monitor Mode APs	Disabled	Disabled	Disabled
Auto Contain Level	1	1	1
Auto Contain Same-SSID	Disabled	Disabled	Disabled
Auto Contain Valid Clients on Rogue AP	Disabled	Disabled	Disabled
Auto Contain Adhoc	Disabled	Disabled	Disabled
Containment Auto-Rate	Enabled	Enabled	Enabled
Validate Clients with CMX	Enabled	Enabled	Enabled
Containment FlexConnect	Enabled	Enabled	Enabled

The **Expiration Timeout for Rogue APs** setting defines the amount of time a rogue AP or client is remembered and stored after the time it was last heard. If it is constantly being heard, the timeout does not hit (until it stops being heard).

One problem with WPA2 or open networks is that management frames are not encrypted or signed in any way. A typical attack vector from malicious devices is to spoof beacons of your APs or spoof your APs' MAC addresses to deauthenticate your clients, for example. We already covered having the NDP frames protected rather than "transparent," but that still does not provide a good layer of protection. Enabling AP authentication uses infrastructure 802.11w (also known as Protected Management Frames, or PMF). Every AP then attaches an AP authentication information element to its management frames, which allows other APs hearing it to validate that the frame is coming from a managed AP. In case a management frame is heard with a source MAC address corresponding to a managed AP but the AP authentication information element is either missing or invalid, an AP impersonation attack alarm is thrown.

Management Frame Protection (MFP) can also be configured on the same page. MFP was the prestandard Cisco feature covering the same gap as PMF, and they are therefore mutually exclusive. PMF is now the de facto standard supported by all AP models, whereas MFP support will deprecate in the future.

Classifying Rogue Access Points

You can create custom classification rules in the **Rogue AP Rules** tab. For example, Figure 7-36 creates a friendly rogue rule where all the APs heard broadcasting the SSID of the neighbor company are classified as Friendly (and therefore mostly ignored).

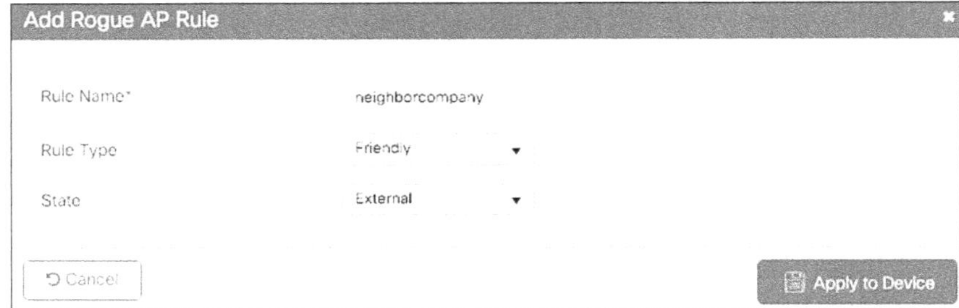

Figure 7-36 *A friendly rogue AP rule*

Internal is the maximum level of trust for a rogue AP. External means the WLC acknowledges the presence of that rogue and does not contain it; the rogue is listed but no alerts are thrown. Alert means actively throwing rogue AP alerts.

Similarly, you could create a rule that classifies as malicious any rogue AP that is heard using any of the managed SSID names and that has a decent RSSI level, as shown in Figure 7-37.

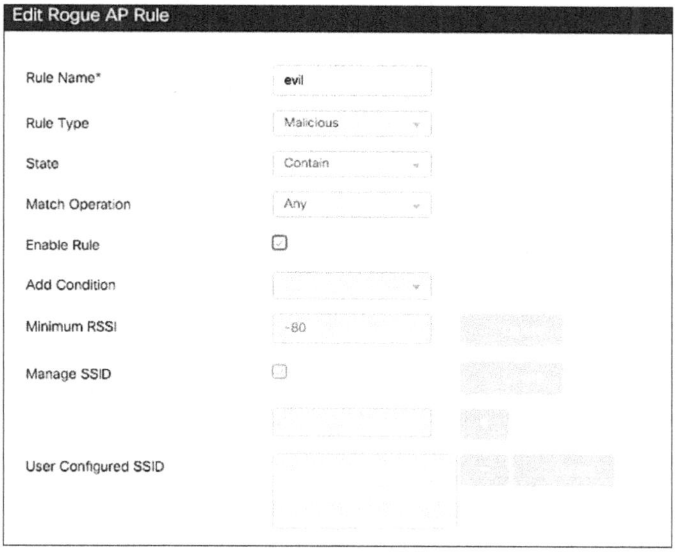

Figure 7-37 *A malicious rogue AP rule*

The action chosen here is to contain it and not simply alert. The WLC actively tries to block this rogue AP (see the later section for more information on containment).

Understanding the Danger of a Rogue Access Point

Attackers have great benefits in placing a rogue access point in the physical area of a network. If the rogue AP advertises the same SSID as the corporate SSID targeted, clients might connect to it inadvertently, and the attacker can collect private data. Authentication and encryption help with this problem but not completely. If a rogue AP is connected to your wired network, it can give attackers access to your wired network from a safe place within the premises but outside of the secured zone. So, detecting whether a rogue AP is connected via wire to the same network is a key element. Finally, rogue APs can simply perform a denial of service against your APs and/or your clients and prevent them from communicating. A combination of all of the previous can be put in place as well: a denial of service at Layer 1 forces the legitimate client to connect to the rogue AP that impersonates the corporate SSID; then the rogue can act as an on-path attacker with the rest of the network and perform various types of attacks.

Containing Rogue Access Points

Containing represents the main mitigation method of rogue APs. It is often misunderstood, and people often have wrong expectations about its effects and impact. There is no way to prevent a rogue AP from transmitting signals. There are no invisible electromagnetic pulses you can send to disable a device remotely and no way you can prevent a foreign device from transmitting RF energy. So what does containment consist of? Managed APs can send deauthentication frames to clients that are trying to connect to the rogue AP, therefore preventing any client from connecting to the rogue and therefore negating its potential malicious impact but not affecting the RF nuisance of the rogue AP that will still beacon.

As usual, things are a bit more complicated in reality. It's important to understand that containment is only an effort to prevent clients from connecting to a rogue AP, but it does not have a 100 percent efficiency. Your APs cannot possibly be using 100 percent of airtime to send deauthentication frames, so the client could still connect very briefly to the rogue, but things also depend on the client behavior. Some clients ignore the deauthentication frame when they see a lot of it, precisely because they assume it's a foreign AP trying to contain them and they persist in trying to connect, whereas other clients might give up trying to connect and instead blacklist the AP (which is the behavior network admins hope for).

Enabling auto containment means that the WLC picks the best access points to contain the rogues that are in the "contain" state (typically, the best is the closest, but if one is very busy, it may pick another one that is less busy). You can decide to limit this activity to only monitor mode APs if you do not want your APs to face any performance impact (spending time off-channel to contain a rogue means less time spent servicing clients). Each AP radio can contain up to three rogues at a given time.

A few facts about rogue containment:

- If you manually contain a rogue, the rogue entry stays in the rogue AP list. If you let a rogue entry expire (because it is not heard anymore), all containment activity related to it stops.
- Deauthentication frames are sent only if rogue clients are detected to be connected to the rogue AP. There is no point in sending deauthentication frames if no clients are connected to the rogue.
- Rogue clients are contained with unicast deauthentication frames targeted at their MAC address.
- Rogue containment cannot happen on DFS channels. It is forbidden to move onto such a channel and to transmit anything until you are certain there are no radars or if a trusted access point is already beaconing on the channels. Cisco APs do not allow containment on DFS channels to avoid violating this rule.
- Only the "alert" rogue state sends the rogue alert to management controllers such as Prime Infrastructure or Cisco DNA Center. Internal or external states do not send any alert.
- It is easy to forget that if rogue containment works for you (that is, it prevents your company clients from associating with rogue APs), it can also work for others. Imagine a neighboring company also having a Cisco wireless network. That company's APs will definitely consider your APs as rogue devices and will prevent any client (including your legitimate clients) from connecting to your access points. For this reason, containing rogues is considered illegal in many countries. Be sure to do your due diligence before configuring any containment activity. Also work with your RF neighbors to mark each other's APs and networks as friendly rogues so that all wireless networks can coexist peacefully.

- On top of this, rogue containment is based solely on sending deauthentication frames and spoofing the rogue AP MAC address. If PMF is enabled on the rogue SSID (which can be set optionally on WPA2 networks but is mandatory on WPA3 SSIDs), Cisco APs are technically unable to contain the rogue AP because they are not able to spoof the rogue AP identity and correctly sign their management frames. Rogue containment therefore works only with open SSIDs or WPA2 networks that do not use PMF. The good news is that configuring PMF on your networks also makes you immune from containment from foreign APs.

The AP join profile allows you to define a few rogue detection and containment settings for groups of APs. As you can see in Figure 7-38, you can enable or disable the detection of rogues but also define the report frequency and the time a rogue has to be seen for it to count as a rogue AP. An interesting setting is the RSSI threshold because it may be interesting to require a certain level of signal to even consider a rogue to represent any concern. If the rogue is below –80 dBm, you could very well decide that the risk it represents for your clients is minimal. The **Rogue Containment Automatic Rate Selection** makes the AP determine the best data rate to use to contain the rogue client based on the received RSSI of that client. The **Auto Containment on FlexConnect Standalone** option makes the FlexConnect APs that lost their links to the WLC keep containing the malicious rogues they are still detecting. The same page allows you to enable aWIPS on the APs, which is covered in the next section.

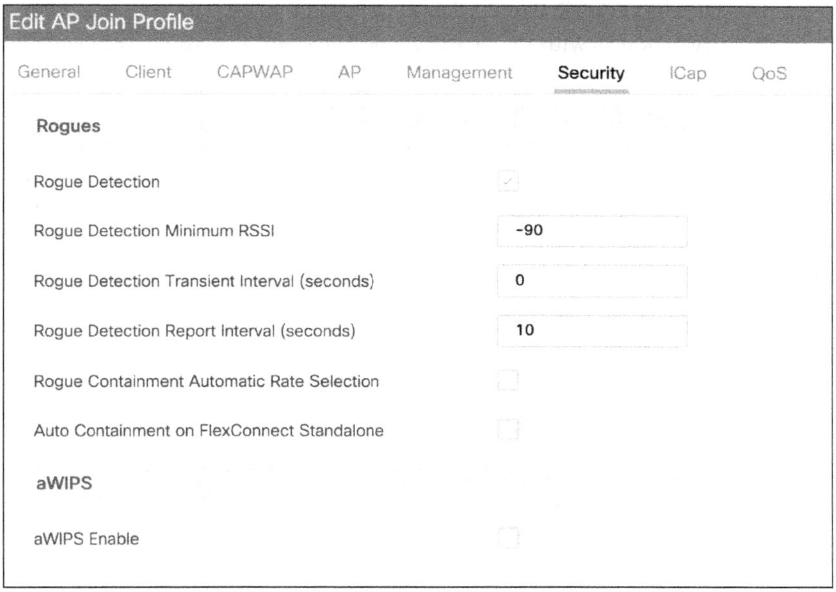

Figure 7-38 *AP join rogue detection settings*

Adaptive WIPS

In the past, the controller was able to detect only low complexity attacks, and external servers were required to correlate attacks over different access points. The Catalyst 9800 has an integrated Adaptive Wireless Intrusion Prevention System (aWIPS), and although DNA Center gives a great dashboard for reporting and managing attacks, no device other than the WLC is required to run the attack detection.

The following attacks were detected as of IOS-XE 17.6:

1. **Authentication flood:** An attacker could be sending 802.11 authentication frames with ever-changing client MAC addresses, therefore filling up the AP client table quickly.
2. **Association request flood and reassociation request flood:** An attacker could be sending abusive association requests to the AP, having the same effect as the previous attack.
3. **Probe request flood:** An attacker could be sending probe requests, forcing the AP to answer nonexistent clients with probe responses and wasting airtime.
4. **Disassociation flood:** An attacker could contain legitimate clients by spoofing disassociation frames from the AP, forcing clients to associate again to regain connectivity.
5. **Broadcast disassociation flood:** An attacker could send disassociation frames with the AP source MAC address with the broadcast destination MAC to disassociate all the clients connected at once.
6. **Deauthentication flood:** Similar to the two previous attacks, sending deauthentication frames spoofed with the AP MAC typically disconnects clients.
7. **Broadcast deauthentication flood:** This flood is similar to the preceding attack, but with a broadcast destination MAC.
8. **EAPOL-Logoff:** An attacker could spoof a client MAC and send an EAPOL-Logoff frame, which terminates the 802.1X authentication status and has the effect of forcing the client to reauthenticate with 802.1X.
9. **Request to send flood:** A wireless device could perform a denial of service if sending back-to-back requests-to-send frames in order to reerve the medium and prevent other devices from transmitting.
10. **Clear to send flood:** Similar to the request to send flood, a clear to send flood prevents other wireless devices from getting a fair chance of accessing the medium.
11. **Fuzzed beacons and probes:** These attacks detect when clients send invalid SSID lengths and fuzzed beacons and probe frames.
12. **PS Poll flood:** A client could be sending too many PS Poll frames to ask for buffered traffic, therefore preventing everyone else from transmitting.

13. **Eapol-start flood:** A client could be sending eapol-start, causing constant restarts of the 802.1X state machine and leaving many sessions opened.
14. **Beacon flood:** An attacker could be sending dummy beacons from nonexistent APs to flood the airtime and make it hard for clients to find the right SSID.
15. **Probe response flood:** An attacker could respond to probe requests on behalf of APs and therefore lure clients into believing wrong settings.
16. **Block ACK flood:** An attacker could be flooding block ACKs to force the client to send retries.
17. **AirDrop:** Apple offers the AirDrop feature, which allows you to send files directly between two devices on a separate channel, which could consume all the airtime on that channel.
18. **Malformed authentication and association frames, invalid MAC OUI (reserved by IEEE) and authentication failure flood:** This alarm includes various malformation conditions of authentication and association frames as well as invalid MAC formats.

Cisco DNA Center has an optional Rogue Management application that gives you rogue management capabilities on top of allowing you to receive aWIPS reports. Click the menu icon and choose **Assurance > Rogue and Awips.** Choose **aWIPS > Enable** to enable aWIPS detection on all the APs.

Cisco DNA Center allows you to configure aWIPS profile to select which alarms you want to focus on. It also correlates all the alarms in an easy-to-read dashboard where you can replay a timeline slider showing the attacks for each time period (the 9800 WLC provides you only with raw alarm counts per AP). Last but not least, Cisco DNA Center also allows you to collect Intelligent Capture forensics of the attack.

On the WLC, you can verify the aWIPS status of an AP with **show awips status <ap Radio MAC>** and list all alarms with **show awips alarm detailed.** The Catalyst 9800 also offers the possibility to send aWIPS alarms (as well as regular rogue alarms) through syslog, which allows for another monitoring tool to receive the information.

Client Exclusion

A basic, yet efficient, method of limiting low-level attacks is to exclude misbehaving clients. Excluding means that for a given period of time (60 seconds by default, but this is configured in the advanced settings of the policy profile), the misbehaving client is completely ignored by the WLC: no data traffic is passed and its association requests are completely ignored. You can configure the conditions for exclusion in **Configuration > Security > Wireless Protection Policies**, as shown in Figure 7-39. They mostly revolve around excessive attempts at associating or authenticating as well as stealing the IP address of another device.

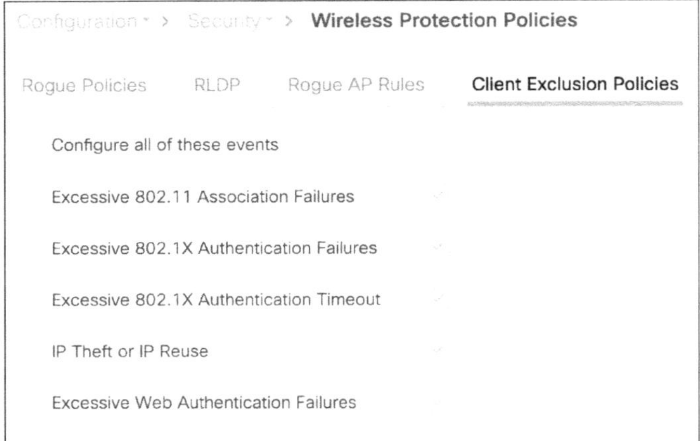

Figure 7-39 *Client exclusion policies on the C9800*

Summary

RF environments can be very challenging, and although 802.11 is a resilient protocol, it cannot operate optimally without a proper design or configuration. Cisco has got your back with a variety of features and constantly improving automated RF tuning. However, it is still critically important to make the right hardware purchasing decision, have good site survey data, mount APs properly, and configure appropiate settings, depending on your environment, to make the best out of your network. This chapter covered the basic antenna concepts and algorithm tuning details you need to have a wireless network running like clockwork.

References

Self-identifying antennas: https://www.cisco.com/c/en/us/td/docs/wireless/antenna/installation/guide/ant2544v4m-r.html

Wi-Fi 6E: https://www.cisco.com/c/en/us/products/wireless/what-is-wi-fi-6e.html

9130 AP deployment guide: https://www.cisco.com/c/dam/en/us/td/docs/wireless/controller/9800/17-3/deployment-guide/c9130-ap-dg.pdf

Chapter 8

Multicast and Multicast Domain Name System (mDNS)

Network traffic is either transmitted from one to one (termed *unicast*), from one to many (termed *multicast*), or one to all (termed *broadcast*). Unicast packets have one source and one destination. However, when you have multiple recipients for the same traffic, unicasting to each recipient would mean sending one copy of the data for each destination. Sending data this way requires the sender to create copies of the data matching to the number of the receivers and results in an increased network load. Multicast forwarding enables a set of sources to send the same traffic to multiple recipients without creating a copy of the packet per recipient. Following are some examples of applications where multicast is used:

- One-way streaming media by online content providers, cable TV operators, educational institutions, and enterprise networks
- Push-to-talk by Vocera badges and music on hold by Cisco Call Manager
- File distribution in campus and commercial networks like operating system images and updates to remote hosts
- Stock tickers and such by financial institutions
- Plug and Play Services like Bonjour

Multicast traffic routing is different compared to unicast traffic routing in that it is almost backward. With unicast routing, the focus is on where the traffic is destined to; with multicast, the attention is more on the source of the traffic. In fact, with multicast traffic, the sender is unaware of the number or the identity of receivers. The sender relies on the recipients subscribing to the multicast stream of interest. A multicast stream is identified by a group IP address that falls in the range of the Class D address space 224.0.0.0 to 239.255.255.255. Some of these addresses are used by well-known protocols (OSPF, BGP, HSRP) and applications, whereas some are available for Internet and private network spaces:

- **224.0.0.0 through 224.0.0.255:** Reserved link-local multicast addresses
- **232.0.0.0 through 232.255.255.255:** Source-specific multicast addresses
- **233.0.0.0 through 233.255.255.255:** GLOP multicast addresses

- **224.0.1.0 through 238.255.255.255:** Globally scoped (Internet-wide) multicast addresses
- **239.0.0.0 through 239.255.255.255:** Administratively scoped (local) multicast addresses

These IP addresses are used only as destination addresses for multicast packets and never as source addresses. As opposed to unicast traffic, the destination does not identify a specific hop or node but rather identifies a content interest group with a constantly changing list of devices interested in that content.

Before we get to wireless multicast, let's briefly review some basic multicast concepts:

- **Internet Group Management Protocol (IGMP):** IGMP is a Layer 3 protocol used by hosts on a local-area network (LAN) to signal their interest in multicast content and hence to join a multicast group; it is also used by a router on the LAN to track multicast groups and the hosts subscribed to them. Each of the three versions of IGMP—version 1, version 2, and version 3—has its own characteristics. Some common IGMP messages include
  - **IGMP join or membership report:** Hosts send these messages to multicast routers to become part of a multicast group.
  - **IGMP leave group:** Hosts send these messages to inform the multicast router they do not want to receive multicast for a specific group.
  - **General membership queries:** A multicast-capable router sends these messages to an entire network to update the multicast group membership for all groups on the network.
  - **Group-specific membership query:** A multicast router sends these messages to a specific group instead of the entire network to update the group's membership.
- **IGMP snooping:** Layer 2 devices like switches and wireless LAN controllers (WLCs) do not, by default, act on IGMP that flows between hosts and routers. IGMP snooping is a feature to enable L2 devices to intercept IGMP packets and process them to have valuable information. By examining these IGMP packets, switches and WLCs note the ports where interested receivers and multicast-capable routers are connected in a snooping table. The snooping table is then leveraged to forward multicast traffic to only the subscribed users and ports and avoid flooding the VLAN to optimize bandwidth consumption.
- **Multicast routing:** Multicast-capable routers on a LAN perform two functions:
  - Manage multicast group membership by listening to IGMP joins from hosts.
  - Create a loop-free path from multicast source(s) to multicast receiver(s) to enable multicast traffic to flow. A loop-free path can be in the form of
  - **Source tree:** A multicast source is at the root of the tree with branches leading to receivers. The tree uses a short path through the network and is referred to as the *shortest path tree (SPT)*. In the case of multiple sources, multiple SPTs exist with

each source at the root of its SPT. On the router, each SPT is represented by an (S,G) entry, where S is the IP address of the source and G represents the multicast group address.

- **Shared tree:** In the network, a single node is chosen as the shared common root for all the multicast groups, known as the rendezvous point. The shared tree is unidirectional where each multicast source sends multicast traffic toward the RP, which then forwards it to the receivers. The shared tree is represented on the router as an (*,G) entry, where * represents all sources and G refers to a multicast group address.

The most common multicast routing protocol is Protocol Independent Multicast (PIM). PIM relies on the unicast routing table made up of static routes or dynamically learned routes to build a tree structure. PIM is available in four flavors: namely, PIM Sparse Mode (PIM-SM), PIM Dense Mode (PIM-DM), bidirectional PIM (Bidir-PIM), and PIM source-specific Multicast (PIM-SSM). The most common flavor of PIM deployed is a hybrid mode of PIM Sparse-Dense Mode.

A wireless deployment relies on L3 multicast routing capabilities of the underlying wired infrastructure to forward multicast traffic. In other words, it is expected that switches or routers serving as a gateway and performing Layer 3 routing for the wired and wireless client subnets are configured with a multicast routing protocol, so #ip pim sparse-dense mode or equivalent is enabled on the L3 interfaces (for example, Switch Virtual Interface [SVI] on an L3 switch).

Plenty of literature is available for the multicast concepts mentioned so far. This chapter focuses only on how these technologies are used in the context of wireless multicast traffic flow.

Figure 8-1 is a simplified representation of what multicast looks like in wireless.

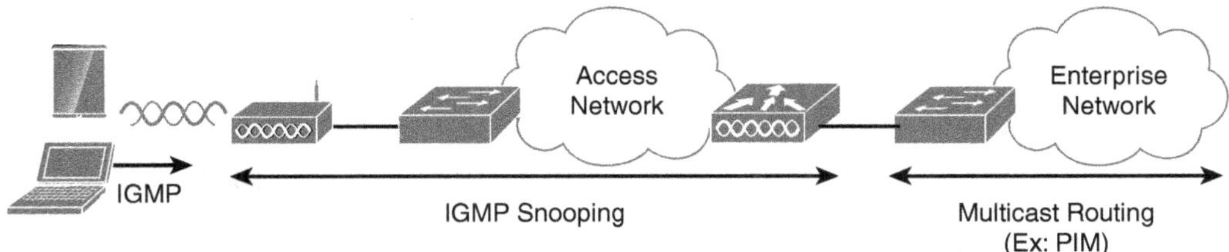

Figure 8-1 *Multicast in wireless*

Broadcast refers to one-to-all transmission within a Layer 2 domain. Broadcasts are typically limited to the subnet and are not routed. In a wireless deployment, similarly to multicastm broadcast forwarding is disabled by default.

This chapter takes a deep dive into

- C9800 multicast configuration and operation
- Reliable multicast or video streaming or media streaming
- Multicast Domain Name System (mDNS)

Wireless Multicast

In a Layer 2 wired network, the default behavior with multicast is to flood to all ports that are part of the destination LAN on the switch. The C9800 defaults to not forwarding multicast to the APs and wireless clients and requires explicit configuration to start forwarding multicasts.

Multicast Packet Flow in Wireless

In a centrally switched wireless deployment, the C9800 oversees data forwarding from and to the wireless clients. In Flex and Fabric deployments, the AP alone owns the multicast forwarding responsibility.

Multicast in a Centralized Wireless Deployment

The wireless multicast packet flow is depicted in Figure 8-2.

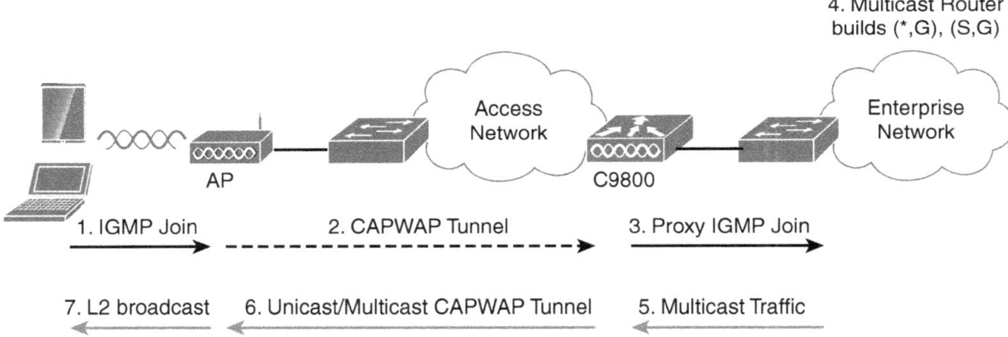

Figure 8-2 *Multicast packet flow in wireless*

As noted by number in the figure,

1. Multicast applications are typically mapped to multicast group IP addresses. Wireless clients running multicast applications send an IGMP Membership Report or IGMP join for the corresponding multicast group IP address to the network.
2. APs receive the IGMP join over radio and transmit it over unicast Control and Provisioning of Wireless Access Point (CAPWAP) tunnels to the C9800.
3. The C9800 runs IGMP snooping, which allows it to intercept these IGMP joins. Upon receipt of an IGMP join, the C9800 performs two actions.
   a. The C9800 creates a multicast group identifier (MGID) based on the source, multicast group IP address, and VLAN (S, G, V) on which the IGMP join was received.
   b. The C9800 sends a proxy IGMP join to the upstream multicast-enabled router sourced from the switch virtual interface (SVI) for the client VLAN on which the IGMP join was received.

4. The upstream switches do not see IGMP joins from each wireless client connected to the WLAN, but they process only the proxy IGMP join from the C9800. They also record the client VLAN SVI IP from the C9800 as the last listener on their local IGMP snooping membership table.
5. When the Layer 3 switch or router designated to perform multicast routing receives a proxy IGMP join, it creates a loop-free path for multicast traffic from the sender to the C9800.
6. When the C9800 receives the multicast traffic from the sender, it encapsulates the multicast traffic in a unicast or multicast CAPWAP header, as configured. The C9800 inserts the MGID to the WLAN mapping derived in step 3a. in the CAPWAP header of the multicast traffic destined to the APs.
7. The AP is responsible for getting multicast packets to all clients in the VLAN. It uses the MGID table updated by the WLC to identify WLANs with active clients subscribed to the multicast group. The AP duplicates a copy to each WLAN with a subscription. The AP sends all multicasts as Layer 2 broadcasts at the highest mandatory data rate using best effort WLAN QoS priority by default or at the multicast data rate configured in the RF profile.

Multicast in Flex

In a FlexConnect deployment, for central switching, the C9800 is still in the path of multicast forwarding and follows the same method as centralized local mode AP deployment. However, for Flex local switching, the C9800 does not receive multicast traffic, and the AP is responsible for bridging the multicast traffic between its wired port and its wireless radios. The C9800 configuration options of enabling multicast globally, IGMP snooping, and multicast mode are not relevant to FlexConnect deployment. The AP does not discriminate between unicast, multicast, and broadcast traffic and bridges them to the upstream access switch that would be responsible for forwarding the traffic to the wired network and to the multicast source. Over the air, the AP sends multicasts as Layer 2 broadcasts at the highest mandatory data rate using Best Effort WLAN QoS priority by default or at the multicast data rate configured in the RF profile.

Multicast in Fabric

In SDA Fabric deployments also, the C9800 is not part of the data path for wireless traffic. The AP forwards the IGMP join from wireless clients over the VXLAN tunnel to the Fabric Edge (FE) switch. In the Fabric, wired and wireless multicast control and data traffic are handled in the same manner. FE transmits multicast traffic to the AP over the same VXLAN tunnel. The AP handling of multicast over the air is exactly the same as in FlexConnect.

How to Configure Multicast on the C9800

To enable multicast forwarding globally on the graphical user interface (GUI), navigate to **Configuration > Services > Multicast** and slide **Global Wireless Multicast Mode** to **Enabled**, as shown in Figure 8-3.

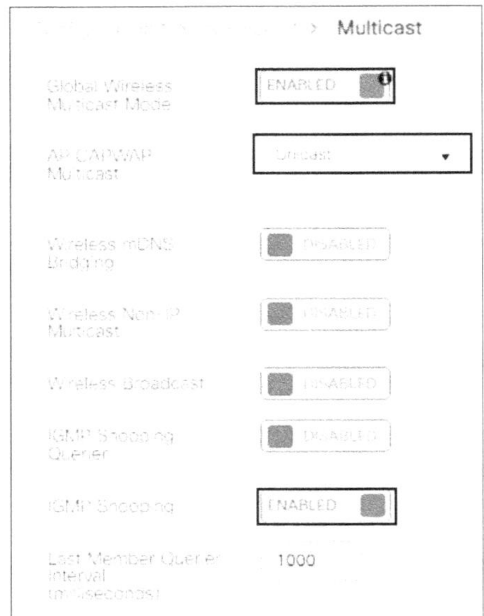

Figure 8-3 *Enabling Multicast and MoU/MoM on the C9800*

You can configure the same settings from the command-line interface (CLI) using

```
C9800(config)#wireless multicast
```

Here are the important configuration aspects to know:

1. **Global Wireless Multicast Mode:** Enabling this option is mandatory for any multicast traffic to be sent or received from wireless clients.

2. **AP CAPWAP Multicast:** This option determines how multicast client traffic is transmitted from the C9800 to the APs and specifically the destination IP address of the encapsulated CAPWAP packet. It supports two modes: Multicast over Unicast (MoU) and Multicast over Multicast (MoM). More details on this in the subsequent section.

3. **Wireless mDNS Bridging (optional):** mDNS bridging refers to the forwarding of the link-local mDNS/Bonjour protocol within an L2 broadcast domain. This option gets set to Enabled when global multicast is enabled but is not required for the mDNS bridging feature because mDNS bridging configuration on the C9800 is managed at the wireless LAN (WLAN) config level.

4. **Wireless Non-IP Multicast (optional):** This config allows a non-IP multicast like the traffic involved in geo-networking, ad hoc vehicular networks, and fleet management to be forwarded to the wireless clients.

5. **Wireless Broadcast:** At Layer 2, broadcasts are flooded to all devices in a given VLAN. However, the C9800 requires this configuration to be explicitly enabled and in conjunction with global wireless multicast to forward broadcasts to wireless clients. This setting applies only to non-ARP/non-DHCPv4/non-DHCPv6 traffic because ARP and DHCP broadcasts are treated in a special way on the C9800.

6. **IGMP Snooping Querier (optional):** The multicast router usually sends out general and group-specific queries in a deployment. However, if a multicast-capable router is not present on network, the C9800 can be configured as the IGMP Snooping Querier to send out queries and keep group membership updated.

7. **IGMP Snooping:** This option enables the C9800 to listen to the IGMP messages sent by wireless clients to identify the hosts as well as multicast groups they are interested in.

8. **Last Member Querier Interval (milliseconds) (optional):** When a host leaves a multicast group, it sends an IGMP leave message. To verify if this host was the last member of the group, the multicast querier sends out group-specific IGMP queries for the duration defined by this Last Member Querier Interval. This timeout is set to 1000 ms by default and applies only when the C9800 is set as the querier. Note that if no response is received for a group-specific query within this 1000 ms timeout, the group is expected to have lost its last member, and mapping pertaining to the multicast group IP is deleted.

IGMP and MLD on the C9800

IGMP is used on IPv4 networks. For IPv6 networks, Multicast Listener Discovery (MLD) is the multicast group membership protocol. The C9800 supports IGMPv1 and MLDv1 with limited support of IGMPv2 and IGMPv3. In addition to IGMP snooping, the C9800 also supports MLD snooping, which allows it to intercept listener queries, listener reports, and listener leave messages exchanged between wireless clients and multicast-capable routers.

IGMP Report Suppression: This option is enabled on the C9800 by default. When multiple hosts are requesting traffic for a multicast group address, the C9800 proxies only the first IGMP report received from all the hosts for that group. The rest of the reports are not transmitted by the C9800. This prevents duplicate reports from arriving at the multicast router.

IGMP Leave: When a client wants to unsubscribe from a multicast group address, it can do so by transmitting an IGMP leave. This traverses the unicast AP CAPWAP tunnel to reach the C9800, which proxies IGMP leave to a multicast-enabled router. The router sends a group-specific multicast and waits 1000 ms to confirm this client is the last receiver for the multicast group. When a timeout occurs, the router prunes traffic to this multicast group. The outgoing interface list on **#show ip mroute** reflects Null. The C9800 and the AP remove the group entry from their corresponding IGMP snooping and MGID tables.

IGMP Snooping: IGMP snooping is enabled by default but comes into effect only when global multicast forwarding is configured. When IGMP snooping is enabled, the C9800 generates a wireless client MAC to multicast group address mapping to keep track of active clients and multicast groups. The C9800 derives Layer 3 MGID from the VLAN on which the wireless clients send IGMP joins and a multicast group IP address for which joins are sent. Once MGID is created, the C9800 creates a replication group for each MGID mapping to all the CAPWAP interfaces that should receive the multicast.

MGID is a 14-bit value that is filled in the 16-bit reserved field of wireless information in the CAPWAP header with the remaining two bits set to zero. Using the CAPWAP header, the C9800 updates the AP MGID table, so it is aware of active wireless multicast clients. The C9800 supports three ranges of MGIDs:

- **1–4095:** Broadcast MGID that represents (S=0,G=0,V). This MGID is allocated for each VLAN for which a WLAN is mapped. The broadcast MGID is the same as the VLAN number.
- **4160–8191:** Multicast MGID that represents the (s,G,V) or (*,G,V) tuple. This MGID is created when a client joins a multicast group. This pool is used to allocate IPv4 multicast MGIDs.
- **8192–(16K-1):** Multicast MGID for IPv6 multicast. This range is used for IPv6 RA packets.

In the case of a Layer 2 multicast or where IGMP snooping is disabled, a broadcast L2 MGID is derived just from the ingress interface and used to update the AP. On the C9800, the L2 MGID value is equal to the VLAN ID. Also, when IGMP snooping is disabled, the WLC does not proxy IGMP, and all the IGMP joins from all the clients are forwarded to the multicast router. The upstream switch reflects the wireless client as the last listener instead of the C9800 client VLAN SVI IP address.

IGMPv3: IGMP snooping on the C9800 snoops IGMPv1, IGMPv2, and IGMPv3 joins by default without any explicit configuration. However, you need to make sure that IGMPv3 is explicitly enabled on all the switches and routers in the path from multicast source to receiver. If you want to use IGMPv3 even for the AP CAPWAP multicast group, explicit configuration is needed.

MLD snooping: MLD snooping is the equivalent of IGMP snooping but for the IPv6 environment. The C9800 uses MLD snooping to build multicast membership lists, which are used to identify IPv6 receivers to forward multicasts to.

CAPWAP Multicast

When an AP registers to a C9800, it forms a unicast CAPWAP tunnel. IGMP packets like IGMP joins sent by a wireless client to subscribe to a multicast stream and IGMP leaves sent by a wireless client to unsubscribe from previously subscribed multicast streams traverse this unicast CAPWAP tunnel in an upstream direction. However, when transmitting the multicast traffic in the downstream direction, from the C9800 to the APs, the C9800 supports both unicast and multicast CAPWAP tunnels. These two modes of operation are named Multicast over Unicast (MoU) and Multicast over Multicast (MoM).

Multicast over Unicast (MoU)

In MoU mode, the C9800 encapsulates a multicast packet for wireless clients into a unicast CAPWAP tunnel destined to the individual AP. The C9800 is responsible for creating as many copies as there are APs. Figure 8-4 shows the replication being done by the C9800 in MoU mode.

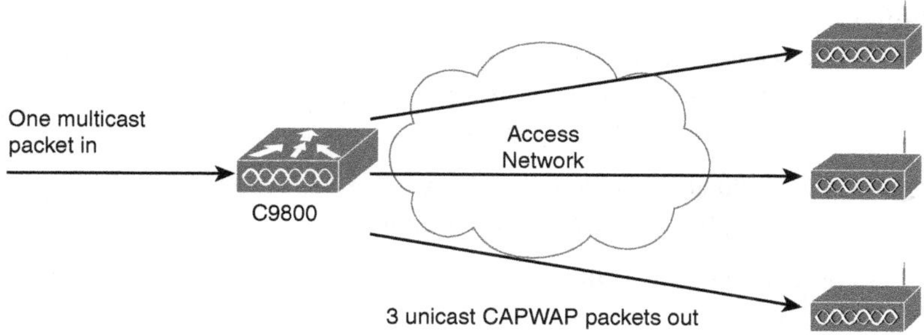

Figure 8-4 *MoU: the C9800 creating copies for each multicast packet*

Duplicating the multicast traffic creates a replication burden on the C9800, leading to a performance impact. Therefore, this mode is supported on only low-scale WLC form factors like the C9800-CL (small template) and C9800-L, where the maximum supported AP count is 1000 or less. Although supported, this mode is not recommended because it increases the traffic load on the network, and it should be used only in deployments where multicast routing is not available in the network between the WLC and the APs. To enable MoU, you only need to enable multicast globally and no additional configuration is needed.

Figure 8-5 shows the packet format for the MoU flow. Figure 8-6 shows a sample snippet of a multicast packet being sent from the C9800 in MoU mode. The highlighted inner header shows a multicast packet sourced from 10.15.1.2 (multicast source) destined to the multicast group IP 234.5.6.14. The outer header is the unicast CAPWAP encapsulation from the C9800's wireless management IP 192.168.1.5 to the AP with IP address 10.5.1.11.

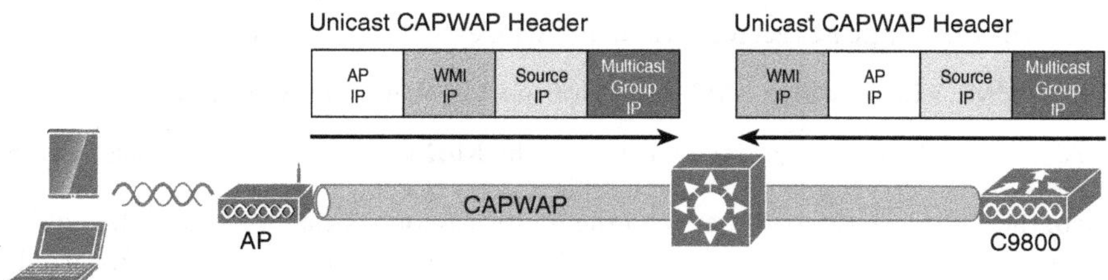

Figure 8-5 *Packet format in MoU mode*

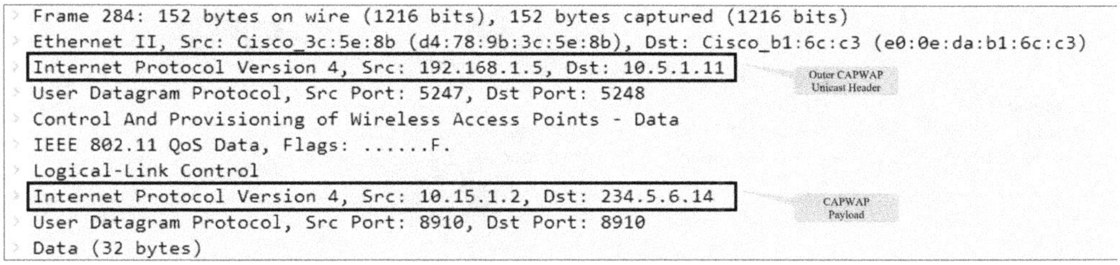

```
Frame 284: 152 bytes on wire (1216 bits), 152 bytes captured (1216 bits)
Ethernet II, Src: Cisco_3c:5e:8b (d4:78:9b:3c:5e:8b), Dst: Cisco_b1:6c:c3 (e0:0e:da:b1:6c:c3)
Internet Protocol Version 4, Src: 192.168.1.5, Dst: 10.5.1.11          Outer CAPWAP Unicast Header
User Datagram Protocol, Src Port: 5247, Dst Port: 5248
Control And Provisioning of Wireless Access Points - Data
IEEE 802.11 QoS Data, Flags: ......F.
Logical-Link Control
Internet Protocol Version 4, Src: 10.15.1.2, Dst: 234.5.6.14          CAPWAP Payload
User Datagram Protocol, Src Port: 8910, Dst Port: 8910
Data (32 bytes)
```

Figure 8-6 *Multicast packet snippet from the C9800 to the AP in MoM mode*

Multicast over Multicast (MoM)

In MoM mode, the C9800 needs to be explicitly configured with a multicast group IP address in the AP CAPWAP multicast group address field. While choosing the CAPWAP multicast group IP, you need to honor these limitations:

1. The IP address should belong to the multicast IP range of 224.0.0.0 to 239.255.255.255, ensuring the IP address is not reserved.
2. The IP address configured on the C9800 should not be used elsewhere on the network for any applications.
3. The IP address should be unique to each C9800 in case of a multi-WLC deployment. Otherwise, duplicate joins are seen at the AP.

The APs send an IGMP join to subscribe to this multicast group. Note that the APs in monitor mode, sniffer mode, or rogue detector mode do not join the multicast group. The C9800 uses IGMPv2 by default for the AP CAPWAP multicast group but can be configured to use IGMPv3. The FlexConnect APs join the multicast group only if they host centrally switched WLANs. Layer 2 switches along the way use IGMP snooping to intercept IGMP joins from the APs and create a membership table of the CAPWAP multicast group to the ports where interested APs are connected. When the APs and the C9800 Wireless Management Interface (WMI) IP address belong to different subnets, CAPWAP multicast flow from the C9800 to the AP relies on a multicast-capable router being present in the deployment. A multicast router needs to have multicast routing configured on all VLANs involved in the MoM flow. This includes

- VLANs of the multicast source, wired and wireless
- VLANs of the multicast receiver, wired and wireless
- C9800 WMI VLAN, which is the source of the MoM AP CAPWAP multicast group
- VLANs of the APs, which are the receiver of the MoM AP CAPWAP multicast group

Only then, the routers in the path between the multicast source, the WLC, and the APs can build the corresponding tree with (*,G) or (S,G) entries for the multicast groups used by client traffic and AP CAPWAP multicast. Figure 8-7 depicts the replication done by the network in MoM for a deployment of three APs.

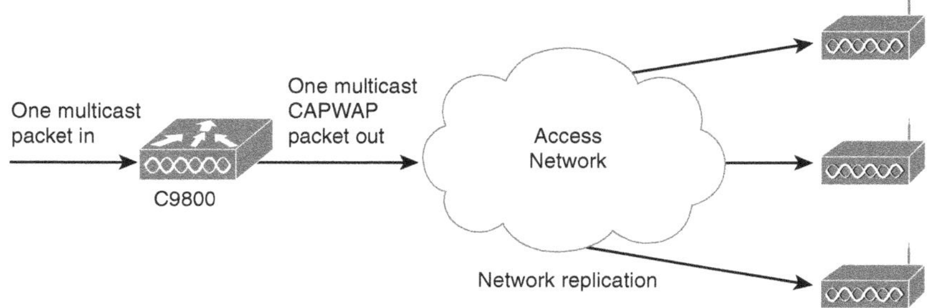

Figure 8-7 *MoM: Network creates copies of multicast packets*

With this control path set up, the C9800 sends only one copy of the multicast application packet wrapped in a CAPWAP multicast header to the AP, as shown in Figure 8-7. The outer CAPWAP header has the source IP set to the wireless management interface IP of the C9800 and destination pointing to the CAPWAP multicast field. The inner packet has the source IP of the device sourcing multicast traffic with the destination being the multicast group address used by the application. This is an asynchronous tunnel with all communication from the APs to the C9800 still encapsulated as unicast CAPWAP traffic, while from the C9800 to the AP, it is multicast encapsulated.

Figure 8-8 shows a packet format for the MoM flow. Figure 8-9 shows an example of a multicast packet being sent from the C9800 to the AP in MoM mode. The highlighted inner header shows a multicast packet sourced from 10.15.1.2 (multicast source) destined to the multicast group IP 234.5.6.13. The outer header is the multicast CAPWAP encapsulation from the C9800's wireless management IP 192.168.1.5 to the AP CAPWAP multicast group IP 239.10.10.10.

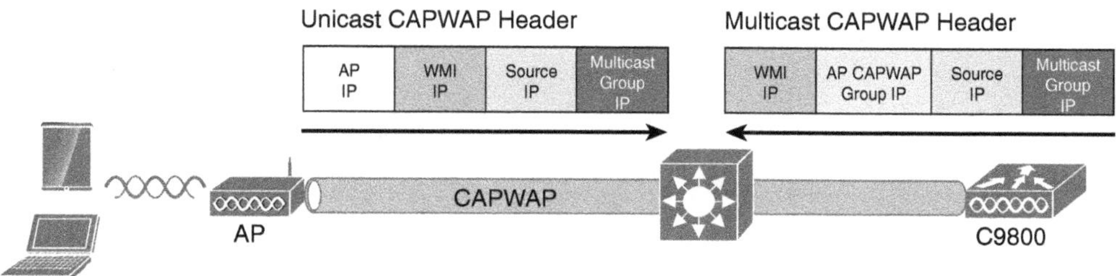

Figure 8-8 *Packet format in MoM*

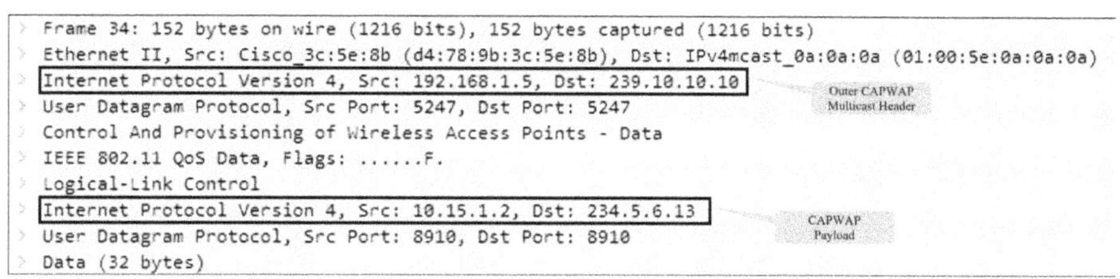

Figure 8-9 *Multicast packet snippet from the C9800 to the AP in MoU mode*

MoM is the only supported multicast mode on large C9800 form factors like the C9800-CL (medium template), C9800-CL (large template), C9800-40, and C9800-80. When multicast is enabled globally on the CLI, it will not take effect until AP CAPWAP Multicast is defined as MoM, and AP CAPWAP Multicast Group IP Address is configured. To configure MoM, navigate to **Configuration > Services > Multicast**, and choose **Multicast** in the AP CAPWAP Multicast drop-down. When Multicast is chosen, the AP CAPWAP IPv4 Multicast Group Address and the AP CAPWAP IPv6 Multicast Group Address fields become visible. In addition, a prompt to define a valid IPv4 or IPv6 multicast IP address is presented, as shown in Figure 8-10. Populate these fields with the group IP address that the APs subscribe to. The corresponding configuration CLI is

```
C9800(config)#wireless multicast x.x.x.x
```

On the C9800, MoM does not work across inter-subnet mobility events such as guest tunneling but does work across Layer 3 roams.

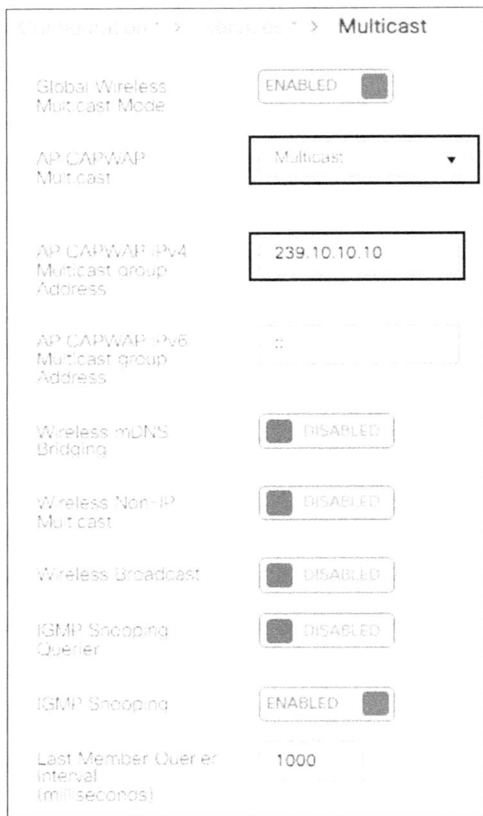

Figure 8-10 *Configuring MoM on the C9800*

When MoM is configured, you can verify the MoM tunnel status and the APs using the MoM tunnel:

```
C9800L#show ap multicast mom
AP Name                              MOM-IP TYPE      MOM-STATUS
-----------------------------------------------------------------
sudha-9130                           IPv4             Up
```

802.11 Multicast

On a wired infrastructure, multicast brings efficiencies, but it was not designed for wireless networks and, over the air, multicast experiences many challenges. When a wireless client sends multicast to the AP, the AP responds with an acknowledgment (ACK), if no errors are found, and then retransmits as multicast. Over the air, the AP sends multicast as broadcast. However, when the AP sends multicast to the client, there is no ACK, and the AP does not retransmit. Packet loss due to signal loss and collisions severely impact multicast.

The AP sends multicast as broadcasts over the air at the highest mandatory data rate. The intent with multicast is to reach all the receivers. With wireless being a collision domain, the clients at the cell edge risk not receiving multicast. To maximize multicast receivers reached, the AP is forced to send at the lowest common denominator settings. Low data rate means more airtime and increases waiting for other stations to transmit. For high-bandwidth applications like video, this leads to poor efficiency and potentially poor video stream quality.

In the presence of power save (PS) clients, the AP buffers all multicasts intended for the clients and sends them at the end of the Delivery Traffic Indication Message (DTIM) interval configured on the C9800. This buffering introduces a delay, which, in the case of video, leads to choppiness. Further, all stations must keep track of this DTIM interval and wake up to receive the packet and discard it if they don't need it. The higher the DTIM interval, the more delay added, but if DTIM is set too low, battery-powered clients must wake up, often depleting their battery life. A more reliable approach to get multicast to most clients would be to convert multicast to unicast, referred to as *video streaming*, as discussed in subsequent sections.

To send multicast out on the air, the APs rely on the MGID table updated by the C9800 to identify the WLAN that has active subscribed clients for the multicast group. Multicast traffic is forwarded only on that specific WLAN. If the WLAN is mapped to multiple VLANs via VLAN grouping or VLAN pooling, the C9800 creates different MGIDs for the same multicast group for different VLANs in the VLAN pool. The router sends as many copies of the multicast packet as VLANs in the pool, which is then transmitted over the air, causing clients to receive duplicate multicast packets. To suppress this duplication and better use the wireless medium, an optimization called VLAN Select is enabled.

Multicast VLAN: One of the VLANs in the VLAN pool mapped to the WLAN is designated as a multicast VLAN. Clients listen to this specific multicast VLAN for the multicast stream. The C9800 creates an MGID based on the multicast group address and the multicast VLAN. If multiple clients on the VLAN pool of the same WLAN are listening to the same multicast stream, only one MGID is created, and only one copy of the multicast packet is sent to the AP and over the air to the client. On the C9800 GUI, navigate to **Configuration > Tags and Profiles > Policy**, select the policy that is mapped to the VLAN pool, and define one of the VLANs in the pool as a multicast VLAN, as shown in Figure 8-11.

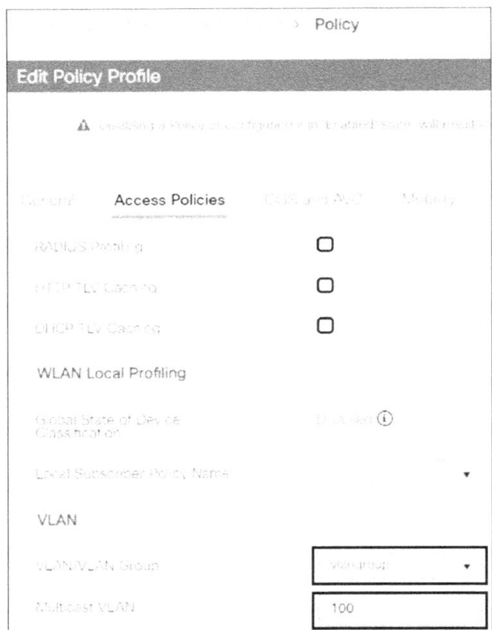

Figure 8-11 *Configuring VLAN Select or multicast VLAN on the C9800 wireless policy profile*

Wireless Broadcast and Non-IP Multicast

Non-IP multicast refers to multicast that is neither IPv4 nor IPv6. Unlike wired networks, the C9800 does not flood broadcast or non-IP multicast to wireless clients by default and requires explicit configuration. Following are some guidelines for this feature:

- Enabling or disabling of non-IPv4 multicast and broadcast forwarding is supported only in central switching deployments.
- In Flex and Fabric deployments, the APs always bridge non-IPv4 multicasts and broadcasts.
- This feature can be enabled and disabled globally or per VLAN.
- The non-IPv4 multicasts and broadcasts are not forwarded on VLANs returned by AAA override and are restricted to clients on the VLAN mapped to the WLAN.
- Non-IPv4 multicast and broadcast forwarding are not supported when a VLAN pool is mapped to the WLAN profile.

By default, all VLANs show broadcast and non-IP multicast enabled when wireless multicast is enabled globally, as shown in Figure 8-12. However, the configuration does not take effect until the individual features of wireless broadcast and wireless non-IP multicast are enabled globally, as shown in Figure 8-13.

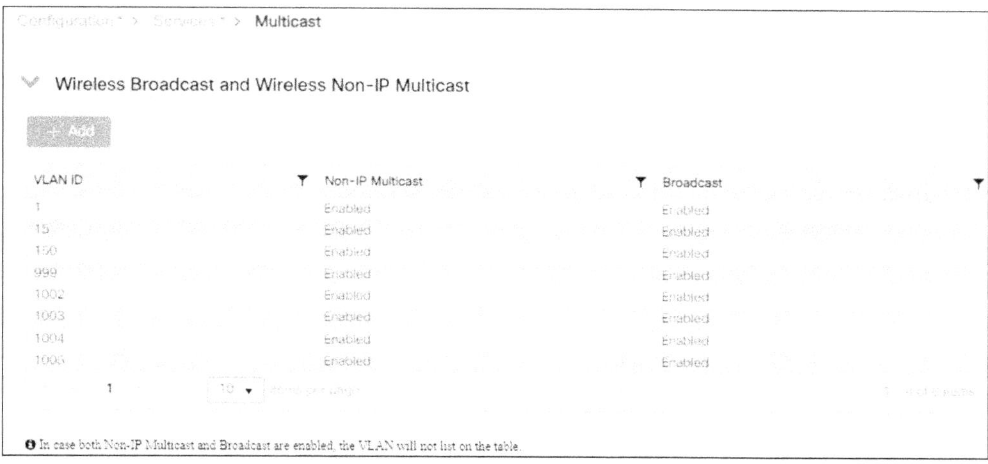

Figure 8-12 *Default settings of wireless broadcast and non-IP multicast per VLAN when wireless multicast is enabled*

Figure 8-13 *Configuring wireless broadcast and non-IP multicast forwarding*

Multicast in Client Roaming Scenarios

On the C9800, when a wireless client is onboarded, an entry is created for the client along with the associated AP, WLAN, and other contextual information. When the client roams, as discussed in Chapter 7, "RF Deployment and Guidelines," the context is maintained to allow for seamless roaming. In the case of multicast, roaming introduces the additional challenge of maintaining group membership across roams, while ensuring the least impact to the multicast application due to traffic loss. Client roaming can be Layer 2 or Layer 3.

Specific to the C9800 architecture, Layer 2 or Layer 3 roaming can occur in three different ways. Consider a client roaming from AP1 to AP2.

- **Intra-WNCd roam:** When a client roams between two APs, AP1 and AP2, on the same site tag and therefore on a single WNCd instance, it is called *Intra-WNCd roam*. When the client roams to AP2, because the same WNCd instance is aware of both APs, no mobility handoff message is needed. WNCd looks up the client MAC address to identify its record, along with (S,G,V) and MGID info, and sends a CAPWAP config update to AP2 to keep traffic flowing.
- **Inter-WNCd roam:** When the clients roam between two APs, AP1 and AP2, mapped to different site tags, which, in turn, map to different WNCd instances within a C9800, the roam is called *Inter-WNCd roam*. When the client roams to AP2, mobility handoff occurs between the two WNCd instances. An IGMP join from the client informs the new WNCd of the continued subscription from the client. The new WNCd looks up the client record for (S,G,V) and MGID to WLAN mapping information and sends a config update to AP2 to keep the multicast traffic flowing. In addition, with MoU, WNCds update the data plane to point to the new CAPWAP interface.
- **Inter-WLC roam:** When the clients roam between two APs, from AP1 to AP2, mapped to different C9800s, it is referred to as an *Inter-WLC roam*. When the client roams to AP2 mapped to the second C9800, Mobile Announce is sent by the second C9800 to verify if any other controller in the mobility group is aware of the client. The first C9800 where AP1 is registered responds with the unicast Mobile Handoff to the second C9800, which includes multicast groups to which the client is subscribed. The second C9800 creates an MGID for this (S,G,V), which, in turn, programs the MGID table on the AP. An entry is created or updated on the data plane of the C9800, so traffic forwarding is handled in hardware.
  - **Layer 2 Roaming:** If the two C9800s have the same client VLAN assigned, the roam is Layer 2. In the case of L2 roaming, the client context is moved to the new C9800 and deleted from the previous C9800. Accordingly, the multicast traffic traverses the new C9800 to get to the client.
  - **Layer 3 Roaming:** If the two C9800s map the client to different VLANs, it is a Layer 3 roam. In L3 roaming, the client context is maintained on the newly roamed-to WLC called the *foreign WLC* and the original roamed-from WLC termed the *anchor WLC*. The unicast traffic is tunneled back to this anchor C9800, which enables the client to maintain the same IP address and security

context and continue to forward traffic. However, multicast traffic does not traverse the mobility tunnel and is, in fact, handled by the foreign C9800. Upon roam, the multicast information about the list of multicast groups that the client has joined is included in the optional mobility handoff message. The foreign C9800 sends out the query to identify whether the client is interested in the multicast stream. When the IGMP join is received from the client, the foreign C9800 sends out the proxy IGMP join upstream to the network, and all multicast traffic for the client is redirected to the foreign C9800.

- **Auto-anchoring:** This is a special L3 roaming scenario that is typically used to anchor the wireless guest clients to a specific WLC, usually located in the demilitarized zone (DMZ) of the deployment. In this case, the intent is to have all guest client traffic terminate at the DMZ WLC and not be able to access any internal devices. However, multicast traffic is not allowed over the mobility tunnel, and consequently, multicast is not supported with auto-anchoring.

Media Stream Feature

MoM optimizes the flow of wireless multicast over the wire from the C9800 to the AP. However, 802.11 multicast still has two fundamental problems that can be addressed via Media Stream or what is also called VideoStream or Multicast Direct:

- **Multicast reliability:** The IEEE 802.11 wireless multicast delivery mechanism is not reliable because there is no ACK mechanism for clients to acknowledge whether they have received a multicast frame sent by the access point. As a result, any lost or corrupted multicast packets are not retransmitted. Wi-Fi networks typically operate with a packet error rate (PER) of 1 percent to 10 percent (due to collisions, fading, and interference), and without acknowledgments/retransmissions, this results in a multicast packet loss rate (PLR) of 1 to 10 percent. But video applications typically require PLR on the order of 0.1 to 0.5 percent. Due to this PLR mismatch, using the native 802.11 multicast delivery mechanism often causes an IP multicast video stream to be unviewable. Media Stream addresses this concern by converting the multicast frames to unicast 802.11 frames over the air, referred to as MC2UC (for Multicast to Unicast). When the frames are sent as unicast, the AP is able to receive ACKs from the clients and to determine when frames need to be retransmitted (in the case of lost or corrupted frames).

- **Video QoS handling:** As discussed in Chapter 9, "Quality of Service (QoS)," the 802.11 WMM provides a mechanism for video traffic to be sent in a special video queue that has higher priority than best effort traffic (but lower than voice). However, for true QoS protection, a resource reservation and admission mechanism must be implemented to make sure the air is not oversubscribed, and that voice, video, and best effort data traffic can all be accommodated. This is particularly true because in the MC2UC approach—when multiple clients are listening to the same multicast stream—the packets must be repeated over the air, thus using more airtime. Resource Reservation Control (RRC) implemented as part of Media Stream addresses this need.

Cell Planning

For deploying voice and video in a wireless network, the most important aspect is cell planning. The cell planning section of the VideoStream deployment guide included in the later "References" section provides some samples of total voice and video calls that can be accommodated with different data rates. In addition to planning for the wireless medium, video brings additional considerations like latency, jitter, and packet loss. Table 8-1 depicts the expectations of different video applications to latency, jitter, throughput, and packet loss. As you can see, live teleconferencing has higher sensitivity to latency and jitter, whereas video on demand has higher tolerance to latency and jitter. High-definition (HD) videoconferencing would invariably have higher throughput requirements than non-HD video applications. In videoconferencing applications, Table 8-2 quantifies network latency, jitter, and packet loss of some common video applications. Note that video applications like conferencing, collaboration, and telepresence require audio and video to sync, and the user experience is dictated by the slower of the two. The references include voice over wireless design guides that can be used to plan coverage cells for the wireless APs.

Table 8-1 *Expected Metrics for Common Video Applications*

Application Name	Latency	Jitter	Throughput	Packet Loss
Teleconferencing	High	High	Low	Medium
HD videoconferencing	High	High	High	High
Video on demand	Low	Low	Medium	Low
Live streaming video	Medium	Medium	Medium	High

Table 8-2 *Metric Values That Can Be Used to Design Video over Wireless Deployment*

Application Name	Latency (ms)	Jitter	Packet Loss
Video collaboration	150	30	1%
Digital signage	200	10	0.05%
TelePresence	150	10	0.05%
Video surveillance	300	10	0.05%

Components of VideoStream

Media Stream helps to deliver reliable and superior quality video over the C9800 wireless deployment. The components involved in Media Stream are stream admission, radio reservation control, multicast to unicast, and quality of service.

Stream admission: Stream admission enables you to specify the media streams that need to be given preferential treatment. It provides the configuration options for you to define characteristics of the media streams, including multicast destination IP addresses, bandwidth required, and resource reservation parameters for the media streams of interest. The bandwidth defined is the guaranteed bandwidth allocated to each client receiving this media stream. This needs to be chosen with care to be as close to the video bit rate as possible. If bandwidth is less than the video bit rate, the video on the wireless client end will be choppy, pixelated, or freeze often. In contrast, if the bandwidth dedicated is too high, the number of clients that can be accommodated will be less and bandwidth is wasted.

Resource Reservation Control (RRC): RRC is the decision-making algorithm for media streaming that manages admission and policy controls for media streams. RRC on the C9800 determines whether a particular client needs to be admitted or denied to the multicast stream. If a client is admitted, RRC also informs the level of service that the stream should receive. RRC uses the following parameters to make these decisions:

1. **Radio resource measurements:** Client Count, Channel Utilization, Current Transmit Rate, Radio Transmit Queue Limit, Radio Transmit Queue Size
2. **Traffic statistics:** Media Bandwidth, Media Latency, Client Link Rate
3. **System configuration:** All radio measurements and traffic statistics are calculated at the AP through continuous data collection and delivered to the C9800. These metrics and statistics are collected on a per-stream, per-client basis. When the C9800 receives the RRC response, it processes metrics and statistics against the configured policy with the intent to prevent oversubscription and makes one of three decisions.
   - **Admit:** If RRC metrics match the policy, the client is admitted to the stream by creating an MGID entry with MC2UC enabled for the client.
   - **Violation:** If RRC metrics violate a policy in some aspect such as channel utilization or client link rate but video bandwidth is available, the client is accommodated to the stream but as a best effort client.
   - **Deny:** If video bandwidth is not available per configured stream requirements, the C9800 offers protection to existing client streams by denying admittance to the new client.

RRC evaluation can be done only initially when a new client subscribes to the stream, or it can be periodic with service adjusted to adapt to the changing radio conditions and client statistics.

4. **Call admission control (CAC):** CAC is a per-radio configuration that guarantees configured media bandwidth and blocks new media users to safeguard existing clients. On the C9800, both voice and video CAC are disabled by default, and configuration options are for media, which includes both voice and video. When

CAC is enabled, maximum radio bandwidth reserved for media is 85 percent by default with the remaining 15 percent serving the best effort data traffic. This reserved 85 percent needs to be carved out to accommodate voice and video clients as needed by the specific deployment.

MC2UC: While the RRC decision marks the client for MC2UC on the C9800, the actual replication and transition of multicast to unicast 802.11 frames occurs at the AP.

Stream prioritization and QoS: On the C9800, video is the only priority allowed for a video stream that meets the policy requirements. In case of violation where bandwidth is available, but all radio and client metrics are not optimal, you can configure the option to allow the media stream but at best effort in lieu of dropping the traffic entirely. The AP delivers on the QoS by placing traffic in the video transmission queue at the radio or best effort queue as applicable.

Figure 8-14 provides the flow diagram for the C9800 video stream admittance and denial process:

1. The C9800 monitors its IGMP snooping database to process IGMP joins from clients and determine multicast IP addresses of the video streams. If the WLAN profile of the client has the media stream enabled and the stream's multicast IP address is defined as a stream of interest on the C9800, the C9800 sends an RRC request to the APs.
2. The AP blocks the video stream and responds to the C9800 with radio and client RRC metrics.
3. The C9800 reviews available bandwidth, stream metrics, required airtime, current load on the radio, and health of the media against the configured policy as part of its RRC calculation. Based on this, the C9800 admits the client to or denies the client from the media stream. Further, the C9800 concludes whether the client will be given video priority or best effort service when admitted.

4a. If the client is admitted, an MGID entry with MC2UC enabled is created and updated to the AP from the C9800.

4b. If the client cannot be accommodated, the client is denied from receiving the media stream.

5. The AP performs replication and converts multicast into an 802.11 unicast frame, which is sent to the client.

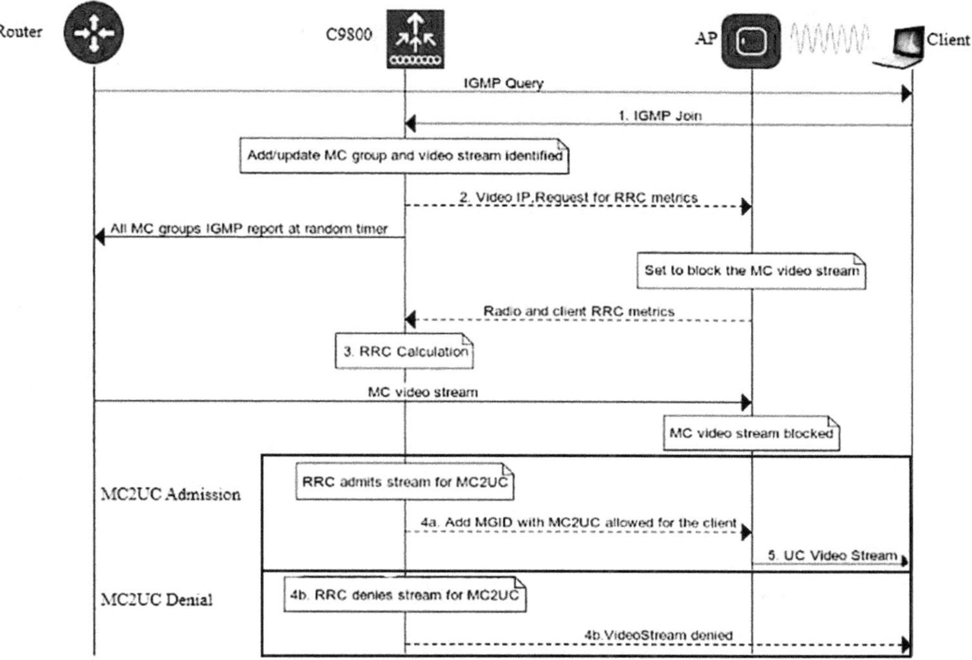

Figure 8-14 *Media Stream packet flow*

How to Configure Media Stream

To enable reliable multicast or streaming on the C9800, you need to perform the following minimum configuration:

- Enable Media Stream globally.
- Define Multicast Stream characteristics and policy.
- Enable Media Stream on the specific radio band.
- Enable Media Stream on the client WLAN profile.

To enable Media Stream globally, navigate to **Configure > Wireless > Media Stream > General**, and choose **Multicast Direct Enable**, as shown in Figure 8-15.

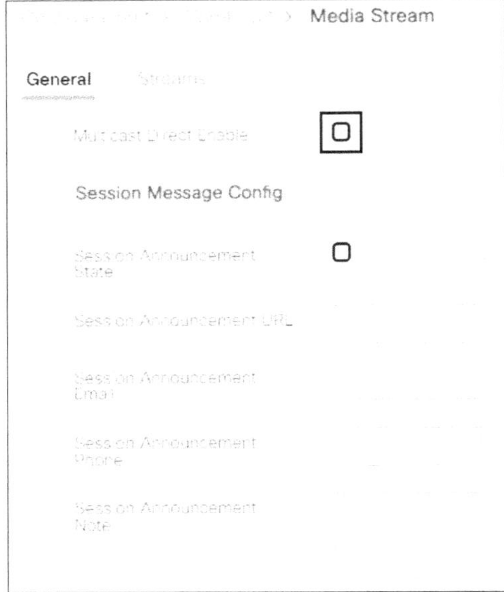

Figure 8-15 *Configuring Media Stream globally*

To define a multicast stream of interest along with the policy configuration, navigate to **Configure > Wireless > Media Stream > Streams**, as shown in Figure 8-16. The relevant configuration fields are

- **Stream Name:** This is a user-friendly name to define the stream.
- **Multicast Destination Start IPv4/IPv6 Address:** This is the first multicast group IP address for the stream if the stream uses a range of multicast IP addresses.
- **Multicast Destination End IPv4/IPv6 Address:** This is the last multicast group IP address for the stream if the stream uses a range of multicast IP addresses.

Note The end address is the same as the starting address if the stream maps to only one multicast IP address.

- **Maximum Expected Bandwidth (Kbps):** This field indicates the dedicated radio bandwidth that needs to be available to accommodate a client to join the stream. The default value is set to 1000 kbps, but the C9800 supports all values from 1 Kbps to 35000 Kbps. Until the configured max expected bandwidth is available, the client is not allowed to receive the media stream.
- **Average Packet Size:** This field should be matched to the average packet size of the multicast stream. The C9800 supports a range of 100 to 1500 bytes with the default set to 1200 bytes.

- **Policy:** This field can be set to Admit, allowing clients to join the media stream if RRC parameters are met, or to Deny to prevent clients from joining the stream.
- **Priority:** RRC Priority takes effect only when there is congestion or contention in the wireless access point. Valid values for the field range from 1 to 8, with 1 considered the lowest RRC priority and 8 as the highest RRC priority. When there is congestion and there are too many media streams, the stream with the highest numerical RRC priority is given precedence over other media streams. The configured media stream has lower priority than voice and higher priority than best effort traffic. All the other multicast traffic is admitted as best effort traffic even though they are marked for QoS for Video priority.
- **QoS:** This field configures the over-the-air QoS class to Video by default and does not allow the field to be modified. After it is permitted, the media stream traffic is placed in the prioritized transmit queue for video on the radio of the AP.
- **Violation:** If the RRC declines the request, the stream can be dropped or fall back to the best effort queue.

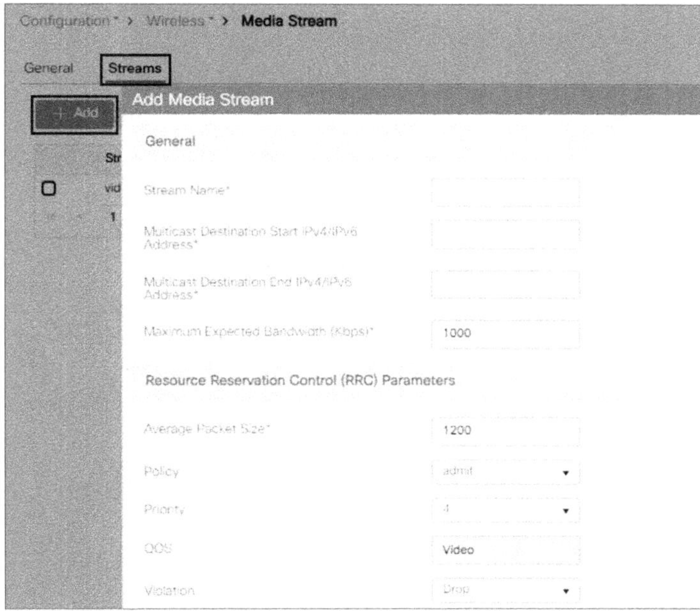

Figure 8-16 *Configuring multicast stream details with their characteristics*

To enable a media stream on the radio, navigate to **Configuration > Radio Configurations > Media Parameters** > and select **Multicast Direct Enable** on each radio. Figure 8-17 shows the same for 5 GHz radio.

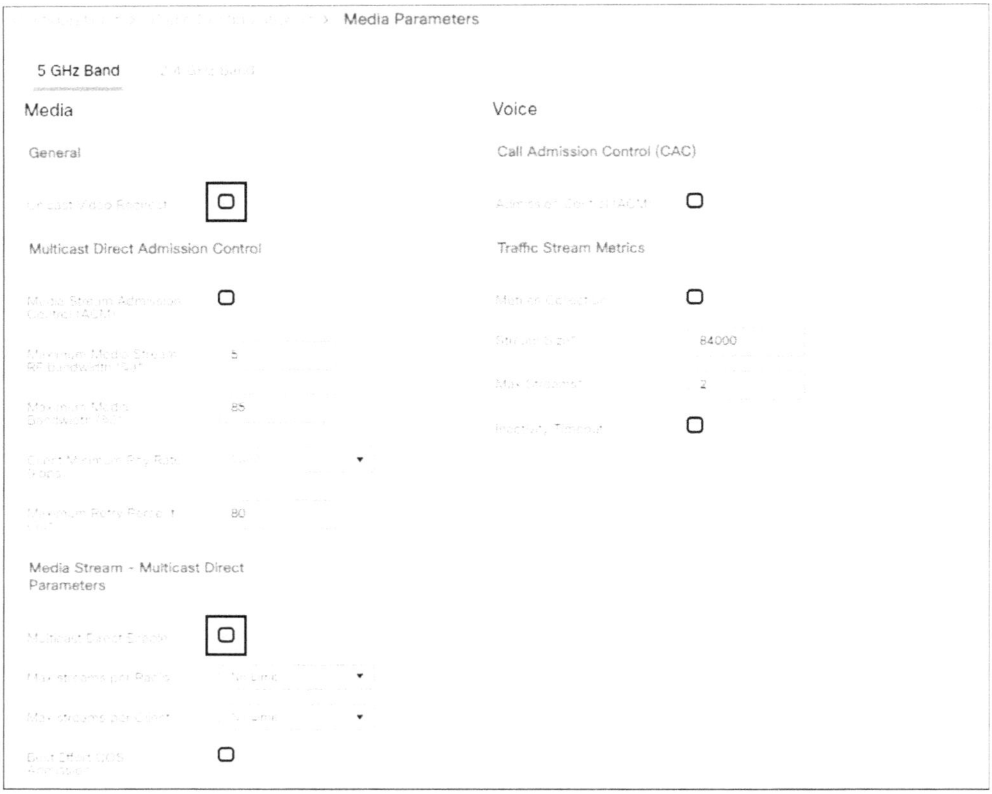

Media Parameters

5 GHz Band

Media

General

Multicast Direct Admission Control

Media Stream - Multicast Direct Parameters

Voice

Call Admission Control (CAC)

Traffic Stream Metrics

Figure 8-17 *Configuring Media Stream on 2.4 GHz and 5 GHz bands*

To configure media parameters on radio, the radio must be disabled. On each radio band, several optional configuration fields provide granular control to you to permit voice and video clients based on bandwidth and ensure quality for both voice and video. You can set these fields:

- **Unicast Video Redirect:** This field enables unicast video traffic flow to wireless clients.
- **Media Stream Admission Control (ACM):** This field enables CAC for the video stream.
- **Maximum Media Stream RF Bandwidth (%):** This field indicates the maximum bandwidth (in percentage to total bandwidth on the AP radio) that is allocated for the media stream traffic of the wireless clients on this radio band. The allowed range is from 5 to 85 percent with the default set to 5 percent. When the client reaches the configured value, the AP rejects new calls on this radio band.

- **Maximum Media Bandwidth (%):** This is the maximum reserved bandwidth for media (voice+video) set to 85 percent by default with 15 percent available for the use of best effort data traffic. The range supported by this field is 5 to 85 percent.
- **Client Minimum Phy Rate (kbps):** This is the lowest rate that a client can drop to on the radio and still join a video stream and defaults to 6000 kbps.
- **Maximum Retry Percent (%):** This field is set to 80 percent by default. The system keeps track of the retries on the radio. If the retries exceed the configured value, the client is not allowed to join the stream.
- **Max Streams per Radio:** No limit is set by default and usually not needed to be modified because CAC and maximum bandwidth limit the client count as needed.
- **Max Streams per Client:** No limit is set by default and does not need to be modified for any deployment.
- **Best Effort QoS Admission:** Clients that do not pass RRC criteria are dropped by default. This field can override this behavior and admit clients at best effort.
- **Voice CAC:** This field enables CAC for voice and helps to determine the voice clients that can be allowed on the radio.

To enable Media Stream on the WLAN profile for the client, navigate to **Configuration > Tags and Profiles > WLAN > Select Client WLAN > Advanced** and select **Media Stream Multicast-Direct**, as shown in Figure 8-18.

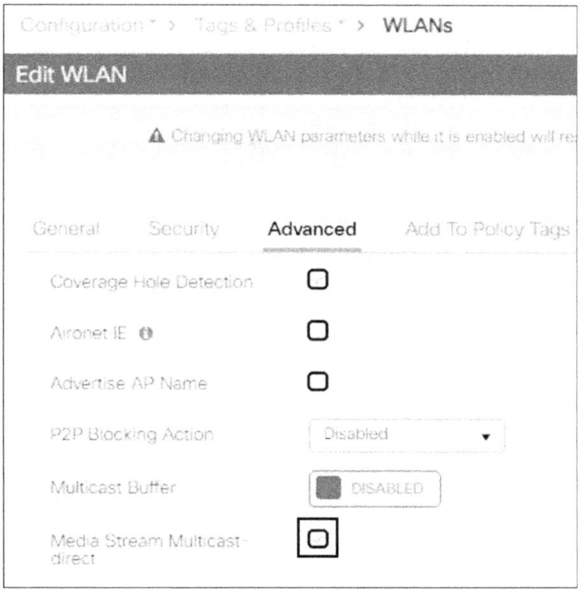

Figure 8-18 *Enabling Media Stream on a client WLAN profile*

To verify that a client passed RRC criteria and has been admitted to the media stream, navigate to **Monitoring > General > Multicast > Media Stream Clients**, as shown in Figure 8-19.

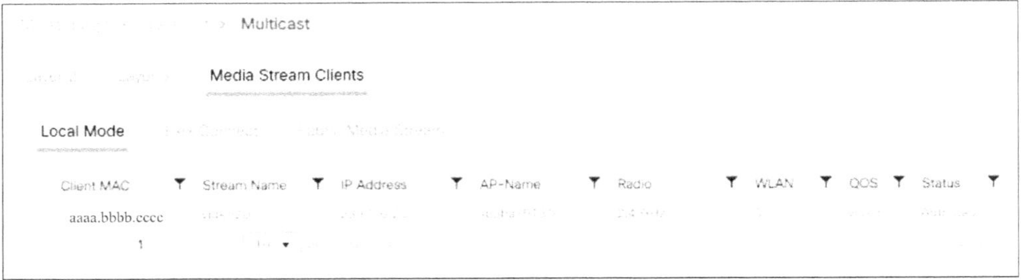

Figure 8-19 *Monitoring a client admitted to the MC2UC Media Stream*

mDNS

Domain Name System (DNS) is the protocol that resolves hostnames to IP addresses. Multicast DNS (mDNS) does the same within a network that does not include a local name server. mDNS is mainly used by Apple's Bonjour protocol, which is a link-local service discovery protocol. Bonjour is responsible for locating service providers and the specific services advertised by these providers within a local-area network. Bonjour is made up of mDNS queries and advertisements and has the following main characteristics:

1. It operates over User Datagram Protocol (UDP) port 5353.
2. Traffic is destined to the link-local reserved multicast IPv4 group address of 224.0.0.251 and IPv6 group address of FF02::FB.
3. The term *link-local* means traffic stays local to a Layer 2 domain and cannot traverse subnets.
4. Time to live (TTL) for mDNS traffic is set to 1, which prevents routers from using multicast routing to send mDNS between subnets.
5. Some services advertised by mDNS devices include
   - Printing services like AirPrint
   - File sharing like iTunes sharing
   - iTunes wireless iDevice syncing
   - AirPlay streaming services like music or video broadcasting or full-screen mirror sharing

Consider an enterprise deployment with Apple TVs in conference rooms and AirPrinters on the floor. These are mDNS service providers for specific AirPlay and AirPrint services. For example, if a user wants to mirror presentations from a MacBook to an Apple TV in the conference room or if a user wants to print to an AirPrint server on the floor, that user sends out an mDNS query looking for these services. Apple TV or AirPrinter would advertise the services over mDNS and respond to the queries. The client device can then reach out to the device offering the service in unicast because it discovered its IP address via mDNS. The mDNS providers advertise their services as TXT, SRV, PTR, and A/AAAA records.

- **PTR:** Pointer Records enable Service Discovery by mapping each type of service to the instances that are advertising the specific service. The format of this record is <Service Type>.<Domain>.
- **SRV:** DNS Service Records specify the host and port for a specific service. The format of SRV records includes a short name of the protocol with a maximum of 14 characters, which are all lowercase letters, digits, and hyphens and starting and ending with a lowercase letter or digit. It also includes transport protocol (_udp or _tcp).
- **TXT:** Text records are optional for providing additional information. A text-based string is recommended to be under 1300 bytes to fit in an Ethernet frame.
- **A/AAAA:** A records resolve the hostname to the IPv4 address, whereas AAAA records resolve the domain name to its corresponding IPv6 address.

On the C9800, mDNS can be configured at the global level as well as at the WLAN level, with WLAN configuration overriding the global configuration. On the global level, the C9800 is either in bridging or gateway mode. At the WLAN level, you can choose between bridging, gateway, and drop.

mDNS Bridging

mDNS bridging enables the mDNS behavior discussed so far, where mDNS service providers and mDNS services are made available to devices within the same subnet. On the C9800, mDNS bridging is enabled by default at the WLAN level. However, this configuration takes effect only when global multicast is enabled. Note that mDNS bridging at the global level gets enabled when wireless multicast is enabled. In bridging mode, mDNS traffic is intended to reach all the devices in the same VLAN. If the packet has both mDNS multicast IP and multicast MAC, it is transmitted over the CAPWAP multicast tunnel from the C9800 to the AP to reach wireless clients and is Layer 2 bridged to wired ports in the VLAN. If the packets have only a multicast MAC destination, they are flooded to all wired and wireless interfaces in the VLAN. Figure 8-20 shows a diagrammatic representation of default behavior of mDNS bridging on the C9800 when wired and wireless clients are in one subnet.

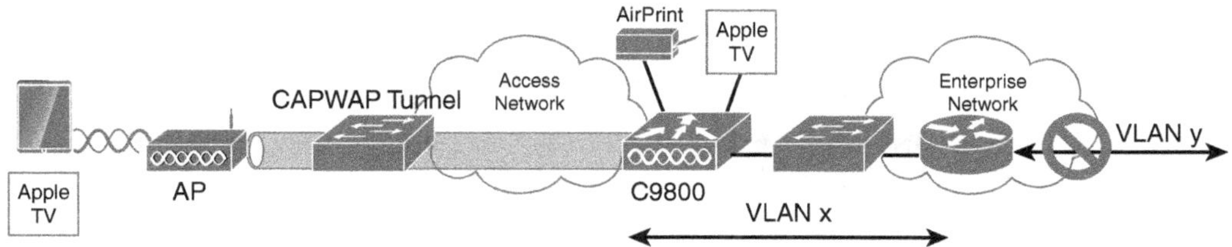

Figure 8-20 *Default mDNS bridging in action*

mDNS Gateway

mDNS gateway is the feature that allows mDNS services to be made available across subnets and helps reduce mDNS chatter with the same subnet. The device configured as the mDNS gateway is responsible for listening to and consuming mDNS advertisements and caching the different mDNS providers, along with their services provided and only advertising the services as requested by interested clients. The mDNS advertisements terminate at the C9800, and other devices and clients do not learn them. The C9800 creates a cache for all services it receives, and it can be viewed by running the CLI **#show mdns-sd cache**. The clients that are interested in the mDNS service send out an mDNS query. The C9800 consumes this query and does not forward it over wire or wireless. The C9800 as mDNS gateway then unicasts the cached service instances to the clients in response to their query. Figure 8-21 illustrates a typical mDNS topology. When the mDNS gateway is enabled:

- The C9800 as mDNS gateway listens to the multicast mDNS and unicast mDNS service instances, both to wired and wireless clients.
- When an mDNS query is received from a wired client, it responds with matching service instances learned from wireless.
- When a query comes from a wireless client, it responds with matching service instances learned from both wired and wireless VLANs.
- For wireless clients, the C9800 unicasts service instances from other VLANs as well as the same VLAN as the wireless client. In other words, bridging and mDNS gateway are mutually exclusive modes.

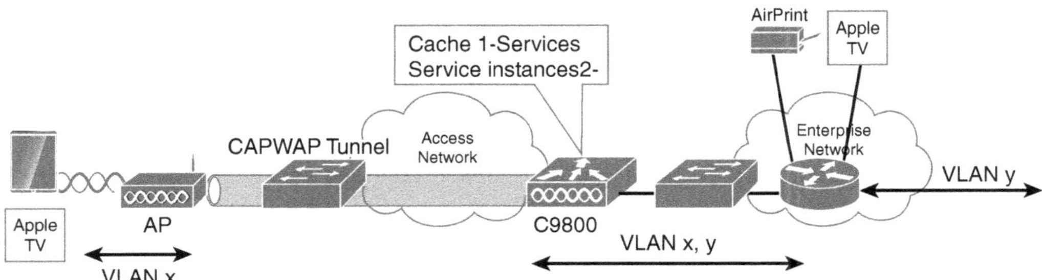

Figure 8-21 *mDNS gateway in action*

How to Configure mDNS Gateway

To configure mDNS gateway mode, navigate to **Configuration > Services > mDNS** and slide **mDNS Gateway** to **Enabled**, as shown in Figure 8-22.

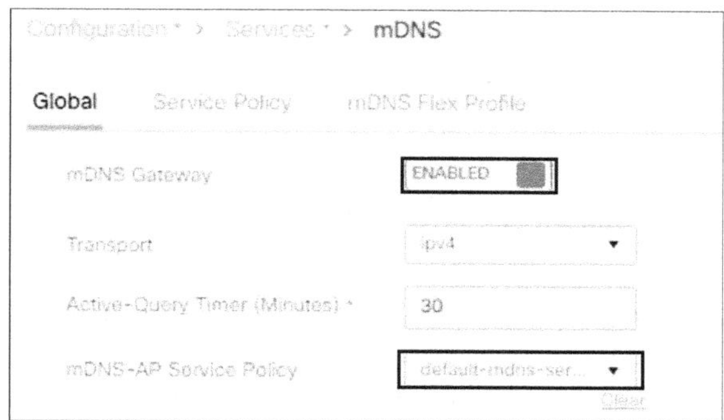

Figure 8-22 *Configuring the C9800 as mDNS gateway globally*

- You can choose an IP transport as IPv4 or IPv6 or both.
- When it learns the services, the C9800 needs to refresh the services periodically to maintain valid services and service providers in its cache. The C9800 uses an active query (AQ) timer to accomplish this. A multicast AQ queries all clients, and it is transmitted in 30-minute intervals by default. This timer can be configured in the range of 1 to 120 minutes. A unicast per service AQ is sent when there are 15–20 minutes of the service TTL left.
- On the C9800, for ease of configuration, most common mDNS service types are prepopulated under a default-service-list that is mapped to default-mDNS-service-policy. This is the service policy that is applied by default at the global and wireless policy profile level when mDNS gateway is enabled. You can view the service types included in the default-service-list via the CLI **#show mdns-sd default-service-list.** Table 8-3 lists the service definitions and service-types for reference. The default-service-list has the same service in both the ingress and egress directions. This list is not editable and does not have any location-based filtering of service instances.

Table 8-3 *Service Definitions and Service Strings in default-service-list*

Service Definition	mDNS Service-Type
Apple-TV	_airplay._tcp.local, _raop._tcp.local
Homesharing	_home-sharing._tcp.local
Printer-ipps	_ipps._tcp.local
Apple-AirPrint	_ipp._tcp.local, _universal._sub._ipp._tcp.local
Google-Chromecast	_googlecast._tcp.local,_googlerpc._tcp.local,_googlezone._tcp.local
Apple-Remote-Login	_sftp-ssh._tcp.local, _ssh._tcp.local
Apple-Screen-share	_rfb._tcp.local
Google-Expeditions	_googexpeditions._tcp.local

Service Definition	mDNS Service-Type
Multifunction-Printer	_fax-ipp._tcp.local, _ipp._tcp.local, _scanner._tcp.local
Apple-Windows-Fileshare	_smb._tcp.local

mDNS Gateway on WLAN

mDNS is configured at both the WLAN and global levels with the WLAN configuration taking precedence over the global configuration. So, even after the mDNS gateway is globally enabled, the WLAN stays at mDNS bridging until the WLAN is explicitly configured. On the C9800 GUI, navigate to **Configuration > Tags and Profiles > WLANs** and select the WLAN of interest; then select **Advanced > mDNS Mode**, as shown in Figure 8-23. Bridging and Gateway mode on WLAN provide the same functionality as the Global configuration. In addition, you have the option to drop all mDNS advertisements and queries on a WLAN by setting the WLAN mDNS Mode to drop.

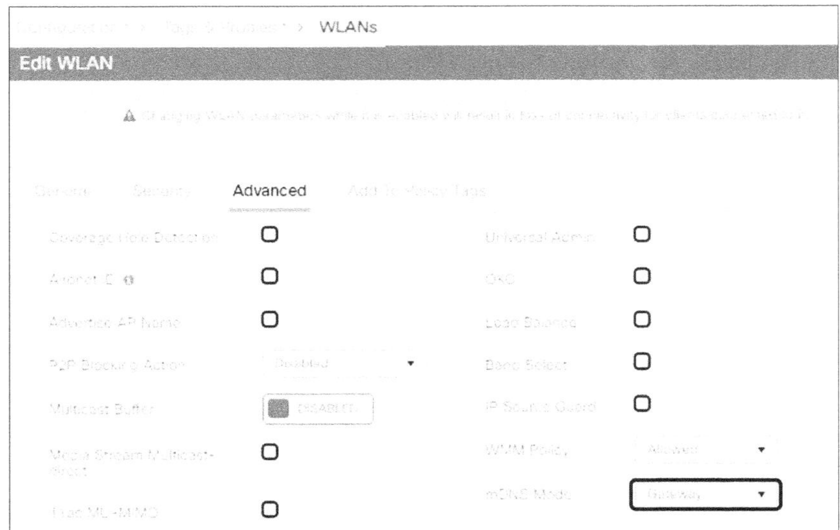

Figure 8-23 *Enabling the MDNS gateway on the WLAN*

The C9800 can also serve as an mDNS gateway for a wired guest deployment. The same mDNS modes are available under the guest-lan profile servicing wired guest users. To configure mDNS mode on guest-lan, navigate to **Configuration > Wireless > Guest LAN**, and select **Wired Guest LAN > General**, and then set **mDNS Mode** to **Gateway** from the default setting of mDNS bridging.

The C9800 can also serve as an mDNS gateway for wired clients connected over a remote LAN (RLAN). To configure mDNS mode on a RLAN, navigate to **Configuration > Tags and Profile > Remote LAN**, select **RLAN Profile > General**, and then set **mDNS Mode** to **Gateway** from the default of mDNS bridging.

mDNS Service Policy on Policy Profile

An mDNS service policy—either default or custom—is applied to the wireless policy profile to filter the wireless mDNS advertisements and queries. On the C9800 GUI, navigate to **Configuration > Tags and Profiles > Policy,** select the policy profile in use, and then select **Advanced > mDNS Service Policy**, as shown in Figure 8-24.

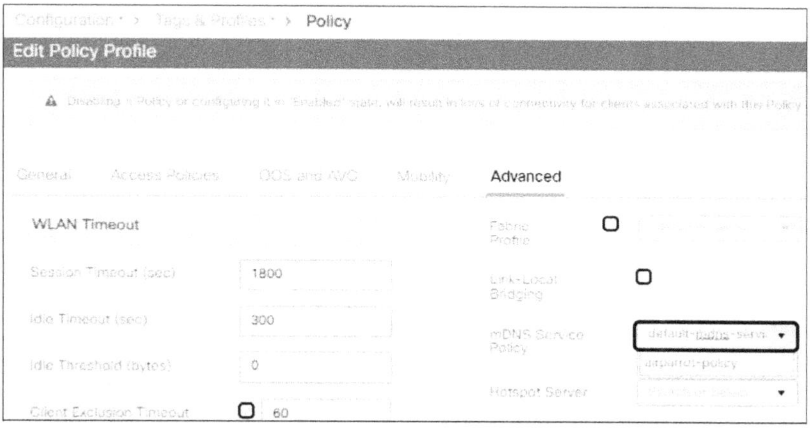

Figure 8-24 *Configuring an mDNS service policy on a wireless policy profile*

By default, this field is populated with default-mDNS-service-policy. The drop-down menu lists all the custom mDNS service policies that are configured on the system. Choose the service policy that matches your filtering intent. Note that the default-mDNS-service-policy is not listed in the drop-down menu. To revert to default-mDNS-service-policy, select **Clear** and then **Update** and **Apply to Device**, as shown in Figure 8-24.

mDNS Service Policy

Besides default-mDNS-service-policy, you have the option to create a customized mDNS service policy. The custom service policy needs to reference a list of mDNS services, whose mDNS queries and advertisements the C9800 will process and unicast to wired and wireless clients. The list of mDNS services can be chosen from a readily available master list of known service definitions mapped to their service strings. This is referred to as the *master service list*. The default-service-list is a subset of most common mDNS services from the master service list. Alternatively, you can create a custom service definition, and the service list can be a combination of predefined and custom service strings. Consider the AirParrot service, which can be used to mirror a screen from a laptop to an mDNS receiver or another laptop running the Reflector app.

1. On the C9800 GUI, navigate to **Configuration > Services > mDNS> Service Policy > Service Definition > +Add.** You can define a user-friendly service definition name and provide the Service Type string **._airparrot._tcp.local** that it maps to. Select the + (plus) sign to add it to the service type strings unique to this service and **Update** and **Apply to Device**, as shown in Figure 8-25.

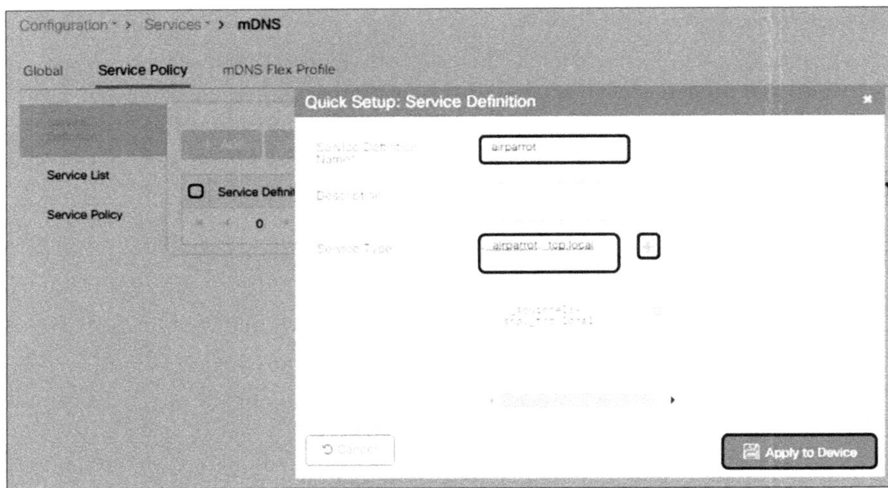

Figure 8-25 *Creating a custom service definition*

2. After defining a custom service, you need to map it to a service list. On the C9800 GUI, navigate to **Configuration > Services > mDNS> Service Policy > Service List > +Add.** Define a service list name of your choice and choose the direction in which the service list will be matched. IN is for ingress or requests, and OUT is for egress or response to the queries, as shown in Figure 8-26.

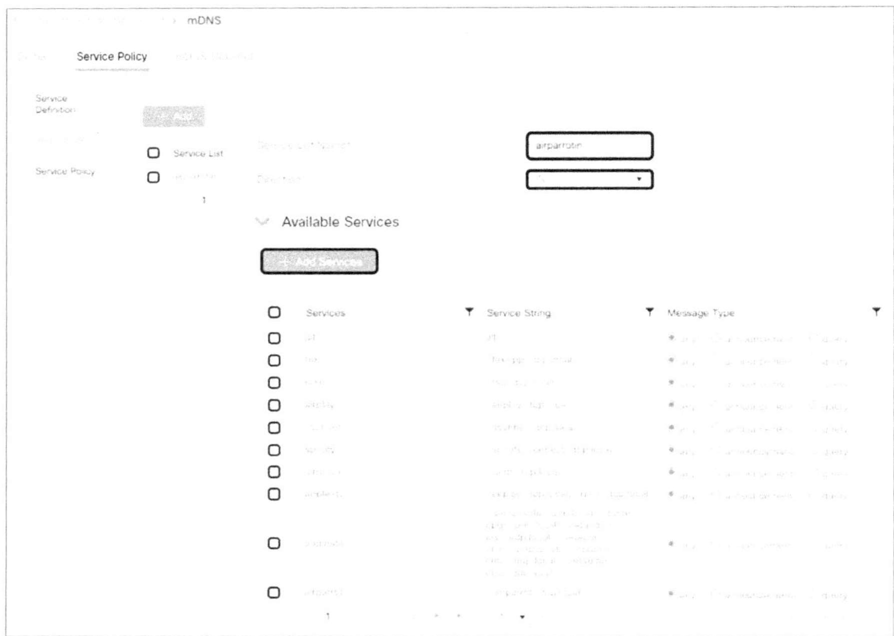

Figure 8-26 *Creating a service list mapping custom service definitions and services from the master service list*

3. Create a service policy matching on the custom service list created in step 2. On the C9800 GUI, navigate to **Configuration > Services > mDNS > Service Policy > Service Policy > +Add.** Define a service policy name and map IN, OUT, or both service lists to the service policy, as shown in Figure 8-27.

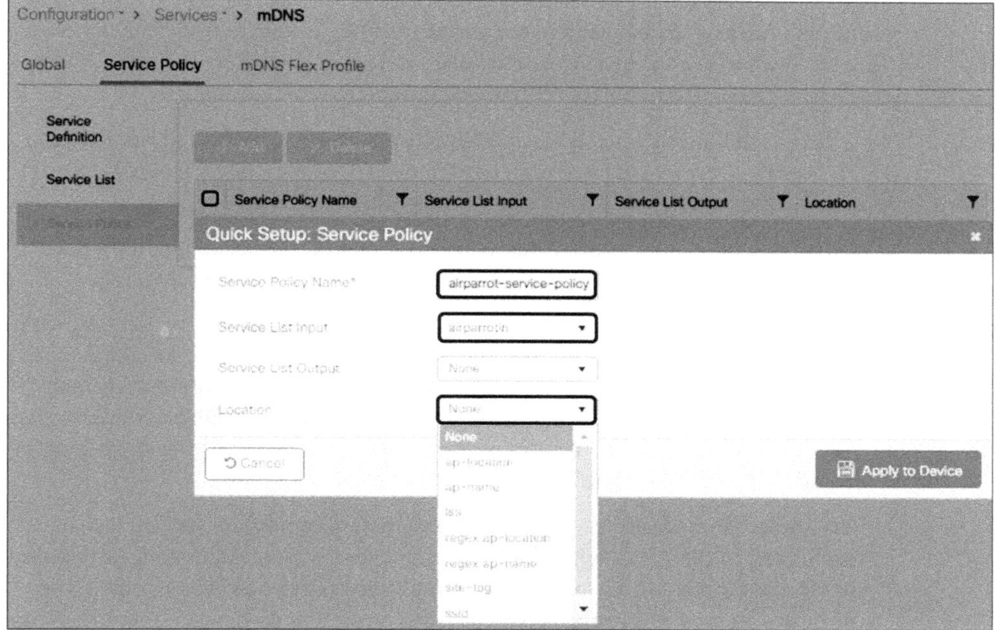

Figure 8-27 *Creating a service policy and mapping service filters*

The service policy filters services learned or query responses. The service list filters it by specific service types. But for each service type, there may be several service instances in a deployment. Consider an example where a user wants to mirror a screen to Apple TV during a meeting. When the user queries for AirPlay service, that user hears about all Apple TVs in the network. It would be cumbersome at the least and impossible at the worst to identify the specific Apple TV belonging to the conference room. To work around this situation, proximity or location is a factor to filter the services. The following location parameters are available to filter:

- **Site tag:** A site tag is used to group together the APs in radio frequency (RF) proximity to be within a roaming domain. When a site tag filter is applied for mDNS, only the service instances learned by the APs that are in the same site tag as the AP of the querying client are included in the response to the client. Note that the mDNS cache also tracks the site tag. Wired service instances do not belong to any site tag and therefore do not honor this filter. All wired service instances are sent in the query response.

- **Location-specific services (LSS):** Depending on the size of a site tag, a user may still see a large number of service providers for a given service. In this case, LSS can be used to limit the number of service providers. When the C9800 creates or updates its cache for wireless mDNS service providers, it also lists and updates the radio MAC address of the AP to which the wireless provider is associated. When LSS is enabled, for any mDNS query from a client, RF neighbors of the AP to which the querying client is associated are pulled from the RRM database. Only the service instances associated with these RFs neighboring the APs are sent to the querying client. LSS filtering applies only to the wireless service instances in the mDNS cache. That is, all wired service instances providing services requested by the query client are always responded to unless otherwise configured.

Note that both LSS and site tags as mDNS filters have been supported since release 16.11. However, the remaining filters listed in this section are available only starting in release 17.x:

- **SSID:** This filter allows only wireless service instances associated with the same SSID to be sent in response, in addition to all the wired service providers.
- **AP Location:** Each AP upon joining a C9800 WLC is assigned, by default, with a location field called default location. Typically, this field is reconfigured with meaningful strings that indicate the physical location of the AP—for example, Bldg5-Floor4-NW. This filter provides only those wireless mDNS services whose AP location matches the location of the AP to which the querying client is associated, in addition to all the wired service providers.
- **AP Name:** With this filter, mDNS response includes only the service instances associated to the AP to which the querying client is associated along with wired service instances. In case of areas with multiple conference rooms, this would be best suited to ensure that your content mirrors the conference room that you are in.
- **Regex AP Location:** In this case, the AP location is matched via regex, and all wireless service instances matching the regex are provided in addition to the wired service instances. For example, for the intent of reaching all wireless providers on the fourth floor of Building 5, where the AP location follows the Bldg5-Floor4-xx convention, the regex match to Bldg5-Floor4 would yield the desired result.
- **Regex AP Name:** With this filter, you can use regex matches on the AP name to filter wireless service instances. For example, if all APs on the fourth floor of Building 5 followed the naming convention of AP-Bldg5-Floor4-xx, the regex match on AP-Bldg5-Floor4 would list all providers on the floor.

mDNS Service Policy on VLAN SVI

The service policies discussed so far filter out only the wireless advertisements and queries. To filter wired mDNS providers, the mDNS service policy should be applied to the VLAN SVI. On the C9800 GUI, navigate to **Configuration > Layer 2 > VLAN > SVI**, select the VLAN that hosts wired providers, then slide **Advanced > Slide mDNS**

Gateway to **Enable**, and finally select the custom mDNS service policy from the drop-down menu, as shown in Figure 8-28. In the drop-down menu, there is an option for None to revert the service policy from the SVI.

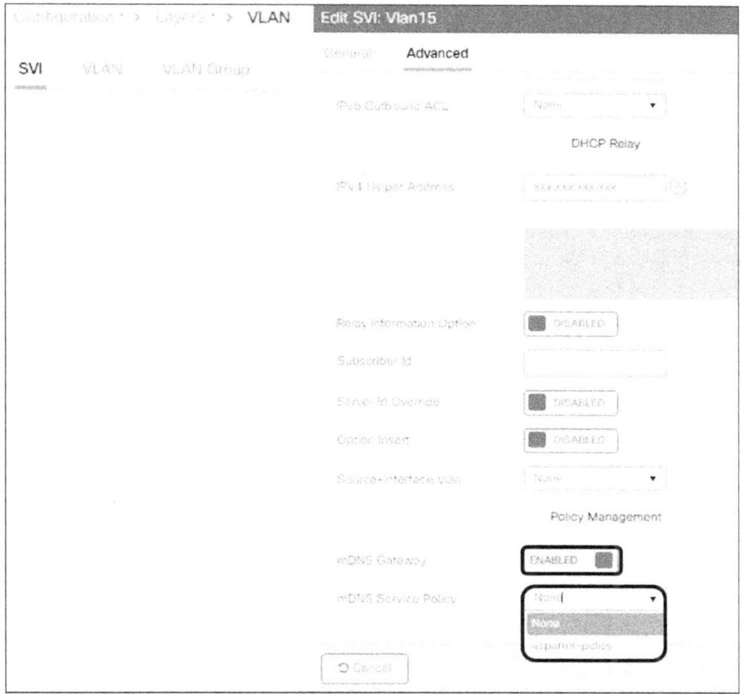

Figure 8-28 *Enabling the mDNS gateway on a VLAN SVI*

mDNS Service Policy via AAA Override

A service policy configured on a wireless profile policy can be overridden by the authentication, authorization, and accounting (AAA) server. For this, the AAA override must be enabled on a policy profile, and the AAA server should be configured to send cisco av-pair mDNS-ServicePolicy-names as part of the client authorization. Note that only the name of a service policy is sent to the C9800. A service policy is expected to be explicitly configured on the C9800 manually.

If service policies are applied in more than one location, as discussed previously, the order of precedence in the AAA override service policy trumps the SVI service policy, which, in turn, trumps the service policy configured on the wireless policy profile.

mDNS-the AP

When the wired mDNS service provider resides in a VLAN not present on the C9800 or on a VLAN that is invisible to the C9800, the mDNS AP is the feature that enables the AP to learn these service providers, if it hears them, and send them to the C9800 over the CAPWAP tunnel. The C9800 caches these service providers and includes them in its mDNS responses to queries from the users. On the C9800, the mDNS AP is supported starting with the 17.x releases and is supported only on the APs that are in local mode or monitor mode. The CLI command to enable the mDNS-AP is **# ap name <AP-Name> mdns-ap {enable|disable} vlan <VLANid>**.

mDNS Gateway in FlexConnect Deployment

The wave2 and Wi-Fi 6 APs can support mDNS gateway functionality in a FlexConnect deployment. All the FlexConnect APs independently snoop and build their mDNS cache but sync their databases with all the APs within the FlexConnect group using Layer 2 multicast designated for mDNS packets. When a wireless client sends a query for mDNS services, the corresponding AP responds with the list of mDNS services and providers. Each FlexConnect AP forwards mDNS service advertisements within the client VLAN itself, which is visible to all the APs within a FlexConnect site tag because they have the same client VLAN mapped. If an AP goes down, all the service providers associated with that AP are invalidated.

Among all FlexConnect APs in a FlexConnect group, only one AP is elected as Master-AP and is responsible for periodically syncing its mDNS database to the C9800. However, if the request comes from a wired client, only the Master-AP responds. Note that in a deployment where all the APs are registered to the C9800, only one Master-AP is present. If this Master-AP goes down, a new one is selected, and the C9800 marks the old Master-AP as a non-Master AP. However, if the Master-AP only goes to a standalone mode, you end up with two Master-APs in same Flex group. Because the mDNS cache is synced, all that this results in is duplicate mDNS responses.

To define an mDNS Flex profile, navigate on the C9800 GUI to **Configuration > Services > mDNS >mDNS Flex Profile**, as shown in Figure 8-29, and apply it under the Flex profile corresponding to the FlexConnect group.

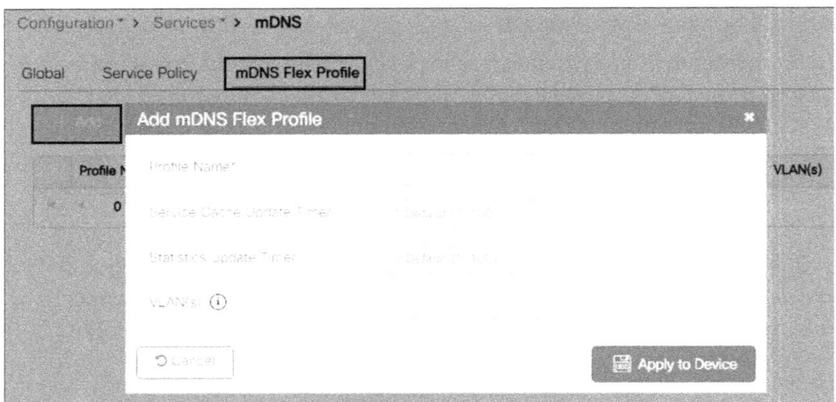

Figure 8-29 *Defining an mDNS Flex profile*

mDNS Gateway with Guest Anchor

mDNS traffic is supported in a Guest anchor deployment wherein guest clients are expected to terminate on the anchor WLC. The mDNS services that are made available to guest clients depend on the mDNS configuration on the foreign and anchor WLC:

1. **mDNS bridging on the foreign WLC and mDNS gateway on Anchor:** In this setup, mDNS queries from a guest client terminate at Anchor, and only the mDNS services cached at Anchor are advertised to the guest client. All mDNS traffic for the guest clients traverses the mobility tunnel from the foreign WLC to the anchor WLC. The different service policy filters that rely on AP name, location, or RF neighbors cannot be used because Anchor has no context for them.

2. **mDNS gateway on foreign and mDNS bridging or mDNS gateway on Anchor:** After the foreign WLC is configured as an mDNS gateway, all mDNS traffic for guest clients terminates at the foreign WLC irrespective of the mDNS mode configured on the anchor WLC. mDNS traffic is not tunneled, and only mDNS providers cached at the foreign WLC are available to the guest clients. All service policy filters are available for this deployment.

In a mixed deployment of guests and non-guest WLANs, starting with release 17.1, the necessary restrictions are applied to ensure the guest users, both wired and wireless, cannot see mDNS providers from non-guest WLANs. The response to mDNS queries from guest clients is provided only by service instances on a guest WLAN and wired guest LAN.

Summary

This chapter describes the configurations and optimizations available on the Catalyst 9800 wireless LAN controller to deliver broadcast, multicast, and mDNS traffic effectively and efficiently on the CAPWAP path as well as over the air. The chapter delves into MoM, multicast VLAN reliable multicast, and mDNS service filtering, all of which seek to reliably transmit time-sensitive and bandwidth-intensive video traffic, which is inherently ill-suited for wireless transmission. All the supported configuration options and design considerations for the local and FlexConnect deployments are detailed.

References

AireOS Media Stream Deployment Guide: https://www.cisco.com/c/en/us/td/docs/wireless/controller/technotes/8-1/VideoStream/b_Cisco_Unified_Wireless_Network_Solution_VideoStream_Deployment_Guide.html#concept_A03AC1E303C241899FBB3896CB0C9168

Mobility Design Guide: https://www.cisco.com/c/en/us/td/docs/wireless/controller/8-5/Enterprise-Mobility-8-5-Design-Guide/Enterprise_Mobility_8-5_Deployment_Guide/Chapter-9.html

Configure VideoStream on 9800: https://www.cisco.com/c/en/us/support/docs/wireless/catalyst-9800-series-wireless-controllers/215859-video-stream-on-catalyst-9800-wireless-c.html

9800 Configuration Guide for Media Stream: https://www.cisco.com/c/en/us/td/docs/wireless/controller/9800/17-6/config-guide/b_wl_17_6_cg/m_ewlc_video-stream.html?bookSearch=true

mDNS technical reference for 9800: https://www.cisco.com/c/dam/en/us/td/docs/wireless/controller/9800/17-3/deployment-guide/c9800-mDNS-technical-guide-rel-17-3.pdf

Chapter 9

Quality of Service (QoS)

Let's start with a simple definition: *quality of service (QoS)* is the measurement of the overall quality and service usability of the network. The factors that define and determine QoS are

- **Bandwidth (or throughput):** The capacity of the network to transmit the maximum amount of data from one point to another in a given amount of time. It's normally measured in megabits per second (Mbps) or gigabits per second (Gbps).
- **Packet loss:** The percentage of packets that are not received (did not arrive or arrived too late and hence discarded). For voice, for example, the default G.729 codec requires packet loss far less than 1 percent to avoid audible problems.
- **Latency (or delay):** The time it takes for a data packet to travel from point A to point B. Each step in the network (links and network devices) adds to the overall latency. The ITU G.114 specification recommends less than 150 milliseconds (ms) one-way end-to-end delay for high-quality real-time traffic such as voice.
- **Jitter:** The variation in delay over time from point to point. If the delay of transmissions varies too widely in a VoIP call, the call quality is greatly degraded. The amount of jitter tolerable on the network is determined by the depth of the jitter buffer on the network equipment in the voice path. The more jitter buffer available, the more the network can reduce the effects of jitter.

QoS for an enterprise network needs to be considered and measured end to end; and like a chain, QoS is as strong as the weakest link.

Quick question: In the typical enterprise network represented in Figure 9-1, what would be the weakest link, in your opinion?

Figure 9-1 *Simplified view on an enterprise network*

If you guessed the wireless portion of the network, you are correct, and there are multiple reasons why. The wireless (Wi-Fi) media is where the following takes place:

- Shift from full-duplex (wired) to half-duplex media (wireless)
- Shift from a dedicated and confined media (wired) to a shared media (wireless)
- Typical downshift in speed (and throughput) when going from wired to wireless

It is critical to understand wireless QoS to guarantee an overall quality of experience for your Wi-Fi clients. The goal of this chapter is to first clarify the main concepts of QoS over 802.11 networks and then help you implement it. This is not a book about wireless QoS in general, instead the main focus is on the Catalyst 9800's QoS implementation and the related design considerations.

Wi-Fi Quality of Service (QoS)

Wireless is evolving, driven by more devices, more connections, and more bandwidth-hungry applications. Present and future Wi-Fi networks will need more wireless capacity to satisfy the requirements of an expanding range of use cases.

A few questions come to mind: With Wi-Fi 6 and Wi-Fi 6E bringing gigabit speed to wireless, is quality of service still needed? Can't we just "throw more bandwidth" at the problem? Is it all about capacity and throughput, or is there more?

These are some of the questions that have been asked when it comes to implementing quality of service. And the answers have not changed: while having more bandwidth and bigger pipes reduces the chances of congestion, this can always happen in any network, and when it happens, you need QoS to properly handle queuing, scheduling, and eventually dropping traffic.

Moreover, it's not only about throughput: services that rely on real-time audio and video, such as voice over IP (VoIP), online gaming, enterprise videoconferencing, and more recently augmented reality/virtual reality (AR/VR), have strong requirements in terms of latency and jitter to deliver a quality experience. Take the example of voice communication: these packets are very small (one audio payload can typically represent 160 bytes), so bandwidth is typically a nonissue. Yet, you need to guarantee that these frames are sent into the air with very low latency (typically 30 ms is the maximum time interval allowed for a Wi-Fi driver to send a real-time voice frame). How can you guarantee this if you don't configure the access point to prioritize these frames' transmission? What if the access point, after receiving a voice packet, goes for off-channel scanning for 50 ms to look for interferences and other information? That would completely compromise the quality of your voice call.

Wi-Fi quality of service is even more relevant because wireless networks are inherently shaky. The radio frequency (RF) environment changes all the time; devices roam continuously (hey, it's wire*less*, better take advantage of it!); and most importantly, wireless is a shared media and the 802.11 protocol overhead makes the links less predictable in terms of capacity. Not to mention that there is always that old device driver that disrupts the performance of your latest shiny fast phone or laptop. So yes, you need to implement Wi-Fi QoS in your network. Here are some basic concepts you need to keep in mind and that will be useful to understand the Catalyst 9800 implementation of QoS.

Wi-Fi (802.11) QoS Fundamentals

Wired networks can rely on dedicated, point-to-point, full-duplex links; wireless is a half-duplex shared medium, where only one device can talk at a given time. Half-duplex environments are susceptible to collisions. For Wi-Fi, there is no way to detect a collision when it happens because you are using the air as the transmit medium: there is no closed loop, like in a wire, and hence every frame must be acknowledged.

Although wired QoS is mostly concerned with managing congestion problems (solved with queuing, shaping, and so on), for Wi-Fi the focus is on reducing the probability of a collision happening and avoiding congestion in the first place. For this, 802.11 networks implement Carrier Sense Multiple Access/Collision Avoidance (CSMA/CA) as the access method, also known as "listen before talk": every device listens for the medium to be free before it starts a transmission.

Quick quiz: In a Wi-Fi network, who gets the higher transmission priority between the access point and the clients?

The right answer is that all would get equal access to the medium because the access method is the same—CSMA/CA. It's through Wi-Fi QoS that traffic belonging to either AP or client would get assigned to one of the access categories (queues) and get higher probability to be transmitted first.

Specifically, for QoS, 802.11n, 802.11ac, and 802.11ax networks must implement Wi-Fi Multimedia (WMM), which is a Wi-Fi Alliance certification that validates a partial implementation of the QoS standard 802.11e. WMM validates that wireless devices implement four access categories (AC): Voice, Video, Best Effort, and Background. It also implements the part of 802.11e amendment called Enhanced Distributed Channel Access (EDCA); this is one implementation of congestion avoidance, where any 802.11 transmitter listens to the medium until the medium is free for the duration of a Distributed Interframe Space (DIFS), also called Arbitrated Interframe Space Number (AIFSN). For each packet, the wireless device also picks up a random number; after the medium has been idle for the duration of the AIFSN matching the AC within which the packet is waiting, the driver counts down from that random number. The range of numbers from where the random number is picked is called the contention window (CW). A random number is picked within 0 and 15, and the device counts down from that number. When the countdown ends (at 0), the corresponding packet is sent.

These values are different for each access category, so higher priority traffic gets more probability of being transmitted first. Figure 9-2 shows the DIFS/AIFSN and random backoff for each AC.

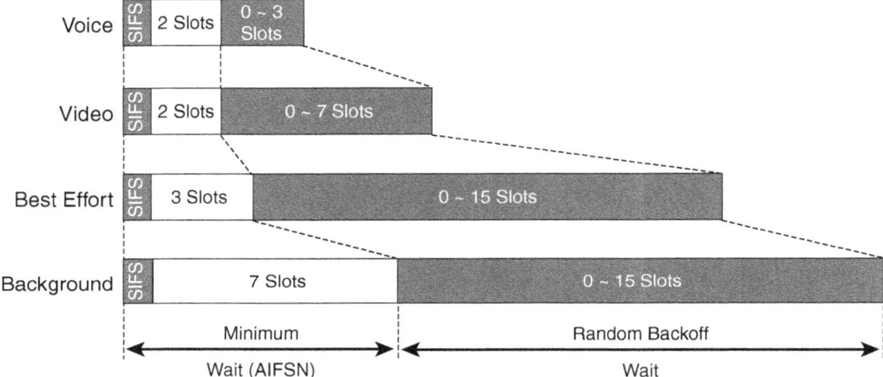

Figure 9-2 *EDCA slots for each access category*

As you can see, AISFN is made of a Short Inter-frame Spacing (SIFS) time interval, common to all the ACs, plus different time slots so that the total AIFSN is different for each access category.

Another way to avoid congestion is to give a transmit opportunity (TXOP) to send a burst and to empty the buffer of packets that each station might have. In each beacon, the AP advertises TXOP values for each of the four queues. Each TXOP is a duration, in units of 32 microseconds. Without TXOP, each station can send only one frame at a time. With TXOP, each station can transmit one frame, then continue transmitting in a burst as long as the transmission is smaller than the TXOP for that queue. Each station would have to wait for SISF between these frames. The 802.11-2016 TXOP values are represented in Figure 9-3.

EDCA/WMM AC	TXOP (μs)	TXOP (Units)
Voice	2080	65
Video	4096	128
Best Effort	2528	79
Background	2528	79

Figure 9-3 *TXOP values for each access category*

There are many other important concepts to fully understand 802.11 QoS: EDCA parameters like contention windows (CWmin and CWmax), Automatic Power Save Delivery (APSD), Traffic Specification (TSPEC), and so on. As stated at the beginning of the chapter, this is not the focus of the book, and there are many publications where you can get more information on these topics. These are the same concepts that you might have learned already with the AireOS-based wireless network; nothing changes with the Catalyst 9800 and IOS-XE–based wireless LAN controllers.

The goal here is to introduce some basic concepts that will help you better understand the QoS implementation on the Catalyst 9800. Let's now consider how the QoS information is carried across the network and introduce two key concepts: the User Priority (UP) in the Layer 2 header of the 802.11 frame and the Differentiated Service Code Point (DSCP) in the IP header, so at Layer 3.

QoS Design

Building a QoS model is the first and fundamental step to determine how to prioritize traffic on your network. Suppose that you have to decide which packet to send first because your wireless medium has a slower data rate than the wired one, or you have to pick which packets to drop at an interface because you cannot sustain the receiving rate. How do you do that? You need a QoS model or blueprint that allows you to classify the traffic in different categories or classes and then apply a different treatment to each.

It's important to say that there is not a "valid for all" QoS model. What you choose depends on what traffic you expect in your network, how granular you want your categorization to be, and of course, what you can do with the classes you define. In networking, the reference QoS model has evolved, starting from four queues to the 12 traffic classes described in IETF RFC 4594:

1. Network Control and Internetwork Control
2. VoIP
3. Broadcast Video
4. Multimedia Conferencing
5. Real-time Interactive
6. Multimedia Streaming
7. Signaling
8. Transactional Data
9. Network Management (OAM)
10. Bulk Data
11. Best Effort
12. Scavenger

As stated earlier, the access point has only four access categories or queues, so the 12 classes model should be used as a reference to classify traffic, and then, of course, you need to map the traffic down to the four queues you have if you really want to utilize this information in your wireless network.

The first thing you need to do is classify and mark the traffic as it starts flowing into your network. You must identify each packet (you see later how this can be done on the Catalyst wireless network) and put a "label" on each identified packet so that you can process it at each and every node of the network. This classification and marking should happen at the edge of the network, so it's one of the primary roles of the AP and wireless LAN controller.

For a Wi-Fi network, you would have a "QoS label" both in the Layer 3 header (IPv4 and IPv6) and in the Layer 2 802.11 header. The Layer 3 QoS label stays within the packet across the network, end to end, and hence expresses a global QoS intent. On the other hand, the Layer 2 information is local to the medium and is removed as you cross between the wired and wireless domains and vice versa: this happens at the AP in FlexConnect mode with local switching or Fabric mode; or at the wireless controller, in local mode or FlexConnect mode with central switching because the client frame is encapsulated in CAPWAP at the AP and tunneled to the WLC. Let's look at the QoS label information and how this is mapped between the Layer 2 and the Layer 3 headers.

UP and DSCP Mapping

When wireless frames are transmitted, a 3-bits QoS value known as the User Priority (UP) is written into the 802.11 frame, as illustrated in Figure 9-4.

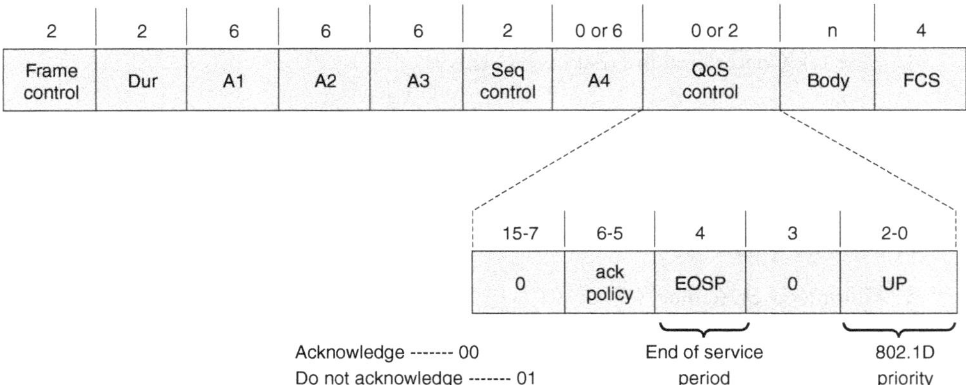

Figure 9-4 *The UP (802.1D) bits in the 802.11 header*

In Figure 9-5, you can see a real traffic example taken from an over-the-air capture of a voice communication. As you can see, the UP value is equal to six, which is the standard value used for voice.

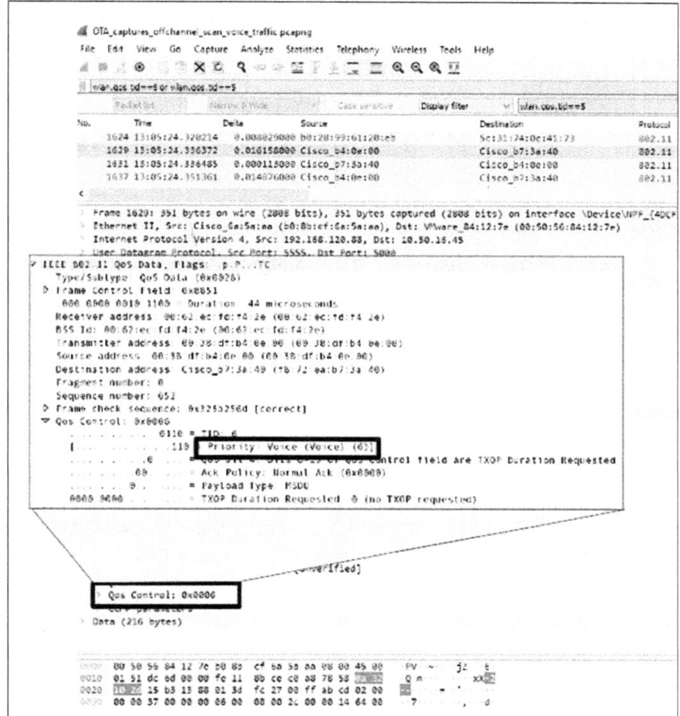

Figure 9-5 *The UP value in a real over-the-air capture*

The UP value is very important for QoS processing at the access point because it's the value used to queue the packet in the appropriate access category following the mapping in the table in Figure 9-6.

Access Category (AC)	UP values
Background (AC_BK)	1,2
Best Effort (AC_BE)	0,3
Video (AC_VI)	4,5
Voice (AC_V0)	6,7

Figure 9-6 *Table with UP to Access Categories mapping*

For example, as shown in Figure 9-7, when the AP receives the traffic from the wired interface, it looks at the QoS information in the original IP packet or in the tunnel header (CAPWAP or VXLAN for Fabric), maps it internally to an UP value (more later on how the mapping is done), and then it uses this value to send the frame to the appropriate queue to the air interface. For each access category, the corresponding EDCA parameters are used to access the medium and transmit the frame into the air.

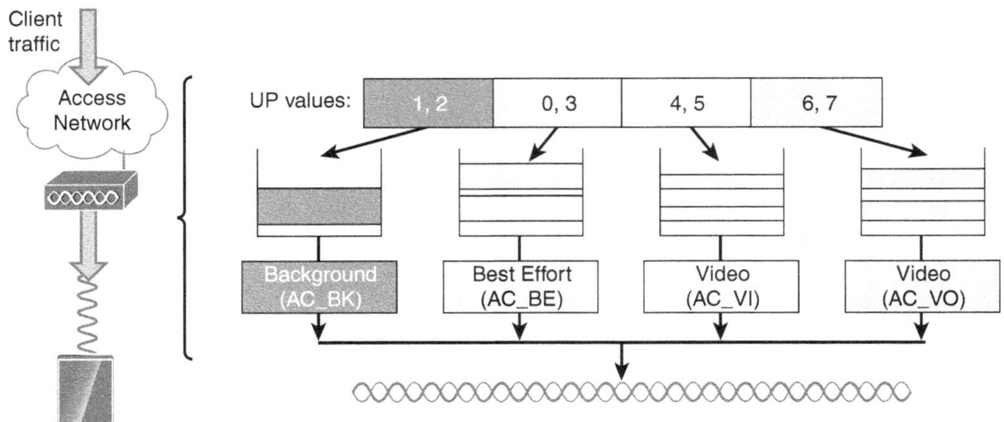

Figure 9-7 *Client UP value to queue mapping at the AP*

At Layer 3, the reference QoS model is the Differentiated Service Model, or DiffServ for short, which is the end-to-end QoS architecture defined in RFC 2474 and RFC 2475.

The QoS label at Layer 3 is carried in the Differentiated Service Code Point (DSCP) bits in the Type of Service (ToS) field in the IPv4 header (Traffic Class field in IPv6), as illustrated in Figure 9-8.

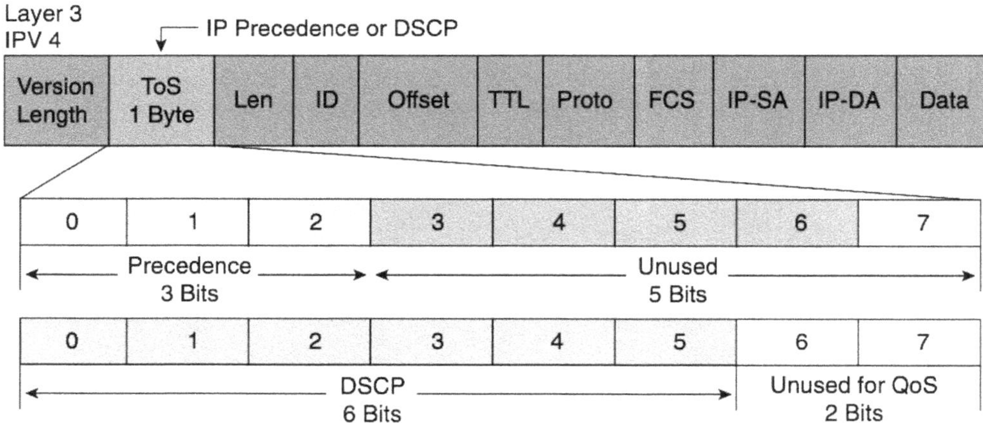

Figure 9-8 *DSCP value in the IPv4 header*

The DSCP is a six-bit value and hence can specify 64 different values (0–63). In reality, the least significant bit is always at 0 (this bit is called MBZ, "must be zero"), and hence, the standard DSCP values should be even numbers (unless the implementation is nonstandard, and the admin uses experimental QoS marking).

The three leftmost bits (class selector) are used to define the class of service, and the convention is that a higher number means that traffic is more sensitive to delay and loss. There are three classes of service:

- **Expedite Forwarding (EF):** For low-loss, low-latency traffic
- **Assured Forwarding (AF):** Different classes for different treatment
- **Default Forwarding (DF):** Typically for best-effort traffic

Within a class, there is a drop probability (DP) that indicates which traffic should be dropped or delayed first. This is represented by two bits, so it means that a drop probability of 11 is higher than a drop probability of 10, and the traffic marked with 11 is dropped before the traffic marked with 10. A representation of the class of service and the drop probability bits is shown in Figure 9-9.

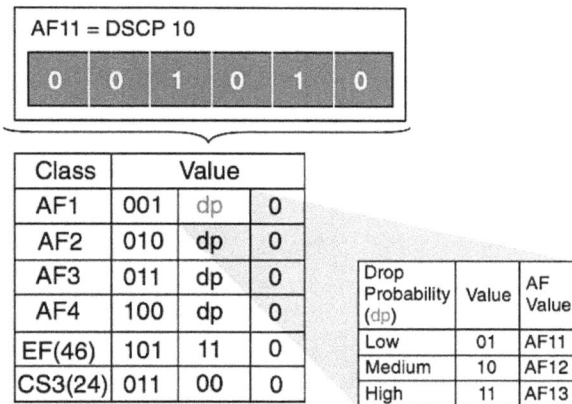

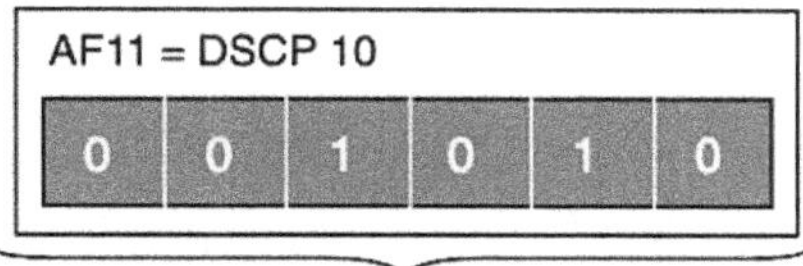

Class	Value		
AF1	001	dp	0
AF2	010	dp	0
AF3	011	dp	0
AF4	100	dp	0
EF(46)	101	11	0
CS3(24)	011	00	0

Drop Probability (dp)	Value	AF Value
Low	01	AF11
Medium	10	AF12
High	11	AF13

AF = Assured Forwarding (DSCP 10 to DSCP 38)
EF = Expedited Forwarding (DSCP 46)
CS = Class Selector. Used to preserve partial backward compatibility with IP precedence.

Figure 9-9 *Class Selector and DP bits*

Let's look at an example. When the class selector is set to 101, traffic belongs to the Expedited Forwarding (EF) class. In this class, for voice traffic, the drop probability bits are set to 11, which gives a resulting tag of 101110, or 46. The EF name is usually used as equivalent to DSCP 46 (101110) for voice traffic, but you should also know that Voice Admit (DSCP = 44) is another value in the same EF class, even if less common.

When there is no drop probability (DP bits 00), the class is called CS and a single number indicates the level. For example, CS3 represents the DSCP value 011 00 0.

In Figure 9-10 you can find a map between the DSCP classes and the relative values in binary, decimal, and hexadecimal. Keep this map handy because it will be useful.

DSCP Class	DSCP (bin)	DSCP (hex)	DSCP (dec)
none	000000	0x00	0
cs1	001000	0x08	8
af11	001010	0x0A	10
af12	001100	0x0C	12
af13	001110	0x0E	14
cs2	010000	0x10	16
af21	010010	0x12	18
af22	010100	0x14	20
af23	010110	0x16	22
cs3	011000	0x18	24
af31	011010	0x1A	26
af32	011100	0x1C	28
af33	011110	0x1E	30
cs4	100000	0x20	32
af41	100010	0x22	34
af42	100100	0x34	36
af43	100110	0x26	38
cs5	100100	0x28	40
ef	101110	0x2E	46
cs6	110000	0x30	48
cs7	111000	0x38	56

Figure 9-10 *DSCP classes and the associated values*

As stated earlier, you should always express your QoS intent using the DSCP values because DSCP information is carried end to end in the network, independently from the local medium. It's for this exact reason that the C9800 implementation of QoS adopts a "trust DSCP" model.

You might also wonder how the 12-class model described in RFC 4594 maps to the corresponding DSCP classes and values. Figure 9-11 illustrates this mapping (here Network and Internetwork Control have been separated; that's why you see 13 rows).

RFC 4594-Based Model	DSCP
Network Control	(CS7)
Internetwork Control	CS6
Voice + DSCP-Admit	EF +44
Broadcast Video	CS5
Multimedia Conferencing	AF4
Real-time Interaction	CS4
Multimedia Streaming	AF3
Signaling	CS3
Transactional Data	AF2
OAM	CS2
Bulk Data	AF1
Scavenger	CS1
Best Effort	DF

Figure 9-11 *DSCP classes and the associated values*

These are the standard QoS classes, and unless you want to use experimental values, you should stick to these values.

DSCP to UP Mapping

Now that you have been introduced to the main concepts, you realize that when you consider QoS at Layer 2 (the Wi-Fi layer) and at Layer 3 (at the network level), you have different information representing your "QoS intent" and you need to map one to the other. The other thing you immediately realize is that when you go from a six-bit value (DSCP field) to three-bit information (UP field), some information will be lost.

Before we describe DSCP to UP mapping, let's define the directions of traffic where QoS policies are applied. As represented in Figure 9-12:

- **Upstream:** Direction of traffic from a wireless source to a wired destination
- **Downstream:** Direction of traffic from a wired source to a wireless destination

Also, if you consider the AP as a reference point and the bridge between the wired and wireless media, these are two other definitions:

- **Ingress:** From a wireless source to a wired destination
- **Egress:** From a wired source to a wireless destination

These definitions are used as a reference in the rest of the chapter when we talk about QoS policies.

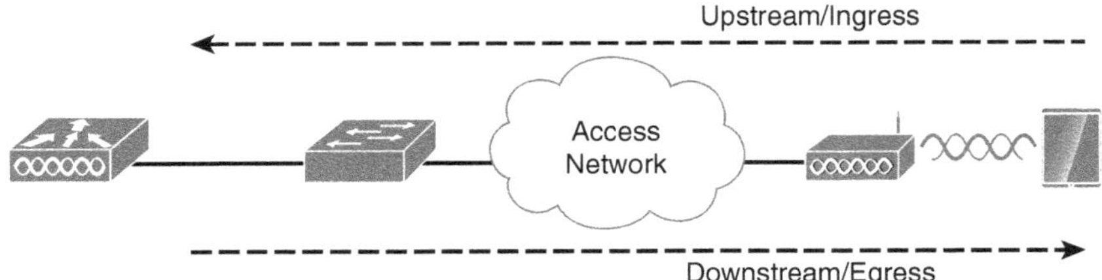

Figure 9-12 *Upstream and downstream directions*

When we talk about the UP and DSCP mapping, it's important to clarify that the C9800 plays the main role of setting this mapping based on the defined policy. This a bit different from the AireOS implementation.

For the downstream direction, in local mode and FlexConnect Central switching deployments, the WLC receives the original IP packet from the wired network. The C9800 is responsible for applying the QoS policy, which results in setting or modifying the original DSCP, mapping it to an UP value in the 802.11e QoS header, which is added to the original packet, and finally in copying the inner DSCP to the outer DSCP value in the CAPWAP header. At the AP, the traffic can be either encapsulated (CAPWAP for local mode and FlexConnect central switching or VXLAN for Fabric mode) or not; when the traffic is encapsulated in CAPWAP, the AP trusts the UP value set by the WLC and embedded in the 802.11e header and uses this value for the internal QoS processing.

If the traffic is encapsulated in VXLAN (Fabric mode), the AP considers the DSCP value set in the external VXLAN IP header; if the traffic is not encapsulated at all (FlexConnect local switching), the AP uses the DSCP in the original packet. In both cases, the DSCP is used to process the QoS policy internally and to derive the UP value that will be used to send the traffic to the right access category queue to schedule it for transmission. The UP value is derived using the table shown in Figure 9-13 in compliance with RFC 8325.

IETF DiffServ Service Class	DSCP name	DSCP value	UP	Access Category
Network Control (Reserved)	CS7	56	0	AC_BE
Network Control or Internet-work	CS6	48	0	AC_BE
Voice (and Voice-admit)	EF	46, 44	6	AC_VO
Signaling	CS5	40	5	AC_VI
Multimedia Conferencing	AF41,AF42, AF43	34, 36, 38	4	AC_VI
Real-Time Interactive	CS4	32	5	AC_VI
Multimedia Streaming	AF31, AF32, AF33	26, 28, 30	4	AC_VI
Broadcast Video	CS3	24	4	AC_VI
Low-Latency Data	AF21, AF22, AF23	18, 20, 22	3	AC_BE
OAM	CS2	16	0	AC_BE
High-Throughput Data	AF11, AF12, AF13	10, 12, 14	2	AC_BK
Standard	DF (default forwarding)	0	0	AC_BE
Low-Priority Data	CS1	8	1	AC_BK

Figure 9-13 *DSCP to UP mapping based on RFC 8325*

Note that CS6 and CS7 are remarked to DSCP zero in compliance with the security recommendations of RFC 8325 because you should not have such network control traffic over a (wireless) access network.

The mapping shown in Figure 9-13 implies that some traffic, within the same access category (AC), will have different UP values. Take, for example, real-time interactive traffic (CS4 > 5) and multimedia streaming (AF31 > 4): both end up in the same Video Access category, but they still receive a different treatment. In fact, Cisco APs have an internal buffering system that prioritizes higher UPs. In other words, although there is no difference between UP 4 and UP 5 as far as serialization goes (the action of effectively sending the frame to the air), the internal buffering mechanisms in Cisco APs process the UP 5 frame in the buffer before the UP 4 frame, thus providing an internal QoS advantage to UP = 5 over UP = 4.

For the Catalyst 9800, the DSCP-to-UP mapping is hard-coded at both the WLC and the AP and cannot be changed. You can see it with the following command on the AP:

```
AP#sh dot11 qos
[...]
Active dscp2dot1p Table Value:
[0]->0 [1]->0 [2]->0 [3]->0 [4]->0 [5]->0 [6]->0 [7]->0
[8]->1 [9]->0 [10]->2 [11]->0 [12]->2 [13]->0 [14]->2 [15]->0
[16]->0 [17]->0 [18]->3 [19]->0 [20]->3 [21]->0 [22]->3 [23]->0
[24]->4 [25]->0 [26]->4 [27]->0 [28]->4 [29]->0 [30]->4 [31]->0
[32]->5 [33]->0 [34]->4 [35]->0 [36]->4 [37]->0 [38]->4 [39]->0
[40]->5 [41]->0 [42]->0 [43]->0 [44]->6 [45]->0 [46]->6 [47]->0
[48]->0 [49]->0 [50]->0 [51]->0 [52]->0 [53]->0 [54]->0 [55]->0
[56]->0 [57]->0 [58]->0 [59]->0 [60]->0 [61]->0 [62]->0 [63]->0
```

Note that the nonstandard DSCP values are mapped to UP =0 for best effort treatment. This again follows the RFC 8325 recommendation specifying that all unused codepoints are recommended to be mapped to 0.

In the upstream direction, traffic is received at the AP directly from the wireless client. If you can trust the client setting, you should configure the AP to use the DSCP value and ignore the UP. However, you may have scenarios where you cannot trust the DSCP, or the client sets the UP only in the dot11 frame. In this case, you need to derive the DSCP value from the UP value. Because you go from a three-bit value (UP) to a six-bit value (DSCP), some hard choices need to be made. You can see the UP-to-DSCP mapping implemented for Catalyst wireless by using the same AP command shown previously, looking for upstream mapping:

```
AP#sh dot11 qos
[...]
DSCP TO DOT1P (UPSTREAM)
Active dscp2dot1p Table Value:
[0]->0 [1]->8 [2]->10 [3]->18 [4]->26 [5]->34 [6]->46 [7]-0
```

This example follows the common RFC 8325–based mapping seen before, with the exception of UP 5, to DSCP 34 or AF 41 because this is commonly used by video applications. Again, the mapping is hard-coded in the AP software, and it's not user configurable.

Wireless Call Admission Control (CAC)

When you talk about QoS and Wi-Fi, chances are you are discussing how to provide a better experience for delay-sensitive applications like voice and that's how the wireless call admission control (CAC) usually comes into the picture. Let's look at what Wi-Fi CAC is and what you should know about it.

CAC enables an access point to maintain controlled quality of service when the wireless LAN experiences congestion. In general, when you have traffic congestion and a particular link in the network gets oversubscribed, queueing, buffering, and packet drop may resolve the congestion. The extra traffic is simply delayed until the link becomes available again, or, if traffic is dropped, the protocol initiates a timeout and requests a retransmission of the information. As you can imagine, this is not a good way to handle congestion if you have a delay and drop sensitive applications like voice.

For applications like voice, it is better to deny access completely than to allow it to later drop and/or delay it. Wireless CAC is a deterministic method to decide whether a voice call can be admitted on the wireless network. The decision is based on whether the required wireless network resources are available to provide suitable QoS for the new call. That's why CAC is about voice traffic only and not any other data applications. Also, don't confuse Wi-Fi CAC with CAC used in the Unified Communication (UC) world. UC CAC is typically implemented on a voice server (like Cisco Unified Communications Manager) and tests end-to-end bandwidth and links performance.

Wireless CAC concerns only the Wi-Fi network, and not the rest of the wired infrastructure. You may encounter two names for this functionality—Wireless CAC and ACM (admission control mandatory)—but they refer to the same thing. ACM is just the term used in the 802.11 standard and WMM.

Traffic Specification (TSPEC) is the protocol used by a QoS-capable wireless client to signal its traffic requirement to the access point for the voice traffic stream (TS) it represents. The TSPEC admission request is sent when the associated client starts a voice call (using an ADDTS action frame) or at association and/or roaming time. In both cases, the TSPEC packet includes the minimum physical rate (PHY) the client device will use, allowing the infrastructure to determine how much the station would consume to send and receive traffic. The client also specifies the QoS marking, which is DSCP = 46 and corresponding UP = 6.

The AP/WLC then decides whether the request is acceptable or not and provides its decision to the client. The client can start the high-priority communication (marking the voice traffic accordingly) only when it receives the approval. This prevents any kind of collision or congestion on the wireless link and thus maintains a good communication quality.

To enable TSPEC on your Catalyst wireless network, you need to

- Enable CAC for the different frequency bands (2.4/5/6 GHz); this is a global setting and automatically enables TSPEC.
- Set the desired maximum RF bandwidth that is allocated for voice traffic (default = 75 percent).
- Set the bandwidth that is reserved for roaming voice clients (default = 6 percent).

The C9800 automatically enables load-based CAC and leverages the settings in the calculation to verify whether resources are available on the channel. In load-based CAC, the access point continuously measures and updates the utilization of the RF channel (that is, the percentage of bandwidth that has been exhausted), channel interference, and the additional calls that the access point can admit. The access point admits a new call only if the channel has enough unused bandwidth to support that call. The C9800 also supports bandwidth-based, or static, CAC, which allows you to specify how much bandwidth or shared medium time is required to accept a new call and in turn enables the access point to determine whether it can accommodate this call. The access point rejects the call, if necessary, to maintain the maximum allowed number of calls with acceptable quality. Load-based CAC is recommended.

Another important setting that you need to consider when configuring CAC is the Metal QoS applied on the Voice SSID; as explained later in this chapter, with a Metal QoS policy, you can configure a ceiling to limit the DSCP and the corresponding UP values allowed on the network. If the client requests a value of UP = 6 in the TSPEC negotiation (which corresponds to DSCP 46 used for voice), but this is not allowed by the QoS policy, the call is denied. That's why you must use Platinum (the highest ceiling) as the Metal QoS policy in the policy profile associated to the voice SSID. One last thing to remember: if you have CAC enabled (global setting), only the clients that perform a TSPEC request can be allowed to send traffic with UP = 6; any other traffic is re-marked to zero.

It's important to understand that QoS is continuously evolving, standards are updated, and new Wi-Fi alliance certifications are issued. It's key to stay up to date. Also, wireless technology itself is evolving. Wi-Fi 6 brings a new set of innovations that are important for QoS: technologies like orthogonal frequency-division multiple access (OFDMA), dual sub-carrier modulation (DCM) uplink, downlink multiple input and multiple output (MIMO), Basic Service Set (BSS) coloring, and spatial reuse play a role to improve the quality of experience for a wireless client on a Cisco network.

The rest of the chapter is dedicated to explaining the Wi-Fi QoS implementation on the Catalyst 9800 wireless LAN controller, the design considerations, and the best practices.

Implementing Wireless QoS on the C9800

To understand the Wi-Fi QoS implementation on the Catalyst 9800, we need to first introduce three important concepts:

- **QoS policy target:** The construct where QoS policies can be applied on the C9800
- **Modular QoS CLI (MQC):** The IOS-XE–based configuration model
- **"Trust DSCP" model:** The QoS architecture model implemented on the C9800

QoS Policy Targets

A policy target is the configuration construct where a QoS policy may be applied. The QoS implementation on the Catalyst 9800 is modular and flexible. The user can decide to configure policies at three different targets: the SSID, client, and port levels. They are represented in Figure 9-14.

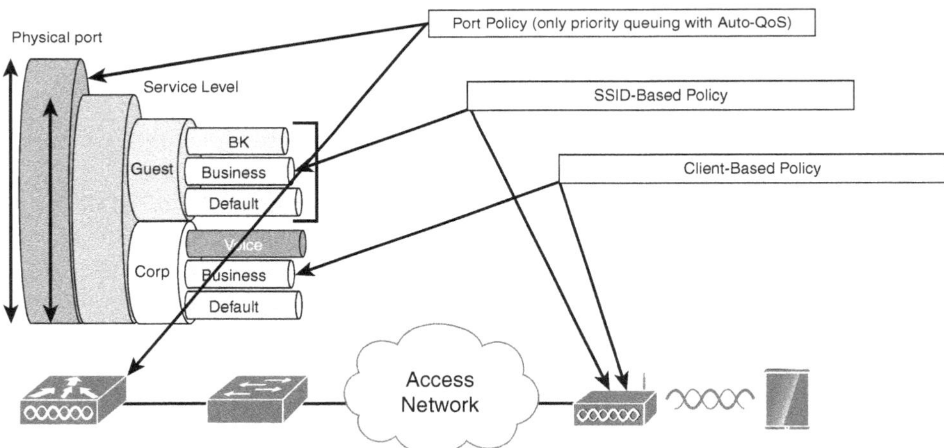

Figure 9-14 *QoS policy targets*

SSID-based policies can be applied to both the upstream and downstream directions. Upstream, the policy is applicable per AP radio, per SSID (or per BSSID). For example, if you decide to rate-limit the traffic on the SSID to, say, 10 Mbps, this policy is applied to all clients on that SSID, and the limit specified is applied per AP on each radio. You can configure policing and marking policies at the SSID level.

Client-based QoS policies are also applicable both to upstream and egress downstream directions. You can configure policing and marking policies at the client level. A client-based QoS policy can also be applied dynamically using AAA override; in this case, the client policy is received directly from the AAA server during the client authorization phase and would override any statically defined policy on the C9800.

Finally, port-based QoS policies can be applied at a physical or at a logical (port-channel) port, and the C9800 supports priority queuing only through Auto QoS configuration (see the "Design and Deployment" section later in this chapter).

Modular QoS CLI

The Catalyst 9800 QoS is based on Modular QoS CLI (MQC). In IOS-XE, MQC is used to implement the Differentiated Service model, the end-to-end network QoS architecture defined in RFC 2474 and RFC 2475. This model is common to routing, switching, and wireless platforms running the IOS-XE operating system.

The main MQC constructs are

- **Class-map:** Classifies traffic
- **Policy-map:** Binds classified traffic to specified actions
- **Service-policy:** Attaches policy-map to a target in a specified direction

Let's refer to a real example to understand how this works together to define and implement a QoS policy: you want to create a policy that matches some critical traffic and use priority queuing to give it a preferred service. Figure 9-15 shows an MQC implementation of the policy.

Classification ACL

```
ip access-list extended AutoQos-4.0-Output-Acl-CAPWAP-C
 10 permit udp any eq 5246 16666 any
```

Class-map definition

```
class-map match-any AutoQos-4.0-Output-CAPWAP-C-Class
 match access-group name AutoQos-4.0-Output-Acl-CAPWAP-C
class-map match-any AutoQos-4.0-Output-Voice-Class
 match dscp ef
```

Policy-map definition

```
policy-map AutoQos-4.0-wlan-Port-Output-Policy
 class AutoQos-4.0-Output-CAPWAP-C-Class
  priority level 1
 class AutoQos-4.0-Output-Voice-Class
  priority level 2
 class class-default
```

Service-policy attachment

```
interface TenGigabitEthernet0/0/0
 service-policy output AutoQos-4.0-wlan-Port-Output-Policy
```

Figure 9-15 *MQC policy*

It's a modular approach, as the name *Modular QoS CLI* implies. First, you need to classify the traffic that you want to consider. This can be done by matching on a user-defined condition (specified with an access-list or a DSCP value) or matching on an application using the Network-Based Application Recognition (NBAR) engine. In the example, the user first defines an ACL to match the CAPWAP control traffic (port UDP = 5246 for CAPWAP between AP and WLC, and UDP = 1666 for Mobility group tunnels). The next step is to define a class-map to match based on the ACL and DSCP = EF for voice traffic. Then the actual policy is defined: for traffic identified as CAPWAP control, assign priority queue one; for voice traffic, use priority queue two. The last step is to apply the policy to the target; in this case, it's the uplink port in the output or egress direction.

Trust DSCP Model

What does it mean that "Trust DSCP" is the QoS model implemented on the Catalyst 9800 and in general on a Catalyst wireless network? To answer this question, you need to first understand the concept of a trust boundary.

The previous section introduced the UP and DSCP values; these are the Layer 2 and Layer 3 labels, respectively, that are inserted in a packet to express your QoS intent. The DSCP value, being at Layer 3, expresses the end-to-end network QoS intent, whereas UP represents the local domain translation of that general intent. To mark the traffic, you would perform some sort of classification first. This classification is more effective as it's done closer to the traffic source. After you do it once, you do not need to perform classification again, and you can just "trust" that label and use it for your QoS policy. The trust boundary represents the point in the network where you start trusting the QoS label on a packet.

For a wireless network, the trust boundary can be either at the access point or at the WLC, depending on the deployment mode, but anyway it's at the access layer. This means that you can trust what you receive from the client (you have control of the devices and their applications). If you cannot trust the end device, you would have to move the trust boundary deeper in your network.

At the trust boundary, when both the Layer 2 and Layer 3 information are available, you can decide to trust one label or the other. Implementing a "trust" DSCP model means that the network device (AP or WLC) will use the DSCP value in the received packet to internally process the packet and apply the related QoS policy. This is the model used in the Catalyst 9800, and Figure 9-16 shows an example for a local mode deployment (the same for Flex central switching), in the downstream direction.

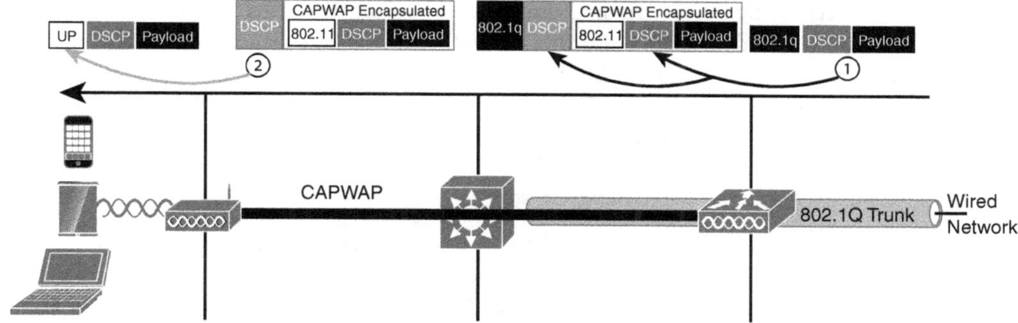

Figure 9-16 *"Trust" DSCP in the downstream direction*

In step 1, the client packet is received at the WLC from the wired network over an 802.1q trunk. The WLC leverages the DSCP value in the original IP packet to apply the user-defined QoS policy; the derived DSCP value is set to both the inner IP and external CAPWAP header. Note that in Figure 9-16, the DSCP is simply copied because there is the assumption that no other QOS policy is applied. The C9800 also maps the derived DSCP to the UP value and sets it in the 802.11e QoS header in CAPWAP before sending the packet out to the AP.

In step 2, the AP leverages the UP value received in the CAPWAP header for internal QoS processing and queuing; the 802.11e UP value is also copied in the egress wireless 802.11 frame to the client, over the air.

In the upstream direction, from the wireless client to the wired network, the trust model is represented in Figure 9-17.

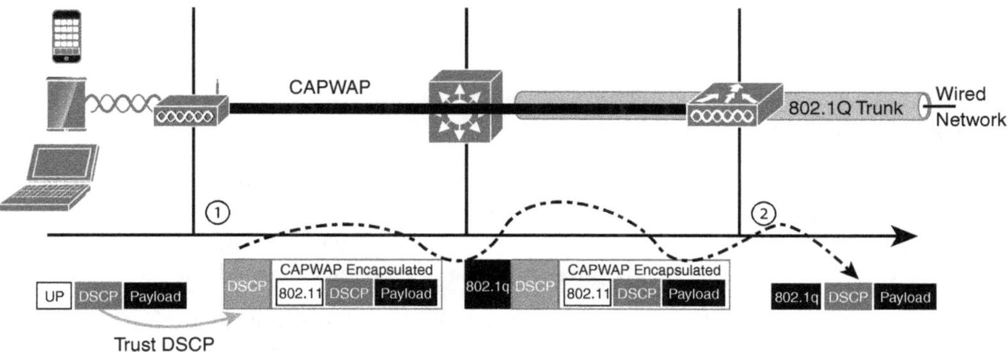

Figure 9-17 *"Trust" DSCP in the upstream direction*

The AP receives the 802.11e frame from the client in step 1. The AP trusts the DSCP value in the original packet for the internal QoS processing. It then maps the derived DSCP value to the outer CAPWAP IP header before sending the packet to the WLC. The inner and original client DSCP value is not touched. As per the "trust DSCP" model, the AP ignores the UP value in the client frame and only considers the DSCP. When the WLC receives the packet, it removes the CAPWAP header, applies any defined QoS policy, and sends the packet to the wired network, step 2.

Similarly for Flex and Fabric modes, the AP only considers and trusts the DSCP value. Upstream, this would always be the DSCP in the original frame received from the wireless client; in the downstream direction, it would be the DSCP in the original header (for Flex local switching) or the DSCP value in the VXLAN header received from the Fabric edge in SDA wireless mode.

Starting with release 17.4, the trust DSCP model is the default mode of QoS operations for the Catalyst 9800. The user may decide to disable the "trust DSCP" and can do so under the AP join profile configuration. In this case, the AP behavior would be different in the upstream direction: the AP takes the UP value from the original client frame and uses the embedded UP-to-DSCP mapping seen in the previous section to derive the DSCP value to copy in the CAPWAP header before sending the packet to the WLC.

Designing and Deploying Catalyst C9800 QoS

Let's now consider how to implement Wi-Fi QoS on the Catalyst 9800. We cover what you need to know to configure QoS successfully and avoid any known limitations at the time of writing. You also learn about QoS profiles and Application Visibility and Control (AVC), how these features work, and how to deploy them. In addition, we introduce Fastlane+, the latest innovation from Cisco and Apple to enhance the user experience on the Cisco wireless network.

QoS Deployment Workflow

To present the steps to implement QoS on C9800, we use the graphical user interface because it provides an intuitive and easy way to deploy a QoS workflow, and it's very popular among customers. You can achieve the same result by leveraging a management system like Cisco DNA Center or Cisco Prime or even the command-line interface.

The first step is to define a Wi-Fi QoS policy. To do this, you connect to the GUI interface and navigate to the **QoS** section under **Configuration > Services > QoS**. A window opens where you can add a new policy, as shown in Figure 9-18.

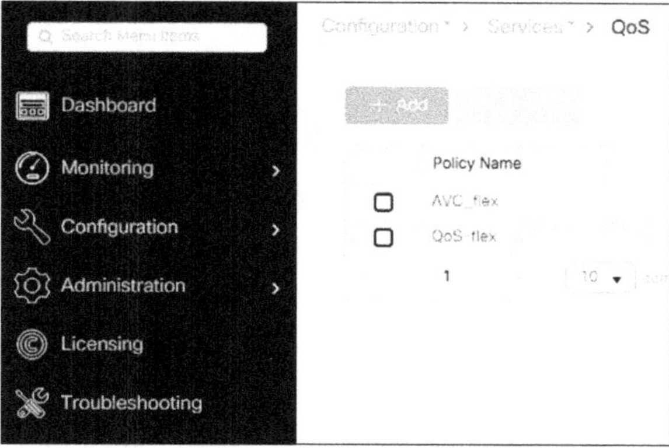

Figure 9-18 *Click Add to start configuring a new QoS policy*

Click **+ Add**, and in the pop-up window that appears, give a name to the new policy (**test** is used in this example...yes, lots of imagination going on here). Then click **+ Add Class-Maps** to define your classification logic, as shown in Figure 9-19

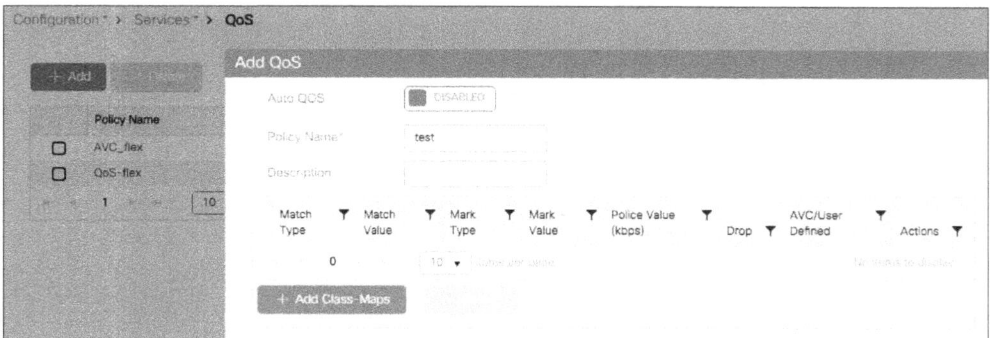

Figure 9-19 *Start adding the classification logic*

In this example, let's assume you want to match ping traffic to drop it and match Facebook to re-mark it down to Scavenger class. When matching traffic, you have two options: user-defined or AVC. For user-defined, you need to further select an ACL that you have previously configured to match the desired traffic. With AVC, you just select the protocol. For ping, you select the related protocol, and then select **Drop** for the action, as shown in Figure 9-20.

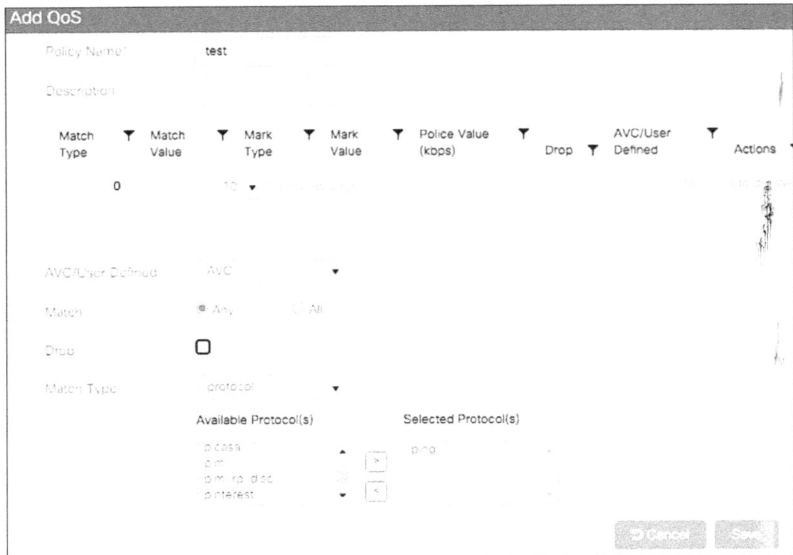

Figure 9-20 *Configuring a policy to match and drop ping traffic*

Important note: if your classification has multiple conditions, you need to select either **Any** or **All** for **Match**, the latter being the default. **Any** is a logical OR: so as long as one condition is met, the related action is taken. If you select **All**, it is a logical AND: all the conditions need to be met for the action to be taken.

Keep this in mind when configuring the policy to drop Facebook traffic in the example: this application has multiple subapplications, such as Facebook-media and Facebook-voice, so it's important to match any of them, as illustrated in Figure 9-21. Also, notice that the action chosen in this case is to mark the traffic with DSCP 8, which corresponds to CS1 or the Scavenger class.

Note At the time of writing, the GUI has a limitation, and you can select multiple match conditions only with AVC (as in Figure 9-21). For user-defined classification, you have to do it in the CLI.

The resulting policy can be seen in Figure 9-22. In the same figure, you can see another interesting capability of the Catalyst 9800: you may decide to apply an action to the best effort class also known as class-default. This class has this name because it collects all traffic that doesn't match any other defined classes. By default, the action is None, which means the traffic is forwarded untouched. But you can use this simple method to make sure that all traffic in class-default is re-marked to DSCP = 0 and sent to the best effort queue. Note that the default behavior is different from AireOS where any new WLAN gets assigned to a "Silver" QoS profile, which re-marks everything greater to DSCP 0 and leaves untouched what goes into the Scavenger class. In the C9800, you can decide to have a similar behavior as AireOS or leave all the traffic untouched.

Add QoS

Match Type	Match Value	Mark Type	Mark Value	Police Value (kbps)	Drop	AVC/User Defined	Actions
protocol	ping	None		8	Enabled	AVC	

AVC/User Defined: AVC

Match: Any / All

Mark Type: DSCP

Mark Value: 8

Drop

Police(kbps): 8 - 10000000

Match Type: protocol

Available Protocol(s): facetime, fasttrack, fasttrack-static, fatserv

Selected Protocol(s): facebook-video, facebook-media, facebook-audio, facebook

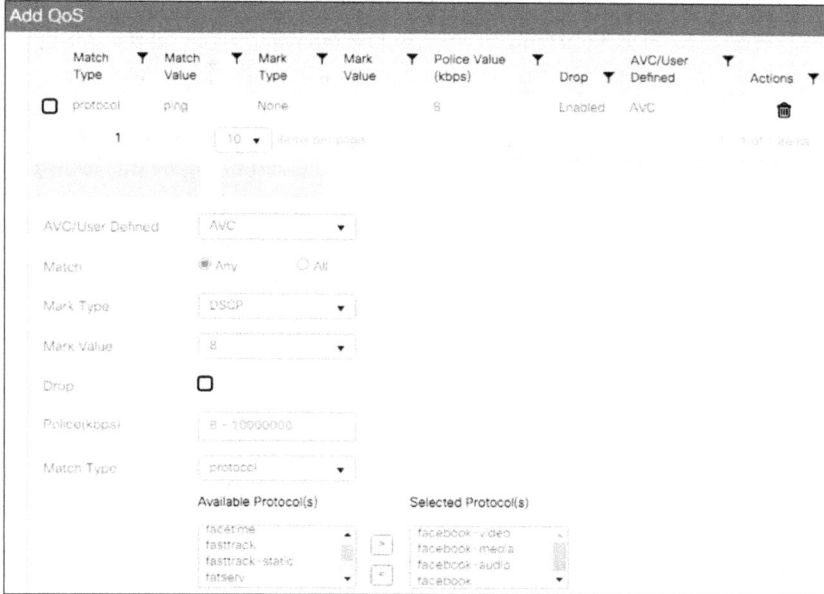

Figure 9-21 *Configuring a policy to match and mark down Facebook traffic*

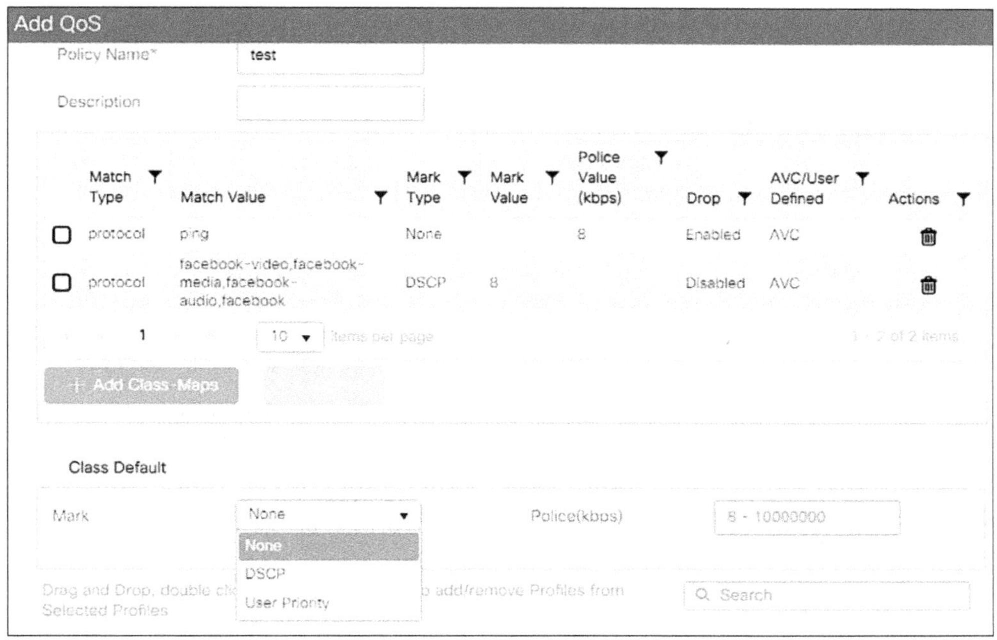

Figure 9-22 *Configuring a policy for class-default*

When defining your QoS policy, it is also important to understand that although the C9800 leverages the same QoS MQC model of switches and routers, not all the settings, filtering, and actions are supported. The reason is that the C9800 is an access platform, mainly working at the Layer 2 level, and therefore, the requirements are different. For example, in a router, when you deal with WAN interfaces and the possible congestion you might have there, you need to have sophisticated queuing and dropping mechanisms. On a wireless controller, these are usually not needed.

It's then important to understand what is supported and what's not on the C9800. Regarding classification, as said before, this can be done using a user-configured rule based on an ACL or matching a DSCP value or using application recognition with the NBAR engine on the WLC or AP (for FlexConnect and Fabric mode). Regarding the possible actions supported, these are only the ones shown in the GUI:

1. **Mark:** Setting the DSCP or UP value
2. **Drop:** Discarding the matched packets
3. **Police:** Rate-limiting the traffic

For example, the C9800 supports policing but not shaping. The two are different, as you probably know from QoS on routers. With policing, when the traffic rate reaches the configured rate, excess traffic is dropped. Traffic shaping retains excess packets in a queue and then schedules the excess for later transmission over increments of time. The result of traffic shaping is a smoothed packet output rate.

For rate-limiting policing on the C9800, you can configure only the average traffic rate or committed information rate (CIR), and no other parameter can be defined. There is no support for bandwidth reservation settings or burst. It is supported to the bidirectional rate limiting (BDLR) policy may be sent from AAA server leveraging AAA override feature. This is supported for both Local and FlexConnect AP mode. For more information on rate-limiting configurations, check the examples in the "References" section.

The Catalyst 9800 supports priority queuing, but with some limitations. Only two priority queues are supported on the physical interfaces, and the dropping mechanism for those is a tail drop. Neither random early detection (RED) nor weighted random early detection (WRED) congestion avoidance techniques are supported. Also, priority queuing is supported when set in the context of auto QoS (see the next section for details).

The final step in your QoS policy workflow would be to assign the policy to the available policy profiles. You can pick one or multiple profiles, and you can choose whether the policy is applied in the ingress direction (from wireless to wired) or in the egress direction (from wired to wireless). In this example, the policy profile called "flex-policy" has been selected, as seen in Figure 9-23.

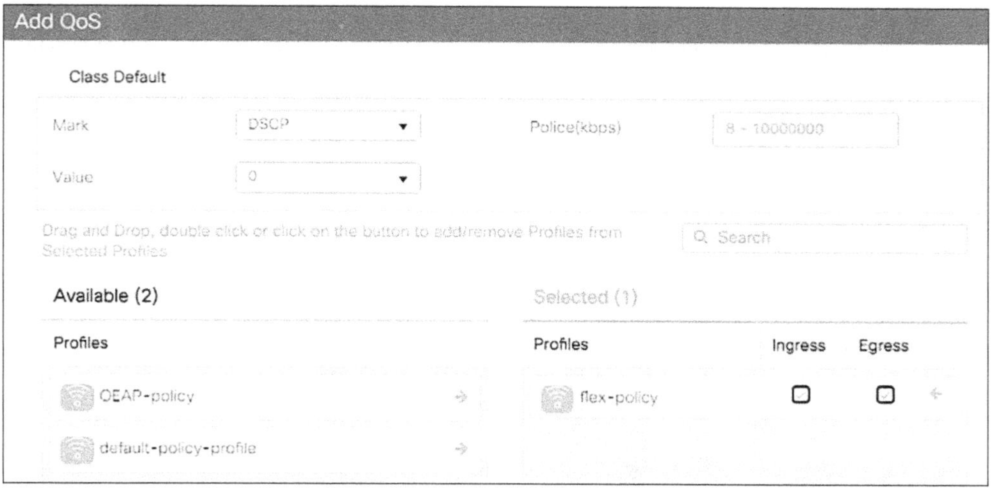

Figure 9-23 *Assigning the policy to the policy profile(s)*

Doing it this way results in the policy being applied at the SSID level, as you can see from the resulting configuration shown in Figure 9-24, which represents the QoS and AVC configuration tab in the policy profile.

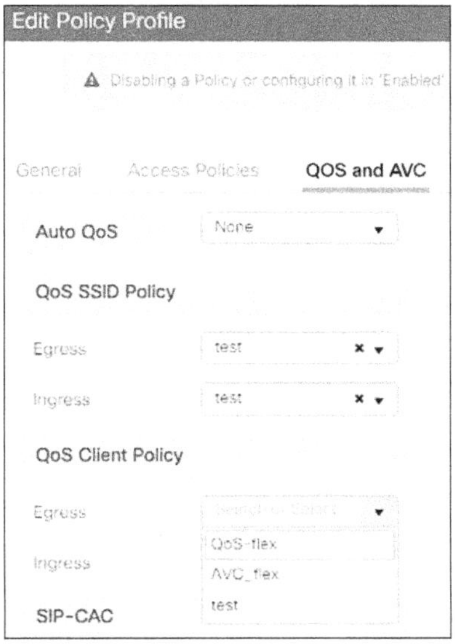

Figure 9-24 *Assigning the policy at SSID or client level, or both*

You may decide to apply the policy only at the SSID level. In this case, the policy gets applied per AP radio to the aggregated traffic for all clients on that SSID. You may decide to apply it instead only at the client level, and in that case, the policy is per client. Lastly, it can be applied at both the SSID and client levels; in this case, the policy is first applied per client, and then the result is applied at the SSID level.

As you can see from Figure 9-24, there is also an auto QoS setting under the policy profile. This can be used by itself or in combination with a custom policy like the one defined in this section. Auto QoS is described in detail in the next section.

What if you want to apply a QoS policy dynamically? For example, you want clients on the same SSID to have a different policy based on the Active Directory (AD) group they belong to and hence based on the authorization profile returned from the Radius server upon client authentication.

You can do this on the C9800 with the QoS policy override feature, which is available for all the AP deployment modes: local, FlexConnect, and Fabric modes. QoS policy override is available only for a client-based policy, and not for SSID-based policies. The way you do it is simple: configure the policy on the C9800 as shown before, but don't assign it to any policy profile. Next, configure the AAA server to return the following cisco attribute-value pair (cisco-av-pair) with the policy name as part of the client authorization:

```
cisco-av-pair = ip:sub-qos-policy-in=MyPolicy
cisco-av-pair = ip:sub-qos-policy-out=MyPolicy
```

Figure 9-25 shows an example of an authorization policy on Cisco Identity Service Engine (ISE), but any AAA server that supports Cisco vendor-specific attributes might be used.

Let's take a look at auto QoS, which represents a simplified way to implement QoS on the C9800.

Auto QoS

You can use the auto QoS feature to simplify the deployment of QoS settings on a Catalyst 9800 wireless network. Auto QoS defines a few reference network designs and enables QoS configurations accordingly so that the wireless network can prioritize specific traffic flows.

The C9800 supports the following auto QoS profiles:

- **Voice:** Sets the recommended QoS policy to correctly mark and prioritize voice at the SSID level and enables CAC
- **Guest:** Sets the recommended QoS policy at the SSID level to mark everything to best effort (DSCP =0)
- **Enterprise:** Sets the recommended QoS policy at the SSID level to mark VoIP data, signaling, multimedia, transaction, bulk data, and Scavenger traffic
- **Fastlane:** Sets specific EDCA parameters to optimize Fastlane clients

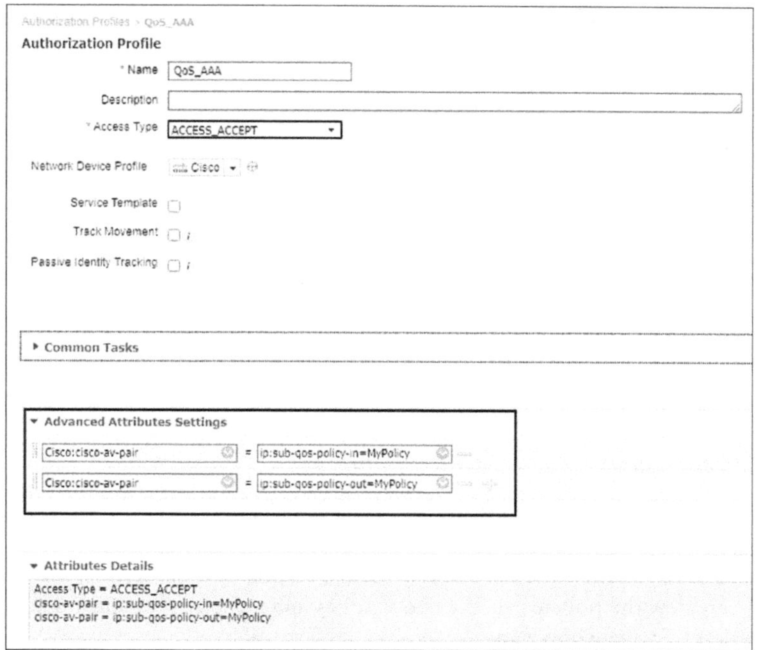

Figure 9-25 *Authorization profile in ISE for AAA QoS override*

For all profiles, queuing is configured on the C9800 egress port for prioritizing voice and CAPWAP traffic. The specific settings for each auto QoS profile are represented in Table 9-1.

Table 9-1 *Auto QoS Settings for Each Profile*

Auto QoS Profile	SSID Ingress	SSID Egress	Port Egress	Radio
Voice	platinum-up	platinum	AutoQos-4.0-wlan-Port-Output-Policy	ACM on
Guest	AutoQos-4.0-wlan-GT-SSID-Output-Policy	AutoQos-4.0-wlan-GT-SSID-Output-Policy	Auto-Qos-4.0-wlan-Port-Output-Policy	N/A
Fastlane	N/A	N/A	AutoQos-4.0-wlan-Port-Output-Policy	Fastlane EDCA
Enterprise	AutoQos-4.0-wlan-ET-SSID-Output-Policy	AutoQos-4.0-wlan-ET-SSID-Output-Policy	Auto-Qos-4.0-wlan-Port-Output-Policy	N/A

The auto QoS profile is configured under the Policy profile, as represented in Figure 9-26. The default setting is None, and no auto QoS setting is applied.

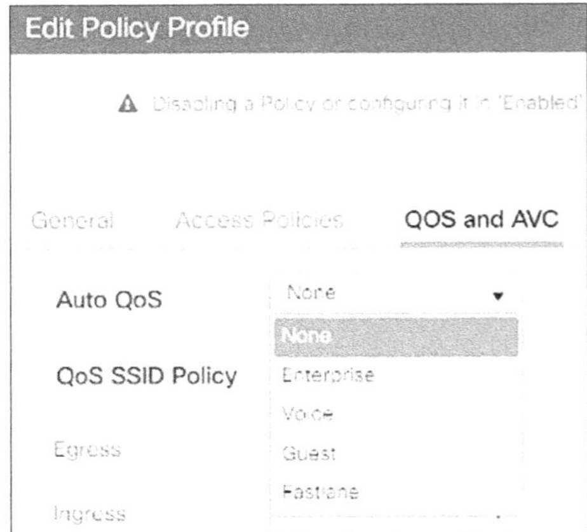

Figure 9-26 *Auto QoS configuration under the Policy profile*

Each profile is a set of predefined policies that apply to different QoS targets (SSID, port, and so on) in IOS-XE. When selected, the associated commands are auto-generated and not displayed for user selection in the GUI. After an auto QoS profile is applied on the Policy profile, you can view the policies via the **show policy-map <name>** command and show the configuration by using **show run**.

Every auto QoS profile has an associated QoS policy called Auto-QoS-4.0-wlan-Port-Output-Policy applied to the uplink port (see Table 9-1). This policy is designed to match voice and CAPWAP traffic (both control and data plane) and use priority queueing so that, in case congestion happens, the most important traffic is prioritized and protected.

The Fastlane profile is optimized for voice and Apple devices. Besides applying the egress port policy, it changes the radio EDCA parameters according to Cisco and Apple best practices for real-time applications. Usually, these settings are also valid for non-Apple clients. To see the EDCA parameters, you can issue the CLI command shown in Example 9-1 directly on the AP.

Example 9-1 *Output of the **show controllers** Command on AP*

```
AP#show controllers dot11Radio <0|1> | begin EDCA
EDCA Config:
====================
L:Local EDCA params C:Cell EDCA params
  AC      Type  CwMin   CwMax Aifs  Txop ACM
 AC_BE     L      4       6    3     0   0
 AC_BK     L      4      10    7     0   0
 AC_VI     L      3       4    1    94   0
 AC_VO     L      2       3    1    47   1
 AC_BE     C      4      10    3    79   0
 AC_BK     C      4      10    7    79   0
 AC_VI     C      3       4    2   128   0
 AC_VO     C      2       3    2    65   1
```

Usually, Radio 0 is the 2.4 GHz radio and 1 is for 5 GHz. As you can see from the command output, each queue has a different value for the EDCA parameters.

The Voice auto QoS profile applies the same policy to the uplink ports as all other profiles; it configures call admission control and applies the Platinum "metal" QoS profile both in the ingress and egress directions at the SSID level. This makes sure that voice packets marked with DSCP = 46 are treated with the highest priority.

QoS Profiles (a.k.a. Metal QoS Profiles)

The Catalyst 9800 provides four preconfigured QoS profiles named Platinum, Gold, Silver, and Bronze; given the names, they also are referred to as *metal QoS policies*.

You might be familiar with these profiles because the same ones are present in AireOS-based wireless controllers. Some differences with the C9800 are highlighted in this section.

The main purpose of the metal QoS profile is to limit the maximum DSCP value allowed on the wireless network for certain traffic, the so-called ceiling value; this also implies a limit to the corresponding 802.11 UP value and hence the mapping to a certain access category (AC) or queue at the AP.

Each profile is designed for a certain type of traffic:

- **Bronze:** Background or lowest-priority traffic, mapped to the access category background (AC_BK)
- **Silver:** Best effort mapped to AC_BE queue
- **Gold:** Video applications, mapped to AC_VI queue
- **Platinum:** Voice applications, mapped to AC_VO queue

It's important to keep in mind, as explained earlier in the "Trust DSCP Model" section, IOS-XE QoS implementation differs from AireOS because C9800 always adds the 802.11 QoS header in the CAPWAP packet to the AP. The AP always trusts the inner UP value and copies it to the 802.11 frame over the air.

Let's look at a metal QoS policy with an example: say you want to make sure that the traffic on a certain SSID would always be mapped to the lowest-priority queue, which is the background one. In this case, you apply the QoS profile Bronze to the SSID, and this automatically limits the traffic to a maximum DSCP value of 8 and hence a corresponding UP value of 1.

Let's now look at the impact of the Bronze QoS profile on the marking and queuing of the downstream (wired to wireless) and upstream (wireless to wired) client traffic. In the example illustrated in Figure 9-27 and Figure 9-28, the AP is in local mode, so the traffic is always centralized at the WLC.

For the downstream direction, the traffic is received at the C9800 from the wired network via the 802.11q trunk to the uplink switch, and it's marked with DSCP 34. Because

34 is greater than the reference "ceiling" value for the configured metal QoS policy (DSCP 8 in this case), both the inner DSCP and the outer CAPWAP DSCP values are marked down to 8 before the packet is transmitted to the AP. The C9800 also maps derived DSCP to the UP value (UP = 1 in this case) and sets it in the 802.11e QoS header in the CAPWAP packet.

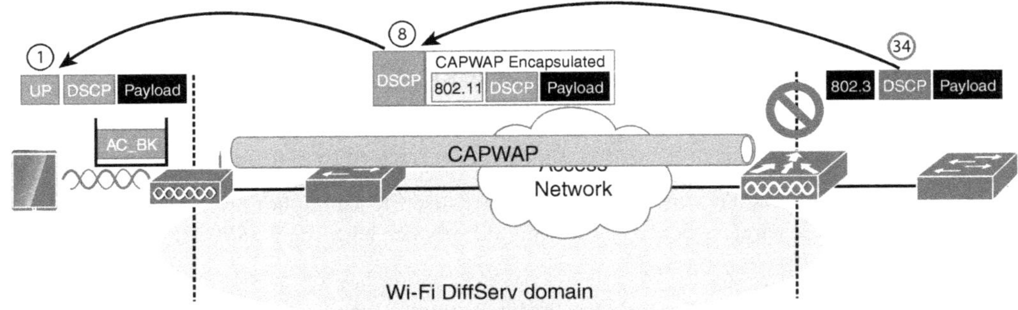

Figure 9-27 *Bronze QoS profile, downstream*

Note another difference in the QoS implementation between IOS-XE and AireOS-based controllers: in the case of the C9800, the inner original packet is also marked down to DSCP 8, whereas in AireOS, the inner is left untouched. This does not change the QoS processing between the WLC and AP because this is the inner packet that is ignored by the intermediate network nodes. The only difference is that the receiving client gets a packet with a different DSCP marking than the original one.

At the AP, only the UP value in the external CAPWAP header is considered. The AP takes the value from the outer header, UP 1 in this case, and then queues the packet for transmission to the background access category.

In the upstream direction, represented in Figure 9-28, the client sends a packet marked with DSCP 34. The UP value is not important here because the AP, by default, trusts the incoming DSCP value. Because DSCP 34 is greater than the ceiling value, as per Bronze policy, the AP marks the outer CAPWAP header with DSCP = 8 before sending the packet to the WLC. This is the same exact behavior as AireOS.

When the C9800 gets the packet, it looks at the original packet's DSCP, and because this is greater than 8, it re-marks the packet using the ceiling value.

This is the main difference between the two systems: with the AireOS QoS implementation, the inner packet is left untouched as it travels across the wireless network domain; with the C9800, the original packet is rewritten to apply the policy at egress.

As a consequence, in this specific example, you would have an asymmetric marking. If you consider the wireless network as your DiffServ domain from a QoS point of view, the incoming packet from the wired network enters the DiffServ domain with a DSCP value of 34 and the return packet egresses the wireless domain with DSCP 8.

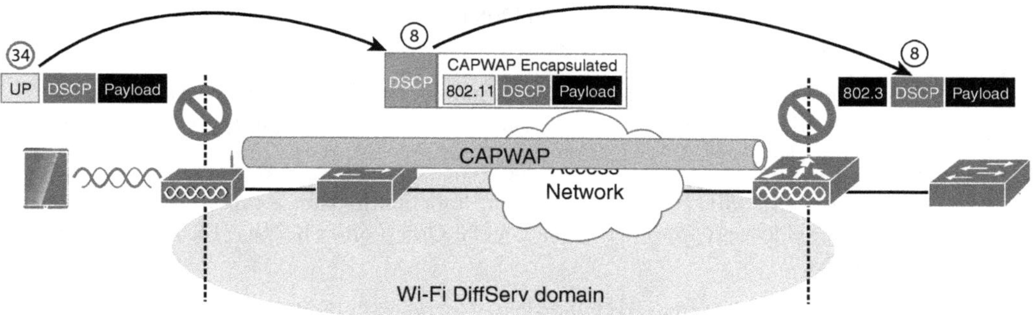

Figure 9-28 *Bronze QoS profile, upstream*

You may be wondering, Is this a problem? Which approach is better? First of all, both AireOS and IOS-XE QoS implementations are compliant with the main architecture RFCs (RFC 2474, RFC 2475, and so on). Second, there is not really a right or wrong way to do things. They are just different, and you need to understand the difference.

To do that, consider that there are two architectural approaches when it comes to implementing QoS over an overlay network:

- Pipe mode tunneling
- Uniform mode tunneling

In pipe mode, the QoS policy is applied only on the outer tunnel header (whatever the encapsulation might be: MPLS, CAPWAP, and so on), and the inner packet is not touched. You can think of AireOS as operating in pipe mode. In uniform mode, the policy is applied at ingress on the outer header and at egress at the inner, so the original packet is indeed modified. You can think of IOS-XE–based C9800 operating in uniform mode. *Pipe* and *uniform mode* are terms often used in the context of MPLS networks, and you can find more information in the "References" section.

Let's now analyze a useful case for the metal QoS implementation of C9800. Imagine you want to re-mark all the traffic from users on a specific SSID to DSCP 0; in other words, you want to "bleach" specific users' traffic; this could be your guest SSID because you don't trust the user behavior. With the C9800 implementation, in uniform mode, as traffic is received on this SSID, not only is the outer header re-marked to DSCP 0 at the AP so that the right queuing can be done in the wired network between AP and WLC, but also the original packet is re-marked to DSCP 0 as it egresses the WLC to the wired network, no matter what the original marking is. This is a truly bleaching behavior that could not be achieved in pipe mode unless you apply another policy at the wired network.

Regarding the implementation of metal QoS profiles on the Catalyst 9800, there are a few things you need to know:

- You can apply a QoS profile in egress and ingress directions separately—for example, Platinum downstream and Silver upstream.

- On the GUI, you can set the metal QoS profile only per SSID. On the CLI, you can also configure it per client target.
- In the C9800, the nonmatching traffic goes in the default class. This applies also to non-WMM traffic (frames without an 11e UP value). You can apply a policy to the default class.
- Per-user and SSID bandwidth contracts (rate limiting) are configurable via separated QoS policies and not directly using metal QoS profiles because they are available in AireOS.
- At the time of writing, the UP-to-DSCP mapping in the C9800 is static, and it cannot be changed. There is an exception: when the WLAN is configured for hotspot 2.0, the standard mandates for the user to be able to set the DSCP to UP mapping that will be advertised by the AP in the beacons. This might change in the future.

For each metal QoS profile, a max DSCP setting is used as a ceiling value and to re-mark traffic. This is shown in Table 9-2.

Table 9-2 *Max DSCP Allowed (Ceiling Value) for Each Metal QoS Profile*

QoS Profile	Max DSCP Allowed
Bronze	8
Silver	0
Gold	34
Platinum	46

Application Visibility and Control (AVC)

AVC stands for Application Visibility and Control. There is a Deep Packet Inspection (DPI) engine in the C9800 wireless controller and in the APs that allows for application identification and related usage analysis. This is the visibility part. Once the traffic is identified, you can apply the control part of AVC to mark or re-mark, rate-limit, and drop unwanted traffic.

How is the application identified? Well-known applications can be recognized by the ports they use. For example, TCP 80 is the signature of web traffic; TCP 443 is HTTPs. This is easy, but what if the traffic is encrypted? What if the application uses random ports? For example, the latter is very common for real-time applications because this helps ensure that two calls on the same site would not use the same port and improves the privacy of the calls by making port prediction more difficult for an eavesdropper.

In these cases, the Catalyst 9800 leverages Network-Based Application Recognition. The NBAR engine combines stateful deep packet inspection with state-of-the-art recognition techniques such as statistical classification, socket caching, service discovery, auto learning, and DNS-AS to identify more than 1,400 applications.

As you can imagine, running DPI doesn't come for free: some packets need to be punted to the NBAR engine for analysis, and this can be CPU intensive. That's a good reason to deploy AVC at your access network: the closer to the edge of the network you identify the traffic, the earlier you can mark the traffic and "trust" the marking across the rest of your network. If you had to perform the classification and marking deeper in the network (on a core router, for example), you would need to analyze a much larger amount of traffic; doing it at the edge scales and optimizes the load. Let's now look at how AVC is implemented in wireless with the Catalyst 9800 in the different deployment modes. For local mode and FlexConnect central switching, as illustrated in Figure 9-29, the traffic is centralized at the WLC, and therefore, you leverage the engine built in the C9800 to perform AVC for both upstream and downstream traffic.

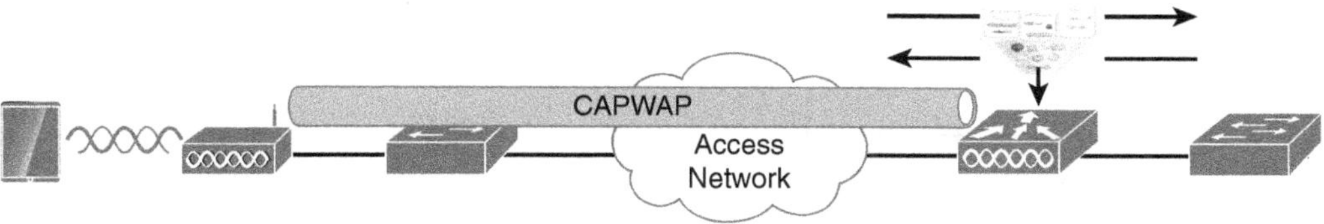

Figure 9-29 *AVC in local mode and Flex central switching is applied at the WLC*

AVC can be applied in a specific direction (upstream or downstream or both) and the "C" in AVC may modify the inner DSCP value, thus influencing the CAPWAP DSCP and wireless UP values; WLC can also drop or rate-limit traffic, as stated before.

In FlexConnect local switching and SDA wireless, the data plane is distributed at the AP and hence the inspection and related actions are applied at the AP itself, as shown in Figure 9-30. Note that the access point doesn't usually run the same NBAR protocol pack as the wireless controller, given the different CPU and memory constraints.

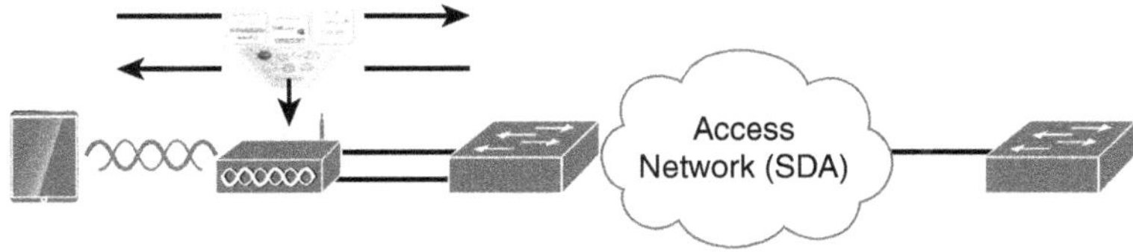

Figure 9-30 *AVC in FlexConnect local switching and SDA wireless is applied at the AP*

Cisco continuously improves and enhances the capability of the NBAR engine to recognize new applications. Starting with release 17.1, C9800 supports dynamic updates for the AVC protocol pack. This means that for a certain IOS-XE release, you can update the C9800's NBAR engine without the need to upgrade the entire software image. Just download the correspondent package file (with the .pack extension) from Cisco Connection Online (CCO) and load it on to the bootflash as you would normally do for any software upgrade. Then click **Download** and **Install**. After that, you can check the engine version in the GUI under **Monitoring > Services > Application Visibility**, as illustrated in Figure 9-31.

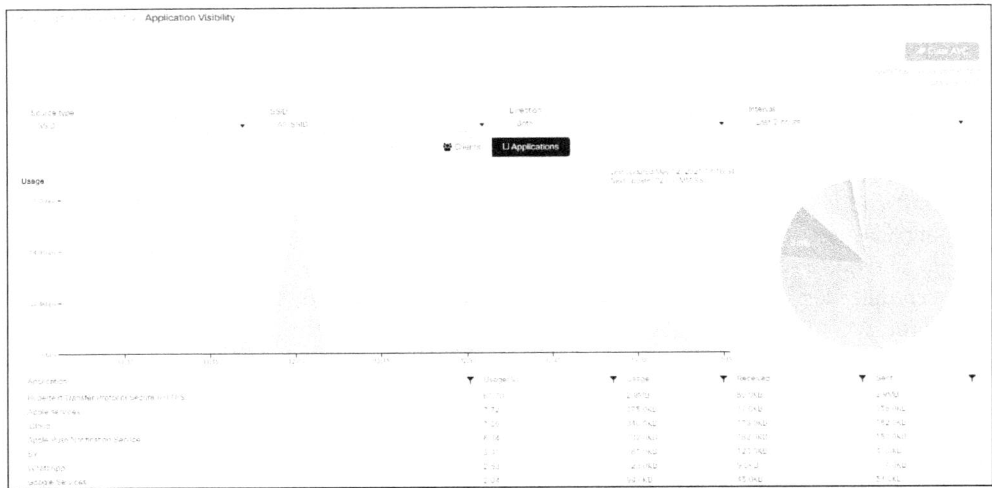

Figure 9-31 *Monitoring AVC*

Note that Cisco might provide multiple updates for a single IOS-XE release, but a certain protocol pack should be used with the corresponding IOS-XE release and later ones. For example, protocol pack 56 was released with IOS-XE 17.3.1; the same pack can be used with any maintenance release of 17.3.x train but also subsequent major releases (such as 17.4 and 17.5). At the time of writing, the dynamic protocol pack update is not supported for FlexConnect APs, so the file is updated only with a major software release.

What if a specific business-critical application is not yet recognized by NBAR? The C9800 implementation of AVC allows you to define your own custom application by specifying a set of parameters: custom IP address for the server, custom ports, DSCP value, the host name, and even the HTTP URL. The host and URL specification strings can take the form of regular expressions to provide more flexibility. At the time of writing, configuring a custom application is available only via the CLI using the commands shown in Example 9-2.

Example 9-2 *The CLI Commands to Configure a Custom NBAR Application*

```
c9800(config)#ip nbar custom ?
  WORD  Name of the User defined Protocol
c9800(config)#ip nbar custom <name of the APP> ?
  <0-255>    BYTE offset of field to search for in payload (DEPRECATED)
  composite  Custom protocols based on multiple underlying protocols
  dns        Custom protocols based on dns parameters
  http       Custom protocols based on http parameters
  ip         Custom protocols based on the IP protocol
  ssl        Custom protocols based on ssl parameters
  transport  Custom protocols based on the transport layer and below
```

The option to define a custom application for AVC is a value-add that only the Catalyst 9800 can offer.

Deployment Verification and Restrictions

In the previous section you learned how to deploy QoS on the C9800 mainly through the GUI. You can use the GUI for a first verification of the settings as well. Here you learn some additional CLI commands for a deeper dive into your QoS deployment.

To verify the policy applied on an AP for a certain SSID, use the following command:

```
C9800#show policy-map interface wireless ssid profile-name <name>
radio type 2GHz|5GHz|6GHz ap name <AP name> input|output
```

A similar command can be used for the client policy verification:

```
C9800#show policy-map interface wireless client mac <MAC> input|output
```

Here's another useful command for client QoS verification:

```
C9800#show wireless client mac <MAC> service-policy input|output
```

To verify the EDCA parameters, you apply this **show** command at the AP:

```
Ap#show controllers dot11Radio 0|1 | begin EDCA
```

At the time of writing, there are some deployment limitations regarding the QoS implementation on the Catalyst 9800 that you should be aware of; these are being tracked and will be fixed in subsequent releases. The following features are not supported in FlexConnect mode with local switching and in Fabric mode:

- In the policy definition, matching traffic using ACLs is not supported. Only matching DSCP and AVC are supported.
- Policy stats on Wave 2 802.11ac and 802.11ax Flex APs.
- SIP Snooping/CAC.
- AVC custom application.

Fastlane+ (Plus)

Cisco has a strong partnership with Apple. The purpose of this collaboration is to codevelop features and solutions that allow Apple clients to work better on a Cisco network. Fastlane+ is the latest example of how Cisco and Apple go beyond the 802.11 standard, in this case 802.11ax, to provide an enhanced user experience for latency-sensitive voice and video applications in a high-density user environment.

Fastlane+ does this by supplementing the 802.11ax Buffer Status Report (BSR) with direct estimation of application requirements indicated by the client. This method significantly reduces BSR polling overhead as compared to the standard uplink OFDMA scheduling and improves uplink scheduling efficiency. Fastlane+ also allows the AP to solicit periodic scheduling feedback from the clients to keep the information updated. Internal testing shows that Fastlane+ has the following benefits in an environment where channel utilization is high:

- Up to 30 percent decrease in uplink transmission latency, which is very important for delay-sensitive applications like voice and video
- Up to 40 percent increase in MOS score for voice calls

Fastlane+ is enabled by default on 802.11ax-capable iPhone and iPad devices running iOS 14 or later. On the infrastructure side, it is currently supported on the Cisco Catalyst 9130 access point and on the C9800 running 17.4.1 release and higher. For more information on Fastlane, check out the frequently asked questions document in the "References" section at the end of the chapter.

Best Practices

Every wireless deployment is different and may require taking into consideration the customer requirements and the specific RF environment to tweak the design accordingly. QoS implementation is no different. Some general best practices can be applied and are hence recommended.

It's a good practice to tune EDCA parameters to optimize your RF environment. Remember that the EDCA profile is set for the whole 6 GHz, 5 GHz, or 2.4 GHz networks, and not per SSID. There is no single recommended setting for all wireless networks, so it is important to test different values. For networks with voice and video traffic, it is a good idea to start with the EDCA to "optimized-video-voice," as shown in Figure 9-32.

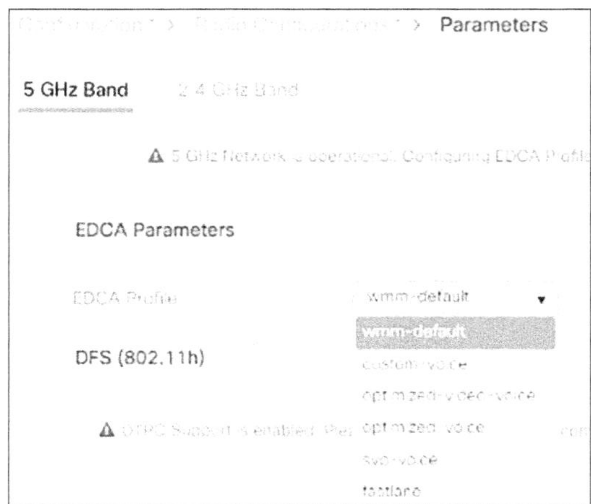

Figure 9-32 *EDCA profile settings for a 5 GHz network*

If you have an SSID with voice traffic, the recommended settings would be to configure the policy profile with

- SSID QoS policy = Platinum
- Auto QoS profile = "Fastlane"

As seen in the previous section, the auto QoS profile Fastlane automatically triggers the following configurations:

- The Client QoS policy is set to Platinum.
- The EDCA parameter is set to Fastlane under **Radio Configurations > Parameters > 5 and 2.4 GHz Bands.**
- The Catalyst 9800's egress priority queuing is set to prioritize voice and CAPWAP traffic, applying the AutoQos-4.0-wlan-Port-Output-Policy service policy.

The settings are shown in Figure 9-33.

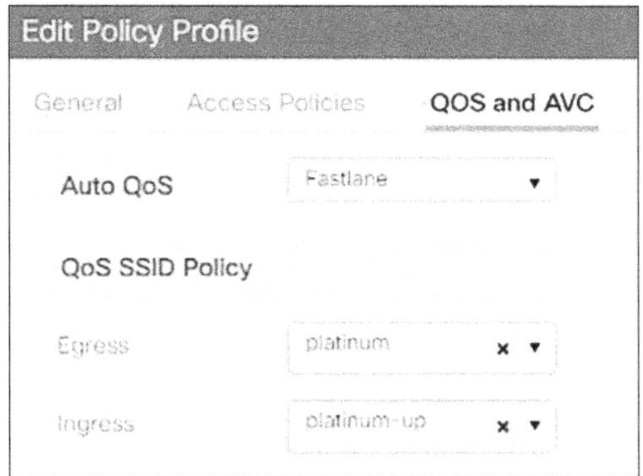

Figure 9-33 *Recommended policy profile settings for an SSID with voice traffic*

In terms of overall wireless network QoS recommendation, you have already learned about the "trust DSCP" setting. Trust DSCP is ON by default in release 17.4.1 and higher. For previous releases, you can turn it on under Join Profile, as shown in Figure 9-34.

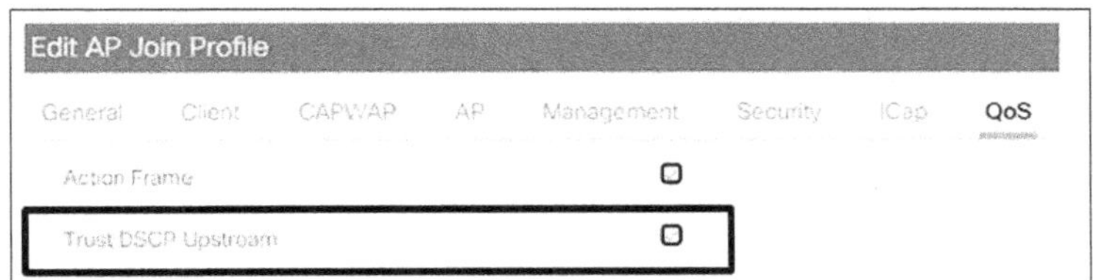

Figure 9-34 *Trust DSCP settings under the AP Join Profile*

A final consideration regarding metal QoS profiles (Platinum, Gold, Silver, Bronze): these preconfigured QoS policies were designed to be most effective for SSIDs with one type of traffic (for example, voice SSID with voice-dedicated handsets). This is rarely true anymore; today most customers deploy a single "enterprise" SSID and have multiple business applications on it, including voice and video. With the Catalyst 9800, you can go beyond metal QoS profiles and leverage the flexibility of IOS-XE and MQC to create your QoS policy using AVC/NBAR, policy maps, or Auto QoS, whichever is more suited for your environment.

Summary

This chapter describes the implementation of Wi-Fi QoS on the Catalyst 9800 wireless LAN controller. The C9800 is based on IOS-XE and leverages Cisco Modular QoS CLI (MQC) to define and implement an end-to-end architecture for QoS. In addition to introducing the Wi-Fi QoS technology, the focus of the chapter is on the design and deployment of the C9800 "trust DSCP" QoS model and on providing some best practices.

References

RFC 4594: https://www.ietf.org/rfc/rfc4594.txt

RFC 8325: https://tools.ietf.org/html/rfc8325

Configuring Call Admission Control (CAC) on C9800: https://www.cisco.com/c/en/us/support/docs/wireless/catalyst-9800-series-wireless-controllers/214863-voice-deployment-on-catalyst-9800-wirele.html

Bidirectional client rate-limiting configuration: http://cs.co/BDRL-QoS-example

QoS tunneling mode: https://www.cisco.com/c/en/us/support/docs/multiprotocol-label-switching-mpls/mpls/47815-diffserv-tunnel.html#summary

Fastlane+: https://www.cisco.com/c/dam/en/us/td/docs/wireless/access_point/9130ax/tech-notes/fastlane-faq.pdf

Chapter 10

C9800 High Availability

High availability (HA) refers to the capability of a network system to maintain uptime and provide service even in case of failure. Typically, this effect is achieved by setting up two or more systems that match capabilities to operate in a load-balanced fashion or in a redundant fashion. Catalyst 9800 wireless LAN controllers—specifically the C9800-L, C9800-40, C9800-80, and C9800-CL on the private cloud (ESXi, KVM and HyperV)—support one-to-one box-to-box redundancy. In other words, only two chassis, in the case of C9800-40, C9800-80, and C9800-L, or two instances, in the case of the C9800-CL, form the redundant pair to provide always-on service. Continuity of service is expected for both planned and unplanned outages.

- Unplanned outages occur due to failures on the WLC or network that prevent the WLC from servicing wireless users. WLC crashes, WLC hardware failures, and WLCs losing network connectivity are examples of unplanned outages. The C9800 provides two solutions to deal with these unplanned outages:
    - **Stateful switchover (SSO):** When two standalone C9800s pair up and sync information to enable one C9800 to take over services from another C9800 seamlessly, without noticeable impact to end users, the process is called *stateful switchover (SSO)*.
    - **N+1 redundancy:** When two or more WLCs operate independently without sharing information, but one or more act as backup to other WLCs, the process is termed *N+1 deployment*. Because there is no sync between the WLCs, service transition between WLCs is stateless, and end users experience some impact. The C9800 supports many-to-one backups up to the capacity of the backup WLC.

- Planned outages typically refer to system or network maintenance that you have planned, for example software management of the WLC or hardware replacement or network maintenance. The C9800 supports the following:
  - **Hot patching:** Software maintenance upgrades (SMUs) are pointed code fixes for service impacting software issues that can be applied on a live system without impacting data plane forwarding or the control plane state.
  - **In-Service Software Upgrade (ISSU):** This term refers to the C9800's capability to allow software code upgrades and downgrades between versions on the C9800 while it continues to service users and traffic.
  - **N+1 Hitless Upgrade:** This term refers to the framework where access points are intelligently migrated, say C9800-1 in a staggered fashion, to an N+1 C9800 chassis or instance that has been upgraded to the intended software version. Then the first C9800, C9800-1 in this example, is upgraded, and APs are migrated back to the C9800-1 seamlessly and with minimal user intervention.

The various options to manage planned outages are covered in further detail in Chapter 15, "Backing Up, Restoring, and Upgrading your C9800." This chapter focuses on unplanned outages where you learn about:

- Steps to form an SSO pair between two C9800s
- Design considerations when deploying a C9800 HA pair
- SSO pair behavior under different network failure conditions
- Impact of SSO on different features on the C9800
- N+1 design and deployment
- HA in Embedded Wireless Controller on AP (EWC-AP) deployments
- HA in Embedded Wireless Controller on Switch (EWC-SW) deployments

SSO Redundancy

The Catalyst 9800 supports a one-to-one box-to-box redundancy model and stateful switchover for access points and clients. *Stateful* means that the system automatically keeps information in sync within the cluster so that when a failure happens, the APs and clients are not affected; there is no network downtime.

Two standalone C9800s form a peer relationship over a dedicated port referred to as a redundancy port (RP). As part of this peer relationship, one of the two C9800s is automatically elected as the active WLC, and a bulk sync of configuration occurs from the active to the standby C9800. After they are peered, both C9800s are accessible via a single IP address, and control plane and data forwarding are centralized at the active

C9800. Access points register their CAPWAP tunnels to the active C9800, and the active C9800 services the clients on these APs. The active C9800 mirrors this operational AP and client database to the standby C9800. In the event of an active C9800 failure caused by a power, box, or network failure, the standby C9800 takes over as the active WLC. AP SSO ensures that APs stay joined and do not have to reinitiate the CAPWAP discovery/join process. Client SSO ensures all the clients that are in the RUN state, meaning the clients that are already authenticated, have an IP address and are forwarding traffic, do not disconnect, and continue data forwarding. Note that the clients that were in the process of joining the active WLC and were in varying states of associating, authenticating, or getting an IP address are dropped at failover and are expected to restart their join process.

The AP and client SSO are supported on the C9800-L, C9800-40, C9800-80, and C9800-CL in all deployment modes (central, flex, and fabric). On Embedded Wireless Controller on Switches, SSO is supported only within a switch stack using stackwise technology and only for APs and clients that are not connected off the Active Catalyst 9K switch. EWC-AP deployments do not support SSO but leverage the Virtual Redundancy Router Protocol (VRRP) to provide resiliency.

Prerequisites

For two C9800s to be SSO peers, they need to meet a few requirements:

- Both C9800s need to be of the same form factor. For example, the C9800-80 can be paired only with another C9800-80, not a C9800-40. By the same definition, the C9800-L-C can pair only with the C9800-L-C, not a C9800-L-F. Even in the case of the C9800-CL, both peers need to be deployed for the same template (small, medium, or large) with the same resources (vCPU, memory, hard disk, vNICs). The commands you can use to verify the two C9800s to be paired are:

  ```
  C9800#show inventory
  C9800-CL#show platform software system all
  ```

- Both C9800s must match any field replaceable units like the SSD drive or Ethernet port adapter (EPA). This requirement is applicable only on the C9800-40 and C9800-80. The commands that include the details of the EPA are:

  ```
  C9800#show inventory
  C9800#show platform
  C9800#show logging process stack_mgr internal to-file
  bootflash:<FILENAME>/log
  ```

  Example 10-1 highlights two C9800-80s—one with an EPA and one without an EPA—being paired and resulting in an error from this HA pairing attempt.

Example 10-1 *HA Pairing Failure Seen When Two C9800-80s with Mismatched EPAs Are Paired*

```
C9800-80-1#show platform
Chassis type: C9800-80-K9
Slot      Type                State                 Insert time (ago)
--------- ------------------- --------------------- -------
0         C9800-80-K9         ok                    00:07:41
0/0       BUILT-IN-6X10G/2X1G ok                    00:04:12
0/1       C9800-2X40GE        ok                    00:04:34

C9800-80-2#show platform
Chassis type: C9800-80-K9
Slot      Type                State                 Insert time (ago)
--------- ------------------- --------------------- -------
0         C9800-80-K9         ok        00:57:34
0/0      BUILT-IN-6X10G/2X1G  ok                    00:54:23
R0        C9800-80-K9         ok, active            00:57:34

C9800#show logging process stack_mgr internal | include MISMATCH
%STACKMGR-1-EPA_MISMATCH: Warning: multi-chassis interfaces are mismatch.
The mis-match may be caused by hardware failure or different EPA types

C9800#show logging process stack_mgr internal | include FRU
Multi-chassis HA: local FRU=D68, remote FRU=not-ready
```

- Both C9800s should run the same software version—both major and minor versions. So, the C9800-80 running 16.x and 17.x cannot be peered. The same applies to a C9800 running 16.12.x and 16.12.y. Even hot patches need to match between C9800s before they can be paired for SSO. Starting with release 17.5, auto-upgrade is supported wherein, if the standby C9800 is running a different version from the active C9800, then the active C9800 auto-upgrades the standby C9800 to the version of the active C9800 by copying over the images. One requirement for auto-upgrade is that the active C9800 should be in Install mode, and the standby C9800 should be either in Install mode or configured to boot in Install mode at the next reload.
- Both C9800s should have the same installation mode, either Install or Bundle, as shown in Example 10-2.

Example 10-2 *Verifying Installation Mode on C9800*

```
C9800#show version | inc Install
Installation mode is INSTALL
```

- In the case of appliances, both the chassis need to use either an Ethernet RJ-45 redundancy port or small form-factor pluggable (SFP) RP and comply with the RP requirements covered in the next section.

When one or more of the requirements listed are not met, an HA pairing fails with a version mismatch indicated by the log message shown in Example 10-3.

Example 10-3 *Error Message When C9800s with Incompatible Installation Modes Are Paired*

```
%BOOT-3-BOOTTIME_INCOMPATIBLE_SW_DETECTED:R0/0:issu_stack: Incompatible
software detected. Details: Active's subpackage boot mode does not match
with member's super boot mode. Please boot switch 2 in subpackage mode
```

Ports and Interfaces

The C9800s host multiple types of physical ports and virtual interfaces, each with their specific functions. The pairing of two C9800s explicitly uses the RP and introduces a Redundancy Management Interface (RMI), which together enable seamless SSO failover.

Redundancy Management Interface (RMI)

A Redundancy Management Interface is a logical communication channel over uplink ports between the active and standby C9800, which serves as an alternate path to the RP link to verify connectivity to the network and to peer C9800. The primary purposes for RMI are

- **Dual-active detection (DAD):** If two C9800s end up in a dual-active state, an RMI link detects it and places one of the C9800s in a recovery state to resolve.
- **Gateway reachability check:** The verification of gateway reachability for the wireless management subnet, from each C9800 in the HA pair, is done via ICMP and ARP requests sourced from the RMI for the respective C9800.
- **Resource health exchange:** RMIs on both C9800s communicate over Transmission Control Protocol (TCP) port 3200 to exchange resource health information.
- **Direct access to standby:** Standby C9800 uses RMI to communicate to external syslog and Network Time Protocol (NTP) servers as well as to provide out-of-band access via Secure Shell (SSH), Hypertext Transfer Protocol Secure (HTTPS), and NETCONF.

RMI is configured in the same subnet as the Wireless Management Interface (WMI) but needs to have a unique IP address. On the active C9800, RMI is represented as a secondary IP address on the WMI with the wireless management IP address defined as the primary IP address. Note that you can configure multiple secondary addresses on WMI via the command-line interface (CLI) without seeing errors, but for HA, only one secondary address is supported. Also, there is no check that prevents to configure RMI in a different subnet than WMI. So, you, as the network administrator, need to take care of it. The subnet mask for the WMI is used as a subnet mask of the RMI.

Standby C9800 cannot use the same WMI IP because the active C9800 has claimed it. However, standby C9800 needs its RMI IP address to be reachable to maintain HA state.

This reachability is achieved by programmatically assigning the RMI IP as the primary IP address of the WMI on the standby C9800. When a standby C9800 takes over as active WLC, the management IP is programmed as the primary IP address and the RMI IP of the local C9800 is reassigned as a secondary IP address. Note that this construct of secondary addressing is relevant only to an IPv4 deployment. When using IPv6 for WMI and RMI, both WMI and RMI are represented as two unique IPv6 addresses.

Redundancy Port (RP)

C9800 appliances like the C9800-40, C9800-80, or C9800L have a dedicated physical port called the redundancy port, and the appliances use the link between RPs of the two C9800s to exchange keepalives to form an HA peer relationship. The RP has to comply with strict requirements.

- The maximum RP link latency allowed is 80 ms round-trip time.
- The maximum transmission unit (MTU) supported on the RP link is 1500 bytes. In other words, jumbo frames cannot be transmitted on the RP link.
- The minimum bandwidth of the RP link is expected to be 60 Mbps.

As shown in Figure 10-1, the RP on the C9800L-Copper or C9800L-Fiber is always an RJ-45 port. However, on the C9800-40 and C9800-80, the RP can be an Ethernet RJ-45 port or a small form-factor pluggable (SFP) transceiver port. Note, if both ports are connected, the SFP port takes priority. Online insertion or removal (OIR) of SFP is not supported. If SFP is added or removed, even if the SFP port is not connected and the Ethernet RJ-45 is serving as the RP interconnect, the C9800 gives priority to SFP and treats the insertion as a physical link event on the RP, triggering redundancy teardown. Only the Cisco SFP modules GLC-LH-SMD and GLC-SX-MMD are supported for the RP, even though a wider variety of SFPs are supported on uplink ports.

Figure 10-1 *Redundancy port on the C9800L-C, C9800L-F, C9800-40, and C9800-80*

You can view details of SFP in use by running the CLI **#show platform hardware slot R0 ha_port sfp idprom on the C9800-40 and C9800-80 platforms.**

On a virtual C9800-CL deployment, a virtual network interface card (vNIC) must be explicitly assigned to function as an RP. Typically, on a C9800-CL, three vNICs are assigned, as shown in Figure 10-2. The breakdown of ports is

- **GigabitEthernet1:** Management (out of band)

- **GigabitEthernet2:** Uplink that forwards CAPWAP and client traffic (trunk)
- **GigabitEthernet3:** Redundancy port

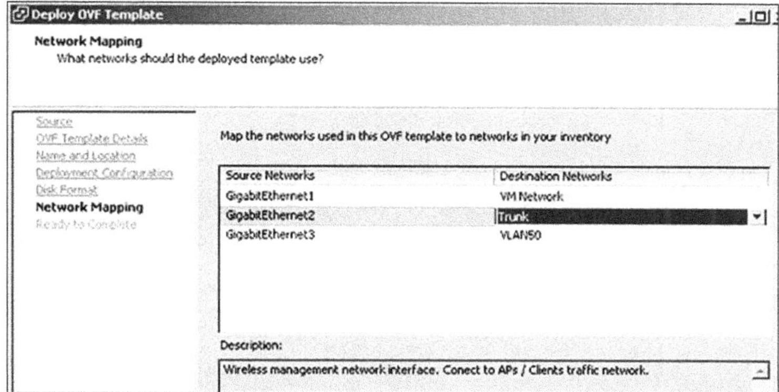

Figure 10-2 *vNIC mapping for a C9800-CL over ESXi*

When the HA pairing is successful, this interface (for example, GigabitEthernet 3) disappears from the configuration, and the RP is directly accessed by the Linux Kernel on the C9800. Starting with release 17.5.1, you can view the status and statistics for the RP, run **ping** tests to validate RP peer reachability, and even do packet captures on the RP using the following commands:

```
C9800#show platform hardware slot R0 ha_port interface stats
C9800#test wireless redundancy rping
C9800#test wireless redundancy packetdump start
C9800#test wireless redundancy packetdump start filter port <0-65535>
C9800#test wireless redundancy packetdump stop
```

On the physical appliance C9800, the RPs on the two chassis can be connected back to back or via a Layer 2 switch, provided the RPs are assigned to a unique VLAN that is not used anywhere else in the deployment. It is recommended that this RP link VLAN be unroutable. Manual configuration of the RP IP address is used only with RP-only SSO, which is available starting with release 16.x. Starting with release 17.1, RP + RMI deployment is recommended, and the RP IP address is auto-derived as 169.254.x.y, where the last two octets match the last two octets of the RMI IP address for a given chassis. You need to make sure that this subnet of 169.254.x.y is not blocked on the RP link via port ACLs or any other means.

When the RP of two chassis are connected, they need to be of the same type. That means the RJ-45 RP on one C9800-40 cannot be connected to an SFP RP on another C9800-40. In the case of the C9800-CL on a private cloud, the RP on one instance can be connected to an RP hosted on the same server or hosted on another server. Figure 10-3 provides some examples of the RP connections.

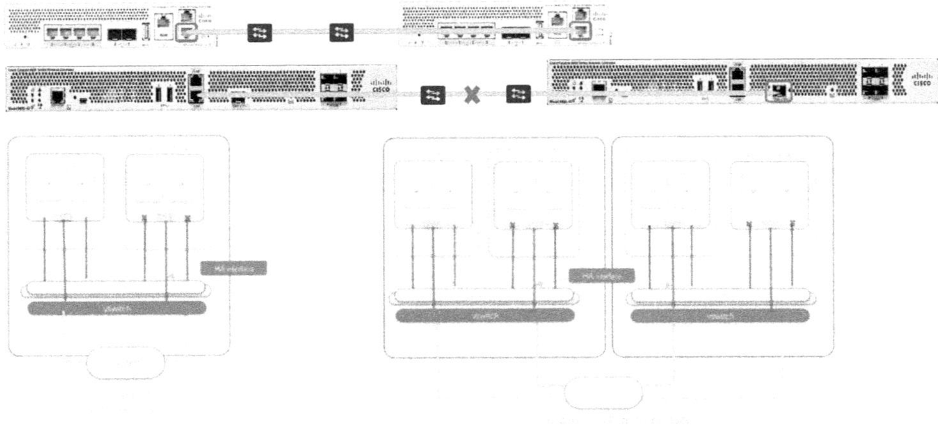

Figure 10-3 *RP-to-RP connectivity between two C9800-L-Cs, two C9800-40s, and two C9800-CLs*

Uplink Ports

On the C9800, IOSd is the process responsible for creating and maintaining port and interface status. IOSd is not HA-aware, meaning that interfaces created on standby are, perennially, in a down state. When an interface state sync occurs from the active, a shadow state is created and maintained by the local chassis, and this shadow state reflects the true synched state from the active.

```
C9800-CL#show interface GigabitEthernet1
GigabitEthernet1 is down, line protocol is down
Shadow state is up, true line protocol is up
```

Console Port

When the HA pairing is done, only the console of the active chassis is accessible by default. You can access the console of the standby chassis via manual configuration on the HA pair, as shown in Example 10-4.

Example 10-4 *Enabling a Standby Console on a C9800 HA Pair*

```
C9800(config)#redundancy
C9800(config-red)#mode sso
C9800(config-red)#main-cpu
C9800(config-r-mc)#standby console enable
```

Out-of-Band Management/Service Port (SP)

When an HA pair forms, the standby service port configuration is synchronized to the configuration on the active chassis and the service port on standby is marked operationally down, thus preventing out-of-band access on the standby WLC.

RP+RMI Supported Topologies

The C9800 HA relies on both RP reachability and the RMI gateway check to validate the chassis health and its ability to service APs and clients. To accomplish this, recommended connections from the RP as well as distribution ports to switch infrastructure are shown in Figure 10-4.

- The C9800 SSO pair is connected to a VSS switch with uplinks from the active and standby C9800 split across the two switches that form the VSS pair.
- The C9800 SSO pair is connected to a VSS switch with uplinks from the active going to one switch in VSS and uplinks from the standby connecting to the other switch of the same VSS.
- The C9800 SSO pair is connected to switches running Hot Standby Routing Protocol (HSRP). The active C9800 is connected to HSRP active, while the standby C9800 is connected to the standby HSRP.

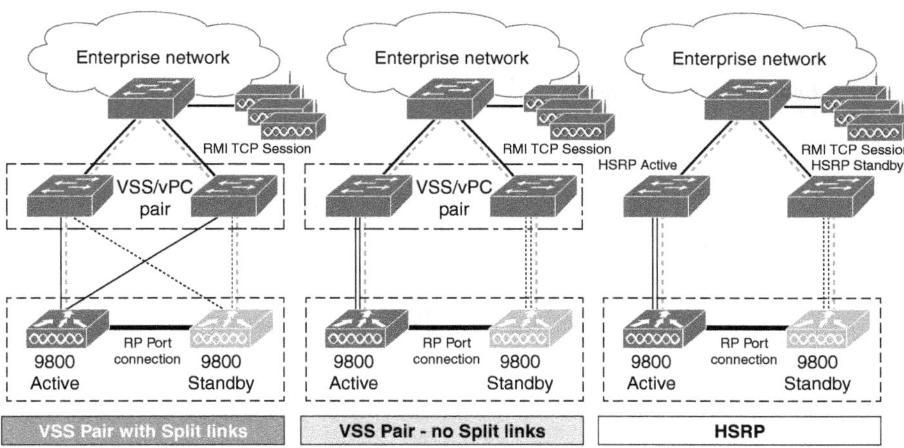

Figure 10-4 *Supported topologies from an RP+RMI C9800 HA pair*

Note that in each of the topologies shown, the RP can be connected back to back, or it can be connected over an L2 switch where switchports belong to a dedicated, nonrouted VLAN. Also, in the topologies, link aggregation group (LAG), discussed later in the chapter, is used to add link resiliency.

Building an RP+RMI HA Pair

Before pairing two C9800s, you need to go through the Day 0 GUI wizard or Day 0 CLI wizard to initialize each chassis, as described in Chapter 1, "Cisco C9800 Series."

You also need to connect redundancy ports and uplink ports as per the supported topologies. In addition, you need to assign the WMI of the two C9800s to be paired in the same subnet and verify reachability. Note that after they are paired, both chassis are accessible via the WMI IP of the chassis elected active, and the WMI IP of the chassis elected standby is available for reuse.

Configuration

When physical connections are in place, to configure redundancy, you can connect to the GUI interface of each chassis to be paired and navigate to **Administration > Device > Redundancy**, as shown in Figure 10-5. Enabling redundancy opens a GUI page that allows you to configure the HA parameters, as shown in Figure 10-6.

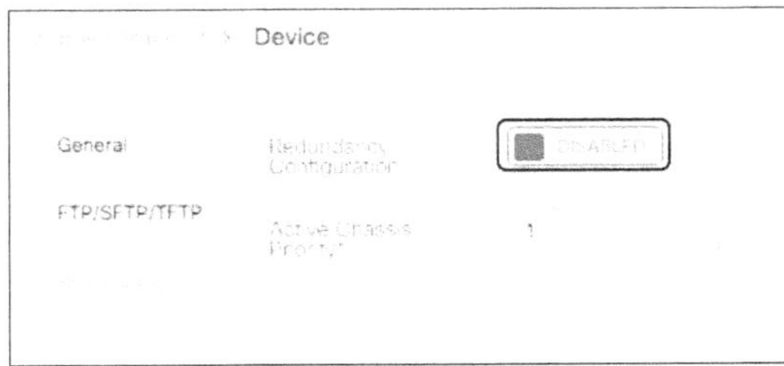

Figure 10-5 *Enabling the HA SSO*

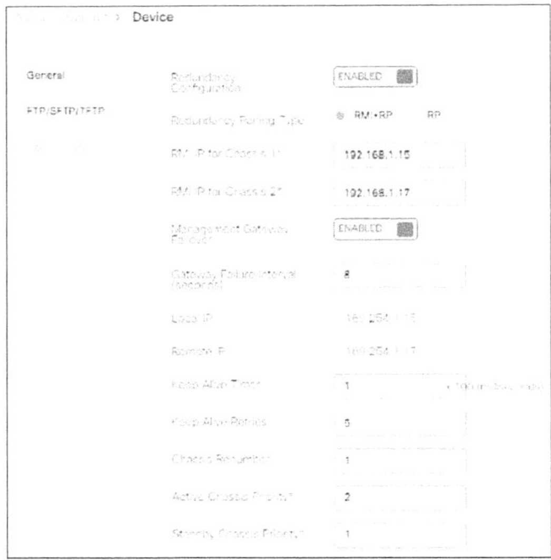

Figure 10-6 *Configuring SSO redundancy on a C9800*

The fields shown in Figure 10-6 can be configured as indicated:

- **Redundancy Pairing Type:** RP+RMI has been available since release 17.1 and is recommended to better handle the network failure conditions. On 16.x releases, the RP-only SSO was the only stateful redundancy supported. RP-only SSO presents severe limitations both in deploying and reacting to network failures, leaving a lot of possibilities to end up with split-brain active-active C9800s.

- **RMI IP Address:** Populate the fields of the RMI IP address for chassis numbered 1 and 2. Note that the RMI IP address is tied to a chassis number irrespective of which chassis is active or standby. The netmask for the RMI is picked up from the WMI netmask:

```
C9800(config)#redun-management interface WMI chassis 1 address
   <Chassis 1_RMI> chassis 2 address <Chassis 2_RMI>
```

- **HA Interface:** This field is available only on a C9800-CL and defines the interface mapped to virtual NIC (vNIC) that will serve as the RP.
- **Management Gateway Failover:** This field enables a default gateway check where Internet Control Message Protocol (ICMP) pings and Address Resolution Protocol (ARP) messages are sent to the default gateway at 1-second intervals to verify reachability. These messages are sent until the Gateway Failover Interval expires. Both active and standby C9800s use the RMI IP address to source these messages. Starting with release 17.4, IPv6 default gateway detection is supported. With IPv6, only ICMP messages are used for the default gateway check.
- **Gateway Failover Interval (seconds):** This field defines the duration for which gateway reachability messages are sent. The default value is 8, but starting with release 17.4, it can be configured for a range of 6–12 seconds. In an IPv4 deployment, a default interval of 8 seconds translates to four ICMP pings followed by four ARP requests. For most deployments, the default value should suffice. If you expect congestion or delay on an uplink from the C9800, you can increase the timer to avoid a reachability failure.

```
C9800(config)#management gateway-failover interval <6-12>
```

- **Local IP:** This refers to the RP IP address auto-generated from the local RMI IP address as 169.254.x.y, where x.y matches the last two octets of the local chassis RMI IP address.
- **Remote IP:** This refers to the RP IP address auto-generated from the remote RMI IP address as 169.254.x.y, where x.y matches the last two octets of the remote chassis RMI IP address.
- **Keepalive Timer:** Active and standby chassis send keepalives to each other over the RP to ensure both are available; this timer defines how often these keepalive messages are sent. Its default value is 100 ms.
- **Keepalive Retries:** This field determines how many keepalive messages are sent before declaring a peer unreachable over the RP. The default value is 5.

Together, the two keepalive fields allow a default peer timeout of 500 ms =100*5, at the end of which the peer is declared unreachable over the RP. Typically, the default value suffices. However, on a C9800CL scale deployment (4kAPs) to leverage vMotion, increasing the HA timers is recommended:

- Peer timeout = 100 ms * 5 on 10GBi vNIC
- Peer timeout = 200 ms * 4 on 1GB vNIC

The following CLIs enable you to view peer timeout and counters for messages exchanged over the RP:

```
C9800#show platform software stack-mgr chassis active R0 peer-timeout
C9800#show platform software stack-mgr chassis standby R0 peer-
timeout

C9800#show platform software stack-mgr chassis active R0 sdp-
counters
C9800#show platform software stack-mgr chassis standby R0 sdp-
counters
```

- **Chassis renumber:** By default, the C9800 chassis number is 1. In an RP+RMI deployment, the RP IP address for each chassis is derived from its RMI IP address. If the chassis number is the same on both chassis, the local RP IP derived is the same, and the HA pairing cannot occur. When you're configuring HA in an RP+RMI deployment, it is mandatory to renumber one of the two chassis to 2. The CLI corresponding to this chassis renumbering is chassis 1 **renumber 2.** After the chassis number is configured, a warning is presented, as shown in Example 10-5, to remind you to reboot the chassis for this renumbering to take effect.

Example 10-5 *Warning to Get You to Reload the Chassis for the Renumbering to Take Effect*

```
C9800#chassis 1 renumber 2


WARNING: Changing the switch number may result in a configuration
change for that switch. The interface configuration associated with
the old switch number will remain as a provisioned configuration. New
Switch Number will be effective after next reboot. Do you want to
continue?[y/n]? [yes]:


C9800#reload
```

- **Active Chassis Priority:** This field is used in determining active-standby roles and allows values of 1–2. You should configure the chassis that you intend to be active with the higher numerical value of 2.
- **Standby Chassis Priority:** This field is used in determining active-standby roles and allows values of 1–2. You should leave the chassis that you intend to be the standby at the default value of 1.

Executing a **write memory** command on the chassis automatically triggers a reload on the standby chassis in the case of the C9800-L, C9800-40, and C9800-80. However, reloading the active C9800 is mandatory so that its RMI can take effect. Note that RMI configuration is reflected on the CLI and GUI even prior to reload and not prompt for reload. Without reloading, some of the features made available via RMI, such as the default gateway reachability check, do not work.

Active-Standby Election Process

When the two chassis boot up, they exchange a chassis discovery message, which includes their capabilities. Each chassis verifies that the other chassis meets prerequisites to ensure a peer relationship can be formed. Then HA role detection occurs.

- For an existing HA pair, the current active continues to be elected as active and no preemption exists.
- For a newly formed pair, chassis priority determines the active. Chassis priority has a default value of 1 and can be configured for values of 1 or 2. If a chassis has been configured with a numerically higher priority of priority 2, it is chosen as the active WLC.
- If both chassis have the same priority, then the chassis with the lowest bootup time is chosen.
- If both chassis have the same priority and same bootup time, then the chassis with the lowest base Ethernet MAC address is chosen as the active WLC. You can see the base Ethernet MAC address by using **#show version | include Base**.

HA Sync

When the election is complete, you have one chassis in the active state and the other chassis in a cold-standby state. Then the configuration is bulk-synched from active to standby. After the startup configuration is synched, the parser on the standby chassis runs through the configuration line by line to build interface state details. This includes syncing the Wireless Management Interface, VLAN Switch Virtual Interface (SVI) IP addresses, service port, and virtual-IP. The only interfaces that remain unique on the HA pair are the RP and RMI interface IP.

After HA is established, privilege level 15 configuration access is available only to the active WLC, and no configuration changes can be made on the standby. All configuration is done on the active and incrementally synched to the standby.

Note Only console access to the standby C9800 is available on 16.x releases, so you need to explicitly configure this access:

```
C9800(config)#redundancy
C9800(config-redun)#mode sso
C9800(config-redun)#main-cpu
C9800(config-r-mc)#standby console enable
```

Starting with release 17.1, SSH, HTTPS, and NETCONF access is available to the standby chassis via RMI. As explained in Chapter 2, "Hardware and Software Architecture of the C9800," and in Chapter 12, "Network Programmability", all the configuration and operational data for 9800 is maintained in databases. For example, OPERATIONAL_DB, CONFIG_DB, and so on are synched to maintain the operational state of APs and clients in the RUN state and ensure SSO. Any certificate imported into the active C9800 also gets synched to the standby C9800. After sync, both C9800s are in an HA pair with the active and hot-standby state. This indicates that in case of active failure, the HA peer is ready to take over in a stateful manner.

HA Formation in Action

When two C9800s are paired, they go through an active standby election followed by a configuration sync to define the roles of the two chassis. On the console session for the two C9800s, the following messages are printed to indicate the different steps in HA formation:

- **Chassis initialization:** When the C9800 chassis is booted up, it comes in standalone mode because it cannot detect a peer, as shown in Figure 10-7. After it is initialized, you can assign the chassis number and chassis priority from the CLI to select a specific C9800 from the two C9800s to take on the role of active WLC.

```
C9800-L-X-K9 platform with 16777216 Kbytes of main memory

File size is 0x000015cf
Located packages.conf
Image size 5583 inode num 26, bks cnt 2 blk size 8*512
#
File size is 0x023f44c5
Located C9800-L-rpboot.17.06.01.SPA.pkg
Image size 37700805 inode num 506916, bks cnt 9205 blk size 8*512
################################################################################################
################################################################################################
################################################################################################
########################################################
Boot image size = 37700805 (0x23f44c5) bytes

ROM:RSA Self Test Passed
ROM:Sha512 Self Test Passed

Package header rev 3 structure detected
Calculating SHA-1 hash...done
validate_package_cs: SHA-1 hash:
    calculated 2f5b2f34:80f4af8a:b3b586c1:d41ca412:7736d66f
    expected   2f5b2f34:80f4af8a:b3b586c1:d41ca412:7736d66f
Validating main package signatures

RSA Signed RELEASE Image Signature Verification Successful.
Image validated
Aug 29 20:43:36.605: %PMAN-3-PROC_EMPTY_EXEC_FILE: R0/0: pvp: Empty executable used for process
bt_logger

Waiting for remote chassis to join
```

Figure 10-7 *Chassis initialization before HA pairing*

- **Active and standby election:** To form an HA pair, the RP is connected between the two C9800s, and necessary RMI configuration is done. Saving the configuration on the C9800 chosen to be the standby triggers a reload, in the case of appliances. However, it is recommended you issue a reload on the active C9800 to ensure all RMI features take effect. In the case of the C9800-CL, a reload is needed after configuring the RMI for the RMI to take effect. As the two C9800s boot up, the election process kicks in to select the active C9800, as shown in Figure 10-8, which in turn detects the peer as standby, as shown in Figure 10-9.

```
*Aug 29 15:10:59.998: %STACKMGR-6-STACK_LINK_CHANGE: Chassis 2 R0/0: stack_mgr: Stack port 1 on
Chassis 2 is up
*Aug 29 15:10:59.998: %STACKMGR-6-STACK_LINK_CHANGE: Chassis 2 R0/0: stack_mgr: Stack port 2 on
Chassis 2 is up
*Aug 29 15:11:00.191: %STACKMGR-6-CHASSIS_ADDED: Chassis 2 R0/0: stack_mgr: Chassis 2 has been added
to the stack.
*Aug 29 15:11:01.468: %PMAN-3-PROC_EMPTY_EXEC_FILE: Chassis 2 R0/0: pvp: Empty executable used for
process bt_logger
*Aug 29 15:11:01.497: %STACKMGR-6-CHASSIS_ADDED: Chassis 2 R0/0: stack_mgr: Chassis 2 has been added
to the stack.
*Aug 29 15:11:03.409: %PMAN-3-PROC_EMPTY_EXEC_FILE: Chassis 2 R0/0: pvp: Empty executable used for
process bt_logger
*Aug 29 15:11:03.497: %STACKMGR-6-CHASSIS_ADDED: Chassis 2 R0/0: stack_mgr: Chassis 2 has been added
to the stack.
*Aug 29 15:11:03.746: %STACKMGR-6-ACTIVE_ELECTED: Chassis 2 R0/0: stack_mgr: Chassis 1 has been
elected ACTIVE.
*Aug 29 15:11:04.390: %PMAN-3-PROC_EMPTY_EXEC_FILE: Chassis 2 R0/0: pvp: Empty executable used for
process bt_logger
*Aug 29 15:11:04.987: %RIF_MGR_FSM-6-RP_LINK_UP: Chassis 2 R0/0: rif_mgr: The RP link is UP.
*Aug 29 15:11:04.987: %STACKMGR-1-DUAL_ACTIVE_CFG_MSG: Chassis 2 R0/0: stack_mgr: Dual Active
Detection link is available now
```

Figure 10-8 *Active election*

```
*Aug 29 20:44:57.925: %RIF_MGR_FSM-6-GW_REACHABLE_ACTIVE: Chassis 1 R0/0: rif_mgr: Gateway reachable
from Active
*Aug 29 20:45:08.545: %IOSXE_REDUNDANCY-6-PEER: Active detected chassis 2 as standby.
*Aug 29 20:45:08.540: %STACKMGR-6-STANDBY_ELECTED: Chassis 1 R0/0: stack_mgr: Chassis 2 has been
elected STANDBY.
```

Figure 10-9 *Standby election*

- **HA config sync:** After the active standby role election, a bulk configuration sync occurs from the active C9800 to the standby C9800 to allow all configuration and operational data to be synched across both C9800s. Note that before the sync, the standby state is cold, as shown in Figure 10-10. Switchover with standby cold does not result in SSO. After the config sync is complete, the standby is moved to the standby-hot state, as shown in Figure 10-11. In the case of any configuration sync failures, explicit messages are printed in the console log. To view the specific configuration that did not sync, you can use the command **show redundancy config-sync failures {bem | mcl | prc}**. The command also includes an option to view historic config sync failures via **show redundancy config-sync historic mcl**.

```
*Aug 29 20:44:07.123: %EWLC_HA_LIB_MESSAGE-6-BULK_SYNC_STATE_INFO: Chassis 1 R0/0: wncmgrd: INFO: Bulk
sync status : COLD
```

Figure 10-10 *Standby comes up cold prior to the bulk sync*

```
*Aug 29 20:45:17.087: %VOICE_HA-7-STATUS: NONE->SSO; SSO mode will not take effect until after a
platform reload.
*Aug 29 20:45:18.589: %REDUNDANCY-5-PEER_MONITOR_EVENT: Active detected a standby insertion
(raw-event=PEER_FOUND(4))

*Aug 29 20:45:18.589: %REDUNDANCY-5-PEER_MONITOR_EVENT: Active detected a standby insertion
(raw-event=PEER_REDUNDANCY_STATE_CHANGE(5))

*Aug 29 20:45:20.149: Syncing vlan database
*Aug 29 20:45:20.165: Vlan Database sync done from bootflash:vlan.dat to stby-bootflash:vlan.dat (616
bytes)
*Aug 29 20:45:30.534: %CRYPTO_ENGINE-4-CSDL_COMPLIANCE_RSA_WEAK_KEYS: RSA keypair
CISCO_IDEVID_SUDI_LEGACY is in violation of Cisco security compliance guidelines and will be rejected
by future releases.
*Aug 29 20:45:51.774: %SMART_LIC-3-COMM_FAILED: Communications failure with the Cisco Smart License
Utility (CSLU) : Unable to resolve server hostname/domain name
*Aug 29 20:47:21.226: %RIF_MGR_FSM-6-RMI_LINK_UP: Chassis 1 R0/0: rif_mgr: The RMI link is UP.
*Aug 29 20:47:23.646: %HA_CONFIG_SYNC-6-BULK_CFGSYNC_SUCCEED: Bulk Sync succeeded
*Aug 29 20:47:23.686: %VOICE_HA-7-STATUS: VOICE HA bulk sync done.
*Aug 29 20:47:24.725: %RF-5-RF_TERMINAL_STATE: Terminal state reached for (SSO)
```

Figure 10-11 *Standby comes up hot after the bulk sync*

SSO Switchover

Switchover on a C9800 HA pair is triggered when the active WLC encounters a service-impacting event. This could be an uncontrolled event such as a power failure on the active C9800 or controlled events like a critical process crash on the active C9800 or manual switchover initiated by the user. In the case of uncontrolled events, the keepalive timeout of 500 ms comes into play, and after the 500 ms have expired, switchover occurs. For system or user-controlled triggers, a fast switchover is executed, which is immediate, without waiting on the timeout. In short, failover is subsecond, and no impact to APs and clients is seen.

In all cases, when the active fails, an HA peer takes over seamlessly if it is in a hot-standby state. After the standby chassis assumes the role of active, it sends out a Gratuitous ARP (GARP) to the network, for each of the interface IP addresses, so the network devices can update the ARP table to map IP addresses to new MAC addresses. The new active WLC also sends out a GARP for each client to ensure traffic for the wireless clients is directed to the new active. When the failed chassis or old active comes back, it boots up in hot-standby mode without preemptively trying to take on the active role. When the active fails, if the HA peer was not in the hot-standby state, both chassis reload and the SSO fails.

Note that a failure on the standby due to a critical process causes only the standby C9800 to reload without impact to the active. When the standby comes up, a bulk sync occurs and then the chassis is in the hot-standby state.

On the C9800, switchover can be manually triggered via the CLI, as shown in Figures 10-12 and 10-13:

```
C9800#redundancy force-switchover
```

```
9800L#redundancy force-switchover
Proceed with switchover to standby RP? [confirm]
    Manual Swact = enabled

Chassis 1 reloading, reason - Non participant detected

*Aug 31 07:26:52.680: RMI-GW-NOTICE: Forced switchover notification received

*Aug 31 07:26:53.782: %RF-5-RF_RELOAD: Self reload. Reason: redundancy force-switchover
Aug 31 07:26:54.805: %PMAN-5-EXITACTION: C0/0: pvp: Process manager is exiting:
Aug 31 07:26:54.842: %PMAN-5-EXITACTION: F0/0: pvp: Process manager is exiting:

*Aug 31 07:26:54.060: %SYS-5-SWITCHOVER: Switchover requested by red_switchover_process. Reason:
redundancy force-switchover.Aug 31 07:27:00.884: %PMAN-5-EXITACTION: R0/0: pvp: Process manager is
exiting: process exit with reload fru code
```

Figure 10-12 *SSO switchover on the active C9800*

```
9800L-stby#Ewlc: triggered dual-active recovery,setting hostname to 9800L, Mode: 4
*Aug 31 07:26:53.881: %REDUNDANCY-3-SWITCHOVER: RP switchover (PEER_NOT_PRESENT)
*Aug 31 07:26:53.881: %REDUNDANCY-3-REDUNDANCY_ALARMS: Unable to assert REDUNDANCY alarm

*Aug 31 07:26:53.881: %REDUNDANCY-3-REDUNDANCY_ALARMS: Unable to assert REDUNDANCY alarm

*Aug 31 07:26:53.881: %REDUNDANCY-3-SWITCHOVER: RP switchover (PEER_DOWN)
*Aug 31 07:26:53.881: %REDUNDANCY-3-SWITCHOVER: RP switchover (PEER_REDUNDANCY_STATE_CHANGE)
*Aug 31 07:26:54.164: SNMPHA-CHKPT: chkpt: msg is NULL

*Aug 31 07:26:54.556: SNMPHA-CHKPT: chkpt: msg is NULL

*Aug 31 07:26:54.676: WLC-HA-Notice: RF Progression event: RF_PROG_ACTIVE_FAST, Switchover triggered
*Aug 31 07:26:54.681: %LINK-3-UPDOWN: Interface Lsmpi0, changed state to up
*Aug 31 07:26:54.681: %LINK-3-UPDOWN: Interface EOBC0, changed state to up
*Aug 31 07:26:54.710: RMI-HAINFRA-INFO: Configured primary IP 192.168.1.5/255.255.255.0 on active(mgmt)
*Aug 31 07:26:54.710: RMI-HAINFRA-INFO: Configured secondary IP 192.168.1.17/255.255.255.0 on
active(mgmt)
*Aug 31 07:26:54.731: %VOICE_HA-2-SWITCHOVER_IND: SWITCHOVER, from STANDBY_HOT to ACTIVE state.
```

Figure 10-13 *SSO switchover on the standby C9800*

System and Network Error Handling

On the C9800 HA pair, if the active C9800 fails, the reason for active failover and the availability of resources like the RP link, RMI, and gateway determine whether the switchover occurs and the state that the two C9800s end up in after the switchover. If resources are not available, the C9800 cannot provide service and then is placed into one of the two supported recovery modes to prevent an active-active scenario. Note that in the recovery mode, ports are administratively down. The two recovery modes are

- **Standby-recovery:** If the gateway becomes unreachable, the standby C9800 goes to standby-recovery. The term *standby* means its state is up to date with the active. But because the gateway is not available, it goes to standby-recovery. The standby cannot take over as active from standby-recovery. When gateway reachability is restored, the C9800 goes to standby without a reload.
- **Active-recovery:** This state occurs when the RP goes down. Active-recovery does not have its internal state in sync with the active. When the RP links come up, active-recovery reloads so that it can come up as standby with bulk sync.

Switchover history captured via **show redundancy switchover history** shows the switchover reason as gateway down in the event of a switchover triggered as a result of the gateway going down.

Table 10-1 depicts the C9800 system failure conditions such as crash and manual switchover that can trigger SSO, along with the outcome, depending on the availability of resources. Table 10-2 lists the network failure conditions that would trigger an SSO failover, along with the failover outcome. In each case, the table provides the status of the RP link and peer reachability through the RMI.

Table10-1 *SSO Behavior on System Failure*

Trigger	RP Link Status	Peer Reachability through RMI	Switchover	Result
Critical Process Crash	Up	Reachable	Yes	Switchover.
Forced Switchover	Up	Reachable	Yes	Switchover.
Critical Process Crash	Up	Unreachable	Yes	Switchover.
Forced Switchover	Up	Unreachable	Yes	Switchover.
Critical Process Crash	Down	Reachable	No	No action; one C9800 is in recovery mode already.
Forced Switchover	Down	Reachable	N/A	No action; one C9800 is in recovery mode already.
Critical Process Crash	Down	Unreachable	No	Double fault; this results in a split brain, where both WLCs are active. After connectivity is restored, active/standby roles are negotiated to keep the active that came up last.
Forced Switchover	Down	Unreachable	N/A	Double fault.

Table 10-2 *SSO Behavior for Network Error Conditions*

RP Link	Peer Reachability through RMI	Gateway from Active	Gateway from Standby	Switchover	Result
Up	Up	Reachable	Reachable	No	No action.

RP Link	Peer Reachability through RMI	Gateway from Active	Gateway from Standby	Switchover	Result
Up	Up	Reachable	Unreachable	No	No action. The standby fails the gateway check and is put in standby-recovery mode. If the RP goes down, the standby (in recovery mode) becomes active.
Up	Up	Unreachable	Reachable	Yes	The gateway reachability message is exchanged over the RMI + RP links. The active reboots so that the standby becomes active.
Up	Up	Unreachable	Unreachable	No	When the active SVI goes down, so does the standby SVI. A switchover is then triggered. If the new active discovers its gateway to be reachable, the system stabilizes in active-standby recovery. Otherwise, switchovers happen in a pingpong fashion.
Up	Down	Reachable	Reachable	No	No action.
Up	Down	Reachable	Unreachable	No	The standby is not ready for SSO in this state because it does not have gateway reachability. The standby goes to recovery mode because LMP messages are exchanged over the RP link also.
Up	Down	Unreachable	Reachable	Yes	A gateway reachability message is exchanged over the RP link also. The active reboots so that the standby becomes active.

RP Link	Peer Reachability through RMI	Gateway from Active	Gateway from Standby	Switchover	Result
Up	Down	Unreachable	Unreachable	No	With this, when the active SVI goes down, so does the standby SVI. A switchover is then triggered. If the new active discovers its gateway to be reachable, the system stabilizes in active-standby recovery. Otherwise, switchovers happen in a pingpong fashion.
Down	Up	Reachable	Reachable	Yes	In all cases, where the RP link is down, irrespective of RMI link status and gateway reachability, switchover occurs, and the standby becomes active with the (old) active going to active-recovery mode. Configuration mode is disabled in active-recovery mode. All interfaces are Admin Down, with the Wireless Management Interface having the RMI IP. The controller in active-recovery reloads to become the standby when the RP link comes up.
Down	Up	Reachable	Unreachable	Yes	The standby becomes active with the (old) active going to active-recovery. Configuration mode is disabled in active-recovery mode. All interfaces are Admin Down, with the Wireless Management Interface having the RMI IP. The controller in active-recovery reloads to become the standby when the RP link comes up.

RP Link	Peer Reachability through RMI	Gateway from Active	Gateway from Standby	Switchover	Result
Down	Up	Unreachable	Reachable	Yes	Standby becomes active with the (old) active going to active-recovery. The configuration mode is disabled in active-recovery mode. All interfaces are Admin Down, with the Wireless Management Interface having the RMI IP. The controller in active-recovery reloads to become standby when the RP link comes up.
Down	Up	Unreachable	Unreachable	Yes	The standby becomes active with the (old) active going to active-recovery. The configuration mode is disabled in active-recovery mode. All interfaces are Admin Down, with the Wireless Management Interface having the RMI IP. The controller in active-recovery reloads to become the standby when the RP link comes up.
Down	Down	Reachable	Reachable	Yes	Double fault; this may result in a network conflict because there are two active controllers. The standby becomes active. The old active also exists. Role negotiation has to happen after the connectivity is restored and keeps the active that came up last.

RP Link	Peer Reachability through RMI	Gateway from Active	Gateway from Standby	Switchover	Result
Down	Down	Reachable	Unreachable	Yes	Double fault; this may result in a network conflict because there are two active controllers. The standby becomes the active. The old active also exists. Role negotiation has to happen after the connectivity is restored and keeps the active that came up last.
Down	Down	Unreachable	Reachable	Yes	Double fault; this may result in a network conflict because there are two active controllers. The standby becomes the active. The old active also exists. Role negotiation has to happen after the connectivity is restored and keeps the active that came up last.

Monitoring HA

The C9800 provides several options, both from GUI and CLI, to monitor the redundancy state of each chassis within the HA pair, the HA timers, the switchover history, and so on. Specific to standby, you can monitor via the active chassis, but there is also a provision to monitor the standby chassis out of band, without going through the active.

Monitoring an HA Pair via the CLI

Several CLIs are available to monitor the standby from the active itself. Alternatively, you can monitor the standby independently as well. Figure 10-14 and Figure 10-15 provide sample output of the **show chassis rmi** command taken from both the active and standby CLI, which enables the pair where chassis number, chassis priority, RP IP, RMI IP, MAC address of the two C9800s, their roles, and state are clearly identified.

```
9800L#show chassis rmi
Chassis/Stack Mac Address : d478.9b3c.5e80 - Local Mac Address
Mac persistency wait time: Indefinite
Local Redundancy Port Type: Twisted Pair
                                                  H/W    Current
Chassis#   Role     Mac Address     Priority Version  State              IP                RMI-IP
-----------------------------------------------------------------------------------------------------------
*1        Active   d478.9b3c.5e80     1       V02      Ready              169.254.1.15       192.168.1.15
 2        Standby  d478.9b3c.5f60     1       V02      Ready              169.254.1.17       192.168.1.17



9800L#show redundancy states
       my state = 13 -ACTIVE
     peer state = 8  -STANDBY HOT
           Mode = Duplex
           Unit = Primary
        Unit ID = 1

Redundancy Mode (Operational) = sso
Redundancy Mode (Configured)  = sso
Redundancy State              = sso
     Maintenance Mode = Disabled
    Manual Swact = enabled
 Communications = Up

   client count = 150
 client_notification_TMR = 30000 milliseconds
          RF debug mask = 0x0
Gateway Monitoring = Enabled
Gateway monitoring interval  = 8 secs
```

Figure 10-14 *HA monitoring from the active C9800*

```
9800L-stby#show chassis rmi
Chassis/Stack Mac Address : d478.9b3c.5e80 - Local Mac Address
Mac persistency wait time: Indefinite
Local Redundancy Port Type: Twisted Pair
                                                  H/W    Current
Chassis#   Role     Mac Address     Priority Version  State              IP                RMI-IP
-----------------------------------------------------------------------------------------------------------
 1        Active   d478.9b3c.5e80     1       V02      Ready              169.254.1.15       192.168.1.15
*2        Standby  d478.9b3c.5f60     1       V02      Ready              169.254.1.17       192.168.1.17



9800L-stby#show redundancy states
       my state = 8  -STANDBY HOT
     peer state = 13 -ACTIVE
           Mode = Duplex
           Unit = Primary
        Unit ID = 2

Redundancy Mode (Operational) = sso
Redundancy Mode (Configured)  = sso
Redundancy State              = sso
     Maintenance Mode = Disabled
    Manual Swact = cannot be initiated from this the standby unit
 Communications = Up

   client count = 150
 client_notification_TMR = 30000 milliseconds
          RF debug mask = 0x0
Gateway Monitoring = Enabled
Gateway monitoring interval  = 8 secs
```

Figure 10-15 *HA monitoring from the standby C9800*

Starting with release 17.3, you can also monitor the health of the standby WLC via SSH to the standby RMI IP address without going through the active. Just follow these steps:

Step 1. Enable SSH on the active C9800 and generate the RSA key.

```
C9800(config)#ip domain name cisco.com
C9800(config)#crypto key generate rsa
The name for the keys will be: C9800.cisco.com
Choose the size of the key modulus in the range of 512 to
4096 for your
  General Purpose Keys. Choosing a key modulus greater than
512 may take
  a few minutes.

How many bits in the modulus [2048]:
% Generating crypto RSA keys in background ...
*Sep 4 08:01:51.458: %CRYPTO_ENGINE-5-KEY_ADDITION: A key
named C9800.cisco.com has been generated or imported by
crypto-engine
```

Step 2. Configure the local AAA for the admin to be authenticated to the standby. If an external RADIUS server is in use, configure the AAA fallback to allow local authentication for SSH access to the standby. Also note that Terminal Access Controller Access Control System (TACACS) is not supported on the standby. Example 10-6 provides a sample configuration to set up AAA authentication.

Example 10-6 *Configuration to Set Up AAA Auth for SSH Access to Standby C9800*

```
C9800(config)#line vty 0 50
C9800(config-line)#password Cisco
C9800(config-line)#authorization exec VTY_ADMIN login
C9800(config-line)#authentication VTY_ADMIN
C9800(config-line)#transport input ssh

C9800(config)#aaa authentication login VTY_ADMIN group AAA_SERVER_GROUP
local
C9800(config)#aaa authorization exec VTY_ADMIN group AAA_SERVER_GROUP
local
C9800(config)#aaa group server radius AAA_SERVER_GROUP server name SRV1

C9800(config)#radius server SRV1
C9800(config-radius-server)#address ipv4 <IP> auth-port 1812 acct-port
1813
C9800(config-radius-server)#key <RADIUS shared secret>
```

Step 3. Verify that the default route for the management VLAN is configured.

```
9800L#show run | include ip route
ip route 0.0.0.0 0.0.0.0 <GW_IP>
```

Step 4. **Login** to the standby via SSH:

```
ssh <username>@<RMI IP>
Password:
```

When you're logged in, the following functional **show** commands are on the standby. Starting with release 17.4, the same monitoring is also available with IPv6:

```
C9800-stby#show environment
C9800-stby#show environment summary
C9800-stby#show ip interface brief
C9800-stby#show processes {cpu | memory | platform | history}
```

Monitoring an HA Pair via the GUI

On the active C9800 GUI, the redundancy information for both active C9800 and standby C9800 can be seen, including uptime, version, MAC address, redundancy role and state, RP IP, and RMI IP address being used by each C9800, as shown in Figure 10-16. In addition, you can also view CPU and Memory utilization of the standby from the active's GUI, as shown in Figure 10-17.

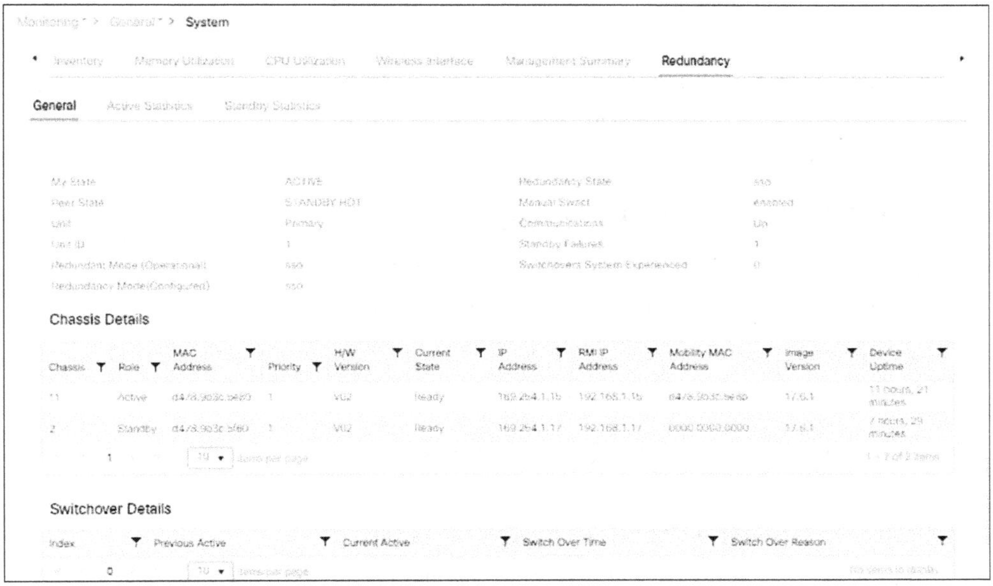

Figure 10-16 *HA state monitoring from the active C9800 GUI*

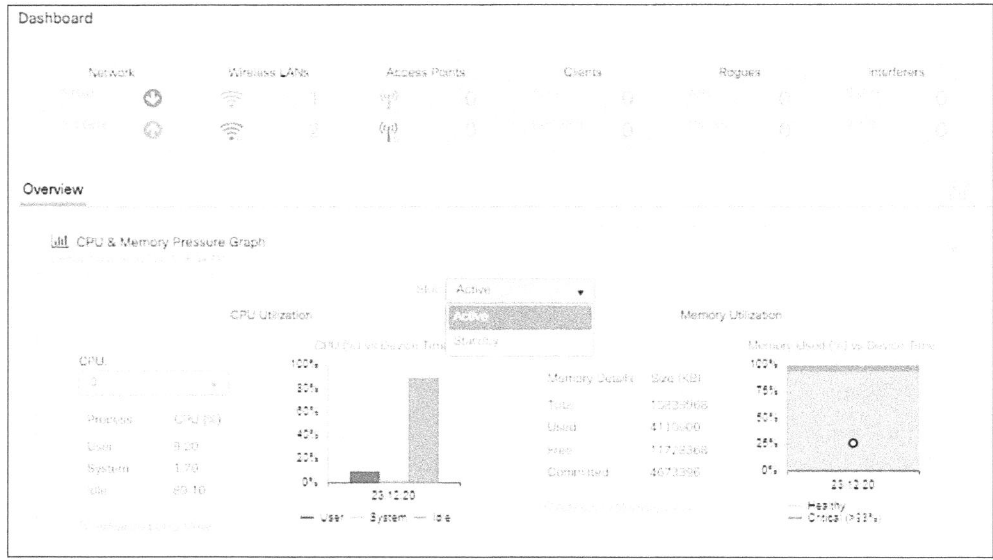

Figure 10-17 *HA active and standby resource monitoring from the active C9800 GUI*

Monitoring an HA Pair via SNMP

Starting with release 17.5, SNMP monitoring on the standby and for standby parameters on the active is supported. The MIBs supported include

- **IF-MIB:** To monitor interface statistics
- **CISCO-PROCESS-MIB:** To monitor CPU/memory information
- **CISCO-LWAPP-HA-MIB:** To monitor HA SSO parameters
- **cLHaPeerHotStandbyEvent:** Object indicates that the peer is in the hot-standby state
- **cLHaBulkSyncCompleteEvent:** Object indicates that the bulk sync is complete

Note that the standby does not support SNMP traps.

Monitoring an HA Pair via Programmatic Interfaces

You also can obtain similar health information on a standby C9800 by using programmatic interfaces like NETCONF, YANG, and RESTCONF. You can connect to the C9800 via SSH over NETCONF port 830 using tools such as YANG Suite to monitor the standby WLC using the operational models shown in the following list. For more information on leveraging programmability to configure and monitor the C9800, see Chapter 12, "Network Programmability."

- **Cisco-IOS-XE-device-hardware-oper.yang:** Provides the serial number of all FRUs and the chassis
- **Cisco-IOS-XE-process-cpu-oper.yang:** Provides CPU utilization averages over intervals of the past 1 minute, 5 minutes, 5 seconds, and per process CPU stats for IOS tasks
- **Cisco-IOS-XE-platform-software-oper.yang:** Provides average CPU utilization of 5-second intervals and allocated memory for the processes
- **Cisco-IOS-XE-process-memory-oper.yang:** Provides per process memory utilization
- **Cisco-IOS-XE-interfaces-oper.yang:** Provides interface operational data including state and stats

RP Only to RP+RMI HA Migration

As mentioned, RMI was only introduced starting with the 17.1 IOS-XE release. However, the C9800 and HA SSO on the C9800 are supported as of release 16.10. In the 16.x releases, HA pairing was done via the RP only. RP reachability was the only factor to trigger switchover and therefore a single point of failure. RP reachability could time out due to a system failure or network failure. In the case of a network failure, the standby C9800 would assume the active role and send out GARP to the network to redirect the AP and client traffic to itself. However, the original active C9800 would still be alive, resulting in a dual-active scenario. Migrating any RP-only HA deployments to RP+RMI HA deployment is recommended. You can do this as follows:

1. Upgrade to the 17.x release as applicable to your environment.
2. Configure the RMI+RP HA using the steps described in the "Configuration" section.
3. Replace the manually configured RP IP address with 169.254.x.y, where x.y is the last two octets in the RMI IP address for each corresponding chassis.
4. Reload so that these new IP addresses of RMI and RP come into effect.

HA Teardown

To disable an HA pairing between two C9800 chassis, connect to the GUI and browse to **Administration > Device > Redundancy** and slide the **Redundancy Configuration** setting to **Disabled**. This results in both C9800s in the pair moving to standalone mode, with the standby C9800 reloading with a Day 0 configuration to prevent the two active C9800s from ending up with a duplicate configuration, including IP addresses. A corresponding warning pop-up appears, as shown in Figure 10-18; you need to confirm this message to proceed with the unpairing.

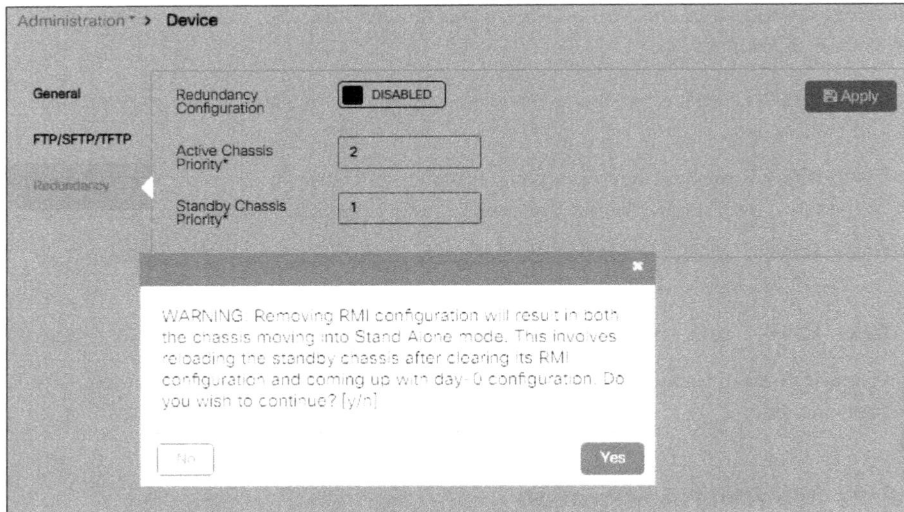

Figure 10-18 *Disabling HA SSO*

The resulting CLI from this action is **no redun-management interface WMI chassis 1 address <Chassis 1_RMI> chassis 2 address <Chassis 2_RMI>**. This command also results in a warning you need to acknowledge before proceeding.

SSO Deployment: Impact on Features

HA SSO impacts all the features such that you would be interested to know which features retain their state information at failover—for example, clients that are in data forwarding mode. Beyond this, some features are more important and show differences in behavior when operating in an HA SSO pair as compared to a standalone deployment.

Mobility (Mobility MAC)

Mobility MAC refers to the MAC address used by the C9800 to establish mobility tunnels with other WLCs. In a standalone setup, a WMI MAC is used as a mobility MAC. However, in an SSO HA deployment, when a failover occurs, the MAC address associated with the WMI changes, causing all mobility tunnels to go down. To avoid this situation, you can configure a mobility MAC that is shared by the two chassis; this solution is recommended. The mobility MAC can be configured on an individual chassis before pairing or on an active C9800 if already paired.

On a C9800 GUI, navigate to **Configuration > Wireless > Mobility: Global Configuration**, as shown in Figure 10-19.

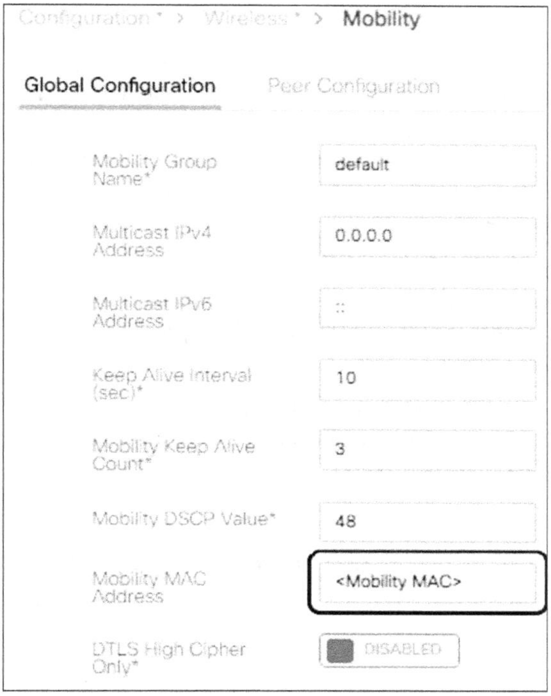

Figure 10-19 *Mobility MAC Configuration*

The corresponding CLI for configuring the mobility MAC is

```
C9800(config)#wireless mobility mac-address <MAC>
```

Link Aggregation Group (LAG)

EtherChannel refers to bundling two or more physical ports to expand the available bandwidth and to provide resiliency in the case of a link failure. On the 16.x release of C9800, only the LAG mode ON setting is supported. LAG negotiation protocols like the Cisco proprietary Port Aggregation Protocol (PagP) or the standard Link Aggregation Control Protocol (LACP) on an SSO C9800 HA pair are supported only starting with release 17.1.

Link Aggregation is supported only on the C9800-L, C9800-40, and C9800-80. The RP+RMI supported topologies shown previously already use multiple uplink ports to connect to the wired infrastructure, which can be bundled in LAG. For PagP/LACP topologies, the following features are *not* supported:

- Auto-LAG
- C9800-CL and EWC-AP
- L3 port channel

Multi-Chassis LAG

Multi-Chassis LAG adds the ability to connect multiple LAG-bundled port channel uplinks from a WLC to separate uplink switches. On the C9800, Multi-Chassis LAG is supported starting with IOS-XE release 17.2.1. This feature provides flexibility in how the WLC is connected to a wired infrastructure and allows traffic to be split on different port channels based on VLANs. For example, guest traffic could egress out a different switch network as compared to enterprise traffic.

Multi-Chassis LAG works on a standalone C9800 as well as an HA pair, as shown in Figure 10-20; it allows the port channel to be configured in the ON mode and use PagP/LACP negotiation. As with LAG, this feature is only supported with physical appliances, including the C9800-L, C9800-40, and C9800-80.

Bundling ports of the same type and speed is recommended. On the C9800-L, this translates to bundling 2.5GB/1GB multigigabyte Ethernet ports separately from the 10G multigigabyte Ethernet ports. Along the same lines, on the C9800, ports on bay 0 and bay 1 are not bundled together typically.

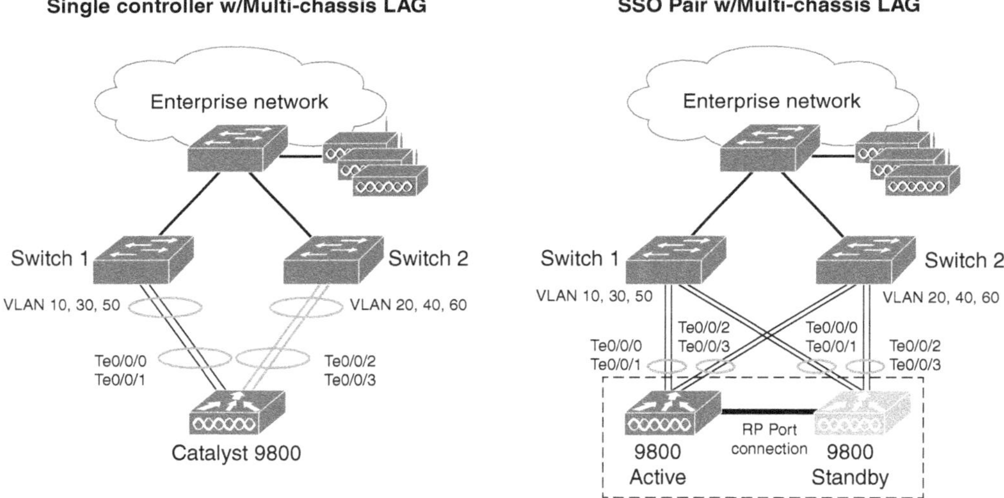

Figure 10-20 *C9800 to switch connections with Multi-Chassis LAG*

N+1 Redundancy

In addition to SSO, the C9800 also supports N+1 redundancy, which is a stateless redundancy. The C9800s involved in N+1 redundancy operate independently and are active at the same time. The two C9800s do not need to be connected over a dedicated RP, and each chassis has a unique IP address that is used to manage and monitor each one on its own. There is no sync for either configuration or operational data.

AP configuration determines which C9800 serves as the primary and which chassis serves as the secondary for that AP. This configuration enables access points to be load-balanced across multiple C9800s. This is not one-to-one box redundancy, meaning one WLC can serve as an N+1 backup for more than one WLC and can be any hardware. The criterion for choosing an N+1 WLC is to ensure that the N+1 WLC hardware can service the AP count and client count for which it is serving as backup. The N+1 WLC can run any software version and any installation mode. However, it is recommended to run the same software version as the rest and in the same installation mode; this way, when APs fail over between C9800s, they do not have to download code and the failover can be faster.

When an AP chooses the WLC for registration, it starts with the primary WLC. If the primary WLC becomes unreachable, the AP tries the secondary and then tertiary WLCs before attempting to join any other WLCs that it would have discovered via other CAPWAP discovery methods like option 43, DNS, or mobility peers. If WLC N is configured as the primary and WLC N+1 is configured as the secondary, the AP registers to the WLC N, and if the WLC N fails either due to system or network failure, the AP fails over to WLC N+1. Because no sync occurs between the WLC N and WLC N+1, the state information of the APs and clients is not maintained at WLC N+1. The failover is service impacting in that the AP must go through a complete CAPWAP discovery and join process to join the N+1 WLC. A local mode AP cannot service clients until it is registered to the N+1 WLC. In a FlexConnect deployment, when the FlexConnect AP switches over from the WLC N to WLC N+1, it goes into standalone CAPWAP mode and continues to service the clients that are already registered and forwarding traffic, and the FlexConnect AP does not accept any new client connections. When the AP registers to the WLC N+1, the AP is moved back to Connected mode and continues serving existing clients and also accepting new clients.

N+1 HA can provide redundancy across geographically separate locations and can be deployed in a data center over a wide-area network (WAN) link to serve as a disaster recovery solution. N+1 HA can be configured in combination with the AP SSO where the primary and/or secondary WLCs are each an SSO pair. You are recommended to have the same configuration in terms of WLANs, profiles, tags, and AP-to-tag mappings on the primary, secondary, and tertiary C9800s. This way, you can ensure the least amount of service downtime in case of APs failing over.

N+1 HA Configuration

On the C9800, the configuration for N+1 is typically done by priming each AP with a primary, secondary, and tertiary WLC name and IP address. To accomplish this, on the C9800 GUI navigate to **Configuration > Wireless > Access Points**, as shown in Figure 10-21.

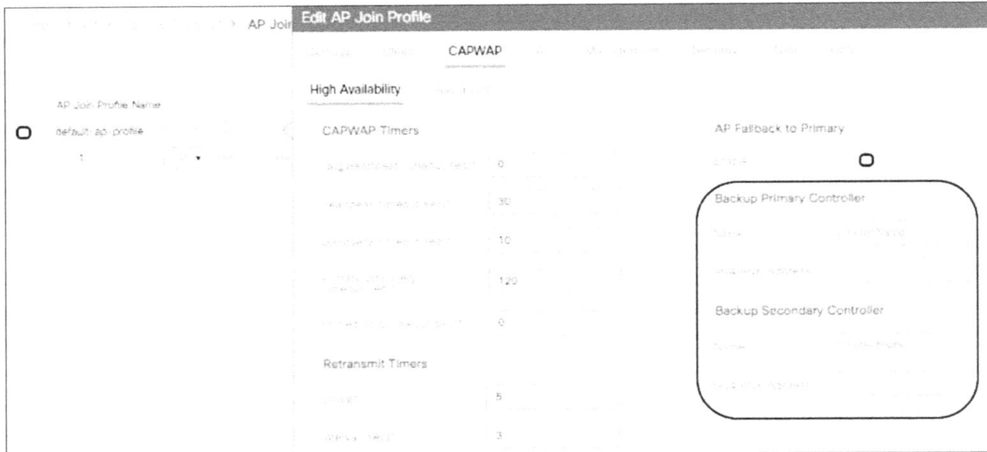

Figure 10-21 *N+1 HA configuration on access points*

The CLI equivalent for this is

```
C9800#ap name <AP_NAME> controller primary <WLC N Name> <WLC N WMI IP>
C9800#ap name<AP_NAME> controller secondary <WLC N Name> <WLC N WMI IP>
C9800#ap name <AP_NAME> controller tertiary <WLC N Name> <WLC N WMI IP>
```

AP Failover Priority is the field where the AP can be assigned the Critical, High, Medium, or Low priority. The term *AP Failover Priority* is a slight misnomer because failover is only one of the situations in which AP priority comes into play. When a C9800 has a large number of APs attempting to join either due to a failover or otherwise, APs with higher priority are given precedence to join ahead at the expense of lower-priority APs. For example, if the N+1 WLC is reaching the upper limit of its AP count and APs from a primary WLC fail over, existing lower-priority APs are dropped to accommodate high-priority APs. The CLI to configure this priority is

```
C9800#ap name <AP_NAME> priority <1-4>
```

In addition to priming the AP, the C9800 AP Join profile supports configuration of backup primary and backup secondary WLCs.

An AP uses heartbeats to validate current WLC reachability. If these heartbeats time out, the AP fails over to the backup WLC candidate list. The AP preference for the WLC to join is in this order: primary, secondary, tertiary, backup primary, and backup secondary.

Configuration on the AP Join Profile

To configure the AP join profile, connect to the C9800 GUI and navigate to **Configuration > Tags and Profile > AP Join: CAPWAP**, as shown in Figure 10-22, and enter the backup primary and backup secondary WLC name and corresponding IPv4 or IPv6 address.

Note that the primary, secondary, and tertiary WLCs defined on the AP take precedence over the backup WLCs configured on the AP join profile. Because there is no configuration sync involved, all configuration needs to be explicitly repeated on the primary, secondary, and tertiary C9800s.

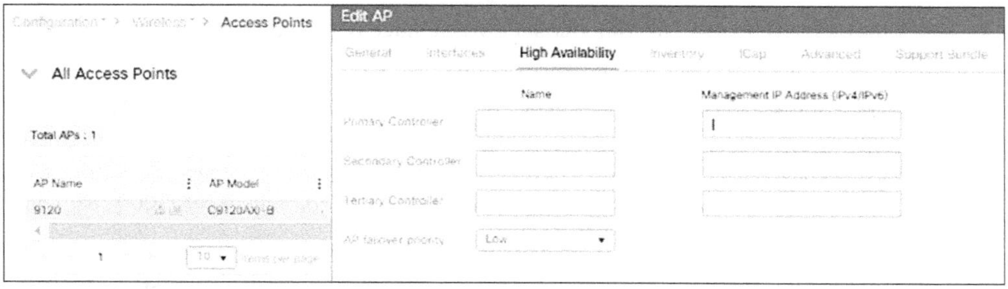

Figure 10-22 *N+1 HA Configuration on AP Join Profile*

The equivalent CLI would be

```
C9800(config-ap-profile)#capwap backup primary <WLC N hostname>
<WLC N WMI ip>
C9800(config-ap-profile)#capwap backup secondary <WLC N+1 hostname>
<WLC N+1 WMI ip>
```

After an AP fails over to the secondary C9800, it continues to send heartbeats to the primary C9800 to check whether the primary C9800 becomes available. If the primary C9800 is up, the default configuration with the AP fallback enabled ensures that the APs move back to the primary C9800. Note that although the AP fallback is defined under the AP join profile, it is not related to the backup primary or backup secondary C9800s. It is only responsible for moving the AP back to its configured primary C9800 if and when that primary C9800 is reachable. Enabling AP fallback is recommended so that you have a planned and intentional distribution of APs across C9800s. You might disable AP fallback when you are doing maintenance work on one C9800 and want to prevent APs from flip-flopping and keep them on the C9800 not under maintenance. To do so, use

```
C9800(config-ap-profile) #capwap fallback
```

CAPWAP Timers

Under the AP profile, you can configure times to extend or delay failover as per network requirements. For most typical deployments, the default values suffice.

- **Fast heartbeat timeout:** Fast heartbeats lower the amount of time it takes to detect Primary WLC Failure. The AP sends fast heartbeats every second, and when the fast heartbeat timer expires, the AP sends three fast echo requests to the WLC. If there is no response, the AP fails over to the WLC in the backup candidate list. This capability is typically more useful in central switching deployments as compared to FlexConnect deployments. The allowed range is 0–10 seconds with the default being disabled /0 seconds.

- **Heartbeat timeout:** This is the time that an AP waits after losing the first heartbeat to the C9800 to check C9800 reachability. A small timeout value reduces the time taken to detect C9800 failure. The allowed value ranges from 1 to 30 seconds with a default value of 30 seconds. For a FlexConnect deployment, this value needs to be at least two times higher than the fast heartbeat timeout.
- **Discovery timeout:** This timeout supports a range of 1–10 seconds with the default set to 10 seconds.
- **Primary discovery timeout:** Primary discovery requests are sent by the AP to the primary C9800 to know when it is online. This timeout comes into play when the AP has failed over to the N+1 WLC, and the N+1 WLC has the AP fallback enabled to get the AP resituated on its primary C9800. The allowed range is 30–3000 seconds with the default set to 120 seconds.
- **Primed join timeout:** This timeout supports a range of 120–43,200 seconds with a default set to disabled/0 seconds.
- **Retransmit count:** This field defines the number of heartbeats retransmitted by the AP to the WLC after the heartbeat timeout. The allowed range is 3–8 with the default set at 5.
- **Retransmit interval:** This field defines the interval at which heartbeats are retransmitted to the WLC by the AP post heartbeat timeout. It supports a range of 2–5 seconds with the default set to 3 seconds.

```
C9800(config-ap-profile)#capwap timers ?
discovery-timeout           Configures AP Discovery Timeout
fast-heartbeat-timeout      Configures fast heartbeat timeout
heartbeat-timeout           Configures heartbeat timeout
primary-discovery-timeout   Configures primary discovery timeout
primed-join-timeout         Configures primed join timeout
```

Preserving AP-to-Tag Mapping across N+1 Failovers

On the C9800, tags (including policy, site, and RF tags) can be mapped to an access point in different ways, referred to as *tag sources*. The sources, in order of priority, are Static, Location, Filter, AP, and default. When tags are mapped to an AP, these tags do not get saved to the AP by default. You can explicitly save the tags to the AP's nonvolatile memory by using the command **ap name <AP_NAME> write tag-config**. Starting with release 17.6, the tags can be saved automatically on all the APs on the C9800 by enabling tag persistency under **Configuration > Tags and Profiles > Tags > AP**. If tags have not been saved to the AP and the AP moves to the N+1 C9800, say from C9800-1, here are the outcomes with the tags:

- If the N+1 C9800 is configured with the same tags and profiles and AP-to-tag mappings including static mappings as C9800-1, the AP registers to the N+1 C9800, receives these tags, and services the same SSIDs as on C9800-1.

- If the N+1 C9800 is not configured with all relevant AP-to-tag mappings using any of the tag sources, the AP joins with default tags.
- Even when the tags are saved on the AP, if the N+1 C9800 does not have the corresponding tags and profiles configured, the AP is dropped to default tags.
- Even when the tags are saved on the AP and a tag is present on the N+1 C9800, if the configuration under the profiles under each tag changes, the APs operate based on these new settings from the N+1 C9800.
- In the case of a FlexConnect deployment, when an AP moves to the Connected state, if the configuration of WLANs or profiles is different on the N+1 WLC compared to the primary, the AP needs to disconnect all clients to make use of the new configuration, and migration is not seamless.

You are recommended to save the tag mapping to the AP to avoid the APs ending up with unintended tags and configuration.

Licensing with N+1

The C9800 uses smart licensing, which maintains the AP Count DNA License. N+1 HA is part of the DNA Essentials Licensing Tier. When an AP fails over to a secondary or a tertiary C9800, the AP MAC is sent as part of an entitlement request to the Cisco Smart Software Manager (CSSM) portal to enable seamless handling of licenses across the N+1 failover. Also, unlike AireOS, there is no HA-SKU on the C9800.

N+1 vs. SSO High Availability

In summary, SSO avoids a service impact, whereas N+1 failover creates downtime. Table 10-3 compares SSO with N+1 on the key functions of failover time, whether a config sync occurs, detection mechanism to trigger failover, topology supports, hardware and software restrictions, and impact to APs and clients.

Table10-3 *Comparison of HA SSO and HA N+1*

Functionality	SSO	N+1
Failover Time	Subseconds for box failover and up to 8 seconds for network GW failure.	45–60 seconds for the rediscovery and join to secondary C9800 with default timers.
Config Sync	Full configuration and AP/client state sync.	No sync of config or AP/client runtime data.
Detection Mechanism	The keepalive timer between active and standby units can be configured between 100 and 1000 ms and the retry count between 5 and 10.	This is based on various timers, including heartbeat timers and discovery request timers.

Functionality	SSO	N+1
L2.L3 Topology	Only L2 is supported between C9800s.	L3 is also supported between WLCs.
Software/ Hardware	Must be the same between the two C9800s.	Can be different.
AP/Client Impact	APs and clients do not disconnect.	APs rejoin and clients need to reassociate and reauthenticate in the case of local mode APs and centrally switched SSIDs on Flex APs.

HA in EWC-AP Deployment

When running as an Embedded Wireless on Catalyst Access Point (EWC-AP), the C9800 runs in a container on top of the AP data plane. When two EWC-APs are deployed, there is no dedicated port over which syncing can occur. So, there is no SSO on EWC-AP. For N+1 redundancy, the EWC-AP uses the Virtual Router Redundancy Protocol (VRRP). With VRRP, active-standby election occurs as follows:

1. Day 0 Active Election:

   - The EWC-AP provides the option for users to configure a preferred controller and, if configured, dictates the EWC-AP among many that will take on the active role.
   - If the user has not configured a preferred controller, the AP of the higher model is chosen as active.
   - If all EWC-APs in the deployment have the same AP model, the EWC-AP with the least number of associated clients is chosen as active.
   - If the associated client count is uniform across EWC-APs of the model, the lowest base Ethernet MAC address is leveraged.

   Note that the standby election does not occur at the same time as the primary EWC-AP.

2. Day 1 Standby Election:

   A standby election comes into play only after all APs have joined the elected active EWC-AP. VRRP is not used to elect standby, but instead the following sequence of steps occurs:

   - If there is an explicit configuration to mark an EWC-AP as the preferred WLC, it assumes the role of the standby.
   - If no configuration is present, the EWC-AP with the highest model number becomes the standby.

- If all EWC-APs are of the same model, the EWC-AP with the least AP join time gets elected as the standby.

There is no support for preemption even with this HA setup.

When an active EWC-AP fails, standby EWC-AP takes over in less than 10 seconds, and another standby election process kicks off. To view the active and standby EWC-APs after the election, on the EWC-AP GUI, navigate to **Configuration > Wireless > Access Points**, as shown in Figure 10-23.

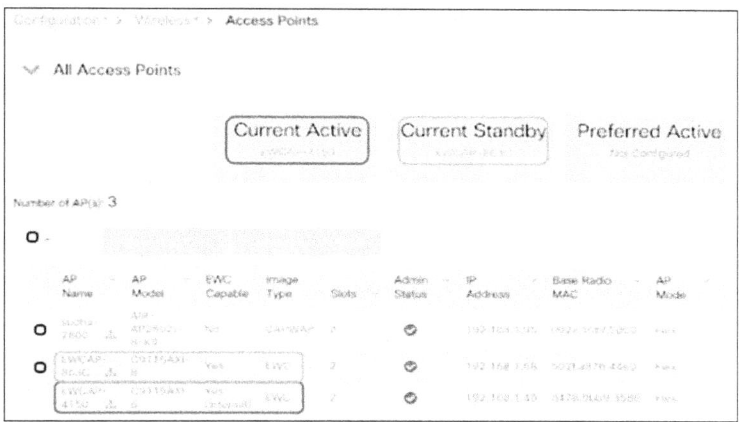

Figure 10-23 *Active and Standby EWC-APs*

HA in EWC-SW Deployment

The Embedded Wireless Controller on Catalyst 9k switches (EWC-SW) relies on stackwise technology using stack ports connected via stacking cables on the backplane of the switch stack to deliver resiliency. See the "References" section for white papers on stackwise architecture for Cat9k switches. Within a switch stack, there is one active and one standby, and the remaining are member switches. The active and standby within the stack sync over stack ports and deliver SSO for APs and clients that are not connected off the failed switch. Across the switch stack, SSO is not possible because there is no sharing of configuration and operational data.

Summary

This chapter describes high availability capabilities of the Catalyst 9800 wireless LAN controller for its various form factors. The C9800 provides stateful switchover to allow subsecond failover for access points and clients. The chapter delves into design and deployment of networks to leverage redundancy features for the C9800.

References

HA SSO deployment guide: https://www.cisco.com/c/dam/en/us/td/docs/wireless/controller/9800/17-6/deployment-guide/c9800-ha-sso-deployment-guide-rel-17-6.pdf

EWC-AP White Paper: https://www.cisco.com/c/en/us/products/collateral/wireless/embedded-wireless-controller-catalyst-access-points/white-paper-c11-743398.html

Catalyst 9300 Stackwise System Architecture White Paper: https://www.cisco.com/c/en/us/products/collateral/switches/catalyst-9300-series-switches/white-paper-c11-741468.html

Chapter 11

Cisco DNA Spaces Integration and IoT

Wireless networking involves a few external services that can be essential for it to work, such as authentication servers or identity databases; we cover them in other chapters. Other types of services can be added to your wireless network to bring more value; they include location-based services, guest user engagement, or analytics of your visitor population. They started as on-premises services on extra appliances but now include cloud connectivity and IoT application integration. This chapter covers these services and how they integrate with the Catalyst 9800.

Value-Added Wireless Services

Many different services sometimes relate very little to each other. Here, we cover the biggest categories of services that may be offered by several Cisco solutions in a different manner or may be unique to a specific solution.

Location Tracking

Global Positioning System (GPS) is a technology that became available to consumers in the 1990s and revolutionized a lot of markets by allowing people to know where they are exactly at any instant. GPS can only work outdoors because it relies on getting a clear signal from several satellites. As the use of smartphones started to explode in the 2010s, end users began to wonder why they could not get their position while being indoors. Getting information about your own location became known as the *blue dot experience* in reference to several apps in which the user is represented on the map by a blue dot (or sometimes a blue arrow). You can have maps but rarely get your current position on one unless a different tracking system than GPS is being used. Cisco has been a pioneer in this domain, starting Wi-Fi location tracking with the 2700 location appliance. Wi-Fi is only the method and frequency range used to track a device, but the location techniques stay similar and revolve around trilateration or triangulation (in the case of *angle of arrival*, or *AoA*): if several sources (for example, access points, or APs) can hear a given device at the same time, a management system that knows the location of the points can compute the device location by assessing the distance between these access points and the device based on the signal strength. There is one major difference (there are many in reality, but let's look at an important one) between Wi-Fi location tracking and the GPS system:

- With GPS, the satellites are constantly transmitting, and client devices are comparing timestamps reported by different satellites to evaluate their position by calculating the time difference of arrival (TDOA). This can be possible because the satellites have fixed coordinates for everyone.
- In Wi-Fi, we rely on the client device transmitting and on the infrastructure to use the strength of the signal to evaluate distance by calculating free space path loss (and possible known obstacles of attenuation). The reason is that the client devices cannot know the location of every access point, and because, until now, there was no standard for location tracking in 802.11 (there is now and it is covered later in this chapter), a common mechanism could not be implemented in all wireless clients.

The GPS is a receive-only system that respects privacy: the satellite operators do not know where you are. Wi-Fi, because it works the other way around, has immediate privacy concerns that caused many wireless client operating systems to take privacy countermeasures. In recent years, smartphone and mobile device vendors started to implement privacy features (such as using a random MAC address) or power-saving techniques that increase the difficulty for a network owner to implement accurate and reliable location tracking based on Wi-Fi. The increase in the number of electronic devices owned also caused another realization: you are tracking a specific device and not a person. A person could have no device, could have a device that is not connected to your infrastructure, or could have several devices.

Accuracy

The accuracy of a location tracking solution refers to the error margin between the calculated position and the real position of the device. There is a possibility of a "glitch" when using signal strength, due to multipath reflection or simply because a device sent frames at a lower power or obstacles were in the way. Accuracy is often expressed like "10 meters accuracy 90 percent of the time" to indicate that 90 percent of the calculations will be accurate within 10 meters of the real position, while the remaining 10 percent can be completely way off. Cisco developed the angle of arrival technique on the 3700 AP with the Hyperlocation module as well as the 4800 AP. Those APs contain an array of antennas that can calculate the angle from which a client signal is coming, which drastically improves the accuracy. Hyperlocation accuracy can go down to 2 meters in a proper deployment. The biggest enemy of Hyperlocation is multipath, and in certain cases, APs that are placed too high or environments that are too metallic prevent you from reaching this level of accuracy.

In general, good accuracy can be met if

- Multipath (for example, mostly metallic reflections) can be kept to a minimum.
- At least three access points can hear the client simultaneously at a good RSSI (stronger than –75 dBm).
- AP placement matches the location guidelines; that is, APs must form a convex hull around the client (the client must be heard by APs that are around it in different directions).
- APs are not placed too high. This does not directly mean that external antenna AP models are out of the picture for location tracking—they can give good results—but the more directional the antenna is, the less accuracy it gives in terms of location tracking if placed vertically. Picture an AP with a directional antenna placed at human height and beaming toward a very long warehouse aisle. This type of tracking works fine because the signal will fade as the client moves away from the AP in the aisle or vice versa. But if you place the AP on the ceiling with this high directionality of the signal, the client on the ground can move around without much impact to the RSSI. When the client moves out of the AP signal beam, its signal fades out very quickly, which does not allow the AP to locate the client accurately. AP placement and antenna selection and orientation are therefore crucial.

Location Update Frequency

Wireless location tracking requires three to four access points to hear a signal from the client at the same time in order to accurately try to calculate its position. That requires a deployment dense enough for several APs to hear the client at a decent signal strength, but that is not the biggest challenge. The problem is that access points are typically on different channels for more efficiency; therefore, they do not hear the same client signal. This is why the basic location tracking method is a probe-based location: when a client probes, it sends probe request frames on several channels more or less at the same time. This gives access points an opportunity to hear a signal from the same client, despite being on different channels. The main drawback that became a bigger concern throughout the years is the fact that wireless clients try to save their battery and thus keep probing activity to a minimum. This means that you might have a refresh of the client position every 30 seconds or every few minutes only. The probing frequency depends on the client adapter but also driver version and even signal strength: clients don't feel the need to probe around if their current signal is good enough.

Reaching a higher frequency requires calculating location based on data frames and that, in turn, requires having APs listening on the same channel at the same time. This capability is enabled by the Cisco FastLocate technology. It typically requires having an extra radio whose job is only scanning: the 3700 can have an extra Wireless Security Module (WSM) module, the 3800 can leverage one of its radios in Flexible Radio Assignment (FRA) to be in monitoring mode, the 4800 has an extra radio dedicated to monitoring and Hyperlocation, and the Catalyst 9120 and 9130 have the RF-ASIC, which is an extra radio dedicated to scanning and monitoring. This strategy requires the wireless LAN controller to synchronize all neighboring APs to scan the same channel at the same time before moving to the next channel: the synchronized off-channel scan ensures that all APs can hear a given client on that channel during the monitoring period. FastLocate allows a refresh rate between 10 and 30 seconds, depending on the number of channels you enabled for scanning. FastLocate would work only with clients that are constantly sending data normally, but the APs can send a Block ACK Request to the specific clients to wake them up and get a signal back, which can have a slight performance impact on the serving radio, though. This technique works only for connected clients; indeed, a client that is passing by and is not connected does not send data frames to your APs.

Presence

Presence is a simpler way to pinpoint the location of a given device. If you do not have the means to do RSSI trilateration (not enough or badly placed APs) or if your business objective is only to track, in which area the device is spending most of its time, Presence checks the APs with the highest RSSI received from the client. Presence tracking requires the client to be connected to the network. The data point that the client is near a given AP for a certain amount of time is already worthy of a certain level of analytics.

The Impact of Privacy MAC Addresses

In that past, probe-based location allowed you, as administrator, to track devices that are not associated to the network because probing is an activity that clients do whether or not they are connected to an SSID. However, the recent push to use privacy MAC addresses (or random MAC addresses), especially for probing, means that the infrastructure can still locate a signal but cannot link it to a given client because that probing uses a MAC address that will not be used anymore by the same device afterward. Even worse, because a given client uses constantly changing MAC addresses to probe, the infrastructure detects it as multiple different clients that briefly passed by, and this drives the number of simultaneously tracked clients through the roof. This is why, in recent years, the tracking of probing clients (that is, those not associated to the network) is disabled by default because it doesn't add any value anymore.

Location Deployment Guidelines

The general accuracy to be expected for RSSI-based location is 5 to 10 meters. When using dedicated RFID tags, you can expect 3 to 5 meters, thanks to the fact that RFID tags are designed to probe a lot on purpose, on all channels to send their location. Hyperlocation can bring the accuracy to 1 to 3 meters.

The update rate revolves around one or two per minute for probe-based location, whereas data frame–based location (FastLocate) can reach a 10-second update interval in typical environments.

Tags—whether RFID tags, BLE tags, or tags that support multiple radios and protocols—offer other features such as a button push that immediately sends a notification, the possibility to advertise the location (by probing) when a movement is detected, or simply a configurable probing interval.

An important concept in RSSI-based tracking is that, considering the signal loses 6 dB every time the distance is doubled (as can be clearly seen in Figure 11-1), when the client is far away, any tiny difference of RSSI could lead to a much larger error margin in distance. Therefore, it is crucial for accuracy to have APs hearing the client at a strong signal (at least –75 dBm on three APs). It is also why the accuracy is not as good when using APs mounted on high ceilings.

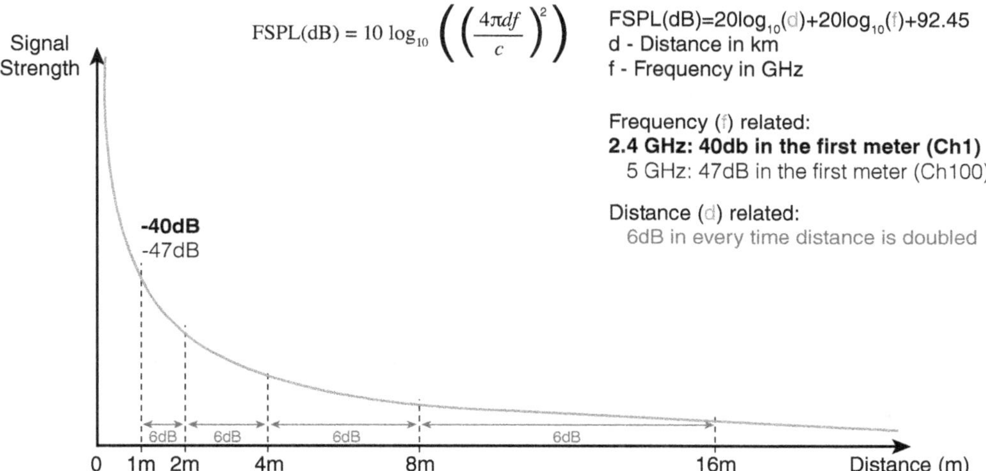
Signal Strength
FSPL(dB) = 10 log10 ((4πdf/c)^2)
FSPL(dB)=20log10(d)+20log10(f)+92.45
d - Distance in km
f - Frequency in GHz
Frequency (f) related:
2.4 GHz: 40db in the first meter (Ch1)
5 GHz: 47dB in the first meter (Ch100)
Distance (d) related:
6dB in every time distance is doubled
-40dB
-47dB
6dB
6dB
6dB
6dB
0
1m
2m
4m
8m
16m
Distance (m)

Figure 11-1 *Free space path loss; a 6 dB decrease every time distance is doubled*

It is also advised to have APs around the client; that is, if you draw a polygon between the APs, the client should be inside it. This setup typically translates into having APs mounted very close to the walls or limits of the building, at least in monitor mode.

Other Technologies

Bluetooth is another technology providing location tracking capabilities since Bluetooth Low Energy (BLE) 4.0 was ratified in the 2010s. Considering Bluetooth is a different protocol altogether, we cover it, including its location tracking capabilities, in the section "Bluetooth and IoT," later in this chapter.

Ultra-Wide Band (UWB) is a new technology that is being implemented in some smartphones but also available in dedicated devices. Due to its very wide bandwidth, UWB can have the benefit of extreme accuracy down to 10 cm but requires a lot of anchors to provide this accuracy.

Wi-Fi has not given its last word as the 802.11-2016 revision of the standard already included location exchanges where a device can get its location from multiple APs by measuring the distance to the APs (using time difference of arrival). The 802.11az is being finalized at the time of this writing and adds a lot more location features that are both beneficial to the client and infrastructure. Time will determine if these amendments get implemented in client devices to provide a real value.

Analytics

Whether RSSI location tracking or Presence is used, the system has a good idea where clients are at all times. If you build a location hierarchy that is meaningful to your business, it then allows you to not only get a breakdown of clients per venue but also the average time spent on each venue, the dwell time in specific areas, and so on.

If you own a shop with accurate location tracking, you can produce a heatmap showing where customers spend most of their time. As a reminder, this capability requires clients to be connected to your network. You do not necessarily need to have an accurate location tracking method in place because the system can rely on presence information and verify to which AP the client is connected. Using presence data rather than location tracking just reduces the accuracy of the geographical data of analytics a bit, but if you are interested in the behavior analysis of your customers at a given branch shop, or the behavior of your devices or employee devices in a given building, presence does the job perfectly. You can see reports of current real-time metrics such as visitor count per location and, how many of those are returning visitors (if they are connected to your network, then, typically even with a private MAC address, at the time of this writing, the same private MAC is reused for some time before being rotated). Cisco DNA Spaces can integrate with Meraki cameras to get a more accurate visitor count (the Wi-Fi–based count actually counts connected devices rather than actual people). A whole set of proximity reports can be produced to do contact tracing and find out who a specific person has been in close contact with, detect clusters of employees, and so on.

Guest Services

Guest services are popular in the hospitality industry or even any enterprise network to allow visitors to easily come in and connect to the network. Guest portals have always been a pain to manage (to update regularly, to support various types and formats of client devices) but also a painful experience for the end users sometimes. Taking the guest portal service to the cloud brings several advantages: no need to maintain the web server, no need to allow firewall rules to allow clients to access the portal page in the DMZ, easy integrations with social networks, and analytics of various sorts are readily available.

Bluetooth and IoT

Although Bluetooth has existed for quite some time, it has drastically gained in popularity because it is included in literally every smartphone device currently, it can be easily integrated in IoT devices, and thanks to the 4.0 and 5.1 revision of the standard, which added a BLE specification, it allows for many new features while keeping the battery consumption to a minimum. Bluetooth and BLE are developed by the Bluetooth Special Interest Group (SIG) and therefore have little connection with the Wi-Fi Alliance. BLE still operates on the 2.4 GHz band like Bluetooth Classic (the name of the original Bluetooth specification) but is oriented toward battery-operated devices and consumes much less energy. The cost of BLE is a much lower throughput (which is typically not a problem for the use cases targeted by BLE). Another big difference is that BLE does not require device pairing necessarily, whereas Bluetooth Classic typically works by pairing two devices (as you are probably used to when connecting your phone to an external Bluetooth speaker, for example). BLE and Bluetooth are incompatible with each other: some devices can support both, but they are two mutually exclusive and different modes of operations. BLE is targeted and mostly adopted by sensor devices or portable devices that require sending small bits of data at a time and preserving their battery. Larger devices like tablets, some phones, and laptops can implement both.

BLE

BLE uses 40 2-MHz-wide channels, in contrast to Bluetooth Classic, which uses 79 1-MHz channels on the same frequency range. Both use the frequency hopping spread spectrum, a technology radically different from 802.11 because it does not focus on a specific channel but keeps hopping between channels to counter interferences. There are 3 advertisement channels and 37 data channels in BLE.

A device can have several roles in BLE:

- The Broadcaster (or advertiser) role advertises Bluetooth packets called *BLE beacons*. This is a role the Cisco AP can have. Three popular beacon formats are supported out of the box and with more formats potentially supported in the future: iBeacon (Apple), Eddystone UID, or Eddystone URL (Google). There can also be custom/proprietary BLE beacon formats (with a static BLE header still). Those beacons are advertised on the specific three advertisement channels. They are broadcasted at fixed intervals (down to 20 ms).
- The Observer role scans the surroundings for BLE tag broadcasts but does not connect to anyone. Considering that this role is included in the "central" role (described next), APs can indirectly act as observers.
- The Central role scans and can also connect. This is a role that the Cisco AP can have.
- The Peripheral role can advertise and also connect.

iBeacon is a popular format of BLE beacon that has configurable parameters called UUID, Major and Minor. Combined, they form a kind of unique identifier. The idea is that you advertise a unique beacon ID in a given location so that the app running on the smartphone can recognize a beacon and act accordingly. (You can leverage the UUID to have a "vendor" identifier, and then the Major and Minor to more granularly identify the specific area/location or purpose of your beacon.) Lots of free BLE scanner apps (an example is depicted in Figure 11-2) are available on smartphones and show the surrounding beacons, their Bluetooth MAC address, and UUID/Major/Minor. Those scanner apps can act as Observer, whereas others act as Central because they can connect to beacons. The Eddystone format uses either a URL or a combination of namespace and ID to achieve the same purpose as iBeacon.

Figure 11-2 *A BLE scanner application running on a smartphone*

To troubleshoot Bluetooth Low Energy, you can use a BLE sniffer, which is basically a USB stick that can be plugged on a laptop and adds the capability to capture BLE traffic and display it in Wireshark or similar packet capture applications.

IoT

Cisco developed an IoT services solution based on Cisco DNA Spaces that can control and leverage the IoT features embedded in the access points. gRPC is the protocol of choice for the APs to report IoT telemetry, such as floor BLE scanning, directly toward the Cisco DNA Spaces connector, which is in charge of talking to the Cisco DNA Spaces cloud. (gRPC is described further in Chapter 13, "Model-Driven Telemetry.") This architecture, depicted in Figure 11-3, presents the advantage of not requiring a direct connection from the APs toward the cloud.

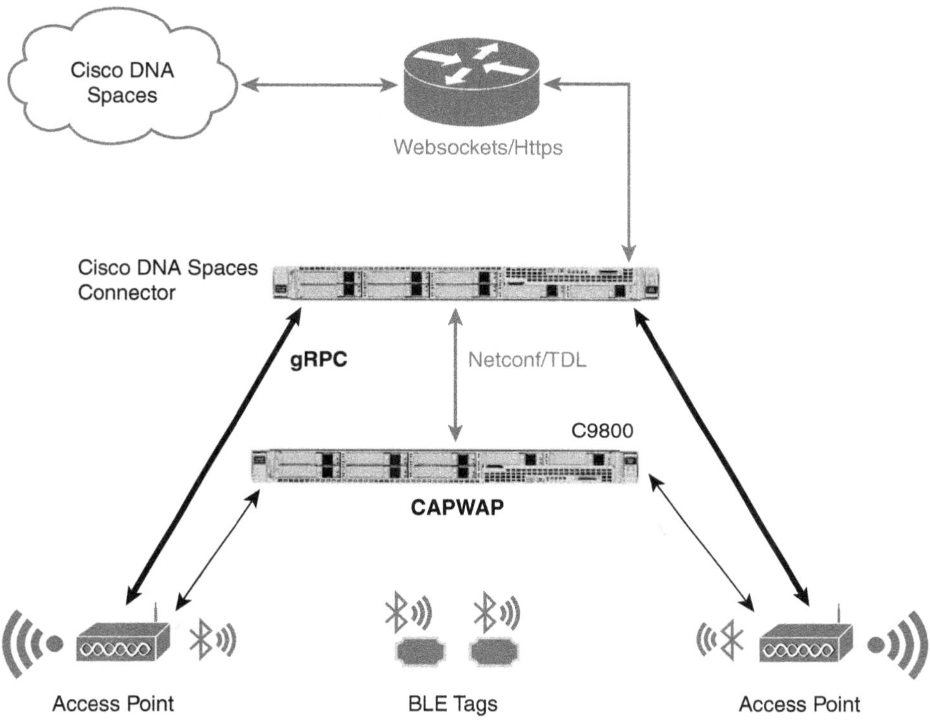

Figure 11-3 *Cisco DNA Spaces Bluetooth telemetry architecture*

There are two modes of operations depending on the AP capabilities:

- The native mode or Base BLE gateway supports basic BLE broadcasting and scanning and is supported on all 802.11ac Wave 2 (that is, x800 series mostly) and later access points (including Catalyst 9100s). Some 802.11ac Wave 2 APs have a native BLE radio, whereas other models like the 2800 or 3800 can leverage a USB BLE radio.
- The IOx mode or Advanced BLE gateway is supported only on Catalyst 9100 and later access points. In this case, the IOx app can configure the embedded BLE radio in the AP as well as configure third-party BLE tags. These IOx apps can be deployed on the APs through Cisco DNA Spaces.

From Cisco DNA Spaces, the APs can be configured as a Central role, where the AP scans for BLE beacons and is able to configure BLE tags, or as a Broadcaster role, where the AP sends BLE beacons in the configured format. Cisco DNA Spaces sends this configuration to the Cisco DNA Spaces connector via HTTP(s), the connector, in turn, configures the appropriate settings on the APs (mostly) and the WLC via NETCONF, and the WLC pushes the configuration to the APs via CAPWAP. Cisco DNA spaces can verify the configuration is active via a feedback from C9800 database because this activity is asynchronous.

You can verify the IoT configuration took place on your devices with a couple of commands. The following command verifies that the APs are configured with an authentication token for streaming the data to the Cisco DNA Spaces connector:

```
WLC#show run | include ap cisco-dna
     ap cisco-dna token 0 <tokened>
```

This next command helps you verify that Bluetooth is enabled on the APs:

```
WLC#show run | include ap dot15
     no ap dot15 shutdown
```

The following command verifies that gRPC is enabled and IOx application hosting is enabled:

```
WLC#show run | begin ap profile default-ap-profile
     apphost
     cisco-dna grpc
```

As of IOS-XE 17.3.2, BLE and Assurance telemetry can coexist at the same time on the AP; however, they remain incompatible with the Intelligent Capture (iCap) feature. Only iCap or BLE can be enabled at a given time. These limitations could be addressed in future software versions.

You can also verify the telemetry subscriptions are active for Cisco DNA Spaces with **show telemetry ietf subscription all**, as shown in Example 11-1. Subscriptions 122 to 127 depict Cisco DNA Spaces telemetry subscriptions.

Example 11-1 *Verifying the TDL Subscription Commands After a C9800 Is Connected to Cisco DNA Spaces*

```
WLC# show telemetry ietf subscription all
  Telemetry subscription brief
  ID              Type        State       Filter type
  --------------------------------------------------------
  122             Configured  Valid       tdl-uri
  123             Configured  Valid       tdl-uri
  124             Configured  Valid       tdl-uri
  125             Configured  Valid       transform-name
  126             Configured  Valid       transform-name
  127             Configured  Valid       tdl-uri
```

The connection can be verified on the APs themselves with **show cloud connector connection detail** and **show cloud connector key access**, as shown in Example 11-2.

Example 11-2 *Verifying the AP Connections to Cisco DNA Spaces*

```
AP# show cloud connector key access
Token Valid : Yes
Token Stats :
        Number of Attempts   : 44
        Number of Failures   : 27
        Last Failure on      : 2020-03-28 02:02:15.649556818 +0000 UTC
  m=+5753.097022576
        Last Failure reason : curl: SSL connect error
        Last Success on      : 2020-04-01 00:48:37.313511596 +0000 UTC
  m=+346934.760976625
        Expiration time      : 2020-04-02 00:48:37 +0000 UTC
Connection Retry Interval : 30

AP# show cloud connector connection detail
Connection State            : READY
Connection Url              :  10.22.243.33:8000
Certificate Available       :  true
Controller Ip               :  10.22.243.31
Stream Setup Interval       :  30
Keepalive Interval          :  30
Last Keepalive Rcvd On      :  2020-04-01 00:32:47.891433113 +0000 UTC
  m=+345985.338898246
Number of Dials                :  2
Number of Tx Pkts              :  2788175
Number of Rx Pkts              :  11341
Number of Dropped Pkts         :  0
Number of Rx Keepalive         :  11341
Number of Tx Keepalive         :  11341
Number of Rx Cfg Request       :  0
Number of Tx AP Cfg Resp       :  0
Number of Tx APP Cfg Resp      :  0
Number of Tx APP state pkts    :  5
Number of Tx APP data pkts     :  2776829
```

Bluetooth Location Tracking

Bluetooth requires a fixed tag (whether physical or virtual) to send a BLE identifier over the air. An application on the phone must be listening to that specific identifier and can calculate the location based on this identifier signal strength. The protocol allows each side to share their power level, and the short range of Bluetooth typically allows for a good accuracy.

The system works the other way around from Wi-Fi, where the client is required to run an application that knows the venue map and layout and listen for specific BLE identifiers present in that venue. Like GPS, the infrastructure is broadcasting, and it's the client doing the trilateration based on the multiple signals it receives.

This system allows for privacy because you only calculate your location yourself in the app when the app is running. Also, every user exercises control whether their device is tracked or not.

Connected Mobile Experiences (CMX)

Connected Mobile Experiences is the successor of the Mobility Services Engine (MSE) appliance and continues its legacy of mostly focusing on location calculations of wireless devices but also a few other services. CMX is a UCS-based appliance on which the CMX apps run on top of a UNIX OS. Its main protocol is the Network Mobility Service Protocol (NMSP), which is a proprietary protocol to talk to switches and wireless controllers. CMX is a software solution that uses location and other intelligence from Cisco wireless infrastructure to generate analytics and deliver relevant services to customers.

CMX is not under a lot of new development, and most of its features have moved to Cisco DNA Spaces. CMX can still be used for a fully on-prem deployment or as a proxy toward Cisco DNA Spaces. It is not the recommended deployment option because not all features are available compared to when using the Cisco DNA Spaces connector. However, it can come in handy if you already own a CMX appliance.

To create the connection between your 9800 WLC and CMX, you need to go in the CMX system page and add NMSP devices. Enter the WLC IP address and SSH credentials. CMX connects to the 9800 and enters the specific configuration to trigger the NMSP connection from the WLC toward CMX.

CMX, like most tracking software, is based on maps. CMX does not offer the capability to edit the maps. Editing involves uploading the floor plans as background, draws walls and obstacles to be the right attenuation level, and so on. CMX can either rely on one-time or regular import of maps from Cisco Prime Infrastructure; anytime you export the Prime Infrastructure map and import it on CMX, it overwrites the existing maps and updates them. With Cisco DNA Center, the map synchronization is in real time. This means that any edits you make to the floor plan or obstacles in Cisco DNA Center nearly immediately are replicated to the maps on CMX. CMX web interfaces stay the source of truth for location tracking and other analytics data, but a good amount is displayed in real time by DNA Center when it comes to client or rogue locations on the map.

Cisco DNA Spaces

Currently, Cisco DNA Spaces uses two different domains: dnaspaces.eu and dnaspaces.io. The first one is specific for residents of the European Union, whereas the other is for international use. Accounts created are specific to the domain they were created on.

Cisco DNA Spaces has several licenses levels.

The SEE license level allows you to

- See behavior metrics (such as visitor behavior).
- Measure location impact. You can measure impact of location behavior on event and layout changes.
- Organize your network in a location hierarchy. You can convert your IT networks to a business taxonomy.
- Search and display through Cloud Detect and Locate. You can search and display devices on a map interface on Cisco DNA Spaces.
- Monitor location analytics. You can filter and analyze behavior based on business criteria.
- Export reports.
- Offer OpenRoaming to your guests. You can enhance the Wi-Fi experience and boost Wi-Fi adoption by seamlessly onboarding guests.
- Obtain camera metrics. You can get insights into the behavior of people within physical spaces using Meraki video cameras.
- Integrate with DNA Center for client and rogue location, Assurance, and Intelligent Capture.

The Extend license level allows you to

- Benefit from on-premises and cloud APIs.
- Access the customer or partner Firehose API, a cloud-first, high-performance, low-latency API.
- Export streaming data.
- Leverage Cisco on Cisco. You can integrate with Cisco Webex applications to improve location services.
- Integrate with the enterprise. You can integrate with enterprise software to correlate with location data.
- Benefit from an app center with advanced analytics, indoor mapping and wayfinding, digital signage, asset management, and many other third-party APPs to address vertical or business specific needs.
- Obtain SLAs and monitoring for APIs for 24×7 API monitoring and support.

The ACT license level allows you to

- Benefit from location-based captive portals.
- Leverage enhanced detect and locate ACT (RSSI location in the cloud, cloud location APIs, cloud location history).

- Use Hyperlocation and benefit from the extra accuracy in location.
- Use profile and segment visitors based on at-location behavior.
- Specify engagement rules. You can trigger notifications to visitors and employees via multiple channels.
- Use indoor IoT services. You can break adoption barriers, lower costs, and unlock IoT use cases and management.
- Use proximity reporting to get insights into clients' location exposure and proximity reports of other clients in the vicinity.
- Use the Asset locator to identify and monitor assets, and soon leverage the IoT explorer and smart workspaces.

DNS Spaces expands the advantages of CMX by being a cloud platform that allows for faster innovation. It also increases the scale at which devices can be tracked and is not restricted by CMX scale limitations. It provides a single interface for third-party integration through the firehose API. The analytics as well as captive portal part can be tailored to the different customer verticals and allow an easier multitenant management. The API allows you to connect different things; for example, the Webex Board can report the number of people detected by the camera in a meeting room to Cisco DNA Spaces so that you can mark a room as effectively busy or not. Other telemetry can also be reported by Cisco devices to DNA Spaces, especially thanks to the Webex-embedded sensors in specific Webex endpoints.

Deployment Modes

Cisco DNA Spaces is a cloud-based product, but giving full access to the Internet from your internal network devices is not always practical or even a wise decision. This is why Cisco offers multiple connection methods to create a link between your Catalyst 9800 and the Cisco DNA Spaces cloud.

Direct Connection

You are able to directly connect the Catalyst 9800 controller to DNA Spaces. However, this setup is recommended for small networks (only one controller and up to 50 APs) or for lab testing purposes. The DNA Spaces connector is heavily recommended for larger scale because it receives the data from all the controllers and APs and compresses/optimizes the Internet bandwidth consumption toward DNA Spaces. A direct connection can consume a lot of resources on larger setups because it sends every update immediately to the cloud without the optimization brought by the connector.

There are two basic steps to ensure direct connection:

- Having a DNS server configured on the 9800:

```
WLC(config)#ip name-server <DNS server-ip>
```

- Making sure the Cisco certificate is trusted:

```
WLC(config)#crypto pki trustpool import url http://www.cisco.com/
security/pki/trs/ios.p7b
```

On DNA Spaces, go to **Setup > Wireless Networks.** Choose your connection method (here direct connection), and choose **View Token.** This gives you a list of steps to connect your WLC to Cisco DNA Spaces. The key points are that you get a token specific to your account, and you also have a server URL specific to your Cisco DNA Spaces account that you need to enter in the WLC configuration.

Cisco DNA Spaces Connector

As explained in the previous section, DNA Spaces Connector is the recommended way to connect to DNA Spaces when having a larger setup or when using all of the features of Cisco DNA Spaces. First of all, it makes your firewall rules simpler by concentrating all the controller telemetry internally in your network before sending this data to the DNA Spaces cloud via TLS in a more bandwidth-efficient manner. You, therefore, do not need to grant Internet access to all your controllers.

The Cisco DNA Spaces Connector is available through a single .OVA container file that offers several deployment options: standard, advanced 1, or advanced 2. These options vary the amount of CPU and memory consumed by the connector VM. Another option is to deploy a single network adapter on the connector VM or two. In the case of a dual network connection, one interface is called the *cloud interface*, and it is the interface that must have access to the Cisco DNA Spaces cloud. The other interface is for connecting to Cisco network devices and needs to have an IP in another subnet and reachability to all your Cisco network devices that you plan on using with Cisco DNA Spaces. The default gateway is configured only on the cloud interface, while the device interface accepts static routes to network device subnets if network devices are in different subnets than the Cisco DNA Spaces connector device interface. The connector VM can be deployed on private clouds like VMware or HyperV but is also available on Amazon Web Services at this time.

When it is deployed, you can access the Cisco DNA Spaces connector WebUI by browsing to the IP of the Cisco DNA Spaces connector. As shown in Figure 11-4, the main settings are in the middle, cloud status and connectivity details are at the bottom, and a banner at the top attracts your attention on specific points. Upon first visit, this screen invites you to set up an HTTP proxy (which is not necessarily needed unless you require one to get Internet access), configure the SALT for anonymizing your data, and configure a token, which is the most essential activity.

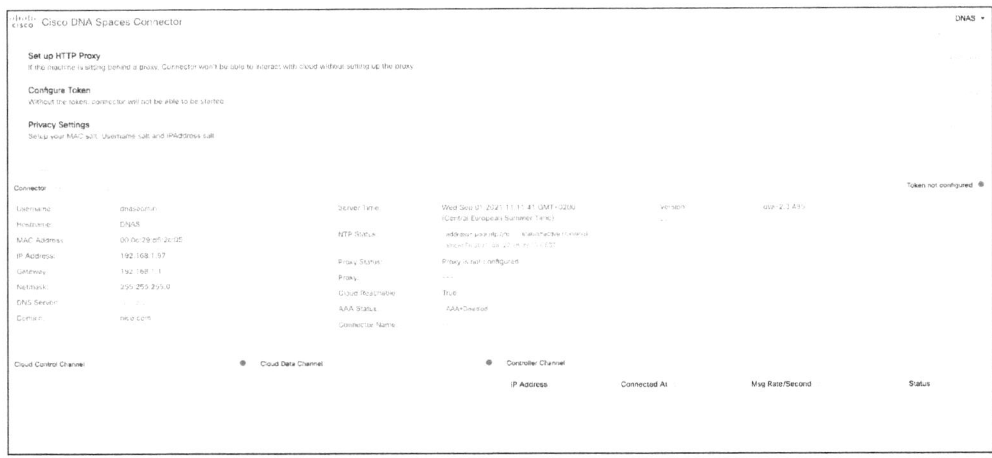

Figure 11-4 *CiscoDNA Spaces connector initial WebUI screen*

To get the token, go to dnaspaces.io or .eu, log in to your account, click the top-left menu, and choose **Setup > Wireless Networks.** A page shows you the most common way to connect your devices: via connector, direct connection, or CMX tethering. In the Connector area, you can create a new token, as seen in Figure 11-5.

DNA Spaces

Connect via Spaces Connector

Spaces Connector is an easy way to get your wireless network connected to Cisco DNA Spaces. No need to upgrade Wireless LAN Controllers or reconfigure your wireless network.

1 **Install Spaces Connector OVA**

Download and install Spaces Connector OVA as a virtual machine.

2 **Configure Spaces Connector**

You will need a token to configure Spaces Connector. You need to connect to https://<your connector IP>/ from a browser to configure the token. You can optionally configure Spaces Connector to connect via HTTPS proxy.

0 / 4 connector(s) active

3 **Add Controllers**

Add and associate controllers to your Cisco DNA Spaces Connector(s)

0 / 1 controller(s) active

4 **Import Maps**

Prime/DNAC map requires in order to work Locate & detect, Asset tracker and IOT services, and proximity Report

19 buildings imported

91 floors imported

5 **Setup location hierarchy**

Once the maps imported, you can add them into location hierarchy

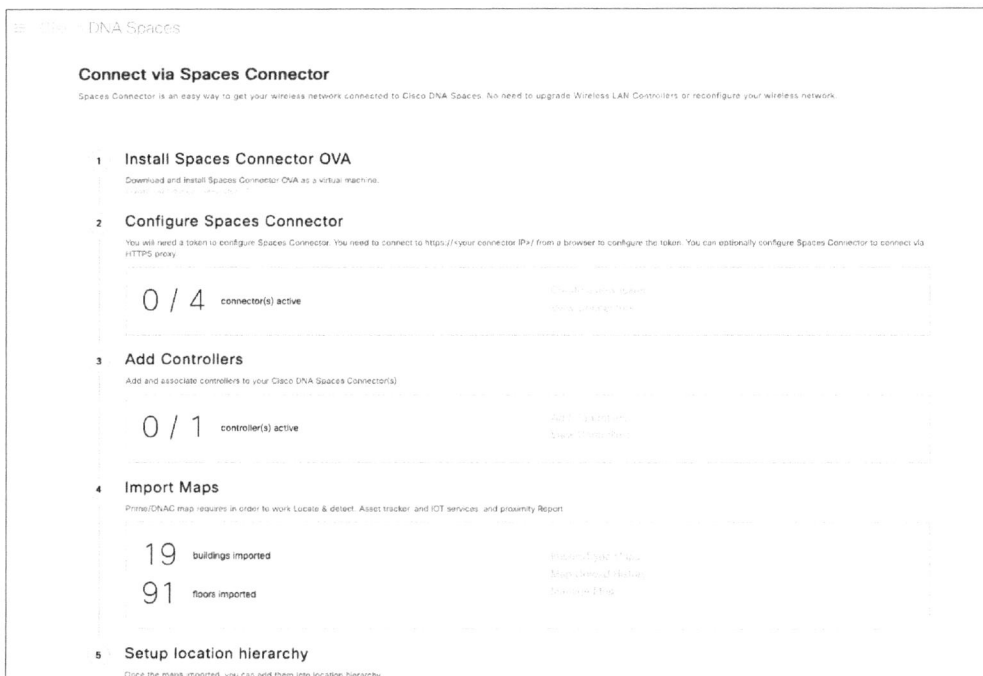

Figure 11-5 *DNA Spaces Setup >Wireless page allowing you to generate tokens*

Copy and paste this token in the Cisco DNA Spaces Connect WebUI on the "setup" link in the banner shown in Figure 11-4. The connector shows the message "downloading images," which is an operation that can take a varying amount of time depending on your Internet connection speed and latency. The connector downloads the latest connector Docker code for itself. After the container code download and token entry, the WebUI, illustrated in Figure 11-6, displays cloud connectivity status details at the bottom of the page.

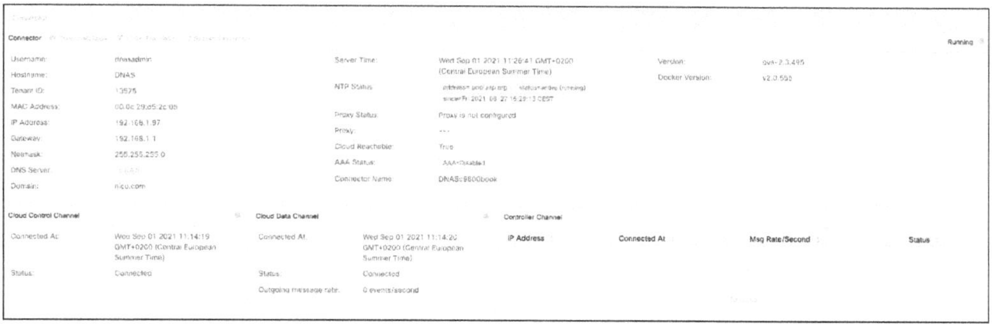

Figure 11-6 *Cisco DNA Spaces WebUI page after successful connection to the cloud*

You can add WLCs to DNA Spaces and the connector by using the DNA Spaces setup page, as in Figure 11-4. Under **Add Controllers**, you can add new controllers or view your current controllers list. To add a controller, as illustrated in Figure 11-7, you need its IP address and NETCONF credentials and need to select the controller type as Catalyst. The DNA Spaces cloud instructs the connector to initiate a local connection to the WLC. The connector then proxies the data to the DNA Spaces cloud.

The bottom of the window, shown in Figure 11-8, shows you the list of CLI commands that DNA Spaces is about to add on your controller, for the controller to trust the DNA Spaces connector for an NMSP session.

When it's done, the WLC should have an NMSP session up and active toward the local DNA Spaces connector IP address. The WLC also should be visible on the DNA Spaces cloud.

If you are going with an IoT/Bluetooth use case, you need to click **Manage Streams** on the three dots next to the connector in DNA Spaces and enable the gRPC tunnel. This action pushes the Cisco DNA Spaces and connector certificate to the WLC for the APs to be able to establish a tunnel back to the controller to stream their IoT data. The end-user guide is a good resource for more information on this topic.

Add Controller

Connector

DNASc9800book

Controller IP

192.168.1.133

Controller Name

lab9800L

Controller Type

Catalyst WLC / Catalyst 98...

Netconf Username

admin

Netconf Password

•••••••• SHOW

Enable Password

•••••••• SHOW

Figure 11-7 *Cisco DNA Spaces connector Add Controller page*

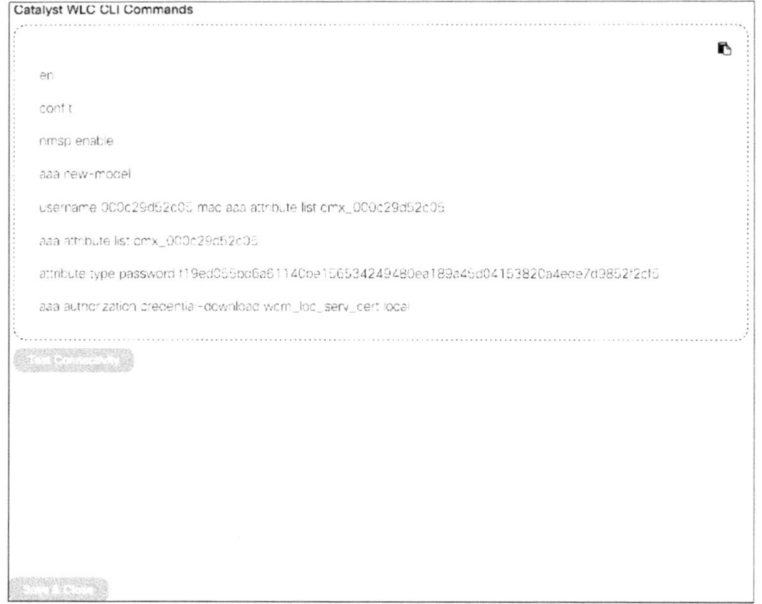

Catalyst WLC CLI Commands
en
conf t
nmsp enable
aaa new-model
username 000c29d52c05 mac aaa attribute list cmx_000c29d52c05
aaa attribute list cmx_000c29d52c05
attribute type password f19ed055bd6a61140be156534249480ea189a45d04153820a4ede7d9852f2cf5
aaa authorization credential-download wcm_loc_serv_cert local

Figure 11-8 *DNA Spaces connector CLI commands overview*

CMX Tethering

CMX can somewhat replace the DNA Spaces connector and act as the proxy toward DNA Spaces. This capability can be a real benefit if you already own a CMX appliance and license and want to put it to good use. It can also be of great value if you use a Hyperlocation tracking system on your CMX and want to get the extra benefits from DNA Spaces. It may not support all features like OpenRoaming or IoT Gateway, which uses a specific container inside the DNA Spaces connector.

Specific Service Examples

The list of services offered by Cisco DNA Spaces keeps increasing constantly, and it's impossible to be always up to date in a book when talking about a cloud offering. Nonetheless, let's review a few of the services offered by Cisco DNA Spaces, like OpenRoaming and IoT/BLE Gateway.

OpenRoaming

Cisco developed OpenRoaming in 2020. Since then, it got handed over to the Wireless Broadband Alliance (WBA) and is therefore not a proprietary solution but an open federation of identity providers that aim at revolutionizing guest wireless access.

What Problem Is OpenRoaming Trying to Solve?

There are several facts around guest wireless access that everyone will agree with:

- Portals can be cumbersome and require lots of end-user efforts to connect.
- It's not easy to figure out which WLAN in a given location is the one you are supposed to connect to and what credentials you are expected to use.
- Guest SSIDs usually have no or poor security. Adding authentication usually reduces the user friendliness and creates an obstacle to have guests connecting to your network.
- Managing guest credentials can be a very time-consuming task.

OpenRoaming solves all of these issues in an elegant manner. It provides the most secure guest WLAN solution (using 802.1X) but at the same time requires no effort from the end-user side, not even tapping or clicking the SSID name, and at the same time, removes the burden of credentials management from the network administrator.

OpenRoaming Architecture

OpenRoaming is built around various actors and roles. One is the access provider, which could be you, the network administrator of a venue desiring to use OpenRoaming. Access providers are venues that can be visited by consumers and where consumers may connect to the network thanks to OpenRoaming credentials.

Identity providers represent all the companies managing user credentials. This could be phone-specific credentials such as a Samsung ID or Google or Apple accounts but also other service providers such as iPass or Boingo that are active in the wireless access domain. This concept establishes that often the venue or access provider will not be managing user credentials and will likely accept users coming from various identity providers. However, it is also possible for a venue to register as an identity provider, which could make sense if you intend for the consumers to use a dedicated app with dedicated credentials maybe linked to loyalty advantages.

The OpenRoaming identity federation is the cloud that connects the access providers (venues around the globe) to the identity providers. This scales the solution because, thanks to a discovery mechanism, millions of networks around the planet can use a constantly growing list of identity providers without any manual configuration.

The system relies on preprovisioned credentials on the mobile (which allows the whole process to be seamless) and PassPoint to be configured on the wireless network (leveraging existing standards and not reinventing the wheel with proprietary protocols). Some smartphone brands have direct support in the OS for OpenRoaming credentials, and the user can enter default credentials when setting up their phone so that those credentials will be used for any OpenRoaming SSID encountered. Other brands have not implemented this capability (yet) and will rely on an app to do so. An OpenRoaming app is available in all application stores and allows you to configure OpenRoaming credentials and install it as a Hotspot 2.0 profile on the phone: this is supported and will work on any phone supporting Hotspot 2.0. On top of this, Cisco provides an SDK that allows you to integrate OpenRoaming profiles to your custom application. This means that a retail store chain, for example, would not require its consumers to install the OpenRoaming app or have their phone properly configured for OpenRoaming. However, they would be able to rely on their existing store app to install the right OpenRoaming profile on the consumer phone so that they will automatically connect when they are present in the store. This is a big step forward and would represent a 100 percent connection rate from users who have the retail store app on their phone because people do not even have to open their phone and select the SSID but are automatically connected. A given client device can have several OpenRoaming profiles configured at the same time. This means that an Android device can have default OpenRoaming credentials set up in the OS itself, but also because a specific retail store app was installed, another specific OpenRoaming profile using that retail chain's credentials is installed and kicks in when they are present in those stores. The bottom line is that end users don't have to worry about anything, and the phone should autoconnect in any case with the best credentials available.

One of the major advantages of OpenRoaming is the ease with which clients can automatically connect to your network. Not only does this increase the user satisfaction by lowering their efforts to connect, but it also means more users are connected to your network at all times (because it is not a manual decision to connect but an automatic one, provided the users have the OpenRoaming profile installed). Because of MAC address randomization, not a lot of useful analytics data can be gathered from passersby anymore, and using OpenRoaming drastically boosts the meaningfulness of your analytics data because it is based on connected users. Being connected also means a device typically gives more frequent location updates. This does not impair the anonymity of users because the network owner still cannot find out who a given device belongs to; however, it does uniquely identify OpenRoaming accounts, and you could retrieve a person's identity if you have the legal right to do so—in case of an investigation, for example.

OpenRoaming Configuration

Assuming you have a Cisco DNA Spaces connector already registered in the cloud as explained previously, go to **Cisco DNA Spaces** main page in the cloud and click the **OpenRoaming** tile. It opens a page that takes you through the steps (see Figure 11-9). You configure an OpenRoaming profile in Cisco DNA Spaces, choose a connector to enable the hotspot container, and push the configuration to WLC(s).

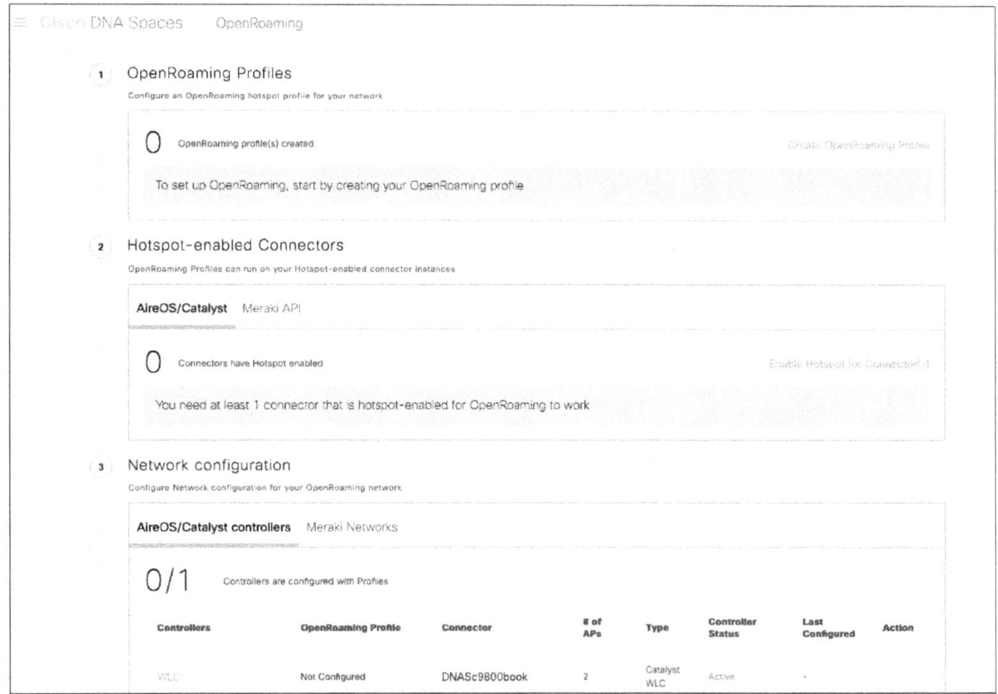

Figure 11-9 *Cisco DNA Spaces OpenRoaming page*

Clicking the OpenRoaming profile creation link takes you to a wizard that helps you define the hotspot policy for your SSID. Illustrated in Figure 11-10, the access policy helps you define who can connect. Having an OpenRoaming SSID does not mean that you have to accept any client on it, although that is a possibility. You could also decide to accept a larger client base but to push clients to use a specific set of credentials if available. This capability is helpful in scenarios where the user may have default

OpenRoaming credentials from the smartphone OS vendor (for example, credentials that were configured in the initial setup of the phone), but you want the client to use the credentials from your loyalty app to promote its use and get more client statistics.

Figure 11-10 *Cisco DNA Spaces OpenRoaming profile access policy page*

On the second page, illustrated in Figure 11-11, you can define the SSID name and basic FT settings. You need to manually configure an 802.1X Enterprise SSID on the WLC at this stage and have it point to the DNA Spaces connector as a RADIUS server. The steps in the wizard simply help tie the Hotspot 2.0/OpenRoaming configuration to the SSID name.

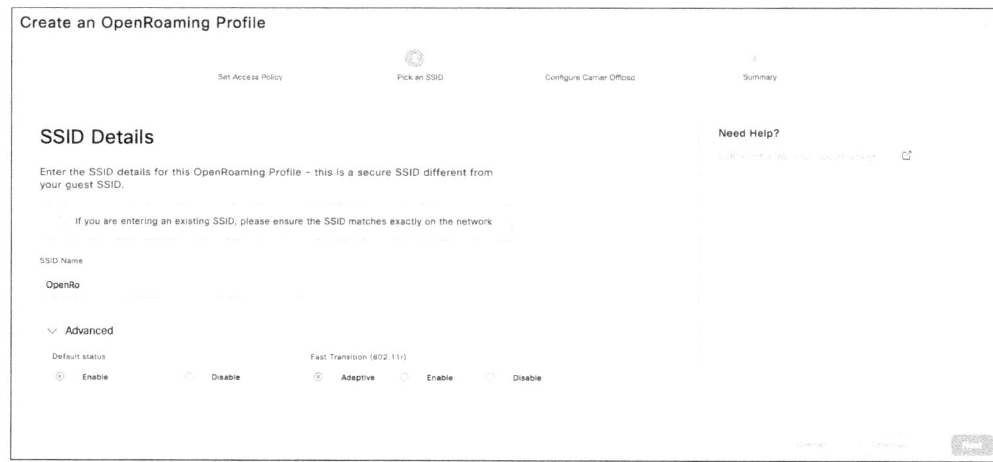

Figure 11-11 *Cisco DNA Spaces OpenRoaming profile SSID configuration page*

The third OpenRoaming profile wizard page, illustrated in Figure 11-12, asks if you want to offload some carrier traffic. This way you can claim the capability to send the client traffic directly to the internal network of a specific mobile provider with whom you have such an agreement.

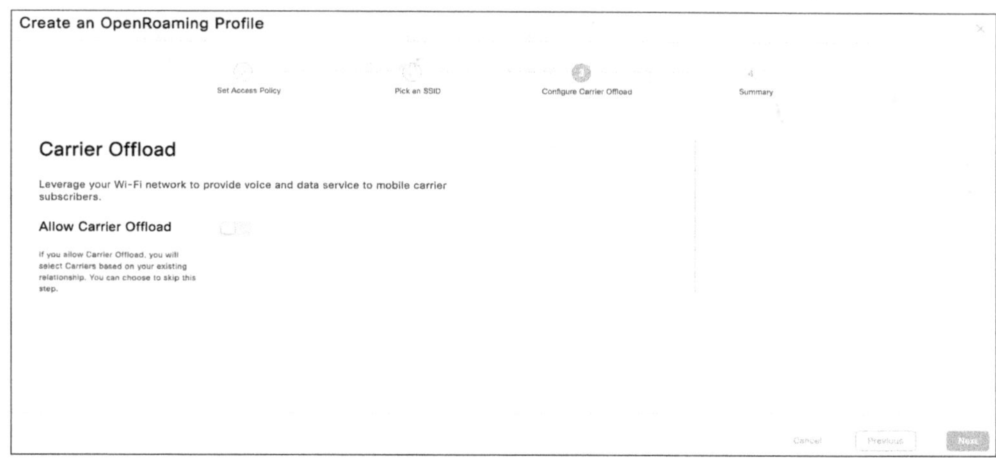

Figure 11-12 *Cisco DNA Spaces OpenRoaming profile carrier offload page*

Finally, a summary page allows you to verify your configuration and submit it.

The second step asks you to choose a Cisco DNA Spaces connector instance, and it installs an extra docker container for the hotspot role on it. If you go back to the Cisco DNA Spaces WebUI, as illustrated in Figure 11-13, you see a new hotspot tab. You have to enter a token again for the hotspot tab to show as fully connected (otherwise the hotspot tab shows but does not indicate cloud connectivity to be working).

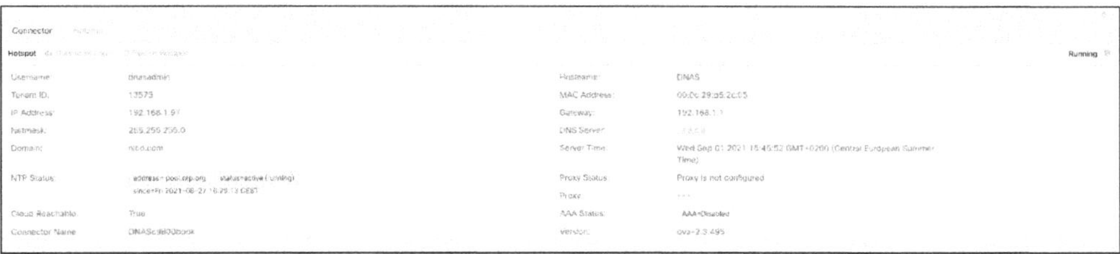

Figure 11-13 *Cisco DNA Spaces home page with the hotspot tab*

Before doing the third step mentioned by Cisco DNA Spaces, it is assumed that you have an SSID ready. Configure an SSID on your WLC for typical WPA Enterprise 802.1X. Configure the RADIUS server to be the Cisco DNA Spaces connector and use the shared secret **radsec**. The WLC sends RADIUS traffic to the Cisco DNA Spaces connector, which acts as a kind of proxy, sending the authentication through RADSEC to the OpenRoaming federation. To make the SSID "OpenRoaming," you need to configure the right 802.11u/Hotspot 2.0 attributes. The WLC has a Hotspot/OpenRoaming page where you can configure them, but Cisco DNA Spaces offers you a wizard that gives you the

CLI commands to enter on your WLC to enable Hotspot 2.0 on your SSID. You need to reference the Hotspot/OpenRoaming profile name in the advanced settings of the policy profile linked to the SSID as well.

The third step on the Cisco DNA Spaces OpenRoaming page asks you to pick one of your WLCs (make sure it is active) and apply the OpenRoaming profile to it, as illustrated in Figure 11-14.

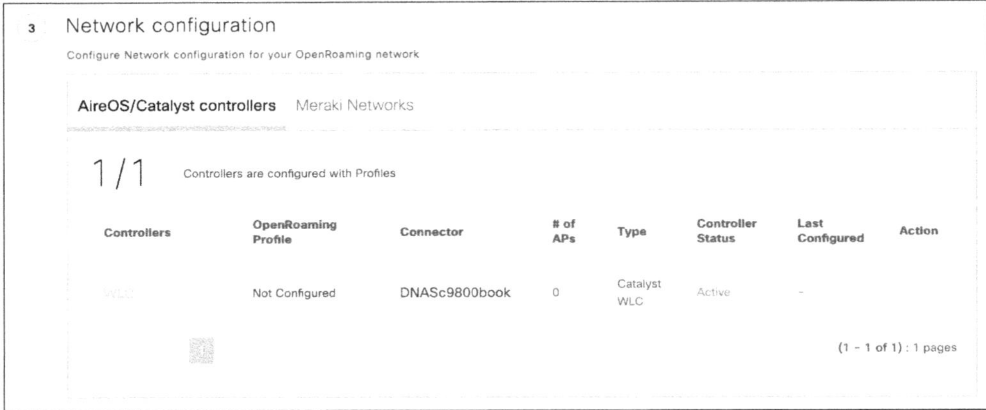

Figure 11-14 *Cisco DNA Spaces network configuration section*

Choose an SSID name (pick the one you configured on your WLC) and a controller to prepare the OpenRoaming configuration. Cisco DNA Spaces takes care of configuring all the 802.11u hotspot parameters that are specific to OpenRoaming, therefore removing any complexity from your hands! You can see an example of those commands pushed in Figure 11-15.

Figure 11-15 *Cisco DNA Spaces controller configuration page shows you the command it pushes*

You can verify that those settings were pushed on the WLC by going to **Configuration > Wireless > Hotspot/OpenRoaming** and verifying that a profile has been configured, as illustrated in Figure 11-16. If not, you can copy and paste the commands recommended by Cisco DNA Spaces yourself in the CLI.

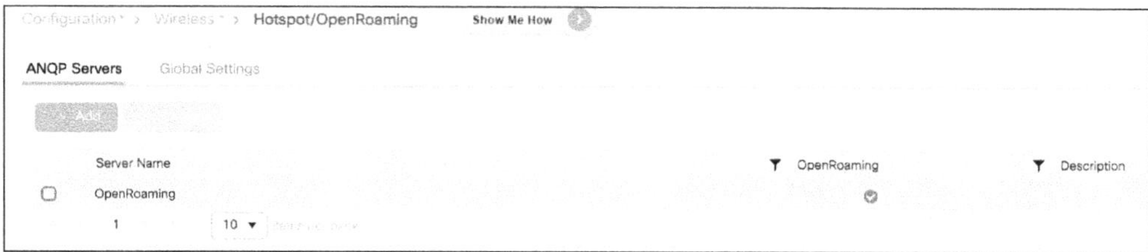

Figure 11-16 *WLC OpenRoaming configuration page*

If you need to change anything, you can change it directly from the WLC OpenRoaming configuration page, as shown in Figure 11-17. Setting the network type to Free Public is the best way to make sure all phones will connect to your network automatically without any restriction. If you set Chargeable Public, your clients have Internet access, but if they try to download an application from their store, they get a notification that they may be charged for the high amount of data (even if you're not charging) as is often the case when connected on LTE. These settings define how the phones behave when connected to the network. The realm configuration defines what EAP types are supported; by default, EAP-TTLS with inner method PAP is configured. This is completely transparent for the user who does not have to enter any credentials or choose a method.

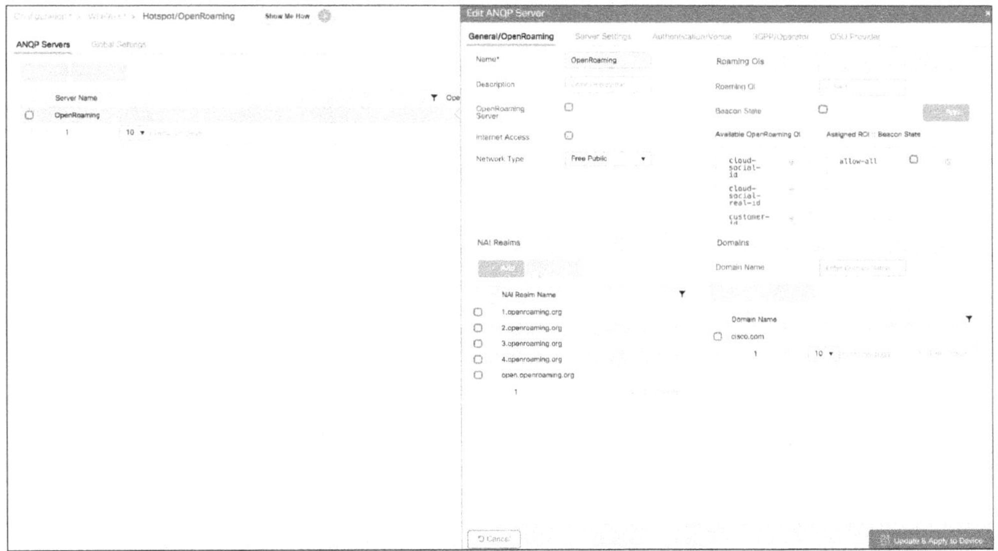

Figure 11-17 *WLC OpenRoaming configuration page*

The Operator tab allows you to define an operator name that will be displayed under the SSID name on the wireless clients. This is convenient if you want to call your SSID OpenRoaming but have a kind of subtitle for your network with your company or venue name. At this stage, if your SSID is broadcasted on access points, any wireless device that is unconnected to any Wi-Fi network automatically connects to your OpenRoaming SSID without user intervention, provided that the users have an OpenRoaming profile either natively on the device or through the OpenRoaming app.

Figure 11-18 illustrates an example of an Apple phone where the OpenRoaming app was installed and the user chose to use the AppleID credentials for OpenRoaming. The phone automatically connected to the SSID named OpenRo with EAP-TTLS. (The name really does not matter; what matters is that it advertises OpenRoaming support in the Hotspot parameters.) The username indicates to the WLC admin that the client used Apple credentials to connect, but the username is obfuscated. It allows you to identify a user uniquely but gives no data with regards to its identity. On the client side, the user did not have to enter any username or credentials of any sort, and the Apple identity was retrieved from the phone.

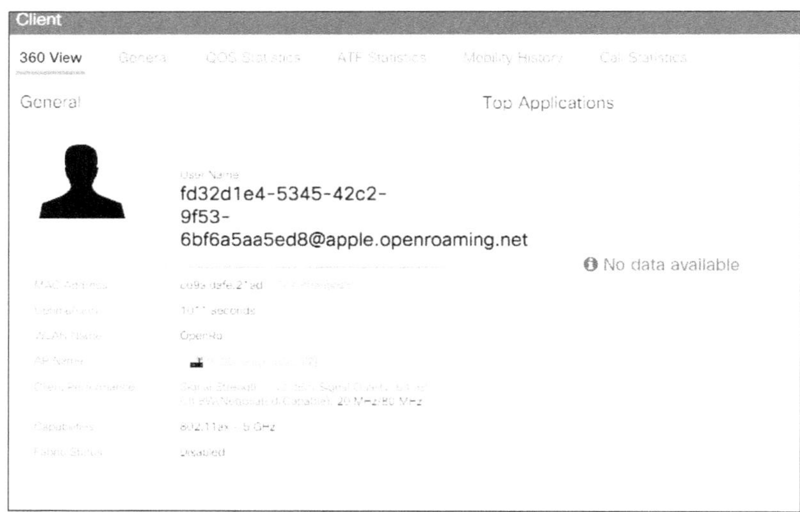

Figure 11-18 *OpenRoaming user details on the 9800 WLC*

Captive Portal

Configuration examples are available on cisco.com, showing you step by step how to configure the integration with the Catalyst 9800 WLC. As a general rule, if you want to understand the workflow, captive portals are a form of external web authentication on the controller: Layer 3 policy web authentication pointing to an external URL.

Pointing to the Cisco DNA Spaces portal URL is an easy step to configure; the only catch might be in allowing access to this portal. As with all external web authentication types, a pre-authentication ACL needs to be in place to allow traffic to the portal to go through unredirected. The Catalyst 9800 webauth parameter map asks you to specify the external portal IP address and automatically adds this IP address to a pre-authentication ACL. This is depicted in Figure 11-19.

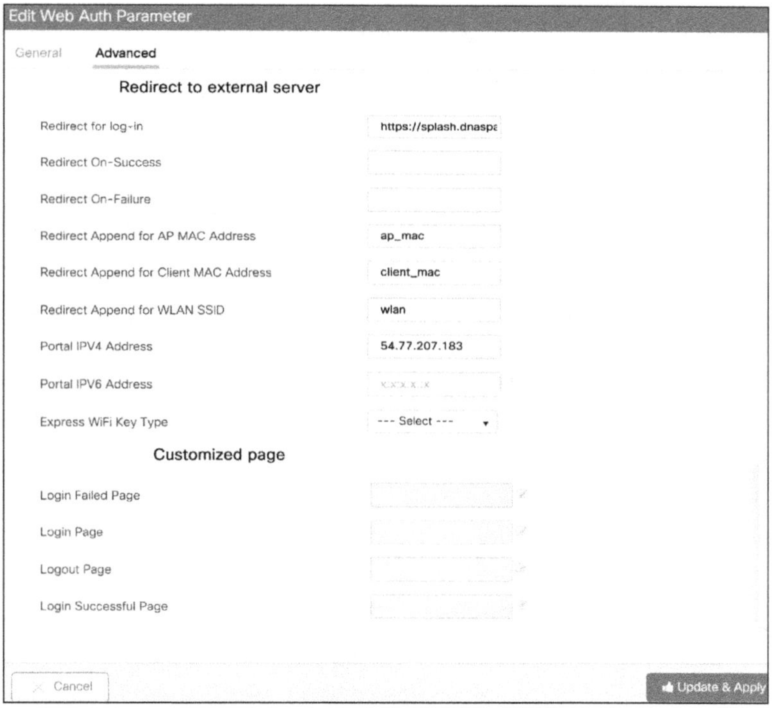

Figure 11-19 *WLC advanced webauth parameter map configuration screen*

Cisco DNA Spaces uses static IP addresses, so this can do the trick if you enter your local Cisco DNA Spaces IP address. However, Cisco DNA Spaces resolves to two different IP addresses in each region right now, and this technique does not allow you to enter the second IP address. The way to solve this issue is to use URL filters, as depicted in Figure 11-20, to allow any IP address that the Cisco DNA Spaces URL resolves to.

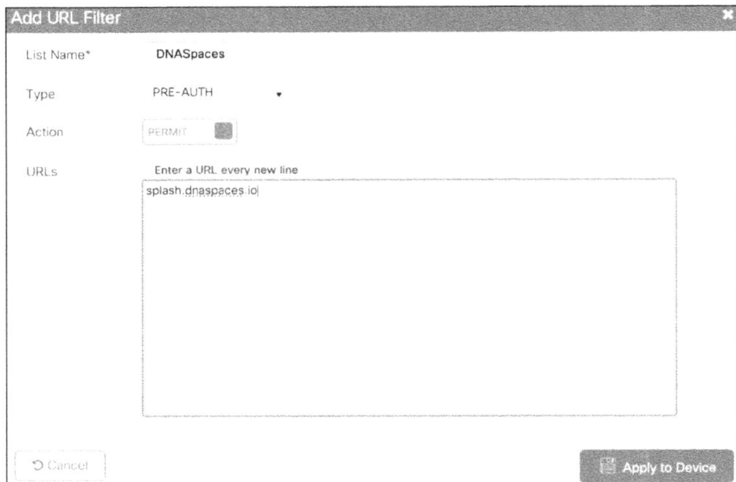

Figure 11-20 *WLC URL filter allows you to define URLs accessible in the pre-auth phase*

Whenever the WLC is not in charge of authenticating the username through RADIUS, the webauth parameter map is of type "consent," which means it is a passthrough. The authentication can be handled by the portal page itself or a third party if you are using a social network login, and the WLC has no idea of the authentication taking place until the process is over. You can also use the DNA Spaces cloud to act as a RADIUS server and define specific usernames and credentials. In this latter case, the parameter map is of type "Webauth," and the WLC intercepts the client credentials and submits them through RADIUS and waits for an access-accept message back.

Advantages of a Portal on Cisco DNA Spaces

If you have a location hierarchy defined in Cisco DNA Spaces, you have the possibility to specify specific portals depending on the AP location, all while retaining a single WLAN and unique Cisco DNA Spaces URL configuration in your controller.

You can configure social network logins such as Facebook, Twitter, or LinkedIn. Customers can enter their social credentials and get authenticated on your network. This allows you to benefit from some analytics based on their public account details.

It is possible to configure integration with an SMS gateway to send the guest credentials via SMS to the visitor.

Cisco DNA Spaces has a few ready-made portal templates that you can customize and that allow for various user experiences:

- Email authentication with data capture
- Inline SMS with password verification and data capture

- Inline social authentication
- SMS with password verification and data capture
- SMS with link verification
- Email authentication
- User agreements

You can customize your portal with ads and promos, videos, a venue map, a feedback field, and much more. You can also configure multiple languages to be available for the same portal, depending on the user browser language setting.

If you're using Cisco DNA Spaces as a RADIUS server, you can also return specific RADIUS attributes such as throughput rate-limiting, depending on the guest user group.

Many options and possibilities are available and best documented in the Cisco DNA Spaces website itself.

Proximity

Cisco DNA Spaces brings a proximity engine, which provides a lot of different reporting about not only the whereabouts of individuals but who they were in contact with. Based on the location data, which can be simple Presence or more or less accurate location coordinates, depending on your design, the engine can report and alert you if a given device was in proximity of another device for an extended (configurable) period of time. (You are tracking devices rather than people, in all fairness.) You can see in which location the device spent most of its time (in case specific devices or people are supposed to stay mostly in a dedicated spot) but also if long contacts were made with other devices (and therefore, often, with other people).

BLE Gateway on Cisco DNA Spaces

From Cisco DNA Spaces, you can configure the APs to either broadcast BLE beacons or to scan for other beacons, but both activities are mutually exclusive currently. When broadcasting BLE beacons, the APs can be configured with up to five beacon profiles provided that the profiles are of the same type (five iBeacon profiles or five Eddystone, no mix allowed). (Each profile advertises a different UUID or URL, for example, allowing you to interoperate with several different smartphone applications.)

Enabling the base BLE gateway for 11ac Wave 2 AP compatibility only allows you to broadcast beacons as well as scans, whereas enabling the advanced BLE gateway for Wi-Fi 6 and later APs only allows you to deploy an IOx APP that can either broadcast beacons or scan. It therefore mostly depends on your AP models.

You can do this in the IoT Services menu in Cisco DNA Spaces, shown in Figure 11-21, where you can see your current AP gateways and add more:

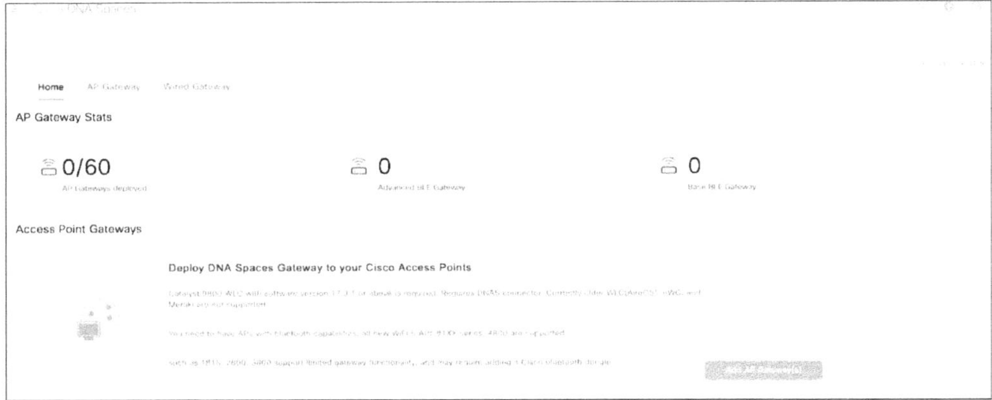

Figure 11-21 *Cisco DNA Spaces IoT Services menu*

If you click **Add**, Cisco DNA Spaces prompts you to decide which type of gateway you want to add; only BLE is available as of now, as seen in Figure 11-22. It also shows you the list of APs it can be enabled on.

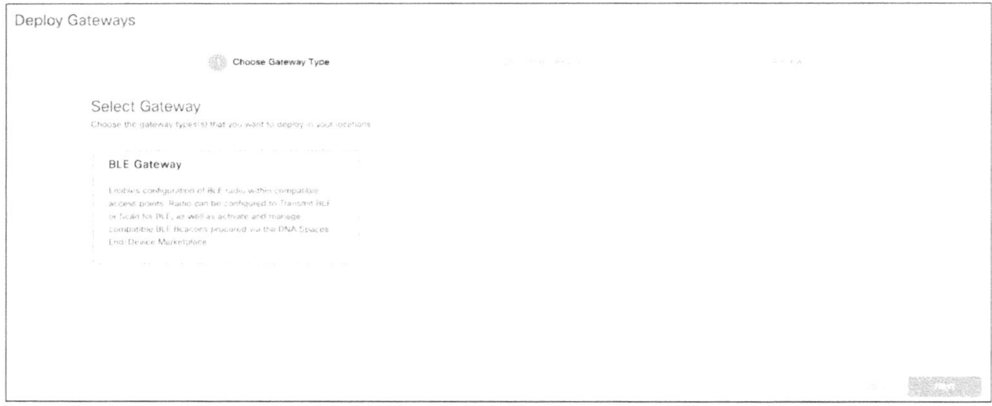

Figure 11-22 *Cisco DNA Spaces BLE gateway configuration*

When the BLE gateway is enabled on a given AP, it is possible to move the AP to scan mode, which stops the transmit mode. The scan mode, illustrated in Figure 11-23, allows each AP to detect surrounding BLE tags. If enough APs (typically four at good signal strength) are detecting a BLE tag present in an area, the BLE tag shows in the Detect and Locate application inside Cisco DNA Spaces on the map.

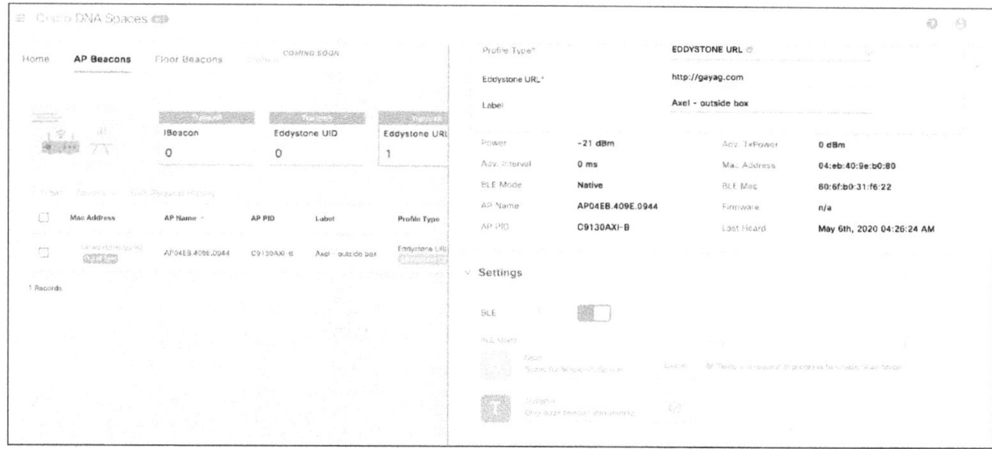

Figure 11-23 *Cisco DNA Spaces allows you to choose the BLE mode of your AP*

It is also possible to enable the advanced gateway mode, where you can deploy an application to the AP, as illustrated in Figure 11-24.

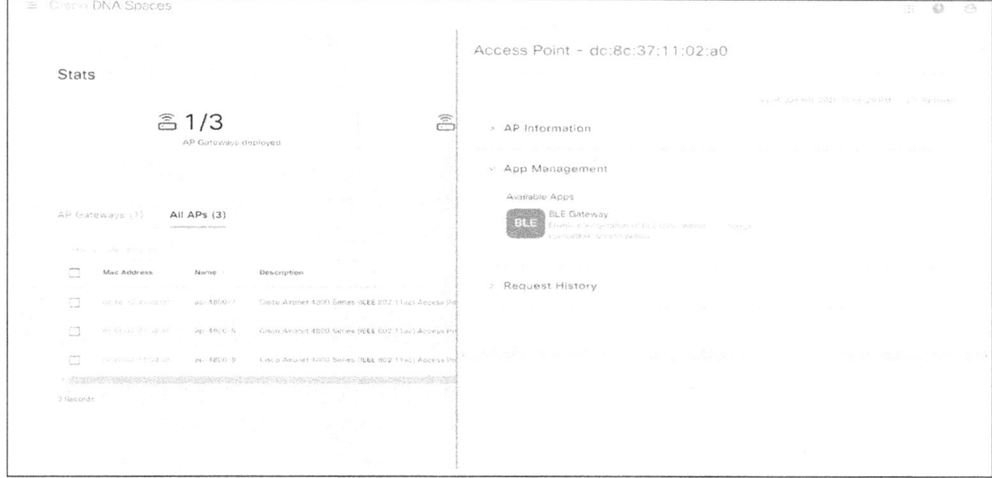

Figure 11-24 *Access Point app management in Cisco DNA Spaces*

The application allows you to connect to interoperable partners, such as Kontakt.io, beacons, and tags, and configure specific settings on these devices. This way you can have more mobile beacons and tags that you place in the relevant location in your venue and configure them centrally from your fixed infrastructure (the Catalyst APs). You can verify in the IoT services dashboard that the IoX app is properly running or check it directly on the access point with the **show iox applications** command. You can then claim the partner beacons in Cisco DNA Spaces by using the order ID that you receive. They are then associated with your tenant.

In the future, the same architecture will allow you to deploy Zigbee gateways on the APs, but that capability is not available at the time of this writing.

Summary

This chapter covers at a high level, the value-added services that the wireless network can provide. Historically, this was provided by the Mobility Solutions Engine (MSE) appliance and then the CMX appliance with the help of the NMSP protocol. Cisco DNA Spaces takes the relay of those solutions and brings many more capabilities. The Detect and Locate section covers the historical location tracking features with all its latest enhancements, the Captive Portal feature of Cisco DNA Spaces gives a new dimension to your web authentication-based WLANs, and OpenRoaming revolutionizes the way you envision a guest network. Sometimes still leveraging NMSP, Cisco DNA Spaces also uses telemetry such as gRPC to allow for new IoT use cases and flexible application hosting directly on the APs. While Cisco DNA Spaces itself is outside the scope of this book, it is important to understand how it interoperates with the WLC and APs, especially through the Cisco DNA Spaces connector, the protocols involved, as well as the possibilities offered. More than just extra services, Cisco DNA Spaces is empowering the standard network of tomorrow with OpenRoaming, for example, which can become your main go-to solution for guest networks rather than "an extra service."

References

DNA spaces installation guide: https://www.cisco.com/c/en/us/td/docs/wireless/cisco-dna-spaces/connector/config/b_connector/m_initialsetup.html

Captive portal configuration example: https://www.cisco.com/c/en/us/support/docs/wireless/dna-spaces/215423-dna-spaces-captive-portal-with-9800-cont.html

DNA Spaces configuration guide: https://www.cisco.com/c/en/us/td/docs/solutions/Enterprise/Mobility/DNA-Spaces/cisco-dna-spaces-config/dnaspaces-configuration-guide/m_preface.html

DNA Spaces direct connect configuration example: https://www.cisco.com/c/en/us/support/docs/wireless/dna-spaces/214668-dna-spaces-direct-connect-configuration.html

DNA Spaces with CMX configuration example: https://www.cisco.com/c/en/us/support/docs/wireless/dna-spaces/214688-dna-spaces-with-cmx-on-prem-configuratio.html

IoT Services guide: https://www.cisco.com/c/en/us/td/docs/wireless/cisco-dna-spaces/iot-services/b_iot_services.pdf

OpenRoaming guide: https://www.cisco.com/c/en/us/td/docs/wireless/cisco-dna-spaces/open-roaming/b-dnas-or-cg/m-config-or.html

Chapter 12

Network Programmability

What Is Network Programmability?

In this chapter, you learn how to deploy and manage the Catalyst 9800 wireless LAN controller using software. You may use open-source software to interact with the wireless LAN controller, and you can also write your own software.

Using this approach, you can tailor the 9800 WLC behavior to your organization's exact needs.

Traditionally, network devices have been configured using the command-line interface, or CLI. This interface was designed with the human in mind and, although it's certainly possible to write software to emulate human interaction, IOS-XE offers solutions better tailored to be used from software. Nevertheless, the CLI continues to be a widely used option to automate device configuration, and there are many tools that help with this task. While not optimal for automation, the CLI has the advantage of being ubiquitous and easy to learn.

Whatever method you use to update the C9800 configuration, the end goal of programmability remains the same: using software to update a device configuration instead of manually updating it.

Why Is Network Programmability Needed?

Networks and, in particular, wireless technologies have evolved rapidly and are now central to business operations. Very often, the primary method to access the network is not a wired connection, but wireless. Users expect to have a wired-like experience when connecting to the network, and it's common to use user-owned devices to connect to the corporate network (this concept is known as bring your own device, or BYOD).

Given the level of integration between the network elements, any change on one of them likely requires matching changes in others.

For example, to create a new SSID, you might need to

- Create the SSID in the wireless controller.
- Update other relevant parts on the wireless controller configuration: tags, policies, VLAN definitions.
- Update the uplink switches to allow the new VLANs to be accepted.
- Update the access point uplink switches to allow new VLANs.
- Update Spanning Tree parameters on some switches.
- Update your RADIUS server with the new parameters.

Making all these changes manually is cumbersome and prone to errors. It also makes troubleshooting hard because tracking changes can be an issue. And, after you complete all these tasks successfully and your new WLAN is active, you need to repeat the entire process each time there's a new change request.

Because humans are bad at repeating tasks, why not just write software to do all those network changes for us? That's exactly the goal of network programmability.

Network programmability can help you during all the deployment phases, as shown in Figure 12-1. This chapter focuses on Day 1 operations, using model-driven programmability. Chapter 13, "Model-Driven Telemetry," focuses on Day 2 operations, using model-driven telemetry.

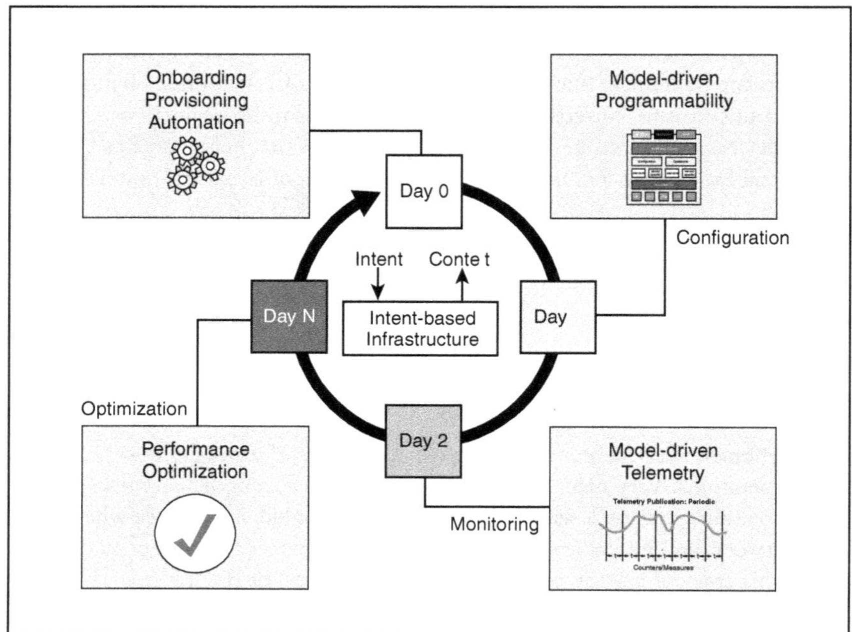

Figure 12-1 *Programmability is used along all deployment phases*

On Day 0, programmability can help with the initial device onboarding and provisioning. Instead of manually copying and pasting initial device configuration, you can write software to automatically push the right bootstrap configuration to the device, with no need for any manual intervention. Based on device serial number or MAC address, your software can dynamically create the initial configuration. This is done through any of the Day 0 protocols supported by IOS-XE, as shown in Figure 12-2:

- ZTP (zero-touch provisioning)
- PXE (pre-boot execution environment)
- PnP (Cisco network Plug and Play)

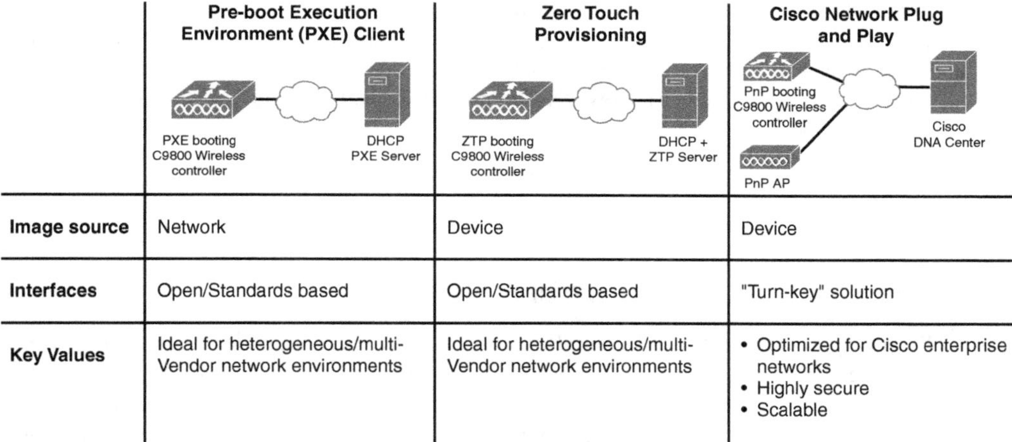

	Pre-boot Execution Environment (PXE) Client	Zero Touch Provisioning	Cisco Network Plug and Play
Image source	Network	Device	Device
Interfaces	Open/Standards based	Open/Standards based	"Turn-key" solution
Key Values	Ideal for heterogeneous/multi-Vendor network environments	Ideal for heterogeneous/multi-Vendor network environments	• Optimized for Cisco enterprise networks • Highly secure • Scalable

Figure 12-2 *Day 0 protocols*

The Day 0 protocols assume the device has an empty configuration (in out-of-the-box condition) and provide a base configuration that includes user and security configuration, IP settings, and management protocol configuration.

On Day 1, after the device has the right bootstrap configuration, you can use model-driven programmability to configure the controller using code, removing the need for manual intervention. Model-driven programmability uses a well-defined data model based in YANG, and protocols such as NETCONF and RESTCONF that configure the controller quickly and reliably.

On Day 2, you can keep track of the network behavior and performance by using model-driven telemetry, which exports key controller metrics to your software using the YANG data model and protocols like gNMI and NETCONF. You can write code that, based on the metrics received, reacts to certain events and modifies the controller configuration as needed.

Is Network Programmability a New Concept?

Given the example in the previous paragraph, you might think network programmability means you have to start writing software. That is not the case; even if you think you're not using programmability in your network, you might already be doing so.

Most likely, your wireless network has elements that are interacting with the wireless controller. You probably use Cisco DNA Center to provide visibility on the network status via DNA Center Assurance, and you might also be using DNA Center or Cisco Prime Infrastructure to do device onboarding and wireless management. Both Cisco DNA Center and Cisco Prime Infrastructure access the configuration of your Catalyst 9800 Wireless LAN Controller and eventually modify it. That is also network programmability!

There are multiple definitions of what network programmability is, but it's currently understood as a set of tools and best practices to deploy, manage, and troubleshoot network devices. It is not a new concept, and it's common to use SNMP or scripts to access the CLI of network devices. However, the protocols used were not optimal, or they were thought to be for human interaction.

This chapter focuses on the modern approach for network programmability, based on YANG data models.

Orchestration of the Entire Network

Both Cisco DNA Center and Cisco Prime Infrastructure have a wider view of the network than the wireless controller itself. Cisco DNA Center has a holistic view of the network, and you, as a network administrator, express how the network as a whole should behave. Cisco DNA Center translates that intent into a specific configuration for each particular device—for example, the wireless controller—and uses programmability to push the configuration in an automated way. By doing so, Cisco DNA Center is able to orchestrate the entire network.

Although both Cisco DNA Center and Cisco Prime Infrastructure use programmability to configure the Cisco C9800 WLC, they use different protocols to do that. In this chapter we focus on the modern protocols like NETCONF and RESTCONF and YANG-based data models. Cisco Prime Infrastructure uses a different approach using SNMP.

Configuration Repeatability

In a production network, even when the network design is frozen, there can be a continuous churn of operational changes. Users are added, access levels are updated, network policies are changed, and new devices are onboarded.

Applying such changes using manual techniques is prone to errors, especially when it needs to be done across multiple network devices, and this is an area where software excels: repeating operations and keeping track of changes.

Using network programmability, when any operational change is needed (for example, a new WLAN needs to be created), you just use a piece of software to do so. That software connects to the network devices, applies the desired changes, and keeps track of the changes.

Idempotency

One important aspect to consider when designing a programmability strategy is what are the possible effects of running a particular change multiple times.

As an example, in your network you might need to control the available bandwidth associated with a particular user. To do so, you decide to write code that increases the bandwidth by 1 MBps.

If, for any reason (network communication issues or faulty software), part of your code is executed several times, you end up increasing the user bandwidth more than expected. This is clearly not what was intended.

When a network programmability solution is designed, it's important to make sure the execution is *idempotent*. That means it can be called one or more times, and the result is equivalent. In this example, the right approach would be writing code that sets the appropriate value, instead of incrementing the existing one.

Imperative vs. Declarative Models

Although your network could be fully automated, in practice there is always a mix of both manual changes and programmability. You might have scripts or other software to perform the most common changes in your network, but in some cases no software is available to cover the particular change you need, or the wireless controller configuration is manually updated to troubleshoot an issue.

Moreover, when the configuration is automated, it is possible there are failures on pushing the configuration due to software errors, network issues, and so on.

This scenario brings a new problem: The actual device configuration might diverge from the expected configuration.

Let's assume you're writing code to manage the set of client VLANs that are allowed on the C9800's uplink trunk interface. You are keeping the list of allowed client VLANs on a file or database, and you want to keep the VLANs on the trunk matching your database.

A first approach could be as depicted in Figure 12-3.

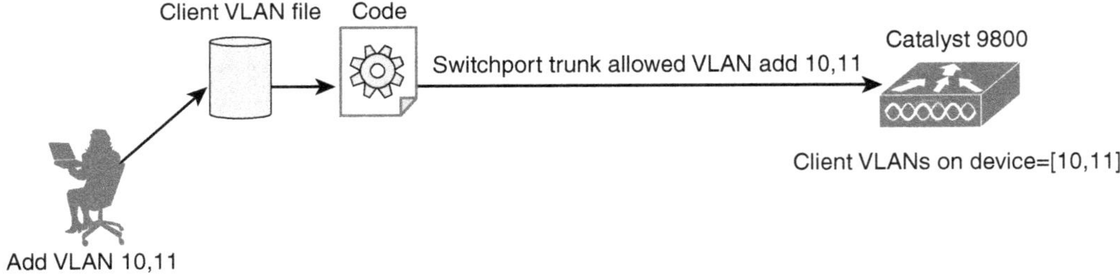

Figure 12-3 *Adding client VLANs using the imperative model*

On this first approach, the code just adds the required VLANs in the allowed list on the interface. If a network administrator logs manually on the device and manually adds more VLANs, there is data inconsistency: the VLANs in the device don't match the ones in the database, as depicted in Figure 12-4.

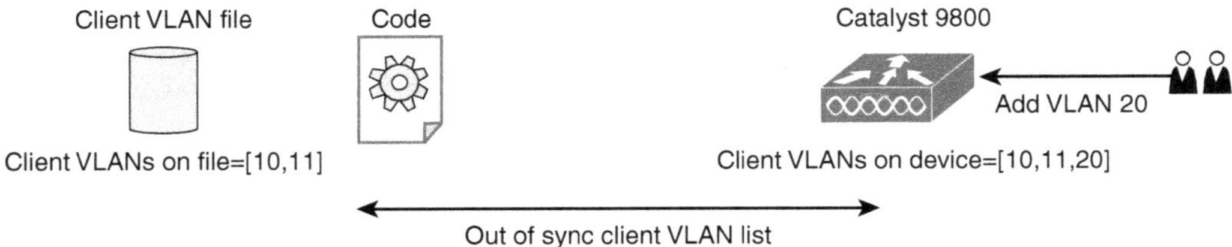

Figure 12-4 *Inconsistency by using imperative models*

This out-of-sync situation can be difficult to detect, and can also happen due to communication errors, as illustrated in Figure 12-5.

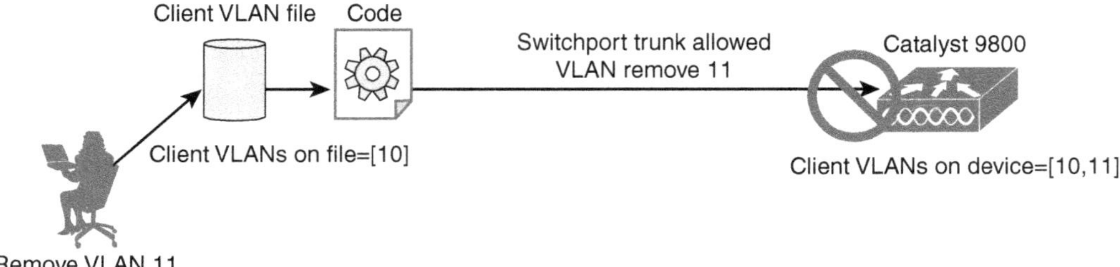

Figure 12-5 *Errors when using imperative models*

Although the code could report an error if this communication error is detected, it's not clear what the administrator can do at that point. Was the VLAN fully removed? There's no way to tell for sure.

The cause of this inconsistency was that your first attempt to automate was done using an *imperative model*, also called an *open-loop system*. Your code was telling the device what to do, but nothing in the system was aware of the actual device status.

A better approach is to use a *declarative model*, also called a *closed-loop system*. With the declarative approach, you don't tell the system what to do. You tell the system the desired state, and the code performs the changes needed to reach that condition.

A declarative, or closed-loop, system works by having the software in a constant loop, as illustrated in Figure 12-6, checking for differences between the desired state and the actual state.

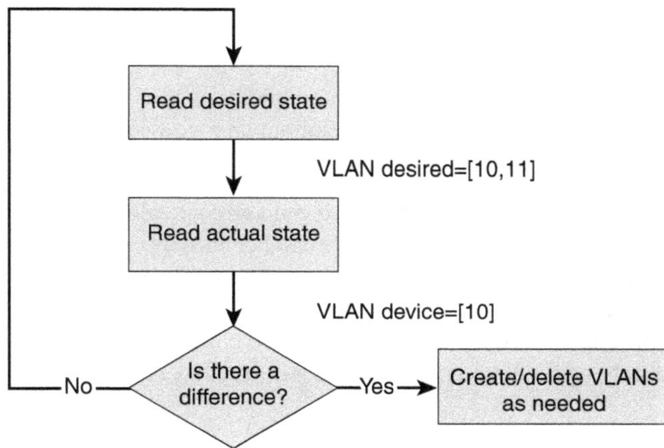

Figure 12-6 *Declarative (closed-loop) model*

Applying this approach to the VLAN example, the system would work as shown in Figure 12-7.

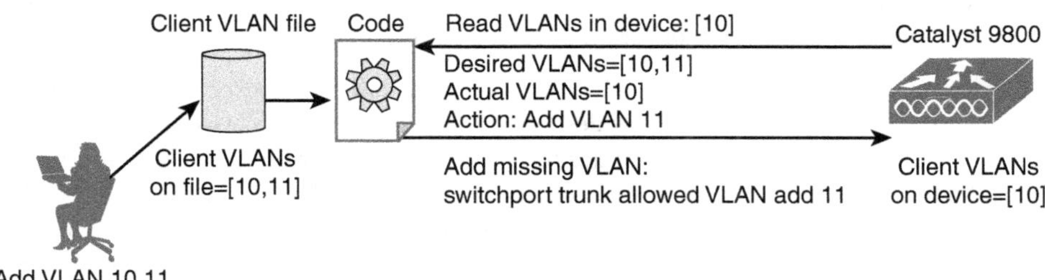

Figure 12-7 *Updating data using declarative models*

If anybody changes the device, the system notices that situation and autocorrects it, as shown in Figure 12-8.

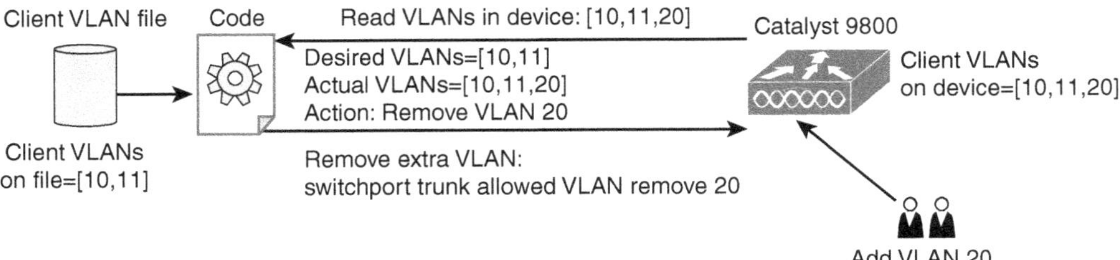

Figure 12-8 *Automatic correction using declarative models*

Infrastructure as Code (IaC)

A core aspect of managing the network devices using programmability is how to store the desired configuration. If you're not using network programmability, you are probably keeping the network documentation on diagrams or spreadsheets.

One advantage of using network programmability is that you can automatically keep in sync the device configuration with the diagrams or spreadsheets. For this approach to be successful, you need to have a versioning strategy for your network documents. You need to keep track of who changed what, and what differences happened between one version and the next. This is the same problem that happens in the software industry, where each program file needs to be accurately tracked, and the software industry has created tools for that purpose.

The purpose of Infrastructure as Code is to manage the infrastructure (networks, connection topology information) using a descriptive model and the same principles and tooling as developers use for the source code. For example, you can keep a list of the WLANs needed in each of the locations in a repository.

With IaC, you would store the WLAN list in a version control system. The fundamental difference with other workflows is that the desired network state is not directly pushed to the wireless controller by a human or script, but instead it is sent to the version control system.

The version control system, when a change is detected, can trigger execution of network scripts that configure the network as described in the new desired network state. Using IaC, the version control system becomes the source of truth for the network configuration. All team members can see it and examine the changes. Changes can also be tested on a test environment before being pushed to production. It is also possible to perform code reviews, where team members need to review and approve the changes before they are pushed to the network. If anything goes wrong, it is also easy to revert to the previous situation.

This capability enhances both repeatability and testability. You consider every network change as a new version that is tested and approved before being released.

When this approach is used along with a declarative model, your network configuration depends only on the changes done using your version control system. If the new release causes any trouble, you are able to revert to the previous state simply by reverting to the previous configuration version, as you would revert to a previous software version.

Network Programmability in the C9800

The goal of network programmability is to share data from the network devices toward applications and back.

To achieve this data transfer in an effective way, you need to define several elements:

- **Data schema:** Describes which data is available and its hierarchy.
- **Encoding:** Describes how the data is represented. For example, it describes whether it is sent in binary or text format, how each value is separated from the other, and so on.
- **Protocol:** Describes the rules defining available operations. For example, how can you ask for data? How do you indicate the data you are interested in?
- **Transport:** The data, represented by the chosen encoding format, needs to be sent over the network. The transport defines how this is done. For example, is it over TCP or UDP? Is there any transport layer encryption? Which ports are used?
- **Application:** This is the code in the computer. It can be a small, self-written application or a full-fledged or vendor-provided application like Cisco DNA Center.

Not all possible combinations of encoding, protocol, and transport are available. Each protocol (NETCONF, RESTCONF, gNMI) defines which encoding and transport are available. This is represented on Figure 12-9.

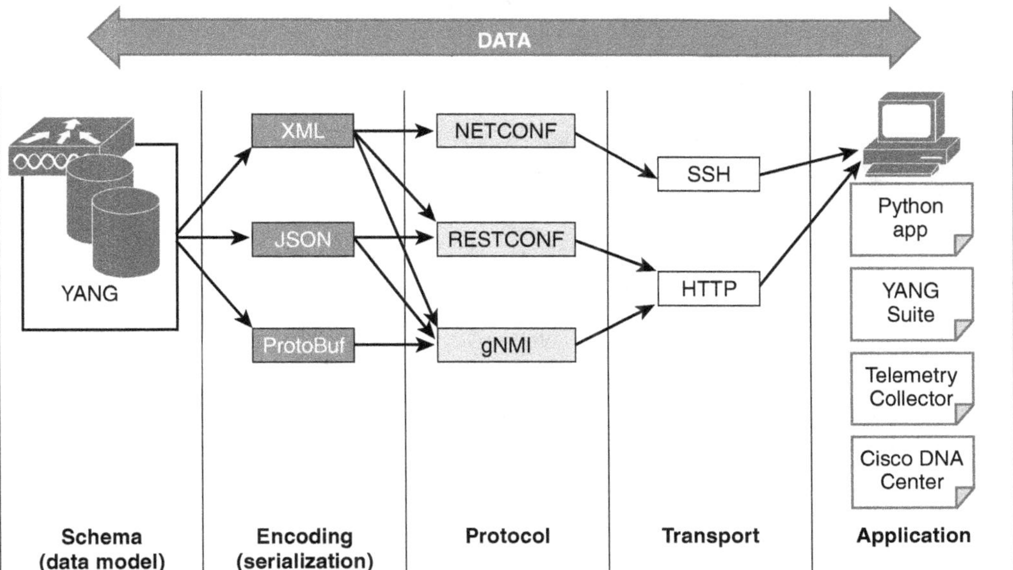

Figure 12-9 *Protocols, encoding, and transport*

Data Models

All the data inside the Catalyst 9800 wireless LAN controller is stored on two internal databases, called the configuration database and operational database.

The configuration database holds the parameters that define how you want the controller to operate (what is commonly called the *device configuration*). In general, the elements in this database allow read and write access.

The operational database holds the wireless controller status. For example, it holds the number of connected clients, uptime, and bytes sent and received on each interface. Because the data in this database represents the controller real-time status, you can read it but not write.

You can access the configuration and operational databases using a variety of protocols:

- NETCONF
- RESTCONF
- gRPC
- WebUI
- SNMP
- CLI

Each protocol is best tailored to some use cases, and part of defining the programmability strategy is to choose the best protocol for the use case. However, at the end of the day, all the protocols just populate the configuration database with the desired values or read them. This is represented in Figure 12-10.

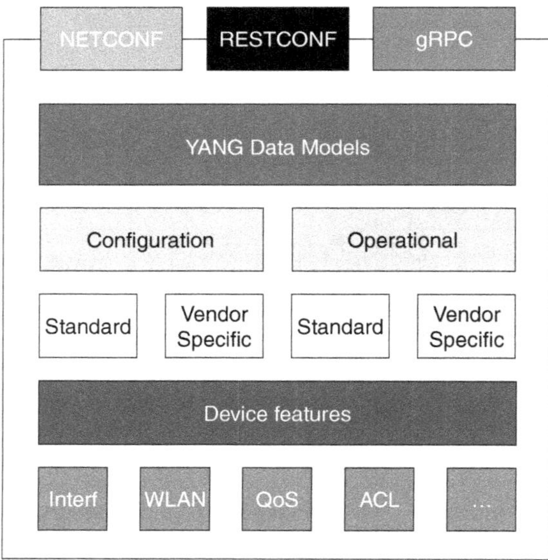

Figure 12-10 *Data model and protocol hierarchy*

To access the configuration or operational databases, you need to define how the data is organized. This data organization is called the *data model*.

Imagine you need to configure the IP address of the wireless management interface of the Catalyst 9800. The data model reflects that the IP address is a property of another element called interface. Also, the data model reflects whether more than one IP address is allowed, which other attributes are part of an interface, and so on. The data model is defined using the YANG language. YANG is defined in RFC 6020 and RFC 7950 and has some key capabilities that make it ideal for this purpose:

- Human readable
- Hierarchical configuration data models
- Structured types, using reusable types and grouping
- Extensibility
- Constraints for configuration validation
- Modularity, using modules and submodules
- Versioning

Although internally the databases have a fixed data model, IOS-XE allows having different *views* of the same data, each view with different data models.

YANG models are defined either by the vendor or by a different organization—for example, IETF, IEEE, or OpenConfig. The same device supports multiple YANG models, even for the same data.

The models defined by the vendor generally have the richest amount of details. However, the model is specific to the vendor and can even change between software releases.

OpenConfig models, on the other hand, are designed to be independent of the underlying platform and compatible between vendors. By using OpenConfig models, you can use the same code to configure devices from multiple vendors, platforms, and software releases.

The best place to see which models are supported on a particular version is the YANG model's GitHub repository at https://github.com/YangModels/yang/tree/master/vendor/cisco/xe.

YANG Data Models

YANG provides a standard way to describe and model the operational and configuration data of network devices. It is not tied to any particular protocol or encoding because it's just a language describing the data model.

YANG models define a hierarchy of nodes. The nodes are organized in modules, containers, groupings, and lists, among other structures.

The module is the base unit of definition in YANG and defines a single data model. Modules include revision numbers, which allow you to manage dependencies from other modules. For example, a module can import a particular revision of another module.

Modules also include other submodules, can be augmented by another data model (using a feature called *augmentation*), and support deviations: a particular release might not support all features contained in the configuration data model or could support them but with different limits. Any application configuring the unsupported feature receives an error. Although this situation could be handled by the application, it would be better for the application to know in advance the feature is not supported.

For example, in the XE switch port YANG module (see https://github.com/YangModels/yang/blob/master/vendor/cisco/xe/1751/Cisco-IOS-XE-switch.yang), the number of secure MAC addresses supported is 4097:

```
list maxcount {
  description
    "Max secure addresses";
  key "max-addresses";
  leaf max-addresses {
    type uint16 {
      range "1..4097";
    }
  }
```

However, on the 9200L platform, only 4096 are supported. To make this limitation explicit, the 9200L platform adds the following deviation (see https://github.com/YangModels/yang/blob/master/vendor/cisco/xe/1751/Cisco-IOS-XE-switch-deviation.yang):

```
deviation "/ios:native/ios:interface/ios:Port-channel/ios:switchport-
config/ios:switchport/ios-sw:port-security-config/ios-sw:port-secu-
rity/ios-sw:maxcount/ios-sw:max-addresses" {
    deviate replace {
      type uint16 {
        range "1..4096";
      }
    }
}
```

Inside a module, as depicted in Figure 12-11, data is organized in

- **Containers:** Nodes in the tree that enclose a set of child elements.
- **Leaves:** Nodes in the tree that contain a value but no child elements.
- **Lists:** Tables containing a set of entries. Each entry can contain leaves, containers, and other elements.

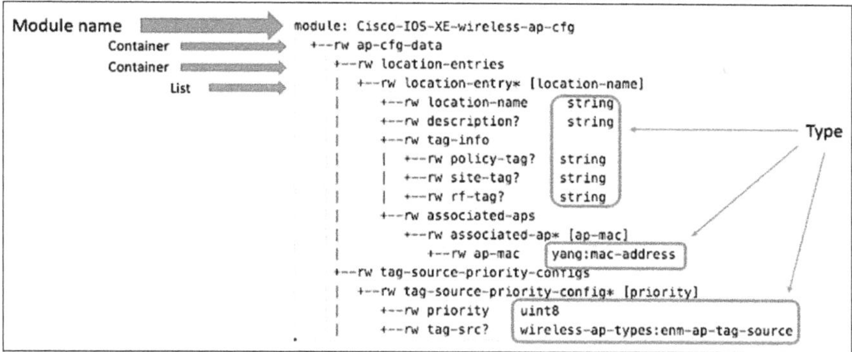

Figure 12-11 *YANG module structure*

Each element has a particular type. YANG defines a few base types (int, binary, boolean, and so on), and additional types are defined using the **typedef** keyword. For example, to describe the AP radio state, IOS-XE defines the type **enm-ewlc-ap-radio-modes** as shown in Example 12-1.

Example 12-1 *YANG Type for AP Radio Modes*

```
typedef enm-ewlc-ap-radio-modes {
  type enumeration {
    enum radio-mode-invalid {
      value 0;
      description
        "Radio mode Invalid";
    }
    enum radio-mode-local {
      value 1;
      description
        "Radio mode Local";
    }
    enum radio-mode-monitor {
      value 2;
      description
        "Radio mode Monitor";
    }
    enum radio-mode-flex-connect {
      value 3;
      description
        "Radio mode FlexConnect";
    }
```

Then on the data model for AP radio, the mode is described as

```
leaf radio-mode {
      type wireless-types:enm-ewlc-ap-radio-modes;
}
```

While the YANG models are human readable, browsing them is tedious because a particular model can be distributed in several files. There are multiple tools that allow you to present YANG models in an easier format, like pyang or YANG Suite.

Encoding Formats

The client and server need to agree on how to represent the values being transmitted. This is the task of the data encoding format. Some common formats are XML, YAML, and JSON. The data is represented as key-value pairs, and the data encoding format describes how to describe both.

XML

The Extensible Markup Language describes the data using text, using tags in a similar way to HTML. The tags create the keys, and the content inside the tags represents the value. An important characteristic of XML is the support of *schemas*: a way to describe the format and possible values of a document. A schema can be described using another XML document, called an XML Schema Definition, or XSD.

XML is the encoding format used by NETCONF. For instance, Example 12-2 shows the YANG model for WLANs, represented as XML.

Example 12-2 *WLAN YANG Model Represented as XML*

```
<wlan-cfg-entries>
    <wlan-cfg-entry>
        <profile-name>test_wlan_1</profile-name>
        <wlan-id>1</wlan-id>
        <wep-key-index>1</wep-key-index>
        <multicast-buffer-value>0</multicast-buffer-value>
        <apf-vap-id-data>
            <ssid>test_wlan_1</ssid>
        </apf-vap-id-data>
    </wlan-cfg-entry>
    <wlan-cfg-entry>
        <profile-name>test_wlan_2</profile-name>
        <wlan-id>2</wlan-id>
        <wep-key-index>1</wep-key-index>
        <multicast-buffer-value>0</multicast-buffer-value>
        <apf-vap-id-data>
```

```
            <ssid>test_wlan_2</ssid>
            <wlan-status>true</wlan-status>
        </apf-vap-id-data>
    </wlan-cfg-entry>
</wlan-cfg-entries>
```

JSON

JavaScript Object Notation is the native format to describe objects in JavaScript. Because JSON is part of a programming language, it's designed to be human readable. It is standardized in RFC 8259.

Each object always has a key and a value, and is described by following a set of simple rules:

- Objects are enclosed in curly braces.
- Keys and values are separated by a colon.
- Keys are always strings (in double quotes).
- Values can be any of the following:
  - Strings
  - Numbers
  - Boolean (true or false)
  - Null
  - Array (a list of values, separated by commas and enclosed in square brackets)
  - Other JSON object

JSON can be used by RESTCONF. For instance, Example 12-3 shows the YANG model for WLANs, represented as JSON.

Example 12-3 *WLAN YANG Model Represented as JSON*

```
{
    "Cisco-IOS-XE-wireless-wlan-cfg:wlan-cfg-entries": {
        "wlan-cfg-entry": [
            {
                "profile-name": "test_wlan_1",
                "wlan-id": 1,
               "wep-key-index": 1,
                "multicast-buffer-value": 0,
```

```
                "apf-vap-id-data": {
                    "ssid": "test_wlan_1"
                }
            },
            {
                "profile-name": "test_wlan_2",
                "wlan-id": 2,
                "wep-key-index": 1,
                "multicast-buffer-value": 0,
                "apf-vap-id-data": {
                    "ssid": "test_wlan_2",
                    "wlan-status": true
                }
            }
        ]
    }
}
```


This object has a key—**"Cisco-IOS-XE-wireless-wlan-cfg:wlan-cfg-entries"**—and the value is another object. The key for this second object is **"wlan-cfg-entry"** and the value is an array. Each of the elements of the array is a WLAN and has a set of keys and values. For the first WLAN, the key **'profile-name'** has the value **'test_wlan_1'**, the key **'wlan-id'** has the value of 1, and so on.

Protobuf

In contrast with JSON and XML, which are text-based protocols, Protobuf is a binary message format. It allows programmers to specify a schema for the data and includes a set of rules and tools to define the messages. The schema is written in a specific language and defines how the data is encoded.

Because it's a binary protocol, Protobuf is very efficient. However, it's not a self-describing protocol. Both sides of the communication need to agree on the meaning of each message before they can understand the message content.

Protocols

The protocol describes the possible operations on a network device: How can you ask for data? How do you indicate the data you are interested in?

Each protocol has different strengths and weaknesses. Depending on the situation, you might prefer to use RESTCONF due to its ease of use and support for JSON, or you might want to use NETCONF because it supports transactions that are needed in your project.

NETCONF

NETCONF is a network management protocol designed with programmability in mind, standardized by the IETF and published as RFC 6241. Applications can use NETCONF APIs to send, receive, and subscribe to full and partial data sets. It uses XML as encoding and generally uses SSH as transport, but other transports could be possible. NETCONF could use any transport protocol that supports a set of basic requirements like being connection oriented and providing authentication, integrity, and confidentiality.

Additionally, RFC 5277 defines an asynchronous notification mechanism, allowing clients (code using NETCONF to connect to the Catalyst 9800) to subscribe to event streams. The client defines the subset of data it is interested in, and the network device sends notifications anytime the data changes.

NETCONF is based on the remote procedure call (RPC) paradigm: the client sends a series of one or more request messages, encoded as XML, and the server replies with a set of XML responses.

NETCONF Capabilities

When a client connects to a device using NETCONF, the first messages that are exchanged are a set of capabilities that both the client and device support. The capabilities include basic NETCONF functionality (whether notifications or validations are supported) and the YANG models supported by the device.

When a NETCONF client connects to a C9800 Wireless LAN Controller, this is an example of the set of capabilities that the controller advertises, enclosed under a **<hello>** element, as shown in Figure 12-12.

Capabilities

YANG modules

```
<capabilities>
  <capability>urn:ietf:params:netconf:base:1.0</capability>
  <capability>urn:ietf:params:netconf:capability:writable-running:1.0</capability>
  <capability>urn:ietf:params:netconf:capability:rollback-on-error:1.0</capability>
  <capability>urn:ietf:params:netconf:capability:notification:1.0</capability>
  <capability>urn:ietf:params:netconf:capability:yang-library:1.0?revision=2016-06-21</capability>
  <capability>http://cisco.com/ns/cisco-xe-deviation?module=cisco-xe-deviation;revision=2016-08-10</capability>
  <capability>http://cisco.com/ns/cisco-xe-routing?module=cisco-xe-unicast-routing;revision=2015-09-11</capability>
...
</capabilities>
```

Figure 12-12 *NETCONF capabilities*

NETCONF Layers

NETCONF can be conceptually partitioned into four layers, as shown in Figure 12-13.

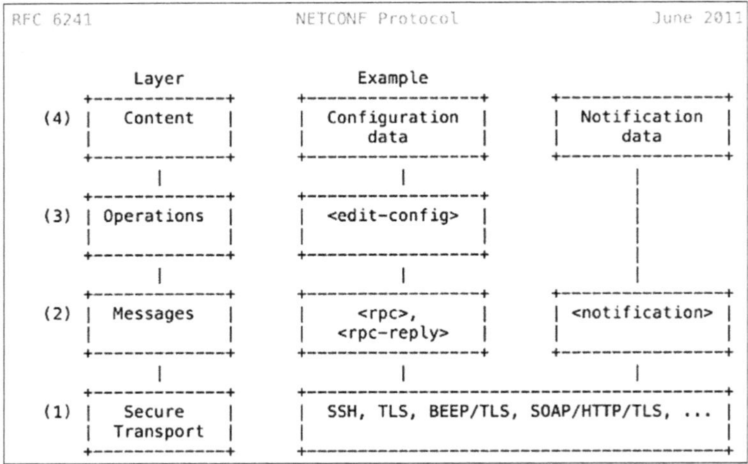

```
RFC 6241                    NETCONF Protocol                   June 2011

              Layer                 Example
         +-------------+      +-----------------+    +----------------+
     (4) |   Content   |      |  Configuration  |    |  Notification  |
         |             |      |      data       |    |      data      |
         +-------------+      +-----------------+    +----------------+
                |                      |                      |
         +-------------+      +-----------------+             |
     (3) | Operations  |      |  <edit-config>  |             |
         |             |      |                 |             |
         +-------------+      +-----------------+             |
                |                      |                      |
         +-------------+      +-----------------+    +----------------+
     (2) |  Messages   |      |     <rpc>,      |    | <notification> |
         |             |      |   <rpc-reply>   |    |                |
         +-------------+      +-----------------+    +----------------+
                |                      |                      |
         +-------------+      +-----------------------------------------+
     (1) |   Secure    |      |  SSH, TLS, BEEP/TLS, SOAP/HTTP/TLS, ... |
         |  Transport  |      |                                         |
         +-------------+      +-----------------------------------------+
```

Figure 12-13 *NETCONF layers*

Content Layer

The content layer contains the actual data content from a query or response. The valid values are defined by the YANG data model.

To specify the data you are interested in, you can use either subtrees or XML XPaths. A full example of a NETCONF data request is shown later in this chapter in Figure 12-20.

Operations Layer

NETCONF provides a base set of operations to manage the device. Using capabilities, a device can also define additional operations supported. The operations defined by NETCONF include

- **get:** Retrieves data (either configuration or operational) from the device.
- **get-config:** Retrieves configuration data.
- **edit-config:** Updates configuration data. The client can request whether the new configuration should be merged or replaced.
- **delete-config:** Deletes the configuration.
- **lock/unlock:** Because multiple clients could be accessing the network device at the same time, a client can lock a set of data, preventing any other session from updating it.
- **commit:** This operation marks the configuration edit as complete. A commit can contain different edits, and the network device needs either to accept all the changes in the commit, or reject all of them, leaving the configuration unchanged.

Messages Layer

The messages layer specifies the type of message: request, reply, notification, and so on.

Secure Transport Layer

The secure transport layer specifies the transport used. SSH is mandatory and most used, but other options are possible.

RESTCONF

RESTCONF is a proposed standard defined in RFC 8040. Similar to NETCONF, it allows you to query or modify information on the network device, but they differ in several key aspects:

- It is based on HTTP. It uses HTTP verbs such as GET, POST, PUT, and DELETE.
- You can choose either XML or JSON as the encoding schema.
- RESTCONF does not support locking or transactions.

HTTP Methods

RESTCONF is a stateless protocol, in contrast with NETCONF. It uses HTTP methods to provide CREATE, READ, UPDATE, and DELETE operations on the device configuration and state, as shown in Table 12-1.

Table 12-1 *HTTP Methods Used in RESTCONF*

HTTP Method	Usage
GET	Read
PATCH	Update
PUT	Create or replace
POST	Create or operations RPCs
DELETE	Delete
HEAD	Header metadata

HTTP Return Codes

For each request, the network device returns a status code to indicate the outcome of the requested operation. HTTP return codes are grouped in five classes:

- **Informational:** 100–199.
- **Success:** 200–299; they indicate the request was successful.
- **Redirect:** 300–399; they indicate the client should retry to a different address.

- **Client error:** 400–499; they indicate an error was caused by some issue on the request (lack of authorization, resource does not exist, and so on).
- **Server error:** 500–599; they indicate an internal error on the network device.

gNMI/gRPC

gNMI is a protocol created by the OpenConfig consortium. It leverages the open-source gRPC framework.

gNMI describes a set of operations (RPCs):

- **Set:** Writes data to the device.
- **Subscribe:** Requests that the device send data. The subscription can be
  - **Stream:** Device pushes data. The device can push the data periodically or based on changes.
  - **Poll:** Client requests data.
  - **Once:** Device pushes data once and closes the connection.
- **Get:** Takes a snapshot of data at a particular time.
- **Capabilities:** Describes supported models and encoding.

gNMI and gRPC are covered in detail in Chapter 13.

Tools to Examine YANG Models

Although the YANG data modeling language is human-readable, there are multiple tools to examine YANG models easily. Two of the most popular ones are pyang and YANG Suite.

pyang

The popular Python utility pyang created by Martin Bjorklund allows you to inspect YANG models. It can be used to validate YANG modules for correctness, to transform YANG modules into other formats, and to write plug-ins to generate code from the modules.

You can download pyang from its GitHub repository: https://github.com/mbj4668/pyang.

Using pyang in a Docker Container

If you prefer not to install pyang on your computer, you can use a Docker container with the tool and dependencies installed by cloning the repository: https://github.com/fsedano/pyang.git.

After you clone the repository, you can start it and clone the repository with the YANG models for the IOS XE software release you're running, as shown in Example 12-4.

Example 12-4 *Using pyang in a Docker Container*

```
$ git clone https://github.com/fsedano/pyang.git
Cloning into 'pyang'...
…
$ cd pyang
$ docker-compose build
Building pyang
…
Successfully built 43a7e7658668
$ docker-compose up -d
$ docker-compose exec pyang bash
root@3d0eeb6db02e:/app#
```

At this point, you are inside the Docker container with pyang ready to be used.

To obtain the YANG models for the wireless controller, you can clone the official Git repository to download them: https://github.com/YangModels/yang.git.

```
root@3d0eeb6db02e:/app# git clone https://github.com/YangModels/yang.git
Cloning into 'yang'...
…
Checking out files: 100% (44617/44617), done.
root@3d0eeb6db02e:/app#
```

And you can examine them by using the pyang **tree** option, as shown in Example 12-5.

Example 12-5 *Examining a YANG Model Using pyang*

```
root@3d0eeb6db02e:/app# cd yang/vendor/cisco/xe/1751
root@3d0eeb6db02e:/app/yang/vendor/cisco/xe/1751# pyang -f tree Cisco-IOS-XE-
  wireless-ap-cfg.yang

module: Cisco-IOS-XE-wireless-ap-cfg
  +--rw ap-cfg-data
     +--rw location-entries
     |  +--rw location-entry* [location-name]
     |     +--rw location-name     string
     |     +--rw description?      string
     |     +--rw tag-info
```

```
 |     |  +--rw policy-tag?   string
 |     |  +--rw site-tag?     string
 |     |  +--rw rf-tag?       string
 |     +--rw associated-aps
 |        +--rw associated-ap* [ap-mac]
 |           +--rw ap-mac    yang:mac-address
 +--rw tag-source-priority-configs
 |  +--rw tag-source-priority-config* [priority]
 |     +--rw priority    uint8
 |     +--rw tag-src?    wireless-ap-types:enm-ap-tag-source
 +--rw ap-filter-configs
 |  +--rw ap-filter-config* [filter-name]
 |     +--rw filter-name        string
 |     +--rw filter-string?     string
 |     +--rw filter-priority?   uint8
 |     +--rw apply-tag-list
 |        +--rw policy-tag?   string
 |        +--rw site-tag?     string
 |        +--rw rf-tag?       string
 o--rw ap-filter-priority-cfg-entries
 |  +--rw ap-filter-priority-cfg-entry* [priority]
 |     +--rw priority       uint8
 |     +--rw filter-name?   string
 +--rw ap-rule-priority-cfg-entries
 |  +--rw ap-rule-priority-cfg-entry* [priority]
 |     +--rw priority       uint32
 |     +--rw filter-name?   string
 +--rw ap-tags
    +--rw ap-tag* [ap-mac]
       +--rw ap-mac        yang:mac-address
       +--rw policy-tag?   string
       +--rw site-tag?     string
       +--rw rf-tag?       string
```

YANG Suite

Another option to examine YANG models is YANG Suite. You can find it at https://github.com/CiscoDevNet/yangsuite.

To use it, you can clone the repository, generate certificates for HTTPS secure connection, and bring it up using Docker. The first time you start it, you need to run **start_yang_suite.sh** to generate the required certificates. For subsequent restarts, you can simply run **docker-compose up -d**. The entire process is shown in Example 12-6. Check the YANG Suite repository README.md file to get the most up-to-date instructions.

Example 12-6 *Running YANG Suite Using Docker Compose*

```
$ git clone https://github.com/CiscoDevNet/yangsuite.git
Cloning into 'yangsuite'...
remote: Enumerating objects: 770, done.
remote: Counting objects: 100% (770/770), done.
remote: Compressing objects: 100% (528/528), done.
remote: Total 770 (delta 306), reused 636 (delta 225), pack-reused 0
Receiving objects: 100% (770/770), 25.23 MiB | 7.34 MiB/s, done.
Resolving deltas: 100% (306/306), done.

$ cd yangsuite/docker
$ ./start_yang_suite.sh
Hello, please setup YANG Suite admin user.
username: admin
password:
confirm password:
email: fran@fransedano.net

Setup test certificates? (y/n): y
################################################################
## Generating self-signed certificates...                     ##
##                                                            ##
## WARNING: Obtain certificates from a trusted authority!     ##
##                                                            ##
## NOTE: Some browsers may still reject these certificates!!  ##
################################################################

Generating a 2048 bit RSA private key
....................................................................................
  ....................................................+++
......+++
writing new private key to 'nginx/nginx-self-signed.key'
-----
You are about to be asked to enter information that will be incorporated
into your certificate request.
What you are about to enter is what is called a Distinguished Name or a DN.
There are quite a few fields but you can leave some blank
For some fields there will be a default value,
If you enter '.', the field will be left blank.
-----
Country Name (2 letter code) []:ES
State or Province Name (full name) []:Madrid
Locality Name (eg, city) []:Madrid
```

```
Organization Name (eg, company) []:Cisco
Organizational Unit Name (eg, section) []:
Common Name (eg, fully qualified host name) []:localhost
Email Address []:
Certificates generated...
Building docker containers...
[+] Building 123.0s (45/45) FINISHED
 => [yangsuite:latest internal] load build definition from dockerfile
 => transferring dockerfile:
 .....
yangsuite_1  | *** WARNING: you are running uWSGI as root !!! (use the --uid flag)
  ***
yangsuite_1  | *** uWSGI is running in multiple interpreter mode ***
yangsuite_1  | spawned uWSGI master process (pid: 35)
yangsuite_1  | spawned uWSGI worker 1 (pid: 39, cores: 1)
yangsuite_1  | spawned uWSGI worker 2 (pid: 40, cores: 1)
yangsuite_1  | spawned uWSGI worker 3 (pid: 41, cores: 1)
yangsuite_1  | spawned uWSGI worker 4 (pid: 42, cores: 1)
yangsuite_1  | spawned uWSGI worker 5 (pid: 43, cores: 1)
```

At this point, YANG Suite is up and running, and you can access it on your browser using https://127.0.0.1:8443.

YANG Suite uses NETCONF to connect to the device, so you need to enable it on the wireless controller before adding it to YANG Suite.

To enable NETCONF, you need to use **netconf-yang** CLI on the wireless controller:

```
WLC(config)#netconf-yang
```

Something to take into consideration is that NETCONF uses SSH as transport, and the authentication uses the default AAA method. If you are using AAA, you need to define the correct authentication and authorization lists for the default method:

```
WLC(config)#aaa authorization exec default local
WLC(config)#aaa authentication login default local
```

When NETCONF is enabled on the controller, you can start using YANG Suite.

The steps to get started with YANG Suite are

1. Add your devices to YANG Suite, using the Device Profiles page.
2. Create a YANG repository. For example, you can create a YANG repository with all the models for a given IOS-XE release.
3. Add modules to the repository, either from Git, a local disk, or the controller using NETCONF. Because YANG modules reference other modules, YANG Explorer warns you about any missing dependency, as shown in Figure 12-14.

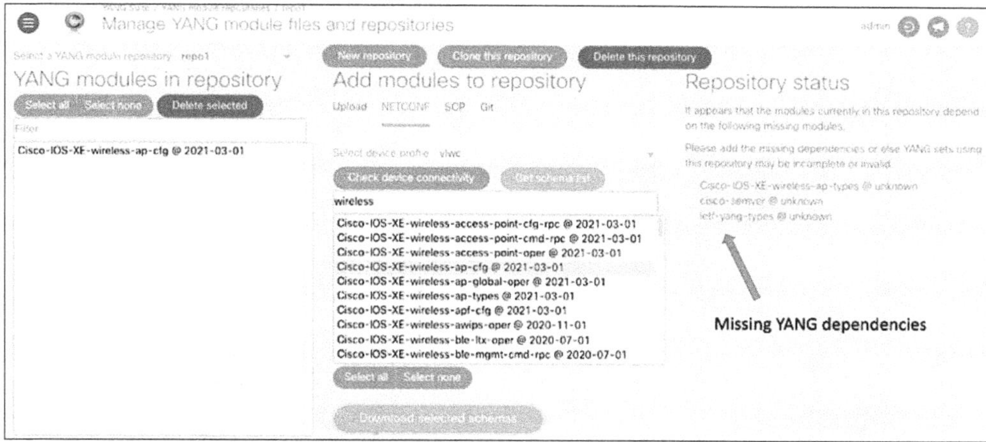

Figure 12-14 *Missing dependencies in a YANG Suite repository*

After you locate and add all the dependencies, you are ready to use the repository, as shown in Figure 12-15.

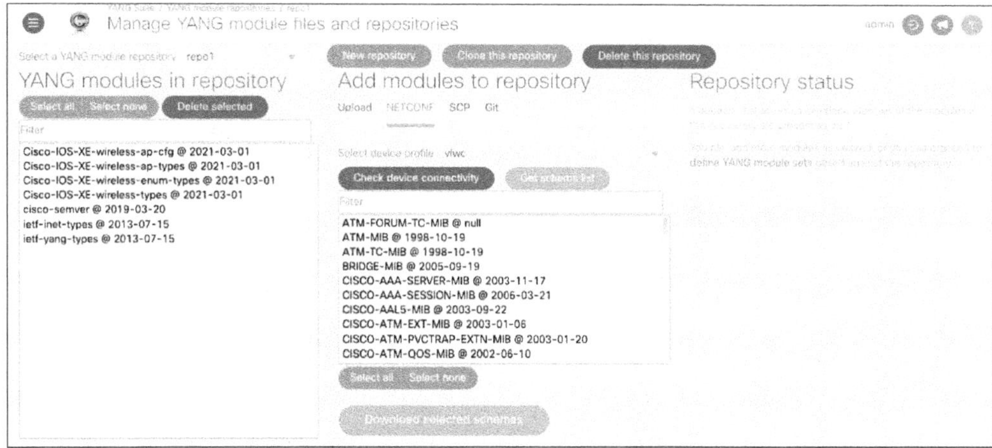

Figure 12-15 *YANG Suite repository ready to be used*

4. Create a module set. This is a subset of modules from a repository. For example, you can create a set with the YANG models regarding access points. Again, YANG Suite warns you of any missing dependency, as you can see in Figure 12-16.

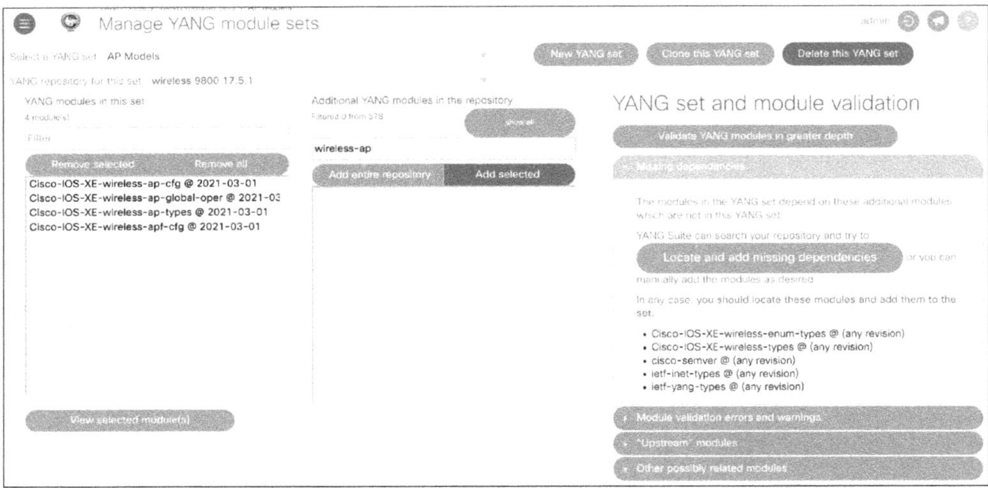

Figure 12-16 *YANG Suite displaying missing dependencies*

5. Explore the models and send and test any calls using NETCONF.

Using the Explore section, you can load any YANG module in the browser and explore it. For example, in Figure 12-17 the ap-tags container from the Cisco-IOS-XE-wireless-ap-cfg YANG model is being explored.

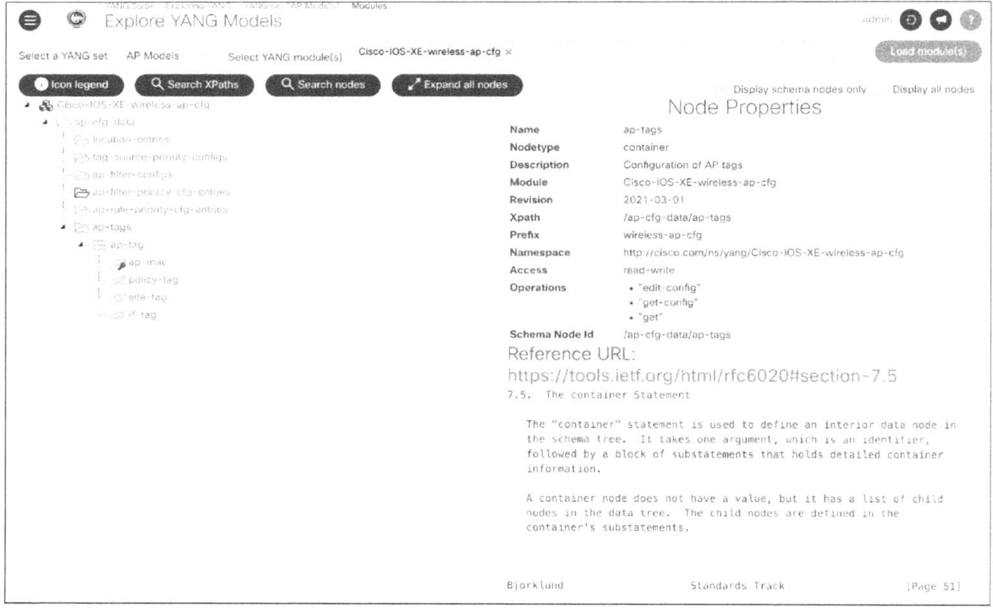

Figure 12-17 *Exploring YANG models*

How to Examine Data Using NETCONF and YANG Suite

YANG Suite also allows you to build and send NETCONF RPCs to the wireless controller. Let's examine the WLANs defined in the controller using NETCONF and YANG Suite.

The first step is to enable NETCONF in the wireless controller, as mentioned before. You can do that using CLI:

```
WLC(config)#netconf-yang
```

NETCONF uses SSH as transport, and it always authenticates using the default AAA methods. If your setup uses AAA, you need to make sure to define the correct authentication and authorization lists for the default method:

```
WLC(config)#aaa authorization exec default local
WLC(config)#aaa authentication login default local
```

The next step is to identify the YANG model where WLANs are available. You can do that by browsing the YANG Git repository at https://github.com/YangModels/yang/tree/master/vendor/cisco/xe or by using a tool like https://yangcatalog.org/yang-search.

When you know which YANG data model contains the information (in this case, use Cisco-IOS-XE-wireless-wlan-cfg), you can load it into YANG Suite.

After loading the module, you can select the device to use and start exploring the nodes in the YANG data model, as shown in Figure 12-18. To access this page, on the Protocols menu, select **NETCONF**.

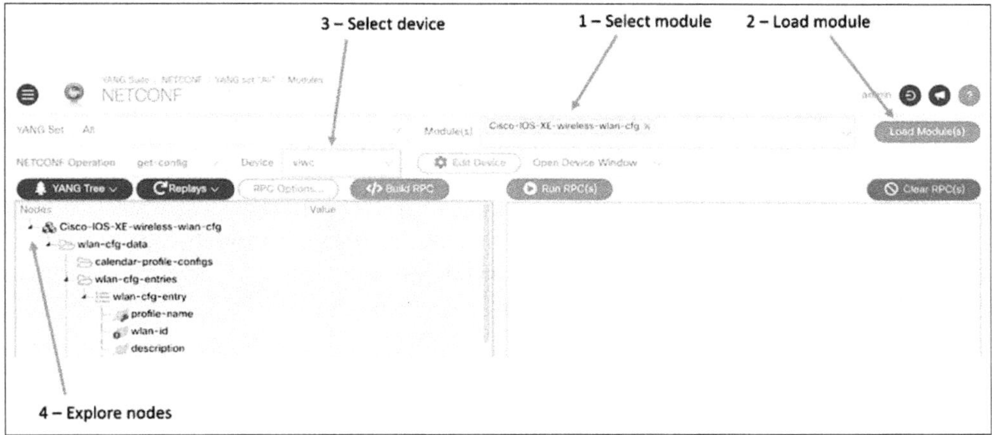

Figure 12-18 *Exploring data in YANG Suite*

The next step is building the RPC to get the data you are interested in. Select the **value** column next to wlan-cfg-entries and click **Build RPC**, as shown in Figure 12-19.

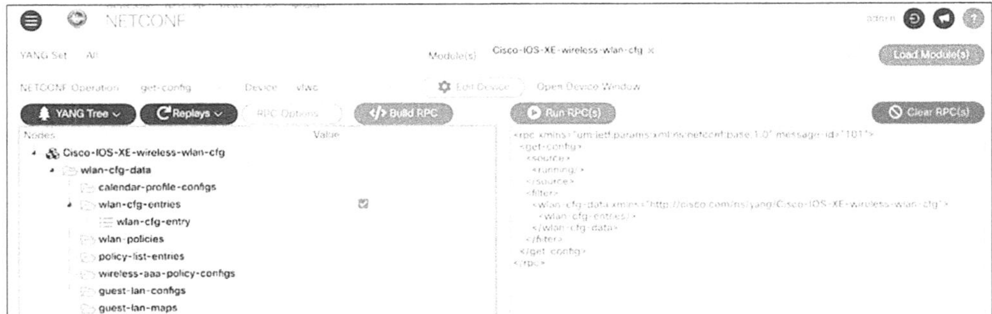

Figure 12-19 *Building RPCs in YANG Suite*

After you build the RPC, you can click **Run RPC(s)** and see the response.

In Figure 12-20 you can see the RPC request.

```
1 ------> <nc:rpc xmlns:nc="urn:ietf:params:xml:ns:netconf:base:1.0" message-id="...">
2 ------>   <nc:get-config>
              <nc:source>
3 ------>       <nc:running/>
              </nc:source>
              <nc:filter>
4 ------>       <wlan-cfg-data xmlns="http://cisco.com/ns/yang/Cisco-IOS-XE-wireless-wlan-cfg">
5 ------>         <wlan-cfg-entries/>
                </wlan-cfg-data>
              </nc:filter>
            </nc:get-config>
          </nc:rpc>
```

Figure 12-20 *NETCONF RPC request*

The request is an RPC call (1), with a **<get-config>** operation (2), requesting the running configuration (3), but asking only for a subset of the data, specified by the filter **wlan-cfg-entries** container (5) from the **wlan-cfg-data** YANG model (4).

Figure 12-21 shows the RPC reply.

The response is an **<rpc-reply>** call (1), including data from the **wlan-cfg-entries** container (3) in the **wlan-cfg-data** YANG model (2). Inside the container there are two entries from the **wlan-cfg-entry** list (4 and 5).

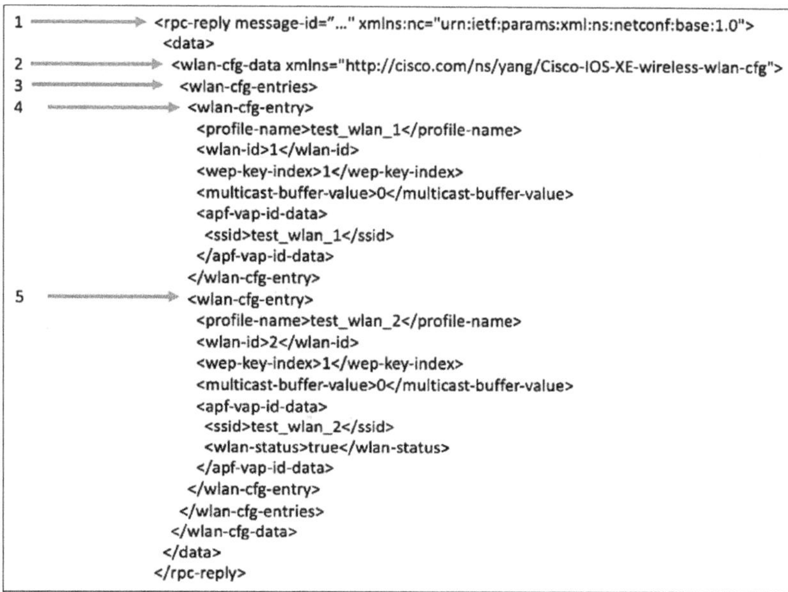

```
1 ----------> <rpc-reply message-id="..." xmlns:nc="urn:ietf:params:xml:ns:netconf:base:1.0">
              <data>
2 ---------->  <wlan-cfg-data xmlns="http://cisco.com/ns/yang/Cisco-IOS-XE-wireless-wlan-cfg">
3 ---------->   <wlan-cfg-entries>
4 ---------->    <wlan-cfg-entry>
                  <profile-name>test_wlan_1</profile-name>
                  <wlan-id>1</wlan-id>
                  <wep-key-index>1</wep-key-index>
                  <multicast-buffer-value>0</multicast-buffer-value>
                  <apf-vap-id-data>
                   <ssid>test_wlan_1</ssid>
                  </apf-vap-id-data>
                 </wlan-cfg-entry>
5 ---------->    <wlan-cfg-entry>
                  <profile-name>test_wlan_2</profile-name>
                  <wlan-id>2</wlan-id>
                  <wep-key-index>1</wep-key-index>
                  <multicast-buffer-value>0</multicast-buffer-value>
                  <apf-vap-id-data>
                   <ssid>test_wlan_2</ssid>
                   <wlan-status>true</wlan-status>
                  </apf-vap-id-data>
                 </wlan-cfg-entry>
                </wlan-cfg-entries>
               </wlan-cfg-data>
              </data>
             </rpc-reply>
```

Figure 12-21 *NETCONF RPC reply*

How to Examine Data Using RESTCONF and POSTMAN

To examine data using RESTCONF, you need a tool to generate HTTP verbs like POST, PUT, DELETE, PATCH, and so on. A browser can send some of them, but to properly use them, you need to use either plug-ins for editors like Visual Studio Code: Thunder, shown in Figure 12-22, or tools like Postman (www.postman.com), shown in Figure 12-23.

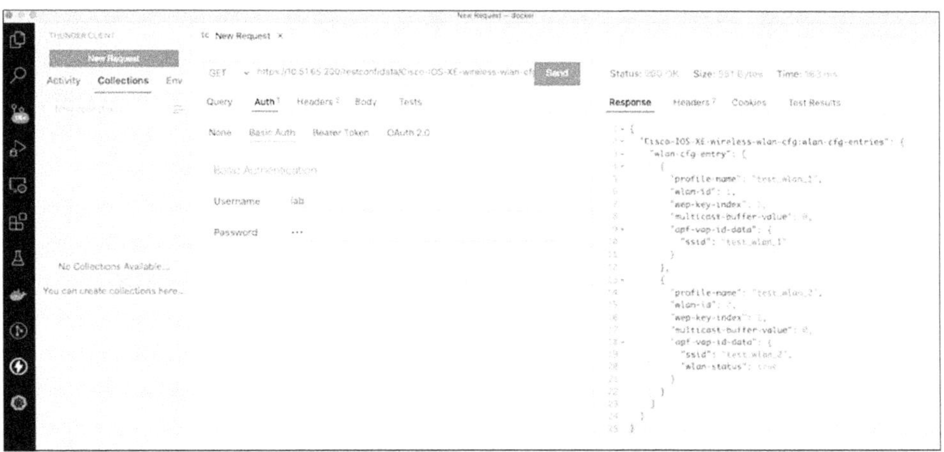

Figure 12-22 *Thunder plug-in for Visual Studio Code*

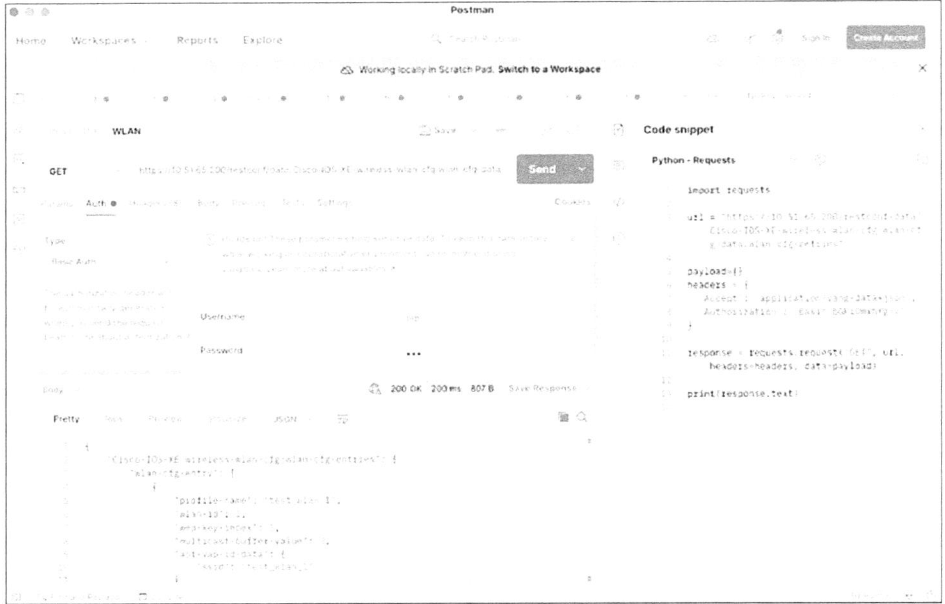

Figure 12-23 *Postman*

Enabling RESTCONF

The first step is to enable RESTCONF on the wireless controller. Because RESTCONF uses HTTPs as the transport, the HTTP server also needs to be enabled. RESTCONF uses the AAA settings in the HTTP server section. If you are using the WebUI on the wireless controller, your HTTP server settings are correct, and you just need to enable RESTCONF:

```
WLC(config)#ip http secure-server
WLC(config)#ip http authentication aaa login-authentication default
WLC(config)#restconf
```

RESTCONF URIs

A key parameter on using RESTCONF is finding the URI you need to access the required data. The URI has the following format:

```
https://<controller-ip>/<root>/<resource>/<data-model>
```

Root

The root is determined by querying the following URI:

```
https://<controller-ip>.well-known/host-meta
```

On the C9800 wireless LAN controller, the root is always **'restconf'**. You can send the well-known query to the controller to verify it:

```
$ curl -k -u lab:lab  https://10.51.65.200/.well-known/host-meta
<XRD xmlns='http://docs.oasis-open.org/ns/xri/xrd-1.0'>
    <Link rel='restconf' href='/restconf'/>
</XRD>
```

Resource

After the root, you need to find the resource to use. This determines the data model and operations available. RESTCONF provides the resources listed in Table 12-2.

Table 12-2 *RESTCONF Resources*

Resource	Description
data	Contains all **data** (configuration and state) resources.
operations	Optional resource, provides access to the data model–specific RPC operations. It is used to trigger operations—for example, clearing MAC address tables, saving configurations, and so on. It is the equivalent of CLI EXEC commands.
yang-library-version	Module date.

To identify the available operations, you can query the root resource URI (the following example was edited for brevity):

```
$ curl -k -u lab:lab  https://10.51.65.200/restconf/operations
<operations xmlns="urn:ietf:params:xml:ns:yang:ietf-restconf">
  <install-ios-xe-rpc:install>
  <install-ios-xe-rpc:activate>
  <install-ios-xe-rpc:install-commit>
  <wireless-access-point-cfg-rpc:set-ap-vlan-tag>
  <wireless-access-point-cfg-rpc:set-ap-monitor-mode-chnl-optimize>
  <cisco-smart-license:register-id-token>
  <cisco-smart-license:de-register>
</operations>
```

Similarly, querying the URI /restconf/data returns the list of YANG modules supported by the controller.

Data Model

The last step to construct the URI is the YANG model tree to query. Let's construct the URI to retrieve the list of WLANs.

As you saw in previous sections, the YANG model that contains the list of WLANs is Cisco-IOS-XE-wireless-wlan-cfg, shown in Figure 12-24.

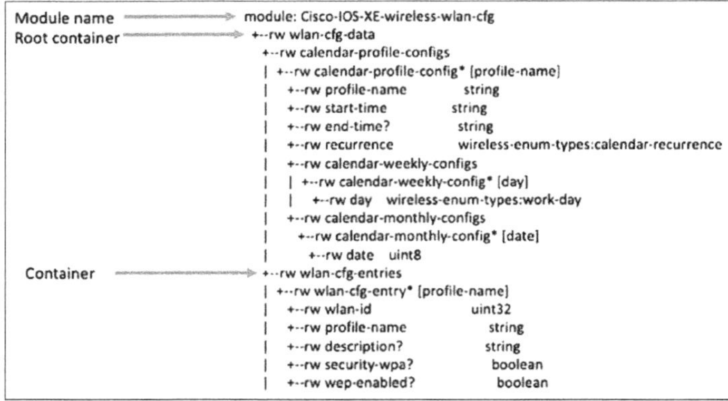
Module name
Root container
Container
module: Cisco-IOS-XE-wireless-wlan-cfg
+--rw wlan-cfg-data
+--rw calendar-profile-configs
| +--rw calendar-profile-config* [profile-name]
| +--rw profile-name string
| +--rw start-time string
| +--rw end-time? string
| +--rw recurrence wireless-enum-types:calendar-recurrence
| +--rw calendar-weekly-configs
| | +--rw calendar-weekly-config* [day]
| | +--rw day wireless-enum-types:work-day
| +--rw calendar-monthly-configs
| +--rw calendar-monthly-config* [date]
| +--rw date uint8
+--rw wlan-cfg-entries
| +--rw wlan-cfg-entry* [profile-name]
| +--rw wlan-id uint32
| +--rw profile-name string
| +--rw description? string
| +--rw security-wpa? boolean
| +--rw wep-enabled? boolean

Figure 12-24 *WLAN YANG Model*

As shown in Figure 12-25, the URI would be https://<controller-ip>/restconf/data/Cisco-IOS-XE-wireless-wlan-cfg:wlan-cfg-data.

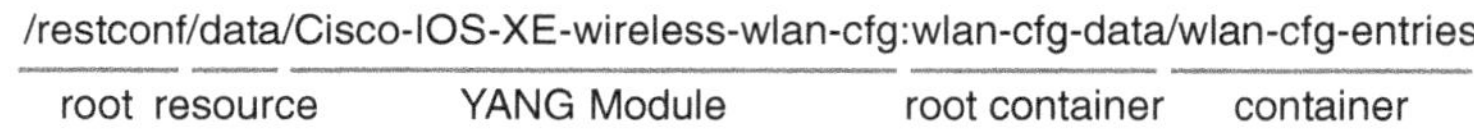
/restconf/data/Cisco-IOS-XE-wireless-wlan-cfg:wlan-cfg-data/wlan-cfg-entries
root
resource
YANG Module
root container
container

Figure 12-25 *Constructing URI from YANG data model*

You can try this request using Postman. To do so, you need to add your username and password in the Authorization tab in Postman, as type Basic Auth, as shown in Figure 12-26.

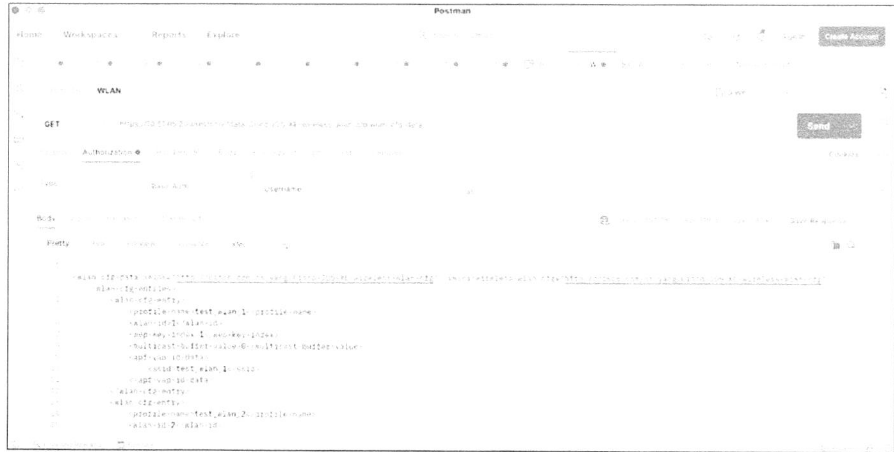
Postman
WLAN
GET
Authorization
Send

Figure 12-26 *Using Postman to retrieve a list of WLANs*

In the Postman output, you can see the response is in XML format. RESTCONF supports both XML and JSON formats. To ask for JSON, you need to add a header to the request specifying **Accept: application/yang-data+json**, as shown in Figure 12-27.

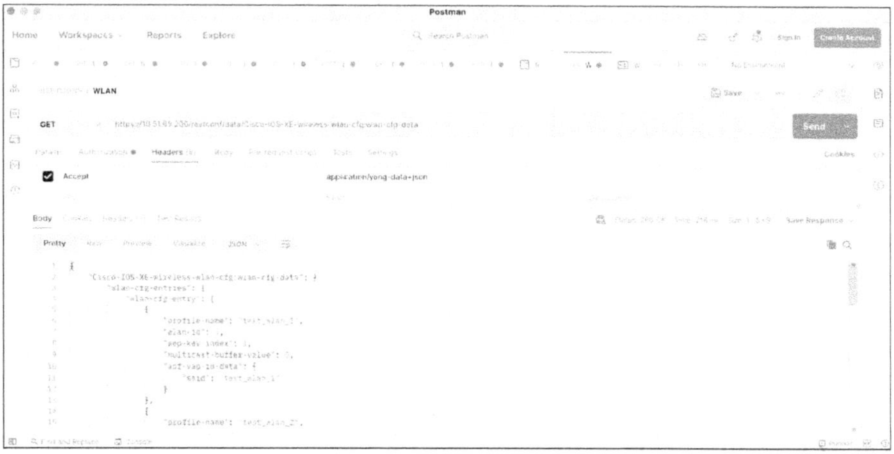

Figure 12-27 *Requesting JSON data in Postman*

You can also use other tools like the command-line utility curl to get this information:

```
$ curl -k -u lab:lab --header 'Accept: application/yang-data+json'
'https://10.51.65.200/restconf/data/Cisco-IOS-XE-wireless-wlan-
cfg:wlan-cfg-data'
{
  "Cisco-IOS-XE-wireless-wlan-cfg:wlan-cfg-data": {
    "wlan-cfg-entries": {
      "wlan-cfg-entry": [
        {
          "profile-name": "test_wlan_1",
          "wlan-id": 1,
          "wep-key-index": 1,
….
```

Searching Data

Instead of retrieving all data, you can search using RESTCONF. The WLANs in the config YANG model are stored in a list, as shown in Figure 12-28.

To search a particular key from a list, you can construct the URI shown in Figure 12-29.

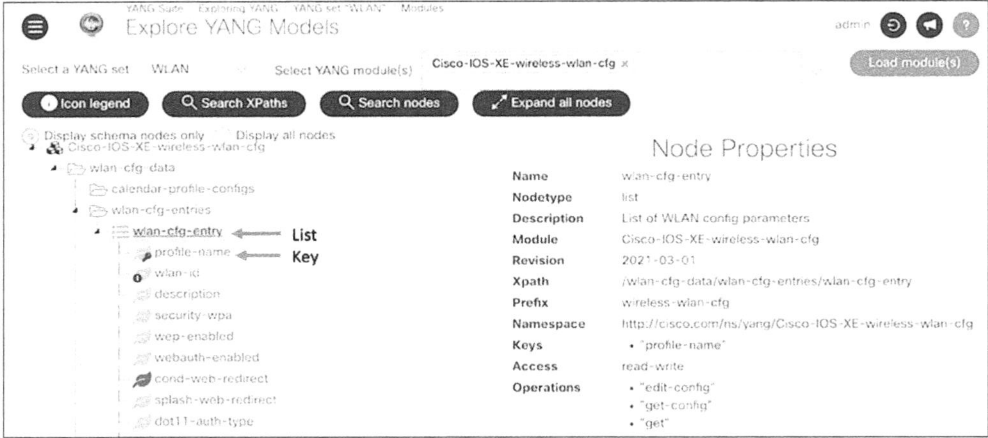

Figure 12-28 *List and key in the WLAN config entry model*

/restconf/data/Cisco-IOS-XE-wireless-wlan-cfg:wlan-cfg-data/wlan-cfg-entries/wlan-cfg-entry=test_wlan_1

root resource YANG Module root container container list search key

Figure 12-29 *Constructing searches in RESTCONF*

Updating the Configuration

So far, the examples in this chapter have used only GET methods to read information. You can also use RESTCONF to create and modify the configuration. Now let's create a new WLAN.

To update the controller configuration, you can use either the PATCH, PUT, or POST method. PUT replaces the entire section of the configuration with the one you send. POST adds a new configuration element, and PATCH modifies an existing element.

First, query the existing WLANs, as shown in Example 12-7.

Example 12-7 *Querying WLANs Using RESTCONF from the Command Line*

```
$ curl -k -u lab:lab --header 'Accept: application/yang-data+json'
  'https://10.51.65.200/restconf/data/Cisco-IOS-XE-wireless-wlan-cfg:wlan-cfg-data/
  wlan-cfg-entries'
{
  "Cisco-IOS-XE-wireless-wlan-cfg:wlan-cfg-entries": {
    "wlan-cfg-entry": [
      {
        "profile-name": "test_wlan_1",
        "wlan-id": 1,
        "apf-vap-id-data": {
```

```
                "ssid": "test_wlan_1"
            }
        }
    ]
  }
}
```

You see there's only one WLAN named **'test_wlan_1'**. To create a new WLAN, you need to add a new element to the array **'wlan-cfg-entry'** as shown on the Postman screen in Figure 12-30. Because you are sending JSON data, you need to add the header **Content-type: application/yang-data+json**, as shown in Figure 12-31.

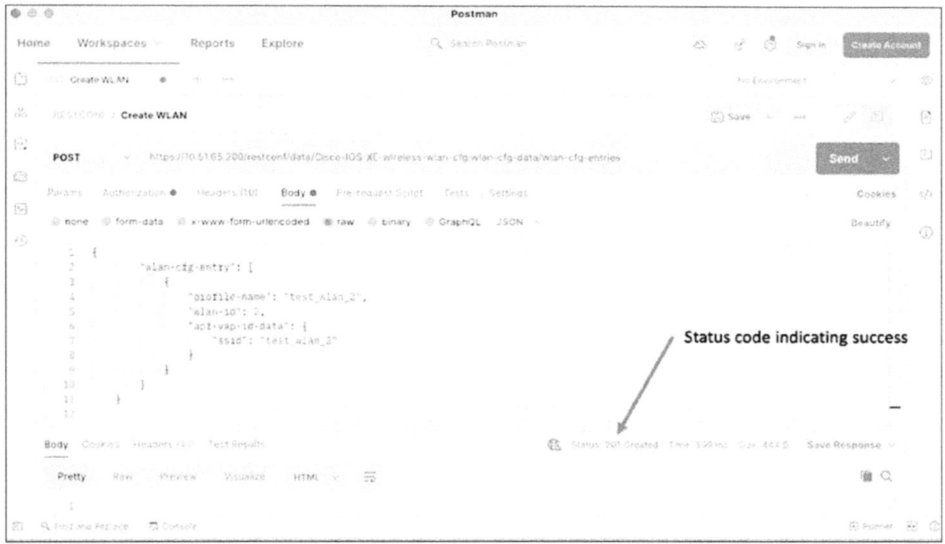

Figure 12-30 *Adding a new WLAN using Postman*

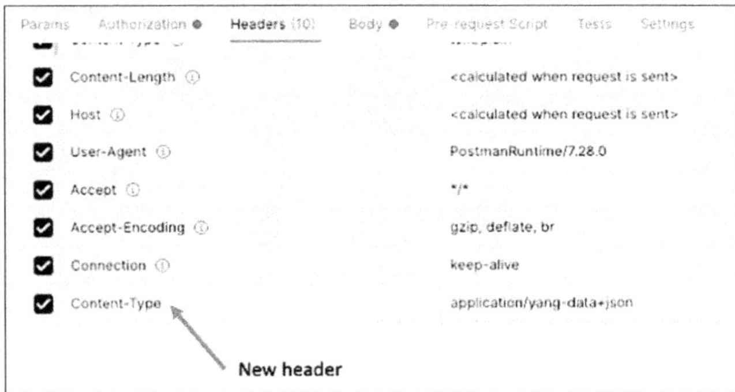

Figure 12-31 *Content-type header*

After sending this request, you can check the WLANs in the controller and see they are created:

```
WLC#show wlan summary
Number of WLANs: 2
ID   Profile Name       SSID                         Status Security
----------------------------------------------------------------------
---
1    test_wlan_1        test_wlan_1          DOWN    [WPA2][802.1x][AES]
2    test_wlan_2        test_wlan_2          DOWN    [WPA2][802.1x][AES]
```

You can also replace the entire WLAN configuration using the PUT method, as shown in Figure 12-32. All the '**wlan-cfg-entry**' list on the YANG model is replaced by the one sent in Postman:

```
WLC#show wlan summary
Number of WLANs: 1
ID   Profile Name     SSID                       Status Security
---------------------------------------------------------------------
5    new_wlan_100     new_wlan_100           DOWN    [WPA2][802.1x][AES]
WLC#
```

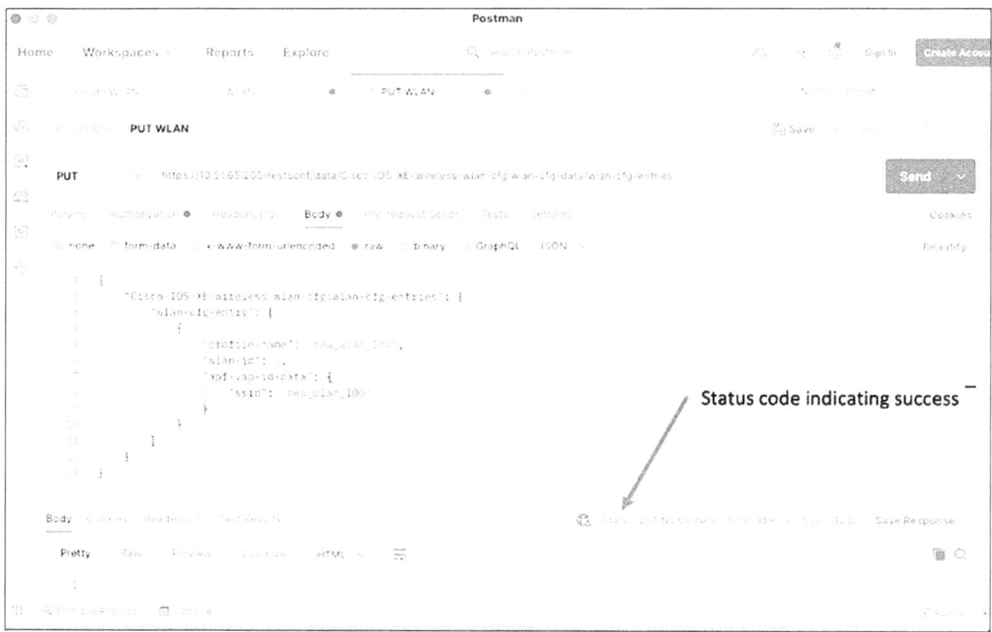

Figure 12-32 *Replacing WLAN configuration using Postman*

Python and Network Programmability

Network programmability allows you to create your own software to customize how your network operates. In this section, we present a few examples of Python code that you can modify to suit your needs.

All the examples are available in the Git repository at https://github.com/fsedano/cisco_9800_programmability.

The examples also include a Docker environment with Python and all the required libraries installed. To start using this environment, you can use **docker-compose** after cloning the repository:

```
$ git clone https://github.com/fsedano/cisco_9800_programmability
$ cd cisco_9800_programmability
$ docker-compose up -d
Creating cisco_9800_programmability_development_1 ... done
$ docker-compose exec development bash
root@218fa72df3de:/app#
```

At this point, you can execute your Python programs at the prompt. The volume named **app** is mapped into the host, so you can also edit the programs on your own computer and execute them inside the container.

Assigning Tags to APs Based on Serial Number

This example uses RESTCONF to update the tag associated with each AP based on the serial number. The program flow is shown in Figure 12-33. This capability is useful when you're deploying a large number of access points. The association between the AP serial number and the AP tag is read from a CSV file, but you can adapt the example to read from any other source, like a corporate database.

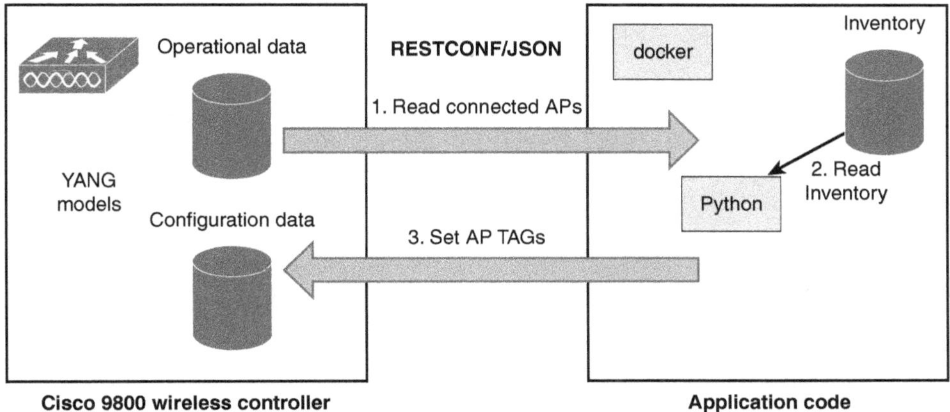

Figure 12-33 *Python application to assign a tag based on the AP serial number*

To execute the example, inside the Docker container, you can change to the directory ap_to_site and execute your program. Make sure to update the inventory list with your AP serial number. To get the AP serial number, you can use the 9800 CLI:

```
WLC#show ap summ
Number of APs: 1
AP Name                                   Slots    AP Model
-------------------------------------------------------------
APCC16.7EDC.27D8                            2      AIR-AP2802E-E-K9
WLC#show ap name APCC16.7EDC.27D8 inventory
NAME: AP2800, DESCR: Cisco Aironet 2800 Series (IEEE 802.11ac
PID: AIR-AP2802E-E-K9, VID: 02, SN: FGL2102AACG
```

With the serial number, update the inventory file, as shown in Figure 12-34.

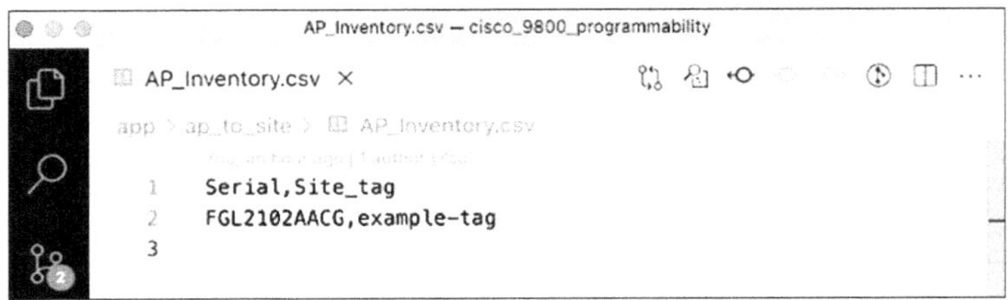

Figure 12-34 *Inventory file*

After the inventory file is updated, you can execute your code, as shown in Example 12-8.

Example 12-8 *Executing a Python Script to Change AP Tags*

```
$ docker-compose exec development bash
root@218fa72df3de:/app# cd ap_to_site

root@218fa72df3de:/app/ap_to_site# python change_ap_tag.py -user lab -password lab
  -wlc_ip <ip_address>

2021-06-27 16:08:59,326 (INFO) Executing method GET on resource Cisco-IOS-XE-wire-
  less-access-point-oper:access-point-oper-data/capwap-data
2021-06-27 16:08:59,822 (INFO) Success!
2021-06-27 16:08:59,822 (INFO) Success! 1 APs joined
2021-06-27 16:08:59,825 (INFO) Creating the site tag example-tag
2021-06-27 16:08:59,825 (INFO) Executing method PATCH on resource Cisco-IOS-XE-wire-
  less-site-cfg:site-cfg-data/site-tag-configs/site-tag-config

2021-06-27 16:09:00,144 (INFO) Success!
2021-06-27 16:09:00,144 (INFO) Changing AP MAC cc:16:7e:dc:27:d8 to have tag exam-
  ple-tag
```

```
2021-06-27 16:09:00,145 (INFO) Executing method PATCH on resource /Cisco-IOS-XE-
  wireless-ap-cfg:ap-cfg-data/ap-tags/ap-tag/
2021-06-27 16:09:00,423 (INFO) Success!
root@218fa72df3de:/app/ap_to_site#
```

Program Structure

The program is divided into three main blocks. The main program is shown in Example 12-9.

Example 12-9 *The change_ap_tag.py Python File*

```
#!/usr/bin/python3
"""
Code to provision AP tags based on spreadsheet data
"""

import logging
from readinventory import Inventory
from c9800 import C9800
import argparse

## Set logging level and format
logging.basicConfig(level=logging.INFO, format="%(asctime)-15s (%(levelname)s)
  %(message)s")

## Parse command line arguments.
parser = argparse.ArgumentParser(description="Utility to update AP tag based on
  serial number")
parser.add_argument('-user', help='c9800 username', required=True)
parser.add_argument('-password', help='c9800 password', required=True)
parser.add_argument('-wlc_ip', help='c9800 IP address', required=True)
args=parser.parse_args()

# Create an object of type C9800 and store it in the wlc variable
wlc = C9800(args.wlc_ip, args.user, args.password)

# Call our C9800 object and ask for the list of joined APs
aps = wlc.get_joined_aps()

# Create an Inventory object, read the file content and store it in the inventory
  variable
inventory = Inventory('AP_Inventory.csv').read()

# We have now the list of joined AP serial numbers (in aps) and the list of APs in
  inventory (inventory)
```

```
# Traverse the joined AP list. It has the following format:
"""
 {
    'FGL2102AACG': {'MAC': 'cc:16:7e:dc:27:d8'},
    'FGL2102AACX': {'MAC': 'cc:16:7e:dc:27:d9'}
}
"""


for ap_serial in aps:
    # We have the serial number now in the ap_serial variable. Check if this serial
  number
    # is in the inventory
    if ap_serial in inventory:
        # If it is in the inventory, update the tag
        ap_inventory = inventory[ap_serial]
        ap_joined_info = aps[ap_serial]
        ap_joined_mac = ap_joined_info['MAC']
        ap_tag = ap_inventory['tag']

        # Create the tag on the 9800
        wlc.create_site_tag(ap_tag)

        # Assign the AP to this tag
        wlc.set_ap_tag(ap_joined_mac, ap_tag)
```

This main program creates two objects: an **'Inventory'** object that contains the serial numbers and tags in the inventory file, and a **'C9800'** object that represents the wireless controller.

Each object has methods to operate. For example, you create a C9800 object and assign it to the **wlc** variable on this line:

```
wlc = C9800(args.wlc_ip, args.user, args.password)
```

And you store the object in the variable called **wlc**. The object has several methods to query the joined APs, another one to set an AP tag, and so on. For example, to query the APs joined in the controller, it calls

```
aps = wlc.get_joined_aps()
```

The main program does not know how the C9800 object does it. The program just asks the object to perform the operation by calling the method.

The next block is the C9800 object, as shown in Example 12-10. The code to implement the C9800 object is in the c9800 directory. It uses RESTCONF to access the controller.

Example 12-10 *The c9800.py Python File*

```
wlc.py

"""
This file implements the C9800 class.

Three methods are available:

-----
set_ap_tag(ap_mac, ap_tag)

This method sets an AP tag to the given AP MAC by calling the REST resource:
/Cisco-IOS-XE-wireless-ap-cfg:ap-cfg-data/ap-tags/ap-tag/

It returns the REST response
-----
get_joined_aps()

This method queries the controller for the list of joined APs using the REST
  resource:
Cisco-IOS-XE-wireless-access-point-oper:access-point-oper-data/capwap-data

It returns a dictionary of joined APs with the following format:

{
    'FGL2115B015': {'MAC': '00:2c:c8:8b:31:b0'},
    'FGL39392819': {'MAC': '00:2c:c8:8a:11:a0'}
}
-----
create_site_tag(ap_tag)

This method creates the site tag using RESTCONF PATCH method

"""
import requests
import urllib3
from requests.auth import HTTPBasicAuth
from requests.exceptions import HTTPError
from netaddr import EUI, mac_unix_expanded
import logging

class C9800:
    def __init__(self, ip, user, password):
        self.controller_ip = ip
        self.controller_user = user
        self.controller_password = password
        self.controller_auth = HTTPBasicAuth(user, password)
        urllib3.disable_warnings(urllib3.exceptions.InsecureRequestWarning)
```

```
        self.ap_list = {}
        self.headers = {
            'Accept': "application/yang-data+json",
            'Content-Type': "application/yang-data+json",
            'cache-control': "no-cache"
        }
        self.baseurl = f"https://{self.controller_ip}/restconf/data/"

    def __execute_REST(self, method, resource, payload=None):
        logging.info(f"Executing method {method} on resource {resource}")
        url = self.baseurl + resource
        response = None
        try:
            response = requests.request(method,
                    url,
                    headers=self.headers,
                    verify=False,
                    auth=self.controller_auth,
                    json=payload)
            response.raise_for_status()
        except HTTPError as http_err:
            logging.error(f'HTTP error occurred: {http_err}')
        except Exception as err:
            logging.exception(f'Other error occurred: {err}')
        else:
            logging.info(f"Success!")
        return response

   def get_site_tags(self):
        resource = "Cisco-IOS-XE-wireless-site-cfg:site-cfg-data/site-tag-configs/
site-tag-config"

        result = self.__execute_REST(method="GET", resource=resource)
        data = result.json()
        logging.info(f"The list of tags is: {data}")
        return data

   def create_site_tag(self, ap_tag):
       logging.info(f"Creating the site tag {ap_tag}")
       resource = "Cisco-IOS-XE-wireless-site-cfg:site-cfg-data/site-tag-configs/
site-tag-config"
        data = {'Cisco-IOS-XE-wireless-site-cfg:site-tag-config': [
                {'site-tag-name': ap_tag, 'is-local-site': False}
                ]
        }
        self.__execute_REST(method="PATCH", resource=resource, payload=data)
    def set_ap_tag(self, ap_mac, ap_tag):
        logging.info(f"Changing AP MAC {ap_mac} to have tag {ap_tag}")
```

```
        payload = {"ap-tag":
            {"ap-mac":ap_mac,
            "site-tag":ap_tag}
        }
        resource = "/Cisco-IOS-XE-wireless-ap-cfg:ap-cfg-data/ap-tags/ap-tag/"
        response = self.__execute_REST(method="PATCH", resource=resource,
payload=payload)
        return response

    def get_joined_aps(self):
        resource = "Cisco-IOS-XE-wireless-access-point-oper:access-point-oper-data/
capwap-data"
        response = self.__execute_REST(method="GET", resource=resource)

        try:
            json_payload = response.json()
            capwap_data = json_payload['Cisco-IOS-XE-wireless-access-point-
oper:capwap-data']
            for entry in capwap_data:
                ethernet_mac = entry["device-detail"]["static-info"]["board-data"]
["wtp-enet-mac"]
                serial = entry["device-detail"]["static-info"]["board-data"]
["wtp-serial-num"]
                MAC = EUI(ethernet_mac, dialect=mac_unix_expanded)
                self.ap_list[serial] = {
                    "MAC":str(MAC)
                }
       except ValueError as err:
            logging.info(f"No data was returned")
        except Exception as err:
            logging.exception(f"Other error: {err}")
        else:
            logging.info(f"Success! {len(self.ap_list)} APs joined. List is
{self.ap_list}")

        return self.ap_list
```

The last block is the Inventory object shown in Example 12-11. The code to implement the Inventory object is in the readinventory directory. It uses a Python library called csv to read from a .csv file and return its contents.

Example 12-11 *The readinventory.py Python Code*

```
import csv

class Inventory:
    def __init__(self, filename):
        self.filename = filename
        self.inventory_data = {}
    def read(self):
```

```
with open("AP_Inventory.csv", encoding='utf-8') as f:
    data = csv.reader(f, delimiter=',')
    next(data, None)
    for row in data:
        if len(row) > 0:
            serial = row[0]
            tag = row[1]
            self.inventory_data[serial] = {
                "tag":tag
            }
    return self.inventory_data
```

Summary

This chapter describes network programmability concepts and the protocols used in modern network programmability, such as NETCONF and RESTCONF. It also describes the YANG data models used in the Cisco Catalyst 9800 Wireless LAN Controller. The chapter also presents examples on how Python can be used to query the Cisco C9800 operational model and how to create site tags using RESTCONF.

References

Zero Touch Provisioning (ZTP): https://www.cisco.com/c/en/us/td/docs/ios-xml/ios/prog/configuration/1612/b_1612_programmability_cg/zero_touch_provisioning.html

RFC 6020: YANG—A Data Modeling Language for the Network Configuration Protocol (NETCONF): https://datatracker.ietf.org/doc/html/rfc6020

RFC 7950: The YANG 1.1 Data Modeling Language: https://datatracker.ietf.org/doc/html/rfc7950

RFC 5277: NETCONF Event Notifications: https://datatracker.ietf.org/doc/html/rfc5277

RFC 8259: The JavaScript Object Notation (JSON) Data Interchange Format: https://datatracker.ietf.org/doc/html/rfc8259

RFC 8040: RESTCONF Protocol: https://datatracker.ietf.org/doc/html/rfc8040

Protocol buffers: https://developers.google.com/protocol-buffers

YANG Suite: https://github.com/CiscoDevNet/yangsuite

Pyang: https://github.com/mbj4668/pyang

Chapter 13

Model-Driven Telemetry

What Is Model-Driven Telemetry?

In Chapter 12, "Network Programmability," you learned how the data in the Catalyst 9800 (C9800) wireless controller is made available for consumption from software and how the data is structured using YANG models. That chapter focused on the configuration data in the C9800. However, there is another set of information that you can retrieve using the same protocols and data models discussed there: the operational data.

This is a set of YANG models that reflect the internal state of the C9800: number of clients connected, number of APs joined, memory usage, and so on.

Although you can access the operational data using the same techniques learned in Chapter 12, operational data presents a specific set of challenges:

- Data Models reflecting operational data are read-only.
- The values will change over time on their own, so you usually want to receive the most up-to-date information from the C9800.
- Due to the dynamic nature of the information, you will be interested in knowing when this information changes. Continuously polling the C9800 would be inefficient.

First let's define what telemetry is. In general, telemetry is a set of automated processes to receive measurements and other operational data from remote entities.

Model-driven telemetry (MDT) uses YANG to model the data and standard protocols like NETCONF, RESTCONF, or gRPC to send it over the network.

In MDT, several roles are defined, as shown in Figure 13-1:

- The **management agent** is the network element creating subscriptions. It sends requests to the C9800 defining the YANG-modeled data of interest and the subscription characteristics (for example, how frequent the data should be sent, which protocol to use, which authentication parameters, and where to send the information).
- The **publisher** is the element sending the telemetry data. In most cases, it is the C9800 that aggregates all information and publishes it (including AP-related information), but for some scenarios, like with Cisco Enterprise IoT solution, it could also be the AP. When the publisher receives the request, it generates a subscription.
- The **collector** is the network element receiving the telemetry data. In some cases (for example, when using Cisco DNA Center), the management agent and the collector could be the same network element.

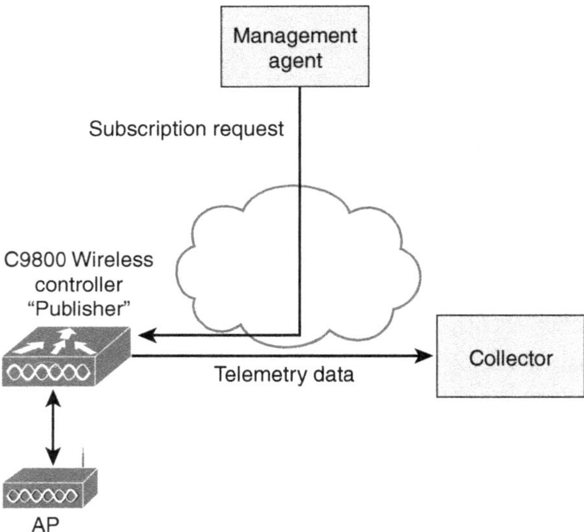

Figure 13-1 *Roles in model-driven telemetry*

The management agent and the collector can use different protocols to communicate with the publisher. For example, the management agent could even use CLI commands to create subscriptions on the publisher, and the publisher could use NETCONF to send the telemetry data.

How to Enable Model-Driven Telemetry

Model-driven telemetry uses NETCONF, RESTCONF, or gNMI protocols. NETCONF must be configured for telemetry to work, even if NETCONF is not used.

NETCONF

To enable NETCONF, use **netconf-yang** CLI in config mode:

```
C9800-telemetry(config)# netconf-yang
```

NETCONF uses the authentication **login** method and authorization **exec** method. You need to configure them to use the local database or any of the supported AAA methods. For example, to use the C9800 local database for authentication and create a username specific for NETCONF, you can use the configuration shown in Example 13-1. NETCONF uses only the default list.

Example 13-1 *Configuring AAA for NETCONF*

```
C9800-telemetry(config)# aaa new-model
C9800-telemetry(config)# aaa authentication login default local
C9800-telemetry(config)# aaa authorization exec default local
C9800-telemetry(config)# username netconfuser privilege 15 secret netconfpass
```

RESTCONF

To enable RESTCONF, you also need to enable the HTTP server, as shown in Example 13-2.

Example 13-2 *Enabling RESTCONF*

```
C9800-telemetry(config)# restconf
C9800-telemetry(config)# ip http secure-server
```

To verify that RESTCONF is enabled, you can use the **show platform software yang-management process** CLI, as shown in Example 13-3.

Example 13-3 *Verifying RESTCONF Is Enabled*

```
C9800-telemetry# show platform software yang-management process
confd            : Running
nesd             : Running
syncfd           : Running
ncsshd           : Running
dmiauthd         : Running
nginx            : Running
ndbmand          : Running
pubd             : Running
gnmib            : Running
```

You can also add the **monitor** keyword after the command, to display further information, as shown in Example 13-4.

Example 13-4 *Retrieving More Details About YANG Processes in C9800*

```
C9800-telemetry# show platform software yang-management process monitor
COMMAND              PID S     VSZ    RSS %CPU %MEM     ELAPSED
confd               3211 S 1916560 171324 5.9  2.1        06:55
confd-startup.s      726 S    6408   5692  0.1  0.0       07:00
confd-startup.s     5790 S    6408   4360  0.0  0.0       06:30
dmiauthd            1263 S 429988 50620  0.1  0.6         06:59
gnmib               1553 S 1154680 60260 5.7  0.7         06:59
ncsshd              2047 S 302128 13864  4.3  0.1         06:58
ncsshd_bp           1910 S 172936 12864  0.0  0.1         06:59
ndbmand              942 S 1640660 188788 0.4  2.3        07:00
nginx               6987 S 106972 13592  0.0  0.1         06:15
nginx               6997 S 107444   7120  0.0  0.0        06:15
nginx               6998 S   57764  6496  0.0  0.0        06:15
pubd               27165 S 2121660 206976 0.4  2.5        07:09
```

gNMI

The gRPC Network Management Interface, or gNMI, is a protocol for configuration manipulation and state retrieval. In the case of the C9800, this data is modeled using YANG. It is more efficient than NETCONF because it uses Protobuf, a compact binary format, instead of the text-based XML used by NETCONF. It provides a set of APIs to Set, Get, and Subscribe data and also allows a third-party device to learn about the C9800 capabilities.

You can enable gNMI in either insecure or secure mode. Insecure mode is appropriate for testing and development, but for production environments, you should use secure mode.

To enable gNMI in secure mode, you need to load certificates into the C9800, either using the CLI or gRPC Network Operations Interface (gNOI) protocol.

gNOI is a suite of services, each corresponding to a set of operations. As of release 17.6, the supported operations are related to certificate management for gNMI.

gNxI is a collection of tools for network management that use gNMI and gNOI protocols.

To enable gNMI in insecure mode, you need to use **gnxi server** CLI, as shown in Example 13-5.

Example 13-5 *Enabling gNMI in Insecure Mode*

```
C9800-telemetry(config)# gnxi
C9800-telemetry(config)# gnxi server
```

To enable gNMI in secure mode, you need to install a certificate and configure a trustpoint. Consult the IOS-XE documentation for your release to do that. After you create the trustpoint, you can enable gNMI in secure mode by using the configuration shown in Example 13-6.

Example 13-6 *Enabling gNMI in Secure Mode*

```
C9800-telemetry(config)# gnxi
C9800-telemetry(config)# gnxi server
C9800-telemetry(config)# gnxi secure-server
C9800-telemetry(config)# gnxi secure-trustpoint <trustpoint_name>
C9800-telemetry(config)# gnxi secure-client-auth
```

By default, secure gNMI uses port 9339 and insecure uses 50052, but you can change them in the configuration if necessary.

To verify the status of the gNMI protocol, you can use the CLI shown in Example 13-7.

Example 13-7 *Verifying the Status of gNMI Protocol*

```
C9800-telemetry# show gnxi state
State              Status
--------------------------------
Enabled            Up
```

Operational Data and KPIs

The Cisco 9800 wireless controller exposes two types of YANG models:

- Configuration data
- Operational data

Each of these types uses a different datastore. A datastore is a fundamental concept binding the YANG data models to network management protocols, such as NETCONF or RESTCONF. Configuration data was the subject of Chapter 12. This datastore keeps all configured values but provides no information on the actual state of the wireless controller. For example, you might have configured one AP using programmability. But has the AP joined the controller? Using which IP address? Are the radios up? What channel is each of the radios using? Operational data can answer those questions.

Figure 13-2 illustrates the configuration and operational YANG models.

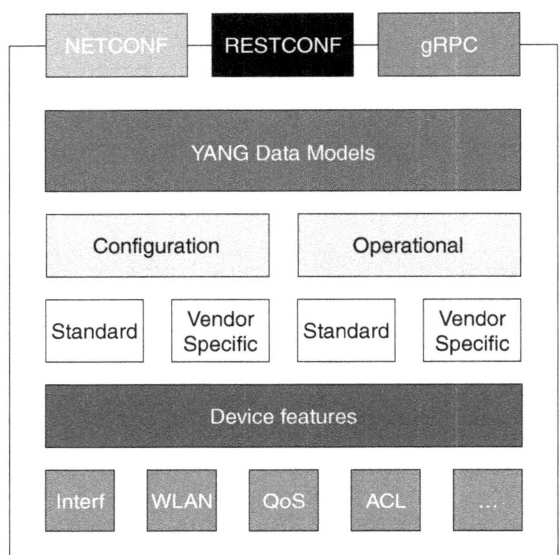

Figure 13-2 *Configuration and operational YANG data models*

You can examine the operational data available with the same tools used for programmability. In Chapter 12, we reviewed some tools like YANG Suite that can be used for that purpose.

The YANG models used for operational data are located at the same place as other YANG models. For example, for the 17.6.1 release of the 9800 wireless controller, you can find the models in GitHub: https://github.com/YangModels/yang/tree/master/vendor/cisco/xe/1761.

Operational and configuration YANG models are similar. The only difference between them is the **config: false** statement on the operational YANG model, as shown on Figure 13-3, corresponding to the **Cisco-IOS-XE-wireless-ap-global-oper** model in IOS-XE 17.6.1.

All operational models are read-only. You can use NETCONF or RESTCONF to read them, but you can't write to them.

Be aware that **config=false** is a property of a YANG container, not of the entire module. Although frequently entire modules contain only operational data, it's also possible to have both operational and configuration trees in the same model. For example, looking at the ietf-interfaces module in YANG Suite, you can see two containers: **interfaces** and **interfaces-state**, as shown in Figure 13-4 and Figure 13-5, respectively.

Interfaces is a configuration container, as indicated with **access: read-write**, and operations include **get** and **edit**, but **interfaces-state** is an operational container, as indicated with **access: read-only** and the only available operation is **get**.

```
    container ap-global-oper-data {
      config false;
      description
        "Root container for AP operational data aggregated across wireless processes";
      container ap-img-predownload-stats {
        presence "ap-img-predownload-stats";
        description
          "AP image predownload stats";
        uses wireless-ap-global-oper:global-ap-stats;
      }
      list ap-join-stats {
        key "wtp-mac";
        description
          "AP join statistics";
        uses wireless-ap-global-oper:st-emltd-ap-stats-info;
      }
    }
  }
```

Indicates this is an operational model

Figure 13-3 *Identifying an operational YANG model*

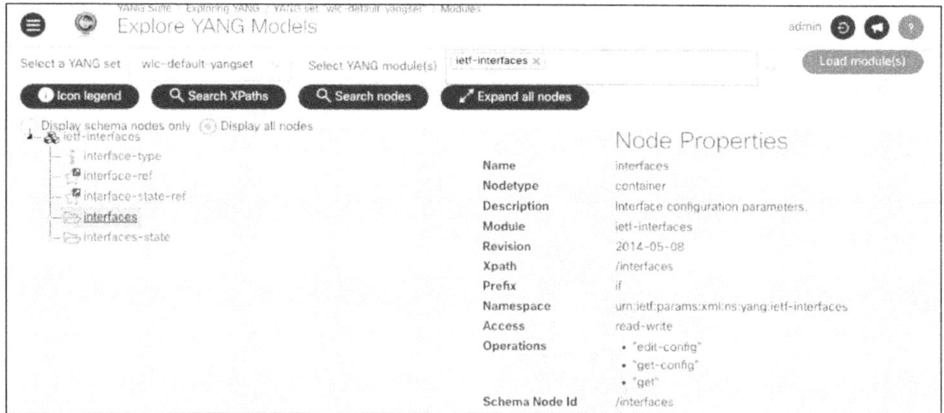

Figure 13-4 *Configuration container for* **interfaces**

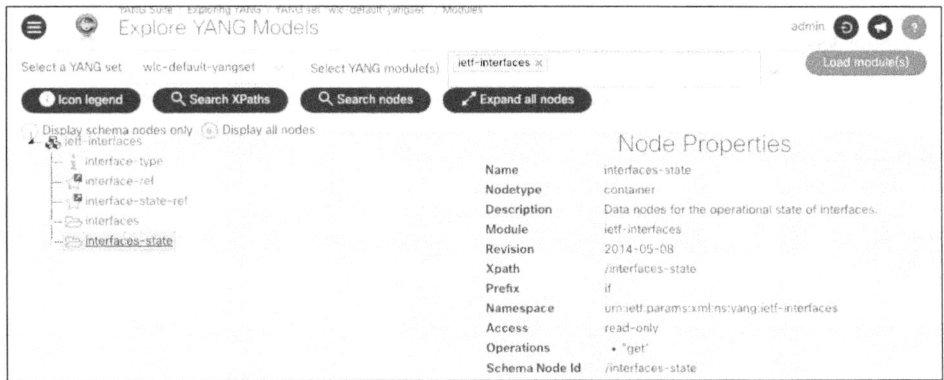

Figure 13-5 *Operational container for* **interfaces-state**

You can use NETCONF to read any of the values, either operational or configuration. As discussed in Chapter 12, all operations in NETCONF are performed using remote procedure calls (RPCs). For example, to read all configured interface names on the wireless controller, you can read from **ietf-interfaces**, container **interfaces** by using the RPC. You can use YANG Suite to generate the RPC for you, as shown in Figure 13-6.

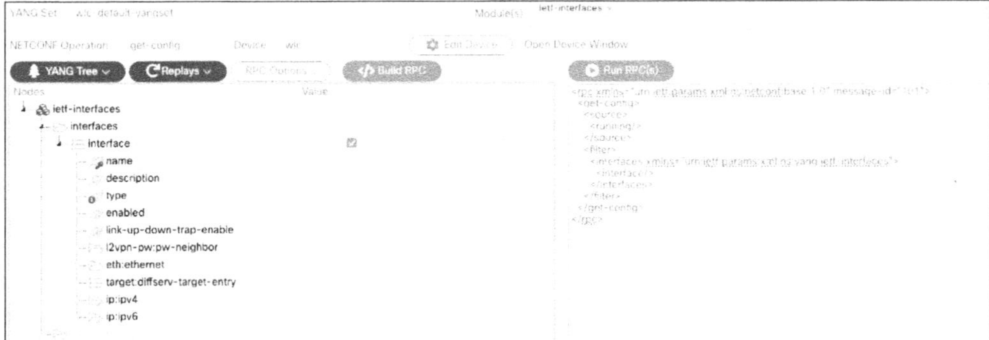

Figure 13-6 *RPC to read configured interfaces using NETCONF*

You can see more details of the NETCONF Query RPC in Figure 13-7.

get-config operation → (points to `<get-config>`)

interfaces container → (points to `<interfaces xmlns=...>`)

```
<rpc xmlns="urn:ietf:params:xml:ns:netconf:base:1.0" message-id="101">
  <get-config>
    <source>
      <running/>
    </source>
    <filter>
      <interfaces xmlns="urn:ietf:params:xml:ns:yang:ietf-interfaces">
        <interface/>
      </interfaces>
    </filter>
  </get-config>
</rpc>
```

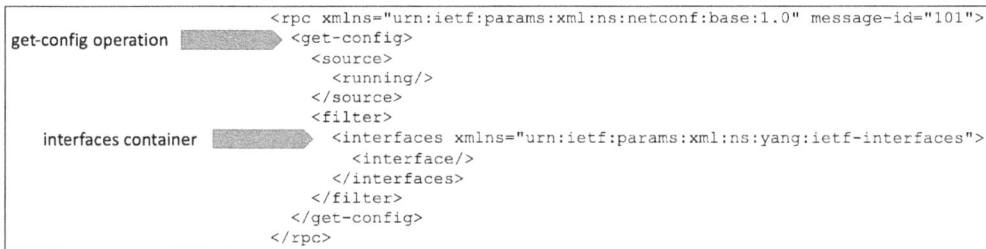

Figure 13-7 *NETCONF query RPC to get interface list*

Example 13-8 shows the NETCONF RPC reply.

Example 13-8 *NETCONF RPC Reply*

```
<?xml version="1.0" ?>
<rpc-reply message-id="urn:uuid:771b3132-b7d5-4140-be34-87f0685a795f" xmlns="urn:
  ietf:params:xml:ns:netconf:base:1.0" xmlns:nc="urn:ietf:params:xml:ns:netconf:
  base:1.0">
  <data>
    <interfaces xmlns="urn:ietf:params:xml:ns:yang:ietf-interfaces">
      <interface>
        <name>GigabitEthernet1</name>
      </interface>
```

```
      <interface>
        <name>GigabitEthernet2</name>
      </interface>
      <interface>
        <name>GigabitEthernet3</name>
      </interface>
      <interface>
        <name>Vlan1</name>
      </interface>
      <interface>
        <name>Vlan20</name>
      </interface>
      <interface>
        <name>Vlan21</name>
      </interface>
    </interfaces>
  </data>
</rpc-reply>
```

To query the operational values from the interface, you use the same mechanism, but querying the operational tree. For example, to get the operational interface names and states, you can use the RPC shown in Figure 13-8.

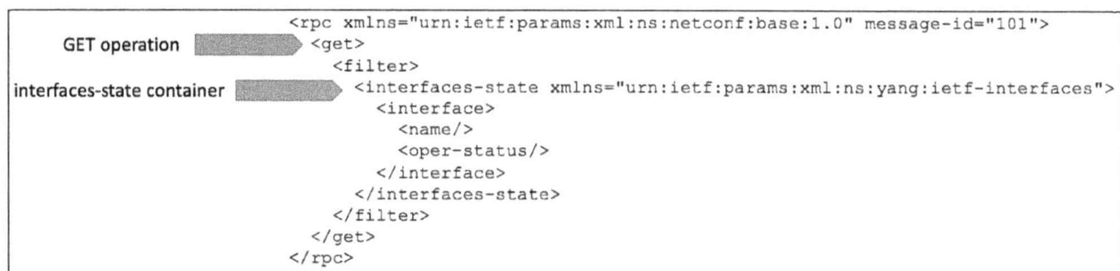

Figure 13-8 *NETCONF RPC to read the interfaces operational state*

Example 13-9 shows the RPC reply for the **get** query to the operational container.

Example 13-9 *RPC Reply for the get Query to the Operational Container*

```
<?xml version="1.0" ?>
<rpc-reply message-id="urn:uuid:7966a6d6-9bc5-472a-bb6c-e487c8383099" xmlns=
  "urn:ietf:params:xml:ns:netconf:base:1.0" xmlns:nc="urn:ietf:params:xml:ns:
  netconf:base:1.0">
  <data>
   <interfaces-state xmlns="urn:ietf:params:xml:ns:yang:ietf-interfaces">
```

```
      <interface>
        <name>GigabitEthernet1</name>
        <oper-status>up</oper-status>
      </interface>
      <interface>
        <name>GigabitEthernet2</name>
        <oper-status>up</oper-status>
      </interface>
      <interface>
        <name>GigabitEthernet3</name>
        <oper-status>up</oper-status>
      </interface>
      <interface>
        <name>Vlan1</name>
        <oper-status>down</oper-status>
      </interface>
      <interface>
        <name>Vlan20</name>
        <oper-status>up</oper-status>
      </interface>
      <interface>
        <name>Vlan21</name>
        <oper-status>up</oper-status>
      </interface>
    </interfaces-state>
  </data>
</rpc-reply>
```

By examining the YANG operational models, you can learn about the operational data available from the C9800 wireless controller. However, not all the data that the controller is exposing might be relevant for your use case.

KPI stands for key performance indicator; KPIs are the set of indicators relevant for your use case, from among all the available ones. Sometimes a KPI can be one of the values the controller provides for you—for example, the number of connected clients. But KPIs can also be derived from a set of metrics. For example, DNA Assurance creates KPIs like "System health" or "Data plane connectivity" by combining data from different operational data sources.

Given the vast amount of data that the Cisco 9800 wireless controller makes available to you by using model-driven telemetry, one of your first tasks is to identify what subset and combination of data is relevant to you. In the "Tools" section later in this chapter, you learn how to use open-source tools to create relevant KPIs from the model-driven telemetry data.

Before going there, let's first examine the different ways to extract information from the C9800: polling and subscribing.

Polling vs. Subscribing

As discussed in the preceding section, while you could periodically poll the controller to retrieve the desired metrics, one of the key advantages of MDT is the ability to subscribe to certain events.

A subscription is used to specify a set of data, when this data is required, which protocol is used to send it, and the destination of the data. You can use different methods to create subscriptions: CLI, NETCONF, RESTCONF, or gNMI.

Each subscription describes one set of data you're interested in. The maximum number of subscriptions varies with the software release. For IOS-XE version 17.6.1, the limit is 100 subscriptions.

Part of the subscription definition is when you are interested in retrieving the value. There are two possibilities: periodic and on-change.

On a *periodic* subscription, you specify how often you need a snapshot of the value. For example, if you are monitoring the traffic on a particular interface, you might want to get this information every 60 seconds, regardless of how much the data has changed during that period.

For the *on-change* subscriptions, the controller sends data as soon as there is any change. For example, you might want to subscribe to the number of APs joined. As soon as this number changes (because an AP disconnected from the controller, for example), the controller notifies the change to the subscription destination. You can see both types in Figure 13-9.

Telemetry Publication: Periodic

Counters/Measures

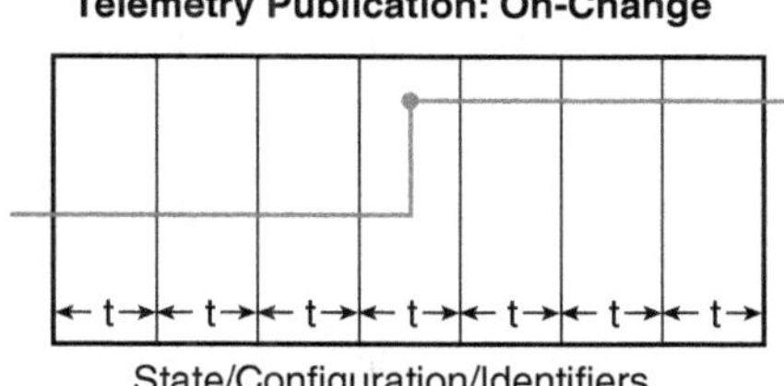

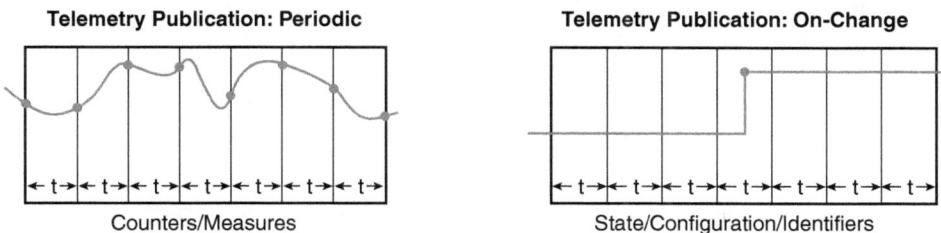

Figure 13-9 *Periodic versus on-change publications*

In the case of an on-change subscription, because the controller notifies the receiver of any change, if multiple changes happen quickly (for example, a network issue causes hundreds of access points to unjoin), the controller has to send a high number of almost simultaneous messages, which could cause performance problems.

To avoid this situation, when you configure an on-change subscription, you could specify a dampening period. This is the interval that must pass before successive update records for the same subscription are generated. After the dampening period passes, the controller sends a single notification with the value at the end of the period. As of release 17.6.1, the only value supported for dampening is 0, indicating dampening is not supported.

Telemetry Streams

A telemetry stream is a set of related events, matching particular criteria. To create a subscription, you define both a stream and a filter.

The stream defines the set of events that can be subscribed to. The filter narrows down the actual data you're interested in. You can find more information about streams in RFC 8639, "Subscriptions to YANG Notifications," and RFC 8641, "Subscription to YANG Notifications for Datastore Updates."

As of release 17.6, the Cisco Catalyst 9800 wireless controller supports two streams:

- Yang-notif-native
- Yang-push

There is an additional stream, native, that is used to send telemetry to Cisco DNA Center, and it is not supported to use with any other product.

Yang-notif-native Stream

The Yang-notif-native stream (defined in RFC 6020) supports any notification where the source is one of the IOS-XE native models. This stream also supports an XPath filter. However, only on-change notifications are supported.

Yang-push Stream

Yang-push stream (based on RFC 8641) adds multiple benefits over Yang-notif-native stream. It contains all data in the configuration and operational databases that is supported in the YANG models. To refine the actual set of information you're interested in, you specify a filter, using XPath. The XPath filter, described in the next section, describes where in the YANG tree the data of interest is, and any filtering you might want to apply to it.

When using Yang-push, you can use either periodic or on-change subscriptions. However, not all YANG subtrees support on-change subscriptions. If you try to create an on-change subscription to a subtree that does not support it, you receive an error.

Periodic subscriptions are hierarchical: if you subscribe to a particular subtree, you receive data from all the leaves under that tree. However, on-change subscriptions are not hierarchical: you receive the data just for the leaf you subscribed to.

This stream is the most common to use unless the device querying the C9800 only supports Yang-notif-native.

How to Identify Subtrees in YANG Models

As described in the previous section, to create subscriptions you need to specify the YANG model subtree you want to subscribe to. YANG relies on XML Path Language (XPath) 1.0 as a notation to specify inter-node references.

Although you can construct the XPath manually by reading the YANG model, YANG Suite provides utilities to identify the XPath to access any of the elements. You construct the XPath from two parts: the module prefix and the element path. The module prefix is identified by the **prefix** statement on the module. You can see this either by looking at the YANG module itself, as shown in Figure 13-10, or by using YANG Suite, as shown in Figure 13-11.

Module prefix

```
module Cisco-IOS-XE-wireless-ap-global-oper {
  yang-version 1;
  namespace "http://cisco.com/ns/yang/Cisco-IOS-XE-wireless-ap-global-oper";
  prefix wireless-ap-global-oper;

  import ietf-yang-types {
    prefix yang;
  }
  import cisco-semver {
    prefix cisco-semver;
  }

  organization
    "Cisco Systems, Inc.";
```

Figure 13-10 *YANG module prefix in the YANG file*

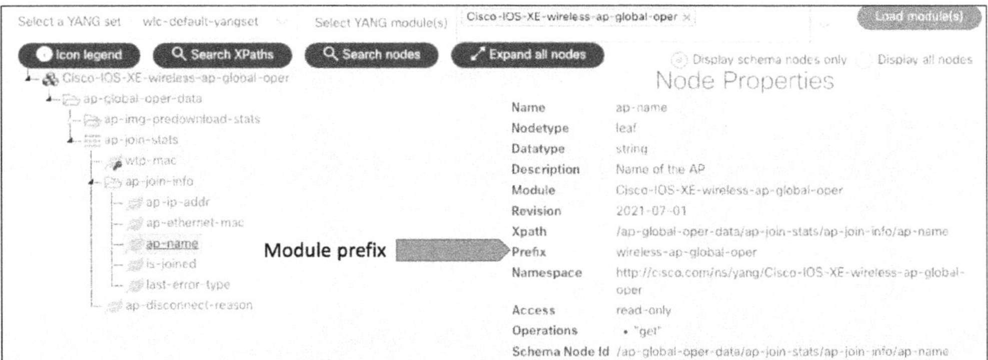

Figure 13-11 *YANG module prefix as shown in YANG Suite*

To get the element XPath, you can use YANG Suite, as shown in Figure 13-12.

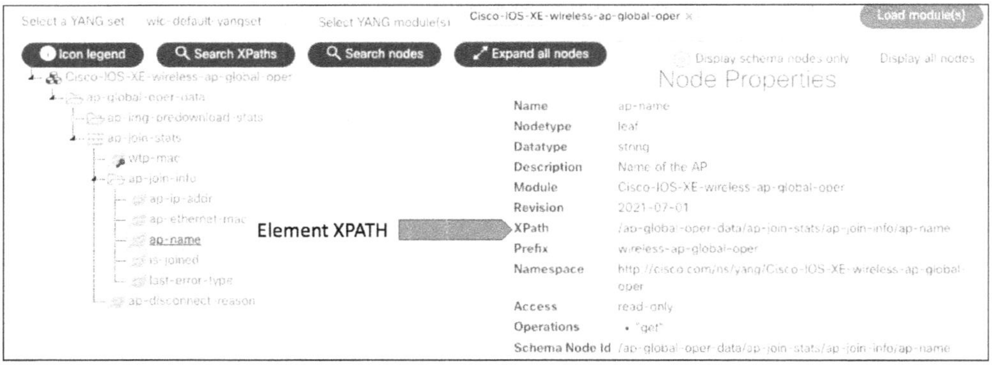
Select a YANG set
wlc-default-yangset
Select YANG module(s)
Cisco-IOS-XE-wireless-ap-global-oper
Load module(s)
Icon legend
Search XPaths
Search nodes
Expand all nodes
Display schema nodes only
Display all nodes
Node Properties
Cisco-IOS-XE-wireless-ap-global-oper
ap-global-oper-data
ap-img-predownload-stats
ap-join-stats
wtp-mac
ap-join-info
ap-ip-addr
ap-ethernet-mac
ap-name
is-joined
last-error-type
ap-disconnect-reason
Element XPATH
Name
ap-name
Nodetype
leaf
Datatype
string
Description
Name of the AP
Module
Cisco-IOS-XE-wireless-ap-global-oper
Revision
2021-07-01
XPath
/ap-global-oper-data/ap-join-stats/ap-join-info/ap-name
Prefix
wireless-ap-global-oper
Namespace
http://cisco.com/ns/yang/Cisco-IOS-XE-wireless-ap-global-oper
Access
read-only
Operations
"get"
Schema Node Id
/ap-global-oper-data/ap-join-stats/ap-join-info/ap-name

Figure 13-12 *Element XPath as shown in YANG Suite*

The final XPath would be as shown in Figure 13-13.

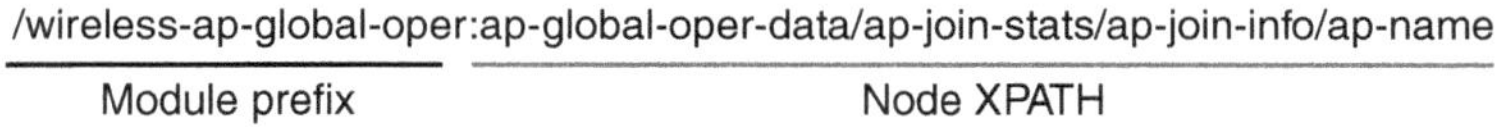
/wireless-ap-global-oper:ap-global-oper-data/ap-join-stats/ap-join-info/ap-name
Module prefix
Node XPATH

Figure 13-13 *Composing a full XPath from the information in YANG Suite*

Dial-out vs. Dial-in

There are two ways to create subscriptions on the C9800: dial-out and dial-in. The terms *dial-out* and *dial-in* refer to which entity is starting the connection.

Dial-out

In Figure 13-14, you can see the flow for a dial-out subscription. First, the management agent connects to the C9800 and creates a subscription, using CLI, NETCONF, RESTCONF, or any other method.

This subscription becomes part of the C9800 configuration. Example 13-10 shows a configuration to create a periodic subscription to the name of joined APs and will send the information using gRPC to the receiver on IP 192.168.20.150 every 500 1/100ths of a second (5 seconds).

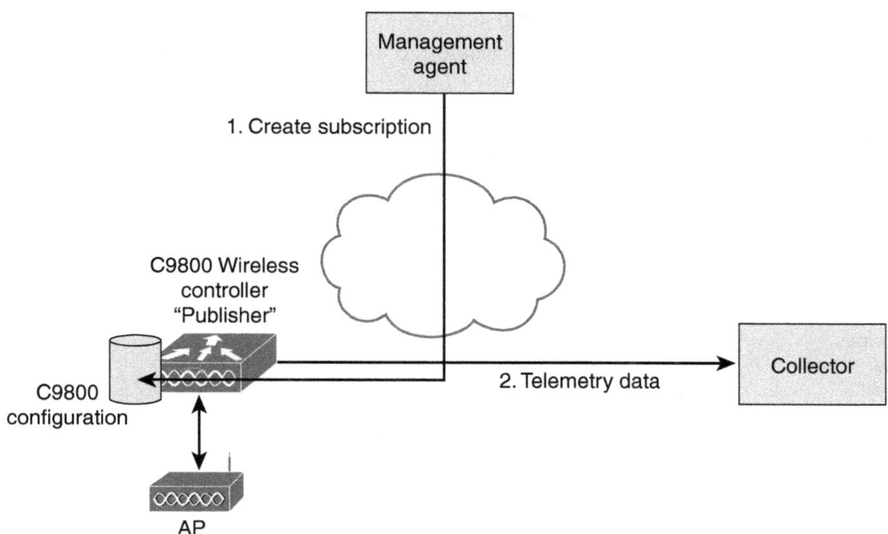

Figure 13-14 *Flow for dial-out telemetry subscriptions*

Example 13-10 *IOS-XE Configuration to Create a Periodic Subscription*

```
telemetry ietf subscription 1
 encoding encode-kvgpb
 filter xpath /wireless-ap-global-oper:ap-global-oper-data/ap-join-stats/ap-join-
  info/ap-name
 source-address 192.168.20.150
 stream yang-push
 update-policy periodic 500
 receiver ip address 192.168.30.188 57000 protocol grpc-tcp
```

The C9800 constantly tries to open a gRPC connection to the indicated IP address to send the telemetry data. Because the C9800 starts the connection process, this mode is called dial-out. The subscription is part of the C9800 configuration, so it stays there until it's manually removed. If there is a Stateful Switch Over (SSO) or reload even, the connection is automatically reestablished.

This is the type of subscription that Cisco DNA Center configures. As described previously, Cisco DNA Center uses the native stream, which is not available for use by other agents.

Dial-in

Instead of having the subscription as part of the C9800 configuration, another option is to have an external agent connecting to it and requesting a subscription on-the-fly. The C9800 then uses this same inbound connection to send the relevant data.

Different agents can connect to the C9800 at the same time and ask for different subscriptions. Each of them receives the ones requested on the same session and, if the session is disconnected, all the relevant telemetry subscriptions are removed.

While dial-out subscriptions are static, stored on the C9800 configuration, dial-in subscriptions are dynamic, created on-the-fly by an incoming connection. Figure 13-15 illustrates the flow for dial-in subscriptions.

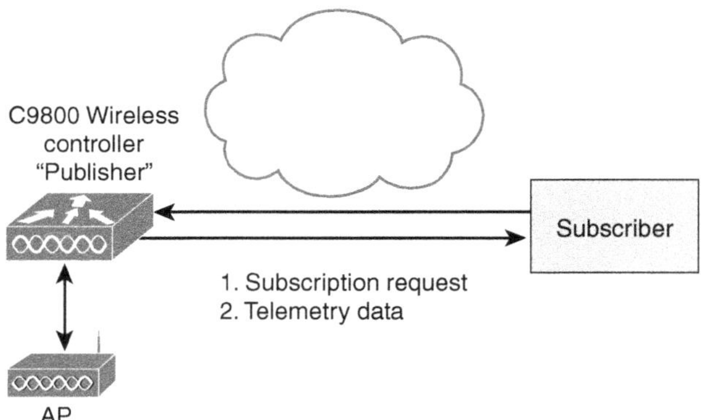

Figure 13-15 *Flow for dial-in telemetry subscriptions*

In Table 13-1, you can see the key differences between dial-out and dial-in MDT.

Table 13-1 *Key Differences Between Dial-out and Dial-in Telemetry Subscriptions*

Dial-Out	Dial-In
Subscriptions are statically configured.	Subscriptions are created dynamically.
Telemetry data is sent to the specified collector.	Telemetry data is sent to the connection initiator (the subscriber).
Subscriptions reconnect to the receiver after a reload or a stateful switchover.	Subscriptions need to be reinitiated after a reload or stateful switchover because sessions are killed on switchover.
Subscription data remains on the C9800 configuration.	There is no change on C9800 configuration, subscriptions are tied to the session that created them and sent on the same session.
Subscription ID is fixed and configured on the device.	Subscription ID is dynamically generated.

Not every protocol used for model-driven telemetry supports dial-in or dial-out. Table 13-2 illustrates the supported combination of protocols, streams, and encodings as per release 17.6. Additionally, on Embedded Wireless Controller in AP (EWC-AP), gRPC is not supported. That implies no dial-out support for Embedded Wireless Controller in AP, as per release 17.6.

Table 13-2 *Supported Protocol/Stream/Encoding Combinations for Both Dial-in and Dial-out Telemetry Subscriptions*

Transport Protocol	NETCONF		gRPC		gNMI	
	Dial-in	Dial-out	Dial-in	Dial-out	Dial-in	Dial-out
Stream						
Yang-push	Yes	No	No	Yes	Yes	No
Yang-notif-native	Yes	No	No	No	No	No
Encoding	XML			Key-value Google Protocol Buffers (kvGPB)	JSON	

Creating Dial-in Subscriptions

To create dial-in subscriptions, you need to use, as per Table 13-2, either NETCONF or gNMI. There are several tools that you can use to connect to the C9800 using either of those protocols. In Figure 13-16, you can see how to use Python code to send a NETCONF subscription.

```
from ncclient import manager
from ncclient.xml_ import to_ele

rpc = """
<establish-subscription xmlns="urn:ietf:params:xml:ns:yang:ietf-event-notifications"
 xmlns:yp="urn:ietf:params:xml:ns:yang:ietf-yang-push">
    <stream>yp:yang-push</stream>
    <yp:xpath-filter>
        /wireless-ap-global-oper:ap-global-oper-data/ap-join-stats/ap-join-info/ap-name
    </yp:xpath-filter>
    <yp:period>500</yp:period>
</establish-subscription>
"""

host = "192.168.20.150"
username = "lab"
password = "lab"

with manager.connect(host=host,
                     port=830,
                     username=username,
                     password=password,
                     timeout=90,
                     hostkey_verify=False) as m:

    response = m.dispatch(to_ele(rpc))
    print("Waiting for notify...")
    while True:
        n = m.take_notification()
        print(n.notification_xml)

```

Figure 13-16 *Python code to create a dial-in telemetry subscription*

This example uses the ncclient library to connect to the C9800 and send a NETCONF RPC. You need to adapt lines 15–17 to the username, password, and IP of the C9800.

The critical piece of information on this example is the RPC you use. You can see it specifies the following:

Stream to use, in line 7:

```
<stream>yp:yang-push</stream>
```

XPath filter, in line 9

```
/wireless-ap-global-oper:ap-global-oper-data/ap-join-stats/
ap-join-info
```

Subscription period, in line 11:

```
<yp:period>500</yp:period>
```

Because you are specifying a period, this is a periodic (instead of on-change) subscription. If you run this code, you see periodic notifications of the joined APs (output edited for clarity). You can see two notifications about the same AP, sent 5 seconds apart from each other, as seen in the **eventTime** field shown in Example 13-11. This notification is sent irrespective of any change happening on the subscribed data. Looking at the notifications, you can see the only change between them is the field **eventTime**.

Example 13-11 *Subscribing to Notifications Using NETCONF*

```
$ python sub.py
Waiting for notify...
<?xml version="1.0" ?>
<notification xmlns="urn:ietf:params:xml:ns:netconf:notification:1.0">
  <eventTime>2022-01-16T11:16:01.29Z</eventTime>
  <push-update xmlns="urn:ietf:params:xml:ns:yang:ietf-yang-push">
    <subscription-id>2147483658</subscription-id>
    <datastore-contents-xml>
      <ap-global-oper-data xmlns="http://cisco.com/ns/yang/Cisco-IOS-XE-wireless-
  ap-global-oper">
        <ap-join-stats>
          <wtp-mac>CC:16:7E:30:59:00</wtp-mac>
          <ap-join-info>
            <ap-ip-addr>192.168.20.72</ap-ip-addr>
            <ap-ethernet-mac>58:AC:78:DE:8D:0E</ap-ethernet-mac>
            <ap-name>AP58AC-78DE-8D0E</ap-name>
           <is-joined>true</is-joined>
            <last-error-type>ap-con-failure-run</last-error-type>
          </ap-join-info>
        </ap-join-stats>
      </ap-global-oper-data>
```

```
    </datastore-contents-xml>
  </push-update>
</notification>

<?xml version="1.0" ?>
<notification xmlns="urn:ietf:params:xml:ns:netconf:notification:1.0">
  <eventTime>2022-01-16T11:16:06.29Z</eventTime>
  <push-update xmlns="urn:ietf:params:xml:ns:yang:ietf-yang-push">
    <subscription-id>2147483658</subscription-id>
    <datastore-contents-xml>
      <ap-global-oper-data xmlns="http://cisco.com/ns/yang/Cisco-IOS-XE-wireless-
  ap-global-oper">
        <ap-join-stats>
          <wtp-mac>CC:16:7E:30:59:00</wtp-mac>
          <ap-join-info>
            <ap-ip-addr>192.168.20.72</ap-ip-addr>
            <ap-ethernet-mac>58:AC:78:DE:8D:0E</ap-ethernet-mac>
            <ap-name>AP58AC-78DE-8D0E</ap-name>
            <is-joined>true</is-joined>
            <last-error-type>ap-con-failure-run</last-error-type>
          </ap-join-info>
        </ap-join-stats>
      </ap-global-oper-data>
    </datastore-contents-xml>
  </push-update>
</notification>
```

While this program is running, you can run **show telemetry ietf subscription all** on the C9800, as shown in Example 13-12.

Example 13-12 *Running show telemetry ietf subscription all*

```
C9800-telemetry# show telemetry ietf subscription all
  Telemetry subscription brief

  ID                Type        State       Filter type
  --------------------------------------------------------
  1                 Configured  Valid       xpath
  102               Configured  Valid       xpath
  2147483672        Dynamic     Valid       xpath
```

In this case, there are three subscriptions: two configured (dial-out) and one dynamic (dial-in). If you stop the Python code, notice that the dynamic subscription disappears.

You can change the RPC to specify a different filter and use on-change notifications. For example, let's build the RPC to subscribe to the VLAN information in the C9800. First, you need to find the data model with the information you're interested in. Because you want to learn about the state of the VLANs, the first step is to find an operational model. You can use YANG Suite to search a model, as shown in Figure 13-17.

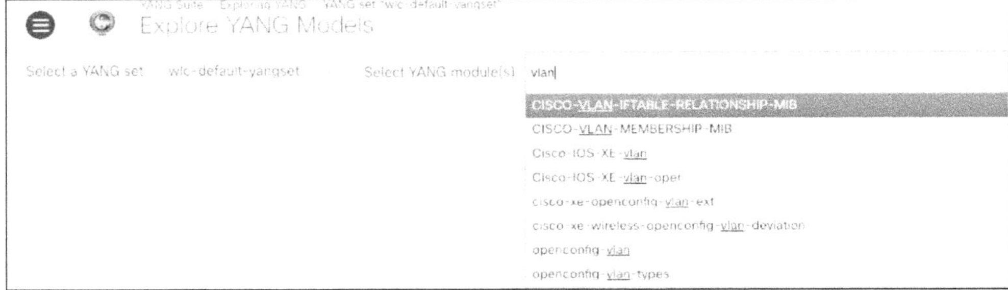

Figure 13-17 *Searching models in YANG Suite*

From the list, you can see a good model would be **Cisco-IOS-XE-vlan-oper**. Select it to get further details, as shown in Figure 13-18.

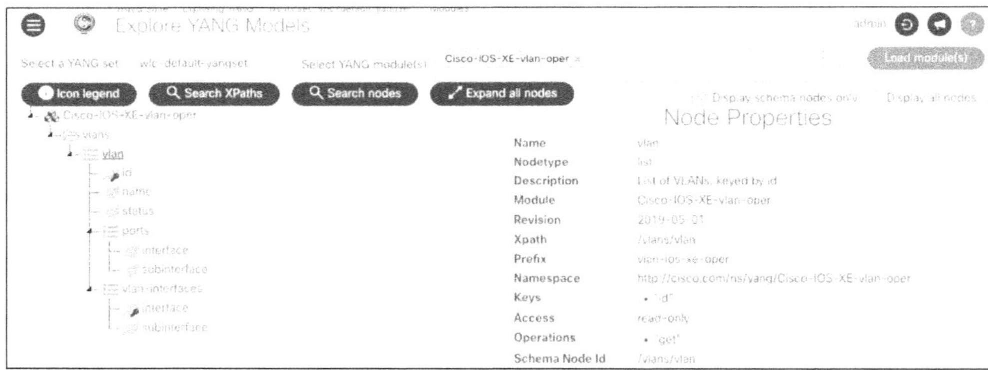

Figure 13-18 *Displaying details of YANG models using YANG Suite*

As described previously, you can use this information to create an XPath filter, by combining the prefix and node XPath:

```
<yp:xpath-filter>
   /vlan-ios-xe-oper:vlans/vlan
</yp:xpath-filter>
```

You can also try to configure an on-change notification. To do that, you need to specify the dampening period, instead of the period. As discussed, only 0 is supported as of release 17.6:

```
<yp:dampening-period>0</yp:dampening-period>
```

With this information, the Python code with the full NETCONF RPC is as shown in Figure 13-19.

```
rpc = """
<establish-subscription xmlns="urn:ietf:params:xml:ns:yang:ietf-event-notifications"
 xmlns:yp="urn:ietf:params:xml:ns:yang:ietf-yang-push">
    <stream>yp:yang-push</stream>
    <yp:xpath-filter>
        /vlan-ios-xe-oper:vlans/vlan
    </yp:xpath-filter>
    <yp:dampening-period>0</yp:dampening-period>
</establish-subscription>
"""
```

Figure 13-19 *Python code with the subscription NETCONF RPC*

If you leave the code running and create some VLANs, notifications start arriving to your Python code.

On initial subscription creation, you get the full VLAN table information (XML edited for clarity), as shown in Example 13-13.

Example 13-13 *Subscribing to VLAN Changes Using NETCONF*

```
$ python notif.py

Waiting for notify...
<?xml version="1.0" encoding="UTF-8"?>
<notification xmlns="urn:ietf:params:xml:ns:netconf:notification:1.0">
   <eventTime>2021-10-12T11:47:34.95Z</eventTime>
   <push-update xmlns="urn:ietf:params:xml:ns:yang:ietf-yang-push">
      <subscription-id>2147483679</subscription-id>
      <datastore-contents-xml>
         <vlans xmlns="http://cisco.com/ns/yang/Cisco-IOS-XE-vlan-oper">
            <vlan>
               <id>20</id>
               <name>VLAN0020</name>
               <status>active</status>
            </vlan>
            <vlan>
               <id>21</id>
               <name>VLAN0021</name>
               <status>active</status>
            </vlan>
            <vlan>
               <id>1</id>
               <name>default</name>
               <status>active</status>
            </vlan>
```

```
          <vlan>
            <id>1005</id>
            <name>trnet-default</name>
            <status>suspend</status>
          </vlan>
        </vlans>
      </datastore-contents-xml>
    </push-update>
</notification>
```

Leave the code running and create a new VLAN on the C9800, as shown in Example 13-14.

Example 13-14 *Creating a New VLAN*

```
C9800-telemetry# conf t
Enter configuration commands, one per line. End with CNTL/Z.
C9800-telemetry(config)# vlan 1234
C9800-telemetry(config-vlan)# end
C9800-telemetry#
```

Immediately, you receive a new notification. Note **operation=merge**, with the value representing the new VLAN, as shown in Example 13-15.

Example 13-15 *Live NETCONF Notification*

```
<?xml version="1.0" encoding="UTF-8"?>
<notification xmlns="urn:ietf:params:xml:ns:netconf:notification:1.0">
   <eventTime>2021-10-12T11:49:52.64Z</eventTime>
   <push-change-update xmlns="urn:ietf:params:xml:ns:yang:ietf-yang-push">
      <subscription-id>2147483679</subscription-id>
      <datastore-changes-xml>
         <yang-patch xmlns="urn:ietf:params:xml:ns:yang:ietf-yang-patch">
            <patch-id>null</patch-id>
            <edit>
               <edit-id>edit1</edit-id>
               <operation>merge</operation>
               <target>/vlans/vlan=1234</target>
               <value>
                  <vlan xmlns="http://cisco.com/ns/yang/Cisco-IOS-XE-vlan-oper">
                     <name>VLAN1234</name>
                     <status>active</status>
                  </vlan>
               </value>
```

```
            </edit>
         </yang-patch>
      </datastore-changes-xml>
   </push-change-update>
</notification>
```

If you delete the VLAN, you receive another notification, as shown in Example 13-16.

Example 13-16 *Deleting a VLAN in the C9800 and Observing the NETCONF Notification*

```
C9800-telemetry# conf t
Enter configuration commands, one per line. End with CNTL/Z.
C9800-telemetry(config)# no vlan 1234
C9800-telemetry(config)# ^Z
C9800-telemetry#

<?xml version="1.0" encoding="UTF-8"?>
<notification xmlns="urn:ietf:params:xml:ns:netconf:notification:1.0">
   <eventTime>2021-10-12T11:51:59.52Z</eventTime>
   <push-change-update xmlns="urn:ietf:params:xml:ns:yang:ietf-yang-push">
      <subscription-id>2147483679</subscription-id>
      <datastore-changes-xml>
         <yang-patch xmlns="urn:ietf:params:xml:ns:yang:ietf-yang-patch">
            <patch-id>null</patch-id>
            <edit>
               <edit-id>edit1</edit-id>
               <operation>delete</operation>
               <target>/vlans/vlan=1234</target>
               <value>
                  <vlan xmlns="http://cisco.com/ns/yang/Cisco-IOS-XE-vlan-oper">
                     <name>VLAN1234</name>
                     <status>active</status>
                  </vlan>
               </value>
            </edit>
         </yang-patch>
      </datastore-changes-xml>
   </push-change-update>
</notification>
```

Note the **operation=delete** setting on this notification.

Not all the models support on-change notification. If you try to create an on-change notification to a model that does not support it, the subscription fails, and you get a message like this on the console:

```
Oct 12 10:00:05.810: %MDT_SUBSCRIPTION-4-NOT_SUPPORTED: Chassis 1
R0/0: pubd: Subscription creation failed (stream yang-push, id none,
client 192.168.30.188:56735): Value 'on-change' not supported for
parameter 'update-trigger' (supported values include 'periodic').
```

Tools

Multiple tools are available to take advantage of model-driven telemetry. We explore two of them: YANG Suite and a set of tools called TIG (Telegraf, Influx, Grafana).

YANG Suite

In the previous chapter, we described how to use YANG Suite to explore YANG models and send RPCs to retrieve data from the C9800 YANG models.

You can also use YANG Suite as a subscriber for dial-out telemetry. For dial-out subscriptions, the C9800 tries to establish a connection to the indicated IP address. That implies that the subscriber (for example, it could be YANG Suite running as a container in your computer) needs to be reachable from the C9800.

In a production environment, firewalls probably make it impossible for the C9800 to reach your computer. However, in testing or learning environments, it could be possible.

If you are running YANG Suite in your computer using Docker, you need to make sure the port you want to use for dial-out telemetry is exposed in Docker, as shown in Figure 13-20.

```
version: '3'
services:
    yangsuite:
      image: yangsuite:latest
      build:
        context: ./yangsuite
      env_file:
        - ./yangsuite/setup.env
      command: /yangsuite/migrate_and_start.sh
      ports:
        - "50052:50052"
        - "50051:50051"
        - "9339:9339"
        - "57344:57344"
        - "57345:57345"
        - "443:443"
```

Figure 13-20 *Ports exposed in a docker-compose file*

In this example, ports 57344 and 57345 (among others) are reachable from the external network. You also need the IP address of the computer where you're running YANG Suite. With this information, you can configure the C9800 for dial-out telemetry, as shown in Example 13-17.

Example 13-17 *Configuring the C9800 to Use Dial-out Telemetry*

```
telemetry ietf subscription 2
 encoding encode-kvgpb
 filter xpath /wireless-ap-global-oper:ap-global-oper-data/ap-join-stats/
  ap-join-info/ap-name
 stream yang-push
 update-policy periodic 500
 receiver ip address 192.168.30.188 57345 protocol grpc-tcp
```

Then you can start the YANG Suite telemetry receiver, and the records appear, as shown in Figure 13-21.

Figure 13-21 *YANG Suite receiving gRPC dial-out telemetry*

TIG (Telegraf, Influx, Grafana)

YANG Suite is a great development and learning tool, but after you have identified the correct YANG models to use for telemetry, you need a different tool.

When using model-driven telemetry, the Catalyst 9800 wireless controller streams metrics based on YANG models. Where are those metrics collected in real-world scenarios?

One of the most common setups is to use a set of tools, like Telegraf, Influx, and Grafana (known as the *TIG stack*). The first one, Telegraf, serves as an aggregator and translator: it collects telemetry data from a variety of sources, and output in a common format, as seen in Figure 13-22. At the time of this writing, there are 218 different input plug-ins and 51 different output plug-ins for Telegraf.

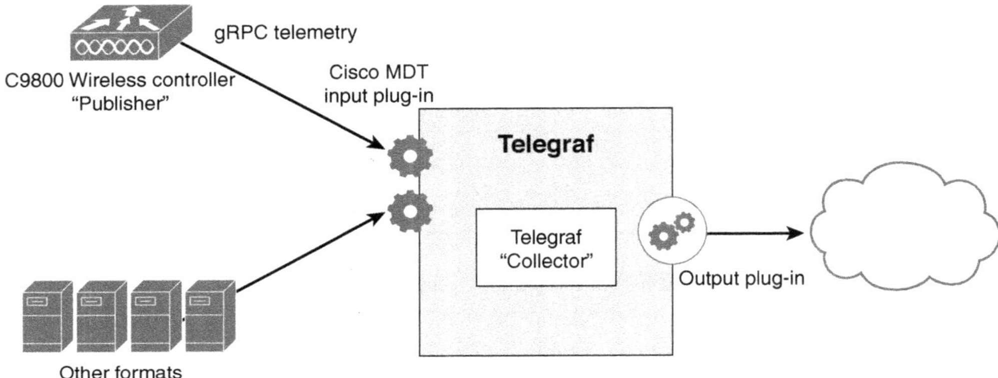

Figure 13-22 *Telegraf serving as an aggregator for telemetry data*

When the data is aggregated and in a common format, it is sent to a special type of database, called a *time-series database (TSDB)*. This type of database specializes in storing time-stamped, repetitive data, like the data produced by telemetry, as shown in Figure 13-23.

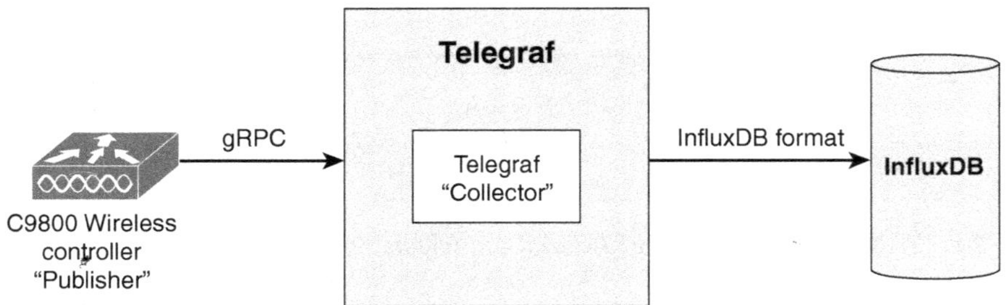

Figure 13-23 *Sending data to InfluxDB*

When the data is in a time-series database, you need a way to visualize the contents. On the TIG stack, that's what Grafana is used for, as shown in Figure 13-24.

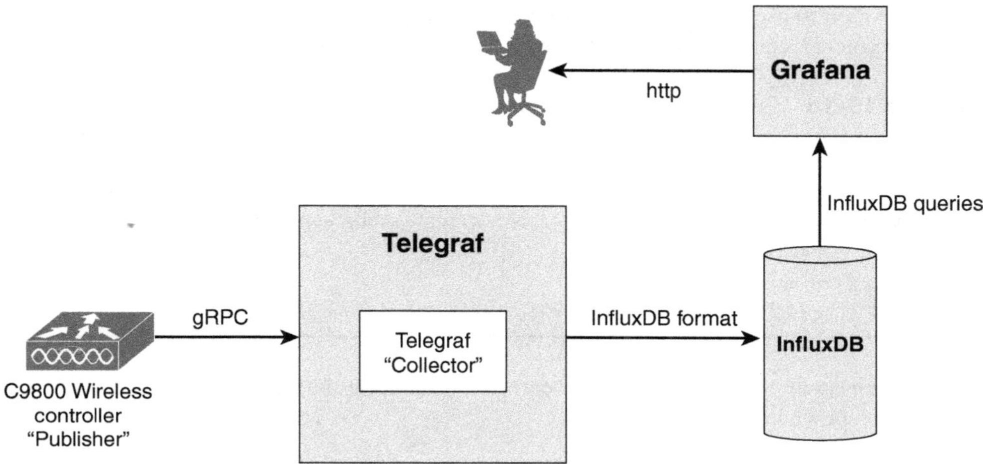

Figure 13-24 *Grafana receiving data from InfluxDB*

Creating a Dashboard

Let's put all the pieces together and build a dashboard to show the number of connected clients over time and also how many of them are in RUN state and how many in AUTH state.

Over the total number, you also want to see how many clients are using a random MAC address. The first step is to identify the metrics:

- Number of clients in RUN state
- Number of clients in AUTH state
- Number of clients using random MAC

Those metrics should be in the client model, and they are operational values. Using YANG Explorer, you can easily identify the model and the XPath, as seen in Figure 13-25.

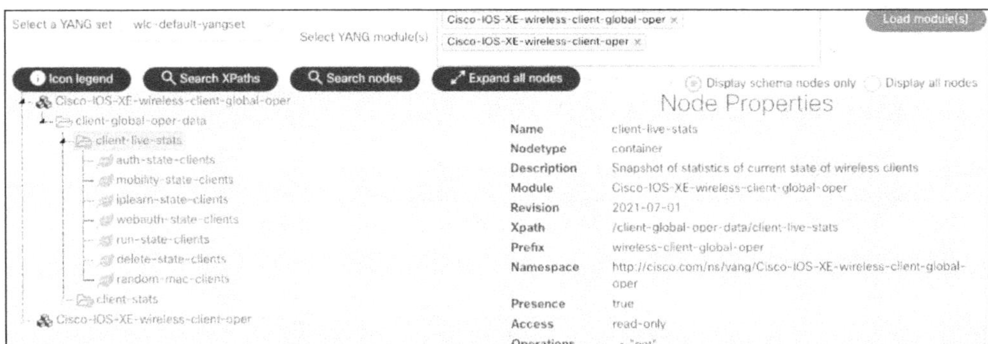

Figure 13-25 *Identifying prefix and XPath using YANG Suite*

The XPath to use is **/wireless-client-global-oper:client-global-oper-data/client-live-stats.** Now you can create the dial-out telemetry, as shown in Example 13-18.

Example 13-18 *Creation of Subscription to Client Live Stats*

```
telemetry ietf subscription 23
 encoding encode-kvgpb
 filter xpath /wireless-client-global-oper:client-global-oper-data/client-live-stats
 stream yang-push
 update-policy periodic 1000
 receiver ip address 192.168.30.188 57000 protocol grpc-tcp
```

You can check whether the XPath is correct by checking the **state** field on the C9800 CLI, as shown in Example 13-19.

Example 13-19 *Checking Status of the Telemetry Subscription*

```
C9800-telemetry# show telemetry ietf subscription 23 detail
Telemetry subscription detail:

  Subscription ID: 23
  Type: Configured
  State: Valid
  Stream: yang-push
  Filter:
    Filter type: xpath
    XPath: /wireless-client-global-oper:client-global-oper-data/client-live-stats
  Update policy:
    Update Trigger: periodic
    Period: 1000
  Encoding: encode-kvgpb
  Source VRF:
  Source Address:
  Notes:
```

You can use the CLI **show telemetry internal subscription all stats** to check the number of updates sent, as shown in Example 13-20.

Example 13-20 *Checking Stats of Telemetry Subscriptions*

```
C9800-telemetry# show telemetry internal subscription all stats
Telemetry subscription stats:

Subscription ID  Connection Info    Msgs Sent  Msgs Drop  Records Sent
---------------- ------------------ ---------- ---------- ------------
23                                          28          0            0
```

On Grafana, you can create a dashboard showing the metrics you're interested in. The first step is to configure the database connection, as shown on Figure 13-26.

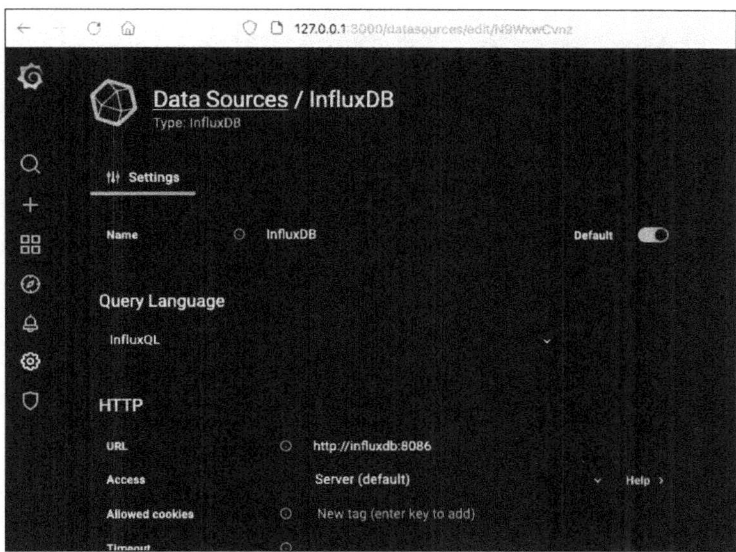

Figure 13-26 *Configuring the database connection in Grafana*

Then you can create the dashboard. Grafana has an interactive option, which you can use to easily select the fields to draw. Clicking on the **measurement** field displays the available measurements (YANG models) in Influx, as seen in Figure 13-27.

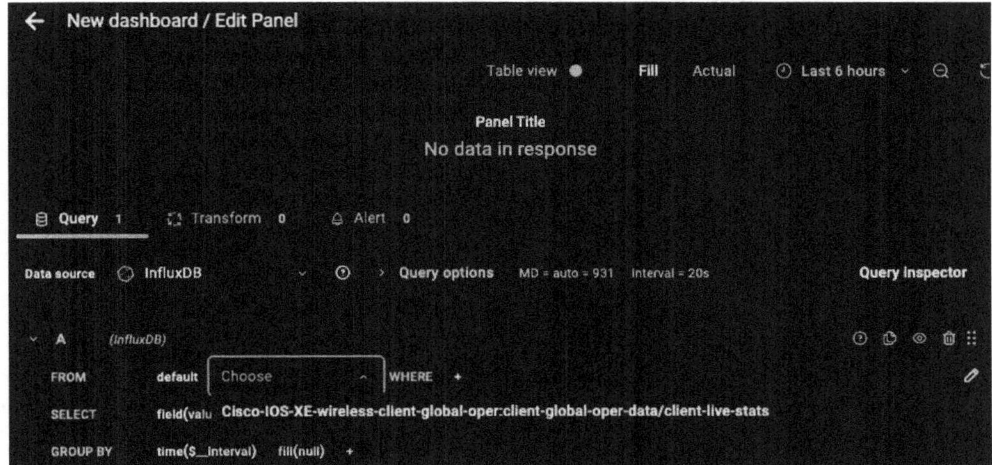

Figure 13-27 *Displaying available measurements in Grafana*

When you select the YANG model, clicking **Field** displays all the available metrics for that model, as seen in Figure 13-28.

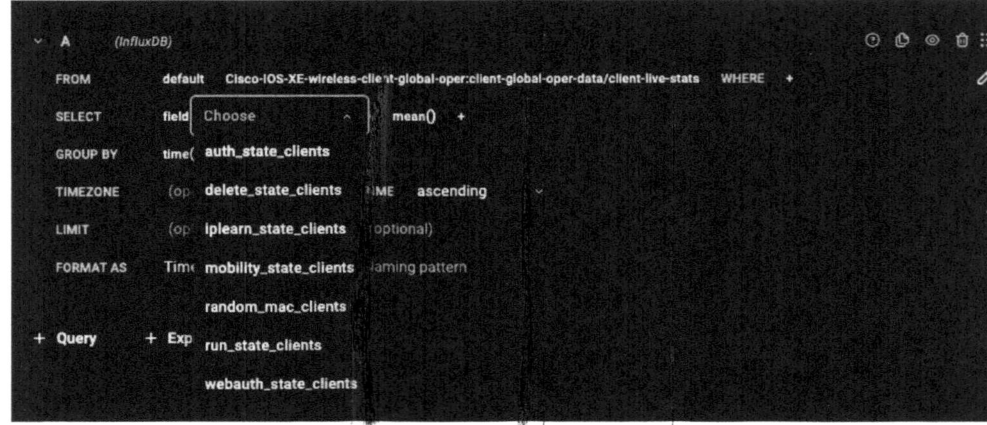

Figure 13-28 *Displaying available metrics in Grafana*

After completing the selections, you can see the metrics over time in the dashboard. Figure 13-29 shows the final selections to create the dashboard:

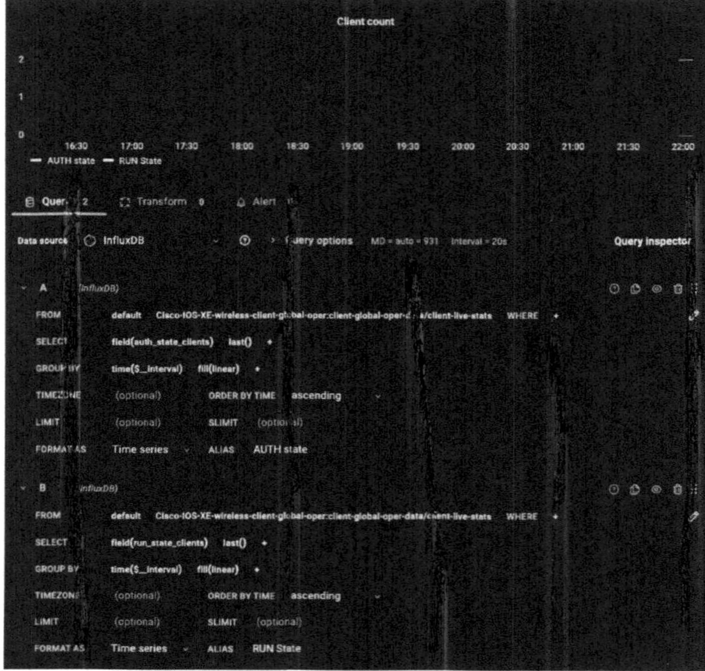

Figure 13-29 *Grafana configuration*

The final dashboard, in Figure 13-30, also shows gauges for total and random MAC clients.

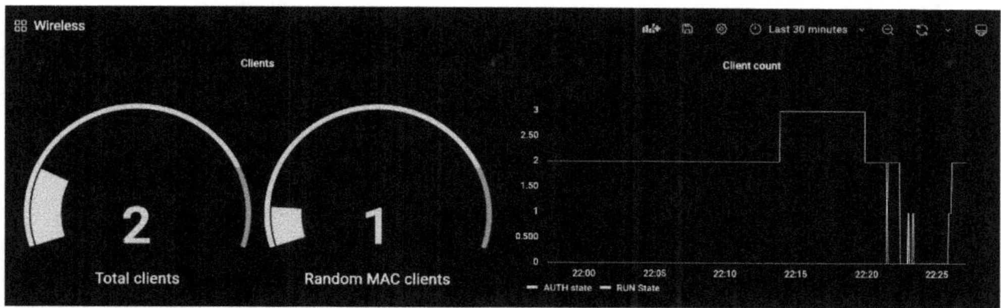

Figure 13-30 *Final Grafana dashboard*

Summary

This chapter describes how you can benefit from model-driven telemetry on the Catalyst 9800 wireless controller and how to integrate the rich data models available with open-source tools.

References

RFC 6020: YANG—A Data Modeling Language for the Network Configuration Protocol (NETCONF): https://datatracker.ietf.org/doc/html/rfc6020

RFC 8639: Subscriptions to YANG Notifications: https://datatracker.ietf.org/doc/html/rfc8639

RFC 8641: Subscription to YANG Notifications for Datastore Updates: https://datatracker.ietf.org/doc/html/rfc8641

YANG Suite: https://developer.cisco.com/yangsuite/

Ncclient Python library: https://github.com/ncclient/ncclient

Repository chapter examples: https://github.com/fsedano/cisco_9800_programmability

Chapter 14

Cisco DNA Center/Assurance Integration

Introduction

The Catalyst 9800 wireless controller brings many improvements with regards to monitoring the wireless network and clients and gives many details that may help you troubleshoot a problem as it is reported. The C9800 also keeps a deep level of always-on logging stored on the box to assist you in troubleshooting after the fact (this capability is discussed more in Chapter 16, "Troubleshooting"). Cisco DNA Center Assurance builds on this and takes network management and monitoring to the next level. If you think about your wireless network, you probably regularly face some of these questions without having an easy answer:

- Is my network actually up and running? Do any parts of the network need to be looked at? Legacy network management applications monitor whether individual network devices are reachable but not really whether or not they are operating at full health.
- Are wireless clients able to connect? Are they connecting in a timely manner? You can see the clients that are connected only when monitoring your wireless controller. You have limited to no visibility on the clients who are not able to connect. You also have little insight as to whether the onboarding time is very slow due to some problem.
- Do the wireless clients have a stable connection? Or are they mostly connected but sometimes getting disconnected and reconnecting shortly after?
- Are wireless clients able to pass traffic? They may be connected but unable to pass any traffic.
- Are individual network services working at their full health (for example, the RADIUS/AAA server is responding in a timely manner, DHCP server is handing over IP addresses promptly to clients, both controllers in a high availability pair are in sync and ready for failover).

Legacy network monitoring tools leverage the SNMP protocol to get information from the network. SNMP is mostly a polling protocol that is very CPU intensive for the device and relies on you choosing the correct SNMP polling interval for the specific information you are tracking. If the interval is too short, you may overwhelm the device; if the interval is too long, you may miss the occurrence of the problem. Real-time updates are possible thanks to SNMP traps, but they do not really solve the efficiency problem. Although Cisco DNA Center keeps SNMP in its toolbelt (because some information is absolutely suitable to be polled at regular intervals), it heavily relies on model-driven telemetry (MDT) to efficiently retrieve the information from the device (stored using YANG model data structures) and efficiently transmit the data (using encoding technologies like JSON, Google Protocol Buffers, and others) so that you have real-time updates about anything happening in the wireless network.

Getting the right data at the right time with the right context enables Cisco DNA Center Assurance to provide actionable insights, because there is more to receiving data from the network devices and providing you with customizable reports. Cisco DNA Center gathers a massive amount of data, stores that data in its database, processes it, and produces business-relevant dashboards and health scores, enabling you to understand the network health in a blink of an eye so that you don't have to go through lengthy, obscure reports. It is moving toward proactive, rather than reactive, troubleshooting, as illustrated in Figure 14-1. As a result, you, as network administrator, can quickly identify whether a critical network service is not running optimally or if certain wireless clients are having a suboptimal experience.

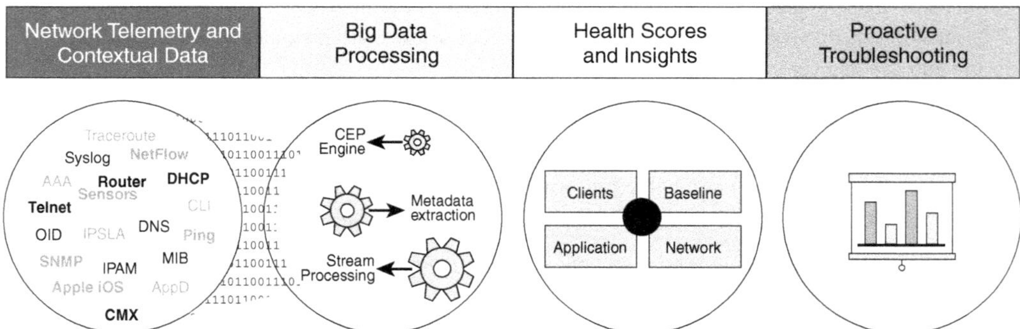

Figure 14-1 *The principles of Cisco DNA Center telemetry and data processing flows*

Knowing there might be a problem before the end users complain to the help desk allows you to resolve the problem before its impact is perceived and hence improve the user experience. This capability is a lifesaver when administrating a network.

Assurance focuses on giving you, the network administrator, visibility into the client devices, the applications, and the network traffic so that you can understand whether your business is running optimally. This is a radically new approach, enabled by various new protocols and tools that are covered in this book.

This chapter covers the basics of Assurance network monitoring and explains how, on top of giving you visibility through health dashboards, Cisco DNA Center also guides you through probable causes for the problem and suggests remediation actions. Therefore, you won't be drowned any longer in thousands of network alarms of various severities and relevance but can use AI-powered network dashboards that point you to the hottest issues in your network and help you solve the problem!

This chapter is based on Cisco DNA Center release 2.2.3.x; please consider this as you refer to functionalities and screenshots shown here.

Cisco DNA Center Assurance Architecture

Figure 14-2 shows a high-level view of Cisco DNA Center architecture highlighting the two major building blocks: the automation and the assurance engines. The DNA Automation part is not covered in detail in this book (to be honest, Cisco DNA Center deserves its own separate book) but is mentioned because it is a key part of the architecture and is required to enable Assurance.

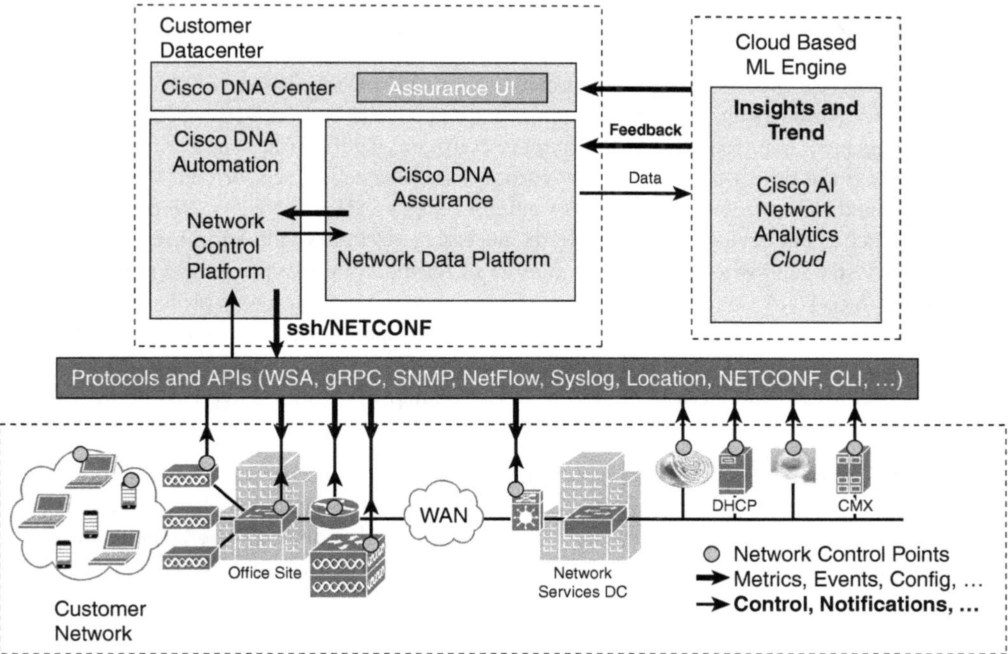

Figure 14-2 *Cisco DNA Center Assurance architecture*

The Cisco DNA Center Automation is referred to as the Network Controller Platform and the Assurance part as the Network Data platform.

Additionally, Cisco provides a cloud-based AI Network Analytics engine, which increases the level of analytics done by Assurance by leveraging machine learning to build a baseline of usage patterns in your network and dynamically identify when any of the performance metrics is abnormal. For example, you may have a complex network with a remote DHCP server that provides an IP address within two or three seconds on busy days, and you might consider this response acceptable. If the onboarding time ever goes significantly above this, the AI Network Analytics immediately identifies it as a severe condition. On the other hand, if you have a very fast DHCP server that always provides an IP address under a second, the same machine learning engine will throw an alarm much sooner if the onboarding time increases due to this factor. The engine is not based on predefined static thresholds but is able to understand what the expected metrics are when your network operates as expected (depending on the load) and can automatically identify any significant deviation from this baseline.

Cisco DNA Center is the overall network controller and management solution encompassing all the components covered in this section and presents everything in a unique web interface or API.

Managing the C9800 with Cisco DNA Center

It is possible and common for customers to use Cisco DNA Center only for Assurance and not Automation, hence relying on something else (a built-in GUI or CLI or another management solution like Cisco Prime) to configure C9800. You would still need to discover the device to load it in the inventory and use Assurance for your C9800 and managed access points. Cisco DNA Center leverages SSH and NETCONF protocols to connect to the device and configure the necessary telemetry subscriptions. This is a mandatory step, and when you see the number of telemetry subscriptions and configurations that Cisco DNA Center pushes on the device, automating this step is probably a good idea anyway. When the network device is in the inventory and managed, you can turn off device controllability if you do not want Cisco DNA Center to modify any further configuration on your network device, which could happen as soon as you assign the device to a site.

At this point, the network device is configured to automatically send the appropriate telemetry data to the Cisco DNA Center appliance. The Assurance collector receives all this data, processes it, and produces the network health indicators, the business insights, and the dashboards you can consult from the web interface or through the API. While the C9800 mostly uses a Cisco proprietary optimized protocol called TDL to send telemetry to the Cisco DNA Center collector both on an on-change and scheduled basis, Assurance leverages many other protocols to reach its goal:

- Wireless Service Assurance is a streaming protocol on top of HTTPS used by AireOS WLCs to provide telemetry.
- gRPC is the telemetry tunnel established between APs and the Cisco DNA Center appliance for features such as IoT management, Cisco Intelligent Capture, or Adaptive Wireless Intrusion Prevention System (aWIPS).

- SNMP is used mostly for legacy devices that do not have NETCONF or TDL telemetry support.
- NetFlow is the key protocol powering the application visibility engine on Cisco DNA Center. It can be sent by the WLC in case of central switching or the APs directly in case of FlexConnect or SDA.
- SYSLOG is a historical protocol implemented by many network devices that provides an easy way to get alerts and notifications.
- SSH is used by Assurance to configure and monitor the device if NETCONF is not enabled (or not configured yet).
- NETCONF is the protocol of choice for grabbing pull-based data (when Cisco DNA Center needs to get a piece of information on demand rather than wait for a telemetry to be sent) as well as configure the network device, when automation is in use.

Figure 14-3 illustrates the Cisco DNA Center inventory page, showing a WLC added and successfully managed. When it is managed, the APs that are joined to that WLC are automatically discovered and appear in the inventory, even if they migrate to another (maybe unmanaged) WLC later.

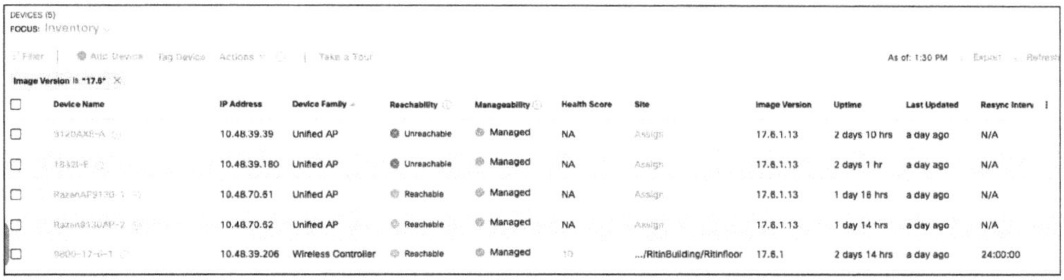

DEVICES (5)
FOCUS: Inventory

As of: 1:30 PM

Image Version is "17.6"

Device Name	IP Address	Device Family	Reachability	Manageability	Health Score	Site	Image Version	Uptime	Last Updated	Resync Interv
9120AXE-A	10.48.39.39	Unified AP	Unreachable	Managed	NA	Assign	17.6.1.13	2 days 10 hrs	a day ago	N/A
[illegible]	10.48.39.180	Unified AP	Unreachable	Managed	NA	Assign	17.6.1.13	2 days 1 hr	a day ago	N/A
RazanAP9130-1	10.48.70.61	Unified AP	Reachable	Managed	NA	Assign	17.6.1.13	1 day 16 hrs	a day ago	N/A
Razan9130AP-2	10.48.70.62	Unified AP	Reachable	Managed	NA	Assign	17.6.1.13	1 day 14 hrs	a day ago	N/A
9800-17-6-1	10.48.39.206	Wireless Controller	Reachable	Managed	10	.../RitinBuilding/Ritinfloor	17.6.1	2 days 14 hrs	a day ago	24:00:00

Figure 14-3 *Cisco DNA Center inventory page*

When the Catalyst 9800 wireless controller is enabled for Assurance, you can use a lot of interesting applications to get visibility on your network and on the clients' performance, and run advanced troubleshooting tasks. Next, let's dig deeper into some of these applications/services in Cisco DNA Center Assurance.

Client 360

The client health dashboard (under the **Assurance > Health > Client** page) shows easy-to-understand widgets indicating the list of wired and wireless clients over time, their health (that is, clients having a bad connectivity experience in proportion to the total count of clients), the average onboarding time (which can quickly help identify a network-side issue...someone said DHCP?), their signal strength and SNR, as well as other statistics. The client RF health metric is a value from 1 to 10 (it could also be null if the client is inactive or so freshly connected that the metric was not calculated yet), where

- 10 means the client is connected with great RSSI and SNR (the threshold is better than –72 dBm of RSSI).
- 7 means the client is connected, but one of the values (RSSI or SNR) is below the threshold.
- 4 means the client is connected, and both values are below the threshold.
- 1 means the client is connected, but onboarding failed,

These numbers can vary further depending on the actual values. For example, Figure 14-4 shows client health of 6 or 9 and shows the RF values that contributed to these numbers.

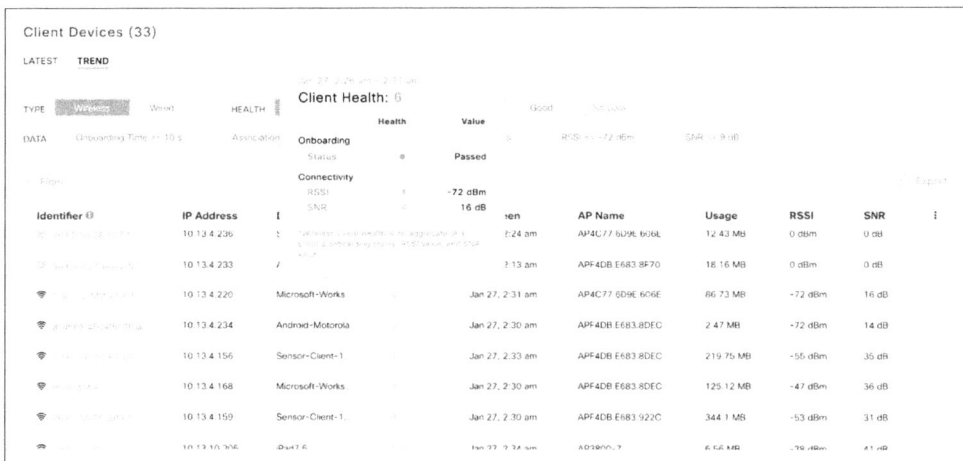

Figure 14-4 *Cisco DNA Center client devices list Assurance page*

The client health page shows the client onboard time, roaming time, and RSSI, and displays the number of clients that match each value, as illustrated in Figure 14-5.

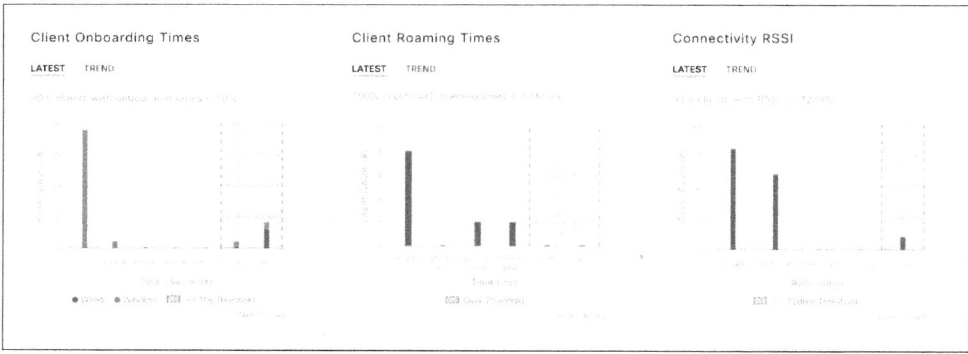

Figure 14-5 *Roaming and onboarding times*

You can drill down in the onboarding times and identify how long the wireless association takes, the DHCP and AAA processes take in general, or for each client. You can also quickly find clients for which onboarding constantly fails and identify why they cannot get onto the network. The client roaming times dashboard helps you to identify possible areas or possible types of clients that don't meet fast roaming standards.

Before the introduction of Cisco DNA Center Assurance, this type of client information was not easily available or would have required you to log in to multiple access points (think of a client roaming, for example) and the wireless controller and run different CLI commands. With client 360, you have this info ready in an easy-to-read dashboard and, most importantly, you have the client history that can extend up to two weeks back. This capability is critical because usually when the user reports an issue, the issue is no longer there, so it's very valuable to be able to go back in time. This is one of the main advantages of Cisco DNA Center Assurance.

AP 360

The overall Network Health page within the Assurance health dashboard gives you an overview of the WLCs, switches, and APs in your network and their overall operational status. If you consider APs, these can be sorted to show the top by utilization or client count to quickly identify the most loaded part of your network. After you click an AP name, Cisco DNA Center takes you to the AP 360 page for the selected access point. This is illustrated in Figure 14-6, which shows a section from the AP 360 general view, and Figure 14-7, which shows the map and comparison view that can be selected on the right of the page.

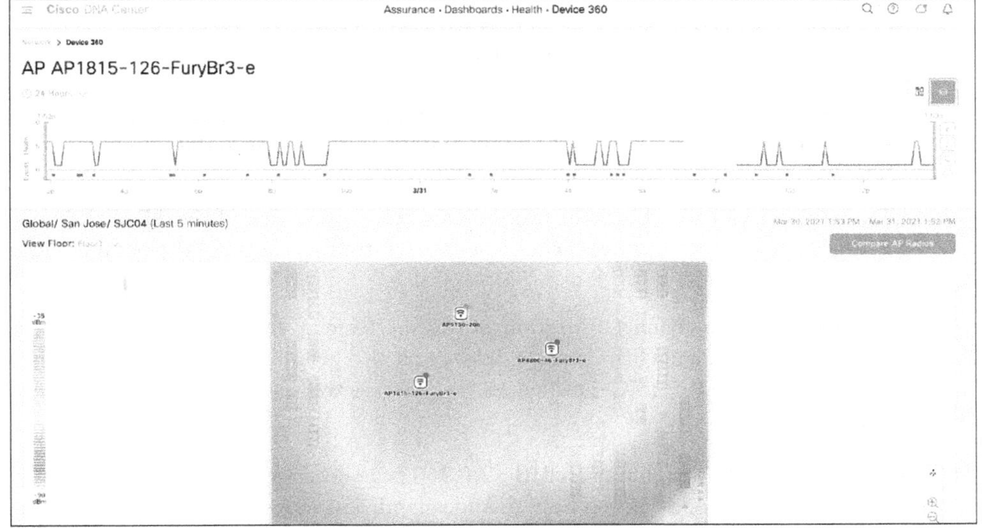

Figure 14-6 *AP 360 page*

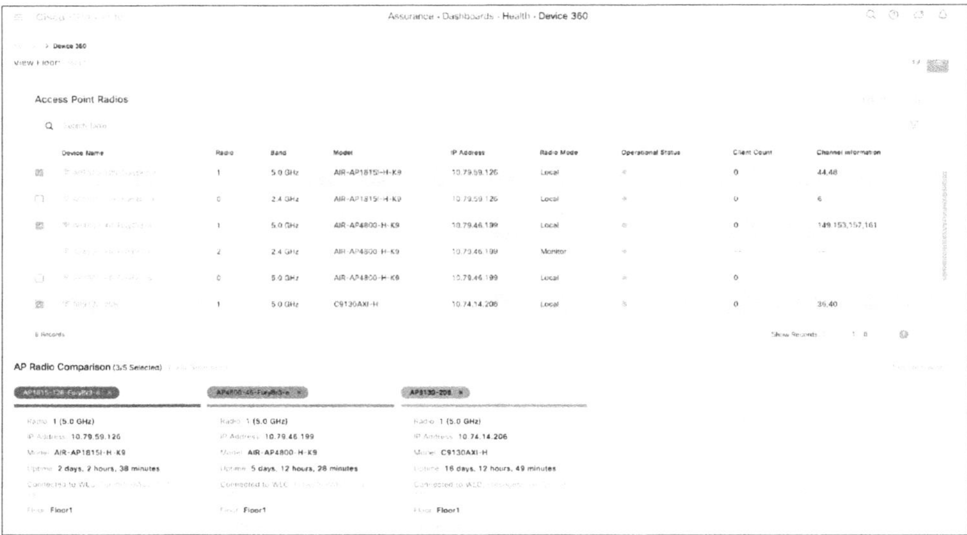

Figure 14-7 *AP radios comparison screen*

This page shows a timeline of a health metric (from 0 to 10) that gives you an idea of the overall health of the AP over time.

A list of issues pertaining to this AP is given and is sorted between resolved and ongoing issues. You can also see a physical neighbor topology diagram, showing the switch where the AP is connected and the WLC where it is registered, as well as a floor map to help you quickly identify the AP's location.

If you scroll to the bottom of the page, you have three tabs to choose from:

- The **Device** tab shows you overall AP information such as model, software version, IP, MAC, name, mode, power status, and uptime. It also shows graphs about AP CPU and memory utilization and WLC connectivity uptime.
- The **Connectivity** tab shows the transmit and receive data rate of the AP over time, airtime utilization, error rate, client count, retries, traffic rate, and Ethernet interface statistics.
- The **RF** tab graphs channel utilization, noise, interference, air quality, Tx power, airtime efficiency, and client latency. Below, you also get a live spectrum view from that AP (if it is an RF-ASIC–enabled access point) as well as a list of rogue APs and clients detected by the AP.

This is a lot of information, but overall, the AP 360 page deserves its name because it gathers key metrics in one place, is easy to understand, and shows at a glance how the AP is operating. There is no need to navigate to many different pages on the WLC WebUI or know CLI commands to run on the APs themselves (not to mention the network infrastructure wired devices) to get this data. On top of this, Assurance allows you to compare up to five AP radios to see whether their KPIs are similar or not. This means you can easily identify if a given problem is specific to an AP, to an area, or to a specific hardware combination.

The Issues section is not a mere 1:1 mapping of what the AP or WLC might send as syslog message. It can aggregate several events and complex conditions, and they come with a clear severity, description, location, time range, and map, as shown in Figure 14-8, so you can immediately determine whether the issue needs your immediate attention.

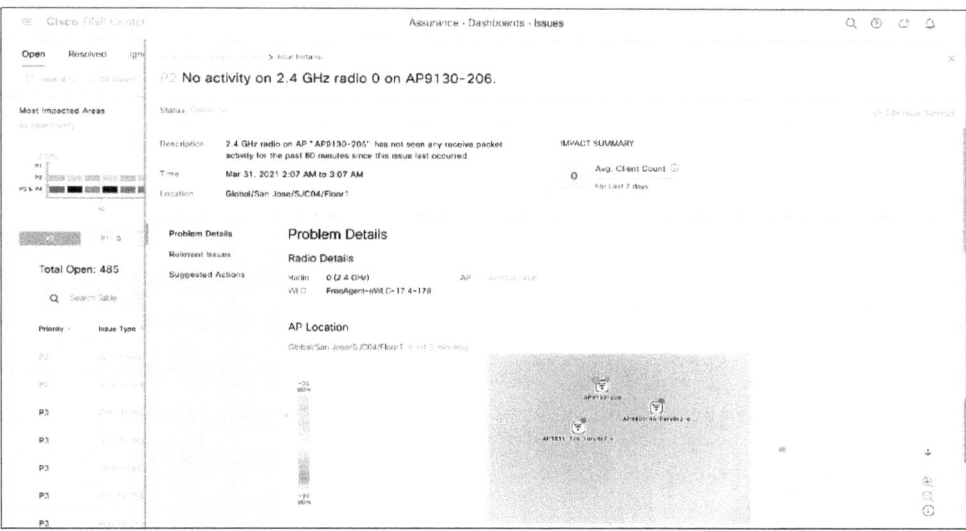

Figure 14-8 *Issues dashboard*

Network Services Analytics

If someone opens a ticket with the title "client cannot connect to the wireless network," the first thing that you must do is to determine if it's a radio frequency (RF) client-related problem or a network-side problem. Network Services Analytics (NSA) provides client, server, and network-side analytics, as well as root cause analysis, for AAA and DHCP services to better troubleshoot the client onboarding process. The Assurance health dashboard now has a Network Services tab just to show you this information.

As shown in Figure 14-9, the page presents the ratio of successful and unsuccessful DHCP and AAA transactions, which can help you get an idea of how the access network is performing.

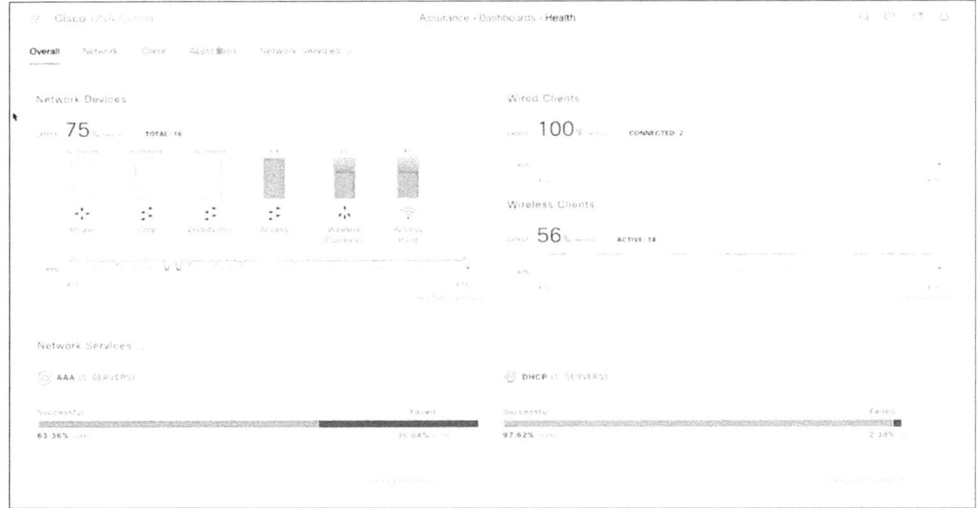

Figure 14-9 *Assurance overview health dashboard*

AAA server latency and transactions can be monitored in more detail on the AAA analysis page. You can view a timeline showing very clearly periods where the AAA process and flow may have failed for an extended period of time, as depicted in Figure 14-10, as well as the AAA servers by WLC, the latency, and failure rate per site, as shown in Figure 14-11.

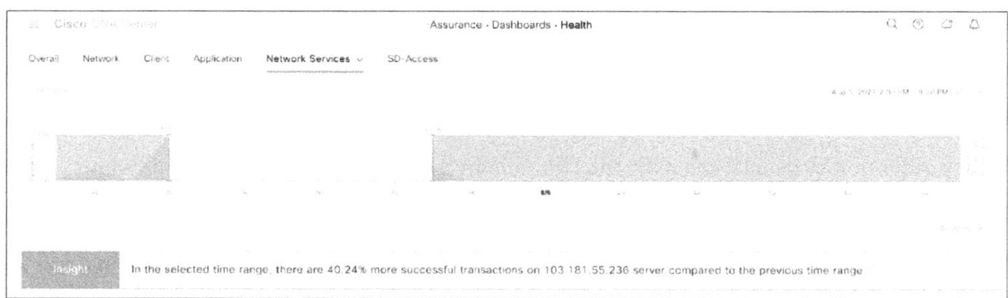

Figure 14-10 *Network services health timeline*

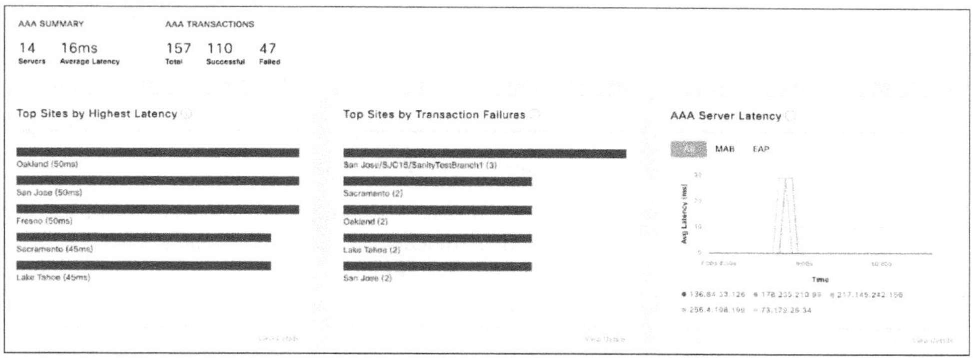

Figure 14-11 *Top AAA sites sorted*

The same type of data is available for the DHCP process of clients.

Device Analytics

When it comes to wireless networks and the quality of user access experience, the wireless client plays a big role: just think of roaming and the fact that roaming is mostly a client decision. In other words, a good wireless access network starts with a good-behaving client. This is the main reason that Cisco has started a close partnership with the main device vendors; started a few years ago with Apple, the partnership is now extended to Samsung and Intel. The idea is first of all to leverage the client to gain insights into its RF point of view of the network; if you think about it, what the client sees can be very different from the access points because they are mounted up in the ceiling, connected to a power source, with powerful antennas, and so on. The other goal is to exploit specific client capabilities to provide a better network connection (think about QoS).

It's also important to understand that these vendor-specific features, when turned on, do not harm the interoperability of the Cisco network with other devices but simply boost the quality of experience for these given client vendors.

Apple Analytics

From its first release, the Catalyst 9800 benefits from the FastLane feature set are as follows:

- Adaptive 802.11r (also called Adaptive FT) allows Apple devices (phones and tablets today, possibly laptops in the future) to use 802.11r for fast and secure roaming even if FT is not configured or advertised on the SSID; this allows for maximum compatibility with older clients that might not be able to connect to an FT-enabled SSID.
- 802.11k and 802.11v are automatically used for Apple clients to enhance their roaming algorithm even if not enabled on the SSID either.

- Finally, FastLane QoS enables WMM UP category usage by selected business applications on client devices.

These features do not penalize other vendors' clients in any way and allow Apple devices to benefit from an enhanced experience. Of course, you can turn on FT and 802.11k and 802.11v for all clients, but you need to make sure they are compatible.

Apple devices also have a method for reporting feedback to the Cisco network infrastructure, which is referred to as Apple Analytics: the client device reports the RF neighboring APs to the infrastructure (other client vendors typically only support the other way around), the reason it last disconnected from the network, and the detailed information about its client model and operating system.

The client health 360 page, depicted in Figure 14-12, has an IOS Analytics tab, which allows you to visualize how the Apple client hears its RF surroundings and the history of its last disconnections. This, for example, gives you a way to troubleshoot situations where it's the client's decision to disconnect from the network rather than the AP disconnecting the client; this simple information was very hard to get unless you did a deep dive into the client device's logs with specialized tools and knowledge.

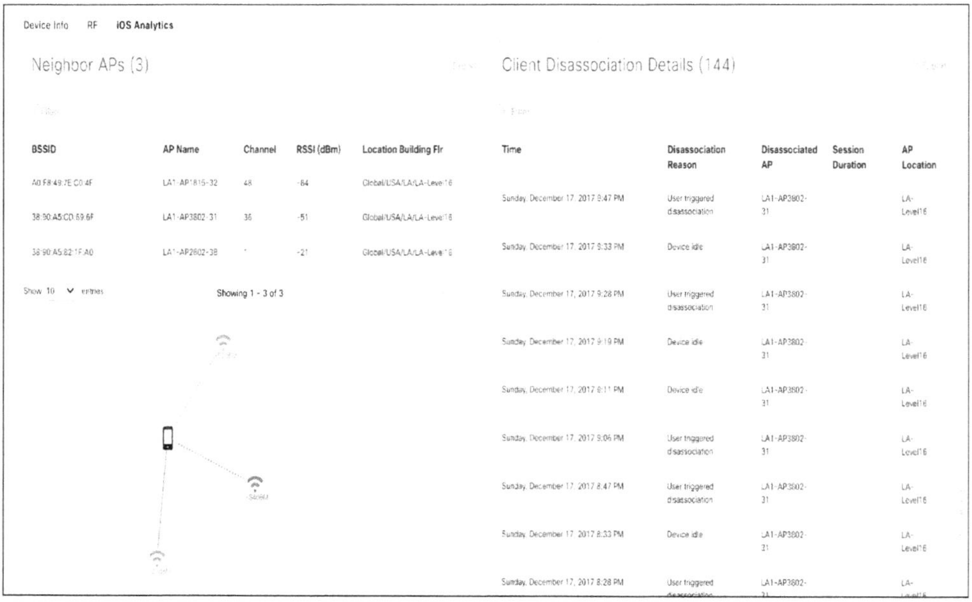

Figure 14-12 *The IOS Analytics tab showing the insights received from an iPhone device*

Samsung Analytics

Starting with the IOS-XE 17.1 release, the C9800 extends the client analytics capabilities and benefits to Samsung clients. The WLAN advanced settings page on the controller allows you to enable Device Analytics support with an easy check box (depicted in Figure 14-13); another check box can be turned on to share data back with the client. The first check box allows the controller (and Cisco DNA Center Assurance client health page) to receive and display the exact Samsung device type and software version, as well as to gain insight with regards to its last disconnection reasons. The second check box shares the WLC and AP hardware and software details with the Samsung client for internal usage by Samsung, so this capability is optional from the Cisco side because it does not add any feature or value for the infrastructure.

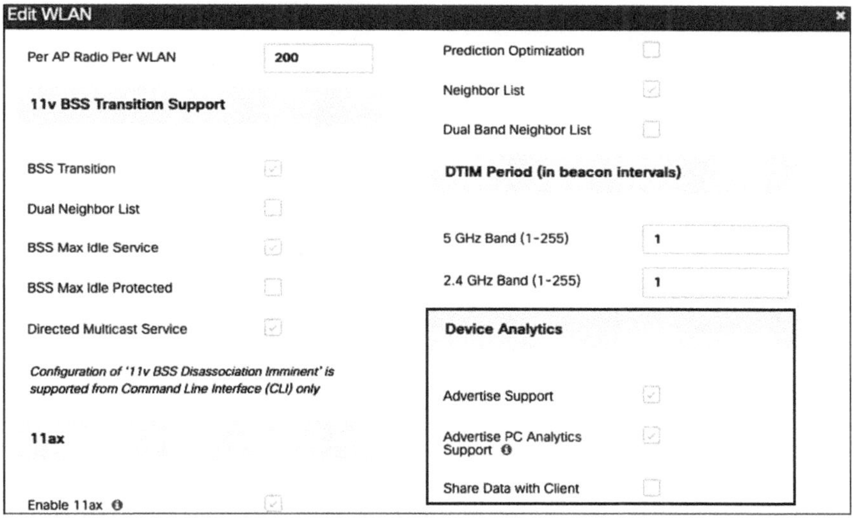

Figure 14-13 *Advanced WLAN settings page, allowing you to enable Device Analytics on the C9800*

Adaptive 802.11r is also supported by higher-end Samsung devices, and they can also benefit from this feature like Apple clients.

Intel Analytics

Starting with the IOS-XE 17.6 release, the Advanced PC Analytics feature (enabled under Device Analytics in the advanced WLAN settings of the C9800 web interface shown in Figure 14-13) enables analytics from Intel devices. Intel chips are popular in enterprise laptops, so this data becomes invaluable for enterprise wireless troubleshooting. It provides you with information on the client laptop model, driver version, adapter capabilities, whether the device is constantly in low RSSI, roaming reasons, 802.11v recommendations, missing beacons information, and failed APs. These details are also visible on the WLC web UI, as depicted in Figure 14-14 and Figure 14-15.

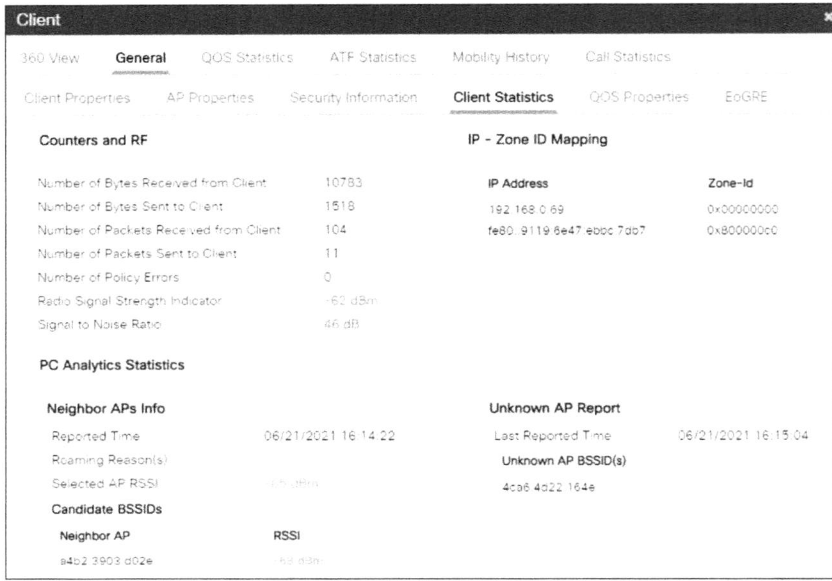

Figure 14-14 *C9800 Client statistics monitoring page*

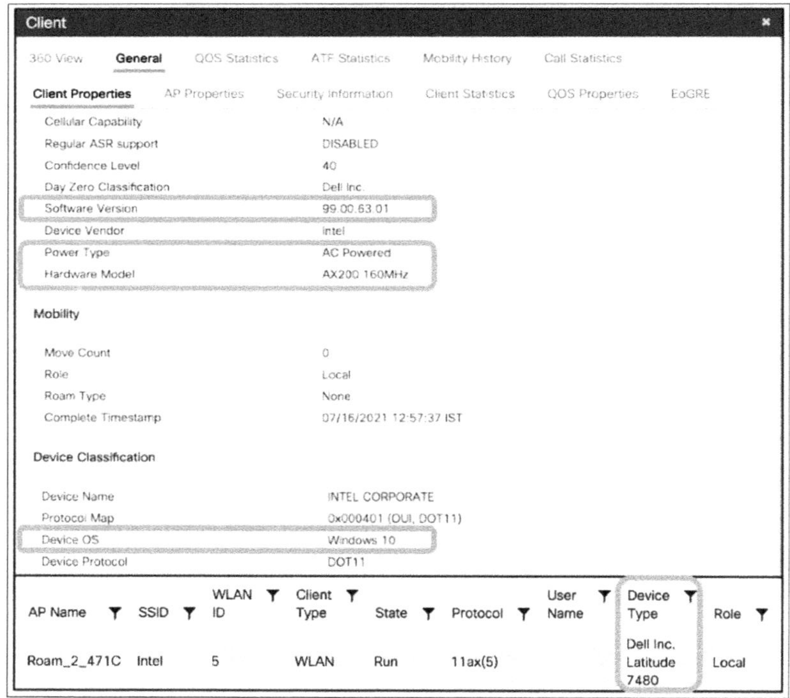

Figure 14-15 *C9800 Client general properties*

Intel Analytics helps you in various scenarios:

- Identifying bad drivers. If the failing clients have the same driver version, it's an immediate hint that a driver update might help to resolve the issue.
- Validating new drivers during a software update rollout.
- Identifying bad hardware. Profiling and network monitoring tools usually provide only operating system information, not the exact device vendor and model type. This can be key information if all the problematic clients are of the same make and model.
- Understanding roaming issues. It's easy to blame a client for bad roaming decisions. However, gaining insights from the client-side perspective can be helpful to understand why a roaming decision was taken, and it may sometimes have to do with your infrastructure too.
- Identifying poor client connectivity. The AP reports the signal strength at which it hears the client, but due to asymmetry in client and AP transmit power, as well as the difference in antenna types and orientations, gaining the signal information from the client perspective, is extremely valuable in troubleshooting RF deployments.
- Identifying misbehaving APs. When an AP is not working properly, that same AP might not be able to feed the information back (depending on the source of the problem). Having the client providing you details on misbehaving APs helps you troubleshoot the network infrastructure side.

Intelligent Capture

If you have tried to troubleshoot an RF-related client problem, you know how difficult the whole process might be. You require specific professional (and rarely free) tools, and you often need to be present physically at the site and reproduce the problem when you're there. If the issue is intermittent, your only solution would be to deploy RF capture tools everywhere on-premises and hope to catch the problem—not really a fun thing to do. Intelligent Capture (iCap) is a set of technical solutions offered by Cisco AP and automated from Cisco DNA Center that come to the rescue. It addresses the challenges in reproducing a wireless issue because it can run proactively and catch the problem as it happens, saving you from being present on site and giving you the real RF visibility you need rather than just hints or high-level metrics.

Capturing wireless client traffic was already possible, to some extent, by issuing specific debug commands on each access point or by converting an access point to sniffer mode, which would require you to dedicate an additional access point or sacrifice a client-serving AP to send the resulting capture streams to your laptop directly. Intelligent Capture takes client troubleshooting and data capturing to a new level of scalability and performance. Figure 14-16 depicts the Intelligent Capture workflow: the Cisco DNA Center appliance configures the WLC to enable Intelligent Capture on all the relevant APs (which ones are relevant is explained later in this section). These APs create a streaming telemetry tunnel (using the gRPC protocol) directly to the Cisco DNA Center and start streaming the captured data in a manner that is both CPU processing and network traffic efficient.

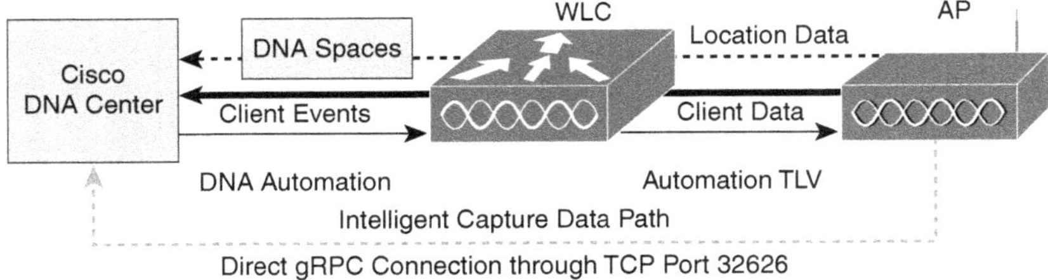

Figure 14-16 *Intelligent Capture workflow*

Intelligent Capture encompasses two types of data: detailed client RF statistics (more detailed and frequent than the airtime utilization statistics the WLC is getting through the CAPWAP tunnel) and/or over-the-air data traffic in a standard PCAP format. The power of Intelligent Capture is that rather than enabling the capture on each individual AP, you can monitor a specific client and have the capture automatically enabled on all the APs that can hear that client in real time as it roams around the premises. Not only is the data collection automatically distributed across APs without any manual effort, but the log result is a single PCAP file stored chronologically by Cisco DNA Center. This makes roaming and onboarding multichannel captures as easy as one click!

Actively monitoring specific clients is not your only choice because Intelligent Capture can automatically capture anomalies, such as client onboarding failure or WIPS attacks, without you specifying a client or time range. You get the data for any problem occurring without having configured any data collection in advance!

Intelligent Capture is supported on all Catalyst 9100 access points (including the small C9105). It is important to highlight some differences in the supported functionalities:

- Only Catalyst 9120 and 9130 series feature the Cisco RF-ASIC, which is needed to extract CleanAir Spectrum captures. Their RF stats are a lot more accurate and detailed than 9105 and 9110 series, which rely on Spectrum Intelligence software capability.
- Only the 9130 series can provide a full client data capture while still serving other clients at the same time, thanks to its third radio and RF ASIC. Other AP models capture only the onboarding process, which includes wireless management and action frames (authentication, association request/response, and so on) and the data frames pertaining to the control plane of the client onboarding process (DHCP, EAP authentication and key exchange, and so on).

The Intelligent Capture set of features also includes pure spectrum analysis. On the AP models that include the RF-ASIC (C9120 and C9130 series), you can monitor the spectrum around the access point while it is still serving clients: you don't need to set the AP to a specific non-client-serving mode!

The WLC contributes to the client troubleshooting process because it sends the client failure events to Cisco DNA Center, which correlates the corresponding anomaly captures coming from the APs at the same time.

The first step is to make sure that you have the Intelligent Capture module installed on your Cisco DNA Center. Go to **System > Software Updates > Installed Apps.** Under the Assurance section, verify that **Automation-Intelligent Capture** is installed.

On Cisco DNA Center, go to **Assurance > Intelligent Capture Settings** (in the Manage section) and then click the **Access Point** tab. This page allows you to select which APs you will enable for client and AP RF stats and/or client anomaly capture, as depicted in Figure 14-17.

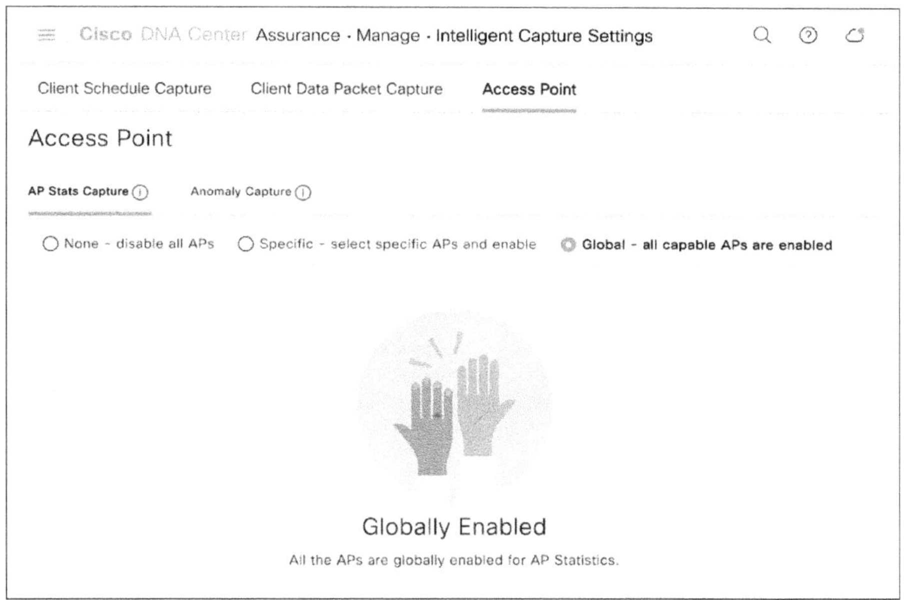

Figure 14-17 *Intelligent Capture access point configuration page*

You can enable Intelligent Capture on specific APs or globally (you can also disable it completely). It's important to note that enabling Intelligent Capture has little performance impact on the APs, and therefore, there are no reasons to disable it except in the case where you have mutually exclusive features being used on the same APs (depending on the version, Intelligent Capture might not be compatible with aWIPS, for example; check the documentation of your exact IOS-XE release).

When going to the client health assurance dashboard and entering the client 360 dashboard by selecting any client, you can get an Intelligent Capture data page. As depicted in Figure 14-18, it shows you the client location on the map, the RF stats pertaining to this, a list of the latest onboarding events with anomalies detected (that is, where the flow breaks in the onboarding), as well as a 14-day time travel bar that allows you to go back in time and check the last two weeks' worth of data for the same client. You can now figure out what exactly occurred even if the end user comes to complain a few days after the issue occurred.

The same page also has links to start collecting a live packet capture (possibly a full data capture if you have 9130 series APs) for the client. The full data capture has the huge advantage of being in clear text even if you are using WPA encryption on your WLAN; the reason is that the capture is taken at the AP radio driver level after decryption. The file can store up to 2 GB of data from the target client.

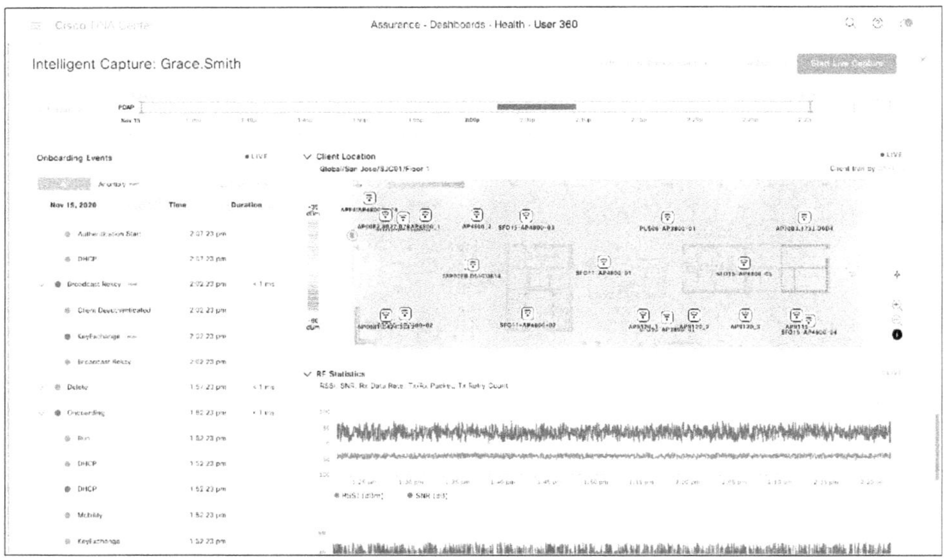

Figure 14-18 *Intelligent Capture client 360 health page*

RF statistics include RSSI and SNR from the client, the corresponding data rate used at that time, the count of packets sent in that reporting window, and the number of retries. The statistics reporting happens on periods of 30 seconds.

You do not need to be a Wireshark expert and have experience looking at packet captures to troubleshoot anomalies either because the interface gives you the problem details in an easy-to-understand chart and identifies the most likely root cause of the problem, as illustrated in Figure 14-19.

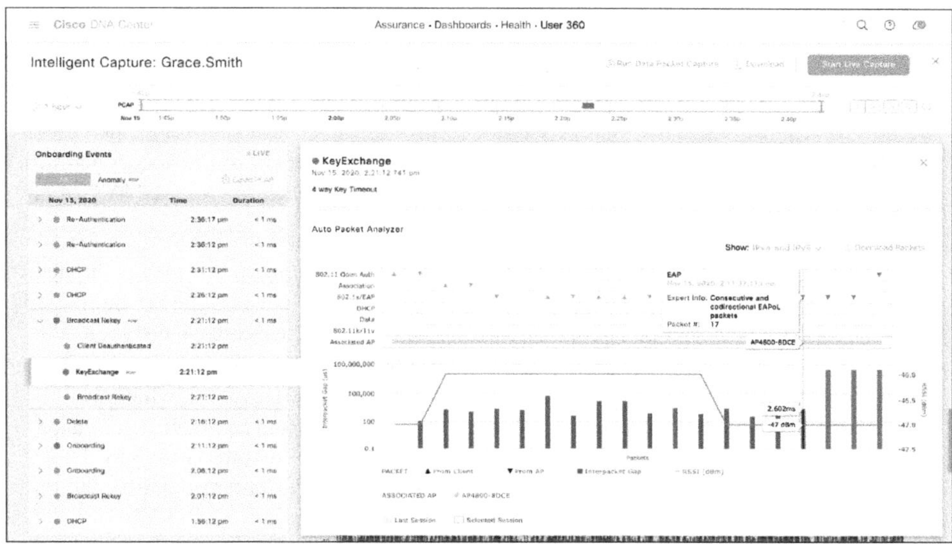

Figure 14-19 *The packet analyzer from the onboarding events list*

The Intelligent Capture page in the AP health dashboard also shows always-on RF stats pertaining to the AP (and not related to specific clients). As shown in Figure 14-20, it shows you the channel utilization over time for each radio, the client count graph, as well as a list of top clients with transmit failed packets per SSID. A timeline allows you again to go back in time and analyze the time range of your choice. You can also see the Tx/Rx error trends, broadcast and multicast count trends (useful to identify some broadcast storms that could impact your performances), the Tx power of the AP, and the noise floor trend.

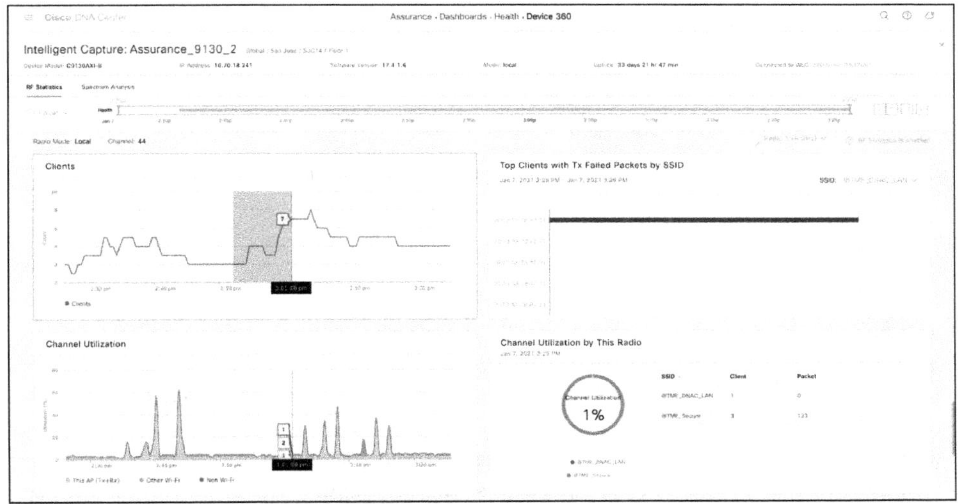

Figure 14-20 *The AP health Intelligent Capture page*

Cisco Active Sensor

Assurance uses all sorts of techniques and tools to get data from the clients and the network infrastructure to understand the client connectivity status. Of course, the best method to achieve this would be to have full control of the client itself. Enter the Cisco 1800S sensor: smaller than any access point, it connects as a client to the access network, sends all sort of client metrics, and runs tests and sends feedback to Cisco DNA Center.

A sensor can be plugged into a power outlet or powered via micro USB; you can decide to connect it fully wirelessly back to Cisco DNA Center. Of course, this means the device needs a wireless backhaul connection so the wireless network needs to be up and passing traffic. This capability can be useful for places where an Ethernet cable is not available or you need to easily move the sensor in different places. The sensor provides data about the quality of the wireless connection or may not provide any data at all, which would imply that the sensor cannot connect, which is useful information by itself. The sensor can also be connected and powered via an Ethernet cable (as illustrated in Figure 14-21). In this case, the sensor is able to feed data back to Cisco DNA Center no matter the state of the wireless network.

Figure 14-21 *The 1800S sensor*

Despite the fact that the Cisco 1800S runs the same operating system (named AP COS) as other APs, it does not do any CAPWAP discovery and therefore does not join any WLC. The sensor is completely managed via Cisco DNA Center: it receives the Cisco DNA Center IP address through DHCP Option 43 (following the PNP guidelines); discovers the Cisco DNA Center through HTTP port 80 initially; and then uses HTTPS for everything else, including receiving its configuration (which includes the test profiles to run, the SSIDs to connect to, the reports to send back), upgrading its image, and sending test data back. This is shown in Figure 14-22.

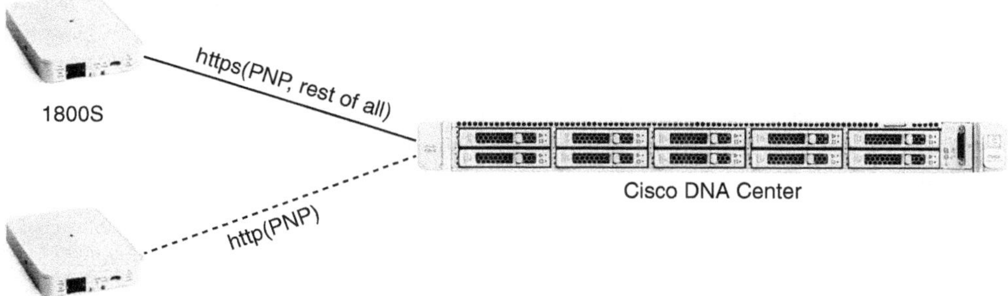

Figure 14-22 *1800S protocol workflow overview*

Sensor Provisioning and Onboarding

When the sensor is coupled with the PoE module, it can receive its power and IP address from the Ethernet connection and then discover Cisco DNA Center as the PNP server. Without an Ethernet connection, the sensor automatically tries to connect to a predefined SSID with the name *CiscoSensorProvisioning*. A custom and more secure backhaul SSID can be configured later, but this is required for out-of-the box sensor provisioning.

The PnP server can be discovered through the option 43 DHCP field with the ASCII format **"5A1D;B2;K4;I<IP address>;J80"**, through DNS discovery through the FQDN pnpserver.cisco.com, or through the Cloud PnP Connect at devicehelper.cisco.com. In all cases, the PNP server should be the Cisco DNA Center appliance. As a last resort, the PNP server can be configured on the sensor through the console (or SSH) command **config dot11 sensor pnp ip** ***DNA-IP-ADDRESS***. It is important to note that the sensor requires a specific console adapter (that is, it does not use the RJ45 format) that can be ordered for free with Cisco.

When it discovers the Cisco DNA Center appliance, the sensor shows up in the **Assurance > Manage > Sensors** page, ready to be claimed. After it is claimed, you can upgrade the image on the sensor (sensors have a version numbering system that looks like Cisco DNA Center itself, but a strict version pairing is not required), change the sensor name, enable or disable SSH on the sensor, and enroll it in a SCEP process to deploy certificates.

Test Suites

You can configure a test suite and deploy it on your sensor: choose an SSID name and security type, enter the SSID credentials (PSK or 802.1X), and select test conditions.

The sensor supports SSIDs with different security settings: Open, WPA2-PSK, web authentication (Central or Local web authentication or even third-party portals are supported with their various username/passcode/password/Acceptable Use Policy page and fields combination); or WPA2-Enterprise with PEAP-mschapv2, EAP-FAST, EAP-TLS, or PEAP-TLS.

A test can validate the onboarding process (association, authentication, and DHCP), the RF conditions (data rate and SNR), the DNS lookup of specific domains, the RADIUS status (direct authentication toward a given server), throughput (using NDT, iperf), and IP SLA, but also application tests like web browsing to specific URL, FTP, email, or ping test to a given host.

Cisco DNA Center keeps a heartbeat to check the connectivity of sensors. After you deploy a new test, Cisco DNA Center sends the new configuration with the next heartbeat message.

The sensor does a fresh radio scan to get the latest neighbor SSID list and then creates a test matrix based on the scan list and the number of allowed target APs and the target APs configured. The sensor runs the tests and then reports to Cisco DNA Center after execution of each test group.

The sensor 360 page on Cisco DNA Center allows you to see all the recent test results for each sensor, the timeline of recent tests, as well as test details, as depicted in Figure 14-23 and Figure 14-24. You can also see a view of your floor areas and a quick view of where physically some tests failed.

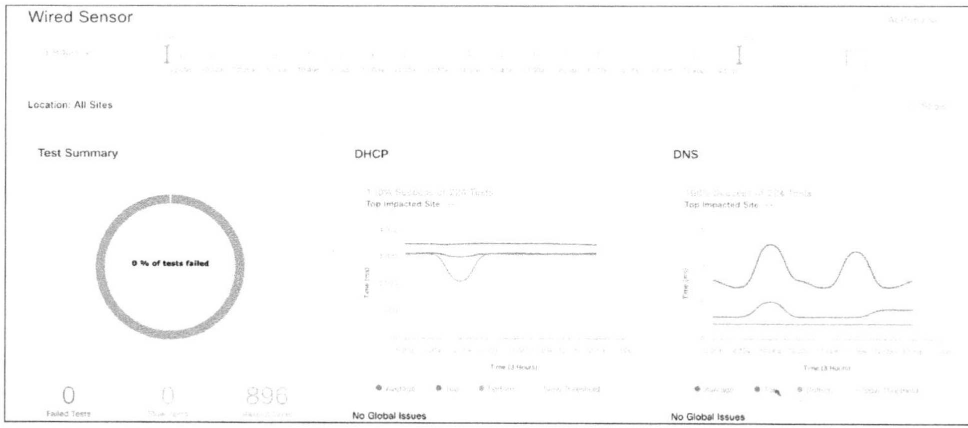

Figure 14-23 *Cisco 1800S test suite dashboard*

Test Results

Find

Test Name	Location	Vlan	Test Types	Test Results
wired_sensor		vlan11	Onboarding Test DHCP Test DNS Test	

Showing 1 of 1

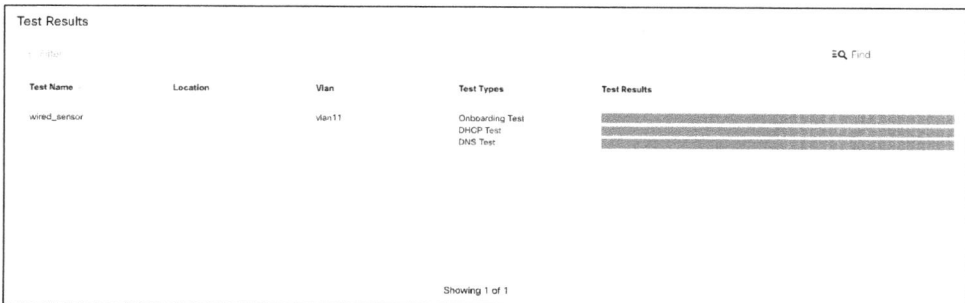

Figure 14-24 *Cisco 1800S test result timeline*

Troubleshooting the Assurance Application

There are two possibilities when suspecting an Assurance problem. Either the data is not being sent at all (or partially) to the Cisco DNA Center appliance, which can be verified on the WLC itself; or the augmentation or correlation of that data is wrong, which would be something you can verify on the Cisco DNA Center appliance side (and this is not covered in this book or section). Verifying that Cisco DNA Center can fully access the controller and is receiving telemetry data is a great first sanity check that will solve most of the issues you could face with Cisco DNA Center Assurance.

Make sure that the following command lines are present in the C9800 configuration:

1. **netconf-yang:** This command enables NETCONF on the WLC, which is required for Cisco DNA Center to push configuration to the controller.
2. **crypto pki trustpoint Cisco DNA Center-CA:** This means that the required certificate has been pushed to C9800 from Cisco DNA Center. This can also be checked via "**show crypto pki trustpoint | begin Cisco DNA Center-CA**".
3. **aaa authorization exec default local:** This is the main **aaa** command that needs to be present on the box. The authorization of the NETCONF access session (that Cisco DNA Center is using) uses the default exec authorization method (no named method can be configured). It should point to the local database initially but can point externally to a AAA server later if needed.

You can check the telemetry connections and their status with the following command:

```
C9800#show telemetry internal connection
Telemetry connections

Index Peer Address         Port  VRF Source Address       State
----- -------------------- ----- --- -------------------- ------
    0 10.48.71.4           25103.    0  10.48.39.206      Active
```

A status of "Active" means telemetry is actively sent. You can either have no entry at all or have an entry stuck in "Connecting" when things are not working as they should.

You can also verify the list of telemetry subscriptions with the **# show telemetry ietf subscription configured** command.

The running configuration can show you what those subscription IDs refer to exactly.

You are not supposed to configure subscriptions yourself through the CLI; it is the responsibility of Cisco DNA Center to enable all the telemetry subscriptions. If they haven't been enabled, it is advised to check how the WLC was added to Cisco DNA Center inventory, if it's reachable and managed, and if NETCONF is working.

You can test whether NETCONF connectivity is working by starting an SSH session to the WLC on port 830. After the authentication prompt, you should then get a big XML dump that corresponds to the NETCONF capabilities of the device. This allows you to verify the reachability on the NETCONF port as well as the authentication phase.

Summary

This chapter gives an overview of how Cisco DNA Center Assurance works and how it can help you in troubleshooting your Catalyst wireless networks. DNA Assurance differs from traditional monitoring and troubleshooting tools: it leverages streaming telemetry to efficiently extract and transmit analytics data from the network devices. It is not a collection of logs and errors difficult to navigate; instead, it brings intelligence and business-relevant metrics and focuses your attention to the real problem at hand, allowing you to find the real root cause rather than chase symptoms. It also helps you to prove that your network is performing fine, which can be essential when some software developers prefer to blame the network rather than their own application.

References

Intelligent Capture Deployment Guide 2.1.1: https://www.cisco.com/c/en/us/products/collateral/cloud-systems-management/dna-center/guide-c07-744044.html

Cisco DNA Assurance User Guide, Release 2.1.2: https://www.cisco.com/c/en/us/td/docs/cloud-systems-management/network-automation-and-management/dna-center-assurance/2-1-2/b_cisco_dna_assurance_2_1_2_ug.html

Cisco DNA Center information: https://www.cisco.com/c/en/us/products/cloud-systems-management/dna-center/index.html?dtid=osscdc000283

Cisco Prime Virtual Network Analysis Module (vNAM) information: https://www.cisco.com/c/en/us/products/cloud-systems-management/prime-virtual-network-analysis-module-vnam/index.html

Cisco CMX Configuration Guide, Release 10.6.0 and Later: https://www.cisco.com/c/en/us/td/docs/wireless/mse/10-6/cmx_config/b_cg_cmx106/getting_started_with_cisco_cmx.html

Intelligent Capture Demo Video: https://www.youtube.com/watch?v=NOO43eMhYVA&t=740s&ab_channel=CiscoWLAN

Chapter 15

Backing Up, Restoring, and Upgrading Your C9800

Backing up the configuration of a device and restoring it are often overlooked topics because they are considered rather trivial. However, many critical questions arise during daily operations, such as "If you restore the configuration on another Catalyst 9800 controller, will everything be restored?" or "When an upgrade takes place, what happens to deprecated commands? What happens to configuration lines pertaining to new features when you downgrade?" This chapter answers these concerns:

- Saving and restoring the configuration for disaster recovery
- Backing up the configuration and restoring it
- Backing up everything for restoring on another controller
- Backing up using Cisco Prime Infrastructure
- Saving the configuration using Cisco DNA Center
- Running IOS-XE in Bundle or Install mode
- Upgrading and downgrading your controller safely
- Upgrading hardware programmables

Saving and Restoring the Configuration for Disaster Recovery

Disasters can vary a lot in severity or shape. You could have a brief power failure in the data center, an unforeseen crash of the controller, or a complete hardware failure that forces you to reconfigure another controller VM or appliance to replace your faulty unit.

Saving the Configuration Changes

The first step is to save all current unsaved changes to the configuration. Similarly to other IOS and IOS-XE devices, the Catalyst 9800 applies any changes you bring to the configuration in real time but does not save them to the starting configuration, and those changes will be lost after a reboot.

IOS-XE devices (and therefore the Catalyst 9800) have two different types of configurations: the running configuration and the startup configuration. The running configuration is modified every time you type a configuration line or click the **Apply** button in the WebUI; it defines the behavior of the controller in real time. The startup configuration represents the configuration loaded upon startup of the controller. Therefore, saving the configuration basically boils down to copying the running configuration to the startup configuration. When that is done, you are certain that the controller will have the same configuration after the next reload as the configuration active at the time you are doing this operation.

```
WLC# write mem
```

Or

```
WLC# copy run start
```

Backing Up the Configuration and Restoring It

So far, you have just saved the configuration to make it persistent. Backing up a configuration simply means saving a copy of it either on the device or externally to that device. If you are about to make configuration changes that you are not totally certain about, you can save the configuration to the bootflash memory of the controller:

```
WLC# copy run bootflash:filename.cfg
```

If you are unhappy with the changes you made, you can return to the configuration checkpoint you saved by restoring the configuration from the saved file. If you copy the saved configuration and overwrite the startup configuration, it loads that saved configuration at the next boot:

```
WLC# copy bootflash:filename.cfg start
```

If you copy the saved configuration to the running configuration, both configurations are merged. This means that all the configuration lines of the saved file get added to the current configuration, but any command line present in the current configuration is not removed if it was not present in the saved one.

Obviously, a much safer and smarter idea is to save your configuration file outside of the controller itself if you want to survive any hardware failure on the controller side (or any human mistakes from one of your pesky fellow network administrators). You can do this by replacing the bootflash path with a FTP/SFTP/TFTP URL or by copying the file manually to an external repository:

```
WLC# copy bootflash:filename.cfg ftp://ftpadminuser@ftpserverip/path
```

After you restore a configuration, everything contained in that text file is restored: IP addresses, WLAN configurations, profiles and tags, static (AP MAC based) and dynamic (using filters) AP-to-tag mappings, QoS/AVC, and other features.

Is this all there is to it? Does saving the running configuration protect you from *any* type of failure? Unfortunately, no. The question is then to define what is not saved with this process:

- AP-specific configuration. Whenever you configure things specific to an AP in the WebUI AP, it is typically saved on the AP flash itself (and not with the controller configuration). Examples are AP hostname, LED flash status, primary/secondary/tertiary WLC configuration, and so on.
- Layer 2 VLAN database (vlan.dat in the flash storage).
- Certificate private keys.
- ROMMON boot variables.
- Customized webauth bundle pages.
- SNMPv3 user credentials.
- The general password encryption key (if you encrypt your configuration).
- Smart Licensing registration token.

Backing Up Everything for Restoring on Another Controller

Saving the running-config to a file somewhere is a pretty safe and complete way to back up your controller.

AP-specific configuration cannot be backed up. This is saved on the APs and will be carried with them until that config is overridden. If you're using configuration templates in Cisco Prime Infrastructure or Cisco DNA Center, you have a safe way of having that configuration stored somewhere and ready to be provisioned again in a couple of mouse clicks.

The vlan.dat file in the bootflash partition contains all the VLANs' information. However, if the VLANs were created through configuration of the controller, they are present in the running-config and startup-config and will be backed up and restored along with your configuration. If using VTP, then your controller would receive the VLANs list anyway when being connected to the switch so there should be little reason to care about backing up that file.

ROMMON boot variables are saved outside of the configuration. They can be listed from the IOS-XE prompt with the **show romvar** command, as shown in Example 15-1. You can save that output to a file somewhere safe. There are few reasons to modify the ROMMON variables apart from specific license boot levels, or the BOOT variable or maybe the console baud rate. In any case, better safe than sorry: copy your "romvars" if you are migrating to a new controller.

Example 15-1 *Listing ROMMON Variables from CLI*

```
9800# show romvar
ROMMON variables:
 SWITCH_NUMBER = 1
 LICENSE_BOOT_LEVEL =
 MCP_STARTUP_TRACEFLAGS = 00000000:00000000
 VXE_INSTANCE_ID = i-045f223a618427cb8
 VXE_INSTANCE_ID_PRI = i-045f223a618427cb8
 RET_2_RTS = 13:25:41 CET Thu Nov 12 2020
 BSI = 0
 RET_2_RCALTS =
 RANDOM_NUM = 615938243
 BOOT = bootflash:packages.conf,12;
```

Your customized webauth bundle web pages (if any) are typically stored somewhere in the WLC **bootflash** or **harddisk** partition. You can spot their location by hunting the **custom-page** statements in your configuration and then copying the HTML files out to a safe place:

```
9800# show run | include custom-page
custom-page login device bootflash:customlogin.html
custom-page failure device bootflash:customfailure.html
```

The Advantage of Backing Up Using the WebUI

If you do the config backup through the WebUI, the config file is not equivalent to what you get with **copy start <filename>**. The WebUI removes any self-signed certificate definition as well as any assignment of certificate to the web server. The reason is that after clearing the configuration or restoring on another box, all the certificate private keys are missing, and therefore, all certificates are unusable. Because a new self-signed certificate can automatically be generated on the new controller, it is best to not include it in the backup and let the controller web server pick the best available certificate then.

A similar problem occurs if you had assigned a third-party signed certificate to the web admin server of the controller. The certificate is backed up with the configuration but is missing its private key and is unusable until you import it again. If the WebUI interface is not accessible after the config is restored, either delete all the certificates (**no crypto pki trustpoint <name>**) so that a new self-signed certificate gets generated or reimport your signed certificate properly.

The Case of Configuration Encryption

All passwords and preshared keys can be shown in clear text in the configuration by default. You can also only store the hash, but those are now reversible and it is sometimes trivial to recover your passwords and keys from a configuration file. This is why it is recommended to encrypt the passwords in the configuration using AES. It is achieved with these commands:

```
key config-key password-encrypt <private key>
password encryption aes
```

Although the latter command is saved in the configuration, the first one isn't. That command defines the AES key that will be used to encrypt all the other passwords and shared keys present in the configuration. It is therefore useless to save that key in the config file, encrypted or not, because it would allow any malicious person to decrypt everything.

Therefore, if you restore an encrypted config file, all your passwords will be missing (IOS admin passwords, guest passwords, WLAN preshared keys, RADIUS shared keys, and so on) after the configuration is restored. It is advised that you reenter the **key config-key password-encrypt <key>** command after the restore and then again paste all your config lines that contain the encrypted keys and passwords. They are now decoded and understood properly by the controller.

Backing Up Using Cisco Prime Infrastructure

On Prime Infrastructure, two jobs can be used to take the backup configurations:

- **Device Config Backup-External**
- **Controller configuration backup**

To take the backup configurations outside of Prime Infrastructure for the 9800 WLC, you need to run the task **Device Config Backup-External.** When the job **Device Config Backup-External** is run, the configurations are sent from Prime to the external server.

If you run the **Controller configuration backup** system job, the configuration is obtained via SSH (through a **show run-config** command) and stored on the Prime Infrastructure TFTP folder no matter what protocol or repository you selected.

However, the config archive feature (which is not a system job but a feature of its own) is also available and can back up the configuration directly on Prime Infrastructure as well.

Config archive takes the running config but also copies the VLAN database on top of it. While archiving the configurations for the 9800 WLC, the WLC executes the **show running-config, show startup-config** commands to take the configurations through SSH and execute the command **copy flash:vlan.dat tftp:** transferring the VLAN database directly to Prime Infrastructure through TFTP.

Rather than backing up the controller configuration, some people prefer to configure everything out of templates on Prime Infrastructure and have those templates saved in config groups. This way, the templates can easily be provisioned again to reconfigure the device. Templates never cover 100 percent of the device configuration, however (one example is the device management IP address).

Backing Up Using Cisco DNA Center

Cisco DNA Center stores your configuration based on your intent, site hierarchy, and overall architecture. Should someone suddenly wipe out the configuration of a WLC, the Cisco DNA Center administrator would simply have to hit the **provision** action and all the configurations would be pushed again (you would have to set up the WLC for basic connectivity and add it to the DNA Center again first).

You might want to take configuration backups for safety's sake but also for logging and accounting of various configuration changes. Cisco DNA Center offers a configuration archive feature that allows you to manually export the configuration of a device.

One option is to go in the **Provision > Inventory > Device** menu and choose the **Export CLI Output** option on the **Configuration** tab, as shown in Figure 15-1.

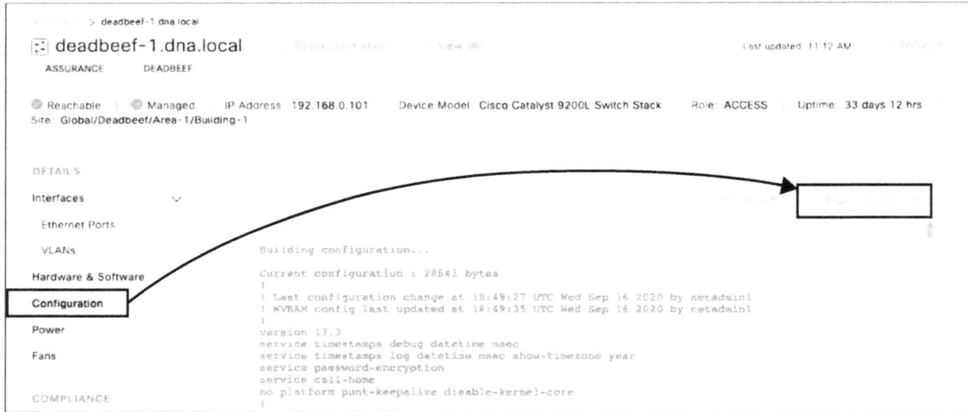

Figure 15-1 *The Export CLI Output allows you to export the configuration of a device*

Another option is to use the configuration archive API call that will export the configuration of a device in a ZIP file. The sequence is as follows:

Step 1. Log in to get a token:

```
POST   https://<_dnac.ip.address_>/dna/system/api/v1/auth/
token
```

Step 2. Get the DEVICE_ID from the device management IP:

```
GET https://<_dnac.ip.address_>/dna/intent/api/v1/
network-device?managementIpAddress=<_device.mgmt.ip.address_>
```

Step 3. Request a configuration export of the specified device:

```
POST https://<_dnac.ip.address_>/dna/intent/api/v1/
network-device-archive/cleartext
```

```
Body example (deviceid obtained from the previous step):
{ "deviceId": [ "2862a8ba-2e2c-4560-8d21-9e6188d862ed",
"e0526635-4c7d-4908-b82b-563c80597d65" ], "password":
"Cisco123#" }
```

Step 4. Check when the export task finishes:

```
GET    https://<_dnac.ip.address_>/api/v1/task/<_task.id_>
Body example (task id obtained from the previous step) :
"taskId": "fa033bb2-81ab-46df-a39c-4455e9410e36"
```

Step 5. Get the file:

```
GET https://<_dnac.ip.address_>/<_additionalStatusURL.
string_>

Body example (additionalstatusURL obtained from previous
step): "additionalStatusURL": "/api/v1/file/4962fff5-33e7-
49f5-a7b1-b7c9731600b3"
```

Step 6. Save the response to a file (see Figure 15-2).

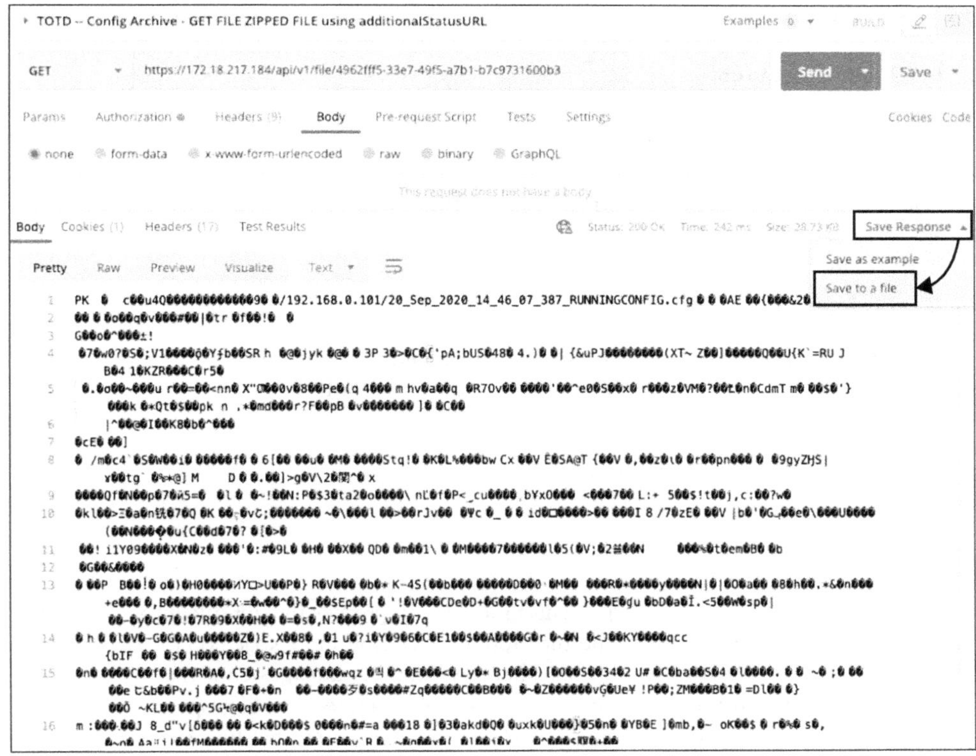

Figure 15-2 *Saving the file containing the configurations through API calls*

Running IOS-XE in Install or Bundle Mode

Every controller you install or purchase shows IOS-XE running in Install mode, and that's the way it should be and the only thing you should care about. However, Embedded Wireless Controller (EWC) on Catalyst 9100 access points runs in Bundle mode only, and that's a mode you may have to go through in case of recovery, so what are the differences?

Bundle Mode

Bundle mode is probably what old-timers (from the days of classic IOS) are most familiar with. You basically set the boot variable to point to a .bin file (the IOS-XE file you downloaded from cisco.com), and the device will boot with that software at the next reload. It is immediately clear that this approach is convenient for recovery operations. Should anything bad happen, you can load an IOS-XE .bin file on a USB drive and quickly boot with it, even from the ROMMON prompt. What is the drawback then? The .bin file must be read and uncompressed in memory when booting, and this is slightly less efficient with regards to memory consumption and boot performance. It also prevents certain features like AP predownload, hot patching, or ISSU to happen because they require Install mode. This is why you should always aim at using Install mode (except on EWC on AP, where it is not available).

```
WLC# show version
(...)
Installation mode is BUNDLE

WLC(config)# boot system bootflash:C9800-CL-universalk9.17.3.1.bin
WLC# wr mem
WLC# show boot
BOOT variable = bootflash:C9800-CL-universalk9.17.3.1.bin,12;
```

On EWC-on-AP, no boot variable is configured, which has the effect that the device will boot in Bundle mode on the first.bin file it finds in the flash:

```
EWC# show boot
BOOT variable does not exist
```

Install Mode

Install mode is the right way to go. This mode uncompresses the .bin file into several .pkg files in the bootflash partition (and therefore it takes a bit more space on the storage). The boot variable needs to point to a file named, by default, **packages.conf** that will point to each .pkg module. If you see your boot variable pointing to a **packages.conf** file instead of a .bin file as illustrated in Example 15-2, you can be certain that the device will boot in Install mode (provided the files are valid and present) at the next reload. This **packages.conf** file is created automatically when doing the install or upgrade process, so you should never have to write it yourself. When recovering a device from ROMMON, it is common practice to first boot in Bundle mode by pointing to a .bin file, and after you have the full IOS-XE running in Bundle mode, then upgrade or migrate to Install mode in an easy step by using the WebUI software upgrade page, for example (staying on the same version but moving from Bundle to Install mode).

Example 15-2 *Verifying the Boot Variable*

```
WLC# show version
(...)
cisco C9800-CL (VXE) processor (revision VXE) with 4794670K/3075K bytes of memory.
(...)
7249919K bytes of virtual hard disk at bootflash:.
Installation mode is INSTALL

WLC# show version
BOOT variable = bootflash:packages.conf,12;
```

Upgrading (and Downgrading) the Controller Safely

There are various ways to upgrade your Catalyst 9800, depending on if you just want to upgrade without caring too much about details or downtime or if you are having several separate controllers or a high availability pair and are looking for zero downtime. Let's go through each possibility.

Standard Upgrade

The **Administration>Software Management>Software Upgrade** WebUI page, shown in Figure 15-3, is the easiest way to upgrade your controller. The Software Upgrade tab allows you to transfer an upgrade .bin file via TFTP, FTP, SFTP, HTTPS, or directly from the controller storage if you copied it beforehand. Cisco highly recommends using secure and fast methods such as SFTP, SCP, or HTTPS. TFTP is extremely slow by default and not suited for large files, whereas FTP is simply unsecure.

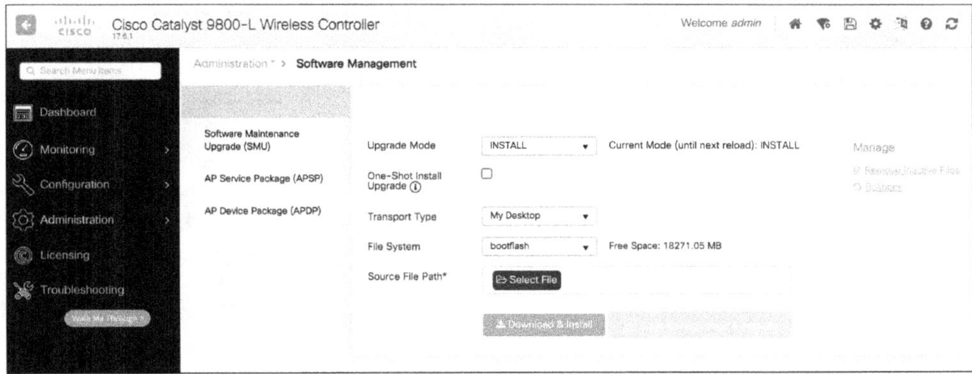

Figure 15-3 *The Catalyst 9800 upgrade page*

The upgrade page allows you to move from Bundle to Install mode if you have been doing a recovery. When you're in Install mode, your upgrades keep the device running in that mode. The page also has a **Remove Inactive Files** option, which is handy to free up some storage space by deleting older and unused IOS-XE images.

The upgrade process is simple: one button downloads and installs the software, and when you're done, you simply have to save and reload to boot in the new version. A **Show Logs** box on the right allows you to see how the install and transfer are going. The upgrade is completely transparent if you have an SSO pair of WLCs: they are both upgraded at the same time.

In reality, the upgrade process has several steps that matter if you are using the CLI or if you are using other upgrade options. CLI details the actual steps a bit further:

```
WLC# install add <filelocation>
```

This command downloads the .bin file, copies it to all the WLC member appliances of the stack, and then uncompresses the file into the .pkg files. After this stage, nothing is really upgraded yet. If you leave the WLC in this state, the new software version shows as an inactive package, and if you reboot it, it keeps using the same software version. The **show install summary** command shows the new software as inactive and the current software as still being the committed software, as shown in Example 15-3.

Example 15-3 *Verifying the Install Status After the New Software Is Downloaded*

```
WLC# show install summary
[ Chassis 1/R0 ] Installed Package(s) Information:
State (St): I - Inactive, U - Activated & Uncommitted,
            C - Activated & Committed, D - Deactivated & Uncommitted
Type  St   Filename/Version
IMG   C    17.05.01.0
IMG   I    17.06.01.0.250

Auto abort timer: inactive
------------------------------------------------------------------------------
```

To finish the upgrade, you need to activate and then commit. The activation triggers a reboot of the WLC, which will come back running the new software version:

```
9800# install activate
```

The **show install summary** command, after reboot, shows the WLC is running the new software version, but this version is still uncommitted (see Example 15-4). It also shows an auto-abort timer of six hours, after which the WLC will revert back to the previous software version if the install is not committed.

Example 15-4 *The New Software Shows as Installed but Uncommitted After the Activation and Reboot*

```
WLC# show install summary
[ Chassis 1/R0 ] Installed Package(s) Information:
State (St): I - Inactive, U - Activated & Uncommitted
            C - Activated & Committed, D - Deactivated & Uncommitted
Type  St   Filename/Version
IMG   U    17.06.01.0.250

Auto abort timer: active , time before rollback - 05:51:36
------------------------------------------------------------------------------
```

When you commit the install, you can see the software version fully installed, as shown in Example 15-5.

Example 15-5 *Final Status of the Installation*

```
9800# install commit
WLC# show install summary
[ Chassis 1/R0 ] Installed Package(s) Information:
State (St): I - Inactive, U - Activated & Uncommitted,
            C - Activated & Committed, D - Deactivated & Uncommitted
Type  St   Filename/Version
IMG   C    17.06.01.0.250

Auto abort timer: inactive
------------------------------------------------------------------------------
```

This procedure also works for other types of upgrades like SMU, APDP, APSP, or wireless packages for switch:

- A software maintenance upgrade (SMU) is a patch that can be provided by the Cisco Technical Assistance Center (TAC) for specific defects or Cisco Product Incident Response Information (PSIRT) where an SMU can be built. It can be hot, requiring no reboot of the WLC to take effect, or cold, requiring a reboot to come into action.
- An APSP (Access point Service Pack) is similar to an SMU, but for AP-specific defects. It reboots APs but does so in a rolling fashion to minimize the impact to the network.
- An AP Device Pack (APDP) is a package that brings support for an AP model not supported by the currently active controller version. It allows for supporting a new AP model without having to upgrade your controller immediately.

In IOS-XE version 17.6 and later, if you upgrade using the web UI, you will have to click **commit** in the upgrade page after the WLC has rebooted in the new software release. Forgetting to do this means the WLC will reboot in the old version 6 hours later.

AP Predownload

Access points have two partitions and therefore can store two software version images, which is extremely handy for rolling back in case of trouble. It also comes in handy during upgrades because you can load the next software release in the backup flash partition while the AP is still operating; the AP then can avoid downloading its new software image after the WLC rebooted (because it already has the new software preloaded) and therefore experience less downtime. The concept of AP predownload consists of upgrading the WLC with the new software version (but not rebooting yet) and transferring the corresponding AP image to the access points. When you reload the WLC, it comes up using the new version, and APs do not have to download the new software but reboot using their second partition already containing the new image. This capability is extremely useful when you have APs behind WAN links where the download process can be both time- and resource-consuming.

To trigger an AP predownload, you need to run the following command just after the **install add** operation and before any activate or commit:

```
WLC# ap image predownload
```

You can verify that the new software was downloaded in the secondary flash slot of the APs with this command:

```
WLC# show ap image
```

You can swap the AP images to force the AP to use the new image after reboot (the command does not reboot the AP). If you don't, the APs will reboot only after trying to join the WLC (after it rebooted with the new version) and realize they are not running the same software but that they already have it on their backup slot (at which point they will swap the image themselves and reboot), as shown in Example 15-6.

Example 15-6 *Swapping AP Images After a Predownload*

```
WLC# ap image swap
WLC# show ap image
Total number of APs  : 1
Number of APs
        Initiated                  : 0
        Downloading                : 0
        Predownloading             : 0
        Completed downloading      : 1
        Completed predownloading   : 1
        Not Supported              : 0
        Failed to Predownload      : 0
        Predownload in progress    : No
AP Name                           Primary Image            Backup Image
Predownload Status    Predownload  Version  Next Retry Time   Retry Count
------------------------------------------------------------------------------------
9120-AP                          17.5.1.12               17.6.1.13
Complete              17.6.1.13             0                  0
```

The swap can save you time if you reboot all the APs at the same time as rebooting the WLC; otherwise, the APs wait for the WLC to power cycle and then reboot when they realize they need to run another software image in order to join it. Don't forget to activate and commit the upgrade after your AP predownload. You cannot trigger an AP predownload after the upgrade was activated.

Efficient Upgrade

If you have a lot of APs behind WAN links that suffer from higher latency, AP image predownload might not save the day yet for you because all APs still have to download the image from the controller at the same time (you could do it by batches as well, but they still all download centrally). Efficient upgrade picks random APs (per AP model) in each site and allows up to three subordinate APs (of the same model, because each model can potentially require a different AP image) to download from the selected primary AP rather than download from the controller. This means that at scale, only 25 percent of your APs would download their image from the controller, while 75 percent would download it from another nearby AP. This drastically reduces WAN link consumption and improves transfer time in case of latency on the WAN.

You can trigger a predownload per site tag with

```
9800# ap image predownload site-tag <tag name> start
```

You can identify which APs have been selected as primaries with

```
9800# show ap master list
```

Note Efficient upgrade works only for FlexConnect APs and only on named site tags other than the default site tag.

Rolling AP Upgrade (for N+1)

The upgrade methods explained so far still involve some downtime because the WLC has to reboot and all APs reboot at the same time. Clients will notice a couple of minutes of loss of wireless services. A rolling AP upgrade solves this problem for scenarios where you have an extra controller, often referred to as the N+1 scenario. The overall concept is to have similar configurations on the controllers (they are not in HA SSO, so the configuration is not automatically replicated) and then upgrade only one controller and move APs progressively to the upgraded controller. This means that the client still has an AP to connect to nearby (provided the AP deployment is dense enough) when the closest AP is rebooting in the new version. That "second nearby AP" reboots and is upgraded after the other AP moves to the upgraded WLC successfully. By upgrading APs in waves of non-neighbor APs, you ensure service is still offered for clients.

The prerequisite is to have a mobility tunnel between your two controllers. They also need to have an identical wireless configuration if you want your APs to keep serving their WLANs when moving between the two WLCs.

Do the **install add** operation on the first WLC. Optionally, do an AP image predownload.

You can then move the APs to the second controller progressively with

```
9800# ap image upgrade destination <wlc2 name> <wlc2 ip>
```

In-Service Software Upgrade (ISSU)

When using a high availability SSO pair of controllers (it is therefore not available on Embedded Wireless Controllers), ISSU is probably the best way to have zero downtime and an easy upgrade. It can be done for complete software upgrades as well as cold SMUs.

The concept is similar to the rolling AP upgrade but has several advantages. The first and biggest is that it is as easy as enabling the ISSU check box in the WebUI upgrade page. The new software is then downloaded to both controllers. AP predownload then happens. Then both controllers upgrade to the new version one after the other (so that APs always stay joined to one of them). The APs are then joined to WLCs running a different release (which is never seen before in Cisco Wireless) and keep operating. The WLC then reboots APs progressively, wave by wave, so that they come back with the new version. Neighboring APs are not rebooted at the same time so that there is always one AP to service clients in a given area, therefore providing zero downtime. The clients do not even notice anything if they support 802.11v because the AP sends them an order to move to another AP before rebooting. If the clients don't support 802.11v, they might be disconnected briefly until they connect to another access point.

Summary

This chapter covers how to save the configuration on the controller and how to back it up to an external location, including management platforms like Cisco Prime Infrastructure and Cisco DNA Center, and how to restore the configurations in case of problems on the device or in case of complete device replacement. It also covers the differences between Install and Bundle modes and why Install mode is the preferred choice in general.

References

Back up the controller configuration with Prime Infrastructure:
https://www.cisco.com/c/en/us/support/docs/cloud-systems-management/prime-infrastructure/216266-prime-infrastructure-backup-of-wireless.html

Chapter 16

Troubleshooting

Installing an operational enterprise wireless deployment involves four steps:

1. Designing and planning
2. Deploying the infrastructure
3. Configuring the network
4. Monitoring and troubleshooting

So far, you have learned about the capabilities and features of the Catalyst 9800 wireless LAN controllers (WLCs), which should contribute to designing, deploying, and configuring your wireless network. Feature-specific troubleshooting and deployment tips are also provided in each chapter, which, in conjunction with the recommended software and best practices guides listed in the "References" sections, should aid in deploying your C9800 successfully. After it's deployed, you need to monitor the network to ensure the good health of the infrastructure and to ensure all the applicable services are delivered to the users at the intended service level and all the applications are performing flawlessly. In general, Cisco recommends the Cisco Digital Network Architecture (DNA) Center as the single pane of glass to manage and monitor the enterprise infrastructure, including wireless deployments. With its raw processing power and intelligent analytics engine, Cisco DNA Center packs the ability to monitor the network end to end to assure the health of the infrastructure and performance of the end clients and applications across the network. Further, Cisco DNA Center Assurance leverages monitoring data collected over days and weeks run against powerful artificial intelligence (AI)–based tools to identify trends and patterns and predict flaws and failures in the deployments. In Chapter 14, "Cisco DNA Center/Assurance Integration," you read about monitoring via Cisco DNA Center Assurance. This chapter focuses on the troubleshooting tools that are native to the C9800 and can be leveraged by traditional or innovative network management systems like Cisco DNA Center.

To deliver the wireless services, the C9800 must interoperate with other network infrastructures like switches, routers, firewalls, and clients with different vendor chipsets and operating systems that may or may not be compatible. In addition, each wireless deployment has its unique properties from an access point (AP) layout and radio frequency (RF) perspective. To address these issues, troubleshooting becomes a necessity. The first step in all troubleshooting is to understand the expected behavior, and so far, this book has detailed how the various features work on the C9800. The next step is to quantify the deviations from the expected behavior and narrow down the deployment conditions and impact of the deviations. To define and triage these deviations and get a view into the inner workings and packet handling of the C9800, you can make use of several capabilities and tools that are available on the C9800.

The C9800 runs a Linux-based operating system called binary OS (BinOS), or more commonly IOS-XE. The Linux kernel is updated with Cisco proprietary components, and the various processes are implemented as Linux daemons. The functionality delivered by the classic Cisco Internetwork Operating System (IOS), which is a large blob of code, has been modularized and segregated into multiple Linux daemons, including the IOS daemon (IOSd) and Session Manager daemon (SMd), among others in BinOS IOS-XE. To this, several new Linux daemon processes are added to deliver wireless functionality, such as the Wireless Network Control daemon (WNCd) to provide Control and Provisioning of Wireless Access Points (CAPWAP) termination, the Radio Resource Management daemon (RRMgrd) to execute RRM operations, and the Rogue daemon (rogued) for rogue AP and client detection. Together, these processes form the code base of the C9800. These central processing unit (CPU) processes deliver the control plane functionality on the C9800, which includes access point onboarding, client onboarding, RF management, and other packets. In addition, the control plane also handles various timers and interprocessor communication (IPC) messages. A tracing infrastructure is provided on the C9800 for each of these processes; it is referred to as *control plane tracing*. Typically, debugging involves enabling certain commands or tools on a system and reproducing symptoms of the problem to see the log messages that help triage the problem. In other words, it is imperative that the problem be repeated. What sets the C9800 apart is its unique capability to generate and store log messages without any user intervention and to do so on an ongoing basis that enables you to look back on the events leading up to a failure incident without needing to reproduce the failure symptom. The failure needs to be reproduced only when a deep-dive analysis is needed. The tracing infrastructure provides several commands to view the generated logs, including several filtering options to customize the view.

After the clients are onboarded, the wireless clients can communicate with each other and with the wired infrastructure, including network devices, servers, and wired clients. All this data forwarding on the C9800 is handled entirely in hardware through dedicated application-specific integrated circuits (ASICs) referred to as the *data plane*. Debugging tools only look at packets that are processed by the CPU and do not have a view into the packets forwarded by the hardware. Packet-capturing and packet-tracing tools are available on the C9800 to monitor each packet from the point it enters the C9800, through its life within the C9800, including all the changes it undergoes based on the features enabled, to the point it egresses out of the C9800. In this chapter, you learn about

- Binary tracing infrastructure on the C9800 and resulting always-on, per-process, and radioactive traces
- Embedded Packet Capture and Packet Tracer
- Other on-the-box and offline tools for the C9800
- Monitoring capabilities on the C9800

Control Plane Tracing

As discussed, *control plane tracing* refers to logging of control flow and data as the various events are handled by the Linux daemons to accomplish the wireless control plane functionality. The C9800 uses a store-and-process approach to generate and view the trace logs as opposed to a display-without-storing approach. There are commands on the C9800 to perform live display of tracelogs, but they tax the system, causing a performance impact; the delay in rendering logs does not give the experience of live logging. Therefore, live debugging is not recommended and not discussed in the scope of this chapter.

Syslog

Traditionally, all IOS-based systems write system logs (syslogs) to notify you of any failures on the system. So, for troubleshooting any Cisco device running IOS, it is a good idea to begin with syslogs. With IOSd being a fundamental building block of IOS-XE, the same is true for the C9800 as well. Note that these syslogs have different logging levels, ranging from critical to error to informational. This topic is not covered in detail here because the C9800 wireless functionality relies more heavily on BinOS processes. By default, the logging process sends log messages to the local system buffer as well as the console. To disable logs being sent to the console, you can run **no logging console** in the configuration mode. The system logging buffer size is set to the minimum allowed 4096 bytes and can be increased by running **logging buffered <4096-2147483647>** in the configuration mode. However, the recommendation is to redirect the log messages to an external syslog server over User Datagram Protocol (UDP) port 514. Most syslog servers have significantly more size than the 4096 bytes allocated for the log messages. This extra size ensures that log messages are not overwritten during any failures. Further, the syslog server can provide a view into the IOS events on the C9800 leading up to the failure.

To view logs in the system buffer, you can navigate on the C9800 graphical user interface (GUI) to **Troubleshooting > Logs > Syslog**, as shown in Figure 16-1, or run the command **show logging** from the command-line interface (CLI) of the C9800. Because syslogs are the first step of troubleshooting, on the C9800, some of the service-impacting BinOS logs generate a copy into the syslog. They include misconfiguration errors, logs pertaining to AP and client disconnects, and resource failures like CPU hog or memory allocation failures. The immediate visibility enables quick remediation and reduced impact. Example 16-1 provide a snippet of syslogs in the system buffer when an AP disjoins due to a regulatory domain mismatch.

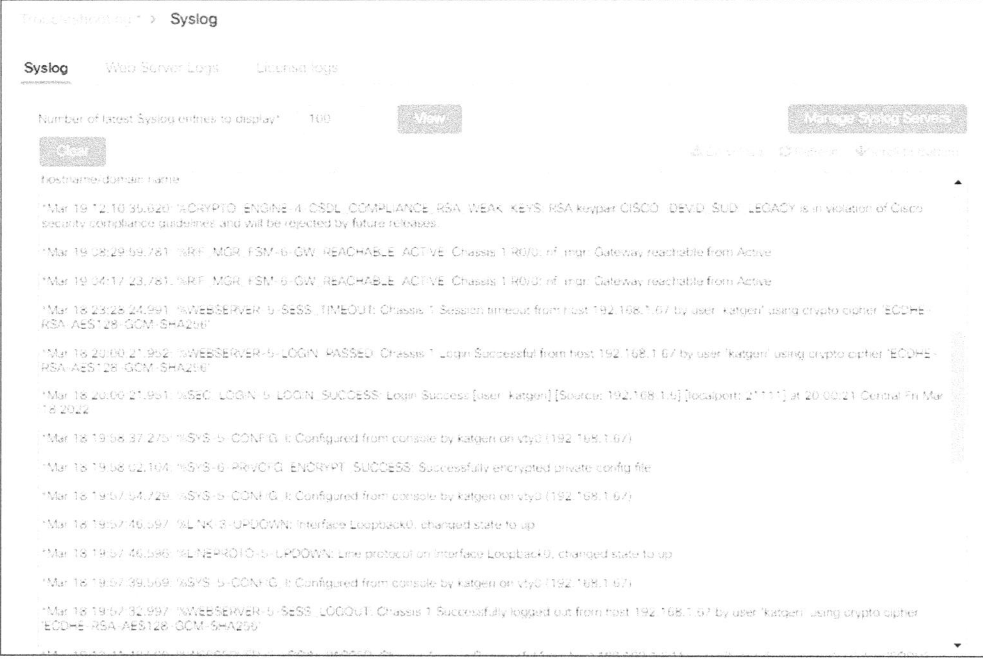

Figure 16-1 *Troubleshooting Dashboard (Syslog)*

Example 16-1 *Syslog Message for AP Join Failed Due to Regulatory Domain*

```
C9800#show logging
*Jan 16 03:36:22.778: %APMGR_TRACE_MESSAGE-3-WLC_GEN_ERR: Chassis 1 R0/0: wncd:
  Error in c064.e423.c640AP will be disconnected. Country code validation failed.
  Country code: US
*Jan 16 03:36:22.778: %CAPWAPAC_SMGR_TRACE_MESSAGE-5-AP_JOIN_DISJOIN: Chas-
  sis 1 R0/0: wncd: AP Event: AP Name: sudha-9120 Mac: c064.e423.c640 Session-IP:
  10.5.1.12[5268] 192.168.1.5[5246] Disjoined Country Code Mismatch
```

In addition to syslog, under **Troubleshooting > Logs**, there are two more tabs that are explained later in the "Per-Process Debugging" section.

Binary Tracing

The fundamental tracing infrastructure used on the C9800 is called *binary tracing* or *btrace*. Each BinOS process running on the C9800 writes to its dedicated file called a *tracelog* at the line rate. In the case of IOSd, only some logical subdivisions in code referred to as modules, like the parser and smart agent used for licensing, use the btrace infrastructure, whereas other modules, like Simple Network Management Protocol (SNMP) and public key infrastructure (PKI), still use legacy debugs. The tracelogs are encoded and written in memory in human-unreadable binary format. The benefit of this is speed and small size footprint: the device can log a lot of things without any performance impact and with minimal storage space usage. When any process tracelog file size reaches a system-defined threshold, the tracelog is compressed into a file with a .tar.gz extension and stored in the hard disk. This sequence of compression and storage of tracelog is referred to as *rotation*. The file size before rotation is 1 megabyte (MB) for most BinOS processes, but WNCd is the exception. You can rotate the trace logs manually with **request platform software trace rotate all**. There are also limits enforced on the total number of compressed tracelogs that can be saved for each BinOS process in the hard disk. Most processes save 25 .tar.gz files, but the WNCd allows a higher number of compressed tracelog files. Table 16-1 lists the maximum tracelog file size before rotation and the maximum number of tracelogs supported by each WNCd for different form factors of the C9800.

Table 16-1 *WNCd Tracelog Maximum File Size and Maximum Count*

	C9800-L	C9800-40	C9800-80	C9800-CL
Maximum WNCd tracelog file size (in MB) before rotation	20	20	20	10
Maximum number of compressed tracelogs per WNCd	350	350	700	400

Note that the listed numbers are per WNCd, and the total number of WNCd tracelogs on a C9800 is derived by multiplying the number of WNCd instances by the maximum number of tracelogs per WNCd. When any process hits its maximum number of files, the oldest file is deleted from the hard disk to make space for the newest file. In the case of Embedded Wireless on Catalyst Access Point (EWC-AP), where there is no hard disk, the second partition (part2) of the flash is used to store the rotated tracelogs. Each btrace log within the tracelog file is written with a specific severity level (inspired by syslog levels) to indicate the impact of the message. This is done to allow progressively finer-grained information to be logged and viewed based on who is consuming the logs and for which purpose. There are 10 such log levels explained in Table 16-2.

Table 16-2 *Btrace Severity Levels and Their Meaning*

Severity Level Name	Severity Level Number	Severity Level Meaning
FATAL		The process is aborted, and it usually results in a crash file or core dump.
EMERGENCY	0	The system is unstable/unusable.
ALERT	1	Action must be taken immediately. Examples: hardware failure, high temperature
CRITICAL	2	A critical event causes loss of essential functions. Examples: no license present, essential component failure, low system memory
ERROR	3	A failure event causes limited loss of functionality without automatic recovery; an unexpected condition may not have immediate impact but have a later impact. Error messages are not expected to be seen during a successful code execution. Examples: error in protocol Finite State Machine (FSM) like AP join, client join; invalid configuration that could not be applied
WARNING	4	A problem could be recovered from automatically, or a condition might end up in loss of functionality if not investigated and addressed. Examples: license limit reached, free disk space below limit, configuration incomplete
NOTICE	5	This is the default logging level for the BinOS processes using btrace; it logs all the significant events at the system and feature level. These messages indicate the module and code to debug in detail to resolve problem symptoms. System-related examples: switchover, process restart, install/upgrade operations Feature-related examples: wireless AP join or leave; client onboarding state transitions
INFORMATIONAL (INFO)	6	There is nonessential information about the system state. This message can also provide additional details about events logged at the NOTICE level. Examples: client connected, switch between radio channels

Severity Level Name	Severity Level Number	Severity Level Meaning
DEBUG	7	This is the first level of code debugging logs; it is used to do focused deep-dive troubleshooting of specific symptoms and code.
VERBOSE	8	This is the second level of code debugging logs. There is potentially extremely verbose output. Examples: track every entry/exit of a state machine, print values that seldom change but provide detailed context
NOISE	15	This message is meant for debugging during development and is compiled out of code posted on cisco.com.

The encoded binary tracelogs need to be decoded and parsed to derive meaningful information about failure triggers and trends. The C9800 has an internal decoder application that decodes the traces into a human-readable string format. Binary trace logs can be decoded on-the-box or offline. To rotate and export all the tracelogs in a single .tar.gz file, you can run **request platform software trace archive**. By default, this command bundles all the tracelogs on the system and saves the resulting .tar.gz into local memory (in this case, the bootflash memory). However, both duration and destination of the archive file can be modified using the suboptions. This activity should not be your default go-to action but can be useful when the support team needs to look at what is happening on the whole box for a longer period rather than looking at more specific events.

- Duration of lookback: You can define the number of days' worth of logs that need to be bundled and exported by using

  ```
  #request platform software trace archive last <1-3650> days
  ```

- Destination for the bundled file: In case of hardware appliances (C9800-40, C9800-80) that have a hard disk, it is recommended you save the archive tracelog to hard disk by using

  ```
  #request platform software trace archive last <1-3650> days target
  {bootflash: harddisk:}
  ```

The CLI tree for triggering the binary-to-text conversion of logs on-the-box is **show logging {process | profile}...**. Considering the maximum .tar.gz file size and maximum number of .tar.gz files that can be stored, the number of lines of logs or the volume of btrace logs on a C9800, even in binary format, is quite high. Decoding only increases the volume of logs. This CLI tree provides several options to restrict the view of the decoded tracelogs to specific context of interest:

- To obtain the decoded per-process logs (in local memory), use **show logging process <process-name>**.

- To obtain the decoded per-process logs (in local memory and harddisk), you need to write the decoded traces to a file on the local file system using **show logging process <process-name> to-file bootflash:<filename>.**
- To identify the process-name, use **show logging process-helper.**
- To obtain the decoded logs across multiple processes, you collate all the relevant tracelogs with profiles using **show logging profile wireless to-file bootflash:<filename>.**
- Typically, when you're triaging a symptom, one failed system like an AP or client is used to drill into details. To obtain the decoded logs filtered to a specific condition, use **show logging {process | profile} {process-name | profile-name} filter {interface | ipv4 | ipv6 | mac | ra | string | uuid} to-file bootflash:<filename>.**
- To obtain the decoded logs up to a specific severity level, use **show logging {process | profile} <process-name | profile-name> level {error | warning | notice| info | debug | verbose} to-file bootflash:<filename>.**
- By default, these commands print the logs only from the last 10 minutes. To gather the decoded logs for a longer duration of time, use **show logging {process | profile} {process-name/profile-name} start last {<duration> | boot} {days | hours | minutes | seconds} to-file bootflash:<filename>.**
- To collect the decoded logs for a specific time window, use **show logging {process | profile} <process-name> start timestamp "MM/DD/YYYY HH:MM:SS" to-file bootflash:<filename>.**

At the end of any of these commands that have the **to-file** option selected, you're left with a file on the C9800 file system bootflash with the file name as defined by **<filename>**. This file can be viewed on the terminal session by running **more bootflash:<filename>.** It can also be exported to an external server via Trivial File Transfer Protocol (TFTP), File Transfer Protocol (FTP), Hypertext Transfer Protocol (HTTP), HyperText Transfer Protocol Secure (HTTPS), and Secure Copy Protocol (SCP) via **copy bootflash:<filename> {tftp: | ftp: | http: | https: | scp:} <filename>.**

The format of each log line within a decoded tracelog is

```
YYYY/MM/DD HH:MM:SS.SSSSSS [PROCESS_NAME] {xx}: [MODULE_NAME][PID]:
(SEVERITY LEVEL): CONDITION TYPE : CONDITION <MESSAGE STRING>
```

- **YYYY/MM/DD:** This field is the date format for the day the log message was written.
- **HH:MM:SS.SSSSSS:** This is the time stamp when the log message was written. Note that the log messages were written in Coordinated Universal Time (UTC) in earlier codes, but starting with release 17.3.1, the tracelogs use the time zone configured on the C9800.
- **PROCESS_NAME:** This is the name of the BinOS Process or IOS that is writing the log.
- **Xx:** This is the internal system–generated identifier.

- **MODULE_NAME:** Each process includes several modules, and at bootup, all modules are set to log messages up to and including the NOTICE level.
- **PID:** This is the process ID of the process writing the log. You can verify it by using **show process** for IOS processes and **show process platform** for BinOS processes.
- **SEVERITY LEVEL:** This is the severity level of the log line.
- **CONDITION TYPE:** This field defines the context or condition to which the tracelogs need to be restricted. The condition types supported by btrace on the C9800 are the interface, IP address (IPv4/IPv6), MAC address, radioactive logs, application context (AppCtx) string, and universally unique identifier (UUID).
  - **UUID:** A UUID on the C9800 is simply a unique identifier generated and assigned internally by IOS-XE to each code execution flow (think about a function being called and all the subfunctions it calls too).
  - **AppCtx:** Application context refers to a set of UUIDs that together provide contextual information at a feature or functionality level. Think of it as a tag affixed to the UUID. Note that a single UUID can be affixed with multiple AppCtx tags and that each AppCtx tag can link to multiple UUIDs.

  Both UUID and AppCtx are constructs internal to the C9800 and not commonly used to filter the decoded tracelogs
- **CONDITION:** This refers to the actual condition of the specific filter type such as IP address or MAC address.

Always-On Tracing

By default, all processes that use btrace always write tracelogs, without any explicit configuration or user intervention, and are therefore referred to as always-on traces. The always-on traces, by default, include tracelogs only up to the NOTICE level. On the C9800, if a failure occurs, these always-on traces are powerful tools to look back at the events leading up to the failure. The tracelogs are constantly being written, rotated, stored, and overwritten upon reaching the maximum supported number. This maximum number of tracelogs is determined by the activity on the system. A steady-state C9800 aims to provide at least 48 hours of lookback. However, on a C9800 operating at scale with high numbers of APs and clients on it, if failure like a large scale or high rate of AP flaps and client drops is seen, there is the potential for logs getting overwritten much more quickly. All serious failures that generate FATAL, EMERGENCY, ALERT, and CRITICAL messages as well as any system-level or feature-level ERRORs can be easily diagnosed. The NOTICE level logs help to isolate the failure and identify any trends of failure being seen, which, in turn, help define the deep-dive debug tracelogs that need to be collected to determine the root cause of the failure. The rotation system ensures the disk and flash never fill up with logs because only a dedicated amount of space is reserved for this purpose.

To view always-on traces, you can use the same CLI tree **show logging...** and its suboptions. The option that is most used is always-on traces filtered to the condition of one or more MAC addresses or IP addresses. Common use cases where this option comes in handy include AP disconnect, AP flaps, client onboarding failures, and mobility tunnel flap. You can view these conditional always-on traces from the GUI of the C9800, as shown in Figure 16-2, by navigating to **Troubleshooting > Radioactive Trace**. Click **+Add** to add the MAC addresses or IP addresses of interest and then click **Generate.** Finally, select the lookback duration and click **Apply to Device** to generate the log file. The lookback duration selected by default is 10 minutes. You can then download this file from the same page.

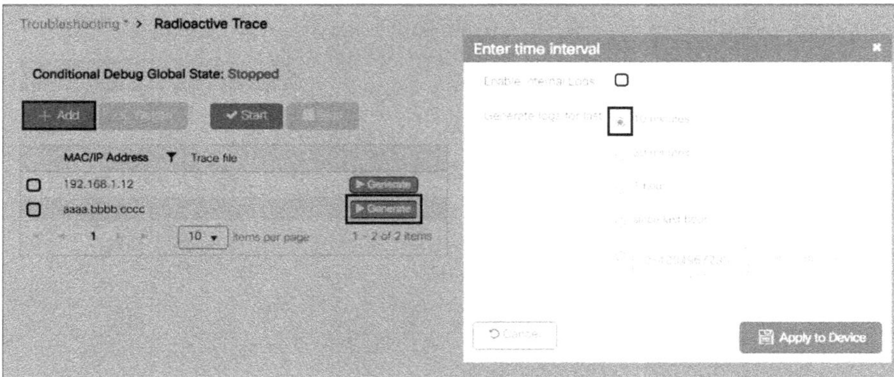

Figure 16-2 *Collecting always-on traces for a client MAC from the C9800 GUI*

The C9800 GUI is designed for ease of use and provides a spanner button on the Monitoring tab besides the common conditions for tracing that enable direct navigation between the Monitoring and Troubleshooting tabs. Figure 16-3 highlights the spanner button provided on the C9800 GUI beside the radio MAC address and Ethernet MAC address of each AP under **Monitoring > Wireless > AP Statistics**, and Figure 16-4 highlights the same for a client MAC address under **Monitoring > Wireless > Clients.** Figure 16-5 shows the prompt that allows you to directly navigate to the Troubleshooting tab.

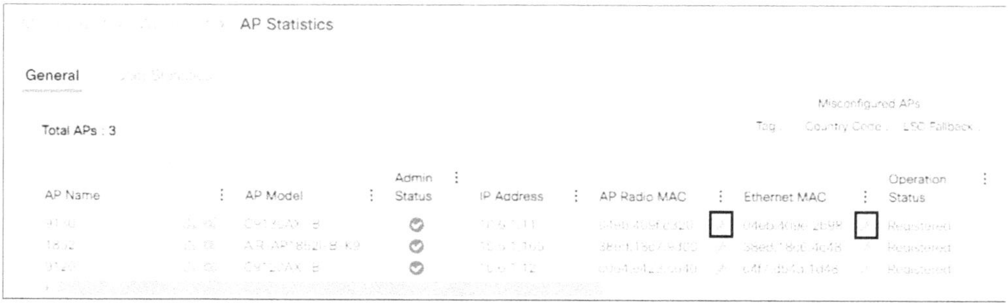

Figure 16-3 *Selecting the AP MAC address for conditional tracing from APs listed under Monitoring*

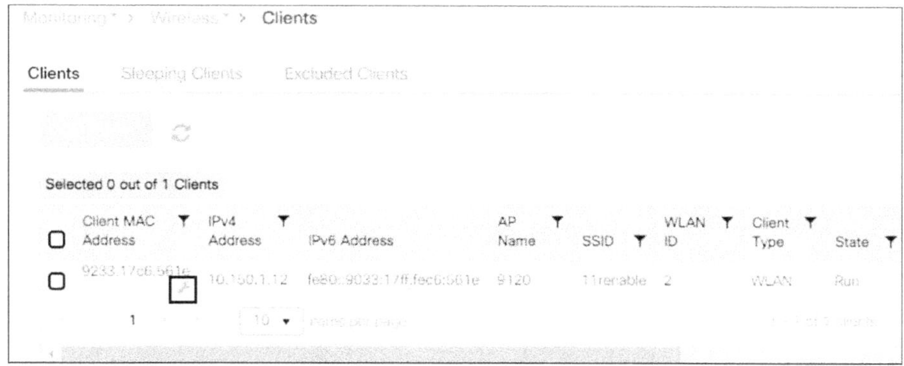

Figure 16-4 *Selecting the client MAC address for conditional tracing from the client listed under Monitoring*

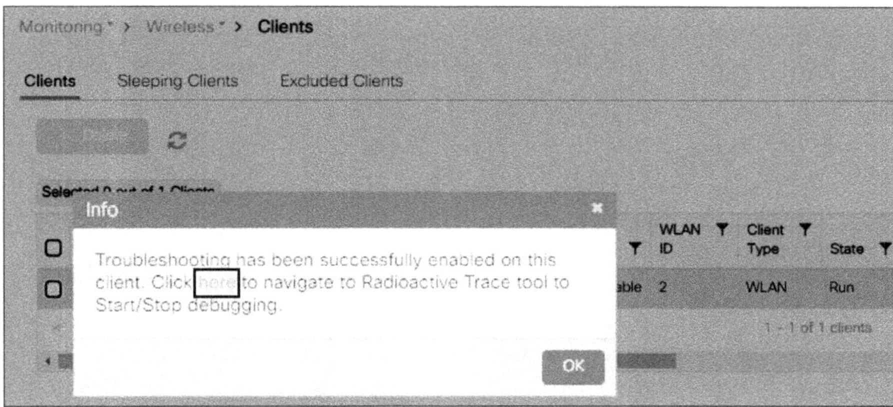

Figure 16-5 *Direct navigation of the C9800 GUI from Monitoring to Troubleshooting*

Example 16-2 provides a log snippet from always-on traces of the AP Ethernet MAC address indicating the tags applied to the AP after AP joins. Example 16-3 shows always-on traces for a successful client onboarding. In each tracelog message, the step within client onboarding, the key takeaway, and the state transition of the client are highlighted. Example 16-4 provides an example of failed client onboarding.

Example 16-2 *Always-On NOTICE Tracelog of the AP Ethernet MAC Address Showing Tags Applied to the AP*

```
C9800#show logging profile wireless filter mac c4f7.d54d.1d48 to-file
  bootflash:c4f7.d54d.1d48snip-APethernetradiomac.log
C9800#more bootflash:c4f7.d54d.1d48-apether.txt | inc Success
{wncd_x_R0-0}{1}: [apmgr-capwap-join] [25038]: (note): MAC: c064.e423.c640  Success-
  fully processed Join request. AP name: 9120, Model: C9120AXI-B, radio slots: 2,
  rlan slots: 0, site tag name: st-central, policy tag name: pt-central-150, rf tag
  name: rf-central-global
{wncd_x_R0-0}{1}: [capwapac-smgr-srvr] [25038]: (note): MAC: c064.e423.c640  Suc-
  cessfully handled Config status request.
```

Example 16-3 *Always-On Trace of Successful Client Onboarding on Dot1x SSID*

```
C9800# show logging profile wireless filter mac 9233.17c6.561e to-file bootflash:
  9233.17c6.561e-clientmac.log
C9800# more bootflash: 9233.17c6.561e-clientmac.log | include ERR
2022/01/16 02:12:54.947167 {wncd_x_R0-0}{1}: [client-orch-sm] [17585]: (note): MAC:
  9233.17c6.561e  Association received. BSSID c064.e423.c64b, WLAN 11renable, Slot 1
  AP c064.e423.c640, sudha-9120
2022/01/16 02:12:54.947375 {wncd_x_R0-0}{1}: [client-orch-state] [17585]: (note):
  MAC: 9233.17c6.561e  Client state transition: S_CO_INIT -> S_CO_ASSOCIATING
2022/01/16 02:12:54.947824 {wncd_x_R0-0}{1}: [dot11] [17585]: (note): MAC:
  9233.17c6.561e  Association success. AID 2, Roaming = False, WGB = False,
  11r = True, 11w = False Fast roam = False
2022/01/16 02:12:54.947932 {wncd_x_R0-0}{1}: [client-orch-state] [17585]: (note):
  MAC: 9233.17c6.561e  Client state transition: S_CO_ASSOCIATING -> S_CO_L2_AUTH_IN_
  PROGRESS
2022/01/16 02:12:54.948023 {wncd_x_R0-0}{1}: [client-auth] [17585]: (note): MAC:
  9233.17c6.561e  ADD MOBILE sent. Client state flags: 0x71  BSSID: MAC: c064.e423.
  c64b capwap IFID: 0x90000005
2022/01/16 02:12:54.950958 {wncd_x_R0-0}{1}: [client-auth] [17585]: (note): MAC:
  9233.17c6.561e  L2 Authentication initiated. method DOT1X, Policy VLAN 150,AAA
  override = 0 , NAC = 0
2022/01/16 02:12:54.951834 {wncd_x_R0-0}{1}: [ewlc-infra-evq] [17585]: (note):
  Authentication Success. Resolved Policy bitmap:11 for client 9233.17c6.561e
2022/01/16 02:12:55.364720 {wncd_x_R0-0}{1}: [client-auth] [17585]: (note): MAC:
  9233.17c6.561e  L2 Authentication Key Exchange Start. Resolved VLAN: 150,
  Audit Session id: 000000000000000D60A8B028
2022/01/16 02:12:55.385353 {wncd_x_R0-0}{1}: [client-keymgmt] [17585]: (note): MAC:
  9233.17c6.561e  EAP Key management successful. AKM:FT-DOT1X Cipher:CCMP WPA
  Version: WPA2
2022/01/16 02:12:55.385480 {wncd_x_R0-0}{1}: [client-orch-sm] [17585]: (note): MAC:
  9233.17c6.561e  Mobility discovery triggered. Client mode: Local
2022/01/16 02:12:55.385483 {wncd_x_R0-0}{1}: [client-orch-state] [17585]: (note):
  MAC: 9233.17c6.561e  Client state transition: S_CO_L2_AUTH_IN_PROGRESS ->
  S_CO_MOBILITY_DISCOVERY_IN_PROGRESS
2022/01/16 02:12:55.386223 {wncd_x_R0-0}{1}: [mm-client] [17585]: (note): MAC:
  9233.17c6.561e Mobility Successful. Roam Type None, Sub Roam Type MM_SUB_ROAM_
  TYPE_NONE, Client IFID: 0xa0000002, Client Role: Local PoA: 0x90000005 PoP: 0x0
```

```
2022/01/16 02:12:55.386466 {wncd_x_R0-0}{1}: [client-auth] [17585]: (note): MAC:
  9233.17c6.561e  ADD MOBILE sent. Client state flags: 0x72  BSSID: MAC: c064.e423.
  c64b capwap IFID: 0x90000005
2022/01/16 02:12:55.386497 {wncd_x_R0-0}{1}: [client-orch-state] [17585]: (note):
  MAC: 9233.17c6.561e  Client state transition: S_CO_MOBILITY_DISCOVERY_IN_PROGRESS
  -> S_CO_DPATH_PLUMB_IN_PROGRESS
2022/01/16 02:12:55.386601 {wncd_x_R0-0}{1}: [dot11] [17585]: (note): MAC:
  9233.17c6.561e  Client datapath entry params - ssid:11renable,slot_id:1 bssid
  ifid: 0x0, radio_ifid: 0x90000003, wlan_ifid: 0xf0400002
2022/01/16 02:12:55.386800 {wncd_x_R0-0}{1}: [dpath_svc] [17585]: (note): MAC:
  9233.17c6.561e Client datapath entry created for ifid 0xa0000002
2022/01/16 02:12:55.386880 {wncd_x_R0-0}{1}: [client-orch-state] [17585]: (note):
  MAC: 9233.17c6.561e  Client state transition: S_CO_DPATH_PLUMB_IN_PROGRESS ->
  S_CO_IP_LEARN_IN_PROGRESS
2022/01/16 02:12:56.326461 {wncd_x_R0-0}{1}: [client-iplearn] [17585]: (ERR): MAC:
  9233.17c6.561e  Failed to build data path client params. Not able to fetch link
  local address
2022/01/16 02:12:57.390380 {wncd_x_R0-0}{1}: [client-iplearn] [17585]: (note): MAC:
  9233.17c6.561e  Client IP learn successful. Method: ARP IP: 10.12.249.101
2022/01/16 02:12:57.390756 {wncd_x_R0-0}{1}: [errmsg] [17585]: (info): %CLIENT_ORCH_
  LOG-6-CLIENT_ADDED_TO_RUN_STATE: R0/0: wncd: Username entry (lxuser) joined with
  ssid (11renable) for device with MAC: 9233.17c6.561e
2022/01/16 02:12:57.390902 {wncd_x_R0-0}{1}: [client-orch-state] [17585]: (note):
  MAC: 9233.17c6.561e  Client state transition: S_CO_IP_LEARN_IN_PROGRESS ->
  S_CO_RUN
2022/01/16 02:12:57.923548 {wncd_x_R0-0}{1}: [client-iplearn] [17585]: (note):
  MAC: 9233.17c6.561e  Client IP learn successful. Method: DHCP IP: 10.150.1.22
```

Example 16-4 *Always-On Trace of Client Authentication Failure Due to Incorrect Password on 802.1X SSID*

```
2022/03/03 11:31:33.699312 {wncd_x_R0-0}{1}: [client-orch-sm] [25038]: (note): MAC:
  9233.17c6.561e  Association received. BSSID c064.e423.c64b, WLAN 11renable,
  Slot 1 AP c064.e423.c640, 9120
2022/03/03 11:31:33.699555 {wncd_x_R0-0}{1}: [client-orch-state] [25038]: (note):
  MAC: 9233.17c6.561e  Client state transition: S_CO_INIT -> S_CO_ASSOCIATING
2022/03/03 11:31:33.721454 {wncd_x_R0-0}{1}: [dot11] [25038]: (note): MAC:
  9233.17c6.561e  Association success. AID 1, Roaming = False, WGB = False, 11r =
  True, 11w = False Fast roam = False
2022/03/03 11:31:33.721592 {wncd_x_R0-0}{1}: [client-orch-state] [25038]: (note):
  MAC: 9233.17c6.561e  Client state transition: S_CO_ASSOCIATING -> S_CO_L2_AUTH_IN_
  PROGRESS
2022/03/03 11:31:33.721704 {wncd_x_R0-0}{1}: [client-auth] [25038]: (note): MAC:
  9233.17c6.561e  ADD MOBILE sent. Client state flags: 0x71  BSSID: MAC: c064.e423.
  c64b capwap IFID: 0x90000004
2022/03/03 11:31:33.725346 {wncd_x_R0-0}{1}: [client-auth] [25038]: (note): MAC:
  9233.17c6.561e  L2 Authentication initiated. method DOT1X, Policy VLAN 150,AAA
  override = 0 , NAC = 0
```

```
2022/03/03 11:31:33.744566 {wncd_x_R0-0}{1}: [ewlc-infra-evq] [25038]: (note):
  Authentication Success. Resolved Policy bitmap:11 for client 9233.17c6.561e
2022/03/03 11:31:34.532757 {wncd_x_R0-0}{1}: [errmsg] [25038]: (note): %DOT1X-5-
  FAIL: R0/0: wncd: Authentication failed for client (9233.17c6.561e) with reason
  (Cred Fail) on Interface capwap_90000004 AuditSessionID 000000000000000B4F8CACE0
  Username: lxuser
```

Per-Process Debugging

For BinOS processes, the NOTICE level always-on traces help with problem isolation, but to determine the root cause of the failure, DEBUG level traces are needed. The DEBUG level for the tracelogs needs to be manually set and can be done on a per-process basis. These per-process debugs are most useful when you are troubleshooting a configuration workflow that is not tied to a session and where behavior of a specific process is of interest. Therefore, typically, there is no context or condition used to debug the individual process daemons. The output of this per-process debugging can be filtered for ease of consumption, but this output is limited to the perspective of the specific process only and does not provide a view into activities of other processes. It is therefore sometimes referred to as *unconditional debugging*. To view the severity level to which the module traces in a process are set, run **show platform software trace level <process-name> chassis active R0**. The CLI command to set a specific process to the DEBUG level is **set platform software trace <process-name> chassis active R0 <module-name> debug.** To run this debug, you need to know the process daemon names and the modules within the processes. Knowing the role of each process and module requires a deep understanding of the C9800 software architecture. To abstract the architecture to some extent and make this debugging user-friendly, process daemon names have been made intuitive to their purpose. For example, mobility is responsible for formation and maintenance of mobility tunnels and thereby generates traces corresponding to the negotiation and formation of the tunnel. Rogued is responsible for rogue detection and reporting and generates the tracelogs related to rogues. You can use the CLI **show logging process-helper <process-name>** to identify the process daemon name. Further, an option called **all-modules** is provided under most processes, so you do not have to choose individual modules and are still able to leverage unconditional debugging to get process level data. Although binary traces are efficient so as to have little impact on the CPU, at the DEBUG level, many modules start logging a lot of information. Therefore, running tracelogs at the DEBUG level contributes to the CPU utilization to some extent. To avoid any inadvertent negative impact, it is especially important to disable debugging after the necessary tracelogs are collected. You can set the individual processes back to NOTICE by running **set platform software trace <process-name> chassis active R0 level <module-name | all-modules> notice**. Alternatively, you can run the CLI **undebug all**, but note that this CLI defaults all the process-specific binary traces to NOTICE, not just the one you're working with at the time. In addition, it also disables any IOS debugs running on the C9800.

All the per-process debugging needs to be triggered and viewed using the CLI. However, you can view per-process tracelogs of two processes on the GUI. On the

C9800, under Troubleshooting > Logs, there are two tabs in addition to Syslog, as seen in Figure 16-1:

- **Web Server Logs:** These are the tracelogs generated by the WebUI process; they provide a view into any obvious web access–related error logs.
- **License Logs:** These are the tracelogs corresponding to the Smart Agent on the IOSd that is responsible for Smart Licensing. This information is not sufficient to determine the root cause but provides a view into any ERROR level logs.

Radioactive Tracing

The per-process debugging works for targeted functions like rrmgrd, rogued, and nmspd. However, when you're troubleshooting a session such as a client or an AP where multiple processes and modules are involved and where only the logs pertaining to a given session are of interest, performing per-process debugging, connecting the dots across processes for a given session while filtering out other sessions would be labor intensive. Consider a medium-sized enterprise deployment, say with 300 APs mapped to each WNCD instance. Per-process debugging would require the following effort and risks:

- Enabling the WNCd to DEBUG without context negatively impacts the CPU utilization of the WNCd process and thereby puts the C9800 at risk.
- The logs resulting from WNCd debugging without context include all APs and clients, and you need to manually filter the logs.
- For any use cases that involve other processes besides the WNCd, you need to collect and manually correlate the decoded tracelogs between processes to define the sequence of events.

To abstract this complexity, radioactive (RA) tracing was created. RA tracing enables you to stitch together a chain of execution for operations of interest across the system at an increased verbosity level. That is, for a given context or condition like a MAC address or IP address, radioactive tracing tracks the condition through all applicable modules and processes and provides concise decoded DEBUG level logs in chronological order. Each execution code flow for binary traces creates a unique UUID. RA tracing tracks these UUIDs for a given context until the code execution completes. For example, when a client sends an association request, the request packet is tracked through the system until the corresponding association response is sent out. Because debugging is done against a specific condition, RA tracing is sometimes referred to as conditional debugging.

The volume of logs generated with the severity level set to DEBUG is at least a hundred times more verbose than those at the NOTICE level. The log lines coded for severity level for DEBUG are written both for general customer consumption and that of Cisco developers involved with developing the features on the C9800. Some of the logs such as internal library calls are useful only for developers and are not needed to diagnose common wireless failure conditions. To limit the DEBUG level logs displayed, the C9800 uses unique system-defined flags to mark and display the tracelogs that are meant for customers from those intended for developers. Note that these flags impact only the tracelog lines displayed and have no impact on tracelogs that are written at the DEBUG level. In other words, all lines defined for severity level DEBUG and below are written to the tracelog .bin files. Note the term *RA trace* is usually used to refer to these customer-friendly DEBUG level tracelog lines, whereas the term *internal RA trace* loosely refers to the more verbose DEBUG tracelog lines intended for developers. With these filters, the number of tracelog lines displayed on the RA trace is about one-tenth of the number of tracelog lines displayed in the internal RA trace.

Unlike always-on traces that do not need to be configured, RA active tracing requires configuration and requires the same failure condition to be intentionally reproduced or allowed to reoccur on its own after the debug has been set. The steps to trigger the RA trace are

1. Define the condition by using **debug platform condition feature wireless {mac | ip | interface | flow-id}**.
2. Start the RA trace by using **debug platform condition start.**
3. Reproduce the issue.
4. Stop the RA trace after the issue has been reproduced by using **debug platform condition stop.**

The wireless use cases that benefit the most from radioactive tracing include client onboarding and roaming, AP join, and mobility tunnel formation, among others. The context in all these use cases is either a MAC address, such as a client MAC address or an AP MAC address, or an IP address, such as a mobility peer IP address. The C9800 GUI supports RA tracing only for these two conditions. From the GUI, the first steps to trigger the RA trace are the same as those shown in Figure 16-2 where you select the condition of interest by navigating to **Troubleshooting > Radioactive Trace > + Add** or through the spanner button, as shown in Figures 16-3, 16-4, and 16-5. However, to get DEBUG level logs versus NOTICE level always-on logs, select **Start.** Then reproduce the use case or failure condition and select **Stop**, as shown in Figure 16-6. Note that the stop puts all relevant tracelogs back to the default NOTICE level.

Figure 16-6 *Starting a radioactive trace on the C9800 GUI*

To view the customer-use RA trace from the GUI, navigate to **Troubleshooting > Radioactive Trace > Generate** for the MAC or IP address for which the RA was triggered. Choose the time duration indicating how far back the log collection should be started and click **Apply to Device.** Then download the resulting file as highlighted in Figure 16-7. This is equivalent to using the CLI **show logging profile wireless start last {<duration> | boot} days | hours | minutes | seconds} <CONDITION-TYPE> <condition> to-file bootflash: <filename>**. To reiterate, the GUI workflow is identical for collecting always-on traces and radioactive traces apart from the Starting debug, which boosts the level of logging to DEBUG for the configuration MAC or IPs until you click **Stop.**

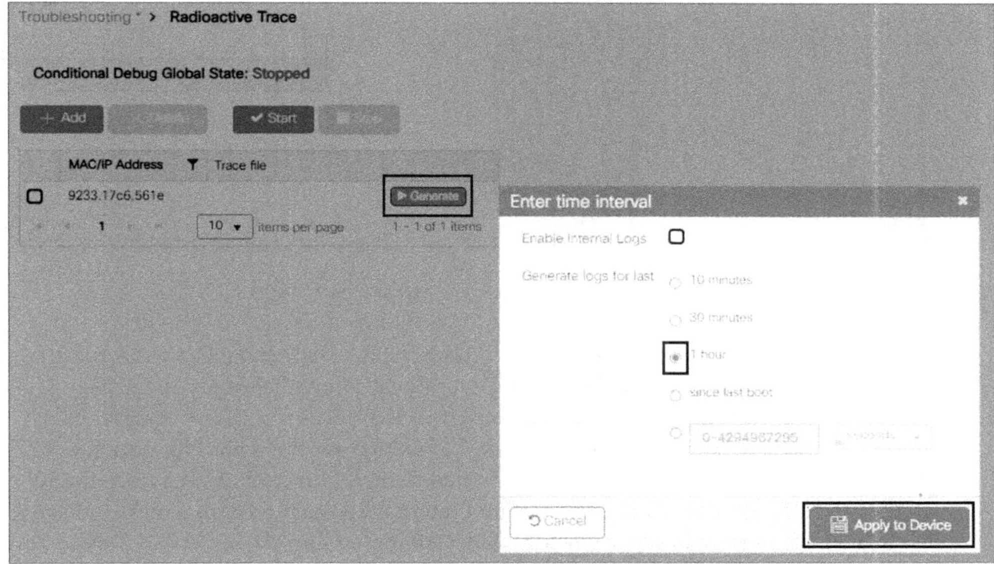

Figure 16-7 *Collecting customer-use radioactive traces from the C9800 GUI*

To view the internal RA trace, navigate to **Troubleshooting > Radioactive Trace > Generate** for the MAC or IP address for which the RA was triggered. Select **Internal.** Then choose the time duration indicating how far back the log collection should be started. Then click **Apply to Device** and download the resulting file, as shown in Figure 16-8. This is equivalent to using the CLI **show logging profile wireless internal start last {<duration> | boot} {days | hours | minutes | seconds} <CONDITION-TYPE> <condition> to-file bootflash: <filename>**. Note that triggering the RA and reproducing the failure condition only once are sufficient to generate and view both the RA trace and internal RA trace.

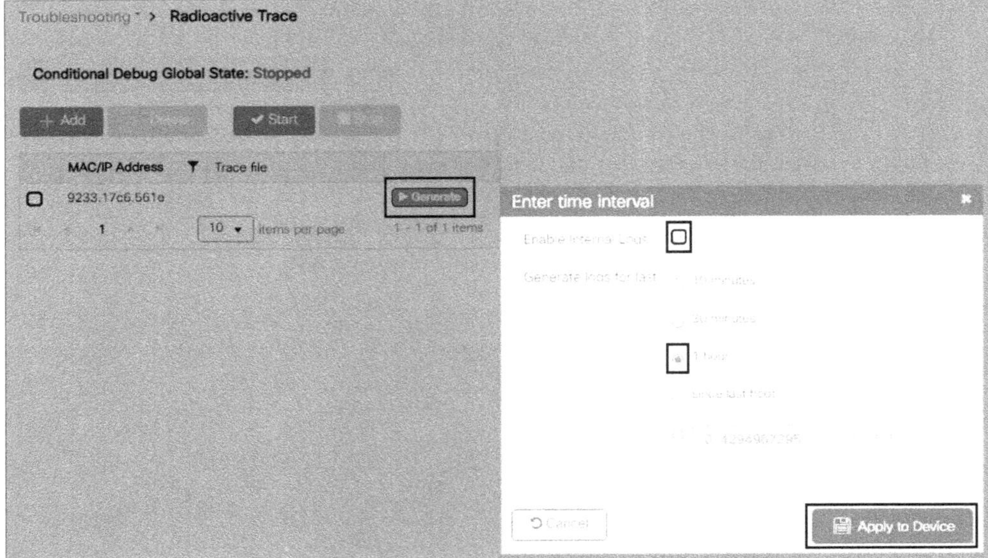

Figure 16-8 *Collecting internal radioactive traces from the C9800 GUI*

Using the GUI is the recommended way to collect the RA trace and internal RA trace. The C9800 is shipped embedded with a TCL script called ra_tracing_tool.tcl, which automates the triggering, generation, and export of RA traces to a file or external File Transfer Protocol (FTP) server into the single CLI **debug wireless {mac | ip} {xxxx.xxxx.xxxx | x.x.x.x}** with additional options. By default, this CLI triggers an RA trace for 30 minutes and stores the resulting RA trace (not internal RA) to bootflash with the filename ra_trace_MAC_aaaabbbbcccc_HHMMSS.XXX_timezone_DayWeek_Month_Day_year.log. The suboptions available with this command include

- **monitor-time:** To run the RA trace for any duration other than the default 30 minutes supported by this CLI, use the monitor-time option **#debug wireless {mac | ip} {xxxx.xxxx.xxxx | x.x.x.x} monitor-time <1-2085978494>.** Note that the duration specified here is in seconds.
- **level:** This option defines the severity level up to which tracelogs need to be written and supports ERROR, WARNING, NOTICE, INFO, DEBUG, and VERBOSE levels.
- **internal:** This option allows you to both trigger and generate only the internal RA trace in one shot. Note that, in this case, you need to use the CLI **show logging profile wireless....**
- **to-file:** This option is used to define the file system (bootflash: or harddisk:) on the box to which the RA trace or internal RA trace needs to be stored.
- **ftp-server:** To export the RA trace to an external FTP server, use **debug wireless {mac | ip} {xxxx.xxxx.xxxx | x.x.x.x} ftp-server <server-IP> <server path to store file>.** If the FTP server is unreachable or packets get dropped in the path from the

C9800 to the FTP server or credentials are incorrect or the credentials do not have write access to the path specified or any such problem, the RA trace output file is not written, and the debug needs to be rerun. To avoid this scenario, using the **to-file** option to save the RA trace output to the local file system is recommended.

For the large volume of logs that get generated on the system for the DEBUG level and the fixed amount of storage available, to maintain a 48-hour lookback, it is mandatory to disable RA traces and go back to the NOTICE level after the problem is reproduced and relevant tracelogs are collected. In general, it is always recommended you undo any configurations introduced for troubleshooting. In the case of the RA trace, it is best to clear any conditions set with **clear platform condition all** to ensure that any future debugging remains unencumbered by stale conditions left behind by the previous debugging.

In summary, for any binary trace-based debugging, there are two aspects:

- Selecting what is written to the tracefile: This must be done before the problem occurs. Always-on tracing on the C9800 ensures some logs are written at all times. For troubleshooting a specific problem, you can bump up the logging level to DEBUG on a process or condition basis.
- Decoding and visualizing the resultant tracelogs: Depending on the use case and need, various filters like level, condition, and process name are available to view tracelogs of interest.

Getting the first step right is critical to ensuring all the necessary tracelogs get written the first time that the problem is reproduced and multiple reoccurrences or reproductions of the problem are not needed. You can then custom-fit the visualization of the logs as needed.

Embedded Packet Capture (EPC)

As the name suggests, Embedded Packet Capture is a packet-capturing tool built into the code of the C9800. When enabled, EPC enables capturing of packets sent and received by the C9800. EPC supports inline, access control list (ACL), and class-map–based packet filters to define traffic to be captured. After they are captured, these packets are stored in a dynamic random-access memory (DRAM) buffer in the IOSd. A summary of the captured packets can be viewed on-the-box. You can also look at each individual captured packet in detail on-the-box. However, a more user-friendly approach is to export the captured packets as a .pcap file to an end device running capture analyzing tools such as Wireshark for offline analysis. To enable this packet capturing, you need explicit configuration. Like most debugging workflows, EPC is run from exec privileged mode, is not visible in the **show run**, and cannot persist through a reload.

EPC is a powerful infrastructure for viewing the contents of packets being processed by the C9800. Packets traversing the data plane or the punt and inject path are copied to the buffer in IOSd. As CPU processing is involved, in theory, there is potential for a large volume of data traffic to overwhelm the CPU and negatively impact the system.

To protect the CPU, the C9800 uses predefined punt policers that can be viewed with **show platform software punt-policer.** Figure 16-9 provides the CLI output filtered to EPC. The default punt policer for EPC is 8738, meaning if the traffic rate is higher than the rate limited by the punt policer, the EPC cannot capture all the packets.

```
C9800#show platform software punt-policer | include Punt|Cause|--|EPC
Per Punt-Cause Policer Configuration and Packet Counters
Punt  Description  Config Rate(pps)   Conform Packets    Dropped Packets     Config Burst(pkts)  Config Alert
Cause               Normal    High     Normal    High     Normal       High    Normal    High      Normal   High
-------------------------------------------------------------------------------------------------------------------
 75     EPC          8738      1000      0         0        0           0       8738      1000      Off      Off
```

Figure 16-9 *Viewing the default EPC punt policer*

Also, the buffer dedicated to EPC needs to be allocated specific memory to prevent any negative impact to the overall C9800. The maximum buffer size for EPC is 100 MB, and unless the buffer is configured to be circular, the packet capturing stops when the buffer is exhausted.

To enable EPC from the C9800 GUI, navigate to **Troubleshooting > Packet Capture > +Add**, as shown in Figure 16-10.

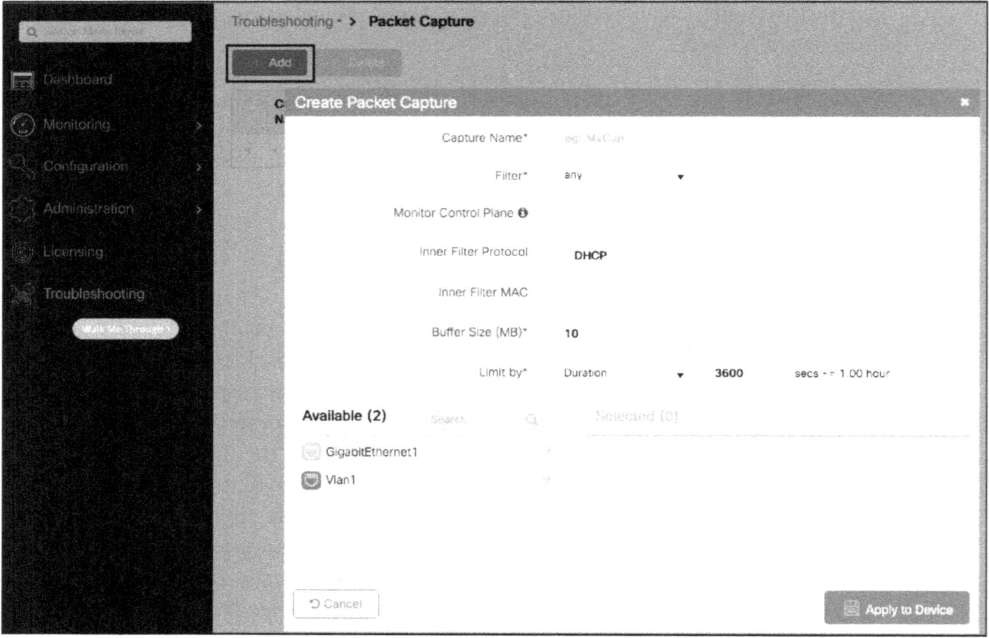

Figure 16-10 *Enabling EPC on the C9800 Troubleshooting dashboard*

The options on the C9800 GUI to enable EPC and the equivalent CLI commands are

- **Capture Name:** In this field, you define the arbitrary name that will be used to reference this capture. The maximum number of characters supported is 8. The corresponding CLI would be **monitor capture <CaptureName>.**

- **Filter:** This field defines the packets of interest that need to be captured. Both the inline filter and ACL options are available. Inline filters for EPC on the C9800 support matching on

  - Any source and destination
  - A specific source or destination IPv4 or IPv6 address
  - An IPv4 subnet for source or destination

  You can also select to capture only Transmission Control Protocol (TCP) or UDP traffic for any source destination combinations listed. Only one match criterion can be included within a single EPC capture.

  If more than one source-destination combination needs to be captured, using an ACL is the best course of action. An extended ACL also provides the granularity of defining the source or destination UDP and TCP ports to further narrow down the capture to the traffic of interest. The filter options available in the GUI translate to the following CLIs:

  ```
  #monitor capture <Capture Name> match any
  #monitor capture <Capture Name> match {ipv4 | ipv6} {any | host
  <source IP address> | <source subnet/netmask>} {any | host <desti-
  nation IP address> | <destination subnet/netmask>}
  #monitor capture <Capture Name> match {ipv4 | ipv6} protocol
  {<0-255 IP protocol number> | tcp | udp} {any | host <source IP
  address> | <source subnet/netmask>} {any | host <destination IP
  address> | <destination subnet/netmask>}
  ```

  A few options that are available only on the CLI are

  ```
  #monitor capture <Capture Name> match mac {<source mac> | any |
  host <source mac>} {<destination mac> | any | host <destination
  mac>}
  #monitor capture <Capture Name> match pktlen-range
  ```

- **Monitor Control Plane:** On the C9800 GUI, when EPC is set up, by default, only data plane traffic from the front panel is selected to be monitored. This is sufficient for troubleshooting data forwarding and packet drops by the C9800. However, it does not give a view of control plane traffic processed by the C9800 CPU. Control plane traffic monitoring requires you to select the **Monitor Control Plane** option, which provides an explicit view into packets punted from the data plane to the control plane. Selecting both the uplink ports and control plane for EPC monitoring provides a view into control plane traffic as it enters the C9800, gets processed by the CPU, and is responded back. Note the EPC capture in this case shows the packets twice: once without any Ethernet header (because the packet is in the control plane) and once with all the proper headers as the packet is entering or exiting the interface.

If you consider where data DTLS is in use, traffic shows up encrypted in all packet captures except where the control plane option is checked, which allows packets to be viewed after decryption. Figure 16-11 shows the EPC capture of client onboarding on dot1x SSID without control plane monitoring selected, with data DTLS in use, and Figure 16-12 shows the EPC of the same client onboarding with control plane monitoring selected.

```
22 2022-03-15 04:29:50.301971 10.5.1.11      192.168.1.5    191 Application Data
23 2022-03-15 04:29:52.336972 10.5.1.11      192.168.1.5    479 Application Data
24 2022-03-15 04:29:52.337979 192.168.1.5    10.5.1.11      411 Application Data
25 2022-03-15 04:29:52.337979 192.168.1.5    10.5.1.11      587 Application Data
26 2022-03-15 04:29:52.340970 10.5.1.11      192.168.1.5    143 Application Data
27 2022-03-15 04:29:52.342969 192.168.1.5    10.5.1.11      155 Application Data
28 2022-03-15 04:29:52.378978 10.5.1.11      192.168.1.5    175 Application Data
29 2022-03-15 04:29:52.446967 10.5.1.11      192.168.1.5    159 Application Data
30 2022-03-15 04:29:52.446967 192.168.1.5    10.5.1.11      155 Application Data
31 2022-03-15 04:29:52.446967 192.168.1.5    10.5.1.11      155 Application Data   encrypted
32 2022-03-15 04:29:52.452964 10.5.1.11      192.168.1.5    303 Application Data
33 2022-03-15 04:29:52.494969 10.5.1.11      192.168.1.5    223 Application Data
```

Figure 16-11 *EPC showing only encrypted packets with data DTLS without control plane monitoring*

```
 80 2022-03-15 04:37:43.681956 92:33:17:c6:56:1e  Cisco_23:c6:4b  419 Association Request, SN=1105, FN=0, Flags=........, SSID=11renable
 90 2022-03-15 04:37:43.723961 92:33:17:c6:56:1e  Cisco_23:c6:4b  107 Response, Identity
 96 2022-03-15 04:37:43.826953 92:33:17:c6:56:1e  Cisco_23:c6:4b  233 Client Hello
102 2022-03-15 04:37:43.968944 92:33:17:c6:56:1e  Cisco_23:c6:4b  102 Response, Protected EAP (EAP-PEAP)
106 2022-03-15 04:37:43.984934 92:33:17:c6:56:1e  Cisco_23:c6:4b  228 Client Key Exchange, Change Cipher Spec, Encrypted Handshake Message
110 2022-03-15 04:37:43.989939 92:33:17:c6:56:1e  Cisco_23:c6:4b  102 Response, Protected EAP (EAP-PEAP)
114 2022-03-15 04:37:43.995935 92:33:17:c6:56:1e  Cisco_23:c6:4b  142 Application Data
118 2022-03-15 04:37:44.000000 92:33:17:c6:56:1e  Cisco_23:c6:4b  144 Application Data
122 2022-03-15 04:37:44.004989 92:33:17:c6:56:1e  Cisco_23:c6:4b  142 Application Data
128 2022-03-15 04:37:44.021986 92:33:17:c6:56:1e  Cisco_23:c6:4b  334 Key (Message 2 of 4)
132 2022-03-15 04:37:44.036985 92:33:17:c6:56:1e  Cisco_23:c6:4b  191 Key (Message 4 of 4)
136 2022-03-15 04:37:44.037992 92:33:17:c6:56:1e  Cisco_23:c6:4b   96 Action, SN=1363, FN=0, Flags=........, SSID=11renable

Frame 106: 228 bytes on wire (1824 bits), 228 bytes captured (1824 bits)
Ethernet II, Src: 00:00:00_00:00:00 (00:00:00:00:00:00), Dst: 00:00:00_00:00:00 (00:00:00:00:00:00)
Internet Protocol Version 4, Src: 10.5.1.11, Dst: 192.168.1.5
User Datagram Protocol, Src Port: 5248, Dst Port: 5247
Control And Provisioning of Wireless Access Points - Data
IEEE 802.11 QoS Data, Flags: .......T
Logical-Link Control
802.1X Authentication
Extensible Authentication Protocol
```

Figure 16-12 *EPC showing decrypted packets even with data DTLS enabled with control plane monitoring*

Another use case where this option is useful is to identify traffic hitting CPU in the case of high CPU troubleshooting. It provides a quick way of identifying or eliminating a misbehaving client or traffic trend, such as an ARP or broadcast storm. Note that the GUI requires an interface to be selected, but the CLI can be used to enable control plane only EPC, as shown in Example 16-5.

Example 16-5 *Enabling Control Plane Only EPC on the C9800 CLI*

```
C9800#monitor capture <Capture Name> clear
C9800#monitor capture <Capture Name> control-plane both
C9800#monitor capture <Capture Name>  match any
C9800#monitor capture <Capture Name>  buffer size 100 circular
C9800#monitor capture <Capture Name> limit pps 1000000
C9800#monitor capture <Capture Name> start
>> wait for CPU to spike
C9800#monitor capture <Capture Name> stop
C9800#monitor capture <Capture Name> export bootflash:<FILENAME>.pcap
```

- **Inner Filter Protocol:** In case of encapsulated traffic like CAPWAP, this option enables EPC to parse the inner packet header protocol field and, if DHCP, to capture the packet as interesting traffic. The CLI is **monitor capture <Capture Name> inner protocol dhcp.**
- **Inner Filter MAC:** All wireless client traffic entering the C9800 is encapsulated in CAPWAP. The EPC that runs before decapsulation occurs needs this option enabled to match the inner packet MAC address to the configured MAC address. This allows traffic to and from that specific wireless client to be captured. The CLI is **monitor capture <Capture Name> inner.**
- **Buffer Size:** By default, the EPC buffer is linear with the buffer size set to 10 megabytes. This option allows buffer size to be increased. The CLI also allows the buffer to be made circular. Note that EPC automatically stops when the buffer is full.

  ```
  #monitor capture <Capture Name> buffer size <1-100MB>
  #monitor capture <Capture Name> buffer circular
  ```

- **Limit By:** This defines the criteria used to limit the EPC capture output. On the GUI, only two options, Duration and Packets, are available. The CLI provides more granular criteria for limiting captured packets:
  - **Duration:** This is the time duration in seconds that the capture should run for. The default value is 3600 seconds, but a range of 1–1,000,000 seconds is supported. The CLI is **monitor capture <Capture Name> limit duration <1-1000000>.**
  - **Packets:** This defines the total number of packets that should be captured. It supports a range of 1–100,000 packets. The CLI is **monitor capture <Capture Name> limit packets <1-100000>.**
  - **Every:** This option enables capturing every *N*th packet. That is, it helps capture a specific packet that appears repeatedly in the same expected sequence. The CLI is **monitor capture <Capture Name> limit every <2-100000>.**
  - **Packet-len:** This option allows packets of a specific length to be captured. The CLI is **monitor capture <Capture Name> limit packet-len <64-9500 (in bytes)>.**

- **PPS:** This option allows you to limit the capture to the maximum packets per second rate. The default pps supported by EPC is 1k. If the traffic rate is higher than 1k pps, EPC cannot capture all the packets. The CLI is **monitor capture <Capture Name> limit pps <1-1000000>.**

- **Available Interfaces:** This field lists all the interfaces on which the capture can be enabled. All the interfaces that have a context and interface identifier on the IOSd support EPC. This includes physical uplink ports on the C9800, any switch virtual interfaces (SVIs) created on the C9800, any Layer 3 port channels configured on the C9800, any tunnel interfaces including Generic Routing Encapsulation (GRE) tunnels, and Ethernet over GRE (EoGRE) tunnels. Note that EPC cannot be run on the out-of-band management service port. The CLI is **monitor capture <Capture Name> interface <Interface-Name>** {**both| in | out**}.

After an EPC is defined, select **Start** to start the capture, as shown in Figure 16-13. The CLI is **monitor capture <Capture Name> start.**

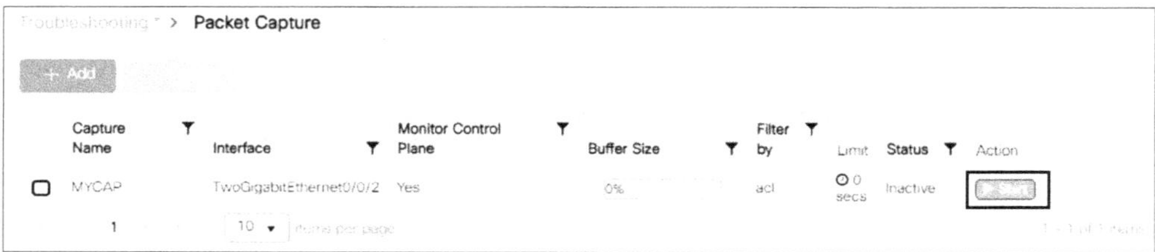

Figure 16-13 *Starting EPC on the C9800 Troubleshooting dashboard*

The status bar on the GUI indicates whether packets are being captured and how much buffer is filling up in percentage. The same can be viewed from the CLI via **show monitor capture <Capture Name>.** This output also provides the state of EPC if it is Active or Inactive, along with the settings that make up the capture definition.

After the packets of interest are captured, you can stop the capture by selecting **Stop**, as shown in Figure 16-14. The corresponding CLI is **monitor capture <Capture Name> stop.** After the capturing stops, the Export tab appears. On it, you can send the capture file to a TFTP server or FTP server or download directly to your desktop via HTTPS for offline analysis or save to the C9800 bootflash for on-the-box perusal. The CLI to export captures is **monitor capture** <**Capture Name**> **export** {*bootflash:*|*ftp:* | *tftp:* | *http:* | *https:* | *rcp:* | *scp:* | *sftp*:}. Instead of executing Stop and Export one after the other, CLI offers a one-shot command that combines both the operations and that CLI is **monitor capture <Capture Name> stop_export.** This option is not available from GUI.

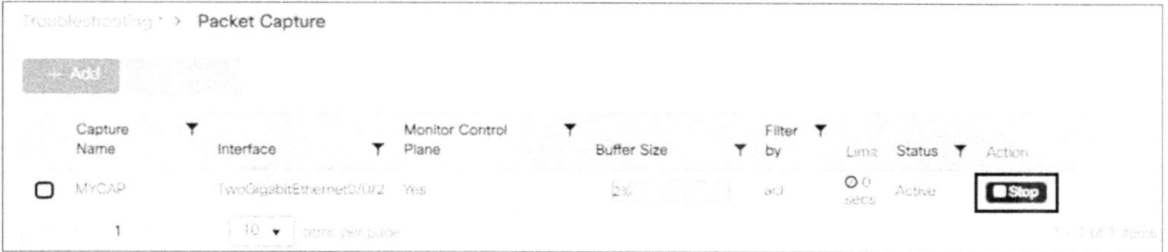

Figure 16-14 *Monitoring progress and stopping EPC capture on the C9800 Troubleshooting dashboard*

To clear the EPC capture from the CLI, run **monitor capture <Capture Name> clear.** For any problems with the EPC infrastructure, such as EPC settings not taking effect or the EPC not capturing traffic of interest, you can triage by using **debug epc capture-point** and **#debug epc provision.**

Packet Tracer

An embedded packet capture is like collecting a sniffer capture on one of the interfaces of the C9800 that gives you a .pcap file. This only shows you the packets as they are entering or exiting the specific interface chosen and does not give you any insight about what is happening to the packet inside the controller. What if there was a way to follow the path of packets inside the controller, to understand why they exit a specific interface, what modifications are applied to the packet, and how long this processing takes? This is what the Packet Tracer allows.

Running the Packet Tracer requires specifying a traffic type through an ACL. Do not expect to be able to trace all the packets like you would with a regular packet capture. It can collect up to 8000 packets before disabling itself to prevent having any throughput toll on the data plane. This means that the Packet Tracer indeed has no noticeable impact on performance, but you need to filter carefully. The C9800 allows you to dive into the path of any of the packets you captured, but it is a manual process. When diving in a packet, the Packet Tracer shows you the path it took inside the device and every feature that looked at the packet. Even better, it gives you the time in nanoseconds spent in each feature. This allows you, for example, to know if a packet was dropped by an ACL, if the CAPWAP fragment re-assembly is taking a long time (due to overfragmentation), or if your quality of service (QoS) policy is causing any latency to the packet processing.

To run the Packet Tracer, start by clearing any preexisting configuration:

```
#clear platform hardware chassis active qfp feature packet-trace
#clear platform condition all
```

Use the following commands to set a buffer of up to 8192 packets, choose an interface, and see the whole packet content of the packets captured:

```
#debug platform condition interface [internal-RP | tenGigE |
port-channel] both
#debug platform packet-trace packet 8192 fia-trace circular
#debug platform packet-trace copy packet both l2 size 256
```

You can start the packet tracing with **debug platform condition start** and stop it with **debug platform condition stop.**

The command **show platform hardware chassis active qfp feature packet-trace summary** shows you a summary of packets captured. Example 16-6 shows you packet-trace summary output filtered for a wireless client MAC address with **debug platform condition mac <CLIENTMAC>.**

Example 16-6 *Snippet from Packet-Trace Summary View for Client Onboarding*

```
C9800#show platform hardware chassis active qfp feature packet-trace summary
Pkt   Input                     Output               State    Reason
0     CAPWAP-IF-0x0090000004    internal0/0/rp:0   PUNT   129 (wls 802.11 Packets to LFTS
1     CAPWAP-IF-0x0090000004    internal0/0/rp:0          PUNT   97   (Packets to LFTS)

9     CAPWAP-IF-0x0090000004    internal0/0/rp:0   PUNT   97   (Packets to LFTS)
10    CAPWAP-IF-0x0090000004    internal0/0/rp:0   PUNT   97   (Packets to LFTS)
11    CAPWAP-IF-0x0090000004    internal0/0/rp:0   PUNT   129 (wls 802.11 Packets to LFTS
12    CAPWAP-IF-0x0090000004    internal0/0/rp:0   PUNT   132 (wls SISF Packets to LFTS)
13    INJ.43                    <none>             CONS   Packet Consumed Silently
14    VLAN-CPPIF-0150           Vl150              DROP   216 (UnconfiguredIpv6Fia)
15    VLAN-CPPIF-0150           Tw0/0/2            FWD
16    CAPWAP-IF-0x0090000004    WLCLIENT-IF-0x00a0000001  DROP   415 (WlsClientError)
17    CAPWAP-IF-0x0090000004    internal0/0/rp:0   PUNT   132 (wls SISF Packets to LFTS)
18    INJ.43                    <none>             CONS   Packet Consumed Silently
19    VLAN-CPPIF-0150           internal0/0/rp:0   PUNT   60   (IP subnet or broadcast pac
20    VLAN-CPPIF-0150           Tw0/0/2            FWD
21    Tw0/0/2                   internal0/0/rp:0   PUNT   132 (wls SISF Packets to LFTS)
22    INJ.43                    Tw0/0/2            FWD

62    CAPWAP-IF-0x0090000004    internal0/0/rp:0   PUNT   132 (wls SISF Packets to LFTS)
```

```
63    INJ.43                     <none>              CONS   Packet Consumed Silently
64    VLAN-CPPIF-0150            internal0/0/rp:0    PUNT   7    (ARP request or response)
65    VLAN-CPPIF-0150            Tw0/0/2             FWD
66    INJ.12                     Tw0/0/2             FWD
67    CAPWAP-IF-0x0090000004     internal0/0/rp:0    PUNT   11   (For-us data)
68    INJ.2                      Tw0/0/2             FWD
69    CAPWAP-IF-0x0090000004     internal0/0/rp:0    PUNT   11   (For-us data)
70    INJ.2                      Tw0/0/2             FWD
71    CAPWAP-IF-0x0090000004     internal0/0/rp:0    PUNT   11   (For-us data)
72    INJ.2                      Tw0/0/2             FWD
73    CAPWAP-IF-0x0090000004     internal0/0/rp:0    PUNT   11   (For-us data)
74    INJ.2                      Tw0/0/2             FWD
75    CAPWAP-IF-0x0090000004     internal0/0/rp:0    PUNT   11   (For-us data)
76    CAPWAP-IF-0x0090000004     Tw0/0/2             FWD
77    CAPWAP-IF-0x0090000004     internal0/0/rp:0    PUNT   132 (wls SISF Packets to LFTS)
78    CAPWAP-IF-0x0090000004     internal0/0/rp:0    PUNT   132 (wls SISF Packets to LFTS)
```

You can then use the command **show platform hardware chassis active qfp feature packet-trace packet <packet_id>** to see the path the packet took and the list of features it encountered.

From Example 16-6, if you focus on packets 67–76, packets 67, 69, 71, 73, and 75 are all similar packets sourced from CAPWAP-IF-0x0090000004 destined to internal0/0/rp:0 0. CAPWAP-IF-0x0090000004, as the name suggests, corresponds to a CAPWAP interface, which means an AP. You can identify specific APs by running **show capwap summary.** Internal0/0/rp refers to the CPU of the C9800, which is why this packet translates to For-us. Packets 68, 70, 72, 74, and 76 are all packets sourced from INJ.2, referring to these packets being injected by the control plane of the C9800 and forwarded out the (FWD) Tw0/0/2 interface. Example 16-7 provides a snippet of the deep-dive view of packet 67 from which you can derive

- Input interface of the packet
- Output interface of the packet
- Whether the packet is punted to the CPU or injected back from the CPU to the data plane (in the case of a PUNT, the specific CPU queue [For-us Data])
- Total time spent to process the packet
- Source and destination IP address and ports of the packet
- Every feature that the packet touches is listed, such as the CAPWAP reassembly; MDNS bridging, which is default enabled on the WLAN profile; destination lookup; and the C9800's decision to consume the packet by the C9800. Even the fact that

the Packet Tracer was enabled (CBUG stands for conditional debugging triggered by the CLI **#debug condition**) with options for fia-trace and copy packet visible in processing of the packet.

- Especially relevant is the time in nanoseconds that each feature process took. When troubleshooting performance issues such as back pressure or drops, the Packet Tracer can tell you exactly which feature is experiencing delays.
- You also have the packet dumped in hex, which can be converted to pcap for ease of parsing.

Example 16-7 *Deep-Dive View of Packet 67 Chosen from Example 16-6*

```
Packet: 67          CBUG ID: 273
Inject Info:
  Packet     : 65534
Summary
  Input      : CAPWAP-IF-0x0090000004
  Output     : internal0/0/rp:0
  State      : PUNT 11  (For-us data)
  Timestamp
    Start    : 113851963623260 ns (03/15/2022 10:19:00.277651 UTC)
    Stop     : 113851963637402 ns (03/15/2022 10:19:00.277665 UTC)
Path Trace
  Feature: IPV4(Input)
    Input        : CAPWAP-IF-0x0090000004
    Output       : <unknown>
    Source       : 10.5.1.11
    Destination : 192.168.1.5
    Protocol     : 17 (UDP)
      SrcPort    : 5248
      DstPort    : 5247
  Feature: CAPWAP_INPUT_REASS_FEATURE
    Entry        : Input - 0x814f43d4
    Input        : CAPWAP-IF-0x0090000004
    Output       : <unknown>
    Lapsed time : 46 ns
  Feature: CAPWAP_TUNNEL_INGRESS_FEATURE
    Entry        : Input - 0x814f1fc4
    Input        : CAPWAP-IF-0x0090000004
    Output       : <unknown>
    Lapsed time : 316 ns
  Feature: CAPWAP_TUNNEL_INGRESS_END_PROCESSING_FEATURE
    Entry        : Input - 0x814f1fc0
    Input        : WLCLIENT-IF-0x00a0000001
```

```
  Output       : <unknown>
  Lapsed time : 74 ns
Feature: LAYER2_IPV4_INPUT_ARL_SANITY
  Entry        : Input - 0x814b3450
  Input        : WLCLIENT-IF-0x00a0000001
  Output       : <unknown>
  Lapsed time : 374 ns
Feature: WLCLIENT_INGRESS_MDNS_BRIDGE
  Entry        : Input - 0x814f2a3c
  Input        : WLCLIENT-IF-0x00a0000001
  Output       : <unknown>
  Lapsed time : 114 ns
Feature: WLCLIENT_INGRESS_IPV4_FWD
  Entry        : Input - 0x814f2a1c
  Input        : WLCLIENT-IF-0x00a0000001
  Output       : <unknown>
  Lapsed time : 142 ns
Feature: SWPORT_VLAN_BRIDGING
  Entry        : Input - 0x814f52b8
  Input        : Vlan150
  Output       : <unknown>
  Lapsed time : 438 ns
Feature: CBUG_INPUT_FIA
  Entry        : Input - 0x814b1474
  Input        : Vlan150
  Output       : <unknown>
  Lapsed time : 62 ns
Feature: DEBUG_COND_INPUT_PKT
  Entry        : Input - 0x814b1490
  Input        : Vlan150
  Output       : <unknown>
  Lapsed time : 64 ns
Feature: IPV4_INPUT_DST_LOOKUP_ISSUE
  Entry        : Input - 0x814ddf0c
  Input        : Vlan150
  Output       : <unknown>
  Lapsed time : 134 ns
Feature: IPV4_INPUT_ARL_SANITY
  Entry        : Input - 0x814b3440
  Input        : Vlan150
 Output        : <unknown>
  Lapsed time : 122 ns
```

```
  Feature: IPV4_INPUT_DST_LOOKUP_CONSUME
    Entry       : Input - 0x814ddf08
    Input       : Vlan150
    Output      : <unknown>
    Lapsed time : 144 ns
|
|
|
Feature: DEBUG_COND_OUTPUT_PKT_EXT
    Entry       : Output - 0x814b149c
    Input       : Vlan150
    Output      : internal0/0/rp:0
    Lapsed time : 42 ns
  Feature: INTERNAL_TRANSMIT_PKT_EXT
    Entry       : Output - 0x814b1f94
    Input       : Vlan150
    Output      : internal0/0/rp:0
    Lapsed time : 2634 ns
Packet Copy In
  450000b2 d0d64000 f811e4a6 0a05010b c0a80105 1480147f 009e0000 00204320
  00000000 04000000 00000000 01880000 c064e423 c6409233 17c6561e d4789b3c
  5e8b6405 0000aaaa 03000000 08004500 00649763 00004001 cbfd0a96 010b0a96
  01020000 1c3f0002 00000000 000006cc 633dabcd abcdabcd abcdabcd abcdabcd
  abcdabcd abcdabcd abcdabcd abcdabcd abcdabcd abcdabcd abcdabcd abcdabcd
  abcdabcd abcdabcd abcdabcd abcdabcd abcd
```

You can export all the packets' content by using **show platform hardware chassis active qfp feature packet-trace packet all decode | redirect bootflash:<FILENAME>.txt**.

Troubleshooting Dashboard

The tracing capabilities discussed so far—syslog, always-on traces, RA traces, and internal RA traces, as well as the EPC capture—are accessible via the C9800 GUI Troubleshooting dashboard. In addition, this dashboard provides a few other tools.

Core Dump and System Report

When a C9800 encounters a software-forced reload, a file with a filename in the format *system-report_YYYYMMDD-HHMMSS-TIMEZONE.tar.gz* is generated to provide the system state at the time of the reload. If the system restart occurred due to a specific process encountering a failure condition, it is termed a *crash*. The process that crashes writes a core dump with a filename in the format *HOSTNAME_PROCESSNAME_PID_YYYYMMDD-HHMMSS-TIMEZONE.core.gz*. The system report, in this case, bundles both the core dump and tracelogs and some other contextual data into a single .tar.

gz for convenience. The destination file system where the system report and core dump are stored depends on the form factor of the C9800. For the C9800-40 and C9800-80, the harddisk: is used, whereas the bootflash: file system holds these files for the C9800-CL and C9800-L. Tools are available within Cisco's internal network to decode and parse through the data in the core dump and system report.

The system report, when unzipped, contains three folders:

1. **bootflash**: This folder includes a reload_info file, which lists the reload reason. This matches the last reload reason in the **show version** of the specific C9800, after the reload.

2. **harddisk**: Note that this is just the folder name and does not imply the harddisk: file system is present on the platform from which the system report was downloaded. This folder includes at least two subfolders called Core and Tracelogs on all C9800s. Some additional files like crash info or pd_info are also included to provide context for the core file for different form factors of the C9800.

   - The Core subfolder contains the core dump for the process that crashed. Typically, processes crash on their own when they encounter unexpected code execution or unhandled signals. If multiple processes crash at the same time, multiple core dump files get generated. A C9800 attempts to recover from a process crash by restarting the process, if the process is indeed restartable. The C9800 generates the core dump to capture the unexpected sequence of functions in the code encountered and the system state at the time, which, in turn, helps to identify why the unexpected sequence of code came into play. This core dump can be downloaded on its own, outside of the system report, by navigating to **Troubleshooting > Core Dump and System Report.**

On the C9800, you can also generate core files that serve as a troubleshooting data point to identify the exact state of a process at a given point in time. This is typically referred to as live core and is available only after service internal is enabled on the C9800 globally. Example 16-8 provides the steps to generate a live core dump.

Example 16-8 *CLI Commands to Generate a Live Core Dump*

```
C9800(config)#service internal
C9800(config)#end
C9800#request platform software process core <PROCESS-NAME> {instance-number}
  chassis active R0
SUCCESS: Core file generated.
C9800#dir bootflash:/core | include core
Directory of bootflash:/core/
416991  -rw-         12602510  Jan 29 2022 21:31:26 +00:00  C9800_1_RP_0_
  wncd_21425_20220129-213122-Central.core.gz
```

   - The Tracelogs subfolder contains all the tracelogs of the extension .bin.gz available in memory and harddisk of the C9800 at the time of the system reload.

These tracelogs for all BinOS processes help you to piece together the sequence of events leading up to the crash or system restart. Therefore, it is prudent to share all the core files and the system reports corresponding to the time of the system reload with a Cisco support representative for the root cause analysis. In fact, it is recommended you share any files generated in bootflash: or harddisk: corresponding to the timestamp of the crash or system restart.

3. **tmp:** This folder usually contains subfolders that provide system context to decode and analyze the files included in the system report. Some common subfolders are

   - **cyan:** *Cyan* is an abbreviation for the platform **C**apabilit**y** **A**bstractio**n.** As the name implies, the purpose of cyan is to abstract the platform-specific capabilities to make IOS-XE reusable across multiple Cisco devices such as switches, routers, and WLCs. The cyan.log is written to this folder and included as part of a system report to identify the platform details such as chassis model, software version, and some registers.

   - **maroon_stats:** Maroon is responsible for memory allocation and management on IOS-XE. This folder provides a map of the memory allocated to the different BinOS processes at the time of the crash. If the crash is triggered due to a memory leak, this file is critical to identify the potential source of the memory leak.

As illustrated in Figure 16-15, the troubleshooting dashboard has a Core Dump and System Report page, where you can easily download these files:

Figure 16-15 *Core Dump and System Report web page that shows the WNCd core and two older system reports*

Debug Bundle

The C9800 provides an option to collect a maximum of five CLI command outputs and package them into a compressed file with the file extension .tar.gz. As illustrated in Figure 16-16, on the C9800 GUI, after selecting **Troubleshooting Dashboard > Debug Bundle**, you can define an arbitrary name for the debug bundle. This name can be up to 25 characters and supports lowercase letters, uppercase letters, hyphens, and underscores (for example, [a–z], [A–Z], -, _). Enter the CLI **show** in the field highlighted and click **+ Add.** Note that **show run** is added to the bundle by default but can be deselected. You also have the option to bundle any web server log, core file, or radioactive trace log that was previously generated. Click **Create Debug Bundle** and download the resulting .tar.gz file.

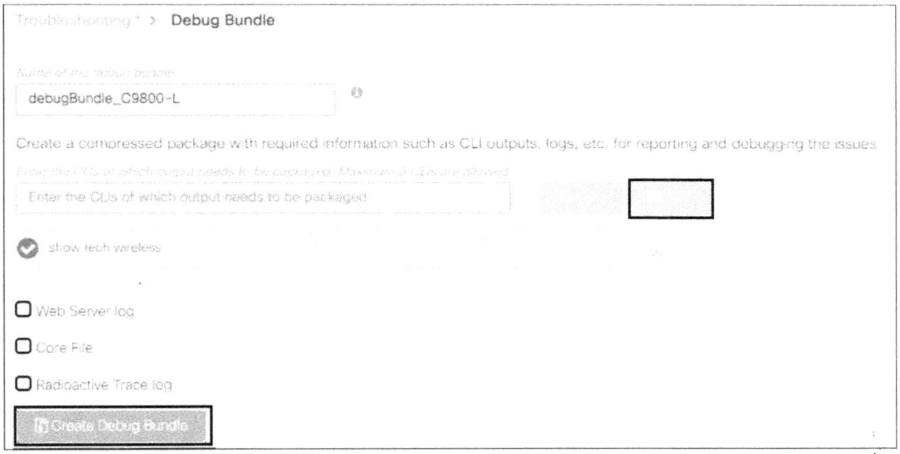

Figure 16-16 *Troubleshooting the dashboard—Debug Bundle*

Ping and Trace Route

The Ping and Trace Route option on the C9800 GUI Troubleshooting dashboard provides the capability to test connectivity from the C9800 to any IP address in the network via an extended ping test. The source of the ping packet can be chosen from any Layer 3 interfaces configured on the C9800. Trace route is a utility used to determine the hop-by-hop L3 path from the C9800 to any destination IP address. Some destination addresses—namely, google.com, yahoo.com, 8.8.8.8, and IPv6 addresses—are prepopulated and can be selected from the drop-down menu, as shown in Figure 16-17.

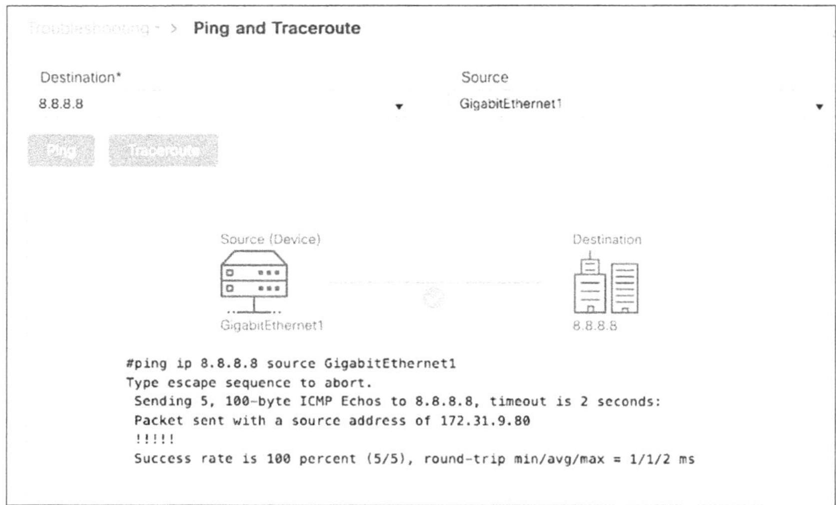

Figure 16-17 *Troubleshooting the dashboard—Ping and Trace Route*

Other On-the-Box Tools on the C9800 GUI

While the Troubleshooting dashboard provides the most-used tools, some additional tools embedded on the C9800 GUI are quite handy.

AireOS Config Translator

The C9800 is the next-generation Catalyst wireless LAN controller. The legacy WLCs run the AireOS operating system. A sizable portion of the C9800 deployment base is the network migrating from AireOS WLCs to the C9800 WLC. To use this tool, the minimum requirement is that the C9800 should have GUI access setup. This can be done by completing the day 0 bootup wizard on the C9800. On the GUI, you can navigate to **Configuration > Services > AireOS Config Translator** as shown in Figure 16-18 and input TFTP back up or the output of **show run-config commands** from an AireOS WLC. The resulting translated configuration file can be directly applied to the C9800. The translator also informs of the commands not applicable to the C9800, the commands that are not supported on the C9800, and the commands that are not yet mapped on the C9800, if any. A cloud version of the same configuration translator is available at https://cway.cisco.com/wlc-config-converter/.

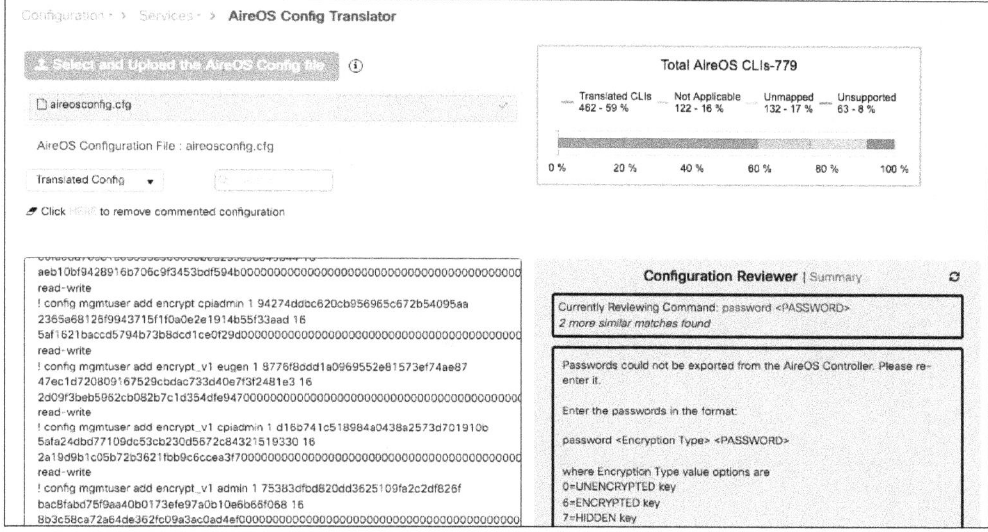

Figure 16-18 *AireOS Config Translator*

Command-Line Interface

The C9800 GUI enables you to run both exec privilege and configuration commands from the C9800 GUI itself. This allows the GUI to operate as a single pane of access into the box without requiring you to switch back and forth between the CLI and GUI. It is particularly useful as a recovery tool when HTTP access is available, but CLI access is blocked due to authentication failure or authorization failure or due to not configuring the enable password. You can access it by navigating to **Administration > Command Line Interface**, as shown in Figure 16-19.

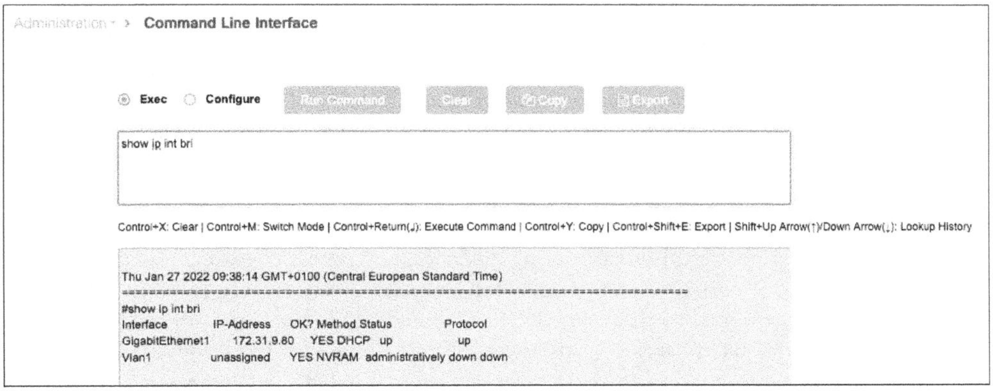

Administration > Command Line Interface

Exec Configure Run Command Clear Copy Export

```
show ip int bri
```

Control+X: Clear | Control+M: Switch Mode | Control+Return(↵): Execute Command | Control+Y: Copy | Control+Shift+E: Export | Shift+Up Arrow(↑)/Down Arrow(↓): Lookup History

```
Thu Jan 27 2022 09:38:14 GMT+0100 (Central European Standard Time)
================================================================================
#show ip int bri
Interface        IP-Address    OK? Method Status                Protocol
GigabitEthernet1 172.31.9.80   YES DHCP   up                    up
Vlan1            unassigned    YES NVRAM  administratively down down
```

Figure 16-19 *Command-line interface on the C9800 GUI*

File Manager

Another tool that is often used on the C9800 GUI is the File Manager. To access it, navigate on the C9800 GUI to **Administration > Management > File Manager.** It provides access to the filesystems on the C9800—namely, bootflash: and harddisk:—so you can download files from the filesystem directly to the desktop that is accessing the C9800 GUI, as shown in Figure 16-20.

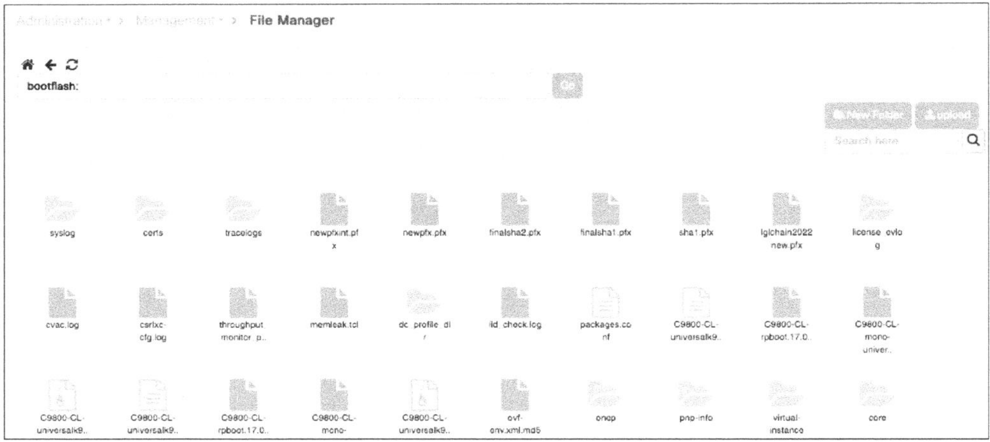

Figure 16-20 *File Manager on the C9800 GUI*

Walk-Me Integrated with the C9800 GUI

The GUI of the C9800 is integrated with the Walk-me digital adoption platform to provide guided workflows for configuring features on the C9800. This tool is especially useful when the configuration of a feature requires multiple steps. For example, configuring a wireless LAN (WLAN) for Local Web Authentication (LWA) requires defining global and custom parameter maps under **Configuration > Security > WebAuth**, and defining an authentication method list under **Configuration > Security > AAA > AAA Authentication.** This method list then needs to get mapped to the WLAN profile under Profiles and Tags. This WLAN profile then is linked to a policy profile via a policy tag under Profiles and Tags and then needs to be applied to an AP. The guided workflow provides a user-friendly experience to get this setup configured successfully. In addition, Walk-me integration has been leveraged to link to documentation from Cisco that provides code guidance, troubleshooting guidance, and access to offline tools that help parse and analyze outputs and logs collected from the C9800. Figure 16-21 illustrates the Walk-me guided workflow that you can click through. Note the Tasks tab on the guided workflow that keeps track of your progress through the guided workflow. Figure 16-22 highlights the integration of the best practices guide listed in the "References" section directly to the C9800 GUI.

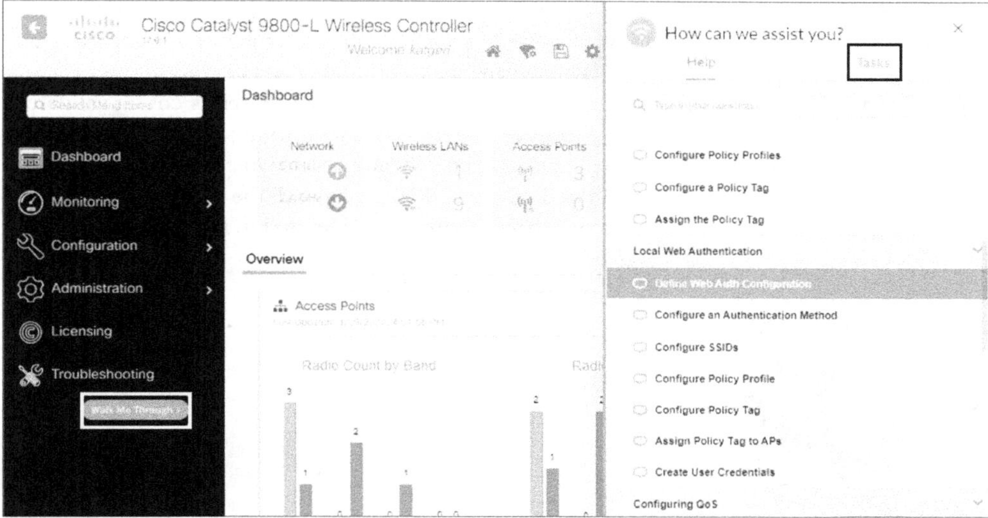

Figure 16-21 *Configuration-guided workflow via Walk-me integration on the C9800 GUI*

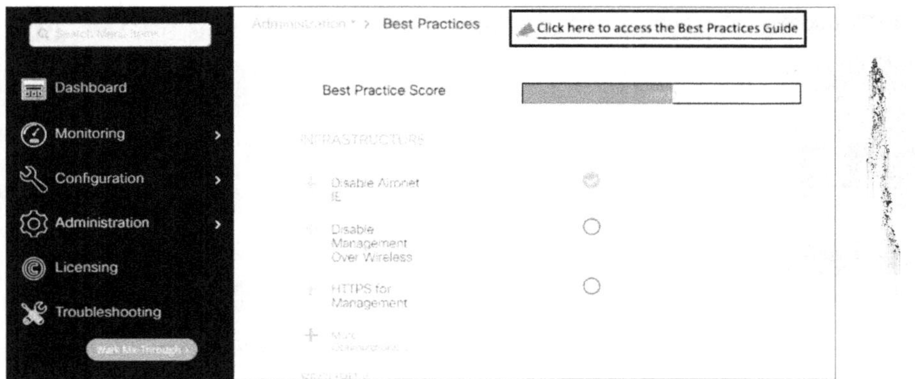

Figure 16-22 *Best practices linked to the C9800 GUI via Walk-me integration*

Configuration Validator

Configuration validation is not an explicit tool available from C9800 GUI. In fact, it is a sort of background task that runs dynamically to prevent misconfigurations on the GUI. Figure 16-23 illustrates an attempt to map a tag to an AP, where the tag has not been defined under Profiles and Tags on the C9800. The GUI calls this out as "Invalid Config" and does not let you apply it to the system, thereby preventing misconfiguration. The

IOS CLI parser is not capable of correlating different lines of configuration and therefore accepts such a misconfiguration. However, on the C9800, configuration validation runs to drop the AP into default-policy-tag and generates a critical syslog alert informing you of the same. This config validation can also be run on demand via **wireless config validate** to parse the configuration and generate syslog messages identifying the misconfiguration. Example 16-9 illustrates how the configuration validator reacts when an AP is mapped to a policy tag that has not been defined on the C9800, along with the log messages generated. Note that the configuration validated by this task is limited to the new configuration model of profiles and tags on the C9800.

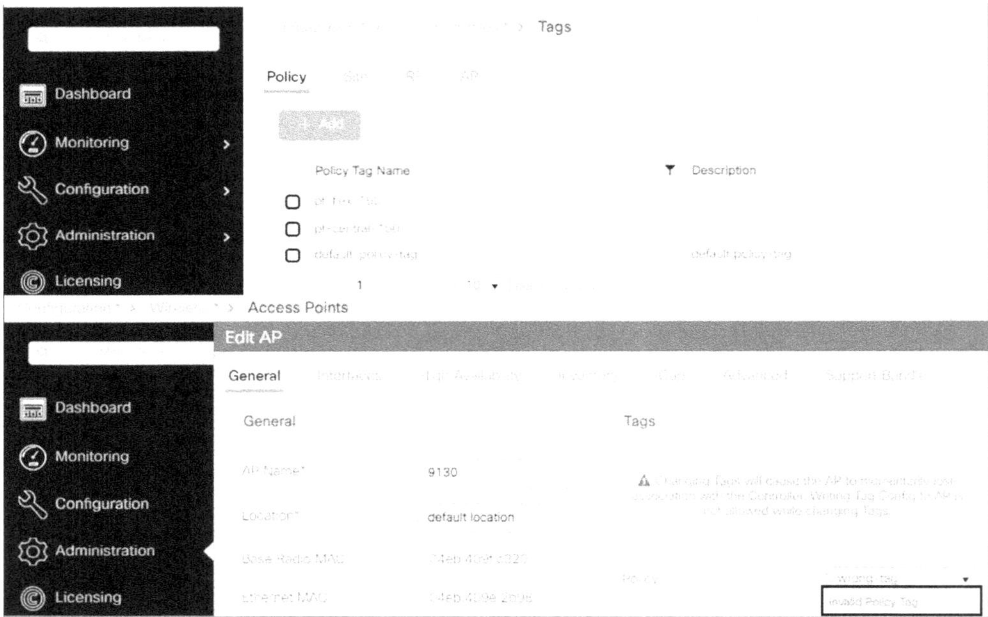

Figure 16-23 *C9800 GUI preventing misconfiguration via configuration validation*

Example 16-9 *C9800 CLI Config Validation and Syslog for Misconfigurations*

```
C9800#show wireless tag policy summary
Number of Policy Tags: 3
Policy Tag Name                     Description
------------------------------------------------------------------------
pt-flex-150
pt-central-150
default-policy-tag                  default policy-tag
```

```
C9800(config)#ap 04eb.409e.2b98
C9800(config-ap-tag)#policy-tag wrong-tag
Associating policy-tag will cause associated AP to reconnect

C9800#show ap tag summary
Number of APs: 3

AP Name  AP Mac  Site Tag Name  Policy Tag Name  RF Tag Name  Misconfigured  Tag Source
------------------------------------------------------------------------------------
9130   04eb.409e.2b98   st-central   default-policy-tag   rf-central-global Yes  Static
1852   38ed.18c6.4c48   default-site-tag   default-policy-tag default-rf-tag  No Default

C9800#show log | include 04eb.409f.c320
*Jan 29 23:51:31.736: %CAPWAPAC_SMGR_TRACE_MESSAGE-5-AP_JOIN_DISJOIN: Chas-
  sis 1 R0/0: wncd: AP Event: AP Name: 9130 Mac: 04eb.409f.c320 Session-IP:
  10.6.1.11[5256] 192.168.1.5[5246] Disjoined Tag modified
*Jan 29 23:52:23.636: %APMGR_TRACE_MESSAGE-2-WLC_APMGR_CRIT_MSG: Chassis 1 R0/0:
  wncd: CRITICAL, 04eb.409f.c320 Configured policy-tag wrong-tag not defined, pick-
  ing default-policy-tag.
*Jan 29 23:52:24.459: %CAPWAPAC_SMGR_TRACE_MESSAGE-5-AP_JOIN_DISJOIN: Chas-
  sis 1 R0/0: wncd: AP Event: AP Name: 9130 Mac: 04eb.409f.c320 Session-IP:
  10.6.1.11[5256] 192.168.1.5[5246] Ethernet MAC: 04eb.409e.2b98 Joined
```

Offline Tools for the C9800

In addition to the tools on-the-box, several out-of-box tools are available to parse the configuration for best practices and RF analysis. They are listed under DevNet resources at https://developer.cisco.com/docs/wireless-troubleshooting-tools/#!wireless-troubleshooting-tools.

Wireless Configuration Convertor

The Wireless Configuration Convertor shown in Figure 16-24 is the cloud version of the AireOS Config Translator. The advantage of the cloud tool over the built-in AireOS Config Translator is that it is not tied to the release of software code on the C9800. This advantage enables the addition or revision of translations more dynamically. In other words, the cloud tool provides more updated translated output than an on-the-box translator between code releases. The cloud tool also includes translation of the configuration across some legacy WLCs.

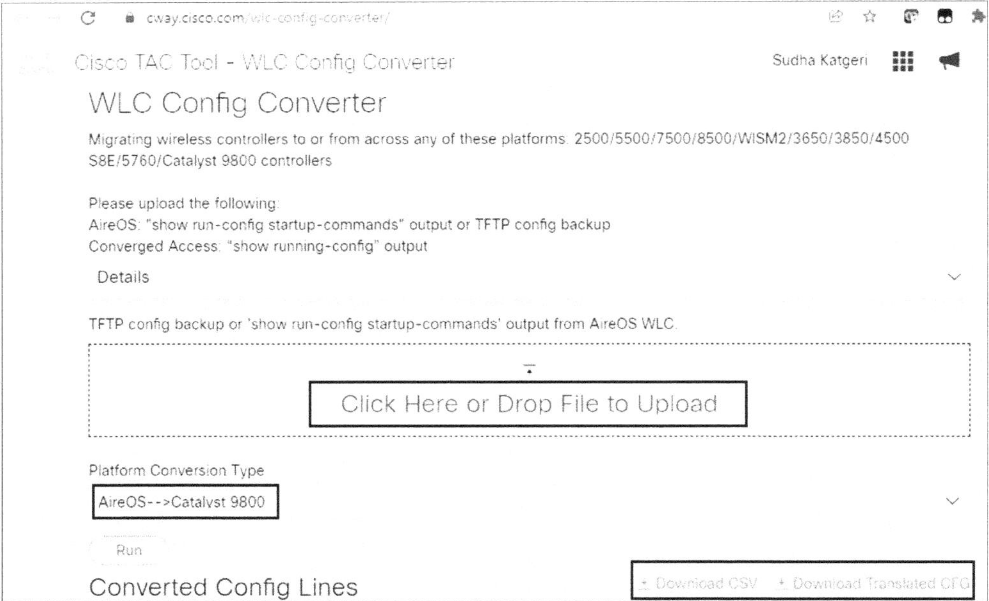

Figure 16-24 *Wireless Config Convertor on cisco.com*

Wireless Config Analyzer

The Wireless Config Analyzer is the generic tool name covering the Wireless Config Analyzer Express (WCAE) and the Wireless LAN Controller Config Analyzer (WLCCA). The WLCCA is a desktop Windows application providing a more graphic experience, but it does not support the C9800 and is no longer getting updates. The WCAE is the cloud version of the tool that is constantly updated to support new controller releases (AireOS or C9800) as well as new checks and verifications and best practices. The WCAE also has a mini-Desktop version you can download for use offline. The analyzer consumes the gigantic output of **show tech wireless** and provides insights covering these checks and features:

- Configuration checks
- RF health analysis
- RF stats summarization
- Upgrade advisor
- Channel stats (per site/flex, covering client, power levels, and rogues)
- Tag/policy usage
- RRM analysis

- Log message summarization
- AP inventory
- RF graph analysis
- Client audits: 8821, iPhone, Drager, and others

It is highly recommended you run this tool regularly on the output of the **show tech wireless** from your controller so that you are aware of any conflicting configuration and aware of the RF environment such as coverage holes, APs under heavy load, APs undergoing frequent channel changes, and so on, that enable proactive remediation to keep the network healthy. Figure 16-25 provides a snapshot of the cloud-based WCAE tool.

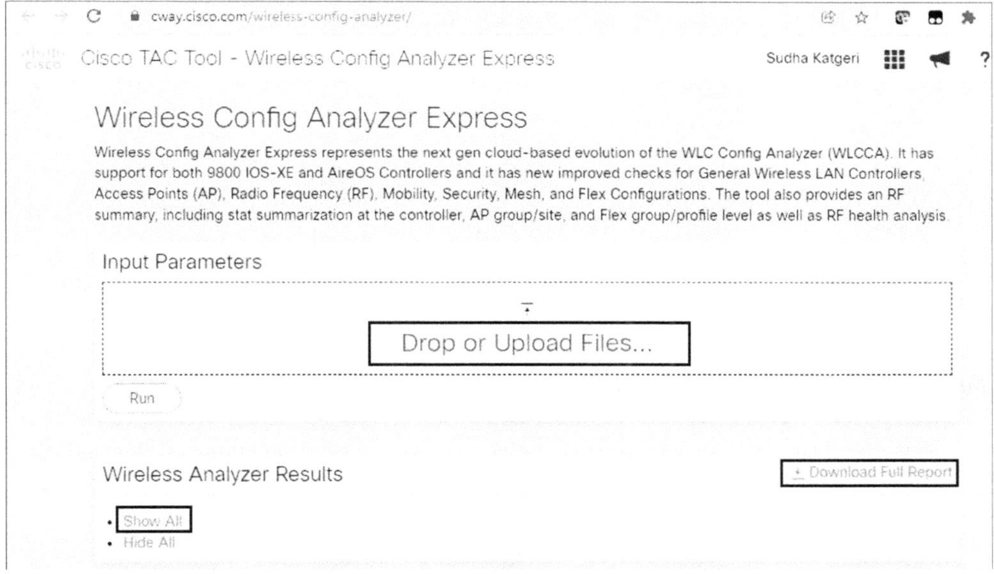

Figure 16-25 *Wireless Config Analyzer Express on cisco.com*

Wireless Debug Analyzer

On the same DevNet home page, the Wireless Debug Analyzer allows you to upload a radioactive trace for assisted analysis. It also works with always-on traces but has much less detailed output and analysis, obviously due to the lower level of logging

As illustrated in Figure 16-26, the tool first splits the output between various connections attempts for each client MAC address present in the file. You can select the client MAC you are visualizing on the top left and have one page for each connection attempt on the top right.

You then can easily see the main state machine changes for the client, for example:

- WPA key exchange messages
- DHCP process
- Moving to RUN state at the end

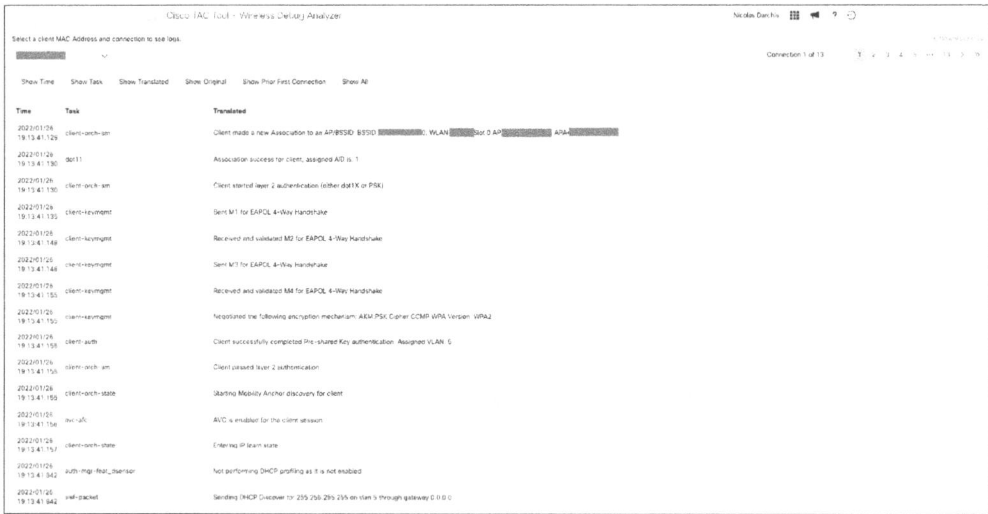

Figure 16-26 *Wireless Debug Analyzer on cisco.com*

This useful tool enables you to parse the radioactive trace for client onboarding and connectivity issues. Note that performance and throughput issues or data forwarding problems are better diagnosed with captures, which are not input to this tool.

Log Advisor

Cisco Log Advisor is not a tool but a web page that acts as a repository for basic outputs to collect and check for various frequent problem scenarios. It is available for many Cisco technologies including wireless. You can access it at logadvisor.cisco.com to learn more.

Health and KPI Monitoring

After the wireless network is deployed, your main responsibility as the network administrator is to monitor the C9800 and wireless infrastructure. This monitoring should ensure the good health of the C9800, including hardware health, system uptime, and resource utilization like CPU and memory, among others. In addition, you need to monitor network service and performance delivered by the C9800 to validate the number of APs connected, if the APs are staying connected, or if the APs are experiencing flaps, the number of clients being serviced, performance being delivered to these clients, and much more. Cisco DNA Center Assurance is the recommended means to monitor the C9800 wireless deployment in conjunction with the wired infrastructure, end clients, and applications. We look at the monitoring tools available locally on the C9800 next.

Dashboard

The Dashboard of the C9800 GUI serves as the single view of important health parameters of the C9800, as shown in Figure 16-27. The key takeaways from this Dashboard include

- Hardware and software running on the C9800
- Temperature reading
- Uptime of the C9800
- Last reload reason
- Redundancy state
- Total AP count
- AP distribution per radio band (2.4 GHz versus 5 GHz)
- AP distribution per AP mode (local, FlexConnect, Bridge, Sniffer)
- Count of successful and failed AP joins
- Active and excluded client count
- Client distribution per device type
- Total number of rogue APs and clients
- Total number of interferers on each band
- Number of WLANs
- CPU utilization, along with top talkers and the usage trend
- Memory utilization, along with trend over time

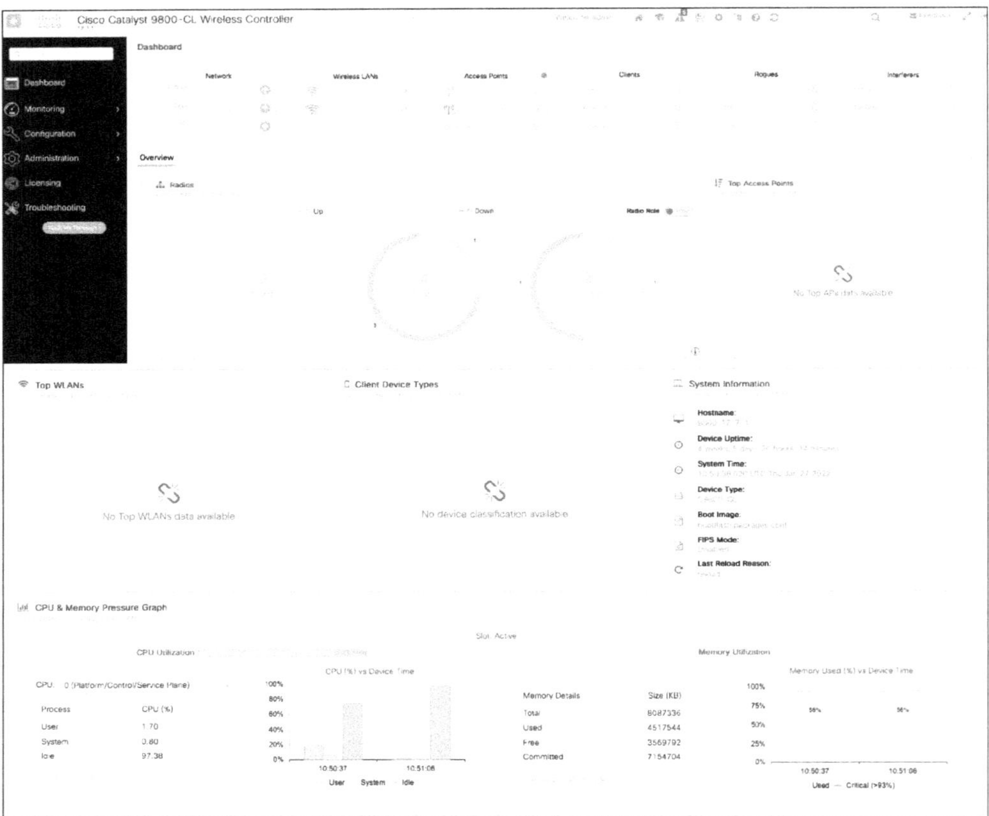

Figure 16-27 *Dashboard on the C9800 GUI in IOS-XE 17.7.1*

Hardware Monitoring

For a C9800, the hardware monitoring is applicable only to the C9800 hardware appliances—namely, the C9800-40, C9800-80, and C9800-L. In the case of the C9800-CL, only a software package is hosted on the guest OS in the private cloud like the kernel-based virtual machine (KVM), VMWare ESXi, or Cisco Enterprise Network Compute System (ENCS), or the public cloud like Google Cloud Platform (GCP), Amazon Web Services (AWS), or Microsoft Azure. The health of the private or public cloud is independent on the C9800. The C9800 does provide a view into the hosting environment from the C9800 exec CLI, as shown in Example 16-10.

Example 16-10 *show platform software system all Shows the Hypervisor Details*

```
C9800#show platform software system all
Controller Details:
=================
VM Template: small
Throughput Profile: low
AP Scale: 1000
Client Scale: 10000
WNCD instances: 1

Processor Details
=================
Number of Processors : 4
Processor : 1 - 4
vendor_id : GenuineIntel
cpu MHz  : 2194.843
cache size : 39424 KB
Crypto Supported : Yes
model name : Intel(R) Xeon(R) Platinum 8276 CPU @ 2.20GHz

Memory Details
==============
Physical Memory : 8106688KB

VNIC Details
============
Name              Mac Address     Driver Name       Status  Platform MTU
GigabitEthernet1  0050.56a5.de91  net_vmxnet3       DOWN    1522
GigabitEthernet2  0050.56a5.66b5  net_vmxnet3       UP      1522
GigabitEthernet3  0050.56a5.dcd5  net_vmxnet3       DOWN    1522

Hypervisor Details
===================
Hypervisor: VMWARE
Manufacturer: VMware, Inc.
Product Name: VMware Virtual Platform
Serial Number: VMware-42 25 5a 81 32 83 ca 70-27 86 0c 27 14 63 aa 16
UUID: 815A2542-8332-70CA-2786-0C271463AA16
image_variant :

Boot Details
===================
Boot mode: BIOS
Bootloader version: 1.1
```

On the physical appliances, you need to monitor multiple components:

- **Environment Health:** The appliances have multiple on-board sensors to monitor the environmental health of the C9800, including the temperature of the cooling air as it moves through the chassis, input and output voltage, and output current, among others. The status conditions returned by the sensors and the monitoring functions are
  - **Normal:** All monitored parameters are normal, that is, within tolerance.
  - **Warning:** The C9800 has exceeded a system defined threshold. The C9800 continues to operate but needs attention and repair to return to normal state.
  - **Critical:** An out-of-tolerance temperature or voltage condition exists. The C9800 continues to operate but is approaching shutdown. Immediate attention is needed to recover.
  - **Shutdown:** The system is shutting down, but before the shutdown occurs, system logs pertaining to monitored parameters are saved to memory that persists after reloading, to help determine the root cause after failure.

  You can view the state of these sensors, as shown in Example 16-11, by using **show environment chassis active r0**.

Example 16-11 *C9800 Sensor States and Readings*

```
C9800#show environment chassis active r0
Sensor List: Environmental Monitoring
 Sensor                   Location         State            Reading
 Vin                      P0               Normal           232 V AC
 Iin                      P0               Normal           2 A
 Vout                     P0               Normal           12 V DC
 Iout                     P0               Normal           34 A
 Temp1                    P0               Normal           21 Celsius
 Temp2                    P0               Normal           27 Celsius
 Temp3                    P0               Normal           28 Celsius
 VDMB1: VX1               R0               Normal           1226 mV
 VDMB1: VX2               R0               Normal           6958 mV
 VDMB1: VX3               R0               Normal           1225 mV
 VDMB1: VX4               R0               Normal           1003 mV
 VDMB1: VP1               R0               Normal           1788 mV
 VDMB1: VP2               R0               Normal           2556 mV
 VDMB1: VP3               R0               Normal           2559 mV
 VDMB1: VP4               R0               Normal           1055 mV
 VDMB1: VH                R0               Normal           11977 mV
 VDMB2: VX2               R0               Normal           4980 mV
 VDMB2: VX3               R0               Normal            855 mV
 VDMB2: VX4               R0               Normal            912 mV
 VDMB2: VX5               R0               Normal           1006 mV
```

```
VDMB2: VP1          R0          Normal          1788 mV
VDMB2: VP2          R0          Normal          3299 mV
VDMB2: VH           R0          Normal          11972 mV
VDMB3: VX1          R0          Normal          970 mV
VDMB3: VX2          R0          Normal          1003 mV
VDMB3: VX5          R0          Normal          5070 mV
VDMB3: VP1          R0          Normal          2489 mV
VDMB3: VP2          R0          Normal          1197 mV
VDMB3: VP3          R0          Normal          1510 mV
VDMB3: VP4          R0          Normal          1511 mV
VDMB3: VH           R0          Normal          11982 mV
Temp: DMB IN        R0          Normal          20 Celsius
Temp: DMB OUT       R0          Normal          30 Celsius
Temp: Yoda 0        R0          Normal          42 Celsius
Temp: Yoda 1        R0          Normal          42 Celsius
Temp: CPU Die       R0          Normal          30 Celsius
Temp: FC FANS       R0          Fan Speed 70%   20 Celsius
VDDC1: VX1          R0          Normal          1003 mV
VDDC1: VX2          R0          Normal          7105 mV
VDDC1: VX3          R0          Normal          949 mV
VDDC1: VP1          R0          Normal          1814 mV
VDDC1: VP2          R0          Normal          2506 mV
VDDC1: VP3          R0          Normal          3312 mV
VDDC1: VH           R0          Normal          11993 mV
VDDC2: VX2          R0          Normal          750 mV
VDDC2: VX3          R0          Normal          749 mV
VDDC2: VX5          R0          Normal          5080 mV
VDDC2: VP1          R0          Normal          1010 mV
VDDC2: VP2          R0          Normal          1006 mV
VDDC2: VP3          R0          Normal          1201 mV
VDDC2: VP4          R0          Normal          1512 mV
VDDC2: VH           R0          Normal          12003 mV
Temp: DDC IN        R0          Normal          20 Celsius
Temp: DDC OUT       R0          Normal          28 Celsius
```

- **Power supply:** The C9800 supports two power supplies for redundancy. With dual power supplies, if one of the power supplies fails, the other power supply suffices to deliver power to the chassis. This allows power supplies to be hot swappable without impacting the C9800. If only one power supply is in use, it is mandatory that the slot be closed with a blank faceplate. The environmental sensors also monitor the power supply, and a status of power supply shutdown indicates that the power supply detected an internal out-of-tolerance overvoltage, overcurrent, or temperature condition and shut itself down.

- **Fans:** The C9800 has a front-to-rear airflow. The C9800-L has two internal fans, the C9800-40 has four, and the C9800-80 has six fans located at the rear of the chassis numbered 0 to 5 from right to left. These fans draw cooling air through the front of the chassis and across internal components to maintain an acceptable operating temperature. If a fan fails, the C9800 continues to operate. The speed of fans adjusts automatically based on chassis inlet temperature to reduce noise and power and increase fan life. You can verify the status of the fans by using **#show platform hardware slot P1 fan status.**

- **Ports and transceivers:** On a C9800 appliance, the uplink or distribution ports that connect to the wired infrastructure are provided by the shared port adapter (SPA). Further, on the C9800-40 and C9800-80, the ports have small form-factor pluggable (SFP) plugged in. The interface and SFP status can be viewed with

  ```
  C9800#show interface <INTF-NAME> <slot/sub-slot/port>
  C9800#show interface <INTF-NAME> <slot/sub-slot/port> transceiver
  detail
  ```

  In case of any issues bringing up a physical link, you can access the SPA module to get status and registers of PHY and optics by using

  ```
  #request platform software console attach chassis <chassis number>
  <slot/subslot>
  !<Press enter> until see prompt "Slot-0-0>"
  Slot0-0>show hw-module subslot 0 status optics 0
  Slot0-0>show hw-module subslot 0 status optics 1
  Slot0-0>show hw-module subslot 0 registers optics 0
  Slot0-0>show hw-module subslot 0 registers optics 1
  Slot0-0>show hw-module subslot 0 status phy 0
  Slot0-0>show hw-module subslot 0 status phy 1
  Slot0-0>show hw-module subslot 0 registers phy 0
  Slot0-0>show hw-module subslot 0 registers phy 1
  ```

  To view the counters, including any packet drops at an interface, run **show interface <IF-NAME>.** Traffic entering any interface on the C9800-40 and C9800-80 is buffered and processed by a port-asic called EZMAN. You can view the statistics and information for EZMAN by running

  ```
  C9800#show platform hardware port <slot/sub-slot/port> ezman info
  C9800#show platform hardware port <slot/sub-slot/port> ezman
  statistics
  ```

- **LEDs:** The LEDs located on the front and rear panels of the C9800 appliance are great visual indicators of the general health of the system. The management LEDs are located on the front panel; their behavior is listed in Table 16-3. On the C9800-80, the front panel also has LEDs for the Ethernet port adapter (EPA) modules that can be inserted; their behavior is listed in Table 16-4. On the C9800-40 and C9800-80, where the power supply is a field replaceable unit (FRU), two LEDs (green and amber) are present on each power supply, and their collective status determines the condition of the power supply, as listed in Table 16-5.

Table 16-3 *C9800-L, C9800-40, and C9800-80 Management LEDs and Their Behavior*

LED	LED Color	Behavior
Power	Green	Remains solid when powered on
System	Green	Remains solid during IOS boot completion
	Blinking green	Remains blinking when IOS booting is in progress
	Amber	Remains solid during system crash
	Blinking amber	Remains blinking during secure boot failure
	Off	Remains off during ROMMON boot
Alarm	Green	Remains solid during ROMMON boot complete
	Blinking green	Remains blinking when system upgrade is in progress
	Amber	Remains solid during ROMMON and system bootup
	Blinking amber	Remains blinking during temperature error and secure boot failure
	Red	Indicates that the system detects critical warnings that can be viewed by **show facility-alarm status**
	Unlit	Remains unlit during normal operation
USB	Green	Indicates the USB port is enabled
	Blinking green	Indicates the USB port is enabled with activity
	Unlit	Remains unlit when the USB port is not enabled
Service port (SP)	Green	Indicates the out-of-band management SP port is enabled
	Blinking green	Indicates activity over the SP port
	Unlit	Indicates the SP port is off
Redundancy port (RP)	Green	Remains solid when HA is active
	Unlit	Remains unlit when HA is disabled
Mini-USB console	Green	Indicates that the mini-USB connector is used as the console
RJ-45 console link	Unlit	Indicates that the RJ-45 connector is not used as the console

LED	LED Color	Behavior
	Blinking green	Indicates that the RJ-45 connector is being used as the console
Built-in ports	Green	Indicates port enabled with a valid Ethernet link
	Blinking green	Indicates port enabled with activity
	Amber	Indicates port enabled with a problem in the Ethernet link
	Unlit	Remains unlit when the port is not enabled
SSD activity	Green	Remains solid during the SSD activity

Table 16-4 *C9800-80 Module LEDs*

LED	Color	Behavior
A/L (Active/Link)	Green	Port is enabled and link is up.
	Amber	Port is enabled but link is down.
	Unlit	Port is not enabled.
Status	Green	Module is ready and operational.
	Amber	Module is powered on and the module is being configured.
	Unlit	Module power is off.

Table 16-5 *C9800-40 and C9800-80 Power Supply LEDs*

Green (OK) LED	Amber (FAIL) LED	Power Supply Condition
OFF	OFF	No AC power to all power supplies
OFF	ON	Power supply failure (includes overvoltage, overcurrent, overtemperature, and fan failure)
OFF	1 Hz blinking	Power supply warning events where power supply continues to operate (high temperature, high power, and slow fan)
1 Hz blinking	OFF	AC present/3/3VSV on (PSU off)
ON	OFF	Power supply ON and OK

In addition to hardware, it is important to monitor the firmware that manages the bootloader—namely, the ROMMON version and the PHY ports on the C9800. In general, it is recommended you run the latest firmware available on cisco.com to avoid any known issues that impact the system. The command that helps to verify the ROMMON version is **show rom-monitor chassis {active | standby} R0.** On the active chassis, the ROMMON version is also printed in the **show version.** You can view the PHY firmware by executing **show platform hardware chassis active qfp datapath pmd ifdev | i FW.** The guide for upgrading the firmware is listed in the "References" section. Note that on the C9800-L, upgrading ROMMON and PHY firmware is mandatory to prevent being locked out of the system or for traffic to blackhole.

Smart Licensing

Starting with IOS-XE release 17.3.2, the C9800 uses Smart Licensing Using Policies (SLUPs), which provide a simplified method of licensing the controller. A C9800 out-of-the-box or after factory reset is not able to connect to a license server and fails the license validation. You get a grace period of 90 days before receiving warnings, but the C9800 keeps operating even if the licenses are expired or do not have enough AP count loaded. Cisco does not want your network to go down simply because of a communication issue between your controller and the licensing server.

To track your license usage, you load all your AP DNA licenses on the licensing portal on cisco.com. Then the licenses are available for use by any of your registered network devices. There are three options for Smart Licensing: you can directly connect your WLC to the Cisco Smart Software Manager (CSSM) in the cloud, you can host a Cisco software on your premises if you want to avoid a direct connection, or you can periodically report the license usage manually if you are in an airgap network.

Direct Connect

The first step is to establish trust between the WLC and Cisco Smart Software Manager using a token obtained from the CSSM portal. This trust binds the WLC to your Smart Account. Example 16-12 shows the commands used to configure a direct connection to CSSM.

Example 16-12 *CLI Commands Required to Establish a Direct Connection to CSSM Licensing*

```
configure terminal
ip http client source-interface <interface>
ip http client secure-trustpoint <TP>
license smart transport smart
license smart url default
exit
write memory
terminal monitor
license smart trust idtoken <token> all force
```

The C9800 uses its HTTP client component to connect and establish a secure connection to CSSM. It is important to verify that the right interface (with connectivity to CSSM) is selected, along with a valid trustpoint (a self-signed certificate is sufficient).

After the token is entered, the C9800 contacts CSSM every 8 hours to report the license usage. If you do not have any APs joined, you might not see any contact toward CSSM because the controller has nothing to report.

On-Premises SSM or CSLU

For some time, Cisco has been offering an on-premises version of the SSM server, formerly called the Satellite license server. Now, Cisco Smart License Utility (CSLU) is a more lightweight application that you can even run on a laptop. Both act as a relay if your controllers do not have direct Internet access and collect the license usage from the controller to report them to the Cisco licensing online. This can be done constantly in real time or periodically (you could decide to run this utility on your laptop every month, for example, to update the license counts from every device at one time).

The procedure is quite similar, but the on-premises SSM does not use a trust token. Example 16-13 shows the key configuration when using a local licensing server such as CSLU or the on-prem SSM.

Example 16-13 *Configuration to Establish Trust Between a C9800 and Local Licensing Server*

```
configure terminal
 ip http client source-interface <interface>
 ip http client secure-trustpoint <TP>
 license smart transport cslu
 license smart url https://<on-prem-ssm-domain>/SmartTransport
 crypto pki trustpoint SLA-TrustPoint
  revocation-check none
 exit
write memory
terminal monitor
```

A significant difference is that you must disable revocation-check on the SLA-Trustpoint. The reason is that a local server typically uses a self-signed certificate, and it is impossible for the C9800 to verify it fully.

Airgap

Some networks have no access to the Internet at all. In that situation, it is still possible to generate a license report file on the C9800, export it through the network or even through a USB drive, and then upload this report out-of-band (from your laptop, for example) to the Smart Licensing portal.

The key point in all these methods is that the C9800 does not stop functioning, and you are in control of how frequently you want to synchronize the number of licenses consumed by the controller. The command **show license air entities summary**, as shown in Example 16-14, helps you identify how many AP licenses were reported as consumed to the licensing server as well as the last and next reporting time.

Example 16-14 *License Consumption on an Air-Gapped C9800*

```
C9800#show license air entities summary
Last license report time........................: 07:38:15.237 UTC Fri Aug 27 2021
Upcoming license report time....................: 15:38:15.972 UTC Fri Aug 27 2021
No. of APs active at last report................: 3
No. of APs newly added with last report.........: 0
No. of APs deleted with last report..............: 0
```

You can remove the trust with the licensing server by using **license smart factory reset**, which removes all licensing configuration and information.

If you want to verify connectivity toward the licensing server, make sure that port 443 is opened toward smartreceiver.cisco.com. For troubleshooting a more complex licensing issue, it is advised you collect the following output:

```
C9800#show license status
C9800#show license summary
C9800#show tech-support license
C9800#show license tech support
C9800#show license air entities summary
```

The licensing deployment guide and troubleshooting example listed in the "References" section provide a view into the various messages and outputs seen in failure conditions.

AP Health Monitoring

The first step in deploying a wireless network with the C9800 is to get APs to register with the C9800. The Day 0 bootup wizard applies the necessary configuration needed to accomplish this task. However, you may need to revise this configuration to meet the needs of the deployment. For example, you might need to manage additional sites from different countries via the same C9800. Following are some of the common problems APs can exhibit:

- **AP crashes:** The AP runs its own operating system, which could crash due to hardware or software failure. This failure can occur both for a brand-new AP out-of-the-box or an AP that is registered and servicing clients. When an AP crashes, it typically generates a crash file that shows the functions in the stack at the time of the crash. This crash file is stored locally on the AP, but in the case of an AP registered to the C9800, the crash file is also saved to the bootflash of the C9800 with the file Extension.crash. Example 16-15 shows how to validate whether any APs have rebooted from the C9800 CLI. Note that the same command output also shows APs that underwent a CAPWAP reset and had to reestablish a tunnel with the C9800. Example 16-16 shows how to locate an AP corresponding crash file on the C9800.

Example 16-15 *AP Uptime and CAPWAP Association Time on the C9800*

```
C9800#show ap uptime
Number of APs: 1

AP Name    Ethernet MAC    Radio MAC       AP Up Time            Association Up Time
-----------------------------------------------------------------------------------
9120       c4f7.d54d.1d48  c064.e423.c640  7 hours 46 minutes 36 seconds
  7 hours 43 minutes 4 seconds
```

Example 16-16 *Locating an AP Crash File on the C9800 bootflash*

```
C9800#dir bootflash:*.crash
61      -rw-          112662  Jan 3 2022 13:16:14 -05:00  <APNAME>_<APMAC>.crash
```

- **AP joins and flaps:** The AP needs to have Layer 3 connectivity and communicate over UDP ports 5246 and 5247 to the C9800 to successfully register. In addition, you need to configure the proper country code aligned to the regulatory domain in which the AP will be deployed on the C9800. Datagram Transport Layer Security (DTLS) is used to encrypt the CAPWAP tunnel via an exchange of certificates between the AP and WLC. This process requires that valid certificates be present on both the AP and C9800. Also, the DTLS version and cipher suites need to match between the AP and C9800 for successful tunnel negotiation. When these conditions are met, the AP joins the C9800. When an AP is registered, it exchanges CAPWAP keepalives with the C9800 periodically to ensure continued connectivity. However, network congestion could cause keepalives to be dropped, which results in APs flapping. Example 16-17 provides the output of the **show wireless stats ap join summary** command that shows a summarized view of join status of all APs on the C9800. It also provides failure reasons for a disjoined AP and the base radio MAC and Ethernet MAC that can be used to trigger an RA trace for deep-dive troubleshooting of a particular AP.

Example 16-17 *Verifying Join Status of the AP on the C9800*

```
C9800#show wireless stats ap join summary
Number of APs: 3

Base MAC        Ethernet MAC    AP Name  IP Address  Status   Last Failure Phase
  Last Disconnect Reason
-------------------------------------------------------------------------------------
002a.10bf.5d00  cc16.7ef0.7500  2802     10.5.1.12   Joined   Join
  DTLS close alert from peer
38ed.18c7.9d00  38ed.18c6.4c48  1852     10.5.1.13   Joined   Image-Download
  Image Download Success
c064.e423.c640  c4f7.d54d.1d48  9120     10.5.1.11   Joined   NA             NA
```

Example 16-18 provides a view into DTLS connections between the AP and WLC, including the DTLS version and cipher suite used in negotiation. This output is particularly useful to identify the ports involved in the case of a teleworker deployment where the AP is behind a router performing Network Address Translation (NAT) or Port Address Translation (PAT).

Example 16-18 *Verifying the DTLS Version, Cipher Suite, and Ports Used for an AP on the C9800*

```
C9800#show wireless dtls connections
APName  Local Port  Peer IP  PeerPort  Version   Ciphersuite
-----------------------------------------------------------------------------------
9120  Capwap_Ctrl  10.6.1.42   5269  DTLSv1.2 TLS_NUM_ECDHE_RSA_WITH_AES_128_GCM_SHA256
1852  Capwap_Ctrl  10.5.1.185 5248  DTLSv1.2 TLS_NUM_ECDHE_RSA_WITH_AES_128_GCM_SHA256
9130  Capwap_Ctrl  10.6.1.41   5248  DTLSv1.2 TLS_NUM_ECDHE_RSA_WITH_AES_128_GCM_SHA256
```

Example 16-19 provides a join history for APs. Each AP can have several entries corresponding to each join and disjoin. This history helps to establish a trend of multiple APs experiencing the same type of failure or if a group of APs experienced a join failure at the same time.

Example 16-19 *Verifying Aggregate AP Join History*

```
C9800#show wireless stats ap history
AP Name    Radio MAC    Event     Time      Recent Disconnect Time    Disconnect Reason
           Disconnect Count
-------------------------------------------------------------------------------
9130  04eb.409f.c320  Joined     01/27/22 19:18:49  NA          NA             NA
9130  04eb.409f.c320  Disjoined  01/27/22 19:06:30  NA  Max Retransmission to AP  1
1852  38ed.18c7.9d00  Joined     01/27/22 19:18:48  NA          NA             NA
1852  8ed.18c7.9d00   Disjoined  01/27/22 19:06:30  NA  Max Retransmission to AP  1
9120  c064.e423.c640  Joined     01/27/22 19:19:53  NA          NA             NA
9120  c064.e423.c640  Disjoined  01/27/22 19:06:30  NA  Max Retransmission to AP  1
```

- **AP stale entry:** When an AP registers, an entry for it needs to be created across multiple software components of the C9800. This is referred to as *plumbing of the AP*. Conversely, if the AP drops from the C9800, either due to network problems or AP failure or any other reason, it needs to be deleted from all the components. On the C9800, this process occurs sequentially where the WNCd updates the Forwarding Manager Router Processor (FMAN-RP), which, in turn, updates the Forwarding Manager Forwarding Processor (FMAN-FP), which then updates the ASIC driver (CPP-Client) to program the data plane. [WNCd > FMAN-RP > FMAN-FP > CPP-

Client > DP]. Each component needs to receive the update from the previous component, create an entry on itself, and then update the next component in the sequence. Any one of these steps can fail at any one of the components, and it is usually an indicator of a software defect. Example 16-20 shows the commands in sequence to view the AP entry at each software component on a HA SSO C9800 pair.

Example 16-20 *AP Plumbing Across WNCd > FMAN-RP > FMAN-FP > CPP-CLIENT > DP on a C9800 Pair*

```
C9800#show ap summary
Number of APs: 1
APName  Slots  AP Model  Ethernet MAC  Radio MAC  Location  Country  IP Address  State
-------------------------------------------------------------------------------------
9120  2  C9120AXI-B  c4f7.d54d.1d48  c064.e423.c640  lab  US  10.5.1.13  Registered

C9800#show platform software capwap chassis active R0
 Tunnel ID  AP MAC          Type          IP               Port
-----------------------------------------------------------------
0x90000005  c064.e423.c640  Data          10.5.1.13        5268

C9800#show platform software capwap chassis standby R0
 Tunnel ID  AP MAC          Type          IP               Port
-----------------------------------------------------------------
0x90000005  c064.e423.c640  Data          10.5.1.13        5268

C9800#show platform software capwap chassis active F0
 Tunnel ID  AP MAC          Type          IP               Port      AOM ID  Status
-------------------------------------------------------------------------------------
0x90000005  c064.e423.c640  Data          10.5.1.13        5268         140  Done

C9800#show platform software capwap chassis standby F0
 Tunnel ID  AP MAC          Type          IP               Port      AOM ID  Status
-------------------------------------------------------------------------------------
0x90000005  c064.e423.c640  Data          10.5.1.13        5268         127  Done

C9800#show platform hardware chassis active qfp feature wireless capwap cpp-client summary
 cpp_if_hdl     pal_if_hdl    AP MAC        Src IP          Dst IP      Dst Port  Tun Type
-------------------------------------------------------------------------------------
    0X24        0X90000005    c064.e423.c640    192.168.1.5    10.5.1.13     5268     DATA
```

```
C9800#show platform hardware chassis standby qfp feature wireless capwap cpp-client summary
cpp_if_hdl   pal_if_hdl  AP MAC            Src IP          Dst IP     Dst Port   Tun Type
-----------------------------------------------------------------------------------------
  0X1A       0X90000005  c064.e423.c640   192.168.1.5     10.5.1.13    5268       DATA

C9800#show platform hardware chassis active qfp feature wireless capwap datapath summary
Vrf Src Port Dst IP          Dst Port Input Uidb Output Uidb Instance Id
--- -------- ------          -------- ---------- ----------- -----------
0   5247     10.5.1.13       5268     65506      65500        3

C9800#show platform hardware chassis standby qfp feature wireless capwap datapath summary
Vrf Src Port Dst IP       Dst Port Input Uidb Output Uidb Instance Id
--- -------- ------       -------- ---------- ----------- -----------
0   5247     10.5.1.13     5268     65516      65510        3
```

Client Health Monitoring

The ultimate purpose of a wireless deployment is to provide wireless connectivity to end-client devices. The type and quality of services delivered to the end users, whether they are static or mobile, both for data and real-time applications, are critical indicators of the overall well-being and performance of the wireless deployment. A few different data points can be used to monitor clients and troubleshoot client connectivity issues:

- **Client statistics:** Client onboarding involves associating a client to an AP, authenticating as configured on the WLAN, determining the mobility state of the client, and getting an IP address on the client so that it can send traffic to the network and other clients. When a client is onboarded, the C9800 creates an entry in its client database. Some key data points need to be monitored:
  - Total number of clients on the C9800
  - The distribution of clients across the RF spectrum
  - The distribution of clients across different states in the onboarding process
  - Time taken by the client in each step of the onboarding process
  - The client delete reason codes when a client entry is deleted from the database
  - Statistics pertaining to client onboarding

  You can use the command **show wireless stats client detail**, which gives a view of most of these data points for all clients on the C9800 in a single shot. Example 16-21 shows a snippet of the output that shows total client count and client distribution across 802.11 protocols.

Example 16-21 *Total Client Count and Client Distribution Across 802.11 Protocols*

```
C9800#show wireless stats client detail
Total Number of Clients : 15146
Protocol Statistics
----------------------------------------
Protocol            Client Count
 802.11b              : 0
 802.11g              : 4
 802.11a              : 0
 802.11n-2.4GHz       : 4270
 802.11n-5 GHz        : 97
 802.11ac             : 10778
 802.11ax-5 GHz       : 0
 802.11ax-2.4 GHz     : 0
```

Example 16-22 continues the same output and lists the client distribution across the different client onboarding states, along with a summary view of client status and client mobility state. This information helps to identify networkwide problems and define the level of impact for these failures.

Example 16-22 *Client Distribution Across Onboarding States*

```
Current client state statistics:
---------------------------------
  Authenticating         : 875
  Mobility               : 17
  IP Learn               : 168
  Webauth Pending        : 319
  Run                    : 13763
  Delete-in-Progress     : 0


Client Summary
-----------------------------
Current Clients : 15148
Excluded Clients: 0
Disabled Clients: 0
Foreign Clients : 1839
Anchor Clients  : 1072
Local Clients   : 11473
Idle Clients    : 6
```

Example 16-23 provides a snippet of the next section of the aggregate client statistics where more than 120 global statistics pertaining to clients are listed. These statistics include the number of client association failures per reason code, statistics for different roam types, client count for Layer 2 and Layer 3 authentication failures, key exchanges involved in the different fast roaming algorithms, and IP addressing issues seen by the aggregate clients on the C9800.

Example 16-23 *Client Distribution Across Onboarding States*

```
Client global statistics:
---------------------------------------------------------------
Total association requests received              : 14001367
Total association attempts                       : 13750538
Total FT/LocalAuth requests                      : 0
Total association failures                       : 1489020
|
Total association drops due to unknown bssid     : 72372
Total association drops due to parse failure     : 2260
|
Total 11r PMKR0-Name mismatch                    : 200921
Total 11r PMKR1-Name mismatch                    : 129521
Total 11r MDID mismatch                          : 160
Total AID allocation failures                    : 14728
Total AID free failures                          : 0
Total Roam Across Policy Profiles                : 289569
Total roam attempts                              : 3830002
Total CCKM roam attempts                         : 0
Total 11r roam attempts                          : 2558631
 …
Total client state starts                        : 4731656
Total client state associated                    : 12168293
Total client state l2auth success                : 5835659
Total client state l2auth failures               : 22375
…
Total add mobiles sent                           : 14539643
Total delete mobiles sent                        : 8437712
…
Total key exchange attempts                      : 1260942
Total broadcast key exchange attempts            : 246611
Total broadcast key exchange failures            : 0
Total eapol key sent                             : 2773241
Total eapol key received                         : 2630538
Total m1 sent                                    : 1260474
Total m3 sent                                    : 1233805
Total m5 sent                                    : 179775
```

```
Total m2 received                                   : 1236587
Total m4 received                                   : 1227492
Total m6 received                                   : 152075
…
Total data path client create success               : 14995015
Total data path client create failed                : 0
```

Example 16-24 provides a snippet of the output that shows average time that clients spent in each onboarding state. It helps to identify whether a particular step in onboarding is experiencing an unusual amount of delay, thereby identifying performance problems in the network. It also provides a good view of variations seen over time.

Example 16-24 *Client Average Time per Onboarding State for Systemic Performance Problems in Client Onboarding*

```
client state statistics:
-------------------------------------------------------
Average Time in Each State (ms)
  Associated State    : 0
  L2 State       : 0
  Mobility State : 0
  IP Learn State : 0
  L3 Auth State   : 0

Average Run State Latency (ms) : 1

Average Run State Latency without user delay (ms) : 0

Latency Distribution (ms)
   1 - 100               : 2732789
   100 - 200             : 136164
   200 - 300             : 215397
   300 - 600             : 301561
   600 - 1000            : 128027
   1000+                 : 628490
```

This output includes many statistics for webauth including aggregate webauth HTTP statistics, client count for the various webauth HTTP response codes, webauth queue counters, as well as webauth backpressure counters that help troubleshoot Secure Socket Layer (SSL) session exhaustion. Aggregate dot1x statistics for clients are also included. The last section of the output is the most useful when troubleshooting client connectivity. When you're troubleshooting client onboarding failures, identifying the client delete reason code is paramount to determine the step within the onboarding process that is failing, to chart out the course of deep-dive troubleshooting. The last section of this super command output lists client count for each client delete reason code. Client deletions initiated by the C9800 or APs or clients themselves, along with client deletions that occur in the usual course of network operations, are all individually listed and counters are provided for each. By capturing and comparing multiple iterations of this trend of client connectivity, you can establish where the failure occurred. Table 16-6 lists some of the client delete reason codes, along with their meaning. The client delete reason code string is printed as is, in the client MAC-based conditional always-on and RA traces, and is used to parse these logs to determine the root cause of client onboarding issues. You also can view the client delete reasons and their corresponding counters by using **show wireless stats client delete reasons.**

Table 16-6 *Client Delete Reason Codes Sample on the C9800*

Client Delete Reason	Initiated by	Explanation
No operation	C9800	Default delete reason. It is seen only when no classified delete reason applies. It is an initialized value in some of the scenarios, such as (1) Add mobile sending; (2) L3 reauthentication process; or (3) L2 authentication success.
WPA key exchange timeout	C9800	This code is triggered when the M1 key retransmit reached the maximum number of retries in a four-way handshake.
Dot11 IE validation failed	C9800	The client deleted as an invalid information element (IE) is seen in a packet from a wireless client.
Dot11 received invalid PMKID in the received RSN IE	C9800	An invalid PMKID is seen in the RSN IE of the dot11 packet sent from the client.
IP theft	C9800	The client is deleted due to IP address theft.
VLAN failure	C9800	The client is deleted due to a policy profile failure/VLAN failure.
L2-Auth connection timeout	C9800	The client is deleted because it did not complete L2 authentication within the timeout period.
IP-LEARN connection timeout	C9800	The client is deleted because IP Learn was not completed within the timeout period.
WPA group key update timeout	C9800	The client is deleted due to the WPA-WPA2 M5 retransmit reaching the maximum number of retries.
Mobility-peer delete	C9800	It is deleted by the remote mobility peer.

Client Delete Reason	Initiated by	Explanation
MAC authentication failure	Informational	This code is triggered when MAC authentication bypass (MAB) fails for the client.
Due to SSID change	Informational	The client is deleted if the client switches the SSID during dot11 packet processing or during mobility failure or during a guest scenario.
Mobility tunnel down	Informational	The client is deleted due to the mobility tunnel going down.
Dot11 max STA	Informational	When AP is catering to the maximum number of allowed clients per radio per WLAN, if a new client tries to associate, the AP rejects the client with this delete reason.
Client user triggered disassociation	Client initiated	This is the reason code seen when the client initiates disassociation by sending a dot11 disassociate packet.
Client EAP timeout	Client initiated	This is the reason code received from the dot11 disassociate packet due to client Extensible Authentication Protocol (EAP) authentication timeout.
AP initiated delete for idle timeout	AP initiated	The AP initiates client deletion with this reason code if the client has been idle for a defined period.
AP initiated delete for reassociation timeout	AP initiated	The FlexConnect AP triggers this type of client deletion if the reassociation timer at the AP expires without receiving a reassociation request from the client.

- **Client plumbing path:** Like AP, when a client is onboarded, its entry needs to be created across multiple software components on the C9800. The plumbing path for a client entry is the same as that of the AP entry—that is, from WNCd > FMAN-RP > FMAN-FP > CPP-client > DP. As before, each component receives an update from the component before it in line, processes the update to create an entry on itself, and pushes out an update to the next component in line. The individual commands to verify this plumbing across an HA C9800 pair are shown in Example 16-25. These outputs can be collected explicitly, but a simple macro that includes all these outputs is made available for ease of use: **show tech-support wireless client mac <CLIENTMAC>**.

Example 16-25 *Client Plumbing Across WNCd > FMAN-RP > FMAN-FP > CPP-CLIENT > DP on a C9800 Pair*

```
C9800#show wireless client summary
Number of Clients: 1
MAC Address     AP Name     Type ID   State          Protocol  Method     Role
-------------------------------------------------------------------------------------
48eb.6260.293e 9120         WLAN 2    Run             11ax(5)  Dot1x      Local
Number of Excluded Clients: 0
C9800#show platform software wireless-client chassis active R0
        ID  MAC Address      WLAN  Client State
----------------------------------------------------
0x9000000e  38ed.18c7.9d07     65  Run
0xa0000001  48eb.6260.293e      2  Run

C9800#show platform software wireless-client chassis standby R0
        ID  MAC Address      WLAN  Client State
----------------------------------------------------
0xa0000001  48eb.6260.293e      2  Run

C9800#show platform software wireless-client chassis active F0
        ID  MAC Address      WLAN  Client State          AOM ID  Status
----------------------------------------------------------------------------
0x9000000e  38ed.18c7.9d07     65  Run                      693  Done
0xa0000001  48eb.6260.293e      2  Run                      703  Done

C9800#show platform software wireless-client chassis standby F0
        ID  MAC Address      WLAN  Client State          AOM ID  Status
----------------------------------------------------------------------------
0xa0000001  48eb.6260.293e      2  Run                      442  Done

C9800#show platform hardware chassis active qfp feature wireless wlclient cpp-client
  summary
Client Type Abbreviations:
  RG - REGULAR BL - BLE
  HL - HALO    LI - LWFL INT
Auth State Abbreviations:
  UK - UNKNOWN IP - LEARN IP IV - INVALID
  L3 - L3 AUTH RN - RUN
Mobility State Abbreviations:
  UK - UNKNOWN IN - INIT
  LC - LOCAL   AN - ANCHOR
  FR - FOREIGN MT - MTE
  IV - INVALID
EoGRE Abbreviations:
  N - NON EOGRE Y - EOGRE
```

```
CPP IF_H   DPIDX       MAC Address     VLAN  CT  MCVL  AS  MS  E    WLAN       POA
-----------------------------------------------------------------------------------
0X3E     0XA0000001  48eb.6260.293e  150   RG   0    RN  LC  N  11renable 0x9000000d

C9800#show platform hardware chassis standby qfp feature wireless wlclient
  cpp-client summary
Client Type Abbreviations:
  RG - REGULAR BL - BLE
  HL - HALO    LI - LWFL INT
Auth State Abbreviations:
  UK - UNKNOWN IP - LEARN IP IV - INVALID
  L3 - L3 AUTH RN - RUN
Mobility State Abbreviations:
  UK - UNKNOWN IN - INIT
  LC - LOCAL   AN - ANCHOR
  FR - FOREIGN MT - MTE
  IV - INVALID
EoGRE Abbreviations:
  N - NON EOGRE Y - EOGRE
CPP IF_H   DPIDX       MAC Address     VLAN  CT  MCVL  AS  MS  E    WLAN       POA
-----------------------------------------------------------------------------------
0X3D     0XA0000001  48eb.6260.293e  150   RG   0    RN  LC  N  11renable 0x9000000d

C9800#show platform hardware chassis active qfp feature wireless wlclient datapath
  summary
Vlan   pal_if_hdl   mac            Input Uidb Output Uidb
------ ------------ -------------- ---------- -----------
150    0xa0000001   48eb.6260.293e 65480      65474

C9800#show platform hardware chassis standby qfp feature wireless wlclient datapath
  summary
Vlan   pal_if_hdl   mac            Input Uidb Output Uidb
------ ------------ -------------- ---------- -----------
150    0xa0000001   48eb.6260.293e 65481      65475
```

CPU Monitoring

On the C9800, multiple CPU cores are present and shared by the Linux kernel and the various BinOS processes. The total number of cores is dependent on the form factor of the C9800. To view the CPU utilization for the BinOS processes in descending order, you can run **show processes cpu platform sorted**, as shown in Example 16-26.To view the usage across all processes of the IOS in descending order, you can run **show processes cpu sorted**, as shown in Example 16-27. The sorting is done based on the 5Sec column, which shows average CPU usage over the last 5 seconds when the command is run. Because the top talkers are of interest, descending order provides the crispest view. To view trends of CPU usage over time for the IOS processes, run **show processes cpu history.** Other options are available in the **show processes cpu CLI tree**, but these two are the most used and are included in both **show tech-support** and **show tech-support wireless.**

Example 16-26 *CPU Utilization Across BinOS Processes on the C9800 (Data plane processing high traffic volume is top talker in this example)*

```
C9800#show proc cpu plat sorted
CPU utilization for five seconds: 7%, one minute: 5%, five minutes: 4%
Core 0: CPU utilization for five seconds: 26%, one minute: 6%, five minutes: 3%
Core 1: CPU utilization for five seconds: 7%, one minute: 2%, five minutes: 2%
Core 2: CPU utilization for five seconds: 6%, one minute: 6%, five minutes: 3%
Core 3: CPU utilization for five seconds: 1%, one minute: 1%, five minutes: 1%
Core 4: CPU utilization for five seconds: 1%, one minute: 1%, five minutes: 1%
Core 5: CPU utilization for five seconds: 2%, one minute: 4%, five minutes: 4%
Core 6: CPU utilization for five seconds: 1%, one minute: 1%, five minutes: 1%
Core 7: CPU utilization for five seconds: 30%, one minute: 33%, five minutes: 33%
Core 8: CPU utilization for five seconds: 0%, one minute: 0%, five minutes: 0%
Core 9: CPU utilization for five seconds: 0%, one minute: 0%, five minutes: 0%
Core 10: CPU utilization for five seconds: 0%, one minute: 0%, five minutes: 0%
   Pid    PPid    5Sec    1Min    5Min  Status        Size  Name
--------------------------------------------------------------------------------
 20934   20918     46%     46%     46%  R           349808  ucode_pkt_PPE0
 29364   29355     23%      8%      1%  S           112360  smand
 22037   21999      1%      1%      1%  S           226436  fman_fp_image
  4039    3929      1%      1%      1%  S          1374336  linux_iosd-imag
```

Example 16-27 *CPU Utilization Across IOSd Threads and Processes on the C9800 (SNMP is top talker in this example)*

```
C9800# show process cpu sorted | exclude 0.00
CPU utilization for five seconds: 52%/4%; one minute: 41%; five minutes: 30%
 PID Runtime(ms)     Invoked      uSecs   5Sec     1Min   5Min   TTY   Process
 335    71086383     2230181      31874   21.99% 16.03% 12.00%   0  SNMP LA Cache pr
 737    63339001    46679321       1356   14.07%  6.46%  4.47%   0  SNMP ENGINE
  94    27713849    36395051        761    8.31%  6.31%  4.65%   0  IOSD ipc task
 358    10182521    21838675        466    2.39%  3.29%  3.58%   0  IGMPSN
 146     2606821    72660954         35    1.11%  1.04%  0.90%   0  IOSXE-RP Punt Se
 739     1510402    56630069         26    0.71%  0.42%  0.28%   0  IP SNMP
 243     1359296    32333604         42    0.55%  0.57%  0.56%   0  IP Input
 740     1028238    27979655         36    0.47%  0.24%  0.17%   0  PDU DISPATCHER
 241     318857     8605731          37    0.07%  0.08%  0.07%   0  IP ARP Adjacency
```

One instance of this output does not provide much information if the CPU usage is normal or indicative of a problem. For example, a WNCd instance that has a high AP and client count mapped to it and is processing a high rate of chatty traffic like mDNS to and from these clients showing a CPU usage of 60 to 70 percent is not in itself a cause for concern. Multiple iterations of the outputs need to be collected and parsed to define the baseline CPU usage for the given C9800. Only a deviation from the baseline needs to be reviewed. Even further, only a spike in CPU utilization may be normal, depending on the process involved. For example, in Example 16-27, SNMP is the top talker, and this would be the case at the instance when the C9800 is polled by an external SNMP Management server. More than likely, the CPU spikes will line up with the rate of polling by the SNMP server. A higher-than-normal average CPU usage across 5 seconds, 1 minute, and 5 minutes is indicative of a definitive problem condition.

After the top talker is identified, tracelogs (in the case of BinOS) and debugs (in the case of IOS threads) provide the most relevant information on the cause of CPU usage. However, both debugs and tracelogs further add to the CPU usage. So, you must be extremely careful in deciding when and to what extent debugs and tracelogs can be enabled. For example, a CPU usage of 90 percent and above for a given process suggests you should avoid enabling additional debugs on that process. On the C9800, generating a live core dump for the process is helpful to diagnose the cause of increased CPU usage without minimal impact.

Typically, control traffic intended for CPU is punted and fed to the control plane processes through Linux sockets, depending on the type of traffic mapped to individual sockets. Processes read and handle packets off these sockets. On the C9800, control plane policers are put in place, by default, to protect the CPU from being overwhelmed by a particular type of control traffic. For example, in the case of an ARP broadcast storm. These policers, like all other default settings, can be viewed by running **show run all.** These policers define normal and high queues for each punt cause, and multiple parameters for both the normal and high queues, including queue depth, packets per second, and so on. To view the threshold values currently configured by these policers, run **show platform hardware chassis active qfp infrastructure punt policer summary.** When traffic to these queues exceeds the defined thresholds, the packets get dropped and, accordingly, the dropped packets counter increments in the corresponding queue. This can also be used to focus troubleshooting to specific traffic and features. These thresholds, as defined by the policers, can be viewed by running **show platform software punt-policer**, as shown in the sample snippet in Example 16-28.

Example 16-28 *CPU Queue Thresholds Defined by the CPU Punt Policer*

```
C9800#show platform software  punt-policer | include Punt|Cause|--|wls
Per Punt-Cause Policer Configuration and Packet Counters

Punt    Description             Config       Conform       Dropped      Config        Config
Cause                           Rate(pps)     Packets       Packets      Burst(pkts)    Alert
                                Normal  High  Normal High   Normal  High Normal  High   Normal
  High
------------------------------------------------------------------------------------------------
129  wls 802.11 Packets to LFTS  10500   550   588    4     0       0   10500   550   Off   Off
130  wls CAPWAP Packets to LFTS  437     45000 0      39014 0       0   437     45000 Off   Off
131  wls MOBILITY Packets to LFTS 437    45000 0      0     0       0   437     45000 Off   Off0
132   wls SISF Packets to LFTS   437     14400 0      19    0       0   437     14400 Off   Off
```

In some cases, the queue maps to multiple types of traffic; for example, traffic destined to SISF includes Dynamic Host Configuration Protocol (DHCP) traffic, Address Resolution Protocol (ARP) traffic, and IP data traffic from clients with unknown IP addresses, all of which enables IP address learning, so mac-IP binding can be stored in a device-tracking database. For troubleshooting high CPU in one or more cores, one of the steps is to quantify the traffic that is hitting the CPU and determine whether the traffic pattern is aberrant. You can view the counters for the aggregate traffic being punted from the data plane to the CPU for control-plane processing by using **show platform hardware chassis active qfp infrastructure punt statistics type per-cause**, as shown in Example 16-29. To view the aggregate wireless traffic being punted to CPU, run **show platform hardware chassis active qfp feature wireless punt statistics**, as shown in Example 16-30.

Example 16-29 *Aggregate Packet Counters Distributed by System-Defined Punt and Inject Causes*

```
Global Per Cause Statistics
  Number of punt causes =   148
  Per Punt Cause Statistics
  Counter ID  Punt Cause Name                   Packets Received   Packets Transmit-
  ted
  ---------------------------------------------------------------------------------
  002         IPv4 Options                        312                    0
  007         ARP request or response             1187479                1187479
  008         Reverse ARP request or response     26                     26
  010         Incomplete adjacency                4976                   4976
  011         For-us data                         519688                 519688
  021         RP<->QFP keepalive                  4015075                4015075
  026         QFP ICMP generated packet           183                    183
  055         For-us control                      886761                 886761
```

```
060         IP subnet or broadcast packet       1526                 1526
097         Packets to LFTS                     1689694              1689694
129         wls 802.11 Packets to LFTS          3290990              3290990
130         wls CAPWAP Packets to LFTS          1737106              1737106


Number of inject causes = 52
Per Inject Cause Statistics

Counter ID  Inject Cause Name                  Packets Received Packets Transmit-
ted
---------------------------------------------------------------------------------
001         L2 control/legacy                        1476           1476
002         QFP destination lookup                   2078383        2072978
005         QFP <->RP keepalive                      4015075        0
007         QFP adjacency-id lookup                  260            260
009         QFP ICMP generated packet                183            181
012         ARP request or response                  6605           6605
042         switch port layer 2 control packet       148705         148705
043         Applications Injecting Pkts using LFTS   1753226        1753226
```

Example 16-30 *Aggregate Wireless Packet Counters Punted to CPU*

```
C9800#show platform hardware chassis active qfp feature wireless punt statistics
CPP Wireless Punt stats:

                                App Tag      Packet Count
                                -------      ------------
         CAPWAP_PKT_TYPE_DOT11_PROBE_REQ          3290990
              CAPWAP_PKT_TYPE_DOT11_MGMT                0
              CAPWAP_PKT_TYPE_DOT11_IAPP          1689694
              CAPWAP_PKT_TYPE_DOT11_RFID                0
               CAPWAP_PKT_TYPE_DOT11_RRM                0
             CAPWAP_PKT_TYPE_DOT11_DOT1X                0
        CAPWAP_PKT_TYPE_CAPWAP_KEEPALIVE           285369
      CAPWAP_PKT_TYPE_MOBILITY_KEEPALIVE                0
            CAPWAP_PKT_TYPE_CAPWAP_CNTRL          1451678
             CAPWAP_PKT_TYPE_CAPWAP_DATA                0
        CAPWAP_PKT_TYPE_CAPWAP_DATA_PAT                59
          CAPWAP_PKT_TYPE_MOBILITY_CNTRL                0
                       WLS_SMD_WEBAUTH                  0
                      SISF_PKT_TYPE_ARP                 0
```

```
               SISF_PKT_TYPE_DHCP              0
              SISF_PKT_TYPE_DHCP6              0
            SISF_PKT_TYPE_IPV6_ND              0
         SISF_PKT_TYPE_DATA_GLEAN              0
      SISF_PKT_TYPE_DATA_GLEAN_V6              0
         SISF_PKT_TYPE_DHCP_RELAY              0
           WLCLIENT_PKT_TYPE_MDNS              0
 CAPWAP_PKT_TYPE_CAPWAP_RESERVED               0
```

Memory Monitoring

On the C9800, three types of memory allocations are involved:

- **IOS memory allocation:** This is the memory allocated within the IOSd process. IOS memory management is self-contained and relies on IOS primitives. It is separate from BinOS memory allocation, both from a monitoring and a troubleshooting perspective.
- **Database memory allocation:** This is specifically the memory allocated for the various databases corresponding to the different BinOS processes on the C9800.
- **BinOS memory allocation:** This refers to the nondatabase memory allocated to the BinOS processes.

All memory allocations on the C9800 code are tagged with a unique number referred to as a *callsite*. This callsite plays a key role in troubleshooting memory allocation failures and memory leaks on the C9800. To identify whether a memory leak is occurring, you need to monitor system-level memory usage at regular intervals to see if there is an increasing trend in used memory and decreasing trend in free memory.

To view the total system memory, run **show platform software process slot chassis active R0 monitor | include Mem.** To view memory usage per process in the descending order of allocated memory, run **show processes memory platform sorted.** Multiple iterations help verify whether the allocated or holding memory on a process keeps increasing. Get the status over the last hour of the callsites by executing **show processes memory platform accounting.** After identifying the process and callsites that are using the most memory from this output, enable these debugs:

```
#debug platform software memory <PROCESS_NAME> chassis active R0 alloc
callsite clear
#debug platform software memory <PROCESS_NAME> chassis active R0 alloc
backtrace start <CALL_SITE> depth 10
#debug platform software memory <PROCESS_NAME> chassis active R0 alloc
callsite start
#debug platform software memory <PROCESS_NAME> chassis active R0
alloc callsite stop
```

After a few minutes of starting the debug and before stopping the debug, collect the output of the command **show platform software memory <PROCESS_NAME>**

chassis active R0 alloc backtrace. Also collect the output of the command **show platform software memory <PROCESS_NAME> chassis active R0 alloc callsite brief.** You can capture a snapshot of the database memory by using **show platform software memory <PROCESS_NAME> chassis active R0 alloc type data (brief | summary | callsite}** and **show platform software memory database <PROCESS_NAME> chassis active R0 {brief | summary | callsite}.**

Besides these commands, a TCL script called memleak.tcl is shipped along with IOS-XE software on the C9800. You can run this script to capture the memory baseline at a specific point in time. The memory is monitored to verify that a leak has occurred. At this point, you can run memleak.tcl a second time to get another memory snapshot. The difference in memory usage between the two snapshots is reported as a probable memory leak.

Data Plane Monitoring

The C9800 uses the Quantum Flow Processor (QFP) as the data plane. The QFP uses a flexible and programmable CPU design that utilizes 40 four-way-threaded CPU cores called packet processing engines (PPEs). All data traffic that ingresses or egresses the C9800 is processed by these PPEs and forwarded to its intended destination or dropped as applicable. The PPEs also serve as a capture point for EPC. To trace a packet through the data plane, you can use the Packet Tracer. However, there are several statistics that you can monitor to validate the health of the data plane.

To view the overall PPE utilization, run **show platform hardware chassis active qfp datapath utilization.** Check Processing: Load (pct) to see the utilization. Anything greater than 92 percent indicates excess load on the data plane. For the global view of all packets dropped by the data plane, run **show platform hardware chassis active qfp statistics drop.** You need to run and compare multiple iterations to identify the drop counters that are incrementing.

Typically, data path troubleshooting comes into the picture when you experience performance and throughput issues and you suspect packets are dropped by the C9800 data plane. The first point where packet drops can occur is at the port level. The first step is to identify whether the MAC address of the source device was learned by the C9800 data plane. The mac-address table maintained by the C9800 data plane can be viewed with the C9800 command **show platform hardware chassis active qfp feature swport datapath mac-table all.** The next step is to ensure the VLAN data has been plumbed to the QFP, which you can view by running **show platform hardware chassis active qfp feature swport client {vlan | mac-table | pm}.** You can view the drop counters across all VLANs by running **show platform hardware chassis active qfp feature swport datapath system statistics.**

If you are specifically troubleshooting an AP disconnect use case, dropped counters for CAPWAP and DTLS would be of the most interest. You can view them by running **show platform hardware chassis active qfp feature wireless capwap datapath statistics drop all.** You can also limit the view of drop counters to a specific AP by running **#show platform hardware chassis active qfp feature wireless capwap datapath mac-address <APradio-mac> statistics**, as shown in Example 16-31. The command to view global DTLS drops is **show platform hardware chassis active qfp feature wireless dtls datapath statistics all.**

Example 16-31 *Sample Output of Datapath Drops for One AP Radio Mac*

```
C9800#show platform hardware chassis active qfp feature wireless capwap datapath mac
  c064.e423.c640 statistics | exclude 0
                                                    Pkts                Bytes
Punt dot11 mgmt                                      2                   515
Inject data keepalive                               688                 59168
```

For all wireless clients, you can view aggregate data plane drops by using **show platform hardware chassis active qfp feature wireless wlclient datapath statistics drop all**, while a per-client view is available when you run **show platform hardware chassis active qfp feature wireless wlclient datapath mac-address <CLIENT-MAC> statistics**, as shown in Example 16-32.

Example 16-32 *Sample Output of Datapath Drops for One Client*

```
C9800#show platform hardware chassis active qfp feature wireless wlclient datapath
  mac 9233.17c6.561e statistics | exclude 0
                                                               Pkts              Bytes
Rx iplearn Drop                                                  2                 228
Rx ipsg invalid v6 Drop                                          1                 134
Punt
Punt sisf dglean v6                                              1                  94
Punt sisf dhcp discovery from wireless                           2                 716
Punt sisf dhcp offer from wired                                  2                 692
Punt sisf dhcp req from wireless                                 4                1462
Punt sisf dhcp ack from wired                                    1                 346
Punt sisf ipv6 nd rs from wireless                               4                 296
Inject
Inject L2 sisf dhcp discovery                                    2                 716
Inject L2 sisf dhcp offer                                        2                 692
Inject L2 sisf dhcp request                                      4                1462
Inject L2 sisf dhcp ack                                          1                 346
Inject L2 sisf ipv6nd rs                                         4                 296
Inject from vlan ipv6nd rs                                       4                 296
```

Summary

Throughout the book, we explained the architecture and features of the C9800 wireless LAN controller. Following the deployment guidelines and best practices laid out in this book as well as the reference documents will enable a successful C9800 deployment.

Cisco DNA Center Assurance provides the best means for network monitoring and assurance including wireless. This chapter explains the various monitoring, debugging, tracing, and packet-capturing features that are native to the C9800 and introduces tools that are available on-the-box and offline that help with data collection and analysis to determine the root cause of problem scenarios. We discuss the most common use cases, such as system monitoring, AP joins, and client connectivity troubleshooting in the context of leveraging these tools and capabilities.

In summary, the sequence of troubleshooting for all control planes follows these steps:

1. Define the problem and deployment details, such as hardware, software version, number, and model of APs and clients.
2. Review syslogs and run **show tech wireless** through the Config Analyzer to validate configuration and RF environment.
3. Collect always-on traces to narrow down the problem.
4. Collect the radioactive trace for specific conditions (MAC address or IP address).
5. Collect other outputs such as state information, statistics, and plumbing data and parse in conjunction with the RA trace to determine the root cause.

Data plane troubleshooting of data loss and throughput issues is better addressed through statistics and packet-capturing and packet-tracing facilities described in the chapter.

References

Recommended IOS-XE releases for 9800: https://www.cisco.com/c/en/us/support/docs/wireless/catalyst-9800-series-wireless-controllers/214749-tac-recommended-ios-xe-builds-for-wirele.html

Best Practices Guide: https://www.cisco.com/c/en/us/products/collateral/wireless/catalyst-9800-series-wireless-controllers/guide-c07-743627.html

Wireless debugging and log collection: https://www.cisco.com/c/en/us/support/docs/wireless/catalyst-9800-series-wireless-controllers/213949-wireless-debugging-and-log-collection-on.html

Upgrading Field Programmable Hardware Devices for Cisco Catalyst 9800 Series Wireless Controllers: https://www.cisco.com/c/en/us/td/docs/wireless/controller/9800/config-guide/b_upgrade_fpga_c9800.html

Smart Licensing troubleshooting: https://www.cisco.com/c/en/us/support/docs/wireless/catalyst-9800-series-wireless-controllers/217348-configure-troubleshoot-catalyst-9800-w.html

Troubleshooting common wireless client connectivity issues: https://www.cisco.com/c/en/us/support/docs/wireless/catalyst-9800-series-wireless-controllers/213970-catalyst-9800-wireless-controllers-commo.html

Quick Start Guide on logs to collect on the C9800 for various scenarios: https://www.cisco.com/c/en/us/support/docs/wireless/catalyst-9800-series-wireless-controllers/215523-quick-start-guide-on-what-logs-and-debug.html

DevNet wireless tools home page: https://developer.cisco.com/docs/wireless-troubleshooting-tools/#!wireless-troubleshooting-tools

Appendix A

Setting Up a Development Environment

Although there are multiple version control systems (Git, Mercurial, Subversion, ClearCase), the most used for network programmability and Infrastructure as Code is Git. Infrastructure as Code, which was discussed in Chapter 12, "Network Programmability," consists of managing network infrastructure using a descriptive model, storing the same versioning systems that development teams use for source code. Git is so common that a new term has been coined. On top of DevOps and NetDevOps, there is also GitOps!

The purpose of Git is to keep track of the changes in a set of files over time, so you can recall any specific version at any point in time.

If the files are text based (such as the source code or text-based configuration files), Git also displays the difference between two versions, allowing you to answer the question: What changed between the configuration of yesterday and today?

To understand how Git works, let's first define some glossary terms:

Repository: A repository is the place where all the files and their history (the changes over time of a particular file) are stored.

Local repository: If a repository is stored in your local computer, it's called a local repository.

Remote repository: If a repository is in another computer, it's called a remote repository. This remote repository can be in any other regular computer, a dedicated server, a site like www.github.com or www.gitlab.com, or even another directory on your own computer.

Cloning: Copying a remote repository to your computer, thus creating a local repository.

Committing: Sending the new version of your files to your local repository.

Pushing: Sending your local repository changes.

Pulling: Retrieving the changes from a remote repository to your local one.

Setting Up Your Development Environment

The best way to get familiar with network programmability concepts is by practicing. In this appendix, you install the basic tools you need to work with programmability: an editor and Git. You complete your environment with additional tools later.

Code Editor

A basic tool you need is a code editor. There are plenty of editors, and each of us has our own preferences. In this book, we use Visual Studio Code, an open-source editor with rich features for code development that exists for all the usual platforms (MacOS, Windows, and Linux). Visual Studio Code has a rich collection of plug-ins that extend the editor and allow you to work directly with network diagrams, Git, or almost any language that you could imagine. You can download Visual Studio Code from https://code.visualstudio.com/download.

After you install Visual Studio Code, go to the Plugins section on the side bar and search for Git. You can choose from several plug-ins, as you can see in Figure A-1. We recommend both GitLens and GitGraph. These plug-ins display changes done to your code over time, who changed it, a graphical representation of your commits, and more.

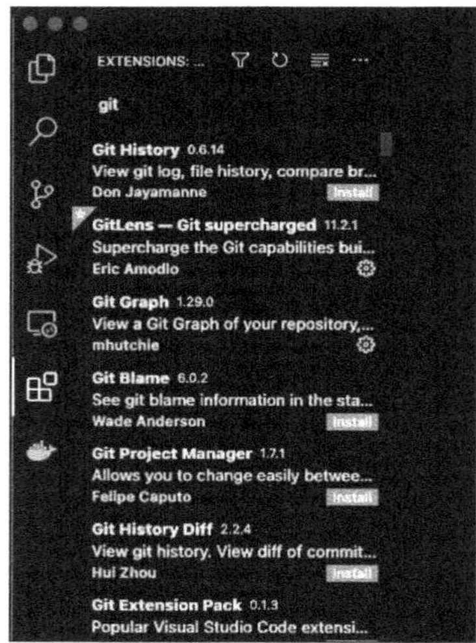

Figure A-1 *Git plug-ins for Visual Studio Code*

Git Installation

You need to install Git on your local computer. Versions are available for MacOS, Windows, and Linux. You can download them from the official Git page: https://git-scm.com/downloads.

The screenshots in this appendix are from MacOS, but Windows or Linux versions are similar.

On top of the official Git downloads, some available GUI tools can make some operations easier. We evaluate some of those tools later.

How to Get Started with Git

In Figure A-2, you can see the general Git workflow. In this section, we go through each of the steps.

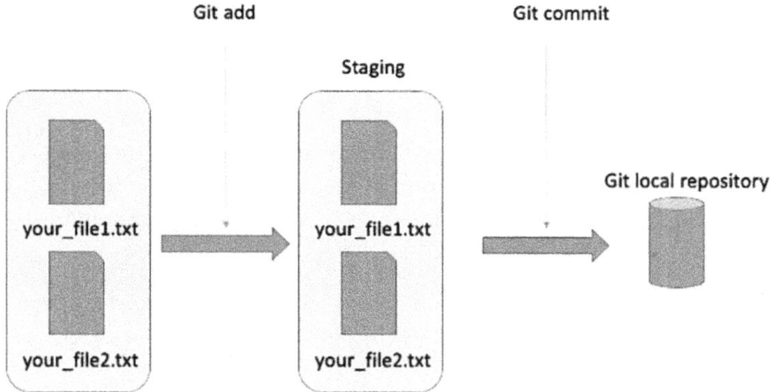

Figure A-2 *Git workflow*

After you install Git and a code editor in your computer, create and enter a new, empty directory:

```
$ mkdir git_learning
$ cd git_learning
```

This directory stores your Git local repository. To create the repository, use the **git init** command, adding a dot as argument to represent the current directory:

```
$ git init .
Initialized empty Git repository in /Users/fsedano/git_learning/.git/
```

Open this directory in your editor and create a text file called **example1.txt**, writing some memorable content inside. This is the first version of the text, as you can see in Figure A-3.

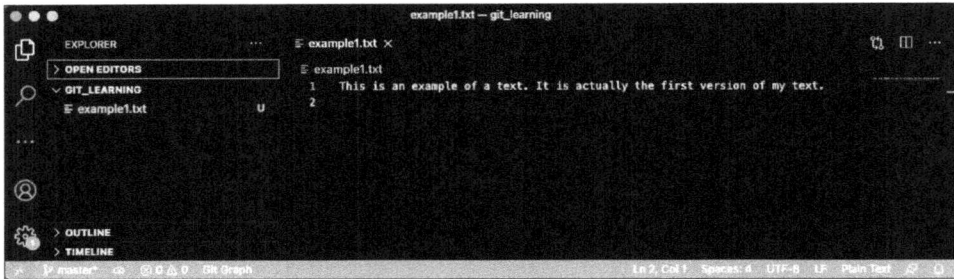

Figure A-3 *Creating first file*

When you are happy with your text, commit it. Committing a set of changes is the process of adding them to your local repository. When your files are in the repository, you can track any changes to them and revert to their original state.

Before committing a change, you need to specify which files will be part of that commit. You might have changed a lot of files in your directory, but probably you want to commit only a few. There might be temporary files, others you created to test something, and so on.

To specify which files need to be committed, Git uses a concept called the *staging area*. Let's look closer at the entire commit process and how the staging area fits into it.

Commit Process in Git

Several steps are required to commit a set of files in Git, as shown in Figure A-4.

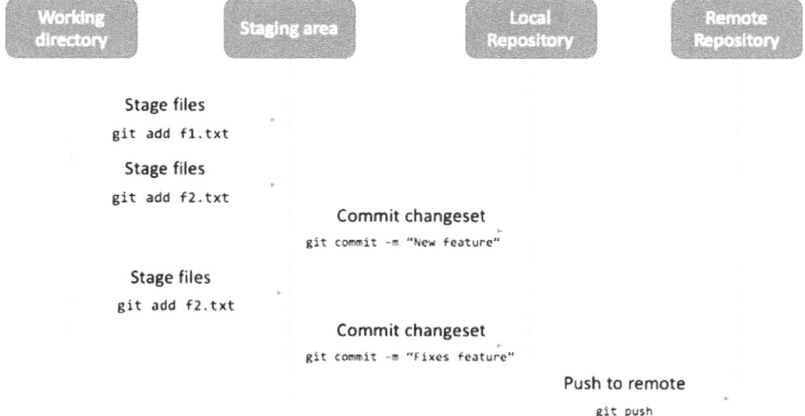

Figure A-4 *Commit process in Git*

To add the files to the staging area, you use the **git add** command, followed by the files or directories you want to add. You can also specify wildcards. To add the sample file to the staging area, you type

```
git add example1.txt
```

Working with Git can be confusing at times. If you want to quickly see what status you're in, a good command to use is **git status**. It shows which files are part of the commit and much more. If you use **git status** after adding the sample file to the staging area using **git add**, you see a screen similar to this:

```
$ git status
On branch master

No commits yet

Changes to be committed:
  (use "git rm --cached <file>..." to unstage)
     new file:   example1.txt
```

Here you can see the file (example1.txt) is going to be committed (so it's part of the staging area) and you've not done any commit yet.

Each commit should be a consistent set. For example, if you're deploying a new feature that needs changes to four configuration files, the commit should have the changes to all four files. The idea is that each commit should bring the system to a known state.

You can keep adding more files and changes to the staging area as many times as you want, and all those changes are part of a single commit.

Let's now change something in the example1.txt file and again run the **git status** command, as you can see in Figure A-5 and Figure A-6.

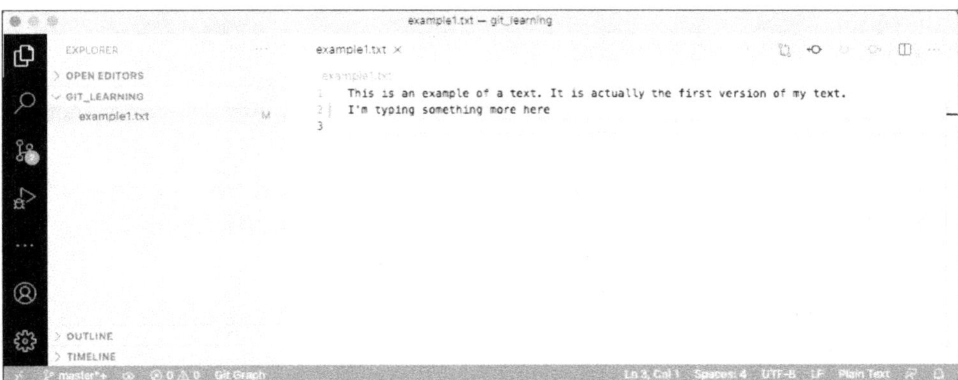

Figure A-5 *Updating the first file*

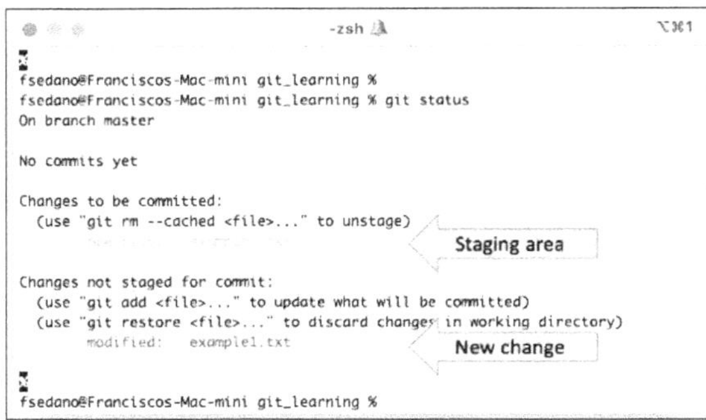
-zsh
fsedano@Franciscos-Mac-mini git_learning %
fsedano@Franciscos-Mac-mini git_learning % git status
On branch master
No commits yet
Changes to be committed:
(use "git rm --cached <file>..." to unstage)
Staging area
Changes not staged for commit:
(use "git add <file>..." to update what will be committed)
(use "git restore <file>..." to discard changes in working directory)
modified: example1.txt
New change
fsedano@Franciscos-Mac-mini git_learning %

Figure A-6 *Git detecting changed files*

By using **git status**, you can see

- A change is already part of the staging area (the previous **git add** you ran).
- Another change happened to the file, but that one is not yet part of the staging (the one you just did).

If you want to see the changes in your file that are not yet part of the staging area, you can use the **git diff** command, as shown in Figure A-7. It displays the differences between the previous version (which Git calls "a") and the new version (which Git calls "b").

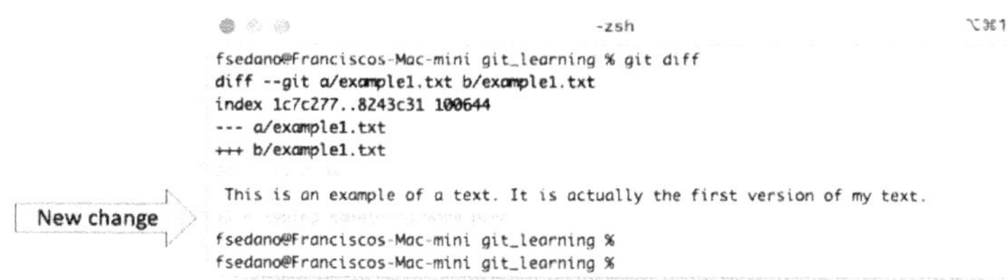
-zsh
fsedano@Franciscos-Mac-mini git_learning % git diff
diff --git a/example1.txt b/example1.txt
index 1c7c277..8243c31 100644
--- a/example1.txt
+++ b/example1.txt
This is an example of a text. It is actually the first version of my text.
New change
fsedano@Franciscos-Mac-mini git_learning %
fsedano@Franciscos-Mac-mini git_learning %

Figure A-7 *Diff between two file versions*

The lines added to the file appear with a + symbol before them. Any deleted line shows with a – symbol.

To see the changes that are inside the staging area, you also use the **git diff** command, but you need to add the **--staged** parameter, as shown in Figure A-8.

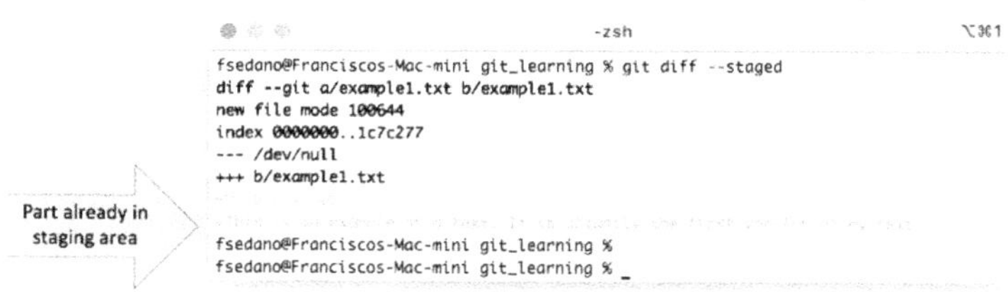
-zsh
fsedano@Franciscos-Mac-mini git_learning % git diff --staged
diff --git a/example1.txt b/example1.txt
new file mode 100644
index 0000000..1c7c277
--- /dev/null
+++ b/example1.txt
Part already in staging area
fsedano@Franciscos-Mac-mini git_learning %
fsedano@Franciscos-Mac-mini git_learning %

Figure A-8 *Git displaying staged files*

Note that there is no mention of the second phrase added here. This figure shows just what's already in the staging area.

To add the new changes to the staging area so that they are also committed, you use the **git add** command again:

```
git add example1.txt
```

Running **git status** again shows the two sentences that are part of the commit.

When you're happy with the changes, you are ready for your first commit. Just run **git commit**, as shown in Figure A-9. You need also to pass a comment as part of the commit. Because this comment will be visible in the commit list, try to be descriptive with it:

```
git commit -m "First commit"
```

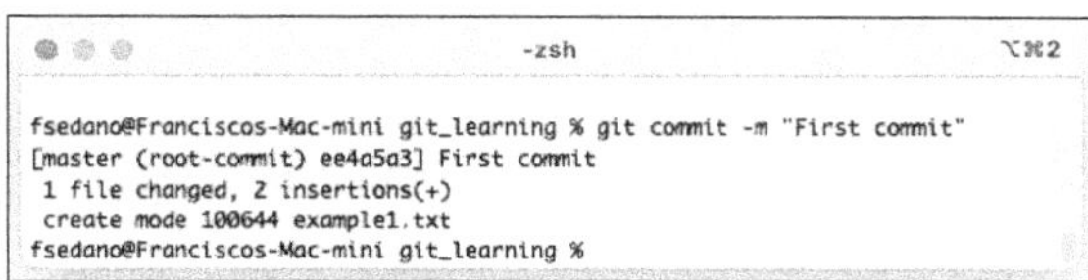

-zsh
⌥⌘2
fsedano@Franciscos-Mac-mini git_learning % git commit -m "First commit"
[master (root-commit) ee4a5a3] First commit
1 file changed, 2 insertions(+)
create mode 100644 example1.txt
fsedano@Franciscos-Mac-mini git_learning %

Figure A-9 *Committing changes*

Examining Repository Activity

So far, you have a local Git repository with your changes. To see the commit log, you could use the CLI **git log**, as shown in Figure A-10:

```
git log
```

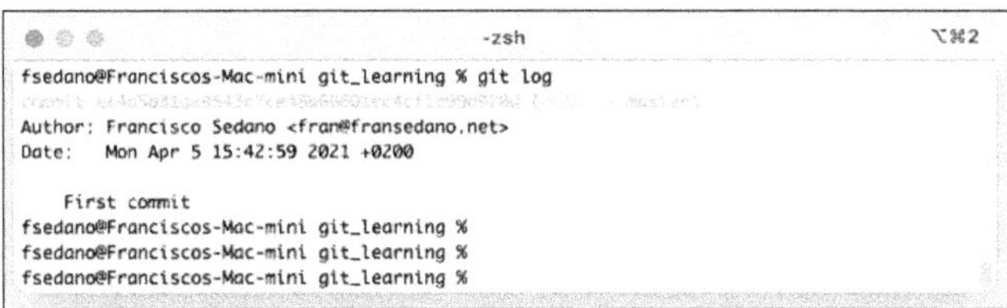

Figure A-10 *Examining Git repository changes*

Alternatively, you could see the activity in the Visual Studio Code interface if you installed the Git plug-in, as shown in Figure A-11.

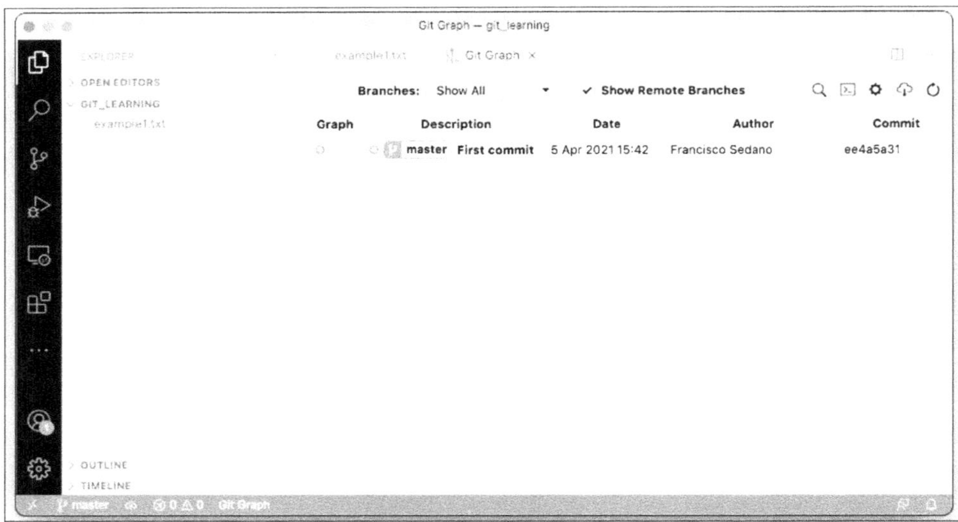

Figure A-11 *Viewing repository activity*

To create more commits, just repeat the process:

- Modify files
- git add
- git commit

If you add some text and changing parts of the existing one, Visual Studio marks the changed or added lines as you type, as shown in Figure A-12.

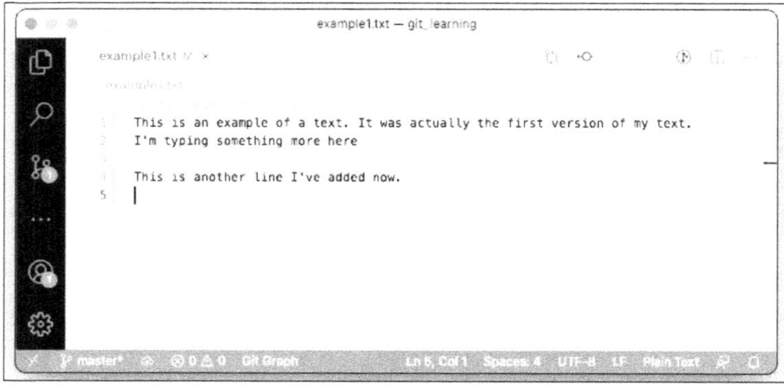

Figure A-12 *Marks in Visual Studio Code*

Now you can commit the changes, as displayed in Figure A-13.

```
-zsh
fsedano@Franciscos-Mac-mini git_learning % git add example1.txt
fsedano@Franciscos-Mac-mini git_learning % git commit -m "Changes to the text"
[master 9b6bcc1] Changes to the text
 1 file changed, 3 insertions(+), 1 deletion(-)
fsedano@Franciscos-Mac-mini git_learning %
```

Figure A-13 *Committing changes*

You can use **git log** again to see the commit history, as displayed in Figure A-14.

```
-zsh
fsedano@Franciscos-Mac-mini git_learning % git log
Author: Francisco Sedano <fran@fransedano.net>
Date:   Mon Apr 5 15:55:09 2021 +0200

    Changes to the text

Author: Francisco Sedano <fran@fransedano.net>
Date:   Mon Apr 5 15:42:59 2021 +0200

    First commit
fsedano@Franciscos-Mac-mini git_learning % _
```

Figure A-14 *Viewing commit history with git log*

Next to each line you see a string with a long hexadecimal number (ee4a5a...). This is the commit hash that you can use to uniquely reference each commit. To see the details of each commit (what was changed, added, or deleted), use **git show** with the commit hash, as shown in Figure A-15:

```
git show <commit hash>
```

-zsh
fsedano@Franciscos-Mac-mini git_learning % git show 9b6bcc19e95fbe99c89d68339ddb0f6fbcfb1b1f
Author: Francisco Sedano <fran@fransedano.net>
Date: Mon Apr 5 15:55:09 2021 +0200
Changes to the text
diff --git a/example1.txt b/example1.txt
index 8243c31..393ffb0 100644
--- a/example1.txt
+++ b/example1.txt
-This is an example of a text. It is actually the first version of my text.
I'm typing something more here
fsedano@Franciscos-Mac-mini git_learning %
fsedano@Franciscos-Mac-mini git_learning %

Figure A-15 *Displaying changes in a commit using the command line*

Alternatively, you can use Visual Studio Code, as shown in the Figure A-16.

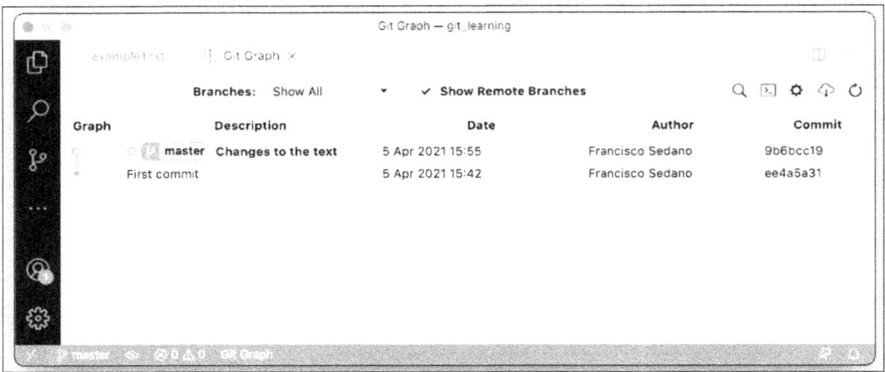

Figure A-16 *Displaying commits using Visual Studio Code*

Using Remote Repositories

The real power of Git comes from its distributed nature. Until now, you have been working in a local repository with your changes. By pushing this repository somewhere else, you can collaborate with other users and avoid any data loss if the contents of your hard drive are lost.

A Git remote repository can be anything from a different directory in your computer to a dedicated remote server. Several online services offer Git repository hosting. Two of the most popular ones are github.com and gitlab.com.

Git has become a common way to distribute software. Instead of uploading your software, configuration files, and so on to a server for others to download, you can just host your Git repository on one of the online services (for example, github.com). That way, anybody can just clone your repository from there.

Those online services offer features on top of just hosting the Git repositories.

For example, you can configure them so that each time you (or somebody else you allow) pushes a commit, a set of scripts is run to validate the changes made. Only if that validation passes, the commit is allowed.

Another useful feature is code review. That provides a discussion board where anybody can comment on the changes that have been done and either approve or reject them.

Next, let's push the newly created local repository to github.com. To get started, you need to create an account on github.com and add keys to authenticate yourself to Git.

Adding SSH Keys to github.com

SSH works via two keys: the private key and the public key. The private key should always stay private. The public key is the one you can freely share and can be used to authenticate you. The public key allows others to verify that you are the owner of your private key.

SSH keys can be created by using **ssh-keygen**:

```
ssh-keygen -t rsa -b 2048
```

You are prompted for an optional password to unlock the private key and a filename to save them.

To add SSH keys to your github.com account, go to the profile settings on the top menu and select **Settings.** In the left menu, you see a section for SSH keys. You can add your public SSH key there, as shown in Figure A-17.

Caution Never share your private key. It must always remain secret.

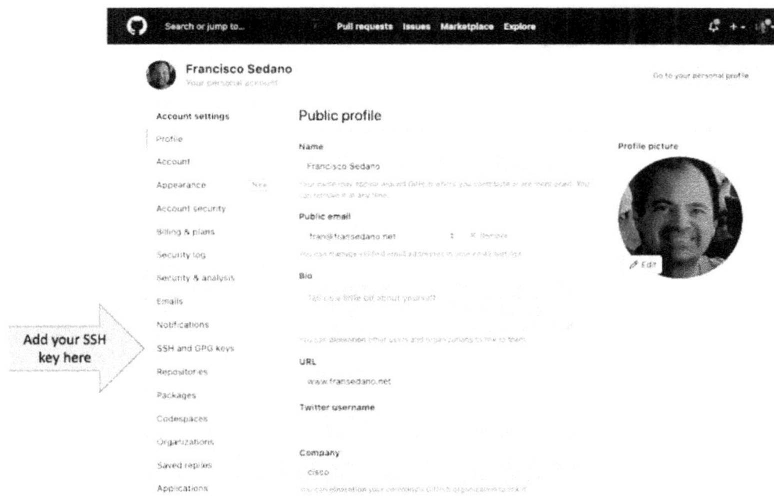

Figure A-17 *Adding SSH keys to GitHub*

When you click there, you can add your SSH keys.

You can now create the repository. At this stage, you can select whether your repository will be public (so anybody can see it) or private. Also, you can select to add some initial files to it. Because you'll be importing an existing repository, don't select any of the options there. You can see the process in Figure A-18.

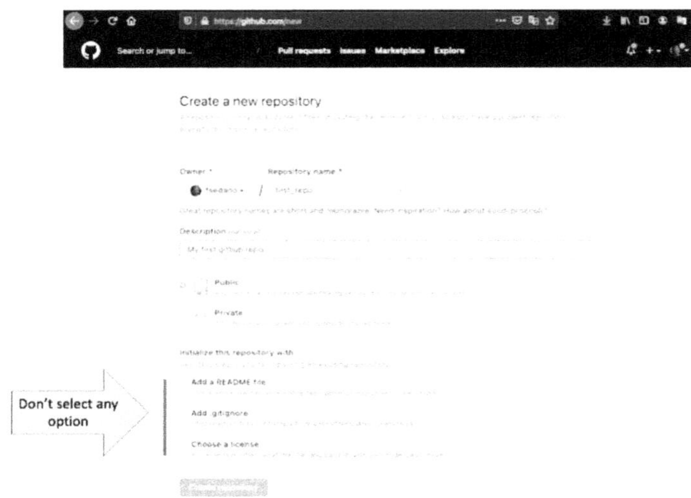

Figure A-18 *Creating a new repository in GitHub*

When you click **Create Repository**, you are brought to a page with some instructions on what to do next, as shown in Figure A-19. For this example, you want this repository to be a remote repository to the one you created in your computer.

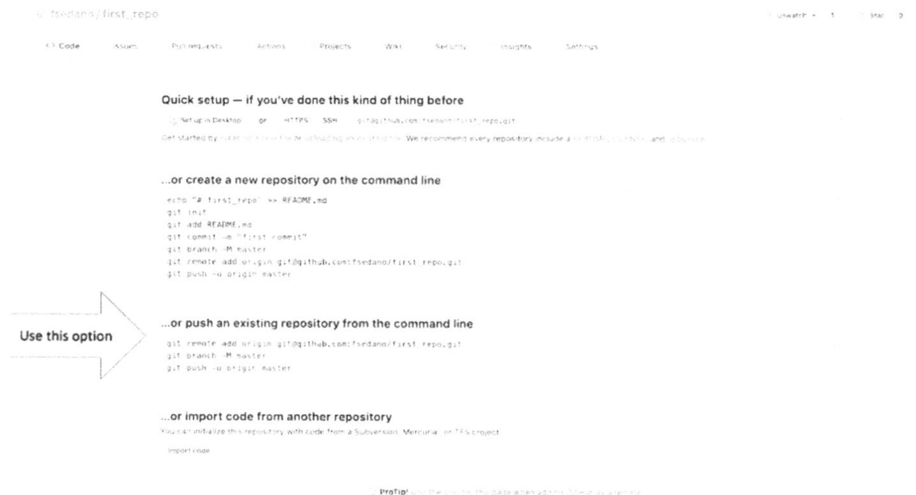

Figure A-19 *New repository in GitHub*

First, you add the GitHub repository you created as a remote repository, with the name **origin**:

```
git remote add origin git@github.com:fsedano/first_repo.git
```

Then you push the entire local repository to GitHub. That includes not only the current files but also all the commit history to the repository:

```
git push -u origin master
```

The local repository and the remote are now clones. Refreshing GitHub's page displays the files in your repository, along with all the story from your commits, as shown in Figure A-20.

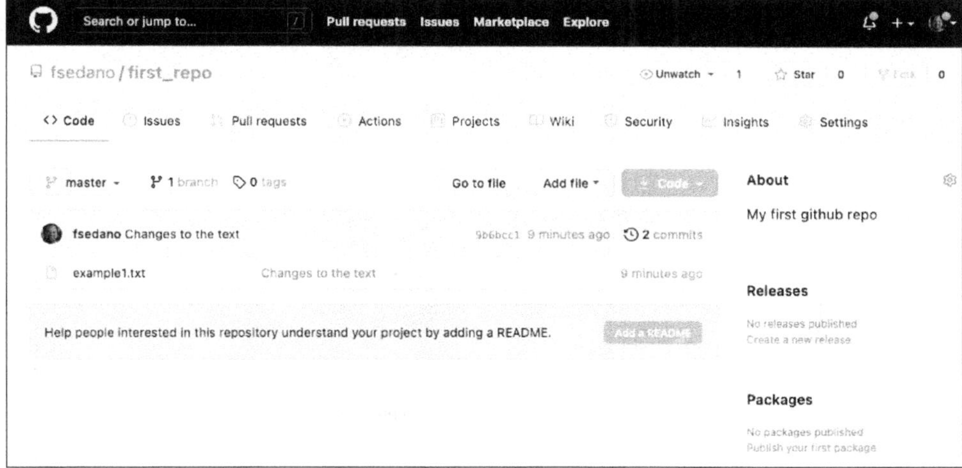

Figure A-20 *Repository after the first commit*

Keeping Repositories in Sync

Because Git is a distributed version control system, it's common to have multiple copies of the repository, as you saw in the previous section. You need a way to synchronize the changes from or to other repositories.

If you want to update your local repository with the most recent changes from another repository, you need to use the **git pull** command, as shown in Figure A-21.

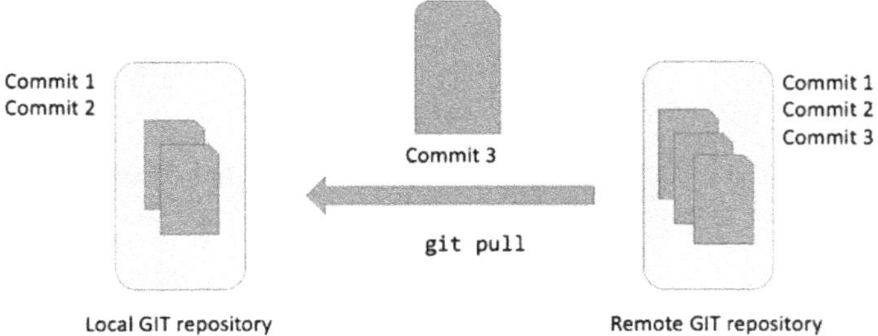

Figure A-21 *Pulling from remote repositories*

Git pull pulls the missing commits from a remote repository and adds them to your local one, making them in sync.

If you have made changes to your local Git repository, and you want to send them to the remote, you need to use the opposite CLI: **git push**. This flow is depicted in Figure A-22.

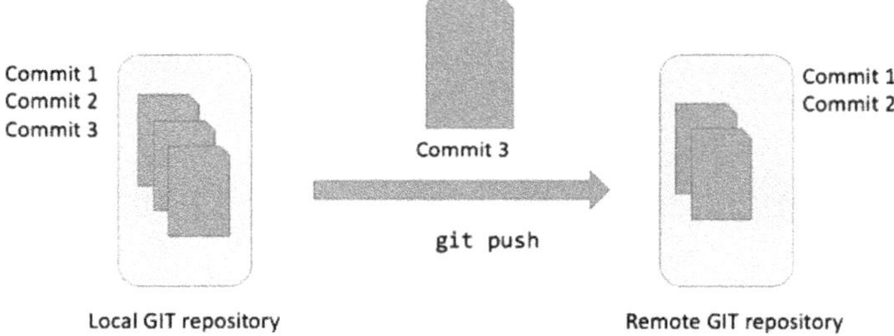

Figure A-22 *Pushing to remote repositories*

Resolving Conflicts

It is possible to have conflicts while pushing or pulling changes from remote repositories. If you have modified a file locally and another commit modified the same file in the remote repository, when you pull it, Git will find a conflict. Figure A-23 shows what happens when you commit and try to push changes to the remote repository in this type of situation. The pull failed because the remote repository had more recent commits, so it asks you to perform a **git pull** before you can push.

```
-zsh
fsedano@Franciscos-Mac-mini cloned_repo % git add example1.txt
fsedano@Franciscos-Mac-mini cloned_repo % git commit -m "Modify a line"
[master 8c16e00] Modify a line
 1 file changed, 1 insertion(+), 1 deletion(-)
fsedano@Franciscos-Mac-mini cloned_repo % git push
To github.com:fsedano/first_repo.git
 ! [rejected]        master -> master (fetch first)
error: failed to push some refs to 'git@github.com:fsedano/first_repo.git'
fsedano@Franciscos-Mac-mini cloned_repo %
```

Figure A-23 *Git message if the local repository lags behind a remote one*

However, when you try to perform **git pull**, it fails because one of the commits in the remote repository modified the same file that the local changes.

Sometimes, Git is able to automatically fix the conflict. However, if the resolution is not trivial, Git asks for your help to solve the conflict. You can see this in the Figure A-24.

```
-zsh
fsedano@Franciscos-Mac-mini cloned_repo % git pull
remote: Enumerating objects: 5, done.
remote: Counting objects: 100% (5/5), done.
remote: Compressing objects: 100% (1/1), done.
remote: Total 3 (delta 1), reused 3 (delta 1), pack-reused 0
Unpacking objects: 100% (3/3), done.
From github.com:fsedano/first_repo
   d1bc3cb..54a39fa  master     -> origin/master
Auto-merging example1.txt
CONFLICT (content): Merge conflict in example1.txt
Automatic merge failed; fix conflicts and then commit the result.
fsedano@Franciscos-Mac-mini cloned_repo %
```

Figure A-24 *Git conflicts*

If you look at your file now, you'll see it has the text from both commits, with a special marking to note which part is coming from which repository, as shown in Figure A-25.

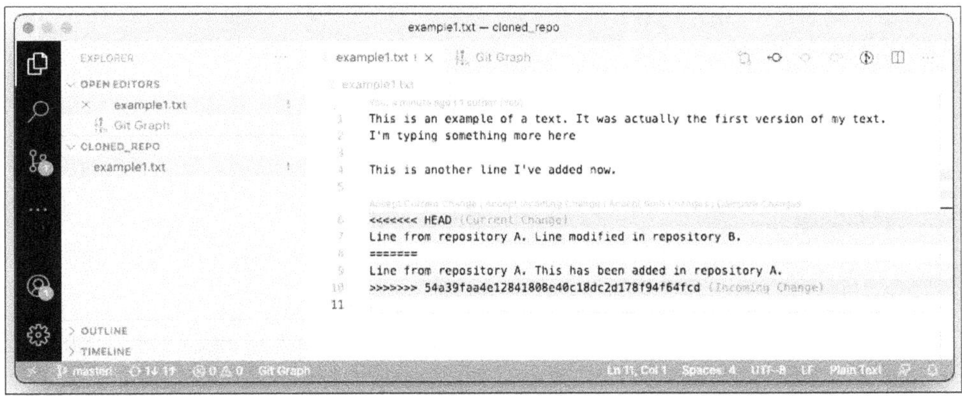

Figure A-25 *Solving Git conflicts*

Because both commits changed the same line, Git cannot solve this conflict automatically. You need to manually edit the file and fix it by merging the changes. **Git status** gives you some hints, as shown in Figure A-26.

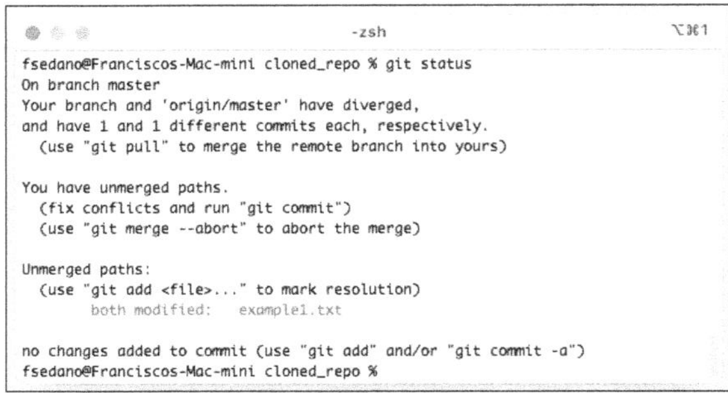

-zsh
fsedano@Franciscos-Mac-mini cloned_repo % git status
On branch master
Your branch and 'origin/master' have diverged,
and have 1 and 1 different commits each, respectively.
(use "git pull" to merge the remote branch into yours)
You have unmerged paths.
(fix conflicts and run "git commit")
(use "git merge --abort" to abort the merge)
Unmerged paths:
(use "git add <file>..." to mark resolution)
both modified: example1.txt
no changes added to commit (use "git add" and/or "git commit -a")
fsedano@Franciscos-Mac-mini cloned_repo %

Figure A-26 *Solving Git conflicts*

For each detected conflict, Git adds the following markers:

```
<<<<<<< HEAD
<the contents from your local repository>
=======
<the contents from the remote repository>
>>>>>>> <commit hash>
```

At this point, you must manually edit the content between <<<<< and >>>>> and remove all markers. When the content is fixed, you need to finish the merge with the **git add** command. Then you can finally push your changes to the remote repository, as shown in Figure A-27.

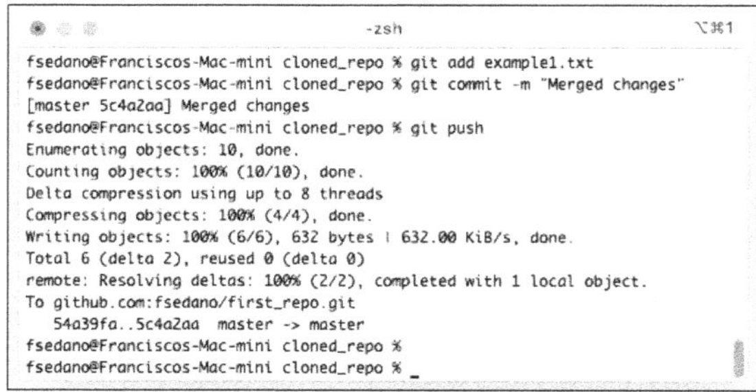

-zsh
fsedano@Franciscos-Mac-mini cloned_repo % git add example1.txt
fsedano@Franciscos-Mac-mini cloned_repo % git commit -m "Merged changes"
[master 5c4a2aa] Merged changes
fsedano@Franciscos-Mac-mini cloned_repo % git push
Enumerating objects: 10, done.
Counting objects: 100% (10/10), done.
Delta compression using up to 8 threads
Compressing objects: 100% (4/4), done.
Writing objects: 100% (6/6), 632 bytes | 632.00 KiB/s, done.
Total 6 (delta 2), reused 0 (delta 0)
remote: Resolving deltas: 100% (2/2), completed with 1 local object.
To github.com:fsedano/first_repo.git
54a39fa..5c4a2aa master -> master
fsedano@Franciscos-Mac-mini cloned_repo %
fsedano@Franciscos-Mac-mini cloned_repo %

Figure A-27 *Git conflict solved*

The Git Graph plug-in on Visual Studio Code gives a great visualization of the entire commit and merge flow, where you see at some point the two repositories diverged, and changes have been successfully merged. You can see it in action in Figure A-28.

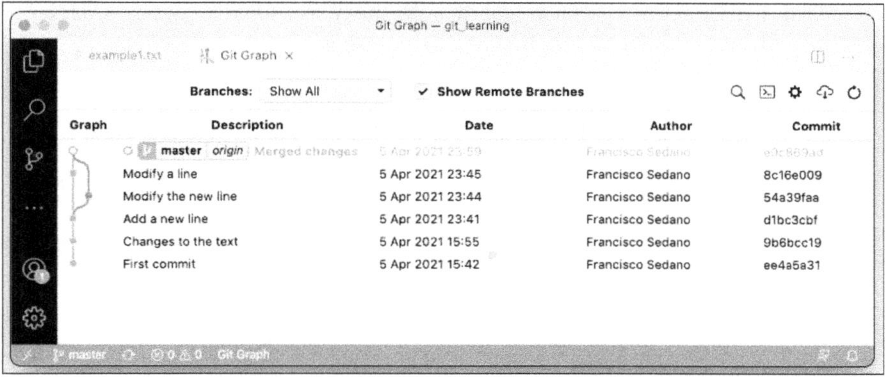

Figure A-28 *Git history flow*

Sharing Your Repository with Other Users

When your repository is cloned in a public site, like GitHub or GitLab, you can easily share it with other users, using the link shown in the Code button, as shown in Figure A-29.

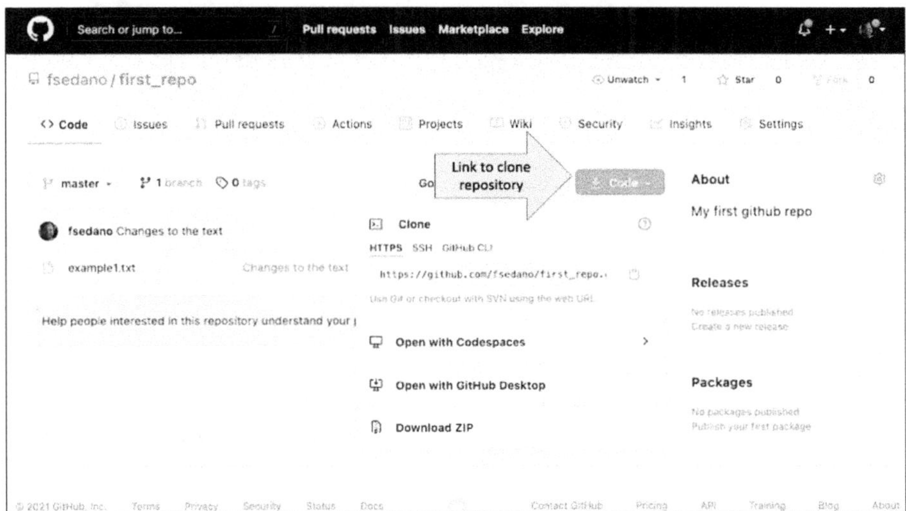

Figure A-29 *Sharing repositories using GitHub*

Anybody with the right permissions can use the link on that page to clone the repository (and all the history) on their computer. You can try it yourself by cloning it to a different directory, as shown in Figure A-30. You always have control over who can access your repository. Public repositories allow anybody to clone them, and private repositories allow you to choose who can see them. You always can choose who can commit to the repository. This can be set when you create the repository, as shown in Figure A-31, or can be changed at any time by using the repository settings menu.

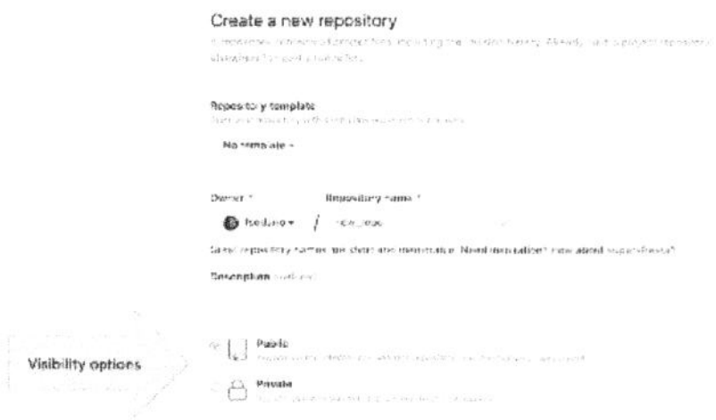
Create a new repository
Visibility options

Figure A-30 *Cloning Git repositories*

-zsh
fsedano@Franciscos-Mac-mini ~ % git clone https://github.com/fsedano/first_repo.git
Cloning into 'first_repo'...
remote: Enumerating objects: 6, done.
remote: Counting objects: 100% (6/6), done.
remote: Compressing objects: 100% (4/4), done.
remote: Total 6 (delta 0), reused 6 (delta 0), pack-reused 0
Unpacking objects: 100% (6/6), done.
fsedano@Franciscos-Mac-mini ~ % cd first_repo
fsedano@Franciscos-Mac-mini first_repo % ls
example1.txt
fsedano@Franciscos-Mac-mini first_repo % git log
Author: Francisco Sedano <fran@fransedano.net>
Date: Mon Apr 5 15:55:09 2021 +0200
Changes to the text
Author: Francisco Sedano <fran@fransedano.net>
Date: Mon Apr 5 15:42:59 2021 +0200
First commit
fsedano@Franciscos-Mac-mini first_repo %

Figure A-31 *Choosing repository visibility at creation time*

Programming Languages

To create your own network automation tools, you use a programming language. To automate networks, you probably want an easy-to-read language, with plenty of available libraries. Python, Ruby, and Go are the most-used languages in network automation, and Python is by far the most popular one.

Python is an interpreted language, which means you don't need to compile it to run it. However, to run a Python program, you need to have Python installed on the computer, along with the required set of libraries.

Having to code everything needed to access your network device would be very tedious and error-prone. Instead, you can use Python libraries that provide all the low-level code needed to access the device.

A huge number of Python libraries are used in network programmability. Some of them provide very low-level access. For example, they might just take care of creating an SSH connection to the network device, and it would be up to you to explicitly code all the interaction needed with the device. This is the case of the paramiko library.

Other libraries provide you with all the building blocks to manage the configuration and status of your device. For example, Cisco's own pyATS library knows how to properly parse the output of the **show** CLIs and how to handle the different CLI modes (enable, config, and so on) by automatically replying to the various CLI prompts. With pyATS, you just need to provide connectivity information to the device and the config you want to push, and it automatically takes care of all the steps needed to send the configuration and verify its operation.

Other options to automate your network using programmability are tools like Ansible or Terraform.

How to Run Code in a Repeatable Way

Regardless of the libraries you choose to use, you need to have them installed in the computer you use to develop your code and in the computer that will be used to run the scripts against the network devices. This can be a challenge:

- Over time, new library (or Python) versions will appear, and old ones might be deprecated. New library versions might require updates to your code so that it keeps working.
- Before running the code, you need to make sure all the libraries are present at the correct version.

You can handle this problem in several ways. The most common ones are using a virtual environment like venv and running your application in a container.

Both venv and containers solve the problem by encapsulating your code in an environment with a fixed set of libraries and dependencies.

Python Virtual Environments

As a general guideline, you should not install Python libraries directly on your computer. You could run into version conflicts and leave your computer in a bad state. Virtual environments are useful to avoid this issue.

venv is a Python version management tool. It allows you to change the Python version used and set virtual environments, each of them with an isolated set of libraries.

When you use venv, you don't install libraries directly on the shared directories on your system. Instead, you

- Create a virtual environment.
- Activate the virtual environment.

- Install the required libraries on it.
- Execute your application inside the virtual environment.

However, venv is itself a Python module. To use venv, you need to install Python in your development machine and in the hosts where you plan to run your code.

The purpose of a Python virtual environment is to isolate the libraries and dependencies you install from the main ones on the system. After you start a virtual environment, any library you add will affect only that virtual environment, preserving how your system works.

To create a virtual environment, use

```
python3 -m venv <new directory>
cd <new directory>
source bin/activate
```

In Figure A-32, you can see an example.

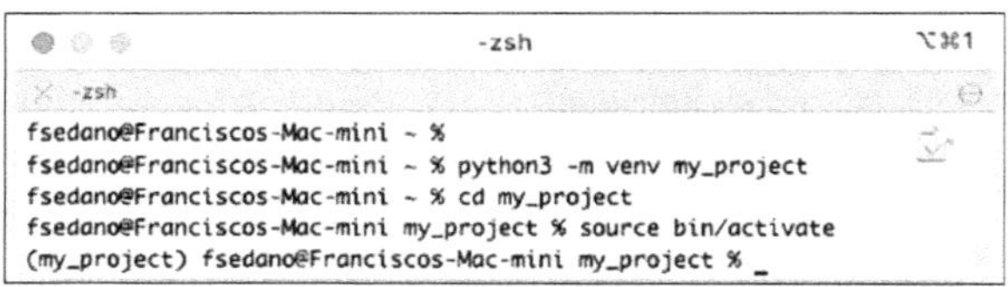

Figure A-32 *Creating a Python virtual environment*

When the virtual environment is active, you see the prompt change to reflect the virtual environment name. From this moment, all Python libraries you install don't affect the global setup on your computer; they affect only the active virtual environment.

To install Python libraries, you use pip. For example, to install a library called requests, you would run

```
pip install requests
```

However, manually installing all required libraries would be prone to errors and tedious, so the convention is to have a file called requirements.txt with all the libraries listed and then reference it by calling

```
pip install -r requirements.txt
```

You can use the main documentation site for more details on venv: https://docs.python.org/3/library/venv.html.

Containers

Another option to run your code in a repeatable way is to use containers. You can use several applications to run containers, Docker being the most popular of them.

A container image consists of all the files, libraries, and parts of the operating system needed to run your programs. For example, you can run your program in an Ubuntu Linux container from your Windows or Mac computer. Regardless of the Windows or Mac version you're using, your container will keep working exactly the same as when you created it.

While a container might look similar to a virtual machine, it is fundamentally different. A virtual machine emulates an entire computer's hardware. It requires you to install an operating system and all required environments for your application to run. This architecture is shown in Figure A-33.

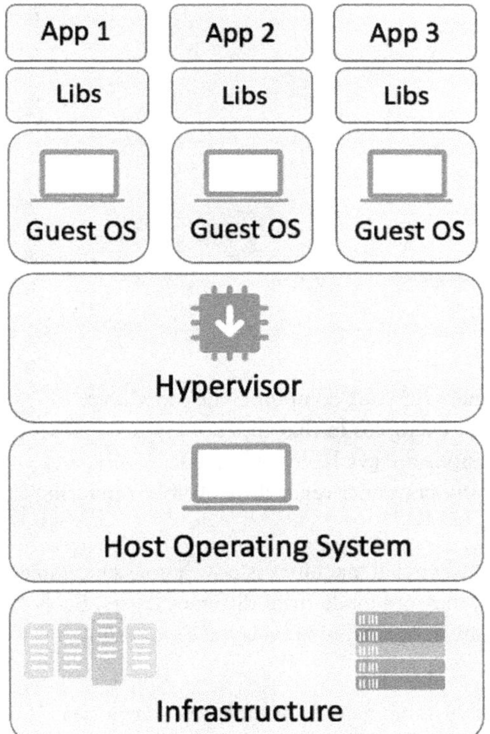

Figure A-33 *Hypervisor architecture*

A virtual machine provides a great level of isolation between applications, but it is difficult to handle, heavy and slow to bring up. Also, re-creating it from scratch is not easy.

Containers, in contrast, follow a different approach. Instead of emulating a whole computer, they run on top of a shared operating system, and they use namespace techniques to provide isolation between applications. You can see this architecture in Figure A-34.

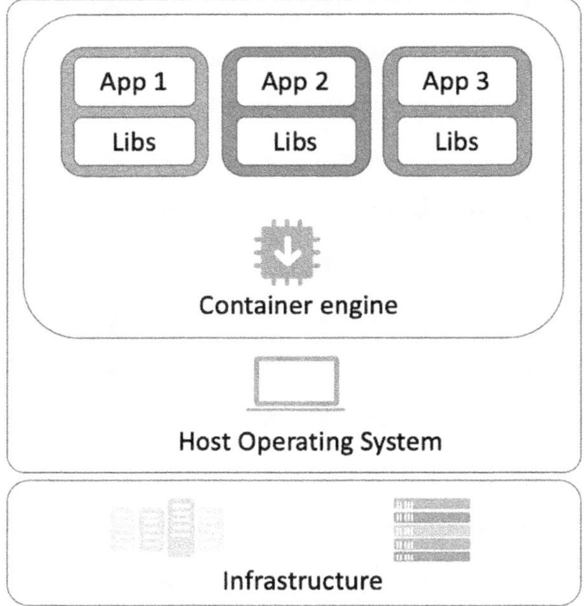

Figure A-34 *Container architecture*

Several applications allow you to run containers on your computer; the most common ones are Docker and Podman. To run the examples in this appendix, you need to install Docker in your computer. Docker versions are available for MacOS, Linux, and Windows, and all of them allow you to run any container regardless of your operating system.

Another key difference between containers and virtual machines is how a container image is structured. Instead of being a single entity, they are made from different layers. Each layer just lists the differences on the filesystem from the previous layer. This is shown in Figure A-35.

The topmost layer in a container image is the writable one. Each time you start a container, a fresh writable layer is created for you. That means that any change you do in the container is stored in that layer, but previous ones are untouched. You can delete, add, or change any file inside the container, but when you re-create it, you go back to a pristine, unmodified copy.

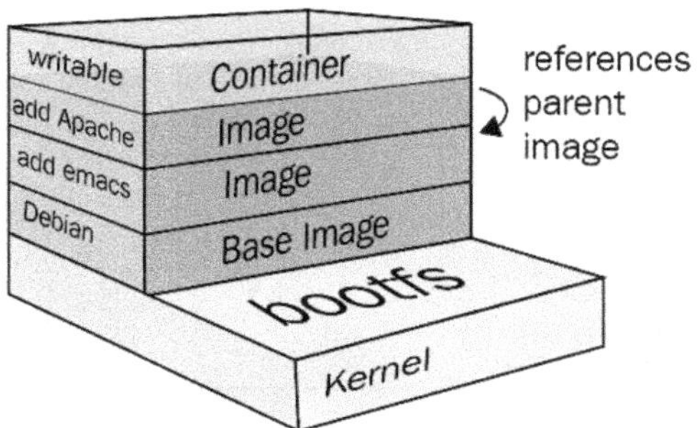

Figure A-35 *Layers in a container image*

The instructions on how to create an image are described in a Dockerfile.

For example, to build an image that has Ubuntu version 18.04 as the base and install Python 3 to it, you would write the following Dockerfile:

```
FROM ubuntu:18.04
RUN apt-get update && apt-get install -y python3
```

Save this file with the name **Dockerfile** and build it with the following command:

```
docker build -t myimage .
```

The dot in the command indicates where the Dockerfile is located—in this case in the same directory where you're running the command (indicated by the dot).

The first time you run the command, it pulls the image called ubuntu:18.04 from the public Docker registry (a repository of container images), and it executes the first Dockerfile instruction on it. In this case, it is a **RUN** instruction, which means execute the argument in the container and save the filesystem state as a new layer. After running the sample Dockerfile, you have an image with the following layers. Docker keeps track of each layer by using a hash (the long hexadecimal number displayed at the end of each step), as shown in Example A-1 and in Figure A-36.

Example A-1 *Docker Uses a Hash to Keep Track of Each Layer*

```
$ docker build -t myimage .

Sending build context to Docker daemon  2.048kB
Step 1/2 : FROM ubuntu:18.04
18.04: Pulling from library/ubuntu
Status: Downloaded newer image for ubuntu:18.04
 ---> 1715a469e5df

Step 2/2 : RUN apt-get update && apt-get install -y python3
 ---> Running in 300539e4f8b4
```

```
Get:1 http://ports.ubuntu.com/ubuntu-ports bionic InRelease [242 kB]
running python post-rtupdate hooks for python3.6...
...
Processing triggers for libc-bin (2.27-3ubuntu1.4) ...
Removing intermediate container 300539e4f8b4
 ---> 2bec8cbaddb0
Successfully built 2bec8cbaddb0
Successfully tagged myimage:latest
```

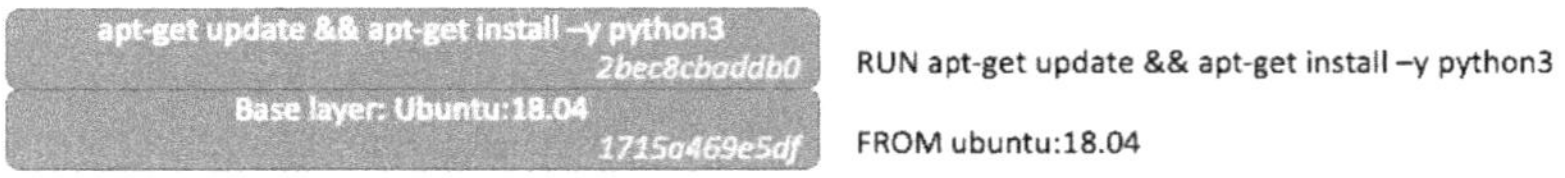

apt-get update && apt-get install –y python3
2bec8cbaddb0
RUN apt-get update && apt-get install –y python3
Base layer: Ubuntu:18.04
1715a469e5df
FROM ubuntu:18.04

Figure A-36 *Layers created by a Dockerfile*

You can add files to the image with the **COPY** command. For example, if you have your Python code saved in your computer as program1.py, to add it to the Docker image, you would use the same Dockerfile as before, just adding the **COPY** command to it:

```
program1.py:

print("Hello world from python inside the container!")


Dockerfile:

FROM ubuntu:18.04
RUN apt-get update && apt-get install -y python3
COPY program1.py /app/
```

When you run the same command again to build the container, you will notice it is much faster than before, because Docker notices it already built the first two layers and just needs to add the third one. You can check that by looking at "using cache" in the output from the **build** command:

```
$ docker build -t myimage .
Sending build context to Docker daemon  3.072kB
Step 1/3 : FROM ubuntu:18.04
 ---> 1715a469e5df
Step 2/3 : RUN apt-get update && apt-get install -y python3
 ---> Using cache
 ---> 2bec8cbaddb0
Step 3/3 : COPY program1.py /app/
 ---> 49b7614d6e34
```

```
Successfully built 49b7614d6e34
Successfully tagged myimage:latest
```

After the image is built, you can use the following **Docker run** command to start it:

```
docker run -it <image name> <command to execute>
```

After running this command, notice the prompt changes to indicate you're inside the container. You can run your program from there:

```
$ docker run -it myimage bash

root@408b73a8bb57:/# python3 /app/program1.py
Hello world from the container
root@408b73a8bb57:/#
```

Remember that any change you do inside the container will affect only the topmost writable layer. For example, if you delete your program inside the container, when you restart it, the program will still be there.

Container images are described by Dockerfiles, and this has several advantages:

- You can build your own copy of a container image just by downloading (or cloning with Git) a Dockerfile, which is a small text file. No need to download huge images.
- The Dockerfile serves as a very good documentation tool. It precisely describes how to build the environment.
- Other users can build upon your work. Your image can be used in a FROM line from other Dockerfiles and add more features to it.

For the examples in the programmability and telemetry chapters, a set of Dockerfiles is provided in GitHub, so you can practice the programmability and telemetry examples in your own setup.

Docker Compose

The Dockerfile provides a great way to define the contents of a container image. However, to properly run an application using containers, you need more than that. You might need to start several containers, each of them with a different image, pass different parameters to each of the containers, set up the rules to allow communication between them, describe how to persist files, and so on.

It's possible to do that by using the **docker** command line and passing all the required arguments there, but it's a tedious process that is prone to errors and it requires manual intervention.

A better option is to use a wrapper for Docker: docker-compose. Docker-compose uses a YAML file to define how to handle your complete environment. By default, it refers to a file called docker-compose.yaml.

The docker-compose.yaml file shown in Example A-2

- Creates a container called app_1 by building a Dockerfile located on the current directory.
- Pulls a database server (mariadb) and a database web admin application from the public docker image repository.
- Sets the root password of the database as mypass, by setting up the environment variable MYSQL_ROOT_PASSWORD.
- Maps port 8080 on the web application so it's visible to the external world as port 9000.

Example A-2 *docker-compose.yaml:*

```
version: '3.1'
services:
    app_1:
        build: .
    db:
        image: mariadb:10.2
        environment:
            MYSQL_ROOT_PASSWORD: mypass
    adminer:
        image: adminer
        ports:
        - 9000:8080
```

To bring up the environment, you can use the **docker-compose up** command. To list the status of the containers, you can use the **docker-compose ps** command. You can see this in Figure A-37.

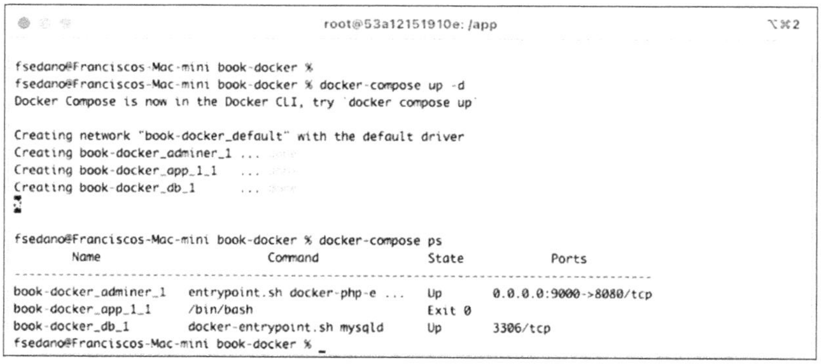

```
root@53a12151910e: /app

fsedano@Franciscos-Mac-mini book-docker %
fsedano@Franciscos-Mac-mini book-docker % docker-compose up -d
Docker Compose is now in the Docker CLI, try `docker compose up`

Creating network "book-docker_default" with the default driver
Creating book-docker_adminer_1 ...
Creating book-docker_app_1_1   ...
Creating book-docker_db_1      ...

fsedano@Franciscos-Mac-mini book-docker % docker-compose ps
        Name                    Command                State          Ports
-------------------------------------------------------------------------------------
book-docker_adminer_1   entrypoint.sh docker-php-e ...   Up       0.0.0.0:9000->8080/tcp
book-docker_app_1_1     /bin/bash                        Exit 0
book-docker_db_1        docker-entrypoint.sh mysqld      Up       3306/tcp
fsedano@Franciscos-Mac-mini book-docker % _
```

Figure A-37 *Checking container status by using docker-compose*

If you now open http://0.0.0.0:9000 in a browser, you access the web application you're running in your container.

To tear down the environment when you're done, use the **docker-compose down** command, and all the containers are stopped.

The workflow described in this appendix greatly simplifies sharing working environments. Now you can share your docker-compose and Dockerfiles on GitHub, and anybody can quickly re-create the same environment you used for your application!

Index

Numerics

A

B

D

J-K

L

M

N

O

P

Q

R

S

T

U

V

W

www.ingramcontent.com/pod-product-compliance
Lightning Source LLC
Chambersburg PA
CBHW080347220326
41598CB00030B/4627

* 9 7 8 0 1 3 7 4 9 2 3 2 9 *